Ocean Noise: From Science to Management

Ocean Noise: From Science to Management

Editors

Michel André
Christine Erbe

MDPI • Basel • Beijing • Wuhan • Barcelona • Belgrade • Manchester • Tokyo • Cluj • Tianjin

Editors
Michel André
Universitat Politècnica de
Catalunya, BarcelonaTech
(UPC)
Spain

Christine Erbe
Curtin University
Australia

Editorial Office
MDPI
St. Alban-Anlage 66
4052 Basel, Switzerland

This is a reprint of articles from the Special Issue published online in the open access journal *Journal of Marine Science and Engineering* (ISSN 2077-1312) (available at: https://www.mdpi.com/journal/jmse/special_issues/ocean_noise).

For citation purposes, cite each article independently as indicated on the article page online and as indicated below:

LastName, A.A.; LastName, B.B.; LastName, C.C. Article Title. *Journal Name* **Year**, *Volume Number*, Page Range.

ISBN 978-3-0365-4377-2 (Hbk)
ISBN 978-3-0365-4378-9 (PDF)

Contents

About the Editors

Michel André

Michel André holds a M.Sc. degree in Engineering (Institut National des Sciences Appliquées, Toulouse, France), a M.Sc. degree in Animal Physiology, a M.Sc. degree in biotechnologies (Université Paul Sabatié, Toulouse, Franc) and a Ph.D. in Bioacoustics (University of Las Palmas de Gran Canaria, Spain). He worked as research assistant at the San Francisco State University, was an intern researcher at the Marine Mammal Center (Marin County, San Francisco), an associated professor at the Universidad de Las Palmas de Gran Canaria. He created and is the Director of the Laboratory of Applied Bioacoustics and is a full Professor at the Technical University of Catalonia, BarcelonaTech (UPC). Michel's interests are underwater sounds and human pressure effects to marine ecosystems, the development of acoustic technology to monitor biodiversity. He is a former Chair of the European Cetacean Society, Founder and President of the Sense of Silence Foundation, Chair of the international OCEANOISE Conference series and a founding member of the European Commission Task Group on Noise (TGNoise, Marine Strategy Framework Directive).

Christine Erbe

Christine Erbe holds an M.Sc. degree in Physics (University of Dortmund, Germany) and a Ph.D. in Geophysics (University of British Columbia, Canada). She worked as a Research Scientist at Fisheries and Oceans Canada, was Director of JASCO Applied Sciences Australia, and after a brief stint in high-school education, returned to academia as Director of the Centre for Marine Science and Technology at Curtin University (Perth, WA, Australia). Christine's interests are underwater sound (biotic, abiotic, and anthropogenic), sound propagation, signal processing, and noise effects on marine fauna. She's a John Curtin Distinguished Professor, Fellow of the Acoustical Society of America, former Chair of the Animal Bioacoustics Technical Committee of the Acoustical Society of America, and former Chair of the international conference series on The Effects of Noise on Aquatic Life.

Preface to "Ocean Noise: From Science to Management"

This book is a collection of articles relevant to ocean noise and its management. Scientific and societal concern about the effects of underwater sound on marine ecosystems is growing. While iconic megafauna was of initial concern, more and more taxa are being included. Some countries have joined in multi-national initiatives to measure, monitor, and mitigate environmental impacts of ocean noise at large, trans-boundary spatial scales. Approaches to regulating ocean noise change as new scientific evidence becomes available, but may also differ by country.

Articles in this collection cover topics such as ocean soundscapes, sources of underwater sound, anthropogenic sound, sounds made by marine fauna, animal sensitivity to sound, the potential effects of sound on marine fauna, sound propagation under water, sound modelling, noise management, and noise impact mitigation.

The idea for this book was born when we had to cancel the OCEANOISE conference two years in a row, due to Covid. The OCEANOISE conference series has provided a platform for the exchange of scientific results, management approaches, research needs, stakeholder concerns, etc. Attendees have represented various sectors, including academia, offshore industry, defence, NGOs, consultants and government regulators.

Michel André and Christine Erbe
Editors

Journal of
Marine Science and Engineering

Article

The Underwater Soundscape at Gulf of Riga Marine-Protected Areas

Muhammad Saladin Prawirasasra *, Mirko Mustonen and Aleksander Klauson

Department of Civil Engineering and Architecture, Tallinn University of Technology (TalTech), 19086 Tallinn, Estonia; mirko.mustonen@taltech.ee (M.M.); aleksander.klauson@taltech.ee (A.K.)
* Correspondence: muhammad.prawirasasra@taltech.ee

Abstract: Passive acoustic monitoring (PAM) is widely used as an initial step towards an assessment of environmental status. In the present study, underwater ambient sound recordings from two monitoring locations in marine-protected areas (MPAs) of the Gulf of Riga were analysed. Both locations belong to the natural habitat of pinnipeds whose vocalisations were detected and analysed. An increase of vocal activity during the mating period in the late winter was revealed, including percussive signallings of grey seals. The ambient sound spectra showed that in the current shallow sea conditions ship traffic noise contributed more in the higher frequency bands. Thus, a 500 Hz one-third octave band was chosen as an indicator frequency band for anthropogenic noise in the monitoring area. It was shown that changes in the soundscape occurring during the freezing period create favourable conditions for ship noise propagation at larger distances. Based on the monitoring data, the environmental risks related to the anthropogenic sound around the monitoring sites were considered as low. However, further analysis showed that for a small percentage of time the ship traffic can cause auditory masking for the ringed seals.

Keywords: passive acoustic monitoring; shallow water; pinnipeds; anthropogenic sound; auditory masking

Citation: Prawirasasra, M.S.; Mustonen, M.; Klauson, A. The Underwater Soundscape at Gulf of Riga Marine-Protected Areas. *J. Mar. Sci. Eng.* **2021**, *9*, 915. https://doi.org/10.3390/jmse9080915

Academic Editors: Michel André and Christine Erbe

Received: 26 July 2021
Accepted: 11 August 2021
Published: 23 August 2021

Publisher's Note: MDPI stays neutral with regard to jurisdictional claims in published maps and institutional affiliations.

1. Introduction

The pressure on marine ecosystems from anthropogenic underwater noise has been recognised as a challenging problem during the last decades. This cross-border issue can be solved only with an international joint effort. The EU Marine Strategy Framework Directive (MSFD) adopted in June 2008 is aiming to achieve the Good Environmental Status (GES) of the European seas [1]. The directive sets qualitative descriptors for GES that list Descriptor 11 as relevant to the energy introduced to the marine environment, including underwater sound. The initial step towards assessing the environmental pressure posed by anthropogenic sound is passive acoustic monitoring (PAM). One-third of octave bands (TOBs) with nominal frequencies of 63 Hz and 125 Hz have been suggested as most relevant to monitor the anthropogenic continuous low-frequency sound in water [2].

Underwater soundscapes are known to manifest spatial and temporal variability [3,4]. According to the types of contributing sources, underwater soundscapes can consist of geophony, biophony, and anthropophony [5]. Geophony includes naturally occurring non-biological sounds such as wind-generated breaking waves [6] and precipitation [7]. Anthropophony includes underwater noise induced by human activities, such as commercial ship traffic [8]. Anthropogenic underwater noise is considered a pollutant that can have long-term adverse effects on marine ecosystems. Potential impacts of continuous underwater noise are the reduction of communication space and auditory masking [9–11] as well as increased stress levels [12]. In the passive acoustic monitoring data, geophony and anthropophony are mixed, but by estimating the wind-dependent natural sound levels, these two components can be separated [13].

Biophony in the Gulf of Riga is typically dominated by pinniped calls [14] but can potentially also include fish vocalisations [15,16]. Underwater vocalisations of pinnipeds are known to play a significant social role in their intraspecies communication [17,18]. Vocalisations can express, for example, aggressive or submissive behaviour. Vocal interaction during the breeding period is very intensive with a high variation of call types and an increased number of calls [17,19]. It has been reported that both grey (*Halichoerus grypus*) and ringed seals (*Phoca hispida*) can use vocalisations as an aid for under-ice orientation during the winter [18,19].

The objective of this study is to provide baseline information on underwater soundscapes at two monitoring locations within the marine-protected areas (MPAs) of the Gulf of Riga. Special focus is made on the detection and identification of pinniped sounds, bearing in mind that elevated detection rates show both the abundance of animals and the importance of the respective marine areas for the pinnipeds. The environmental pressure from anthropogenic underwater noise and its potential risks are also addressed. Quantification of the proportion of the anthropogenic sound makes it possible to draw some conclusions about the current environmental status of the monitoring sites.

2. Materials and Methods

2.1. Underwater Sound Monitoring Locations

Underwater sound monitoring was conducted in two monitoring locations situated in the Natura 2000 MPAs of the Gulf of Riga. These locations are further referenced as Kihnu and Moonsund and their respective positions are shown in Figure 1. The monitoring took place from 2018 to 2019, lasting 9 months in Kihnu and 6 months in Moonsund.

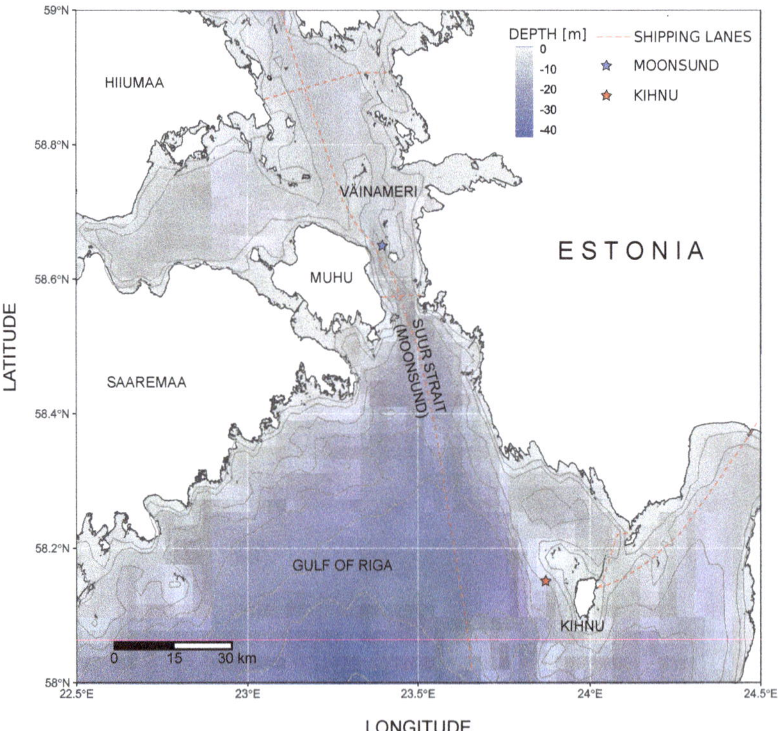

Figure 1. Sound monitoring locations in Kihnu (58.149° N, 23.873° E) and Moonsund (58.651° N, 23.393° E), marked by asterisks.

The Kihnu monitoring location is relatively far from shipping lanes, while the Moonsund monitoring location is close to the local shipping lane which is known to be moderately active in summer and closed for navigation during the winter period. About 8 km to the south, a busy regular ferry line is operating year round between the mainland and Muhu Island.

Moonsund is also known as an important migration route for ringed seals from their haul-outs at the islets in the Väinameri to the feeding grounds in the south of the Gulf of Riga [20,21]. During the winter, the monitored areas are often covered by ice, making it an attractive breeding ground for seals.

2.2. Underwater Acoustic Monitoring Equipment

Autonomous recorders by two different manufacturers were used for the ambient sound measurements. One was the SM2M [22] by Wildlife Acoustics, Inc. with a sampling frequency of 32 kHz and standard HTI hydrophone. The other recorder was the SoundTrap ST500 [23] by Ocean Instruments with a sampling frequency of 36 kHz and equipped with a standard hydrophone.

SM2M recorders were used in the Moonsund location during the whole monitoring period. In Kihnu, the sound was recorded with SM2M in summer and with ST500 during the second monitoring period, extending from autumn to early spring.

Figure 2 shows the rig designs for the two autonomous recorders. The output of both recorders was 16 bit WAV format sound files that were processed using 20 s time-averaging and a rectangular window function without overlap in order to follow the Life+ BIAS project signal processing standard [24]. The sound processing in the study was performed using PAMGuide software [25].

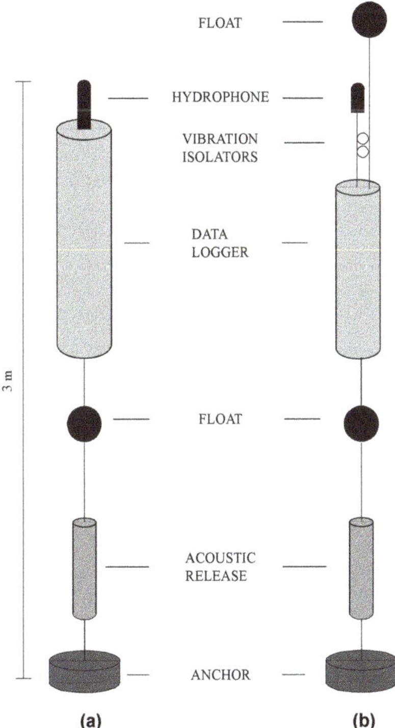

Figure 2. Mooring setups for (**a**) SM2M and (**b**) ST500 data loggers.

2.3. Detection of Pinniped Calls

At the monitoring sites, the bulk of the biological sounds were produced by grey and ringed seals, whose vocalisations were detected and identified in the wake of the results of numerous bioacoustics studies [17,18,26–30]. The identified calls were analysed and their patterns, including frequency ranges and call durations, were entered into the band limited energy detector (BLED) [31] for a subsequent search for similar patterns in the recorded data. Instead of detecting patterns, BLED detects events based on energy exceeding a threshold value in a selected frequency band for a specified time.

2.4. Ship Traffic Data

For characterisation of the ship traffic, the automatic identification system (AIS) data around the monitoring locations were analysed. Figure 3 depicts the AIS-based average daily number of ships by their types, passing within a 10 km radius from the sound monitoring locations. It can be seen that the overall ship traffic density in Kihnu is very low. In Moonsund, most of the distant ship traffic is caused by the ferry line. Pleasure boats appear mostly in the summer season, and their number is likely to be underestimated since not all of them are equipped with AIS transceivers. In some cases, the noise emissions from pleasure boats can dominate the soundscape in the coastal waters [32], and therefore, their contribution should be properly addressed.

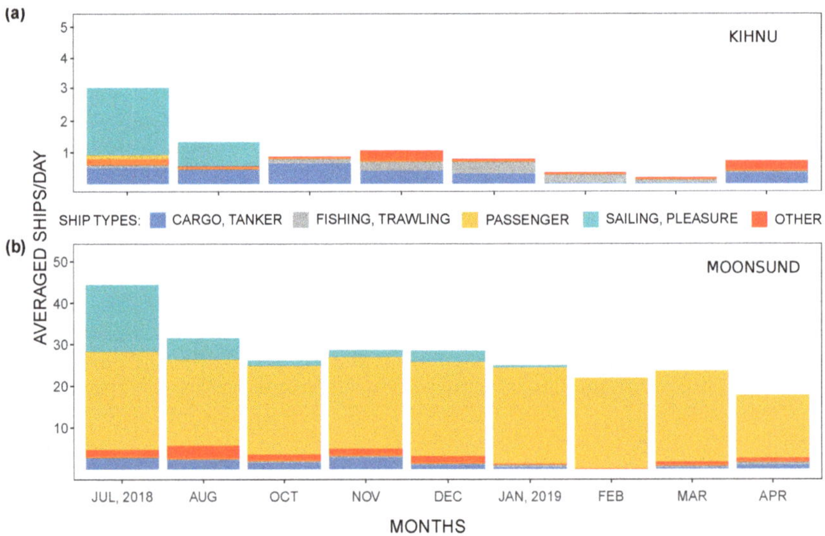

Figure 3. Monthly averaged numbers of ships per day based on automatic identification system (AIS) position reports within 10 km radius from the sound monitoring locations: (**a**) Kihnu and (**b**) Moonsund.

To assess the factual contribution of shipborne noise, acoustic detection was used. As the ship approaches, the ambient sound level increases, and its excess over the background noise level can be calculated by assuming that the running minimum of broadband (10 Hz–1 kHz) received level (RL) is a reasonably good proxy for the background noise [33]. In this study, the window for the running minimum was set to 3 h. The detection threshold was 3 dB for low sea states (under the ice in the wintertime) and 6 dB for other seasons.

3. Results

3.1. Biological Sound Detection

Out of the two monitoring locations, recordings from Kihnu were very rich in biological sounds. During the monitoring period, around 37,000 seal calls were detected in Kihnu.

In contrast, Moonsund recordings contained fewer biological sounds, having only around 1400 detected seal calls. Although the seal's vocal repertoire is quite rich, we have focused only on the most frequent call types. Thus, the moan, guttural rup (rup) and guttural rupe (rupe) were taken into account for the grey seal and bark and yelp for the ringed seal.

3.1.1. Grey Seal Vocalisations

The most frequently detected grey seal call was the rup, making up 41% of all the grey seal call detections. This was followed by the moan at 32%, the rupe at 25%, and the percussive signalling (clap) at 2%. Almost all of the grey seal calls (98%) were detected in the recordings from Kihnu.

Figure 4 shows the spectrograms of the recorded grey seal calls. The moan (Figure 4a) is a low-frequency call that can last up to a few seconds. The rup (Figure 4b) is characterised by a sharp upsweep that lasts for less than 0.5 s. Most of the detected rups appeared in pairs. The rupe (Figure 4c) has a sharp upsweep similar to the rup that is followed by a longer-lasting downsweep. The rupe call sounded similar to the bark and yelp type calls of the ringed seal. Additionally, a recent article [34] described the behaviour of the grey seals where they used percussive signalling by repeatedly clapping their forelimbs. These clap sounds were also detected in our recordings and are shown in Figure 4d.

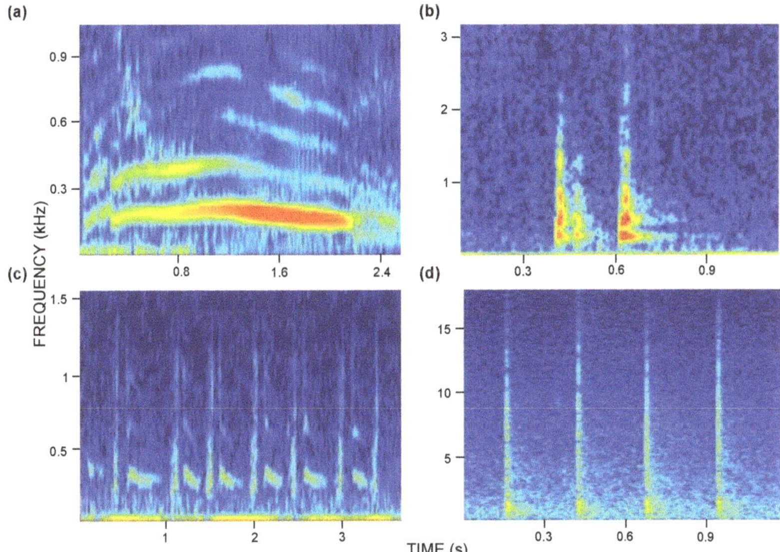

Figure 4. Underwater sounds produced by grey seals: (**a**) moan, (**b**) rups, (**c**) series of rupes, and (**d**) forelimb clapping.

3.1.2. Ringed Seal Vocalisations

Ringed seals are known to vocalise less frequently than their grey counterparts [18,19]. Around 600 ringed seal calls were detected, mostly in the Moonsund site. Ringed seal yelps and barks were detected in almost equal proportions.

Spectrograms of the yelps and barks are shown in Figure 5. The yelp (Figure 5a) is a sweeping tonal sound at 500–600 Hz that lacks harmonics. By contrast, the bark sound (Figure 5b,c) has a lower frequency range and contains several harmonics. In addition, scratching sounds that were attributed to the digging of breathing holes into the ice are presented in Figure 5d.

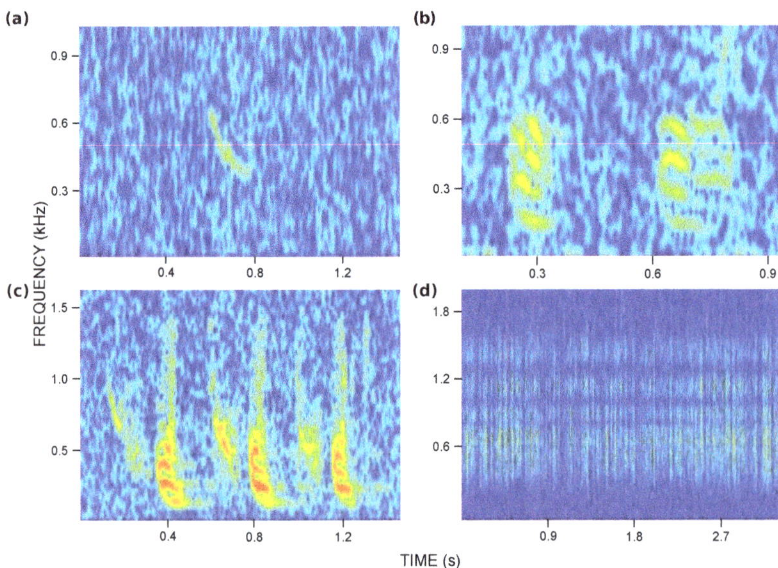

Figure 5. Underwater sounds produced by ringed seals: (**a**) yelp, (**b**) bark, (**c**) alternating series of yelps and barks, and (**d**) digging of a hole in the ice.

The BLED code used in this study has shown a good detection efficiency in the case of the grey seals calls but had less success with the ringed seal calls, probably because of their lower signal-to-noise ratio. The performance evaluation of the automatic detections was made by collecting 100 sample recordings of each call type. The performance characteristics of the detector by call type are shown in Table 1. It can be seen that the sensitivity of the detector is higher for calls with a shorter duration such as the rupe and rup. On the other hand, longer-lasting calls such as moans are often undetected presumably because of their highly variable durations and frequency content. As a result [35], the total number of moans is likely to be underestimated in the case of automatic detection. In contrast, the detected number of rup calls is better predicted as their sensitivity for detections reached 70%.

Table 1. Band limited energy detector (BLED) performance for finding specific types of grey seals' calls.

Types of Calls	TP	FP	FN	TN	Accuracy	Sensitivity
Moan	11	0	89	300	77.75%	11%
Rupe	26	5	74	295	80.20%	26%
Clap	19	4	81	296	78.77%	19%
Rup	70	67	30	233	75.75%	70%

TP = true positive, FP = false positive, FN = false negative and TN = true negative.

3.2. Ship Traffic Noise

Both monitoring sites are located in a very shallow sea area with a maximum depth of 16 m. The low-frequency cutoff [36], corresponding to the average depth (11 m) in the region, is around 60 Hz. Nevertheless, pleasure boats usually radiate underwater noise at higher frequencies. Therefore, the MSFD indicator frequency bands are not well suited for the assessment of the environmental pressure by anthropogenic sound in these shallow watered monitoring sites. A typical spectrogram of recorded sound from two detected vessels is shown in Figure 6.

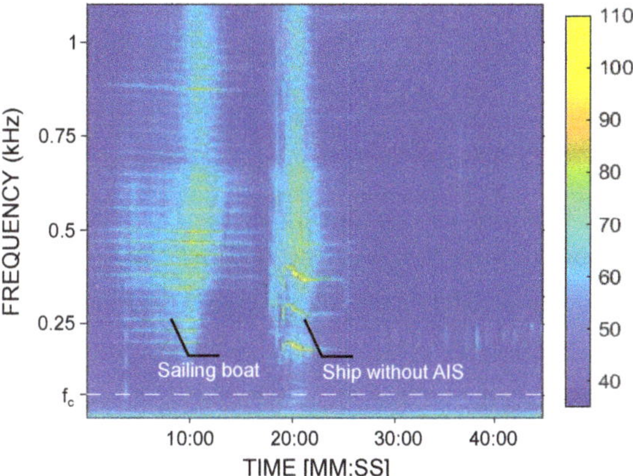

Figure 6. Spectrograms of two vessels recorded in Moonsund during the summer period. The dashed line shows the estimated cutoff frequency f_c = 60 Hz below which ship-radiated sound does not propagate. The first vessel is an AIS-equipped sailing boat and the second is an unknown boat without AIS transmissions.

To select an indicator frequency for the ship traffic noise in the region, the TOB ambient noise spectra for all time intervals containing ship noise were computed and analysed (Figure 7). It can be seen that the MSFD indicator frequency bands are demonstrating quite low levels. Based on the average spectrum, 500 Hz TOB was selected as a more relevant indicator for the environmental pressure posed by shipborne underwater noise. Although the higher TOBs also show higher levels, they were not chosen as they potentially contain an increasingly significant portion of natural ambient noise.

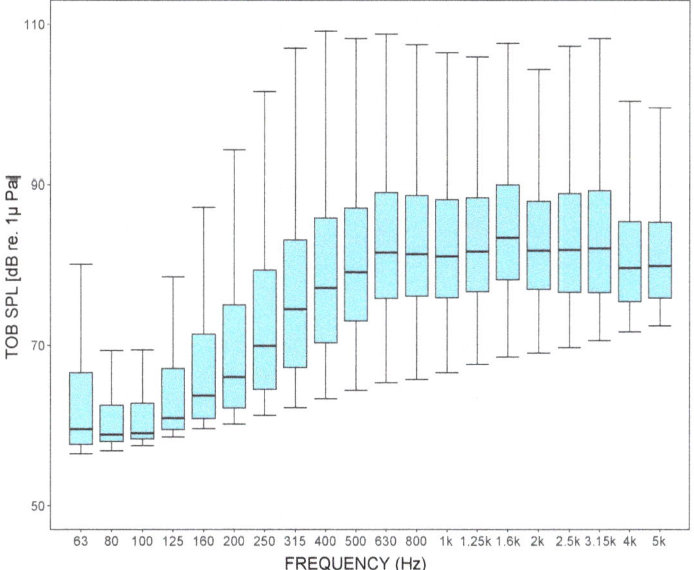

Figure 7. Boxplot of one-third of octave band (TOB) received level (RL) ship noise for the period July 2018 with outliers removed. The lower and upper hinges show the exceedance levels L75 and L25, while medians are shown by the middle lines. The upper and lower whiskers indicate the 1.5 times difference of exceedance levels L25 and L75.

3.3. Underwater Sound Propagation under the Ice

The alteration from an agitated sea surface to a frozen one changes considerably the underwater soundscape. Under ice cover, the natural ambient sound level lowers considerably. The water temperature and salinity near the sea surface change also, thus creating a positive gradient in the sound speed profile, which in turn causes upward refraction of the sound [37]. As a result, the sound rays from distant shipping interact less with the sea bottom and propagate further due to the smaller propagation loss.

Such favourable sound propagation conditions were observed in the winter period when the detection range of ship noise increased considerably. Obviously, the lower ambient sound levels also improved the signal-to-noise ratio, yet distant shipping was never detected outside the freezing period, even at low sea states.

Figure 8 shows the time series of the ambient sound level (500 Hz TOB), the wind speed, the ice concentration, and the number of acoustic ship detections. The sound pressure level (SPL) shows a clear correlation with wind speed. It can be seen that a longer range for detection appeared with the formation of the ice cover. The longest detection range attained was 10 km when the ice concentration was 82%, and wind speed was 2.7 m/s. Furthermore, a gap in long-range detections can be seen during periods of strong winds.

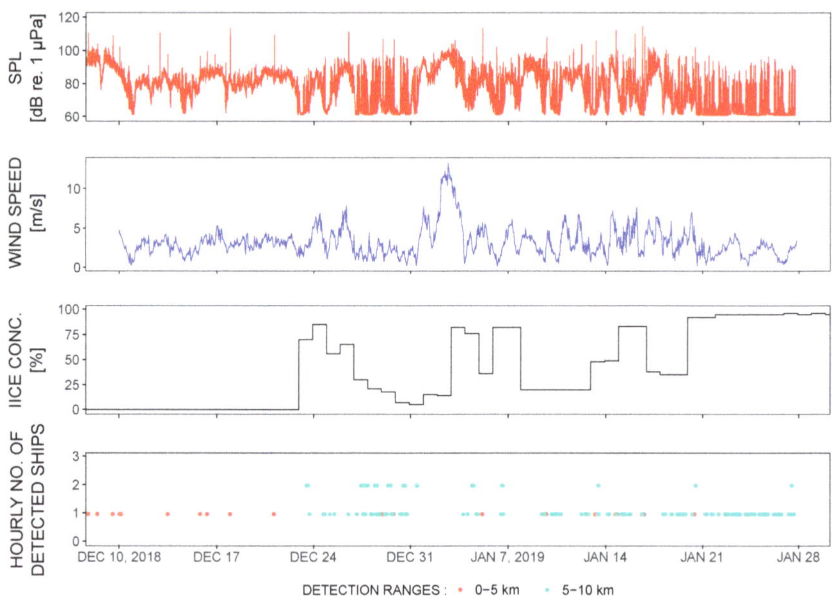

Figure 8. Time series of 500 Hz TOB sound pressure level (SPL), wind speed, ice concentration, and detections of AIS-equipped ships grouped in two categories according to their detection ranges. SPL correlates with the wind speed, and long-range detections start with the appearance of ice cover.

3.4. Ambient Sound Analysis

An overview of the ambient sound levels in both monitoring locations is presented in Figure 9 as monthly estimated probability density functions (PDF) in the 500 Hz TOB. For each violin plot, the surface area equals unity, and the abscissa of the plot shows the relative likelihood of the occurrence of every SPL value displayed on the vertical axis. The key statistical measures of the arithmetic mean and the exceedance levels L95 and L05 are also shown in the violin plots.

Figure 9a presents the monthly estimated PDF of the SPLs recorded in the Kihnu location. The monthly arithmetic means vary from 75 dB in April to 84 dB in August. Most of the monthly PDFs were negatively skewed, which was an indication of natural sound domination [4].

The monthly PDFs of the SPLs from the Moonsund location (Figure 9b) have lower mean values but higher exceedance levels of L05 corresponding to louder and less frequently occurring events that can be caused, for example, by close passing ships. According to the PDFs, the anthropogenic sound was not dominant in the Moonsund location. However, it was more prevalent than in Kihnu (Figure 9a). In January, PDF was particularly skewed so that the mode of the levels was only slightly above the self-noise level of the recorder. Such low levels were due to the presence of ice, which is known to drastically decrease the agitation of the sea surface.

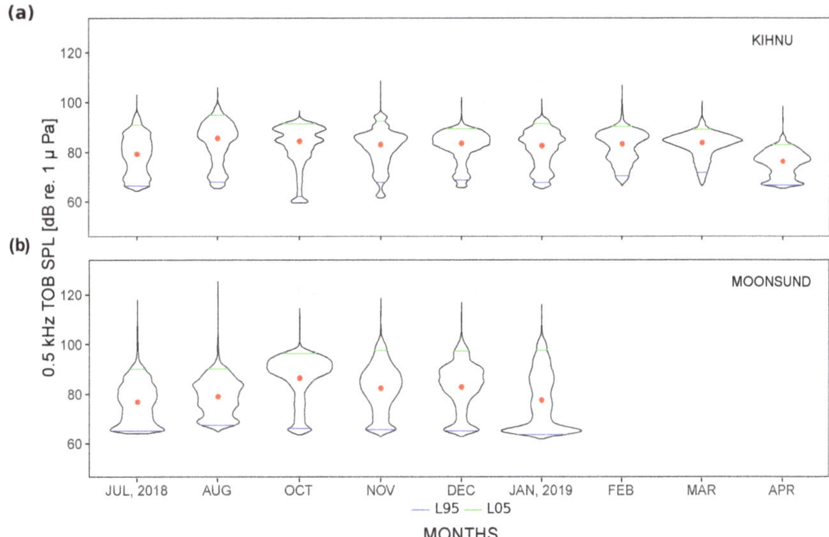

Figure 9. The monthly estimated probability density functions (PDFs) in 500 Hz TOB SPLs: at (**a**) Kihnu and (**b**) Moonsund monitoring location. Red points mark the arithmetic mean values. Blue and green horizontal lines inside the violin plots mark the exceedance levels L95 and L05, respectively.

3.5. Analysis of Co-Occurrence of Ship Traffic Noise and Pinniped Calls

Next, the focus was on the time intervals with overlapping anthropogenic noise and pinniped calls, in order to evaluate the risk of masking the pinnipeds' communication. For the ship traffic, the AIS data, along with the acoustic detection, were used. Comparisons of the hourly detection rates of pinniped calls and the estimated number of ships in a week are depicted in Figures 10 and 11.

3.5.1. Kihnu Monitoring Location

As shown in Figure 10a, in the Kihnu monitoring location, two periods of major biological activity can be observed. During the summer months, the most frequent call type was the moan of the grey seal (Section 3.1). Starting from February, there were numerous detections of the rupe, rup, and forelimb claps. The peak grey seal call detection rate reached 106 calls per hour in March. This drastic increase of detection rates happened during their main mating period, which starts in February and lasts until March [38]. Based on the rates, it can be concluded that the Kihnu monitoring location is an important site for both non-breeding and breeding periods of grey seals. In contrast, almost none of the ringed seal calls were detected in this location.

The Kihnu location has very sparse ship traffic, with only some pleasure and fishing boats each day that appear mainly during the summer months (Figure 10b). Thus, the co-occurrence of biological and anthropogenic sound in this location was extremely rare and, with regard to continuous anthropogenic sound, it can be stated that the Kihnu MPA has a good environmental status.

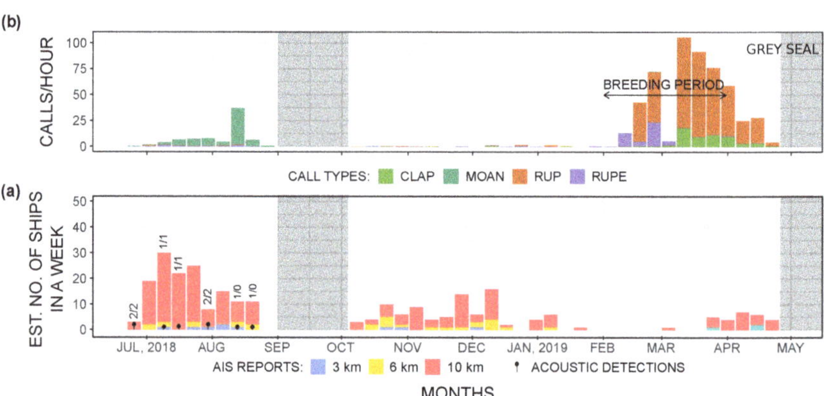

Figure 10. Biophony of (**a**) detected grey seal calls throughout the year. The number of detections increases significantly during the breeding period. The coloured bar charts in (**b**) show the number of ships and their respective ranges based on AIS position reports. The total numbers of the acoustically detected ships are shown by the lollipop chart. Numerical values of the detections written over the number of detected ships without AIS can be seen above the bars. All data are presented on a weekly basis.

3.5.2. Moonsund Monitoring Location

In Moonsund, both ringed and grey seal calls were detected but at much lower rates. Similar to the Kihnu location, the moan of grey seals was the most frequent of the call types in the summer period (Figure 11a). The detection of ringed seals' vocalisation was rare and was mainly found in recordings from the winter period (Figure 11b). It should be noted that the monitoring did not cover the mating periods of the ringed (February or March) [39] and grey seals.

The bar chart in Figure 11c presents the AIS-based number of ships, and the lollipop chart shows the number of acoustic detections. Over 700 ship passages were revealed by the acoustic detection during the whole monitoring period. As expected, the summer months were the busiest, with more than 400 detections. Around one-third of them were ships without AIS. The number of ships drastically decreased during the autumn, resulting in 44 recorded events only. The number of detections started to increase in the winter and specifically during the freezing periods, with weekly detections being constantly above 40 ships in the last three weeks of monitoring. According to the high rates of ringed seal calls and ship traffic (Figure 11b,c), co-occurrences between them were likely to happen.

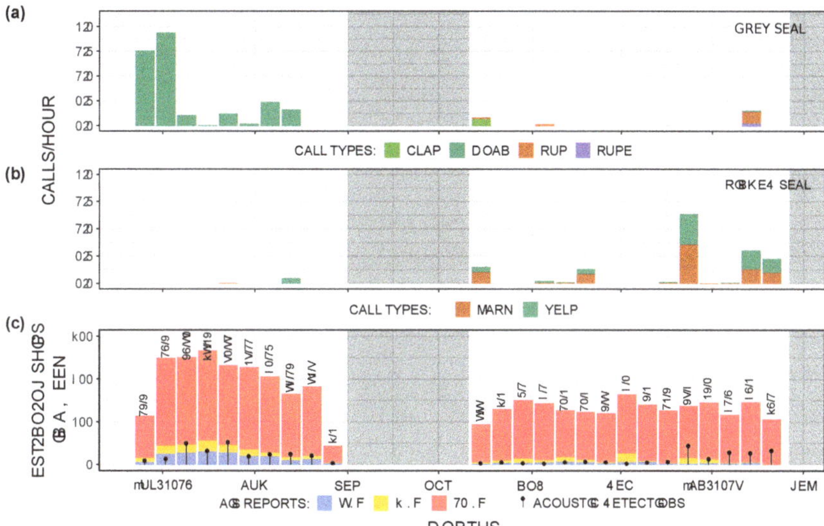

Figure 11. Biophony of (**a**) grey seals and (**b**) ringed seals calls that are detected in the Moonsund monitoring location. The high detection rates of the two pinniped species occurred during two separate seasons. Shown in (**c**) are the coloured bar charts of the number of ships along with their respective distances based on AIS data. The total numbers of detections are shown with a lollipop chart along with the labels. The lollipop chart reveals that the ship noise was detected throughout the year and most frequently during the summer. The rates also start to increase in winter due to the extension of sound propagation ranges. Furthermore, the number of detection relates to no-AIS ships present in the labels. All data is presented on a weekly basis.

3.6. Assessment of the Auditory Masking Potential of the Ringed Seal Calls

According to the monthly PDFs of the SPLs recorded in Moonsund (Figure 9b), the ranges of RLs were within the suggested criteria for not causing the pinnipeds strong disturbance [40]. As a result, injuries to pinniped hearing from continuous anthropogenic noise were very improbable. Thus, as a sudden impact of continuous noise in the monitoring areas, auditory masking was considered.

Auditory masking is defined as "the process by which the threshold of hearing for one sound is raised by the presence of another (masking) sound; and the amount by which the threshold of hearing for one sound is raised by the presence of another (masking) sound, expressed in decibel" [41]. The masking potential is estimated following the steps of the power spectrum model with a critical ratio (CR), as proposed in [11]. The CR is defined as the minimum span of the SPL of an audible tone against a white noise background. Both the hearing capacity (audiogram) and the CR were taken from documented ringed seals hearing tests [42]. In this study, the CR for single intermediate tones was approximated by linear regression.

As it was shown in Section 3.5.2, biological and anthropogenic sounds can occur simultaneously in the Moonsund monitoring location. During the freezing period, the noise from the ferry line propagates over larger distances and can mask the communication signals of the ringed seals. To estimate the masking potential, two case studies were performed, with results shown in Figure 12. It can be seen that the frequency ranges of the ringed seal calls (yelp and bark) and ship noise overlap. For simplicity, only the frequencies with the highest RL were chosen. Figure 12a,b depict the spectrogram and spectrum level plots in the case of the ambient sound level being less than one CR below the audiogram. In this case, the detection of a signal is audiogram limited, and bark with 14 dB excess over the ambient sound was likely to be detected by other seals in the vicinity of the hydrophone.

Figure 12c,d shows the second case study where the yelp signal has 17 dB excess over the ambient sound. By contrast with the previous example, the gap between the ambient sound and audiogram level is less than one CR. Thus, the detection of a signal is limited by ambient sound (ship noise) level. As the yelp is less than one CR above the ambient sound level, it would probably be undetected by the seals close to the hydrophone position.

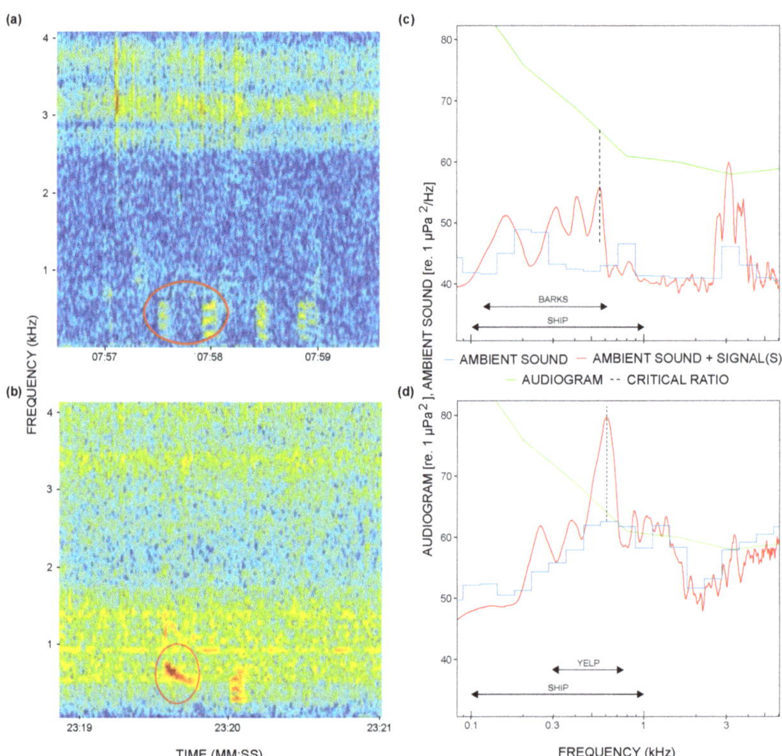

Figure 12. Masking potential estimation of the co-occurrence cases of ringed seal calls with anthropogenic sound from distant shipping. Vocalisations are marked on the spectrograms by red ovals (**a**) bark and (**c**) yelp. On the right side, the spectral overlaps of two masking events are shown. In (**b**), hearing is audiogram limited, and masking does not occur. In (**d**), hearing is shipping-noise limited, and masking is likely. Blue lines show the mean-square sound pressure spectral density level of the ambient sound averaged in TOBs.

The estimated potential masking occurrences due to ship noise are summarised in Table 2. Although the number of co-occurrences of ringed seal vocalisations was quite small, compared to the total numbers of seal detections, a considerable number of them (13 out of 17) have the potential of being masked by the ship traffic noise.

From the above examples, one can deduce that a suitable measure for assessing masking potential is the excess of anthropogenic sound over the natural ambient sound. Even though the source levels and distances to the receivers are unknown, the averaged values of the detected biological signals and natural ambient sound can be compared to evaluate an average excess leading to a potential for auditory masking. To assess the masking potential, we focused on the frequency band that is important for anthropogenic sound (500 Hz TOB) in the area and the CR interpolated value for the pure tone of 500 Hz. The exceedance level L90 was used for representing the natural sound level.

Table 2. Summary of the auditory masking analysis.

Dates/Sea-Ice Concentrations (%)	Call Types	Number of Signals			Ships/CPA (km)
		Co-Occurrences with Ship Noise	Masked (Incl. Ambient Sound Limited)	Not Masked (Incl. Ambient Sound Limited)	
2018-12-29/18	Yelp(s)	1	1(1)	0	Ferry/8.5
2018-12-30/7	Bark(s)	4	0	4(0)	Ferry/8.9
2019-01-10/20	Yelp(s)	1	1(1)	0	Ferry/8.3
	Bark(s)	1	1(1)	0	
2019-01-19/35	Yelp(s)	10	10(10)	0	Ferry/8.5

The result of this analysis is shown in Figure 13. The TOB averaged mean-square sound pressure spectral density level of around 400 RLs of ringed seal calls was compared with the natural ambient sound spectrum for the time period when the calls occurred, as well as with the audiogram. It can be seen that, in the case of an average situation, the signal reception was audiogram limited in the absence of anthropogenic sound. However, 12 dB of an excess over the ambient sound level at 500 Hz would lead to a situation where masking could happen. At such a critical excess level, the anthropogenic sound would raise the ambient sound to the level where it would be just one CR below the RL of the signal, by which its reception could start to be hindered.

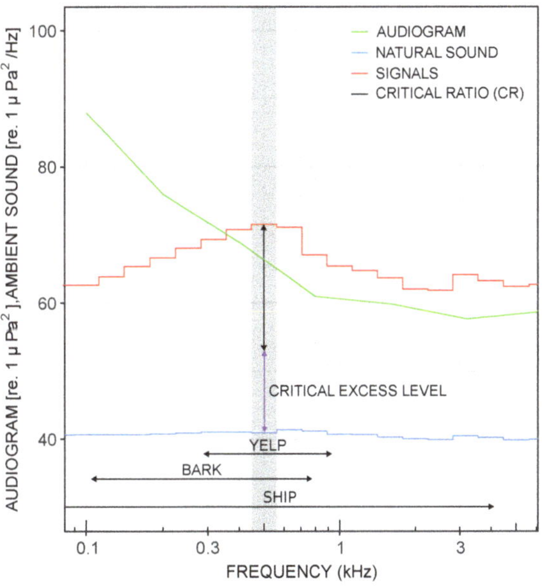

Figure 13. A 12 dB elevation of the ambient sound, called the critical excess level (purple line), can initiate ringed seal call masking. Averaged TOB mean-square sound pressure spectral density level of natural ambient sound and seal calls are shown in blue and red lines. The audiogram of the ringed seals is shown with a green line. The grey area shows the 500 Hz TOB.

3.7. Proportion Estimates for the Anthropogenic Sound

Previous studies [10,43] have proposed key metrics for assessing the proportion of anthropogenic underwater sound using their relative sound levels. For instance, the signal excess is defined as the difference between the RL and the detection threshold. In this study, the signal excess was specified as the difference between the RL and the estimated natural ambient sound level. The natural ambient sound was estimated by calculating the exceedance level L90 [44] for time periods without anthropogenic sound.

Figure 14 shows sound excess level PDFs for 500 Hz TOB recorded in Moonsund for two selected weeks in summer (23–29 July 2018) and winter (24–30 December 2018). In summer (Figure 14a), the range of sound excess varied between −3 and 48 dB, with the major portion of sound excess being slightly above 0 dB for 50% of the time. A higher excess level of 7 dB over the natural sound level occurred for only 10% of the time.

In winter (Figure 14b), the sound excess distribution was practically the same as in summer but with slightly higher levels. Considering the proposed critical excess level of 12 dB, it can be stated that ringed seals were at risk of communication masking by the elevated ambient sound for 8% of the time or approximately for 13 h in a week.

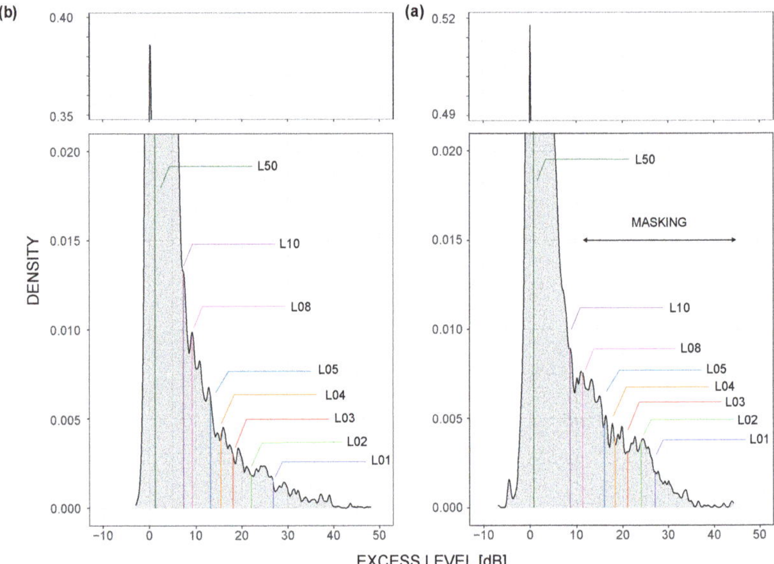

Figure 14. Weekly PDFs of 500 Hz TOB sound excess levels recorded in Moonsund for two periods in (**a**) summer and (**b**) winter. The risk of masking occurs less than 8% of the time when it exceeds the estimated natural ambient sound by 12 dB. The excess level of 0 dB means that the RL coincided with the estimated natural ambient sound level.

4. Conclusions

The underwater ambient sounds from two sound monitoring locations in the MPAs of the Gulf of Riga were analysed. The analysis of the PAM data revealed the presence of both anthropophony and biophony in the soundscape. That offered a possibility to assess the temporal and spectral overlaps of these components and, in particular, to assess the potential for auditory masking of the pinniped calls by anthropogenic noise.

The detection and identification of biological sounds in the PAM data revealed the presence of grey and ringed seals in the vicinity of the monitoring sites. Various types of seal vocalisations were detected. For the grey seals, mainly the guttural rupe, rup, together with forelimb claps in the breeding period and moan in other time periods, were recorded. Even though ringed seals vocalise less than grey seals, their acoustical presence was revealed in Moonsund. The bark and yelp of ringed seals were recorded throughout the monitoring periods. The highest detection rates were found with the formation of the ice cover.

Acoustic detection of shipping noise confirmed the very low shipping activity in the Kihnu location, where the soundscape is largely dominated by natural sounds. Slightly higher shipping activity in Moonsund contributed also to the anthropophony of the soundscape. In summer, the main sources of the anthropogenic noise were pleasure boats, and in winter, distant ferry boats. Long-range detection of the ferries was made possible by the presence of ice cover. The under-ice upward sound refraction and low ambient noise level significantly reduced the propagation loss, thus making the detection of ship noise possible at distances of up to 10 km. Therefore, the effects of ice cover should be considered when assessing the impact of anthropogenic sound on the shallow sea marine environment.

Analysis of the recorded ship spectra in Moonsund showed that they contribute more noise in frequencies higher than 63 Hz or 125 Hz TOBs. Consequently, the 500 Hz TOB was chosen as an indicator frequency band for the anthropogenic noise in the monitoring area. The excess level higher than 12 dB within this frequency band can lead to communication masking for the ringed seal. However, even during the "noisiest" weeks, this risk of masking occurred for a quite small fraction of the time (8%). Based on this assessment, the environmental risks related to the anthropogenic sound around the monitoring sites can be considered as low.

Author Contributions: Conceptualisation, M.S.P. and A.K.; methodology, M.S.P. and A.K.; investigation, M.S.P. and A.K.; data curation, M.M. and A.K.; writing—original draft preparation, M.S.P.; writing—review and editing, M.S.P., M.M., and A.K.; supervision, A.K.; funding acquisition, A.K. All authors have read and agreed to the published version of the manuscript.

Funding: Support from the Estonian Environmental Investment Centre (KIK) is gratefully acknowledged.

Institutional Review Board Statement: Not applicable.

Informed Consent Statement: Not applicable.

Data Availability Statement: Not applicable.

Acknowledgments: Alar Siht from the Transport Administration kindly provided the AIS data. We express our thanks to marine biologists Mart Jüssi and Ivar Jüssi from NGO Pro Mare for sharing their vast knowledge about seal habitats.

Conflicts of Interest: The authors declare no conflict of interest.

References

1.	Directive, M.S.F. Directive 2008/56/EC of the European Parliament and of the Council of 17 June 2008 establishing a framework for community action in the field of marine environmental policy. *Off. J. Eur. Union L* **2008**, *164*, 19–40.
2.	Decision, E.C. Decision 2017/848/EC of the European Commission of 17 May 2017 laying down criteria and methodological standards on good environmental status of marine waters and specifications and standardised methods for monitoring and assessment, and repealing. *Off. J. Eur. Union L* **2017**, *125*, 43–74.
3.	Haver, S.M.; Fournet, M.E.H.; Dziak, R.P.; Gabriele, C.; Gedamke, J.; Hatch, L.T.; Haxel, J.; Heppell, S.A.; McKenna, M.F.; Mellinger, D.K.; et al. Comparing the Underwater Soundscapes of Four U.S. National Parks and Marine Sanctuaries. *Front. Mar. Sci.* **2019**, *6*, 500. [CrossRef]
4.	Mustonen, M.; Klauson, A.; Andersson, M.; Clorennec, D.; Folegot, T.; Koza, R.; Pajala, J.; Persson, L.; Tegowski, J.; Tougaard, J.; et al. Spatial and Temporal Variability of Ambient Underwater Sound in the Baltic Sea. *Sci. Rep.* **2019**, *9*. [CrossRef] [PubMed]
5.	Pijanowski, B.C.; Villanueva-Rivera, L.J.; Dumyahn, S.L.; Farina, A.; Krause, B.L.; Napoletano, B.M.; Gage, S.H.; Pieretti, N. Soundscape Ecology: The Science of Sound in the Landscape. *BioScience* **2011**, *61*, 203–216. [CrossRef]
6.	Haxel, J.H.; Dziak, R.P.; Matsumoto, H. Observations of shallow water marine ambient sound: The low frequency underwater soundscape of the central Oregon coast. *J. Acoust. Soc. Am.* **2013**, *133*, 2586–2596. [CrossRef] [PubMed]
7.	Ma, B.B.; Nystuen, J.A.; Lien, R.C. Prediction of underwater sound levels from rain and wind. *J. Acoust. Soc. Am.* **2005**, *117*, 3555–3565. [CrossRef] [PubMed]
8.	Bittencourt, L.; Barbosa, M.; Bisi, T.L.; Lailson-Brito, J.; Azevedo, A.F. Anthropogenic noise influences on marine soundscape variability across coastal areas. *Mar. Pollut. Bull.* **2020**, *160*, 111648. [CrossRef]
9.	Payne, R.; Webb, D. Orientation by means of long range acoustic signaling in baleen whales. *Ann. N. Y. Acad. Sci.* **1971**, *188*, 110–141. [CrossRef] [PubMed]
10.	Clark, C.W.; Ellison, W.T.; Southall, B.L.; Hatch, L.; Van Parijs, S.M.; Frankel, A.; Ponirakis, D. Acoustic masking in marine ecosystems: Intuitions, analysis, and implication. *Mar. Ecol. Prog. Ser.* **2009**, *395*, 201–222. [CrossRef]

11. Erbe, C.; Reichmuth, C.; Cunningham, K.; Lucke, K.; Dooling, R. Communication masking in marine mammals: A review and research strategy. *Mar. Pollut. Bull.* **2016**, *103*, 15–38. [CrossRef]

12. Williams, R.; Wright, A.; Ashe, E.; Blight, L.; Canessa, R.; Clark, C.; Cullis-Suzuki, S.; Dakin, D.; Erbe, C.; Hammond, P.; et al. Impacts of anthropogenic noise on marine life: Publication patterns, new discoveries, and future directions in research and management. *Ocean Coast Manag.* **2015**, *115*, 17–24. [CrossRef]

13. Mustonen, M.; Klauson, A.; Folégot, T.; Clorennec, D. Natural sound estimation in shallow water near shipping lanes. *J. Acoust. Soc. Am.* **2020**, *147*, EL177–EL183. [CrossRef] [PubMed]

14. Prawirasasra, M.S.; Mustonen, M.; Klauson, A. Underwater monitoring of pinniped vocalizations in the Gulf of Riga. *Proc. Meet. Acoust.* **2019**, *37*, 070015. [CrossRef]

15. Wahlberg, M.; Westerberg, H. Sounds produced by herring (*Clupea harengus*) bubble release. *Aquat. Living Resour.* **2003**, *16*, 271–275. [CrossRef]

16. Vester, H.I.; Folkow, L.P.; Blix, A.S. Click sounds produced by cod (*Gadus morhua*). *J. Acoust. Soc. Am.* **2004**, *115*, 914–919. [CrossRef]

17. Mizuguchi, D.; Tsunokawa, M.; Kawamoto, M.; Kohshima, S. Underwater vocalizations and associated behavior in captive ringed seals (*Pusa hispida*). *Polar Biol.* **2016**, *39*, 659–669. [CrossRef]

18. Asselin, S.; Hammill, M.O.; Barrette, C. Underwater vocalizations of ice breeding grey seals. *Can. J. Zool.* **1993**, *71*, 2211–2219. [CrossRef]

19. Jones, J.M.; Thayre, B.J.; Roth, E.H.; Sia, M.M.I.; Merculief, K.; Jackson, C.; Zeller, C.; Clare, M.; Bacon, A.; Weaver, S.; et al. Ringed, Bearded, and Ribbon Seal Vocalizations North of Barrow, Alaska: Seasonal Presence and Relationship with Sea Ice. *Arctic* **2014**, *67*, 203–222. [CrossRef]

20. Oksanen, S.M.; Niemi, M.; Ahola, M.P.; Kunnasranta, M. Identifying foraging habitats of Baltic ringed seals using movement data. *Mov. Ecol.* **2015**, *3*, 33. [CrossRef]

21. Härkönen, T.; Stenman, O.; Jüssi, M.; Jüssi, I.; Sagitov, R.; Verevkin, M. Population size and distribution of the Baltic ringed seal (*Phoca hispida botnica*). *NAMMCO Annu. Rep.* **1998**, *1*, 167. [CrossRef]

22. Wildlife Acoustics. User Manual Supplement. Available online: https://www.wildlifeacoustics.com/uploads/user-guides/SM2M-User-Manual.pdf (accessed on 1 August 2019).

23. Ocean Instruments. SoundTrap ST500 User Guide. Available online: http://www.oceaninstruments.co.nz/wp-content/uploads/2018/03/ST500-User-Guide.pdf (accessed on 1 August 2019).

24. Betke, K.; Folegot, T.; Matuschek, R.; Pajala, J.; Persson, L.; Tegowski, J.; Tougaard, J.; Wahlberg, M. BIAS Standards for Signal Processing. Aims, Processes and Recommendations, Amended Version. Available online: https://biasproject.files.wordpress.com/2016/01/bias_sigproc_standards_v5_final.pdf (accessed on 15 September 2019).

25. Merchant, N.D.; Fristrup, K.M.; Johnson, M.P.; Tyack, P.L.; Witt, M.J.; Blondel, P.; Parks, S.E. Measuring acoustic habitats. *Methods Ecol. Evol.* **2015**, *6*, 257–265. [CrossRef]

26. Stirling, I. Vocalization in the Ringed Seal (*Phoca hispida*). *Can. J. Fish. Aquat.* **1973**, *30*, 1592–1594. [CrossRef]

27. Rautio, A.; Niemi, M.; Kunnasranta, M.; Holopainen, I.J.; Hyvärinen, H. Vocal repertoire of the Saimaa ringed seal (*Phoca hispida saimensis*) during the breeding season. *Mar. Mamm. Sci.* **2009**, *25*, 920–930. [CrossRef]

28. Schusterman, R.J.; Balliet, R.F.; St. John, S. Vocal displays under water by the gray seal, the harbor seal, and the stellar sea lion. *Psychon. Sci.* **1970**, *18*, 303–305. [CrossRef]

29. McCulloch, S. The Vocal Behaviour of the Grey Seal (*Halichoerus grypus*). Ph.D. Thesis, University of St. Andrews, St. Andrews, Scotland, UK, 2000.

30. Gallus, A.; Dähne, M.; Verfuß, U.K.; Bräger, S.; Adler, S.; Siebert, U.; Benke, H. Use of static passive acoustic monitoring to assess the status of the 'Critically Endangered' Baltic harbour porpoise in German waters. *Endanger. Species Res.* **2012**, *18*, 265–278. [CrossRef]

31. K. Lisa Yang Center for Conservation Bioacoustics. *Raven Pro: Interactive Sound Analysis Software (Version 1.5) [Computer software]*; The Cornell Lab. of Ornithology: Ithaca, NY, USA, 2014.

32. Hermannsen, L.; Mikkelsen, L.; Tougaard, J.; Beedholm, K.; Johnson, M.; Madsen, P.T. Recreational vessels without Automatic Identification System (AIS) dominate anthropogenic noise contributions to a shallow water soundscape. *Sci. Rep.* **2019**, *9*, 15477. [CrossRef]

33. Merchant, N.D.; Witt, M.J.; Blondel, P.; Godley, B.J.; Smith, G.H. Assessing sound exposure from shipping in coastal waters using a single hydrophone and Automatic Identification System (AIS) data. *Mar. Pollut. Bull.* **2012**, *64*, 1320–1329. [CrossRef] [PubMed]

34. Hocking, D.P.; Burville, B.; Parker, W.M.G.; Evans, A.R.; Park, T.; Marx, F.G. Percussive underwater signaling in wild gray seals. *Mar. Mamm. Sci.* **2020**, *36*, 728–732. [CrossRef]

35. Aide, T.M.; Corrada-Bravo, C.; Campos-Cerqueira, M.; Milan, C.; Vega, G.; Alvarez, R. Real-time bioacoustics monitoring and automated species identification. *PeerJ* **2013**, *1*, e103. [CrossRef]

36. Urick, R.J. *Principles of Underwater Sound*, 3rd ed.; McGraw-Hill Publishing Company: Westport, CT, USA, 1983.

37. Jensen, F.B.; Kuperman, W.A.; Porter, M.B.; Schmidt, H. *Computational Ocean Acoustics*; Springer: New York, NY, USA, 2011. [CrossRef]

38. Hook, O.; Johnels, A.G.; Matthews, L.H. The breeding and distribution of the grey seal (*Halichoerus grypus*) in the Baltic Sea, with observations on other seals of the area. *Proc. R. Soc. Lond.* **1972**, *182*, 37–58. [CrossRef]

39. Helle, E. *Hylkeiden Elämää. [Seal Life]*; Kirjayhtymä: Helsinki, Finland, 1983. (In Finnish)
40. Southall, B.L.; Bowles, A.E.; Ellison, W.T.; Finneran, J.J.; Gentry, R.L.; Greene, C.R., Jr.; Kastak, D.; Ketten, D.R.; Miller, J.H.; Nachtigall, P.E.; et al. Marine mammal noise exposure criteria: Initial scientific recommendations. *Aquat. Mamm.* **2007**, *33*, 411. [CrossRef]
41. American National Standards Institute. *Bioacoustical Terminology (ANSI S3.20—1995, R 2008)*; Acoustical Society of America: New York, NY, USA, 2008.
42. Sills, J.M.; Southall, L.B.; Reichmuth, C. Amphibious hearing in ringed seals (*Pusa hispida*): Underwater audiograms, aerial audiograms and critical ratio measurements. *J. Exp. Biol* **2015**, *218*, 2250–2259. [CrossRef] [PubMed]
43. Ellison, W.; Southall, B.; Clark, C.; Frankel, A. A new context-based approach to assess marine mammal behavioral responses to anthropogenic sounds. *Conserv Biol.* **2012**, *26*, 21–28. [CrossRef]
44. Gervaise, C.; Simard, Y.; Roy, N.; Kinda, B.; Ménard, N. Shipping noise in whale habitat: Characteristics, sources, budget, and impact on belugas in Saguenay–St. Lawrence Marine Park hub. *J. Acoust. Soc. Am.* **2012**, *132*, 76–89. [CrossRef]

Journal of
Marine Science and Engineering

Article

Hindcasting Soundscapes before and during the COVID-19 Pandemic in Selected Areas of the North Sea and the Adriatic Sea

Hüseyin Özkan Sertlek

Department of Hydraulic Engineering, Section of Offshore Engineering, Delft University of Technology, Stevinweg 1, 2628 CN Delft, The Netherlands; h.o.sertlek@tudelft.nl

Abstract: The national measures in several European countries during the COVID-19 pandemic also affected offshore human activities, including shipping. In this work, the temporal and spatial variations of shipping sound are calculated for the years before and during the pandemic in selected shallow water test areas from the Southern North Sea and the Adriatic Sea. First, the monthly sound pressure level maps of ships and wind between 2017 and 2020 are calculated for frequencies between 100 Hz to 10 kHz. Next, the monthly changes in these maps are compared. The asymptotic approximation of the hybrid flux-mode propagation model reduces the computational requirements for sound mapping simulations and facilitates the production of a large number of sound maps for different months, depths, frequencies, and ship categories. After the strictest COVID-19 measures were applied in April 2020, the largest decline was observed for the fishing, passenger and recreational ships. Although the changes in the number of fishing vessels are large, their contribution to the soundscape is minor due to their low source level. In both test areas, the spatial exceedance levels and acoustic energies were decreased in 2020 compared to the average of the previous three years.

Keywords: ship noise; sound mapping; acoustic propagation

Citation: Sertlek, H.Ö. Hindcasting Soundscapes before and during the COVID-19 Pandemic in Selected Areas of the North Sea and the Adriatic Sea. *J. Mar. Sci. Eng.* **2021**, *9*, 702. https://doi.org/10.3390/jmse9070702

Academic Editors: Michel André and Christine Erbe

Received: 30 May 2021
Accepted: 22 June 2021
Published: 26 June 2021

Publisher's Note: MDPI stays neutral with regard to jurisdictional claims in published maps and institutional affiliations.

1. Introduction

Anthropogenic underwater sound creates potential risk for marine life with its possible effects on communication, prey–predator relations, behavioral changes, and temporary and permanent effects on hearing [1–3]. Underwater noise international regulations [4–6] aim to avoid the potential impact of the underwater sound caused by anthropogenic activities. Sound maps are a cost-effective tool to evaluate and monitor the underwater sound characteristics over large areas, as they do not require large-scale measurements [7–9]. The North Sea and Adriatic Sea involve busy traffic lanes of various ship types [10]. Thus, it is crucial to monitor [11,12] and predict underwater sound, which is sensitive to temporal and spatial variations of the sound sources [13].

The offshore human activities' temporal and spatial patterns were affected by the precautions against the spread of the COVID-19 due to the radical changes in transportation, tourism, fisheries, cargo, and constructions. According to March et al. [14], the number of ships and shipping densities at sea based on AIS data of the global ocean decreased after the WHO Pandemic Declaration on 11 March 2020 [15]. Based on statistics of port calls in the EU, the average decline over the first 49 weeks of 2020 compared to 2019 is 12.3 %, with the highest decline in chemical tankers, cruise, and passenger ships, especially around Croatia, Iceland, Slovenia, and Spain [16]. Another study also confirmed that the passenger ships were the most affected category during the pandemic in 2020, followed by the container ships [17]. The changes in the shipping traffic naturally led to changes in the shipping sound. The sound pressure levels around the Port of Vancouver decreased 1.5 dB due to the reduction in shipping traffic [18]. On the Oregon coast, the measurement showed the sound pressure level at 63 Hz third-octave band reduced in the spring of 2020 by about

1.6 dB compared to the prior five years [19]. In addition to the COVID-19 pandemic, other factors such as the oil crisis and Brexit in 2020 could influence the cargo and tanker shipping densities in the North Sea [14]. Because the ships are the most significant contributor to the North Sea underwater soundscape [13], a detailed investigation of shipping sound is essential for the pandemic-related changes in the shipping density.

In this work, the shipping sound is investigated before and during the pandemic for the selected shallow water test areas from the North Sea and the Adriatic Sea. First, the monthly averaged shipping densities [10] are investigated for the chosen years from January 2017 to December 2020. The shipping densities are used as an input to calculate the average source level depending on the number of ships per unit area. Next, the sound propagation is modeled with the asymptotic approach of hybrid mode-flux theory, which has comparable accuracy to adiabatic mode theory (within 1–1.5 dB) without requiring long computational time [20]. The adiabatic mode theory is usually used for the slowly varying bathymetry and environmental parameters. The computational efficiency of the proposed method allows the generation of a large number of sound maps with detailed spatial and spectral resolution without long computational times. Finally, the selected test areas' monthly shipping and wind sound maps are calculated for 21 center decidecade frequencies from 100 Hz to 10 kHz. The comparison between these maps provides an insight into the temporal changes in the shipping sound before and during the pandemic. In Section 2, the model inputs such as the shipping density, bathymetry and source level are introduced. Next, the proposed underwater propagation model and its advantages for sound mapping are described in Section 3. Based on the model inputs and propagation model, sound maps for the selected test areas are calculated and compared before and during the pandemic in Section 4. Finally, the conclusions are discussed in Section 5.

2. Model Inputs: Shipping Densities and Bathymetry in the Test Areas

For the investigation of changes in sound pressure levels during the pandemic, two test areas were chosen. The first test area involves the entire Dutch, Belgian and German Exclusive Economic Zones in the Southern North Sea, and part of the Danish, British and French Exclusive Economic Zones. The second area is around the Gulf of Venice in the Adriatic Sea. These areas were chosen because of their high shipping density and shallow water depth. These two test areas and their bathymetries are shown in Figure 1 [21]. The ETRS89-Extended/LAEA Europe coordinate system (EPSG:3035) was used to calculate the compatibility with the EAA datasets [22].

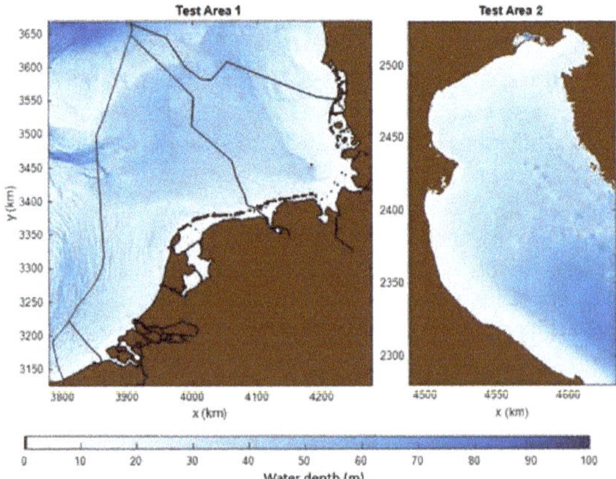

Figure 1. The selected test areas and their bathymetries. Test area 1 (the Dutch, Belgian and German Exclusive Economic Zones in the Southern North Sea) and test area 2 (The Gulf of Venice). Test area 1 partly involves the Danish, French and English Exclusive Zones.

Shipping density maps based on AIS data provide the temporal and spatial distribution for various ship categories. The number of ships in the world fleet has increased in previous years [23,24] and is expected to increase in the future [25]. However, the number of ships is affected due to the various measures in EU countries during the pandemic. First, we analyzed the total number of ships three years earlier than the pandemic (in 2017, 2018 and 2019) and during the pandemic in 2020. Figure 2 shows the monthly variation of the total number of ships from the different categories in test areas 1 and 2.

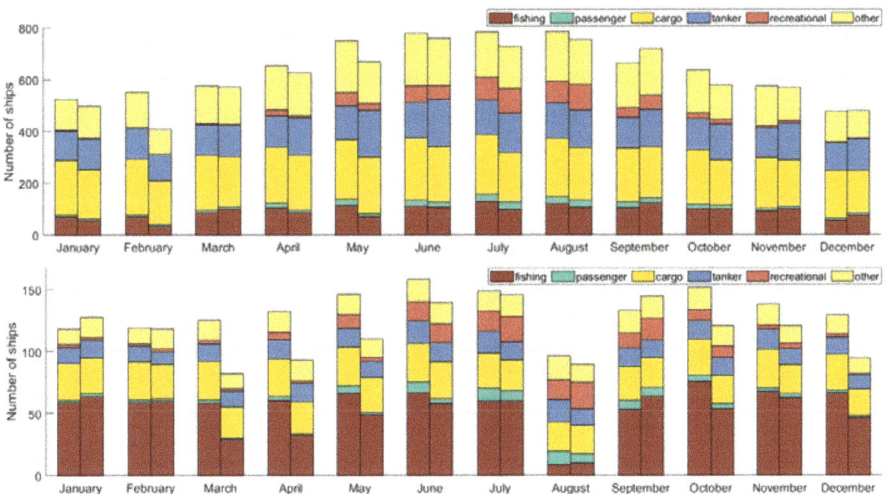

Figure 2. The total number of ships in test area 1 (**upper**) and test area 2 (**lower**). The left bar of each month is for the 3-year average (from 2017 to 2019). The right bar of each month is for 2020. The different colors in the bars correspond to different ship categories.

The composition of ship densities differs between the test areas (Figure 2). Cargo ships and fishing vessels form the largest proportion in test areas 1 and 2, respectively. The smallest number of ships in test area 1 is observed in February 2020, with a noticeable decline in the tanker, cargo and fishing ships. This seems related to high wind speed due to the storms Ciara [26] and Dennis [27], which affected shipping traffic in the North Sea. In area 2, the smallest number of ships is observed in March 2020. Furthermore, fishing is prohibited in some areas during August, which leads to a smaller number of fishing vessels in that month [28]. As it can be noticed from the comparisons between test areas, monthly variations of the number of ships are different during the pandemic. This difference could be related to the composition of shipping densities, size of the areas and local measures against the pandemic in 2020. In test area 2, Italy was one the most affected countries and applied the first strict measures against the pandemic. In test area 2, a noticeable decline in the fishing vessels (which have the largest proportion of the number of ships in test area 2) compared to the prior years was observed in March 2020. Test area 1 is relatively larger than test area 2. Furthermore, test area 1 involves busy traffic lanes in the exclusive economic zones of different countries such as the Netherlands, Germany, Belgium, France, Denmark and the United Kingdom. Furthermore, the cargo and tanker ships with the largest proportion of ships in test area 1 have a lower proportional decline than the fishing vessels due to the pandemic. Thus, the impact of the national measures was not easily visible, as observed in test area 2.

The wind speed corresponding to the same months is shown in Figure 3. The wind speed dataset is based on the measurements that are freely accessible from the data portal of the Royal Netherlands Meteorological Institute (KNMI) [29] and were extrapolated over the sea by the analysis tools in the same data portal.

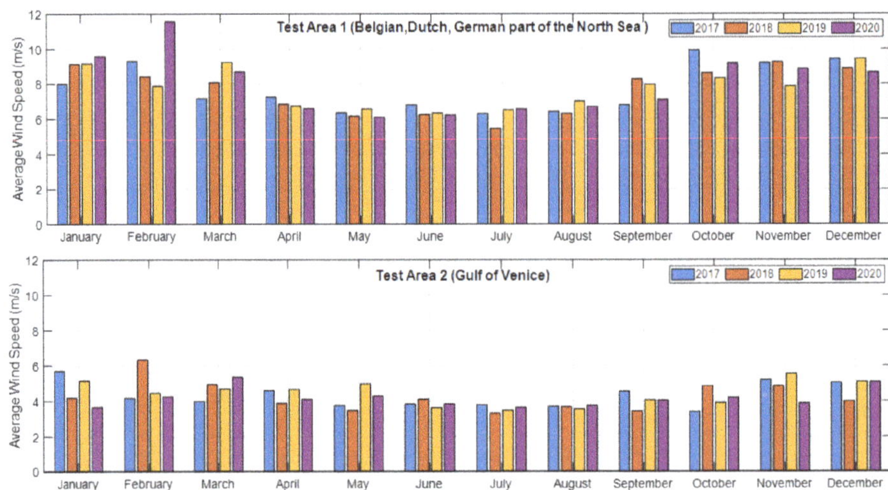

Figure 3. Monthly variation of average wind speed in test areas 1 and 2 for 2017, 2018, 2019 and 2020.

The average wind speed in test area 1 was very high (11.8 m/s) during February 2020. These extreme weather conditions affected the shipping traffic for the corresponding month. The wind speed's seasonal change can be seen in Figure 3, as low wind speed in the summer and high wind speed in the winter months.

When the strictest measures were taken in April 2020 by the European countries, the number of ships and their spatial distribution were changed. Figure 4 shows the shipping density maps of April (3-year average) and April 2020 [10]. The 3-year average is the average of the shipping densities' in April 2017, April 2018 and April 2019.

When we compared the April average of shipping density, the monthly averaged density of all ships decreased by approximately 8% and 13% in test areas 1 and 2, respectively. The differences between the selected ship types are visualized in Figure 5.

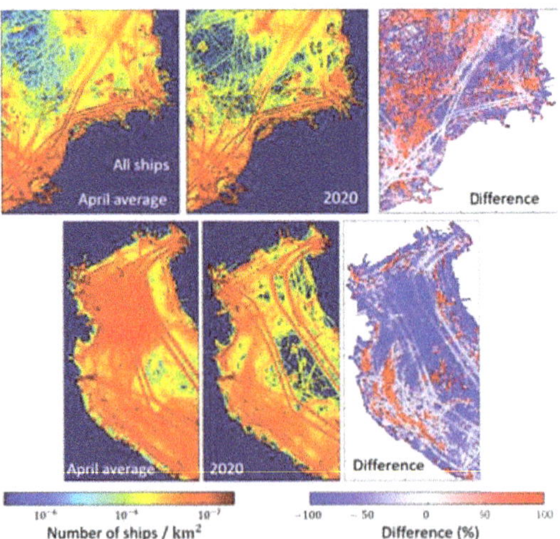

Figure 4. The shipping densities (number of ships per km^2) of the test area 1 (**upper**) and test area 2 (**below**) for April Average (over April 2017, 2018 and 2019) and April 2020.

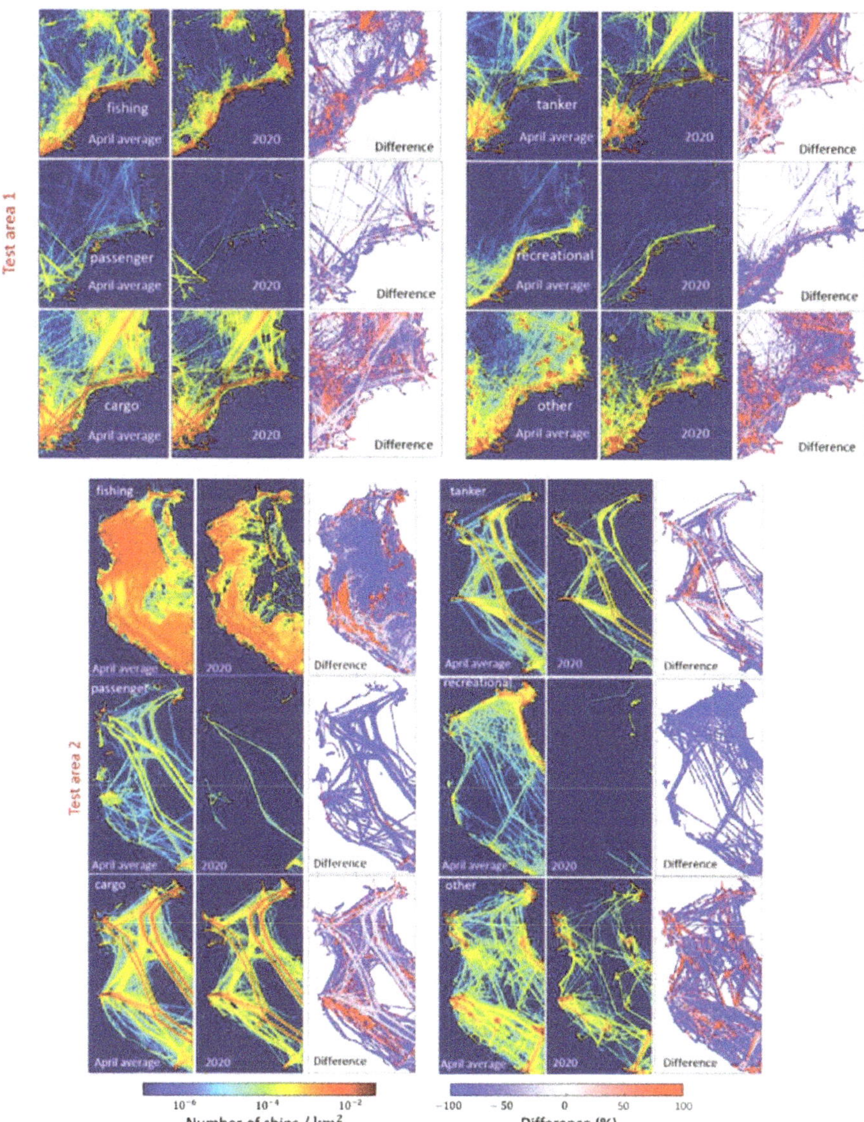

Figure 5. The shipping densities (number of ships per km^2) of the test area 1 (**upper**) and test area 2 (**below**) for April Average (over April 2017, 2018 and 2019) and April 2020.

Comparing the shipping densities of April Average and April 2020, each category has a different change during the pandemic. The fishing, passenger and recreational ships decreased most during April 2020 in both test areas. The total number of tankers slightly increased, related to increasing crude oil exports during the record falls in oil prices [14]. Shipping density-based sound maps can provide practical information related to the ship sound in the test areas. For more accurate sound maps, AIS-based sound maps can be calculated, including ship speed and length information. The AIS-based sound maps should be repeated for each time-snapshot to have an insight into the monthly and annual variation of the ship noise by requiring large computational time, disk space and memory. Alternatively, sound maps based on shipping density input require less computation. The balance between accuracy and computational load should be decided depending on the problem type. In this work, the long-term changes (from 2017 to 2020) are analyzed. Thus, the shipping density-based maps are calculated for practical reasons. In the shipping density dataset of EMODNET used here, all ships with AIS signals are included. However, some stationary ships with an active AIS such as passenger ships or service vessels may stand still for several hours close to the platforms and ports. In the calculations, these ships are first identified by comparing the available AIS data from the previous years. The locations of the stationary ships around the ports and platforms are estimated from these comparisons. For instance, some of the passenger ships waiting around the ports instead of their regular routes are assumed to be stationary based on these comparisons. The locations where the stationary ships are expected are excluded from the shipping densities. There may be more stationary ships far from the ports, platforms, and seashores that cannot be easily identified without the speed information; thus, the AIS-based sound maps that consider the speed and length of the vessels could provide more detailed results. However, AIS-based sound maps require more computation than shipping density-based sound maps and when a long-term trend is investigated, as in this study, over a large area.

The average source level of the ships was modified according to the number of ships in the ship source location; that information is obtained from the shipping density maps. A source-level that takes into account ship speeds, type and lengths [30] was used based on the measurements during the Joint North Sea Monitoring Program (JOMOPANS) [11]. In Figure 6, the source level for the different ship categories is shown. The average speed and length values were used for the different categories since the used shipping density dataset does not have these details.

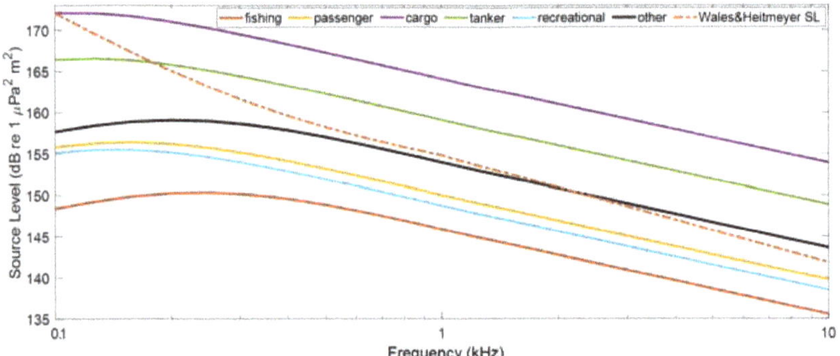

Figure 6. Source level of different ship categories based on the formulas from MacGillivray and de Jong [30] and Wales and Heitmeyer [31].

Due to the use of different source levels in each category, the contribution of the different ship categories to the underwater sound pressure levels are different. Some of the previous works of sound maps [13] often uses the Wales and Heitmeyer source level [31], which is based on the measurements of 272 merchant vessels in the Mediterranean Sea and Eastern Atlantic Ocean over 7 years (from 1985 to 1992). However, the Wales and Heitmeyer formula neglects the vessel speed and length, which can be critical for the sound mapping simulation. For instance, there is a large difference between the Wales–Heitmeyer source level and that of fishing ships, which may lead to misleading results if we would use the Wales and Heitmeyer formula for all ship categories, as described in Appendix B, especially for test area 2.

3. Sound Propagation Modeling

The method for modeling the underwater sound propagation can be chosen depending on the frequency range, water depth and environmental conditions (such as sediment type, sound speed profiles, etc.). For large-scale problems, the number of sources and receivers also plays a critical role in choosing a convenient model to find a balance between significant computational expenses and the accuracy. To overcome computational resource problems, a hybrid method based on normal mode and flux theories [20], called SOPRANO (Sound Propagation Algorithm for Noise Mapping), brings together the accuracy of the adiabatic range dependent normal mode and the speed of Weston's flux theories for shallow water propagation problems [32]. SOPRANO considers bathymetric variations and range-dependent sediment properties. The accuracy of SOPRANO was verified against a detailed multi-model comparison based on the propagation loss calculations of various methods (adiabatic mode theory, coupled modes, ray tracing, parabolic equation, and flux theory) [20] and compared with the measurements for shipping [33] and explosions [34]. These benchmark studies showed that the propagation loss (PL) for a variable sediment type and bathymetry is similar in accuracy to the adiabatic mode theory. SOPRANO is freely available for research purposes from the underwater acoustic simulation tools webpage of TU Delft [35]. Despite the computational benefits of SOPRANO, sound mapping simulations can sometimes require a more practical model than SOPRANO. For these simulations, alternatively, an asymptotic approximation of SOPRANO's is preferred for fast calculations in the mode-stripping region, especially at high frequencies when the mode-contribution is relatively small. M-SOPRANO is another implementation of hybrid mode-flux formulation, where the solution of flux integral is replaced with its asymptotic approximation (as described in Equation (13) in Sertlek et al. [20]). The formulation is described in detail in Appendix A. In Figure 7, propagation loss (PL) calculations of different methods are compared for a realistic North Sea water depth profile. Results obtained by SOPRANO, M-SOPRANO and KRAKENC with the adiabatic approximation were considered. Comparisons were made for 250 Hz, 1 kHz and 2.5 kHz, to a maximum range of 100 km. The depth and range resolutions were chosen as 0.1 and 5 m, respectively, while the source depth is 6 m.

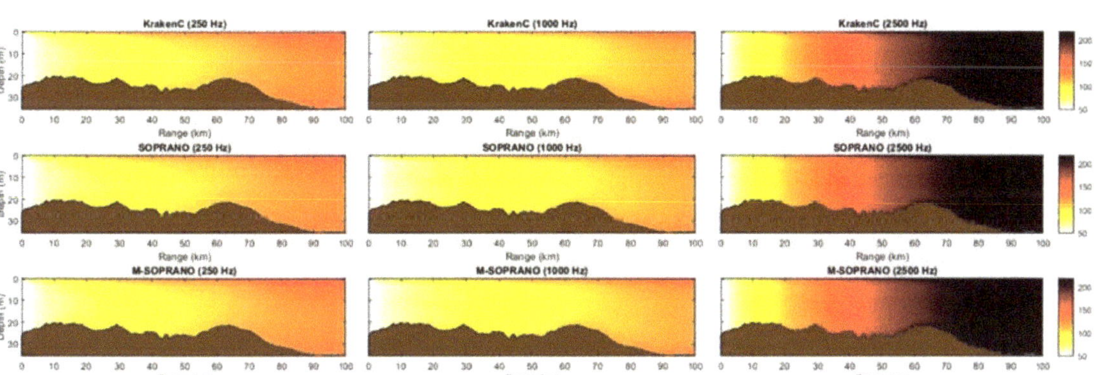

Figure 7. Comparisons for the propagation loss (dB re 1 m^2) at 250 Hz, 1 kHz and 2.5 kHz with KRAKENC (**upper**), SOPRANO (**middle**) and M-SOPRANO (**lower**).

The differences between KRAKENC and SOPRANO are less than 0.9 dB, except for the receiver close to the seabed for the selected bathymetry. KRAKENC uses a stair-step approximation, which differs from the piecewise linear approximation of SOPRANO. Thus, the differences between KRAKENC and SOPRANO results increase (1.8 dB for 0.5 m above the sediment) when the receiver points are close to the seabed. The differences between SOPRANO and M-SOPRANO were investigated with systematic tests, including various frequencies, receiver depths and bathymetry slices. Figure 8 shows propagation loss vs. frequency comparisons for specified ranges between 500 m and 100 km. Mean-square sound pressure is averaged over the water depth (dz = 0.01 m).

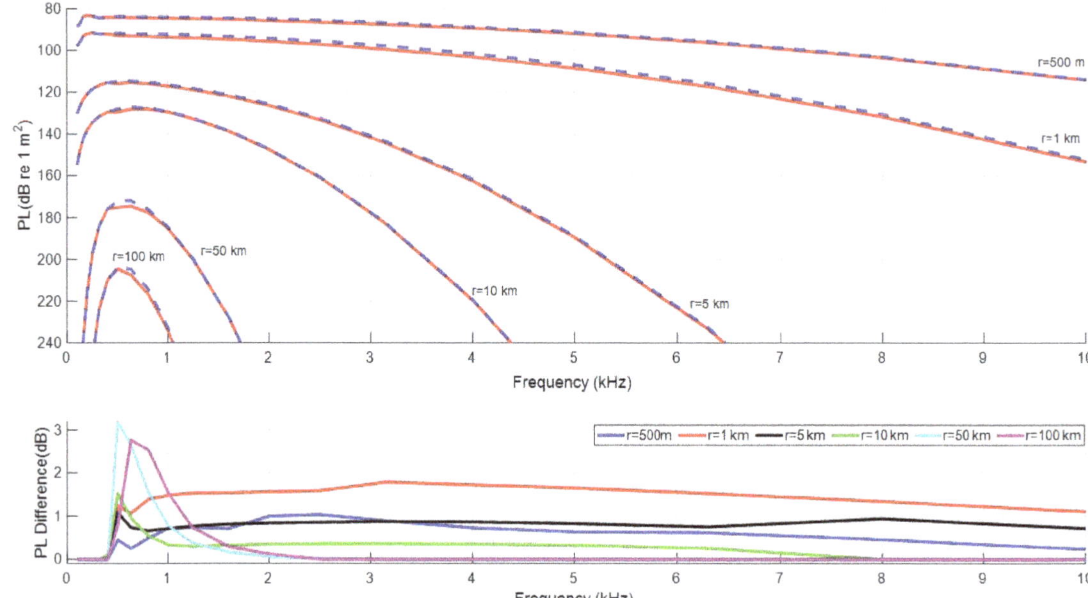

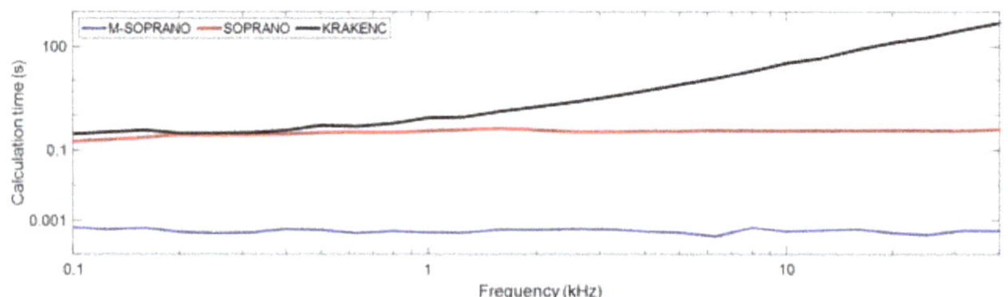

Figure 8. PL vs. frequency comparisons for the different distances over a range-dependent bathymetry. The dashed blue curve is M-Soprano. The red curve is Soprano. The lower figure shows the differences between SOPRANO and M-SOPRANO at different distances, i.e., 500 m, 1 km, 5 km, 10 km, 50 km and 100 km. PL Difference is $PL_{M-SOPRANO} - PL_{SOPRANO}$.

SOPRANO and M-SOPRANO both use the same mode sum up to 400 Hz, which makes the difference zero in this frequency range. The asymptotic approach over-predicts the propagation loss up to 3.2 dB for frequencies below 1 kHz. For increasing range and frequency, the differences are less than 1 dB after a few kilometers. Based on the convergence tests, the total number of the discrete modes in the mode sum was chosen as M = 10. When more discrete modes propagate, the asymptotic solution of mode-flux integral is used. Figure 9 shows calculation times at the same bathymetry slice (Figure 7).

Figure 9. Comparisons for the calculation time of KRAKENC, SOPRANO and M-SOPRANO for an arbitrary bathymetry.

Figure 9 shows that SOPRANO and M-SOPRANO's calculation times do not depend on frequency, while KRAKENC does. M-SOPRANO is approximately 720 times faster than SOPRANO for all frequencies. Although the flux-integral of SOPRANO is equal to zero up to 400 Hz, this integral is still numerically evaluated. Thus, the calculation times are different up to 400 Hz. For 10 kHz, M-SOPRANO is approximately 70,000 times faster than KRAKENC for the selected case. The use of M-SOPRANO decreases the computational times further with an accuracy within 1 dB above 1 kHz for the shallow water bathymetry

(20–25 m). For low frequencies (<1 kHz) and short ranges (<5 km), an exact solution of mode-flux integral or mode-theory can be used.

4. Sound Maps of Ship and Wind

In this section, sound pressure level (SPL) maps are calculated based on monthly shipping densities between January 2017 and December 2020. The calculations are performed for the center frequencies of decidecade bands (100 Hz to 10 kHz) [36]. Although the source level formula of ships is described at lower frequencies than 100 Hz, some part of the frequency range is usually below the cut-off frequency in the test areas. The mode sum part of the hybrid solution uses only discrete modes above the cut-off frequency. The calculation of propagation loss in a range-dependent waveguide below the cut-off frequency requires more detailed propagation modeling, considering the effect of the evanescent field, which is not included in this paper. The asymptotic solution of mode-flux theory (M-SOPRANO) makes it possible to create a large number of different sound maps based on various source distributions, which is implemented for the large-scale problem with a high spatial and spectral resolution. A total of 3780 different monthly sound maps were calculated for the selected 5 ship categories (fishing, passenger, tanker, cargo, all ships), 4 years, 12 months and 21 frequencies. Each sound map contains 268832 point sources for test area 1 and 35,074 point sources for test area 2, located at the center of receiver cells. The receiver grid cell has a 1×1 km resolution, corresponding to a spatial observation window of 1 km^2. Propagation loss was calculated at 10 different receiver depths for the selected radial slices over the test area's bathymetry with 100 m range steps to include the bathymetric changes accurately. For each source, 120 radial slices were used. This selected resolution requires 32.3 million and 4.2 million radial slices for test areas 1 and 2, respectively. Mean-square sound pressure was spatially averaged over receiver cell and receiver depths, as described by Equations (1) and (2) of Sertlek et al. [13]. Figures 10 and 11 show the monthly averaged shipping sound maps of test areas 1 and 2. The month average maps are averaged from 2017 to 2019 as a pre-pandemic reference map. Adding wind-generated sound pressure levels facilitates realistic results at high frequency. First, monthly wind sound maps were calculated based on the wind speed data [29] and added to the ship sound maps. Next, monthly shipping and wind sound maps were compared for the different years. Wind generated sound was modeled using the analytical approach of Ainslie et al. [37], which assumes a constant water depth for each source cells. This constant water depth was calculated as an average water depth of each source cell. Based on the wind speed data [29], wind source is described as a sheet dipole source [38 40].

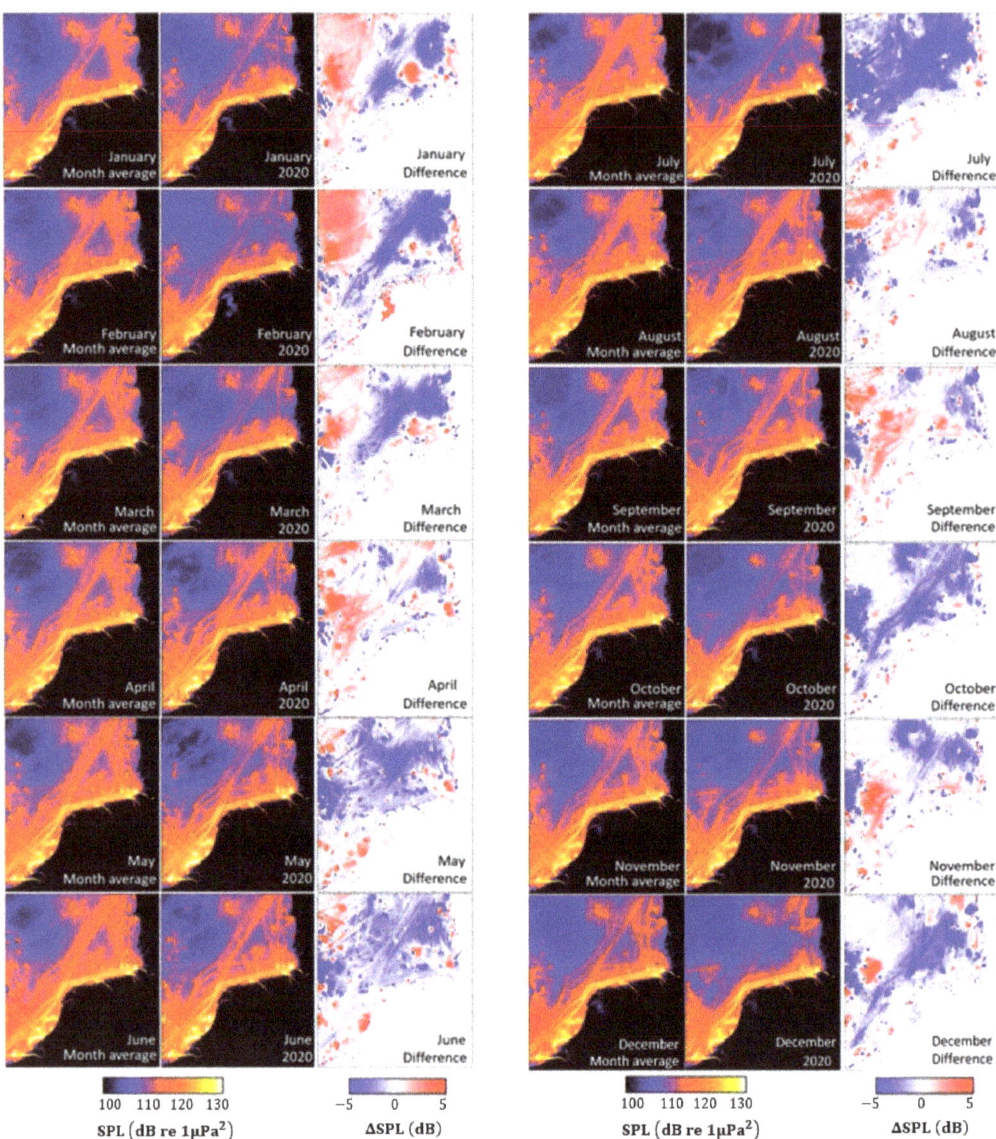

Figure 10. The monthly averaged sound maps before (as a month-average of 2017 to 2019 sound maps) and during the pandemic (2020) in test area 1. The frequency band is from 100 Hz to 10 kHz.

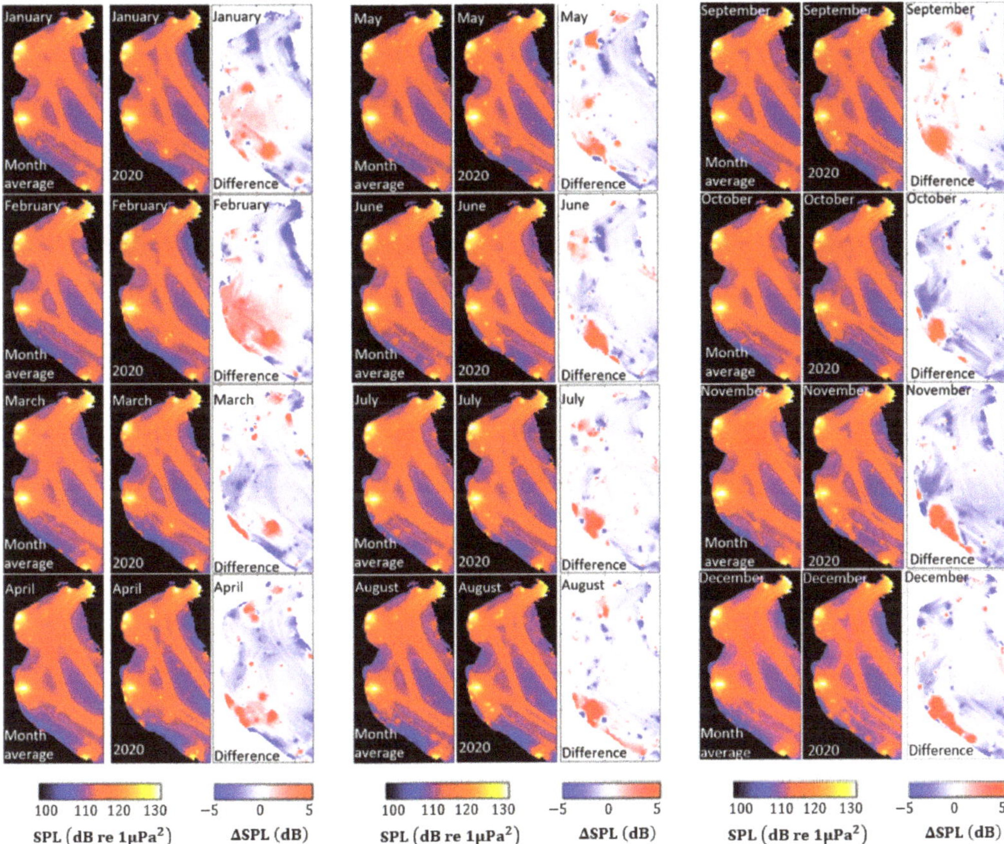

Figure 11. The monthly averaged sound maps before (as a month-average of 2017 to 2019 sound maps) and during the pandemic (2020) in test area 2. The frequency band is from 100 Hz to 10 kHz.

The smallest and largest variations are observed in the Belgian and German parts of the North Sea, respectively. The UK, Danish and French parts of the North Sea are not completely included in this study. The changes are related to the number of ships in the different categories and included ship lanes in the selected areas. The Belgian North Sea, which is smaller than the Dutch and German parts of the North Sea, involves the busy main ship lanes (Figure 2). When the strictest COVID-19 measures were applied during April 2020, the sound maps were investigated in detail for the selected ship categories of the fishing, passenger, tanker, cargo, recreational and sum of all ship categories. Figures 12 and 13 show the April average's shipping and wind sound maps (from 2017 to 2019) and 2020. The most significant decline was observed for the passenger, recreational and fishing ships when the strictest measures of COVID-19 were announced during April 2020. The sound from the tankers increased at the north of test areas 1 and 2.

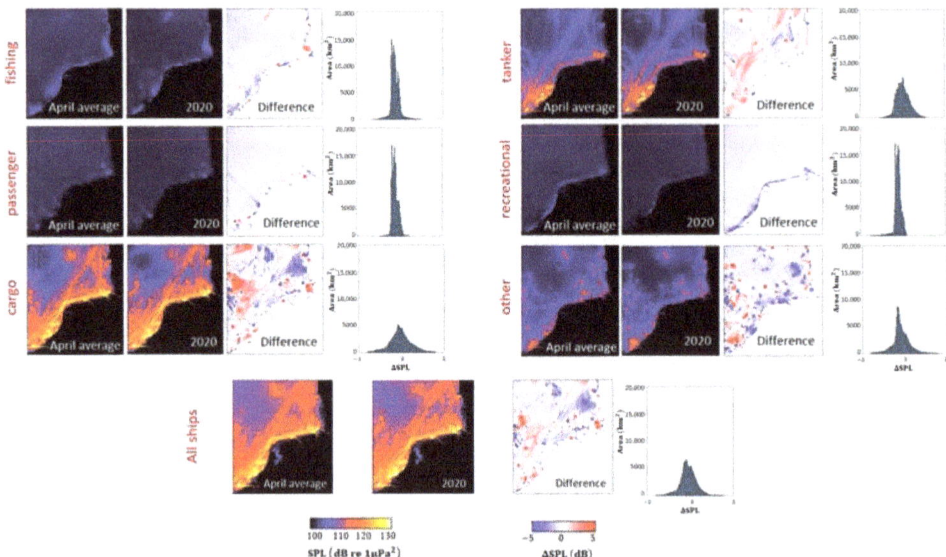

Figure 12. Comparisons for the sound maps of April Average and April 2020 for 100 Hz to 10 kHz in test area 1. The histograms show the amount of the difference in SPL per km^2 for test area 1.

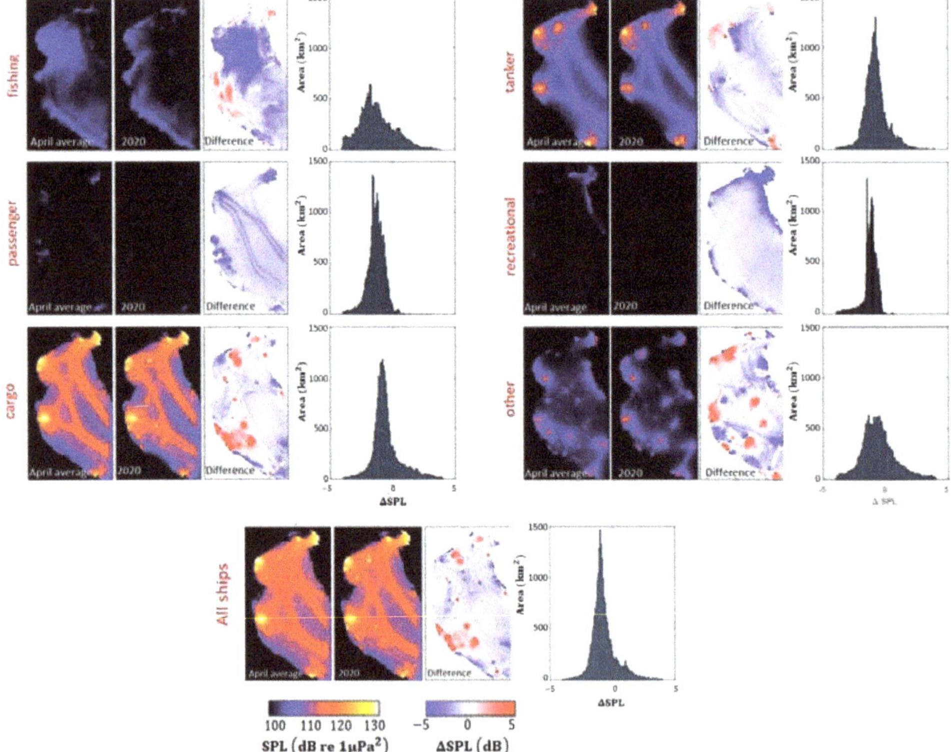

Figure 13. Comparisons for the sound maps of April Average and April 2020 for 100 Hz to 10 kHz in test area 2. The histograms show the amount of the difference in SPL per km^2 for test area 2.

Figures 12 and 13 show that the contribution of each ship category is different based on their numbers and source levels. The contribution from fishing vessels is low despite the large number of shipping vessels (Figure 2). This is because the source level of fishing ships is the lowest between the selected categories in this work (Figure 6).

4.1. Comparison of the Exceedance Levels Based on the Spatial Distribution

Sound maps are calculated for each month for all ship categories. To quantify SPL changes before and during the pandemic, the spatial exceedance levels with the selected percentiles were compared. The ADEON terminology standard defines the N percent spatial exceedance level as "mean-square sound pressure level that is exceeded for N % of the space in a specified spatial analysis window" [41]. A 50% exceedance level corresponds to the median value of SPL. The 10% and 90% exceedance levels help to identify the contribution of very low and high SPL in the sound map, respectively. (The same method as Sertlek et al. [13] but with $25\times$ higher spatial resolution in this work). The monthly variation of the 10%, 50% and 90% exceedance levels before and during the pandemic is shown in Figure 14. The spatial analysis window size is 1 km^2, equal to the receiver grid size of the sound maps.

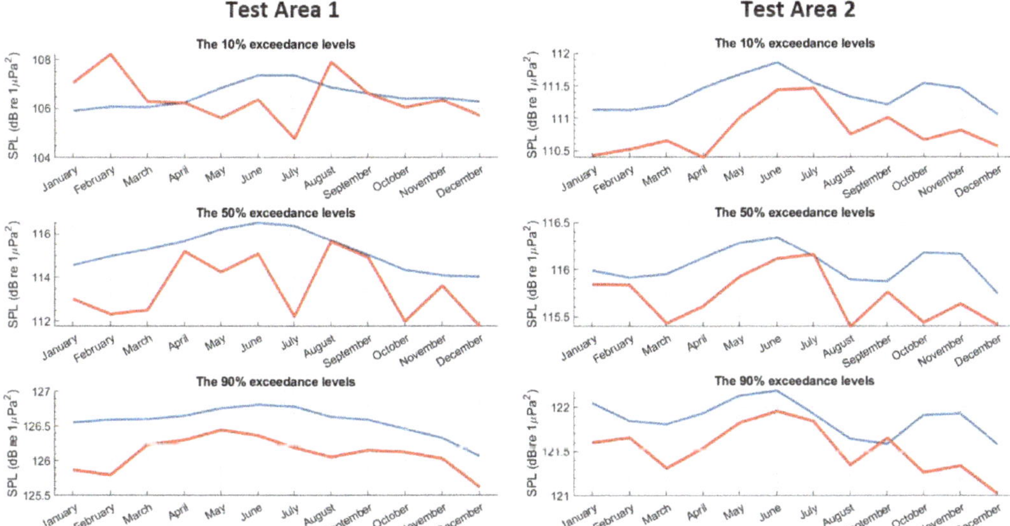

Figure 14. Comparisons for the spatial exceedance levels for 100 Hz to 10 kHz decidecade frequency bands in the test areas. The blue curve is the 2017–2019 average for each month. The red curve is the spatial exceedance levels of 2020.

These comparisons show that the 2020 spatial exceedance levels are lower than the 3-year monthly average for most of the areas. The 3-year monthly average is smoother because it is calculated as an average of three years' curves (2017, 2018 and 2019). The difference is more significant during the winter months, when the pandemic started. Furthermore, in the same period, the oil crisis affected the number of tankers around Europe. With the strictest measures in April 2020, the shipping sound stays lower than in previous years, as seen from the 50% spatial exceedance levels.

4.2. Comparison of Total Acoustic Energies

The monthly total acoustic energy can help quantify the spectral and temporal differences. The energy density can help to compare the shipping sound in the different areas. Figure 15 shows the energy densities (left *y* axis) and energy (right *y* axis) for both areas for

100 Hz to 10 kHz. The observed maximum and minimum total energies are also marked for the month average bars.

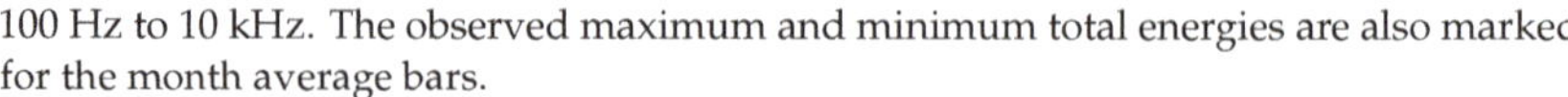

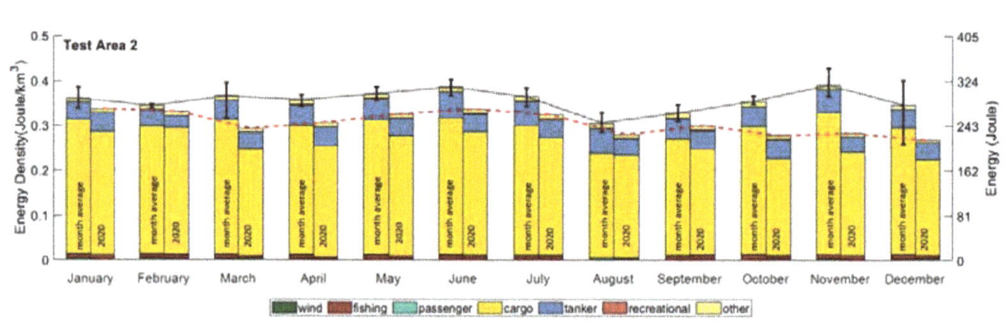

Figure 15. Comparisons for the difference in monthly averaged total acoustic energy (right *y*-axis) and energy density (left *y*-axis) for 100 Hz to 10 kHz decidecade frequency bands. The dotted black curve is based on the month-average of energies between 2017 and 2019. The vertical bars indicate the minimum and maximum energies between 2017 and 2019. The dashed-red curve is the energy density during 2020. The bars show total and individual energy densities of selected ship categories and wind for the month average (left bars) and 2020 (right bars).

In test area 1, the acoustic energy differences were largest during the storms Ciara and Dennis [26,27], in February 2020. July 2020 also has less acoustic energy due to the decrease in the cargo vessels. A decrease in the fishing and passenger ships have a relatively minor contribution to the total acoustic energy. The increasing trend of the shipping sound slows down in contrast to the expected trend before the pandemic [25]. In Test area 2, the most significant decreases in acoustic energy were seen in March and November 2020. During summer, the shipping sound increased with the increasing mobility and relaxed pandemic regulations. The monthly-averaged energy in 2020 is lower than the 3-year month averages for all months in both test areas. For many months, the sound energy densities are even lower than the minimum from the previous 3 years (shown by black bars in Figure 15). The annual average energy density for test area 1 is larger than for test area 2. However, the difference in energy density during 2020 is larger in test area 2 than the test area 1. The month-specific local differences in the total energy can be observed as it was experienced during February 2020 due to the storms in the North Sea.

For the calculations of the sound maps and energies, the source levels of the different ship categories are considered, as mentioned in Section 2. If the Wales and Heitmeyer source level [31] model was used, thus applying the same source level for all ship categories, the contribution from the fishing, recreational and passenger vessels would be overestimated. Thus, it is important to use the category-specific source levels to avoid misleading results. In Appendix B, the sound maps and acoustic energies based on the source levels of Wales and Heitmeyer are compared with the proposed sound maps based on the source levels from MacGillivray and de Jong [30].

The spectral variation of acoustic energy density for different ship categories and wind are compared in Figure 16. The wind and ship energy densities were calculated based on the mathematical models described in the previous sections.

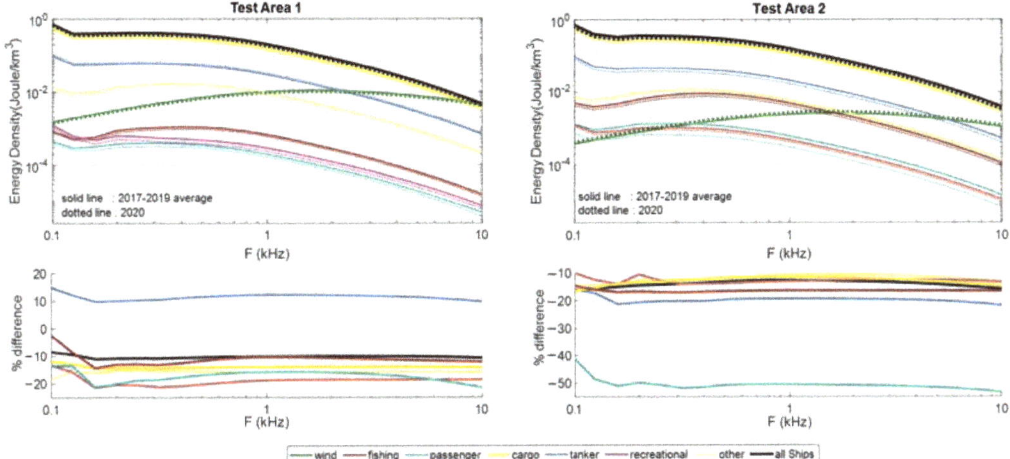

Figure 16. Comparisons of the energy densities of different ship categories and wind for test area 1 (**upper left**) and test area 2 (**upper right**). The percentage differences for each ship categories are shown for test area 1 (**lower left**) and test area 2 (**lower right**).

In test area 1, the contribution of wind-generated acoustic energy density is lower than the total acoustic energy density of the ships up to 10 kHz. The largest and lowest contributions to acoustic energy density are from the tankers and passenger ships, respectively. The contribution of the wind noise is larger than fishing, passenger, and recreational ships. Although the contribution of tankers increased in test area 1, the contribution of all ships (black curve in Figure 16) declined about 10 % during the pandemic. In test area 2, the cargo and tanker categories are responsible from the largest contributions, respectively. The contribution of the fishing vessels are noticeable. The passenger ships have the largest decline in the energy density during the pandemic in 2020 in both test areas. The wind-generated acoustic energy density is lower in test area 2 than test area 1.

5. Conclusions and Discussion

The composition of the shipping densities from the various ship categories vary by time and area. This variation led to changes in the underwater sound pressure levels. This work analyzed the monthly changes in the shipping sound maps, the spatial exceedance levels and acoustic energies in the selected shallow water test areas from the Southern North Sea (test area 1) and the Adriatic Sea (test area 2), which had high shipping densities before the pandemic. Sound propagation was modeled with the asymptotic solution of the mode-flux integral when the propagating modes are larger than 10. The used propagation model had similar accuracy to the adiabatic mode theory, with errors less than 1 dB for ranges exceeding 1 km when multiple modes propagated in the selected test areas. This propagation method enabled the practical calculation of a large number of sound maps to analyze the temporal and spatial variations in detail over a large frequency band without requiring a long computation times.

The sound pressure levels due to the shipping activities decreased during the pandemic in 2020. The most significant fractional decrease in the total number of ships was observed for the passenger ships and fishing ships in both test areas. The differences between the monthly-averaged ship sound maps of 2020 and previous years were analyzed based on the 10%, 50% and 90% spatial exceedance levels. The 50% spatial exceedance level decreased up to 1.6 dB in test area 1 (includes entire Belgian, Dutch, German and partly the UK, French and Danish parts of the North Sea) and 0.7 dB in test area 2 (includes the Gulf of Venice).

The energy density is also a significant metric to compare the acoustic energies in different areas. In test area 1, the most significant differences in the acoustic energy density were observed in February and July 2020 as 0.06 J/km^3 (14.2 % decline) and 0.05 J/km^3 (11.1% decline). In test area 2, the change in the total energy density was around 0.05 J/km^3 in March 2020, and it increases to 0.1 J/km^3 in November. These changes correspond to a 13.5% and 25.6% decline in the total acoustic energy, respectively.

In both test areas, cargo vessels made the largest contribution to the total acoustic energy due to their high source level and large number (as shown by Figures 2 and 6). The wind-generated acoustic energy was larger than the contribution of passenger, recreational and fishing vessels in the examined frequency band of 100 Hz–10 kHz in test area 1. Although the fishing vessels had large numbers, their contribution to the total acoustic energy was low. This showed the significance of the use of the convenient source-level formula for sound mapping.

The noise measurements from the noise monitoring programs in the test areas (JO-MOPANS and SOUNDSCAPE) could be compared with the proposed model results. These comparisons between model and measurements could provide detailed validation for the spatial and temporal variation of sound pressure and improve the confidence in the simulations. In addition, the collected AIS dataset during these programs can help estimate the source level of each ship based on the actual ship speed and length, which can be used to create AIS-based sound maps for a specific time snapshot.

The biological relevance of the change in the sound pressure levels can be investigated by comparing with the fish and marine mammal distribution maps to analyze the potential impact of changed sound pressure levels on marine life during the same time period. In principle, the frequency weighting of these maps can help predict the possible impact on the marine animal's hearing and behavior. A few dB decline in the sound pressure level can significantly extend the communication range of animals. These effects should be investigated as a multidisciplinary work between acousticians and biologists. The natural (such as rain, lightning, etc.) and other anthropogenic offshore activities (seismic airguns, pile driving, explosion, etc.) can be added to the proposed sound maps for the same time period to calculate the total soundscape of the selected areas.

Funding: This research received no external funding.

Acknowledgments: The author thanks Michael A. Ainslie for his extensive review, stimulating discussions and detailed comments, to Apostolos Tsouvalas for his review, encouragement, and support of this work, and to the EMODNET data portal, which makes the shipping density and bathymetry dataset freely available.

Conflicts of Interest: The author declares no conflict of interest.

Appendix A

The sound pressure level (SPL) is

$$SPL = SL - PL \tag{A1}$$

The calculation of source level (SL) is based on the formula of MacGillivray and de Jong [30]. The propagation loss is defined as

$$PL = 10log(F^{-1}/1m^2)dB \tag{A2}$$

where F is propagation factor [40]. Based the incoherent normal mode sum,

$$F(r,z_r,z_0) = \sum_{n=1}^{N} \psi_n^2(z_0,0)\psi_n^2(z_r,r)R_n(r) \tag{A3}$$

where $\psi_n(z_r, r)$ and $R_n(r)$ are horizontal and vertical eigenfunctions, respectively. Sertlek et al. [20] replaces this discrete mode sum with a continuous angle integral, which leads to

$$F(r, z_r, z_0) = \frac{2}{rh(r)} \int_0^{\pi/2} (1 - W(z_r, z_0, \theta_0)) exp(h^2(0)sin^2(\theta_0) \int_0^r \frac{ln(V(\theta'))}{h^3(r')sin\theta' dr'} d\theta_0 \quad (A4)$$

where $h(r)$ is the water depth at range r, z_r is receiver depth, z_0 is source depth, $V(\theta)$ is the reflection coefficient of the seabed , θ_0 and θ' are the mode grazing angles at source and integrant range r'. $W(z_r, z, \theta_0)$ represents the source and receiver depth weighting function (given by Equation (6) of Sertlek et al. [20]). For an exponential reflection coefficient, the angle integral can be analytically solved, and its asymptotic solution for the long ranges can be written as

$$F(r, z_r, z_0) = r^{-3/2} \sqrt{\frac{\pi}{\eta h_{eff}}} \left(1 - e^{-2\phi_0^2 k_w^2 z_0^2} - e^{-2\phi_0^2 k_w^2 z_r^2 \frac{D^2(0)}{D^2(r)}} + \frac{e^{-2\phi_0^2 k_w^2 \zeta_-^2} + e^{-2\phi_0^2 k_w^2 \zeta_+^2}}{2} \right) \quad (A5)$$

where $\phi_0 = \sqrt{\frac{h^2(r)}{2\eta h_{eff}}}$ and η is the sediment absorption coefficient, h_{eff} is the effective water depth (as described in Sertlek et al. [20]), η is the sediment absorption coefficient, k_w is the wave number of the water layer, z_0 is the source depth, z_r is the receiver depth and $D(r)$ is the wave-shifted water depth.

Appendix B

Determining realistic source properties is critical for the sound mapping simulations. The Wales and Heitmeyer source level is widely used to represent the source level of ships in many underwater acoustic applications. The Wales and Heitmeyer formula is given as a function of frequency and compared with the measurements. MacGillivray and de Jong's formulation [30] uses frequency and the vessel speed and length to calculate the source level for the different ship categories. If one compares two formulas, a significant difference can be observed for some ship categories such as fishing, passenger and cargo, as it was compared with the measurements in [30,42]. In this Appendix, the sound maps are calculated based on both two formulations. The obtained sound maps and differences are shown in Figure A1.

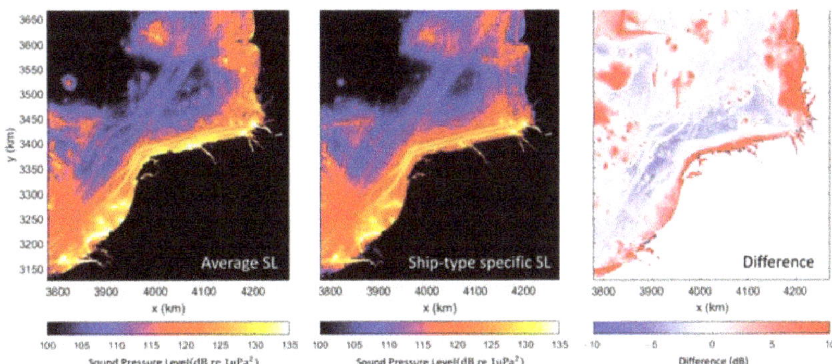

Figure A1. Shipping sound maps for April in test area 1 based on Wales and Heitmeyer (**left**), and ship-type specific (**middle**) source levels are shown for 100 Hz to 10 kHz frequency bands. The differences (dB) between these maps are shown (**right**).

When the Wales and Heitmeyer source level is used, the sound pressure levels are overestimated at some areas, especially with the contribution of the fishing ships, and underestimated at the locations of cargo ships. The use of Wales and Heitmeyer formulations would lead to different total energy comparisons because of the assumption that all ship categories have identical source levels, as shown by Figure A2.

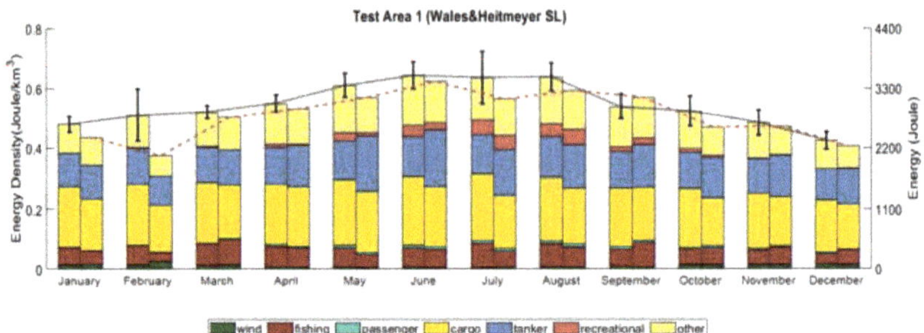

Figure A2. Total acoustic energy (right *y*-axis) and energy density (left *y*-axis) based on Wales and Heitmeyer's source level formula for 100 Hz to 10 kHz frequency bands. The black curve is the month average between 2017 and 2019, indicating the minimum and maximum energies. The red dashed curve is the energy density (and energy) for 2020. The bars show total and individual energy densities of selected ship categories and wind for the month average (left bars) and 2020 (right bars). The frequency range is from 100 Hz to 10 kHz.

In Figure A2, the modeled energy and energy densities are higher than in Figure 15. The total energy in September 2020 is even higher than the energy in the September average, which conflicts with the calculated results in this paper. Thus, the Wales and Heitmeyer source level should be carefully used to estimate the sound pressure level depending on the composition of the shipping density with various ship types. AIS-based sound maps can be preferred to increase the accuracy, including the exact shipping and length. However, depending on the size of the area and the number of ships, shipping density maps can be used for practical reasons.

References

1. Slabbekoorn, H.; Bouton, N.; Opzeel, I.; Coers, A.; ten Cate, C.; Popper, A.N. A noisy spring: The impact of globally rising underwater sound levels on fish. *Trends Ecol. Evol.* **2010**, *25*, 419–427. [CrossRef] [PubMed]
2. Rolland, R.M.; Parks, S.E.; Hunt, K.E.; Castellote, M.; Corkeron, P.J.; Nowacek, D.P.; Wasser, S.K.; Kraus, S.D. Evidence that ship noise increases stress in right whales. *Proc. Biol. Sci.* **2012**, *279*, 2363–2368. [CrossRef] [PubMed]
3. Nowacek, D.P.; Johnson MPand Tyack, P.L. North Atlantic right whales (Eubalaena glacialis) ignore ships but respond to alerting stimuli. *Proc. R. Soc. Lond.* **2004**, *271*, 227–231. [CrossRef] [PubMed]
4. Lucke, K.; Winter, E.; Lam, F.P.; Scowcroft, G.; Hawkins, A.; Popper, A.N. International harmonization of approaches to define underwater noise exposure criteria. *J. Acoust. Soc. Am.* **2014**, *135*, 2404. [CrossRef]
5. EU 2008, Directive 2008/56/EC of the European Parliament and of the Council of 17 June 2008 Establishing a Framework for Community Action in the Field of Marine Environmental Policy (Marine Strategy Framework Directive). 2015. Available online: http://eur-lex.europa.eu/legal-content/en/txt/?uri=celex:32008l0056 (accessed on 20 May 2021).
6. Andersson, M.H.; Andersson, S.; Ahlsén, J.; Andersson, B.L.; Hammar, J.; Persson, L.K.G.; Pihl, J.; Sigray, P.; Wikström, A. A framework for regulating underwater noise during pile driving. In *A Technical Vindval Report*; Swedish Environmental Protection Agency: Stockholm, Sweden, 2016.
7. Colin, M.E.G.D.; Ainslie, M.A.; Binnerts, B.; de Jong, C.A.F.; Karasalo, I.; Östberg, M.; Sertlek, H.Ö.; Folegot, T.; Clorennec, D. Ship Noise Mapping in The North Sea. In Proceedings of the Underwater Acoustic Conference, Rhodos, Greece, 22–27 June 2014 .
8. Dekeling, R.P.; Tasker, M.L.; Van der Graaf, A.J.; Ainslie, M.A.; Andersson, M.H.; André, M.; Borsani, J.F.; Bresing, K.; Castellote, M.; Cronin D.; et al. *Monitoring Guidance for Underwater Noise in European Seas. Part I: Executive Summary*; J.V. JRC Scientific and Policy Report EUR 26557 EN; Publications Office of the European Union: Luxembourg 2014.
9. Hatch, L.T.; Fristrup, K.M. No barrier at the boundaries: Implementing regional frameworks for noise management in protected natural areas. *Mar. Ecol. Prog. Ser.* **2009**, *395*, 223–244. [CrossRef]
10. EMODNET Human Activities. Available online: https://www.emodnet-humanactivities.eu (accessed on 20 May 2021).
11. JOMOPANS. Available online: https://northsearegion.eu/jomopans/ (accessed on 28 May 2021).
12. SOUNDSCAPE. Available online: https://www.italy-croatia.eu/web/soundscape (accessed on 20 June 2021).

13. Sertlek, H.Ö.; Ainslie, M.A.; Slabbekoorn, H.; ten Cate, C.J. Source-specific sound mapping: Spatial, temporal and spectral distribution of sound in the Dutch North Sea. *Environ. Pollut.* **2019**, *247*, 1143–1157. [CrossRef] [PubMed]
14. March, D.; Metcalfe, K.; Tintoré, J.; Godley, B.J. Tracking the global reduction of marine traffic during the COVID-19 pandemic. *Nat. Commun.* **2021**, *12*, 2415. [CrossRef] [PubMed]
15. WHO. Available online: https://www.euro.who.int/en/health-topics/health-emergencies/coronavirus-covid-19/news/news/2020/3/who-announces-covid-19-outbreak-a-pandemic (accessed on 2 March 2021)
16. EMSA, Covid-19 Impact on Shipping. Available online: http://www.emsa.europa.eu/newsroom/latest-news/item/4395-may-2021-covid-19-impact-on-shipping-report.html (accessed on 28 May 2021).
17. Millefiori, L.M.; Braca, P.; Zissis, D.; Spiliopoulos, G.; Marano, S.; Willett, P.K.; Carniel, S. COVID-19 Impact on Global Maritime Mobility, PREPRINT (Version 1) Available at Research Square. Available online: https://www.researchsquare.com/article/rs-100286/v1 (accessed on 19 June 2021).
18. Thomson, D.J.M.; Barclay, D.R. Real-time observations of the impact of COVID-19 on underwater noise. *J. Acoust. Soc. Am.* **2020**, *147*, 3390. [CrossRef] [PubMed]
19. Dahl, P.H.; Dall'Osto, D.R.; Harrington, M.J. Trends in low-frequency underwater noise off the Oregon coast and impacts of COVID-19 pandemic. *J. Acoust. Soc. Am.* **2021**, *149*, 4073. [CrossRef]
20. Sertlek, H.Ö.; Ainslie, M.A.; Heaney, K. Analytical and Numerical Propagation Loss Predictions for Gradually Range-Dependent Isospeed Waveguides. *IEEE J. Ocean. Eng.* **2019**, *44*, 1240–1252. [CrossRef]
21. EmodnetBathymetry. Available online: https://www.emodnet-bathymetry.eu (accessed on 20 May 2021).
22. The EAA reference Grid. Available online: https://www.eea.europa.eu/data-and-maps/data/eea-reference-grids-1/about-the-eea-reference-grid/ (accessed on 20 May 2021).
23. Hildebrand, J.A. Anthropogenic and natural sources of ambient noise in the ocean. *Mar. Ecol. Prog. Ser.* **2009**, *395*, 5–20. [CrossRef]
24. Ainslie, M.A.; Andrew, R.K.; Howe, B.M.; Mercer, J.A. Temperature-driven seasonal and longer term changes in spatially averaged deep ocean ambient sound at frequencies 63–125 Hz. *J. Acoust. Soc. Am.* **2021**, *149*, 2531. [CrossRef] [PubMed]
25. Sertlek, H.O.; Slabbekoorn, H.; Michael, A.A. The contribution of shipping sound to the dutch underwater soundscape: Past, present, future. *Proc. Mtgs. Acoust.* **2019**, *37*, 070010.
26. Storm Wiki 1. Available online: https://en.wikipedia.org/wiki/Storm_Ciara (accessed on 20 May 2021).
27. Storm Wiki 2. Available online: https://en.wikipedia.org/wiki/Storm_Dennis (accessed on 20 May 2021).
28. OECD Report. Available online: https://www.oecd.org/eu/34431541.pdf (accessed on 20 May 2021).
29. The Royal Netherlands Meteorological Institute (KNMI). Available online: https://climexp.knmi.nl (accessed on 20 May 2021).
30. MacGillivray, A.; de Jong, C. A Reference Spectrum Model for Estimating Source Levels of Marine Shipping Based on Automated Identification System Data. *J. Mar. Sci. Eng.* **2021**, *9*, 369. [CrossRef]
31. Wales, S.C.; Heitmeyer, R.M. An ensemble source spectra model for merchant ship-radiated noise. *J. Acoust. Soc. Am.* **2002**, *111*, 1211–1231. [CrossRef] [PubMed]
32. Sertlek, H.Ö.; Ainslie, M.A. A depth-dependent formula for shallow water propagation. *J. Acoust. Soc. Am.* **2014**, *136*, 573–582. [CrossRef] [PubMed]
33. Sertlek, H.Ö. Aria of the Dutch North Sea. Ph.D. Thesis, University of Leiden, Leiden, The Netherlands, 2016.
34. Salomons, E.M.; Binnerts, B.; Betke, K.; von Benda-Beckmann, A.M. 2021, Noise of underwater explosions in the North Sea. A comparison of experimental data and model predictions. *J. Acoust. Soc. Am.* **2021**, *149*, 1878–1888. [CrossRef] [PubMed]
35. TU Delft Underwater Acoustic Simulation Tools. Available online: ua.citg.tudelft.nl (accessed on 20 May 2021).
36. ISO 18405. Available online: https://www.iso.org/standard/62406.html (accessed on 20 May 2021).
37. Ainslie, M.A.; Harrison, C.H.; Zampolli, M. An analytical solution for signal, background and signal to background ratio for a low frequency active sonar in a Pekeris waveguide satisfying Lambert's rule. In Proceedings of the 4th International Conference and Exhibition on Underwater Acoustic Measurements, Kos, Greece, 20–24 June 2011; pp. 491–498.
38. Kuperman, W.A.; Ferla, M.C. A shallow water experiment to determine the source spectrum level of wind-generated noise. *J. Acoust. Soc. Am.* **1985**, *77*, 2067–2073. [CrossRef]
39. *APL-UW High-Frequency Ocean Environmental Acoustic Models Handbook*; Applied physics Laboratory, University of Washington: Seattle, WA, USA, 1994.
40. Ainslie, M.A. *Principles of Sonar Performance Modeling*; Springer: Berlin/Heidelberg, Germany; Praxis Publishing: Chichester, UK, 2010; pp. 32–58.
41. Ainslie, M.A.; de Jong, C.A.F.; Martin, S.B.; Miksis-Olds, J.L.; Warren, J.D.; Heaney, K.D.; Hillis, C.A.; MacGillivray, A.O. *Project Dictionary: Terminology Standard. Document 02075, Version 1, Technical Report by JASCO Applied Sciences for ADEON*; University of New Hampshire: Durham, NC, USA, 2020.
42. Jansen, E.; de Jong, C.A.F. Experimental Assessment of Underwater Acoustic Source Levels of Different Ship Types. *IEEE J. Ocean Eng.* **2017**, *42*, 439–448. [CrossRef]

Journal of
*Marine Science
and Engineering*

Article

It Often Howls More than It Chugs: Wind versus Ship Noise Under Water in Australia's Maritime Regions

Christine Erbe [1,*], Renee P. Schoeman [1], David Peel [2] and Joshua N. Smith [3]

[1] Centre for Marine Science and Technology, Curtin University, Perth, WA 6102, Australia; renee.p.schoeman@gmail.com
[2] Data 61, CSIRO, CSIRO Marine Laboratories, Hobart, TAS 7004, Australia; david.peel@data61.csiro.au
[3] Centre for Sustainable Aquatic Ecosystems, Harry Butler Institute, Murdoch University, Perth, WA 6150, Australia; joshua.smith@uqconnect.edu.au
* Correspondence: c.erbe@curtin.edu.au

Abstract: Marine soundscapes consist of cumulative contributions by diverse sources of sound grouped into: physical (e.g., wind), biological (e.g., fish), and anthropogenic (e.g., shipping)—each with unique spatial, temporal, and frequency characteristics. In terms of anthropophony, shipping has been found to be the greatest (ubiquitous and continuous) contributor of low-frequency underwater noise in several northern hemisphere soundscapes. Our aim was to develop a model for ship noise in Australian waters, which could be used by industry and government to manage marine zones, their usage, stressors, and potential impacts. We also modelled wind noise under water to provide context to the contribution of ship noise. The models were validated with underwater recordings from 25 sites. As expected, there was good congruence when shipping or wind were the dominant sources. However, there was less agreement when other anthropogenic or biological sources were present (i.e., primarily marine seismic surveying and whales). Off Australia, pristine marine soundscapes (based on the dominance of natural, biological and physical sound) remain, in particular, near offshore reefs and islands. Strong wind noise dominates along the southern Australian coast. Underwater shipping noise dominates only in certain areas, along the eastern seaboard and on the northwest shelf, close to shipping lanes.

Keywords: marine soundscape; ship noise; wind noise; whale song; fish chorus; Australian EEZ

Citation: Erbe, C.; Schoeman, R.P.; Peel, D.; Smith, J.N. It Often Howls More than It Chugs: Wind versus Ship Noise Under Water in Australia's Maritime Regions. *J. Mar. Sci. Eng.* **2021**, *9*, 472. https://doi.org/10.3390/jmse9050472

Academic Editor: Michele Viviani

Received: 30 March 2021
Accepted: 25 April 2021
Published: 27 April 2021

Publisher's Note: MDPI stays neutral with regard to jurisdictional claims in published maps and institutional affiliations.

1. Introduction

The oceans abound with natural physical sounds (from wind, rain, polar ice, and seismic activity), biological sounds (from crustaceans, fishes, and marine mammals), and anthropogenic sounds (from transport, construction, offshore exploration, and mining). Soundscapes naturally change over time because of temporal cycles in weather (e.g., cyclones and annual monsoon [1,2]) and animal behaviour (e.g., diurnal foraging patterns, lunar spawning, seasonal mating, and annual migration [3–6]). However, in many habitats, soundscapes further change with patterns of human presence (e.g., temporary construction or summer recreation [7]) and some have changed steadily over time with increasing intensity of anthropophony (e.g., due to shipping [8]).

In 1996, the European Commission identified air-borne noise as one of the main terrestrial environmental issues in Europe, having been neglected compared to chemical pollution [9]. Subsequently, the Commission enacted sound mapping as an important step to assess and manage sound exposure levels in urban areas [10]. A little later, the issue of underwater ocean noise received similar attention, being declared a pollutant, and with underwater sound monitoring and mapping being suggested [11]. Nowadays, underwater noise footprints of individual anthropogenic operations are commonly mapped for environmental impact assessments (e.g., [12–14]). Longer-term, large-scale marine sound mapping has focussed on ship noise [15–17], but may also include other sources, such as seismic airguns and explosives [18].

Shipping is a global contributor to ocean noise and, over the past five decades, has caused a steady increase in underwater low-frequency (10–100 Hz) ambient sound levels in many marine regions [19–24]. This is of concern, because ship noise causes behavioural and acoustic responses, auditory masking, and stress in marine animals [25–28]. Hence, various studies have mapped ship noise and overlain the resulting maps with marine habitat maps to identify areas of concern (hotspots; high animal density and high noise) [29] and areas of opportunity (high animal density and low noise) [30] for marine spatial planning.

A problem with ship noise maps is that they often lack validation against in situ measurements. These maps may have several sources of error in the ship positions and routes, source spectra and levels, sound propagation models, and hydro- and geoacoustic parameters required by the models. As well, the spatial (depth and range) and temporal grid over which the models operate introduces uncertainty. In fact, lack of knowledge on the physical environment (i.e., hydroacoustic parameters of the water and geoacoustic parameters of the upper seafloor) is often the limiting factor in sound propagation model accuracy [14,31]. Model validation is essential to confirm accuracy and to support the use of a sound map for management decisions [32].

Finally, underwater anthropogenic noise needs to be put into context. How does it compare to natural, pervasive noise as from wind? Sertlek et al. [18] found that shipping inserted the greatest amount of acoustic energy into the Dutch North Sea and far exceeded that of wind. Similarly, Farcas et al. [32] showed that ship noise exceeded wind noise under water near major ports and shipping lanes, and around industrial sites in the Northeast Atlantic. However, southern hemisphere oceans have a reputation of being less impacted by anthropogenic sounds, largely due to a lower ship density [33]. Thus, wind may supersede ship noise in parts of the southern oceans. Here, we model underwater sound in the Australian Exclusive Economic Zone (EEZ) from ships and wind over a 6-month austral winter period, and validate the model with 6-month recordings from 25 stations. We chose the winter months as this is the peak of baleen whale presence (e.g., [34–36]). The aim is to enable a better understanding of where shipping noise is likely to be a significant contributor to the marine soundscape and thus, a potential stressor to marine life.

2. Materials and Methods

In a nutshell, we used ship tracks from Automatic Identification System (AIS) logs and underwater source spectra from the literature. On a 5 km × 5 km grid over the Australian EEZ, we modelled underwater sound propagation from all source cells (i.e., cells that contained ships) to all surrounding receiver cells over a 100 km radius. We then integrated sound exposure over the austral winter. The computational effort was managed by (1) splitting the EEZ into 28 acoustic zones, in which sound propagates in similar ways [37], hence, where a similar model may be set up, and (2) using a neural network to cluster all source-receiver transects within a zone into 64 groups of bathymetry transects, and modelling sound propagation only for cluster centroids. An overview of the process step-by-step is given in Appendix A. Wind noise was not propagated, but simply computed based on hourly wind speed data in each cell.

All GIS analysis was done with a combination of ArcMap (version 10.5, ESRI, Redlands, CA, USA) and R (Version 4.03, R Core Team, Vienna, Austria). Noise modelling and validation was done in MATLAB (Version 2020b, The MathWorks Inc., Natick, MA, USA). We commenced with a GIS layer of the Australian marine bathymetry, gridded to 5 km × 5 km [38].

2.1. Ship Data

Data on ship type, size, position, and speed were obtained from Automatic Identification System (AIS) logs managed by the Australian Maritime Safety Authority (AMSA). AIS data were extracted for the period 1 April 2015–30 September 2015. Ships were grouped into five classes based on their length only (and not by type or function; e.g., tanker versus passenger ferry): <25 m, $\geq$25–<50 m, $\geq$50–<100 m, $\geq$100–<200 m, and $\geq$200 m. The regularity of the AIS location reporting depends on location and time, and the data we used was subsampled to provide locations, at most, every 5 min. From this data, ship tracks were interpolated by dead reckoning to intervals of 60 s if two successive AIS positions met criteria based on the time between the polls and the straightness of the vessel's path (as per Appendix B in [39]). For each ship track, the time spent in each grid cell was computed, and time was summed over all ships belonging to the same class, yielding a grid of cumulative time spent in each cell over the 6-month period, by ship class.

Underwater ship noise source spectra were taken from the Research Ambient Noise Directionality (RANDI) model [40] and integrated into full-octave bands: $\geq$10–<20 Hz, $\geq$20–<40 Hz, $\geq$40–<80 Hz, $\geq$80–<160 Hz, $\geq$160–<320 Hz, $\geq$320–<640 Hz, $\geq$640–<1280 Hz, and $\geq$1280–<2560 Hz (Figure 1a). The broadband (10 Hz–2.6 kHz) source levels were: 148, 160, 172, 187, and 193 dB re 1 µPa m, for the five classes, respectively.

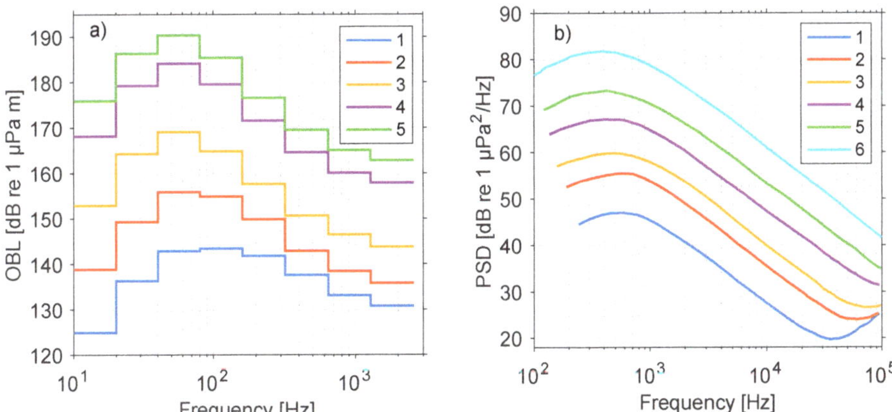

Figure 1. (**a**) Underwater ship noise source levels as full-octave band levels (OBL) for ships of lengths <25 m (Class 1), $\geq$25–<50 m (Class 2), $\geq$50–<100 m (Class 3), $\geq$100–<200 m (Class 4), and $\geq$200 m (Class 5); (**b**) Power spectral density levels (PSD) of wind noise under water at wind speeds 1–3 kn (Beaufort 1; curve 1), 4–6 kn (Beaufort 2; curve 2), 7–10 kn (Beaufort 3; curve 3), 11–21 kn (Beaufort 4–5; curve 4), 22–47 kn (Beaufort 6–9; curve 5), and $\geq$48 kn ($\geq$ Beaufort 10; curve 6).

2.2. Wind Data

Hourly data on surface wind speed (10 m altitude) were obtained over a similar 6-month period (1 April 2012–30 September 2012) from the Bureau of Meteorology and CSIRO [41], based on the NCEP Climate Forecast System [42]. The data varied in spatial resolution (4, 10, and 24 arcminute grids), which we projected and re-sampled to a 5 km × 5 km UTM grid. Over all grid cells, wind speed varied between 0.5 and 30 m/s (i.e., 1–58 kn). Wind speeds were binned to match the sea states represented in the 'Wenz curves' [43] and noise spectra were assigned to each wind speed bin (Figure 1b). The 'Wenz curves' were converted to linear power spectral density, then integrated over frequency, before applying $10\log_{10}$ to yield broadband mean-square sound pressure levels. Expressed as root-mean-square sound pressure levels, the associations became: 79 dB re 1 µPa for $\geq$1–<4 kn, 87 dB re 1 µPa for $\geq$4–<7 kn, 92 dB re 1 µPa for $\geq$7–<11 kn, 99 dB re 1 µPa for $\geq$11–<22 kn, 105 dB re 1 µPa for $\geq$22–<48 kn, and 113 dB re 1 µPa for $\geq$48 kn.

2.3. Acoustic Zones

The Australian EEZ had previously been broken up into 28 'acoustic zones' (Figure 2), whereby each zone was characterised by a unique set of hydroacoustic parameters of the water (i.e., sound speed profiles), geoacoustic parameters of the seafloor (i.e., thickness of the sediment layer, density, compressional sound speed and attenuation, and shear sound speed and attenuation), and bathymetric parameters (i.e., water depth and slope) [37]. The idea was to set up one sound propagation environment for each zone and then model all of the ships in that zone.

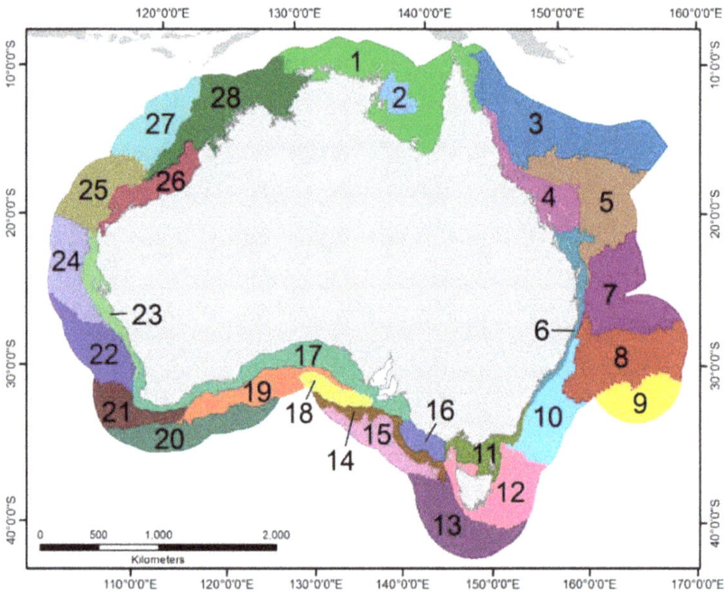

Figure 2. Map of the marine acoustic zones of the Australian EEZ [37].

2.4. Source-Receiver Transects

Within each zone, all of the grid cells that had a cumulative ship time > 0 (no matter the ship class) were identified. From each of these 'source cells', 36 radials were cast at 10-degree intervals. The bathymetry was extracted along each of these radials in 5 km steps out to a maximum range of 100 km, yielding thirty-six 100 km source-receiver transects around each source. A bathymetry reading at 2.6 km range from the centre of each source cell was added, representing the mean distance between two random points inside a square (i.e., 0.5214 times the edge length [44]). If a transect hit land, all subsequent bathymetry samples were set to 'not a number' along this 100 km transect. If a source sat near a zone boundary, then the 100 km transects were extracted with bathymetry from the neighbouring zone or from a 100 km buffer around the outside of the EEZ.

All of the transects from all of the source cells (of all ship classes) within a zone were passed to an unsupervised Kohonen neural network (i.e., a self-organizing map, SOM) with 900 neurons [45] (also see [13], where this SOM was previously used to cluster bathymetry transects). The neural net sorted the transects into 900 groups, based on their bathymetric shape. Further grouping was achieved by k-means clustering allowing for 64 clusters [46]. Cluster centroids were computed as the arithmetic mean of all transects within one cluster (see examples for Zone 16 in Figure 3). Sound propagation was modelled along each of the 64 centroids within one zone.

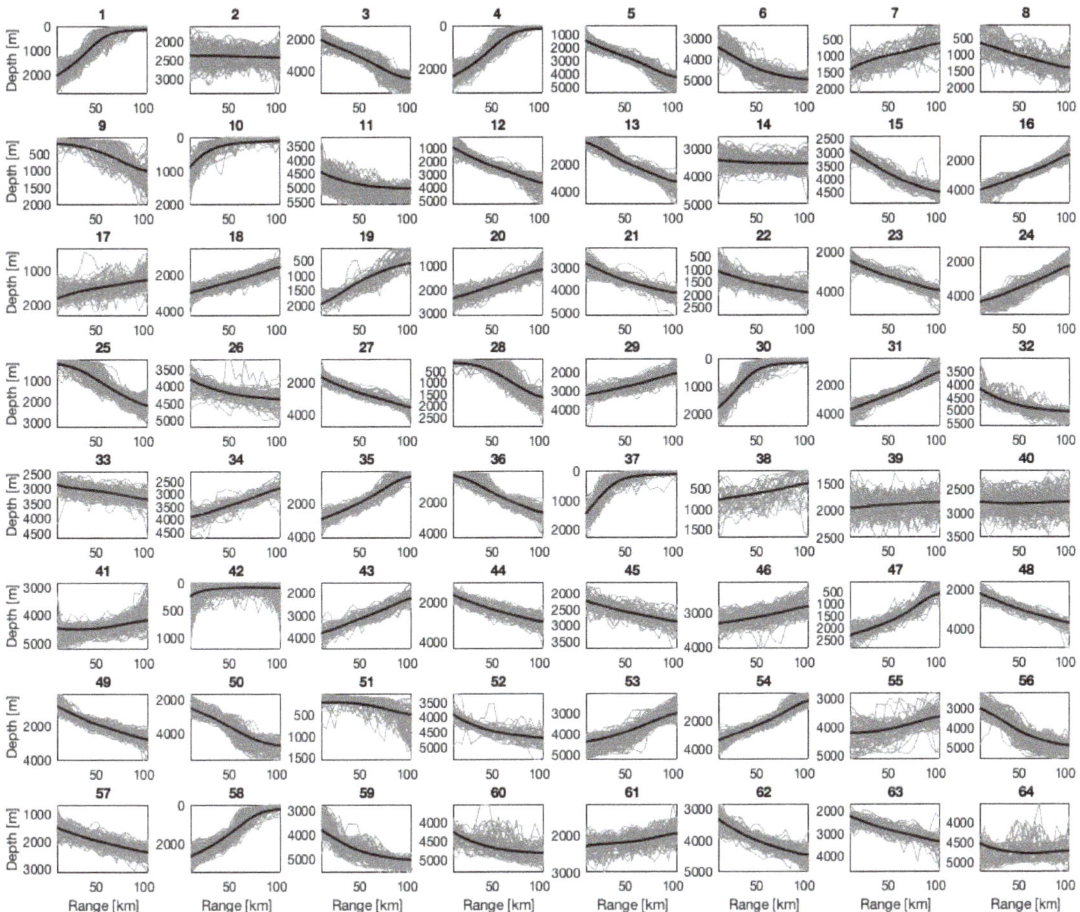

Figure 3. Graphs of all 98,532 source-receiver transects of Zone 16 plotted by cluster (grey), with centroid bathymetries shown in black. X-axes are range (km) and Y-axes are depth below the sea surface (m). Note the changing Y-ranges.

2.5. Sound Propagation Model

Sound propagation over each centroid bathymetry was modelled with RAMGeo in AcTUP V2.8 [47] (https://cmst.curtin.edu.au/products/underwater/ accessed on: 27 March 2021) based on zone-specific acoustic environments consisting of three layers: the water column, an unconsolidated surface sediment layer, and a consolidated calcarenite sediment layer as a half space. Water column parameters included the zone's mean sound speed profile [37] and water density profile. Representative temperature and salinity data were extracted from the World Ocean Atlas [48,49] to calculate water densities based on the UNESCO formula for sea water density [50]. Unconsolidated surface sediment layers throughout the EEZ comprised predominantly fine material (silt-sand) with sufficiently low shear wave speeds (<250 m/s) to allow modelling as a fluid. Hence, unconsolidated surface sediment parameters only included the zone-specific layer thickness, compressional sound speed, compressional wave attenuation, and density [37]. Surface sediment layer thickness was estimated as 0.5 m for zones within the sediment-starved carbonate platform [51]. Surface sediment thickness in the remaining zones appears variable [52–57], and so was modelled as 2 m.

In contrast to the surface sediment layer, calcarenite acts as an elastic material with a shear sound speed of 1400 m/s resulting in important propagation effects [51]. While RAMGeo is a parabolic equation for fluid seabeds, reasonable results can be obtained by using an equivalent fluid approximation with reflection coefficients representative of the elastic model [58]. The procedure to find an equivalent fluid approximation for each zone included (a) creating an environment with calcarenite as an elastic material (see [51] for geoacoustic properties), (b) creating an environment with calcarenite as a fluid layer starting with a compressional sound speed of 1250 m/s and a compressional wave attenuation of 4.5 dB/λ_c, (c) calculating the reflection coefficients for each modelled frequency for both environments with the reflection coefficient model BOUNCE [59], and (d) adjusting the fluid layer parameters until a representative equivalent had been reached (Figure 4).

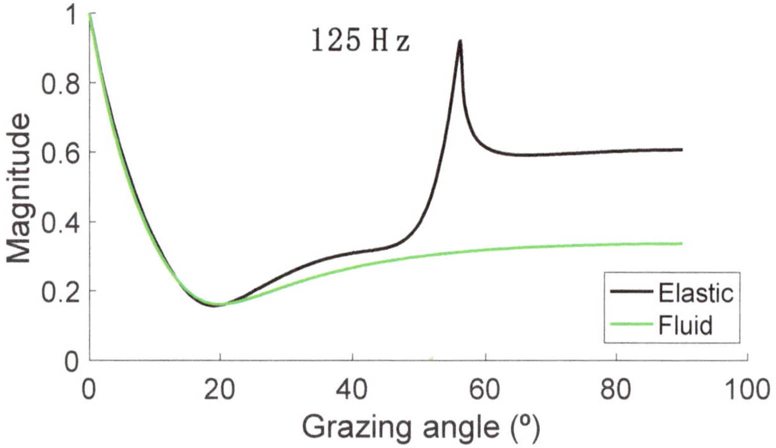

Figure 4. Example of the reflection coefficients of an elastic model (blue line) and a representative fluid model (green line) calculated with BOUNCE at 125 Hz.

RAMGeo modelled sound propagation loss for the centre frequencies of eight full-octave bands between 10 and 2000 Hz over a range of 100 km and up to 7100 m depth, which is more than the maximum water depth (6388 m) of the EEZ. The depth and range resolutions were 10 m. The source depth for all ships was chosen as 5 m below the sea surface. An example RAMGeo output at 640 Hz for the 64 centroid bathymetries of Zone 16 is shown in Figure 5. The bathymetry itself is just visible as a black line, below which propagation loss was greatest. The colours vary from 60 dB propagation loss (dark red) to 110 dB (dark blue). Several patterns are obvious: Convergence zones appeared over all deep bathymetries, leading to low propagation loss (i.e., high received levels) near the sea surface about 60 km from the source (i.e., clusters 6, 11, 26, 32, 41, 52, 53, 55, 59, 60, and 64). Over upwards-sloping bathymetries, propagation loss was greatest (e.g., clusters 1, 4, 19, 35, 37, 54, and 58). Over downwards-sloping bathymetries, sound may reflect into the deep-ocean sound channel, which has an axis at about 1 km depth off Australia. Once inside the channel, sound may propagate over vast ranges at very little additional loss because of no further seafloor (and to a lesser extent, sea surface) interactions (i.e., clusters 12, 13, 25, 36, and 49). Finally, RAMGeo does not include frequency-dependent absorption and so this additional loss was applied outside of and after RAMGeo [60].

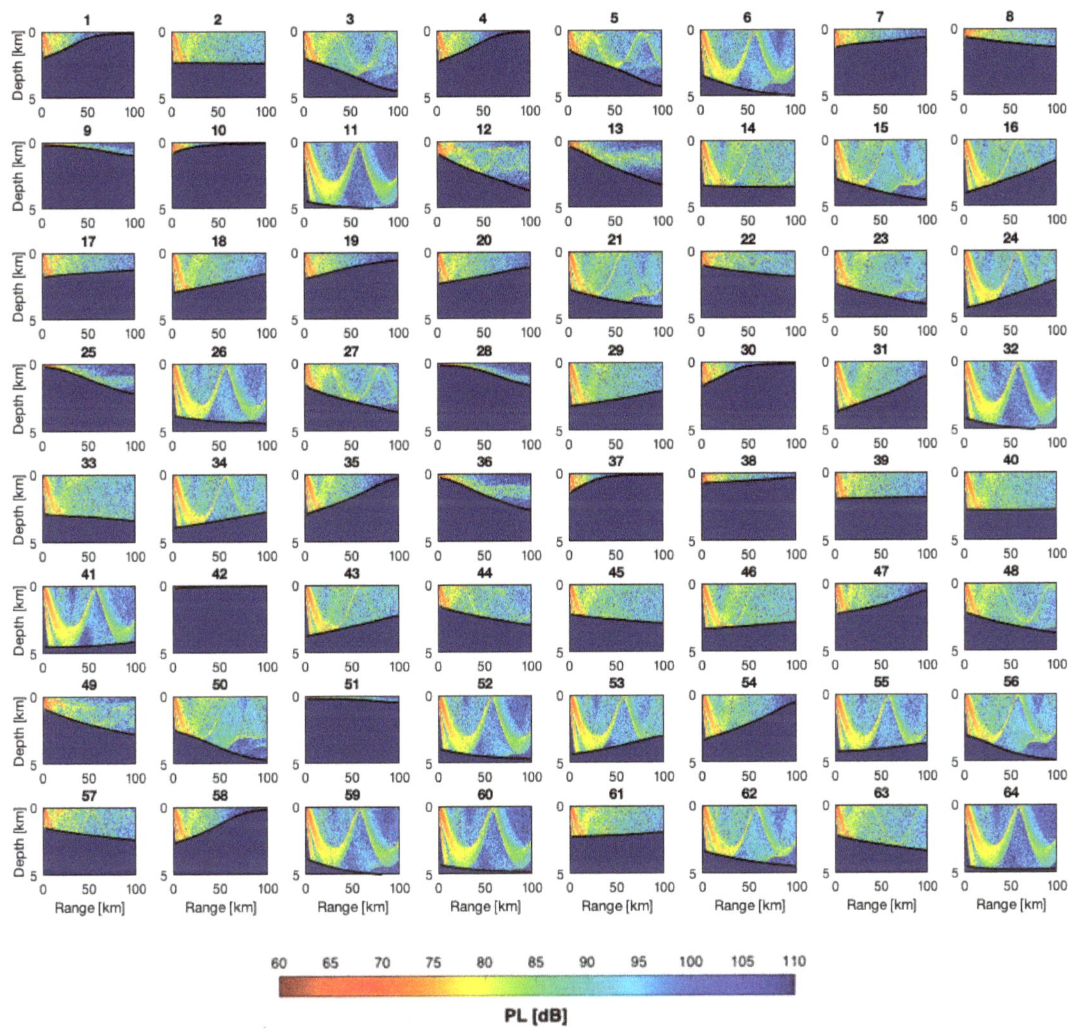

Figure 5. Plots of propagation loss (PL) as a function of range and depth along the 64 centroid bathymetries from Zone 16, for a frequency of 640 Hz. The darkest red corresponds to 60 dB and the darkest blue to 110 dB. X-axes are range (km) and Y-axes are depth below the sea surface (km); both are scaled linearly.

2.6. Accumulation of Received Levels

Within each zone, one ship class was treated at a time. The source cells corresponding to one ship class were found, thirty-six 100 km radials were cast at 10-degree intervals around each source cell, and bathymetry was extracted along each radial and sampled in 5 km steps; the mean distance between a ship and a receiver of 2.6 km within the source cell was inserted at the beginning. In other words, source cells were assigned a received level at 2.6 km. Then, stepping through the source cells for this ship class in this zone, for each of the 36 radial transects, the best matching centroid bathymetry was found. In fact, as the SOM had been trained with all source-receiver transects from all ship classes in this zone, finding the 'best matching centroid' reduced to simply looking up into which cluster this transect originally went. Then, for each frequency modelled, propagation loss (*PL*) along this centroid was recovered and subtracted from the corresponding octave band source level (*SL*) plus the cumulative time in dB re 1 s (with *T* in units of second) of this ship class in this source cell, yielding received levels (*RL*):

$$RL = SL + 10 \log_{10}(T) - PL$$

In this equation, *SL* is a number, the octave band source level expressed as a mean-square pressure level [dB re 1 μPa^2] at the modelled frequency. The duration term is also a number [dB re 1 s]. *PL* [dB], however, is a matrix with values in depth and range. Therefore, *RL* is a matrix containing received sound exposure levels (*SEL*) [dB re 1 μPa^{2}s] as a function of range and depth. There was one such *RL* matrix for each frequency.

To change from the polar coordinates (in which sound propagation was modelled) to the Cartesian grid of the EEZ, *RL* was interpolated to the 5 km grid of the EEZ, at each depth. Including frequency as an additional dimension, this yielded a 4-dimensional matrix of longitude, latitude, depth, and frequency, covering the entire EEZ. The matrix was populated cumulatively by summing sound exposure (i.e., in linear, not logarithmic terms) over all 36 transects about each source cell, and then over all source cells, before taking 10 $\log_{10}$ again to yield cumulative sound exposure levels (*C-SEL*).

2.7. Ship Noise Map

Broadband sound exposure levels were computed by summing sound exposure over frequency, thereby reducing the matrix to three dimensions, then converting to dB. A further reduction to two dimensions was achieved by finding the maximum sound exposure level over the top 200 m, representing the 'worst case' of exposure for animals that dive over this depth [61]. One such map is presented for each ship class, as well as cumulatively over all five classes. These maps of cumulative sound exposure level were accumulated over six months encompassing the austral winter. They can be read as average mean-square sound pressure level (*SPL*) maps by subtracting the 6-month duration in dB re 1 s:

$$SPL = C\text{-}SEL - 10 \log_{10}(183 \text{ d} \times 24 \text{ h/d} \times 60 \text{ min/h} \times 60 \text{ s/min/s}) = C\text{-}SEL - 72 \text{ dB re 1 s}$$

2.8. Wind Map

The wind map was produced by converting the hourly root-mean square sound pressure levels to linear mean-square sound pressures, then integrating over time, and converting back to decibel. 10 $\log_{10}(3600)$ was added to account for the number of seconds per hour, yielding cumulative sound exposure levels from wind at each cell over the 6-month winter period.

2.9. Comparison between Ship and Wind Noise

For comparison, the cumulative sound exposure levels from wind were subtracted from those of ships (summed over all classes) in every grid cell, and plotted, to show in which geographic regions one dominated over the other. We also added the modelled sound exposures from ships and wind, then converted to decibel, to plot the combined ambient noise exposure levels over the 6-month period.

2.10. Validation

Archival underwater acoustic recordings from the northwest, west, south, and southeast of Australia were used in an attempt to validate the modelled noise maps. These data were collected by autonomous recorders [62] deployed over winter between 2006 and 2017. All recorders had been moored on the seafloor, and sampled at 6 kHz, 5 min every 15 min. Most of these datasets were collected while the passive acoustic observatories of Australia's Integrated Marine Observing System (IMOS) were operational and are thus available from the Australian Ocean Data Network (AODN) (https://acoustic.aodn.org.au/acoustic/ accessed on: 15 March 2021).

Long-term spectral averages (LTSA) were computed in 5 min windows and integrated over frequency (10–2000 Hz) and time (1 April–30 September) to yield cumulative sound exposure. LTSAs were visualised in the software CHORUS [63] to provide an overview of the soundscape and its contributors over multiple weeks to months at a time. Spectrograms with a resolution of 1 s (50% overlap) were computed to zoom into any 5 min sample when the sound sources were not immediately identifiable in the LTSAs. Power spectral density percentile plots (in which the nth percentile gives the power spectral density level exceeded n% of the time, at each frequency) helped identify the dominant contributors to the winter soundscapes [64].

3. Results

Australia-wide maps of ship noise C-SEL (by class) over the period 1 April 2015–30 September 2015 are shown in Figure 6. Cumulative sound exposure levels over all classes are also plotted. Wind noise C-SEL, the level difference between ship noise C-SEL and wind noise C-SEL, and the combined C-SEL of ship and wind noise are shown in Figure 7. Hyperlinks to the data can be found in the Data Availability section.

Validation

Modelled C-SELs of ship and wind noise are compared to measured C-SELs from the validation datasets in Table 1. There is good agreement (to within 3 dB) between model and measurement noise levels at sites 9 (NW Shelf, WA, Australia), 16 (Bremer Canyon, WA, Australia), and 25 (Tuncurry, NSW, Australia). The former two were dominated by strong wind, the latter by ships. Figure 8A shows almost continuous strong wind at the Bremer Canyon site, a faint Antarctic blue whale (*Balaenoptera musculus intermedia*) chorus throughout winter, peaking in May, and distant passes of ships. The cumulative energy from wind dominates and is the reason for the good agreement between model and measurement. Figure 8B shows briefer periods of strong wind off Tuncurry and a distant Antarctic blue whale chorus. The dominant feature of this soundscape were numerous passes of ships at close range, and this is the reason for the good agreement between model and measurement at this site.

Disagreement between model and measurement noise levels at other sites was due to unaccounted, additional, non-targeted noise contributions to the soundscape: marine animals and industrial operations. Figure 9 provides an overview of the biological contributors to the soundscape. The stereotypical sounds of Omura's whales (*Balaenoptera omurai*); Antarctic blue whales; pygmy blue whales (*Balaenoptera musculus brevicauda*); the unidentified source of the spot call, fin whales (*Balaenoptera physalus*); dwarf minke whales (*Balaenoptera acutorostrata*), and humpback whales (*Megaptera novaeangliae*) have been well described in the literature; as have Australian fish choruses [65–67]. These animals dominated the winter soundscapes near islands and reefs (sites 1, 4, 6, 9, 12, 14, 15, 17 and 18). Examples of soundscapes almost free from ships but noisy with animals are shown in Figures 10 and 11. Examples of soundscapes affected by anthropogenic noise are shown in Figure 12. At the time of recording, seismic surveying was the most common anthropogenic source that we did not model.

We were able to determine C-SEL variability over time at nearby sites. Winter recordings at sites 17 (2016) and 18 (2017) differed in C-SEL by 1 dB; these sites were only 80 m apart. Similarly, sites 7 (2006) and 8 (2010) were 4 km apart and the C-SEL differed by 1 dB. Moreover, sites 14 (2014) and 15 (2016) were 4 km apart and the C-SEL differed by 1 dB, showing good consistency over 1–4 years at nearby sites. Sites 19–24 were all within 3 km of each other. Recordings were from 2012, 2014, 2015, 2015, 2016, and 2017, respectively. The two simultaneous sets differed by 1 dB in measured C-SEL. The 2014 set had the lowest C-SEL with 179 dB re 1 $\mu Pa^2 s$, and one of the 2015 sets had the highest C-SEL at 185 dB re 1 $\mu Pa^2 s$, indicating the level of variability that may be expected from such in situ recordings over multiple years. There was no linear trend.

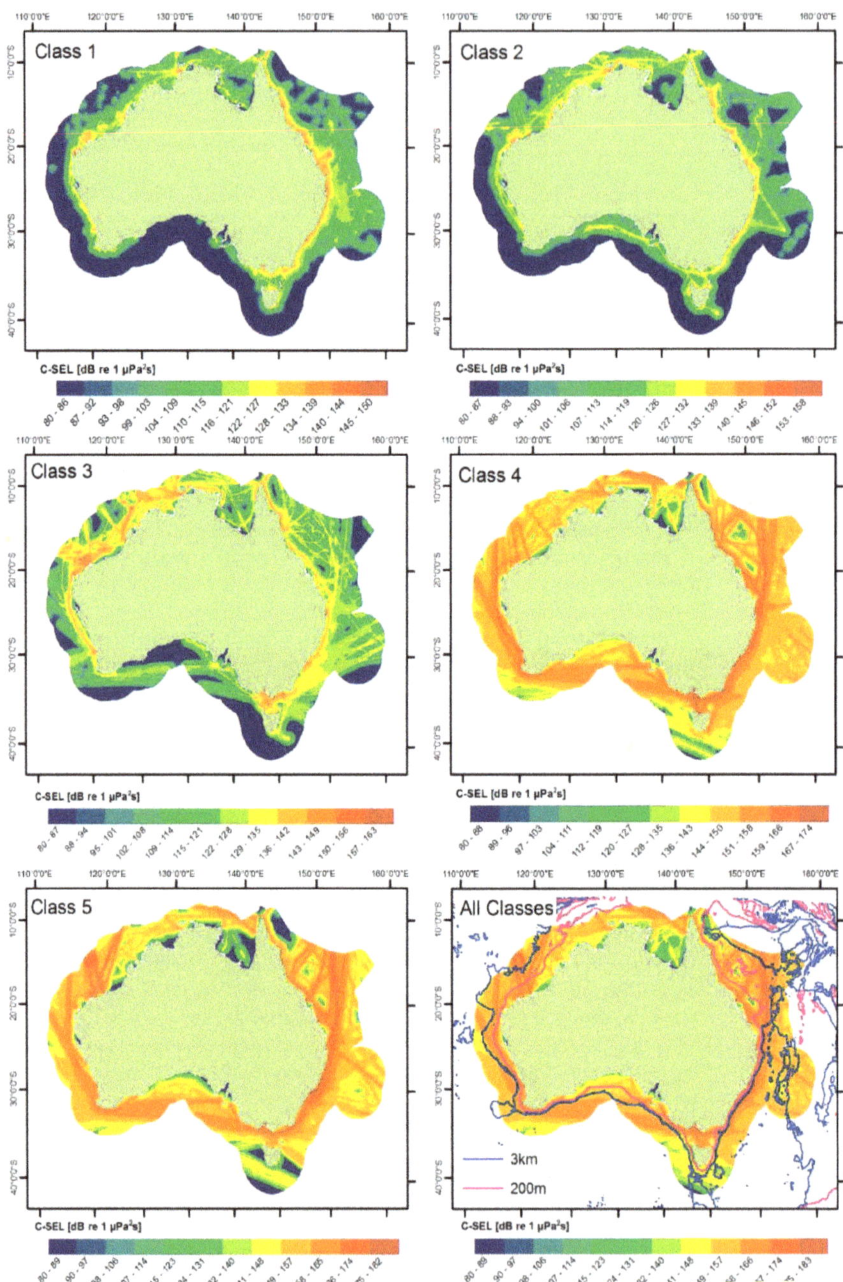

Figure 6. Maps of cumulative sound exposure levels (C-SEL) from shipping in the Australian EEZ, by ship class, and cumulatively over all classes. Maximum received C-SEL over the top 200 m of water were picked, representing a 'worst case' for animals that dive within this depth. Sound exposure was accumulated over 183 days (1 April 2015–30 September 2015). Levels can be converted to average mean-square sound pressure levels by subtracting 72 dB re 1 s. Note that the colour bars all start at 80 dB but the highest levels differ, reflective of the peak C-SEL for each class. The final map also shows 200 m and 3 km bathymetry contours.

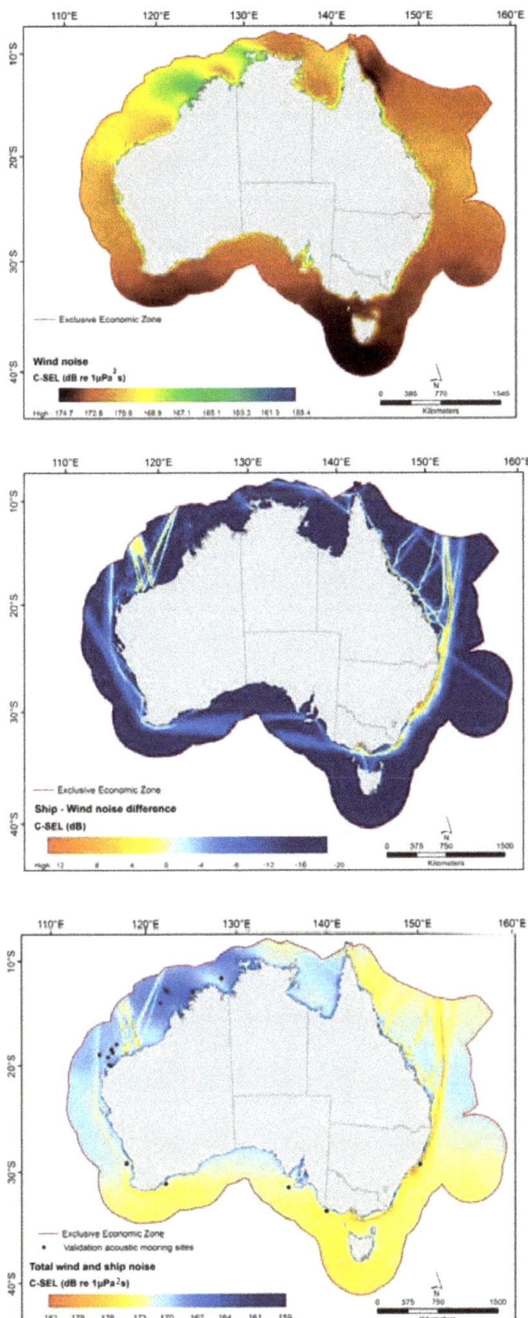

Figure 7. Maps of modelled wind noise within Australia's EEZ during winter 2012 (1 April–30 September: **top**), ship C-SEL less wind C-SEL (**middle**), and C-SEL from ships and wind combined (**bottom**). The black dots identify underwater recording stations used for validation. To convert to mean sound pressure level, subtract 72 dB re 1 s.

Table 1. Site ID, location, year, measured C-SEL and SPL, modelled C-SEL from ships, wind, and combined, C-SEL difference between model and measurement, and rough description of the soundscape.

ID	Location	Longitude	Latitude	Winter	C-SEL [dB re 1 μPa^{2}s] Measured	SPL [dB re 1 μPa] Measured	Ship C-SEL [dB re 1 μPa^{2}s] Modelled	Wind C-SEL [dB re 1 μPa^{2}s] Modelled	Ship + Wind C-SEL [dB re 1 μPa^{2}s] Modelled	C-SEL Difference Measured-Modelled [dB]	Notes
1	Bonaparte Gulf WA	128.2	−13.1	2012	174	102	144	170	170	4	dominated by Omura's whale chorus throughout winter; some fish choruses; strong wind periods; some ships; distant seismic survey
2	NW Shelf	121.9	−14.1	2008	192	120	140	167	167	25	dominated by industrial noise at the time
3	NW Shelf	122.2	−14.3	2008	184	112	154	167	167	17	dominated by 3 seismic surveys at different ranges covering entire winter
4	NW Shelf	124.9	−14.4	2007	176	104	148	166	166	10	dominated by fish choruses, very little anthropophony; pristine
5	NW Shelf	121.3	−15.5	2013	172	100	151	168	168	4	3 seismic surveys overlapping in time at different ranges; Omura's whales and humpback whales

Table 1. *Cont.*

ID	Location	Longitude	Latitude	Winter	C-SEL [dB re 1 μPa²s] Measured	SPL [dB re 1 μPa] Measured	Ship C-SEL [dB re 1 μPa²s] Modelled	Wind C-SEL [dB re 1 μPa²s] Modelled	Ship + Wind C-SEL [dB re 1 μPa²s] Modelled	C-SEL Difference Measured-Modelled [dB]	Notes
6	NW Shelf	115.9	−19.4	2013	177	105	155	170	170	7	strong humpback whale song
7	NW Shelf	115.2	−19.9	2006	183	111	157	170	170	13	dominated by seismic surveys all winter
8	NW Shelf	115.3	−19.9	2010	184	112	160	170	170	14	strong industrial noise throughout
9	NW Shelf	115.4	−20.2	2010	172	100	160	170	170	2	several strong fish choruses and periods of strong wind; pristine
10	NW Shelf	113.9	−20.2	2012	184	112	155	169	170	14	a lot of industrial noise and seismic survey
11	NW Shelf	114.8	−20.6	2010	186	114	153	169	170	16	dominated by industrial noise and seismic surveys, near and far
12	NW Shelf	114.8	−21.4	2010	180	108	157	169	169	11	dominated by humpback whale and fish choruses, also dwarf minke chorus; pristine
13	NW Shelf	115.0	−21.5	2010	177	105	157	168	168	9	industrial noise, fish choruses throughout, humpback whales from 1 Aug.

Table 1. *Cont.*

ID	Location	Longitude	Latitude	Winter	C-SEL [dB re 1 µPa²s] Measured	SPL [dB re 1 µPa] Measured	Ship C-SEL [dB re 1 µPa²s] Modelled	Wind C-SEL [dB re 1 µPa²s] Modelled	Ship + Wind C-SEL [dB re 1 µPa²s] Modelled	C-SEL Difference Measured-Modelled [dB]	Notes
14	Perth Can yon WA	115.0	−31.8	2014	180	108	164	172	172	8	dominated by pygmy blue whale chorus, also strong fish chorus throughout; fin whales in June; spot call in June–July; humpback whales in Sept.
15	Perth Can yon WA	115.0	−31.9	2016	179	107	164	172	172	7	dominated by pygmy blue whale chorus, also strong fish chorus throughout; fin whales in June; spot call in June–July; humpback whales in Sept.
16	Bremer Canyon WA	119.6	−34.7	2015	175	103	158	172	172	3	quiet soundscape with blue whale and spot call chorus and wind; distant shipping only noticeable <5% of the time

Table 1. *Cont.*

ID	Location	Longitude	Latitude	Winter	C-SEL [dB re 1 $\mu Pa^2 s$] Measured	SPL [dB re 1 μPa] Measured	Ship C-SEL [dB re 1 $\mu Pa^2 s$] Modelled	Wind C-SEL [dB re 1 $\mu Pa^2 s$] Modelled	Ship + Wind C-SEL [dB re 1 $\mu Pa^2 s$] Modelled	C-SEL Difference Measured-Modelled [dB]	Notes
17	Kangaroo Isl. SA	135.9	−36.1	2016	180	108	156	173	173	7	dominated by Antarctic blue whale chorus; spot calls; fish; strong wind; very few ships
18	Kangaroo Isl. SA	135.9	−36.1	2017	179	107	156	173	173	6	dominated by choruses of Antarctic blue whales, pygmy blue whales, spot calls, and fish
19	Portland VIC	141.2	−38.5	2012	181	109	167	173	174	7	strong wind and ships; Antarctic blue whale chorus entire winter; strong spot call in Aug.
20	Portland VIC	141.2	−38.5	2014	179	107	167	173	174	5	3 overlapping whale choruses (Antarctic blue, pygmy blue, spot call); wind and ships
21	Portland VIC	141.2	−38.5	2015	184	112	167	173	174	10	strong wind; strong fish; Antarctic blue whale chorus for nearly entire winter in the ship noise band; spot call in July–Aug.

Table 1. *Cont.*

ID	Location	Longitude	Latitude	Winter	C-SEL [dB re 1 μPa²s] Measured	SPL [dB re 1 μPa] Measured	Ship C-SEL [dB re 1 μPa²s] Modelled	Wind C-SEL [dB re 1 μPa²s] Modelled	Ship + Wind C-SEL [dB re 1 μPa²s] Modelled	C-SEL Difference Measured-Modelled [dB]	Notes
22	Portland VIC	141.2	−38.5	2015	185	113	167	173	174	11	Antarctic blue whale chorus entire winter; spot call; strong fish; strong wind and ships
23	Portland VIC	141.2	−38.5	2016	180	108	167	173	174	6	broad ship noise hump at 50 Hz; choruses of Antarctic blue whale; pygmy blue whale; spot call, and fish
24	Portland VIC	141.2	−38.5	2017	181	109	167	173	174	7	Antarctic blue whale chorus and fish all winter; some strong pygmy blue whales; many ships
25	Tuncurry NSW	152.9	−32.3	2016	181	109	177	172	178	3	full of ships; blue whale choruses in the background

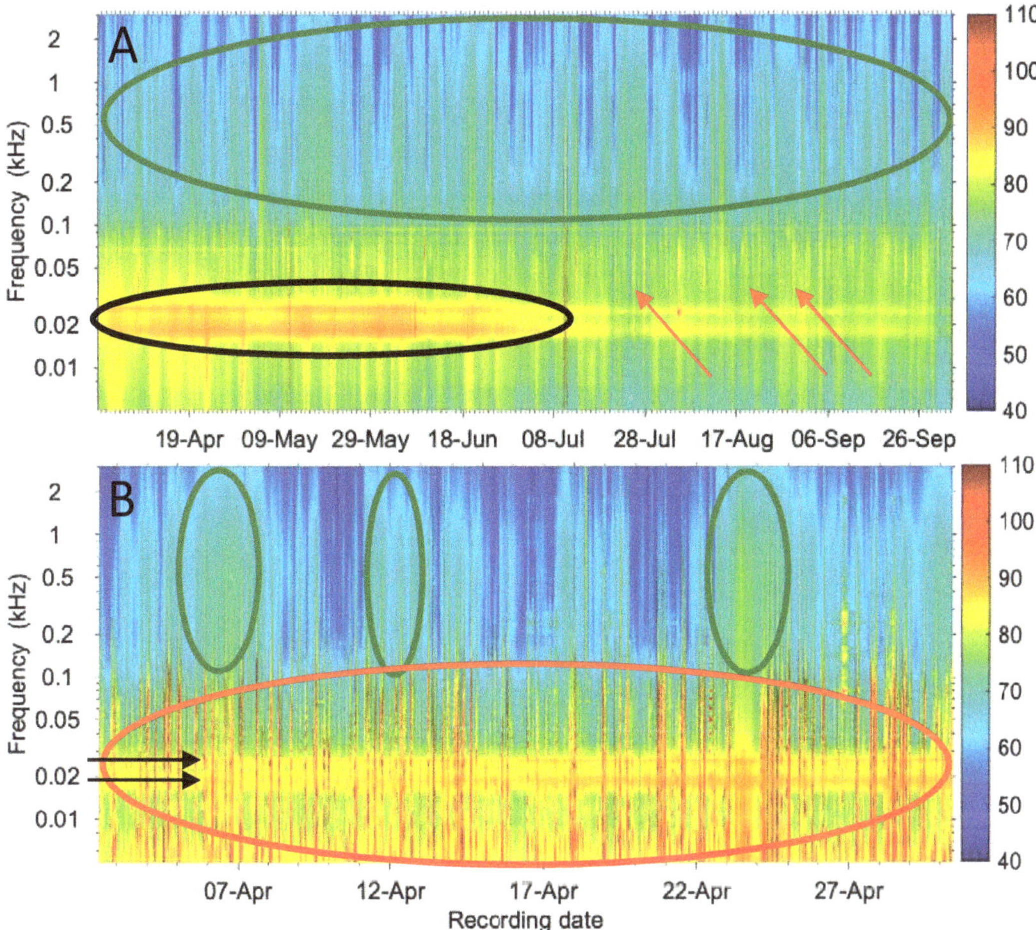

Figure 8. LTSAs [dB re 1 μPa^2/Hz] near (**A**) the Bremer Canyon, WA, site 16, and (**B**) Tuncurry, NSW, site 25. The contributions from ships, wind, and Antarctic blue whales (*Balaenoptera musculus intermedia*) are marked in red, green, and black, respectively. Only a few ships are marked in (**A**).

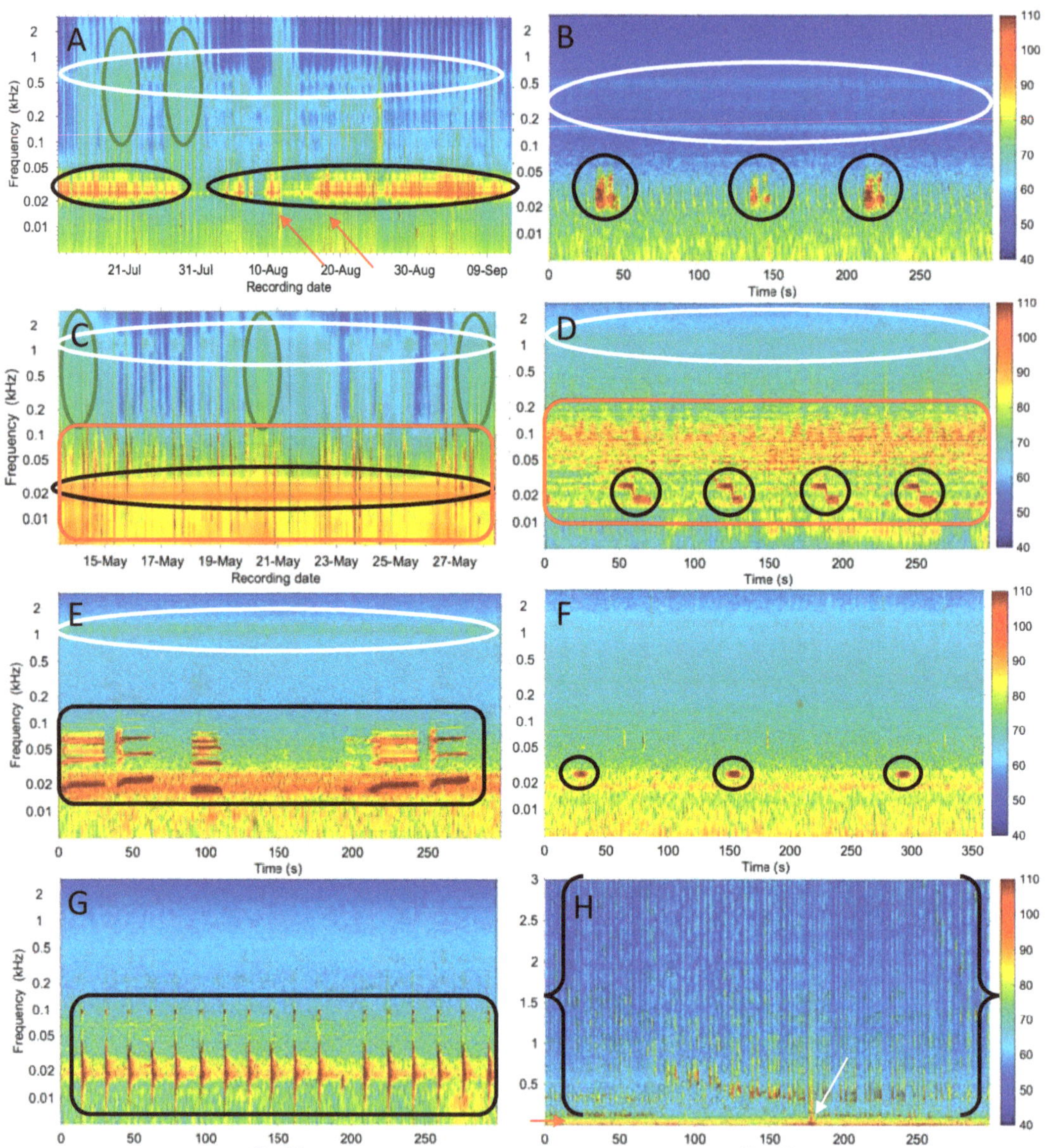

Figure 9. LTSAs and spectrograms [dB re 1 μPa^2/Hz] showing (**A**) an Omura's whale chorus and evening fish chorus at site 1; (**B**) Three Omura's whale calls and a fish chorus at site 5 (this specific example is free from ships and wind); (**C**) An Antarctic blue whale chorus and evening fish chorus at site 22; (**D**) Four Antarctic blue whale Z-calls and a fish chorus at site 23; (**E**) Pygmy blue whale song in front of the Antarctic blue whale chorus and a fish chorus at site 24; (**F**) Three spot calls at site 19; (**G**) Fin whale song at site 14; and (**H**) Humpback whale song at site 6. Note the changing x- and y-scales. All panels but H use a logarithmic y-scale. H uses a linear y-scale to stress the great bandwidth of humpback whale song (100 Hz–>3 kHz) in comparison to the narrow bandwidth of ship noise in this example (<100 Hz), resulting in humpback whales dominating the C-SEL after integration over frequency. An animal (fish?) biting on the hydrophone is marked by the white arrow. Sound from whales, fish, ships, and wind are marked in black, white, red, and green, respectively.

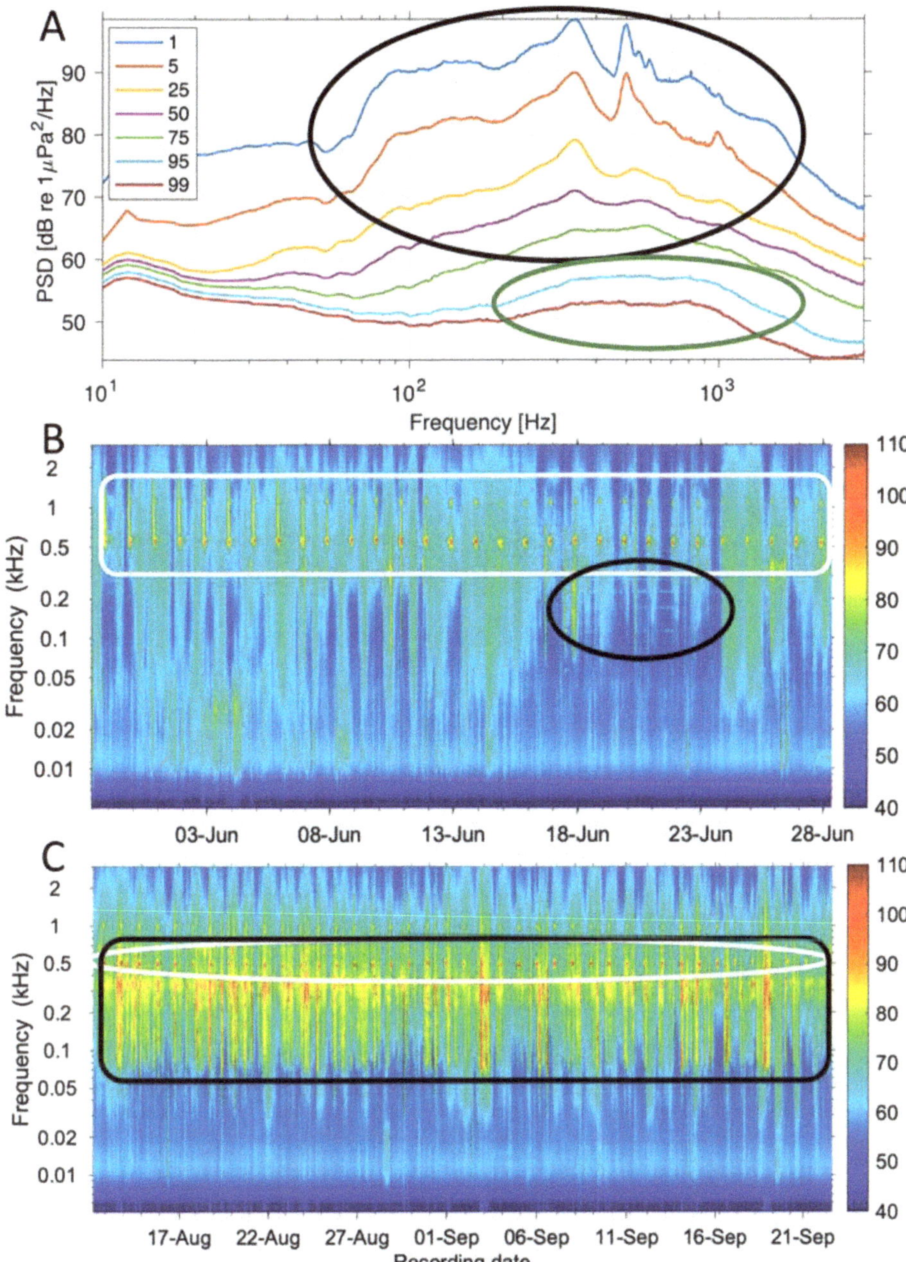

Figure 10. Winter soundscape at site 12. (**A**) Power spectral density percentiles showing domination by humpback whales from late June and fishes throughout. The curves follow the shape of the biological spectra 75% of the time (within black ellipse). The characteristic shape of wind is only seen in the absence of whales (lowest two percentiles within green ellipse); (**B**) LTSA of an evening fish chorus (within white box). A distant dwarf minke whale chorus (thin horizontal lines inside black ellipse) occurred in June–July. (**C**) LTSA showing humpback whales (within black box) and fish (within white ellipse).

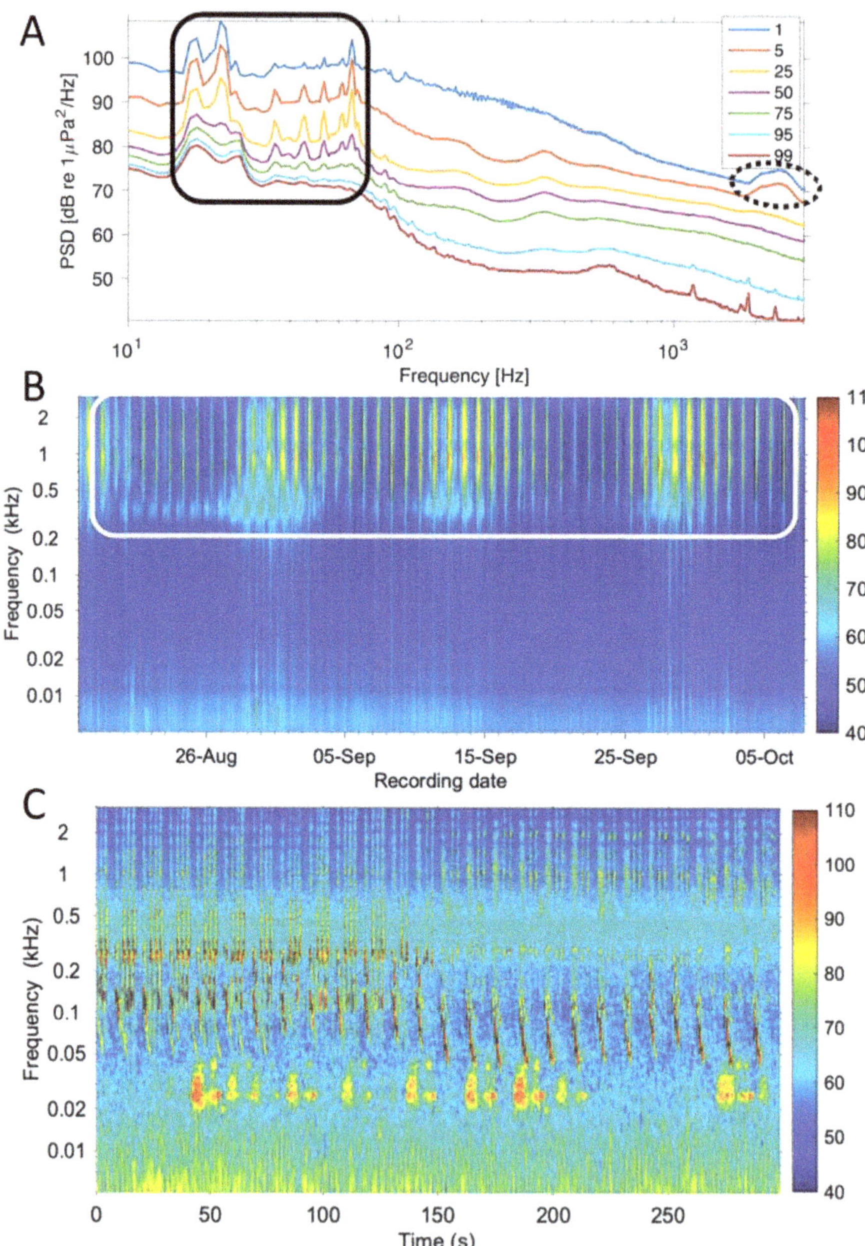

Figure 11. (**A**) Power spectral density percentiles of the winter soundscape at site 15 dominated by pygmy blue whales (within black box). The characteristic spectral shape of a fish chorus is seen at 2–3 kHz in the 1st and 5th percentiles (dotted); (**B**) LTSA of the pristine soundscape at site 4 exhibiting multiple fish choruses at night (within white box), whose intensities vary with the phase of the moon; (**C**) Spectrogram of the soundscape at site 1 showing at least two simultaneous whale species (Omura's whales at 20–50 Hz and one other at 50–3000 Hz, uncertain) and a fish chorus at 300–500 Hz.

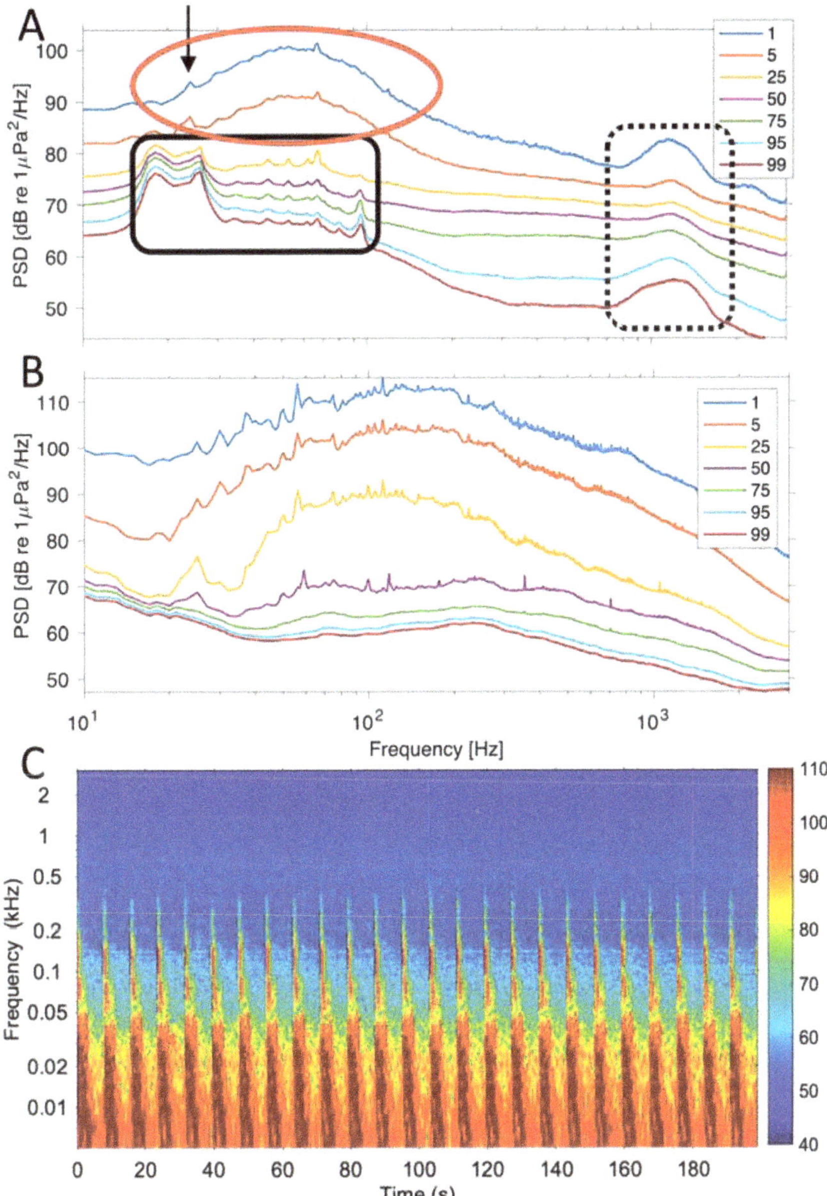

Figure 12. (**A**) Power spectral density percentiles of the winter soundscape at site 23 showing that pygmy blue whales (black box) were present the entire 6 months (because the spectral shape of their song is seen even in the 99th percentile, meaning it did not become quieter than this). However, the strongest sound in this soundscape came from ships (identified by the broad and smooth spectral hump between 20 Hz and 200 Hz; red ellipse). The spot call was also strong at this site (marked by the black arrow). The fish chorus at 800–2000 Hz (dotted box) was present the entire time as well; (**B**) Power spectral density percentiles from site 2 being entirely dominated by broadband industrial noise of unknown origin at this time; (**C**) Spectrogram of a strong seismic survey temporarily present at site 3; no other sounds were visible.

4. Discussion

The aim of our project was to develop a model for underwater ship noise in the Australian EEZ that could be used by industry and government to manage marine zones, their usage, stressors, and potential impacts. To put ship noise into context, we also modelled natural noise from wind under water. The models are based on numerous assumptions and involve a lot of averaging in space and time, leading to uncertainty. We therefore validated the models as a whole by comparing modelled sound exposure levels to measured underwater sound levels from 25 acoustic data sets collected over a 12-year span. Agreement was good when the underwater soundscape mostly contained the two sources modelled: ships and wind. Agreement was poorer when sound sources were missed (i.e., not modelled): seismic surveying, whales, and fishes.

Ship presence and movement were based on AIS data from the winter of 2015. Ships logged their positions at irregular time intervals, requiring that we interpolate between successive logs. We applied criteria for speed and direction continuity before straight-line interpolation, and where these were not met, we accepted holes in tracks, leading to an underestimation of ship time in the corresponding cells. We further had very few vessels in the smallest class (<25 m), as these mostly private recreational vessels do not log AIS positions. We therefore clearly underestimated their underwater noise contribution, in particular to coastal soundscapes. In addition, we did not take into account ships just outside of the EEZ and so underestimated noise levels near the EEZ boundary. Given that most AIS data were available for the larger and noisier vessels, we chose a monopole source depth corresponding to larger vessels (5 m) and applied this to all vessels in the model, for simplicity. The introduced uncertainty in modelled received levels is perhaps greater in winter (which we modelled) than summer, given all of our sound speed profiles exhibited a shallow surface duct of variable depth. Accounting for different source depths for the different vessel classes would require modelling sound propagation over the 64 cluster centroids in each zone multiple times, which we did not do, but could be done to improve accuracy. This might be desirable for more localised applications and modelling over smaller areas than the entire EEZ (e.g., regional seismic surveys or coastal developments). Placing the monopole at deeper depth than the propeller depth of small vessels during sound propagation modelling will likely enhance long-range received levels of the smaller, hence quieter, vessels, which are possibly underrepresented in the AIS data, meaning the errors do not add but work in reverse. Finally, the source levels produced by the RANDI model fall within the broadband quartiles reported recently [68]; however, the spectral shapes might differ. MacGillivray and de Jong [69] very recently showed that the RANDI model overpredicted source power spectral density below ~250 Hz for bulk carriers, vehicle carriers, tankers, container ships, and cruise ships, yet underpredicted source power spectral density above ~250 Hz. This might lead to differential errors in different regions (deep versus shallow water), depending on the efficiency with which sound below and above 250 Hz propagates in each environment. Other studies reported RANDI to overestimate [70,71] or underestimate, particularly above 200 Hz [72]. Underprediction of source levels by the RANDI model might be more common for the smallest vessels, in particular those with powerful motors, such as whale-watching boats and tugs [69,73–76].

In terms of the underwater sound propagation model used, the most common source of uncertainty is a lack of data on the seafloor composition and thus, acoustic properties. We used typical values from [51], but geoacoustic properties may vary from place to place. Hydroacoustic data (i.e., temperature, salinity, and sound speed profiles) were missing in some coastal zones and thus required spatial extrapolation. The equivalent fluid model applied is only approximate up to grazing angles of 50° and thus, more accurate for long-range propagation modelling. Modelling sound propagation only along bathymetry cluster centroids, instead of every source-receiver transect, introduced additional uncertainty. However, with a median water depth of 1809 m for all source cells in the entire EEZ, deviations of individual bathymetries from centroid bathymetries are likely to affect modelled received levels more in shallow and coastal rather than offshore waters. While deviations in bathymetry from cluster centroids may change the pattern of constructive and destructive interference and thus yield rather variable received levels at specific locations in space and depth, there will not be a consistent bias in modelled received levels. Modelling along centroids will lead to both over- and underprediction, depending on range, depth, and frequency. These effects will be important on a fine spatial scale, but disappear on a coarse grid. Finally, the received level depends greatly on receiver depth. We chose to plot maximum received levels over the top 200 m, corresponding to the water layer in which most baleen whales travel. Any receiver depth (or depth range) may, of course, be picked from the model results, corresponding to specific animal depths.

The wind model we used was based on the classic review done by Wenz [43]. Other models, such as the Cato model [77] extend to lower frequencies and thus, yield higher levels (up to 2 dB) in high sea states. The Cato model would reduce the model-versus-measurement difference (i.e., improve the wind noise prediction) at the wind-dominated sites (9, 16).

The map of underwater ship noise was based on AIS data from the year 2015, the map of underwater wind noise was based on wind data from 2012, and the in situ measurements were from various years (2006–2017). For a fine-scale model (i.e., small grid size), the exact positions and types of vessels would matter and therefore, validation with measurements from different years might be less successful. However, on a coarse grid, fine-scale variability averages out. For the ship noise map to differ by 3 dB, twice the number of ships (i.e., twice the power) would be needed. We showed close agreement in measured levels over consecutive years at the same sites, except when strong temporary sources occurred in some sets (e.g., industrial exploration) or when more variable, biological sources dominated in some years.

The geographic grid size chosen for the model might affect the received levels in some cells and change the ship-to-wind noise ratio. We modelled on a 5 km $\times$ 5 km grid, and so the source cells were assigned a received level at 2.6 km range. A 2.5 km $\times$ 2.5 km grid would have a mean receiver range of 1.3 km. If ships are evenly distributed within a 5 km $\times$ 5 km cell, then halving the grid size will increase received levels within the source cells by $20 \log_{10}(2) = 6$ dB. The time spent in the source cell, however, will decrease by a factor 4, or, $10 \log_{10}(4) = 6$ dB, making up for the increase in received level (i.e., decrease in propagation range and thus, propagation loss). If ships are unevenly distributed within the larger grid cell, then changing to a finer grid will yield a net increase in modelled received noise levels within source cells. In comparison, the modelled noise levels from wind, being a sheet (rather than monopole) source, will not vary with grid size as wind speed changes on a much larger spatial scale offshore. Therefore, in areas where shipping lanes are well-defined and narrow (<5 km wide), ship noise levels may exceed wind noise levels by more than modelled in this article.

Based on our model and its 25-point validation, the Australian EEZ has a higher proportion of natural underwater noise from wind over ship noise than the North Sea and likely other northern hemisphere oceans [18,32]. Part of the Australian marine soundscape appears pristine, if pristine is defined as an absence of anthropogenic noise and a richness of biological noise (see also [78]). We have shown that accurate models of the Australian marine soundscape must include biological sources (i.e., primarily whales and fishes). Natural biological and physical noise ought to be considered in management frameworks to provide context (e.g., for noise management in the Southern Ocean [79]).

Our recommendations for future work include the establishment of a databank of Australian ship source spectra as started by [80], which will allow replacing the RANDI model with monopole source spectra from actual measurements. We have shown that other anthropogenic noise sources cannot be excluded in areas and years where these dominate and their contribution to the marine noise budget should be assessed. Comparing long-term cumulative sound exposure might not be the quantity most useful to managers. Instead, sound energy could be integrated over much shorter time frames and maps of % time above certain management thresholds be plotted [81], which is likely more relevant to biological receptors than an annual or seasonal integral or average. The different sound sources have different acoustic features (e.g., ship and wind noise are continuous, while seismic surveying and pile driving are pulsed) and bioacoustic impact is likely driven by different acoustic quantities (e.g., sound exposure versus peak pressure [82]). Therefore, different quantities will have to be mapped for different types of impact. Moreover, these sources exhibit fundamentally different sound radiation fields, where an underwater explosion is a monopole, a ship is a dipole, pile driving a line source, and wind a sheet source, requiring different modelling approaches.

Supplementary Materials: The following are available online at https://www.mdpi.com/article/10.3390/jmse9050472/s1.

Author Contributions: Conceptualization, C.E., J.N.S. and D.P.; methodology, C.E.; software, C.E. and R.P.S.; formal analysis, C.E. and R.P.S.; data curation, C.E., D.P., J.N.S. and R.P.S.; writing—original draft preparation, C.E., R.P.S., D.P. and J.N.S.; writing—review and editing, C.E., R.P.S., D.P. and J.N.S.; visualization, C.E., D.P., J.N.S. and R.P.S.; project administration, D.P.; funding acquisition, D.P., J.N.S. and C.E. All authors have read and agreed to the published version of the manuscript.

Funding: This work was funded as part of the Australian National Environmental Science Programme, Marine Biodiversity Hub, under Project E2.

Institutional Review Board Statement: Not applicable.

Informed Consent Statement: Not applicable.

Data Availability Statement: A spreadsheet with the acoustic parameters that characterise each zone (i.e., mean winter sound speed profile; mean water depth and slope; sediment thickness; and compressional sound speed, shear sound speed, compressional absorption coefficient, shear absorption coefficient, and mean seafloor density) is available for download, as is a shape file of the spatially separated 28 acoustic zones (see https://tinyurl.com/3webp3pn). A spreadsheet with the sound speed profiles, water density profiles, and geoacoustic properties of the seafloor is available as Supplementary Material. Maps and data of cumulative sound exposure levels (C-SEL) from shipping over all ship classes in the Australian EEZ corresponding to Figure 6 can be found at Seamap Australia (https://seamapaustralia.org/; https://tinyurl.com/ahbs6nwr) and the AODN (https://catalogue.aodn.org.au/geonetwork/srv/eng/metadata.show?uuid=480847b4-b692-4112-89ff-0dcef75e3b84). Map and data of modelled wind noise from Figure 7 can also be found at Seamap Australia (https://tinyurl.com/a477j3b9) and the AODN (https://catalogue.aodn.org.au/geonetwork/srv/eng/metadata.show?uuid=0d3c7edc-463a-4fa0-8039-4d5a779035c3). All sites last accessed on: 30 March 2021.

Acknowledgments: We sincerely thank Robert D. McCauley for providing the sea noise data sets used for validation.

Conflicts of Interest: The authors declare no conflict of interest.

Appendix A

Step-by-step process of modelling ship noise:

1. Beginning with a GIS layer of the Australian marine bathymetry.
2. Add layers to the grid with ship positions, grouped by ship size (i.e., ship length), yielding one layer per ship class.
3. Split the EEZ grid into 28 previously determined acoustic zones.
4. For each zone:
   a. Find all grid cells that contain ships of any class, cast 36,100 km radials in 10-degree intervals, and extract bathymetry along the radials.
   b. Cluster all extracted bathymetries (over all radials around all cells with ships) with a neural network and subsequent k-means into 64 clusters.
   c. Compute sound propagation along each cluster centroid, for the centre frequencies of adjacent octave bands.
   d. For each ship size class:

     i.   Find the cells that contain ships of this class (source cells), cast 36,100 km radials in 10-degree intervals, and extract bathymetry along the radials.

     ii.   For each source cell:

- For each radial:

  - Look up into which cluster this radial went;
  - For each frequency:

    - Retrieve propagation loss as a function of range and depth.
    - Add octave band source level for this ship class.
    - Add cumulative time that a ship of this class spent in this source cell to yield sound exposure level as a function of range and depth.
    - Regrid from polar to Cartesian coordinates.

     iii.   Accumulate sound exposure over all radials and source cells to yield a 4-d matrix of cumulative sound exposure level as a function of longitude, latitude, depth, and frequency for each ship class.

    e.   Accumulate sound exposure over all ship classes.

5. Accumulate this 4-d matrix over all zones, EEZ-wide.
6. Sum over frequency to yield a 3-d matrix of cumulative sound exposure level as a function of longitude, latitude, and depth.
7. Pick the maximum cumulative sound exposure level over depth to yield a 2-d map of cumulative sound exposure level versus longitude and latitude.

References

1. Mahanty, M.M.; Sanjana, M.C.; Latha, G.; Raguraman, G. An investigation on the fluctuation and variability of ambient noise in shallow waters of south west Bay of Bengal. *Indian J. Geo Mar. Sci.* **2014**, *43*, 747–753.
2. Haver, S.M.; Fournet, M.E.H.; Dziak, R.P.; Gabriele, C.; Gedamke, J.; Hatch, L.T.; Haxel, J.; Heppell, S.A.; McKenna, M.F.; Mellinger, D.K.; et al. Comparing the underwater soundscapes of four U.S. National Parks and Marine Sanctuaries. *Front. Mar. Sci.* **2019**, *6*. [CrossRef]
3. Gage, S.H.; Axel, A.C. Visualization of temporal change in soundscape power of a Michigan lake habitat over a 4-year period. *Ecol. Inform.* **2014**, *21*, 100–109. [CrossRef]
4. Caruso, F.; Alonge, G.; Bellia, G.; De Domenico, E.; Grammauta, R.; Larosa, G.; Mazzola, S.; Riccobene, G.; Pavan, G.; Papale, E.; et al. Long-term monitoring of dolphin biosonar activity in deep pelagic water of the Mediterranean Sea. *Sci. Rep.* **2017**, *7*, 4321. [CrossRef]
5. Erbe, C.; Verma, A.; McCauley, R.; Gavrilov, A.; Parnum, I. The marine soundscape of the Perth Canyon. *Prog. Oceanogr.* **2015**, *137*, 38–51. [CrossRef]
6. McWilliam, J.N.; McCauley, R.D.; Erbe, C.; Parsons, M.J.G. Patterns of biophonic periodicity on coral reefs in the Great Barrier Reef. *Sci. Rep.* **2017**, *7*, 17459. [CrossRef]
7. Marley, S.A.; Erbe, C.; Salgado Kent, C.P.; Parsons, M.J.G.; Parnum, I.M. Spatial and temporal variation in the acoustic habitat of bottlenose dolphins (*Tursiops aduncus*) within a highly urbanised estuary. *Front. Mar. Sci.* **2017**, *4*, 197. [CrossRef]
8. McDonald, M.A.; Hildebrand, J.A.; Wiggins, S.M.; Ross, D. A 50 year comparison of ambient ocean noise near San Clemente Island: A bathymetrically complex coastal region off Southern California. *J. Acoust. Soc. Am.* **2008**, *124*, 1985–1992. [CrossRef]
9. European Commission. Future Noise Policy. Publications Office of the EU 1996, COM_1996_0540_FIN, Green Paper. Available online: https://op.europa.eu/en/publication-detail/-/publication/8d243fb5-ec92-4eee-aac0-0ab194b9d4f3/language-en (accessed on 15 March 2021).
10. European Commission. Directive 2002/49/EC of the European Parliament and of the Council relating to the assessment and management of environmental noise. *Off. J. Eur. Communities* **2002**, *L 189*, 12–25. Available online: https://eur-lex.europa.eu/legal-content/EN/TXT/PDF/?uri=CELEX:32002L0049&from=EN (accessed on 15 March 2021).
11. van der Graaf, A.J.; Ainslie, M.A.; Andre, M.; Brensing, K.; Dalen, J.; Dekeling, R.P.A.; Robinson, S.M.; Tasker, M.L.; Thomsen, F.; Werner, S. *European Marine Strategy Framework Directive—Good Environmental Status (MSFD GES): Report of the Technical Subgroup on Underwater Noise and Other Forms of Energy*; TSG Noise & Milieu Ltd.: Brussels, Belgium, 2012.
12. Erbe, C. Underwater noise from pile driving in Moreton Bay, Qld. *Acoust. Aust.* **2009**, *37*, 87–92.
13. Erbe, C.; King, A.R. Modelling cumulative sound exposure around marine seismic surveys. *J. Acoust. Soc. Am.* **2009**, *125*, 2443–2451. [CrossRef]

14. Farcas, A.; Thompson, P.M.; Merchant, N.D. Underwater noise modelling for environmental impact assessment. *Environ. Impact Assess. Rev.* **2016**, *57*, 114–122. [CrossRef]
15. Aulanier, F.; Simard, Y.; Roy, N.; Gervaise, C.; Bandet, M. Effects of shipping on marine acoustic habitats in Canadian Arctic estimated via probabilistic modeling and mapping. *Mar. Pollut. Bull.* **2017**, *125*, 115–131. [CrossRef] [PubMed]
16. Erbe, C.; MacGillivray, A.O.; Williams, R. Mapping cumulative noise from shipping to inform marine spatial planning. *J. Acoust. Soc. Am.* **2012**, *132*, EL423–EL428. [CrossRef] [PubMed]
17. Jalkanen, J.P.; Johansson, L.; Liefvendahl, M.; Bensow, R.; Sigray, P.; Östberg, M.; Karasalo, I.; Andersson, M.; Peltonen, H.; Pajala, J. Modelling of ships as a source of underwater noise. *Ocean Sci.* **2018**, *14*, 1373–1383. [CrossRef]
18. Sertlek, H.Ö.; Slabbekoorn, H.; ten Cate, C.; Ainslie, M.A. Source specific sound mapping: Spatial, temporal and spectral distribution of sound in the Dutch North Sea. *Environ. Pollut.* **2019**. [CrossRef] [PubMed]
19. Andrew, R.K.; Howe, B.M.; Mercer, J.A. Long-time trends in ship traffic noise for four sites off the North American West Coast. *J. Acoust. Soc. Am.* **2011**, *129*, 642–651. [CrossRef]
20. Andrew, R.; Bruce, M.H.; James, A.M. Ocean ambient sound: Comparing the 1960s with the 1990s for a receiver off the California coast. *Acoust. Res. Lett. Online* **2002**, *3*, 65–70. [CrossRef]
21. McDonald, M.A.; Hildebrand, J.A.; Wiggins, S.M. Increases in deep ocean ambient noise in the Northeast Pacific west of San Nicolas Island, California. *J. Acoust. Soc. Am.* **2006**, *120*, 711–718. [CrossRef] [PubMed]
22. Chapman, N.R.; Price, A. Low frequency deep ocean ambient noise trend in the Northeast Pacific Ocean. *J. Acoust. Soc. Am.* **2011**, *129*, EL161–EL165. [CrossRef] [PubMed]
23. Miksis-Olds, J.L.; Bradley, D.L.; Niu, X.M. Decadal trends in Indian Ocean ambient sound. *J. Acoust. Soc. Am.* **2013**, *134*, 3464–3475. [CrossRef]
24. Miksis-Olds, J.L.; Nichols, S.M. Is low frequency ocean sound increasing globally? *J. Acoust. Soc. Am.* **2016**, *139*, 501–511. [CrossRef]
25. Erbe, C.; Marley, S.; Schoeman, R.; Smith, J.N.; Trigg, L.; Embling, C.B. The effects of ship noise on marine mammals—A review. *Front. Mar. Sci.* **2019**, *6*, 606. [CrossRef]
26. Ferrari, M.C.O.; McCormick, M.I.; Meekan, M.G.; Simpson, S.D.; Nedelec, S.L.; Chivers, D.P. School is out on noisy reefs: The effect of boat noise on predator learning and survival of juvenile coral reef fishes. *Proc. Biol. Sci.* **2018**, *285*. [CrossRef] [PubMed]
27. Kusku, H. Acoustic sound–induced stress response of Nile tilapia (*Oreochromis niloticus*) to long-term underwater sound transmissions of urban and shipping noises. *Environ. Sci. Pollut. Res.* **2020**. [CrossRef]
28. Williams, R.; Cholewiak, D.; Clark, C.W.; Erbe, C.; George, J.C.C.; Lacy, R.C.; Leaper, R.; Moore, S.E.; New, L.; Parsons, E.C.M.; et al. Chronic ocean noise and cetacean population models. *J. Cetacean Res. Manag.* **2020**, *21*, 85–94. [CrossRef]
29. Erbe, C.; Williams, R.; Sandilands, D.; Ashe, E. Identifying modelled ship noise hotspots for marine mammals of Canada's Pacific region. *PLoS ONE* **2014**, *9*, e89820. [CrossRef] [PubMed]
30. Williams, R.; Erbe, C.; Ashe, E.; Clark, C.W. Quiet(er) marine protected areas. *Mar. Pollut. Bull.* **2015**, *100*, 154–161. [CrossRef]
31. Jensen, F.B.; Kuperman, W.A.; Porter, M.B.; Schmidt, H. *Computational Ocean Acoustics*, 2nd ed.; Springer: New York, NY, USA, 2011.
32. Farcas, A.; Powell, C.F.; Brookes, K.L.; Merchant, N.D. Validated shipping noise maps of the Northeast Atlantic. *Sci. Total Environ.* **2020**, *735*, 139509. [CrossRef]
33. Wu, L.; Xu, Y.; Wang, Q.; Wang, F.; Xu, Z. Mapping global shipping density from AIS data. *J. Navig.* **2017**, *70*, 67–81. [CrossRef]
34. McCauley, R.D.; Gavrilov, A.N.; Jolliffe, C.D.; Ward, R.; Gill, P.C. Pygmy blue and Antarctic blue whale presence, distribution and population parameters in southern Australia based on passive acoustics. *Deep Sea Res. Part II Top. Stud. Oceanogr.* **2018**. [CrossRef]
35. Dawbin, W.H. The seasonal migratory cycle of humpback whales. In *Whales, Dolphins and Porpoises*; Norris, K.S., Ed.; University of California Press: Berkeley, CA, USA, 1966; pp. 145–170.
36. Bannister, J. Status of southern right whales (*Eubalaena australis*) off Australia. *J. Cetacean Res. Manag.* **2020**, 103–110. [CrossRef]
37. Erbe, C.; Peel, D.; Smith, J.N.; Schoeman, R.P. Marine acoustic zones of Australia. *J. Mar. Sci. Eng.* **2021**, *9*, 340. [CrossRef]
38. Whiteway, T.G. *Australian Bathymetry and Topography Grid*; 2009/21; Geoscience Australia: Canberra, Australia, 2009.
39. Peel, D.; Erbe, C.; Smith, J.N.; Parsons, M.J.G.; Duncan, A.J.; Schoeman, R.P.; Meekan, M. *Characterising Anthropogenic Underwater Noise to Improve Understanding and Management of Acoustic Impacts to Marine Wildlife*; CSIRO: Hobart, Australia, 2021.
40. Breeding, J.E.; Pflug, L.A.; Bradley, M.; Herbert, M.; Wooten, M. *RANDI 3.1 User's Guide*; Naval Research Laboratory: Washington, DC, USA, 1994.
41. Durrant, T.; Hemer, M.; Trenham, C.; Greenslade, D. *CAWCR Wave Hindcast Extension Jan 2011–May 2013. v7*; CSIRO Service Collection: Canberra, Australia, 2013.
42. Saha, S.; Moorthi, S.; Wu, X.; Wang, J.; Nadiga, S.; Tripp, P.; Behringer, D.; Hou, Y.-T.; Chuang, H.-y.; Iredell, M.; et al. The NCEP Climate Forecast System Version 2. *J. Clim.* **2014**, *27*, 2185–2208. [CrossRef]
43. Wenz, G.M. Acoustic ambient noise in the ocean: Spectra and sources. *J. Acoust. Soc. Am.* **1962**, *34*, 1936–1956. [CrossRef]
44. Mathai, A.; Moschopoulos, P.; Pederzoli, G. Random points associated with rectangles. *Rend. Circ. Mat. Palermo* **1999**, *48*, 163–190. [CrossRef]
45. Vesanto, J.; Himberg, J.; Alhoniemi, E.; Parhankangas, J. *SOM Toolbox for Matlab 5*; Helsinki University of Technology: Helsinki, Finland, 2000.

46. Vesanto, J.; Alhoniemi, E. Clustering of the self-organizing map. *IEEE Trans. Neural Netw.* **2000**, *11*, 586–600. [CrossRef] [PubMed]
47. Duncan, A.; Maggi, A.L. A consistent, user friendly interface for running a variety of underwater acoustic propagation codes. In Proceedings of the Acoustics 2006, Christchurch, New Zealand, 20–22 November 2006.
48. Locarnini, R.A.; Mishonov, A.V.; Baranova, O.K.; Boyer, T.P.; Zweng, M.M.; Garcia, H.E.; Reagan, J.R.; Seidov, D.; Weathers, K.; Paver, C.R.; et al. *World Ocean Atlas 2018, Volume 1: Temperature*; National Oceanic and Atmospheric Administration: Washington, DC, USA, 2018.
49. Zweng, M.M.; Reagan, J.R.; Seidov, D.; Boyer, T.P.; Locarnini, R.A.; Garcia, H.E.; Mishonov, A.V.; Baranova, O.K.; Weathers, K.; Paver, C.R.; et al. *World Ocean Atlas 2018, Volume 2: Salinity*; National Oceanic and Atmospheric Administration: Washington, DC, USA, 2018.
50. Fofonoff, N.P.; Millard, R.C., Jr. *Algorithms for the Computation of Fundamental Properties of Seawater*; UNESCO Technical Papers in Marine Sciences; UNESCO: Paris, France, 1983; Volume 44.
51. Duncan, A.; Gavrilov, A.; Li, F. Acoustic propagation over limestone seabeds. In Proceedings of the Acoustics 2009, Adelaide, Australia, 23–25 November 2009.
52. Torgersen, T.; Jones, M.R.; Stephens, A.W.; Searle, D.E.; Ullman, W.J. Late Quaternary hydrological changes in the Gulf of Carpentaria. *Nature* **1985**, *313*, 785–787. [CrossRef]
53. Jones, M.R.; Torgersen, T. Late Quaternary evolution of Lake Carpentaria on the Australia-New Guinea continental shelf. *Aust. J. Earth Sci.* **1988**, *35*, 313–324. [CrossRef]
54. Roy, P.S.; Cowell, P.J.; Ferland, M.A.; Thom, B.G. Wave-dominated coasts. In *Coastal Evolution: Late Quaternary Shoreline Morphodynamics*; Woodroffe, C.D., Carter, R.W.G., Eds.; Cambridge University Press: Cambridge, UK, 1995; pp. 121–186. [CrossRef]
55. Heap, A.; Daniell, J.; Mazen, D.; Harris, P.; Sbaffi, L.; Fellows, M.; Passlow, V. *Geomorphology and Sedimentology of the Northern Marine Planning Area of Australia: Review and Synthesis of Relevant Literature in Support of Regional Marine Planning*; 2004/11; Geoscience Australia: Canberra, Australia, 2004.
56. Carter, R.M.; Larcombe, P.; Dye, J.E.; Gagan, M.K.; Johnson, D.P. Long-shelf sediment transport and storm-bed formation by Cyclone Winifred, central Great Barrier Reef, Australia. *Mar. Geol.* **2009**, *267*, 101–113. [CrossRef]
57. Harris, P.T.; Heap, A.D. Cyclone-induced net sediment transport pathway on the continental shelf of tropical Australia inferred from reef talus deposits. *Cont. Shelf Res.* **2009**, *29*, 2011–2019. [CrossRef]
58. Koessler, M.W. An equivalent fluid representation of a layered elastic seafloor for acoustic propagation modelling. In Proceedings of the Acoustics 2017, Perth, Australia, 19–22 November 2017.
59. Porter, M.B. *The KRAKEN Normal Mode Program*; NRL/MR/5120-92-6920; Naval Research Laboratory: Washington, DC, USA, 1992.
60. Fisher, F.H.; Simmons, V.P. Sound absorption in sea water. *J. Acoust. Soc. Am.* **1977**, *62*, 558–564. [CrossRef]
61. Schreer, J.F.; Kovacs, K.M. Allometry of diving capacity in air-breathing vertebrates. *Can. J. Zool.* **1997**, *75*, 339–358. [CrossRef]
62. McCauley, R.D.; Thomas, F.; Parsons, M.J.G.; Erbe, C.; Cato, D.; Duncan, A.J.; Gavrilov, A.N.; Parnum, I.M.; Salgado-Kent, C. Developing an underwater sound recorder. *Acoust. Aust.* **2017**, *45*, 301–311. [CrossRef]
63. Gavrilov, A.N.; Parsons, M.J.G. A Matlab tool for the characterisation of recorded underwater sound (CHORUS). *Acoust. Aust.* **2014**, *42*, 190–196.
64. Erbe, C.; McCauley, R.; Gavrilov, A.; Madhusudhana, S.; Verma, A. The underwater soundscape around Australia. In Proceedings of the Acoustics 2016, Brisbane, Australia, 9–11 November 2016.
65. Erbe, C.; Dunlop, R.; Jenner, K.C.S.; Jenner, M.-N.M.; McCauley, R.D.; Parnum, I.; Parsons, M.; Rogers, T.; Salgado-Kent, C. Review of underwater and in-air sounds emitted by Australian and Antarctic marine mammals. *Acoust. Aust.* **2017**, *45*, 179–241. [CrossRef]
66. McWilliam, J.N.; McCauley, R.D.; Erbe, C.; Parsons, M.J.G. Soundscape diversity in the Great Barrier Reef: Lizard Island, a case study. *Bioacoustics* **2018**, *27*, 295–311. [CrossRef]
67. Ward, R.; Gavrilov, A.N.; McCauley, R.D. "Spot" call: A common sound from an unidentified great whale in Australian temperate waters. *J. Acoust. Soc. Am.* **2017**, *142*, EL231–EL236. [CrossRef]
68. MacGillivray, A.O.; Li, Z.; Hannay, D.E.; Trounce, K.B.; Robinson, O.M. Slowing deep-sea commercial vessels reduces underwater radiated noise. *J. Acoust. Soc. Am.* **2019**, *146*, 340–351. [CrossRef] [PubMed]
69. MacGillivray, A.; de Jong, C. A reference spectrum model for estimating source levels of marine shipping based on Automated Identification System data. *J. Mar. Sci. Eng.* **2021**, *9*, 369. [CrossRef]
70. Chion, C.; Lagrois, D.; Dupras, J.; Turgeon, S.; McQuinn, I.H.; Michaud, R.; Ménard, N.; Parrott, L. Underwater acoustic impacts of shipping management measures: Results from a social-ecological model of boat and whale movements in the St. Lawrence River Estuary (Canada). *Ecol. Model.* **2017**, *354*, 72–87. [CrossRef]
71. Jiang, P.; Lin, J.; Sun, J.; Yi, X.; Shan, Y. Source spectrum model for merchant ship radiated noise in the Yellow Sea of China. *Ocean Eng.* **2020**, *216*, 107607. [CrossRef]
72. Simard, Y.; Roy, N.; Gervaise, C.; Giard, S. Analysis and modeling of 255 source levels of merchant ships from an acoustic observatory along St. Lawrence Seaway. *J. Acoust. Soc. Am.* **2016**, *140*, 2002–2018. [CrossRef] [PubMed]
73. Erbe, C.; Liong, S.; Koessler, M.W.; Duncan, A.J.; Gourlay, T. Underwater sound of rigid-hulled inflatable boats. *J. Acoust. Soc. Am.* **2016**, *139*, EL223–EL227. [CrossRef] [PubMed]

74. Kipple, B.; Gabriele, C. *Glacier Bay Watercraft Noise*; NSWCCD-71-TR-2003/522; Naval Surface Warfare Center: Bremerton, WA, USA, 2003.
75. Kipple, B.; Gabriele, C. Underwater noise from skiffs to ships. In Proceedings of the Fourth Glacier Bay Science Symposium, Juneau, AK, USA, 26–28 October 2004; Piatt, J.F., Gende, S.M., Eds.; U.S. Geological Survey Scientific Investigations Report 2007-5047: Juneau, AL, USA, 2007; pp. 172–175.
76. Gervaise, C.; Simard, Y.; Roy, N.; Kinda, B.; Menard, N. Shipping noise in whale habitat: Characteristics, sources, budget, and impact on belugas in Saguenay–St. Lawrence Marine Park hub. *J. Acoust. Soc. Am.* **2012**, *132*, 76–89. [CrossRef]
77. Cato, D.H. Ocean ambient noise: Its measurement and its significance to marine animals. In Proceedings of the Institute of Acoustics—Underwater Noise Measurement, Impact and Mitigation, Southampton, UK, 14–15 October 2008; pp. 1–9.
78. Marley, S.A.; Salgado Kent, C.P.; Erbe, C.; Thiele, D. A tale of two soundscapes: Comparing the acoustic characteristics of urban versus pristine coastal dolphin habitats in Western Australia. *Acoust. Aust.* **2017**, *45*, 159–178. [CrossRef]
79. Erbe, C.; Dähne, M.; Gordon, J.; Herata, H.; Houser, D.S.; Koschinski, S.; Leaper, R.; McCauley, R.; Miller, B.; Müller, M.; et al. Managing the effects of noise from ship traffic, seismic surveying and construction on marine mammals in Antarctica. *Front. Mar. Sci.* **2019**. [CrossRef]
80. Erbe, C.; Duncan, A.; Peel, D.; Smith, J.N. *Underwater Noise Signatures of Ships in Australian Waters*; Centre for Marine Science and Technology, Curtin University: Hobart, Australia, 2020.
81. Merchant, N.D.; Faulkner, R.C.; Martinez, R. Marine noise budgets in practice. *Conserv. Lett.* **2018**, *11*, e12420. [CrossRef]
82. Southall, B.L.; Finneran, J.J.; Reichmuth, C.; Nachtigall, P.E.; Ketten, D.R.; Bowles, A.E.; Ellison, W.T.; Nowacek, D.P.; Tyack, P.L. Marine mammal noise exposure criteria: Updated scientific recommendations for residual hearing effects. *Aquat. Mamm.* **2019**, *45*, 125–232. [CrossRef]

Journal of
*Marine Science
and Engineering*

Review

A Review and Meta-Analysis of Underwater Noise Radiated by Small (<25 m Length) Vessels

Miles J. G. Parsons [1],*, Christine Erbe [2], Mark G. Meekan [1] and Sylvia K. Parsons [2]

1   Australian Institute of Marine Science, Perth, WA 6009, Australia; m.meekan@aims.gov.au
2   Centre for Marine Science & Technology, Curtin University, Bentley, WA 6102, Australia;
     c.erbe@curtin.edu.au (C.E.); sylvia.osterrieder@gmail.com (S.K.P.)
*   Correspondence: m.parsons@aims.gov.au; Tel.: +61-8-6369-4053

Abstract: Managing the impacts of vessel noise on marine fauna requires identifying vessel numbers, movement, behaviour, and acoustic signatures. However, coastal and inland waters are predominantly used by 'small' (<25 m-long) vessels, for which there is a paucity of data on acoustic output. We reviewed published literature to construct a dataset (1719 datapoints) of broadband source levels (SLs) from 17 studies, for 11 'Vessel Types'. After consolidating recordings that had associated information on factors that may affect SL estimates, data from seven studies remained (1355 datapoints) for statistical modelling. We applied a Generalized Additive Mixed Model to assess factors (six continuous and five categorical predictor variables) contributing to reported SLs for four Vessel Types. Estimated SLs increased through 'Electric', 'Skiff', 'Sailing', 'Monohull', 'RHIB', 'Catamaran', 'Fishing', 'Landing Craft',' Tug', 'Military' to 'Cargo' Vessel Types, ranging between 130 and 195 dB re 1μPa m across all Vessel Types and >29 dB range within individual Vessel Types. The most parsimonious model (22.7% deviance explained) included 'Speed' and 'Closest Point of Approach' (CPA) which displayed non-linear, though generally positive, relationships with SL. Similar to large vessels, regulation of speed can reduce SLs and vessel noise impacts (with consideration for additional exposure time from travelling at slower speeds). However, the relationship between speed and SLs in planing hull and semi-displacement vessels can be non-linear. The effect of CPA on estimated SL is likely a combination of propagation losses in the shallow study locations, often-neglected surface interactions, different methodologies, and that the louder Vessel Types were often recorded at greater CPAs. Significant effort is still required to fully understand SL variability, however, the International Standards Organisation's highest reporting criteria for SLs requires water depths that often only occur offshore, beyond the safe operating range of small vessels. Additionally, accurate determination of monopole SLs in shallow water is complicated, requiring significant geophysical information along the signal path. We suggest the development of appropriate shallow-water criteria to complete these measurements using affected SLs and a comprehensive study including comparable deep- and shallow-water measures.

Keywords: small vessel source levels; acoustic techniques; hydrophone-based observations

Citation: Parsons, M.J.G.; Erbe, C.; Meekan, M.G.; Parsons, S.K. A Review and Meta-Analysis of Underwater Noise Radiated by Small (<25 m Length) Vessels. *J. Mar. Sci. Eng.* **2021**, *9*, 827. https://doi.org/10.3390/jmse9080827

Academic Editor: Michele Viviani

Received: 13 July 2021
Accepted: 27 July 2021
Published: 30 July 2021

Publisher's Note: MDPI stays neutral with regard to jurisdictional claims in published maps and institutional affiliations.

1. Introduction

It is globally accepted that anthropogenic noise is an acute and chronic stressor of marine taxa that can alter marine soundscapes on multiple temporal, frequency and spatial scales, disrupting the acoustic characteristics of ecosystems [1,2]. The sound emitted by vessels under power (particularly large ships; [3]) is a key component of this noise and the size of the commercial shipping fleet has been growing year on year [4]. Vessel noise has been shown to increase stress levels, alter behaviour, displace individuals and mask communication in marine mammals, fish, and invertebrates at multiple life stages, potentially leading to population consequences, e.g., [5–17]. Thus, minimising the exposure of marine fauna to vessel noise is likely to have significant benefits for the global marine ecosystem.

Modelling of instantaneous and cumulative sound exposure levels at any given location requires knowledge of the number of vessels, the level of noise produced by each and understanding their different operations. There are several potential sources of noise emitted by individual vessels, but oscillating bubbles produced by sheet and vortex cavitation near the turning propeller dominate the acoustic output, emitting acoustic energy at frequencies related to diameter range of the bubbles [18–22]. These levels can be broadly estimated from a selection of engineering coefficients including the cavitation number (a function of size, speed, blade number, and depth of the propeller as well as environmental factors), block coefficient (a function of vessel length, breadth, and draft), and Admiralty coefficient (a function of vessel power, speed, and displacement) [18]. However, these values are rarely, if ever, reported in bioacoustics literature and for broader application, it may be more prudent to consider the individual contributing characteristics (e.g., power, displacement and speed), rather than the coefficients themselves (e.g., Admiralty coefficient). In addition, although cavitation contributes significant energy to the broadband level, it is not the sole source of noise, otherwise electric vessels would offer limited noise reduction.

Significant effort has been invested in characterising the sounds of large ships and understanding the potential drivers of source levels see [6,19–22], which have been principally related predominantly to the size and speed of the vessel in the bioacoustics literature [19]. There are, however, little data publicly available on the source levels of small (classed here as <25 m length) vessels and no comprehensive analysis linking the different vessel types within this size class to understand how different vessel characteristics contribute to the acoustic signature.

Anthropogenic noise around offshore shipping lanes is dominated by large (>25 m) commercial ships, whereas shallower coastal waters and inland waterways also host to many smaller vessels that are more variable in speed and highly mobile. These small vessels often traverse spatially controlled areas (e.g., vessel channels) and can have very shallow drafts (<1 m) so they can operate in waters as shallow as <2 m depth, positioning them close to sessile, sedentary or site-attached taxa [11,23,24]. Despite multiple sources of anthropogenic noise in these waters, such as personal watercraft (jetskis), planes, helicopters, unmanned aerial vehicles (drones), swimmers, surfers, scuba divers, and increasingly, remotely operated underwater vehicles, etc., e.g., [25–29], the daytime soundscapes of this habitat are often dominated by the sounds of small vessels, e.g., [30,31].

Small vessels have many applications and designs vary significantly from planing to displacement hulls, single or multiple hulls, using one or more inboard or outboard engines of varying power. Flat (planing) hulls are designed to quickly rise out of the water with increasing speed until the boats skim along the surface at high speed, rather than pushing the water aside (e.g., skiffs, rigid hull inflatable boats {RHIBS}). Displacement hulls are round-bottomed, ploughing through the water and buoyed by the water the hull displaces (e.g., landing craft, cargo vessels). Semi-displacement hulls, such as fishing vessels and mono- or multi-hull cruisers, are a combination of the two hull types and displace water at low speeds, but able to semi-plane at cruising speeds. Thus, their acoustic signature and broadband source level may vary significantly with vessel design and operation.

2. Definitions and Scope

This study reviews the available published literature on noise emitted by small vessels and conducts a meta-analysis of data from studies that provide a source level estimate, matched with additional information about factors that contribute to variance in the acoustic output. We have considered three types of measure to quantify the broadband acoustic levels at a nominal 1 m range:

1. Radiated noise levels (abbreviation: RNL; symbol: L_{RN}), defined as the level of the product of the distance d from a ship reference point of a sound source and the far-field root mean-square sound pressure p_{rms} at that distance for a specified reference value. The RNL is computed as:

$$L_{RN} = 20 \, \log_{10}\left(\frac{p_{rms}}{p_0}\right) + 20 \, \log_{10}\left(\frac{d}{d_0}\right) \tag{1}$$

2. Monopole source levels (abbreviation: MSL; symbol: L_{SL}), defined as the mean-square sound pressure level at a distance of 1 m from a hypothetical monopole source, placed in a hypothetical infinite lossless medium. The MSL is determined by adding the propagation loss N_{PL} to the mean-square sound pressure level $L_p{}^2$ measured at some range d:

$$L_{SL} = N_{PL}(d) + L_p{}^2(d) \tag{2}$$

The propagation loss N_{PL} is determined through modelling (e.g., a parabolic equation model), accounting for the effects of the local environment at the time.

3. Environment-affected source levels (abbreviation: ASL; symbol: L_{ASL}), defined as the mean-square sound pressure level at a distance of 1 m in a natural environment (i.e., with existing surface boundaries) and thus affected by the local environment at the time. The ASL is determined by linear regression of the mean-square sound pressure level measured at a series of ranges.

The first two of these measures are defined in the International Standards Organisation (ISO) standards 17208 (parts 1 and 2) to measure and report noise levels of vessels in deep water [32,33]. However, the highest standard of ISO criteria to measure radiated noise of vessels requires hydrophones to be positioned at vertical incidence angles of 15°, 30°, and 45° to the vessel at a minimum (slant) range of 100 m (i.e., 70 m water depth) in waters depths of a minimum of 150 m [32,33]. Such minimum depth and 45° elevation requirements cannot be achieved in many coastal situations and might involve potentially unsafe operations for some small vessels as they would require travelling significant distances offshore. Additionally, without accurate knowledge of the geophysical characteristics of the seafloor, propagation losses modelled during the estimation of MSLs can be misleading, particularly at low (tens to low hundreds of hertz) frequencies [31]. As a result, some reports provide ASLs determined from linear regression of numerous recordings taken at multiple ranges from a vessel to empirically model the dipole source signal, as it is affected by the shallow waters and interactions with water and seabed surfaces, e.g., [31,34].

3. Objectives

The aim of our study was to improve our understanding of driving factors and variance in noise emission by small vessels, which we achieved through the following objectives:

1. Identify studies that provide broadband source level estimates for individual small vessels, together with appropriate vessel specifications, operations and environmental data.
2. Access supplementary data from these studies (i.e., additional recordings of vessels that were available but not included in the publication, but meet the same standards of quality as those reported).
3. Collate data from (1) and (2) to identify potentially influential common factors that were reported alongside the estimated source levels and produce a dataset from which a statistical model can be developed to identify their contribution to variance in those estimates. This is not a model to predict noise from small vessels, but to tease out the drivers of variation in the available reported data.
4. Compare and characterise the inter-study variability in methodologies and reported source levels.
5. Review and describe the various factors that are known to contribute to variations in the vessel spectra and estimated source levels and, where data are available, quantify their contribution to the variance.

Completing these objectives will reconcile all the informative available data previously reported on small vessel source levels in the bioacoustics literature. It will provide a better understanding into the factors driving estimated source levels recorded in real-world conditions and assist in setting regulations that may limit the levels of noise to which nearby aquatic fauna are subjected. It will also provide a qualitative assessment of the impacts the use of different sites, equipment and methodologies have on the estimation of acoustic signatures.

4. Materials and Methods

To provide an element of continuity, we have followed similar methodologies for gathering data and statistical analysis as Chion et al. [19] in their evaluation of large commercial vessels. To produce a dataset that encompassed as many individual estimates of source levels of small vessels as possible we have:

1. Conducted a literature review using Google Scholar and Scopus to search for papers that included "vessel" or "boat" AND "source level" OR "sound signature" OR "acoustic signature" OR "noise signature" OR "radiated noise" in the title or keywords;
2. Assessed citations found within these publications to see if they met the required criteria for inclusion in the dataset (described below) even though they were missed in the original search;
3. Included data from publications that provided a broadband source level estimate in the dataset;
4. Only included reported values of individual vessel passes and not aggregated assessments of the source levels in the dataset for statistical modelling;
5. Examined previous reports authored by the investigators of this study that met the criteria below for additional recordings that were not included in the original publication, but were collected under the same standard and protocols as those in the publication. Available data were added to the overall dataset.

The estimated source levels for all samples in the dataset were grouped into 11 vessel types (Table 1) and logged with all pertinent information provided in the report. Specifically, intrinsic (originating from the vessel's own static and dynamic characteristics) and extrinsic (data collection, types of propagation model applied, and the environment affecting the recordings) factors [19] that have been previously shown to contribute to vessel source spectra were recorded (Table 2). An initial assessment found that several of these factors (from here called 'predictor' variables) were rarely reported and were discarded from consideration in statistical modelling, thus only recordings that provided information on the bulk of the remaining factors were retained.

This dataset comprised 1719 recordings of individual passes from 224 vessels, with 1 to 349 passes of any individual vessel. The remaining intrinsic and extrinsic factors were defined as either categorical predictor variables (including Vessel type (11 levels), Hull type (3 levels), Vessel ID, Propagation model (3 levels), and Reference (8 levels—each individual study)) or continuous predictor variables (Vessel length, Speed, Engine power, Water depth, Hydrophone depth, closest point of approach (CPA), and CPA/Hydrophone depth) [35].

Speed has been reported as a significant predictor of variation in ship source level, e.g., [19,21,36]. Therefore, as an initial test, linear regression of speed and source level of the entire dataset were investigated in the form of:

$$SL = C_v \times \log_{10} \frac{v}{v_r} + k \tag{3}$$

where SL is the source level (RNL, MSL or ASL, dB re 1 μPa m), Cv is the velocity coefficient, v is the vessel velocity (kn), v_r is a reference velocity (1 kn) and k is the intercept.

Table 1. Categories of small vessels and their descriptions included in the study.

Vessel Category	Vessel Type	Description	Hull Type	Lengths in Dataset (m)
1	Skiff	Single person or small crew, flat bottom, a pointed bow, and a square stern with tiller steering. Typically <10 m length	Planing	4.3–10
2	RHIB	Rigid hull inflatable boat. Typically, centre console, occasionally stern tiller <10 m length	Planing	5–9.5
3	Monohull	Single hull craft without further categorisation. Changes from displacement to planing condition as speed increases. Speed at which change between ploughing and planing occurs is vessel specific.	Semi-displacement	3.4–24
4	Tug	Small, powerful vessel designed particularly to tow or push barges and large ships.	Displacement	15–25
5	Landing craft	Any naval craft designed for conveying troops and equipment from a transport to a beach in an amphibious assault. Typically includes flattened front.	Displacement	6.4–7.9
6	Catamaran	Equivalent of a monohull vessels, but with two separated slender hulls	Semi-displacement	8.2–25
7	Fishing boat	Commercial vessel designed with sufficient power to trawl or pull nets. Changes from displacement to planing hull as speed increases. Speed at which change between ploughing and planing occurs is vessel specific.	Semi-displacement	10.4–25
8	Sailing boat	Monohulled vessel	Displacement	9–19.8
9	Electric	Any vessel powered by an electric motor, often limited in speed. Data recorded at ploughing speeds.	Displacement	10
10	Cargo	Monohull vessel used to transfer cargo. No further information provided.	Displacement	24–25
11	Military	Monohull vessel used for naval procedures. No further information provided.	Displacement	22

An issue with multi-variate analysis is the potential for correlation between predictor variables [35], of which there are several in Table 2 that could intuitively be related. Visual data exploration was conducted to display outliers and aid the testing of collinearity of covariates based on variance inflation factors (VIFs). To avoid including collinear variables, covariates with a VIF > 3 were not added to the model [35,37]. During this assessment, the following observations and actions were made:

- Vessel Type, Length and Hull Type displayed collinearity. Source level displayed a relationship with speed that appeared to vary with vessel type. Further, hull type was reduced to two levels after accounting for missing values. As a result, vessel type was chosen over length and hull type.
- Water Depth and Hydrophone Depth were highly collinear. Water Depth provided a greater range of values to describe variation than Hydrophone Depth. Additionally, the Closest Point of Approach (CPA) was often significantly large with respect to the Hydrophone Depth, and therefore provided a variable that describes the impact of the largest interference pattern, the Lloyd's mirror effect, on source level that could otherwise be explained by hydrophone depth. Therefore, Water Depth was retained for the model and Hydrophone Depth was discarded.
- Propagation Model showed collinearity with Vessel Type, Water Depth and Engine Power and was not included in the model.

Table 2. Factors contributing to variations in source level and justifications for consideration in modelling in this study.

Factor		Justification	Example Citation	Considered
Intrinsic characteristics				
Vessel type		Vessels designs are task-specific, thus type can impact acoustic signature	[21]	Yes
Hull type		Affects the amount of water displaced and therefore the power required to propel the vessel	[38]	Yes
Vessel length		Larger vessels require additional power (and thus noise) to move through water	[22]	Yes
Beam			[3]	No
Engine power		Larger powered engines are associated with increased noise levels	[19]	Yes
Engine type		Whether the engines are inboard or outboard may affect the position of the exhaust and noise radiated into the water	[19]	No
Waterline angle		Angle of the hull to the waterline affects the propeller load and vessel drag, thus power and noise	[31]	No
Draft		Affects the amount of water displaced by vessel and thus power required	[34]	No
Propeller	Size	Propeller size, number of blades, gear ratio and propeller revolution rate dictate the frequency of acoustic tones	[22,38]	No
	No. blades			No
	Speed (rpm)			No
	Gear ratio			No
	Blade angle			No
Propeller (source) depth		Source depth affects signal propagation. However, in small vessels there is minimal difference in propeller depth between vessels	[31]	No
Onboard machinery		Provides additional sources of noise. The number, size, power and condition of this machinery affects the noise emitted	[39,40]	No
Speed		Source level shown to increase with increasing speed in large vessels	[22]	Yes
Displacement		Increased load and resulting displacement requires additional power to attain the same speed	[34]	No
Extrinsic factors				
Water depth		Water depth affects the propagation of an acoustic signal	[41]	Yes
Closest point of approach (CPA)		The relationship between range and hydrophone depth can affect the Lloyd mirror pattern and therefore the received level and estimated noise levels	[19]	Yes
Hydrophone depth			[19]	Yes
Type of propagation model applied		Whether a monopole or dipole source is considered and whether this is considered in the propagation model (RNL or MSL) affects the source estimate.	[19]	Yes
Model source depth		Position of the source in the propagation model affects the received levels and may not always exactly reflect the actual propeller depth	[31]	No

Table 2. *Cont.*

Factor	Justification	Example Citation	Considered
Hydrophone type and system frequency response	Sensitivity is often considered flat over all frequencies even though calibrations are often only conducted at a single frequency (typically 250 Hz). If frequency response is poor at low frequencies, energy is underestimated and thus so is the source level	[42]	No *
Frequency band	If vessel energy is broadband and includes low-frequency energy, an increase in the lower frequency of the reported band can omit energy from the estimate	[43]	No †
Environment conditions	Sea states/currents affect power required to achieve given speed and direction	[22]	No

* Although this information is often presented, details of frequency-dependent calibrations are often not given. † Frequency band is usually reported but has an interactive effect with type of propagation model used, hydrophone depth, water depth and CPA, and comprises two boundaries (high and low frequencies) as continuous variables and would have been complex to include in the statistical model.

Engine Power data displayed imbalance amongst Vessel Types (VIF = 3), therefore, the validity and interpretation of any models that included both variables was assessed with care. We applied generalised additive mixed models (GAMMs) with a Gaussian distribution and identity link function for continuous data to investigate the factors driving estimated source levels of small vessels. GAMMs were chosen to enable the inclusion of smoothing splines (smoothers) in the model, as data exploration revealed a non-linear effect of $\log_{10}(speed)$, which was supported by similar observations found in previous reports, e.g., [31,38]. GAMMs also allow the inclusion of random effects to account for dependencies in the data [35]. Such a model allows investigation of the factors influencing the reported source level. However, as there are intrinsic and extrinsic factors driving the variance in estimated levels, the model was not designed to predict noise from vessels.

For inclusion in the model, the estimated source level required associated data on all chosen variables, which was not always available. Further, each tested level within the model required >20 data points to provide enough information [44] and so those levels of variables with <20 data points were not included. To allow the function to calculate the random intercept, a_v, on more than a single observation, and to avoid an unbalanced random effect, we also decided to only include vessels that had been recorded at least three times. As a result, the following data points and resulting categories were removed from the dataset:

- Engine Power and/or CPA was not reported for 226 data points, which were excluded, removing all data from the Military and Cargo Vessel Types.
- Fishing and Tug Vessel Types only contained 16 and 8 measurements, respectively, that had data on Engine Power, whereas Landing Craft originally only included 8 data points, thus these Vessel Types were not included in the model.
- Data exploration showed that the Vessel Speed and CPA data collected for Electric and Sailing vessels were skewed (e.g., only one electric vessel, recorded multiple times at one speed and one CPA). Therefore, these two levels, with <23 data points each, did not contain enough information to produce sensible smoothers and were not included in the model.
- Several studies reported only one or two measurements for individual vessels (e.g., Veirs and Veirs [3]) and had to be removed.
- In the remaining datasets, a total of 64 data points were present for vessels that were recorded less than three times and were not included in the final dataset.
- Additionally, four CPA outliers, which appeared to be influential data points, were also removed, two from Monohulls and two from RHIB vessel types.

The resulting full model comprised Vessel Type, Speed, CPA, Water Depth and Engine Power as fixed effects. As Speed and CPA displayed non-linear effects on source level during data exploration, these two factors were included as variable effects with smoothing splines (Table 3). Preliminary plots also suggested that the effects of Speed on source level were dependent on Vessel Type. Therefore, the smoothers for speed were added to the model separately for each Vessel Type to investigate these individual relationships, i.e., allowing different speed profiles for each Vessel Type. Vessel ID and Reference were added as random effects to account for dependencies in the data, i.e., to model the correlation between all source levels reported for the same vessels, as well as different vessel types within particular studies. This was designed to account for the risk that observations from an individual vessel (or reference) may be more similar to each other than when compared to observations from different vessels (or references), in part due to the varying sample sizes per vessel and reference. Models were fitted with the *gamm4* package in R [45], as it allows the inclusion of nested random effects.

Table 3. Available variables and their measured or calculated units to include into the model, including available sample sizes and number of categorical levels gathered from published data (raw data only included), and sample sizes and number of categorical levels included in the final model.

Variable	Data Type	Overall Sample Size (No. Levels)	Available No. Levels after Data Exploration
Vessel ID	Categorical	1719 (224)	49
Vessel Type	Categorical	1719 (11)	4
Hull Type	Categorical	1719 (3)	2
Length	Continuous	1719	
Speed	Continuous	1715	
Engine Power	Continuous	1519	
Water depth	Continuous	1714	
Hydrophone depth	Continuous	1714	
Closest Point of Approach	Continuous	1689	
Propagation Model	Categorical	1719 (3)	3
Reference	Categorical	1719 (15)	8

4.1. Model Validation

Model construction and validation started with the most complex model [37,46,47], which included all selected and available independent effects that may have driven variations in source level estimates, based on prior knowledge. Pearson residuals were plotted versus fitted values to validate the model, ensuring no obvious pattern was left unaccounted for. Pearson residuals were also plotted against each covariate in the model and each available covariate not in the model to ensure the model assumption of homogeneity was valid [37,47]. Generalised additive models (GAMs) were also applied to the Pearson residuals, used as the response variable, to check for nonlinear patterns left in each of the continuous covariates. Therefore, each continuous covariate was added as a smoothing function, which was then checked for its significance. A non-significant smoother means that no nonlinear pattern is left in the residuals [47].

Following the full model that included all chosen covariates, non-significant covariates ($p > 0.05$) were removed in a stepwise approach, with the least significant variable first. Each subsequent model was refitted and validated after each adjustment. Models were compared using Akaike's information criterion (AIC), where a reduction by >2 units represents an improved model [48]. All analyses and corresponding figures were produced using R version 3.6.3 [49] run through RStudio Version 1.2.5042—© 2009–2020 RStudio.

4.2. Within Study Assessments

4.2.1. Applied Frequency Band

Chion et al. [19] and McKenna et al. [21] highlighted the high levels of low-frequency (<100 Hz) noise emitted by large vessels, which has also been observed in some recordings of small vessels [38,50]. Therefore, reports that do not include these frequencies may exclude portions of the vessel's acoustic signal. To quantify the impact of the frequency band reported, we looked at 95 example recordings from Wladichuk et al. [51] that included one-third octave band levels between 10 Hz and 63 kHz for each estimate and closely met the ISO guidelines for RNL and MSL estimation. We compared the full broadband RNL estimates for each vessel and speed with estimates for the same recording with the energy from one-third octave bands of centre frequencies between 10 Hz and 100 Hz incrementally removed from the estimate.

4.2.2. Propagation Model and Equipment Configuration

Four studies provided sufficient information to directly compare their estimates of MSL and RNL for each vessel pass. Wladichuk et al. [50,51] and Erbe [36] made recordings of vessel passing in ≈70 m of water, while Parsons et al. [52] and Parsons and Meekan [31]) reported on recordings in 4 and 8 m depths of water, respectively. Data for individual passes was accessible for all studies. Wladichuk et al. [51] also provided one-third octave band levels for both the MSL and RNL estimates, which allowed a comparison of the two approaches at the one-third octave, as well as broadband level.

Finally, we also compared the estimates reported by Erbe [36], who recorded the same vessel passes using two hydrophones positioned at two of either 5 m, 10 m and, on one occasion, 25 m water depth. In that study, MSLs were calculated using a ray tracing propagation model down to 100 Hz and RNLs were provided for each measurement.

5. Results

5.1. Characterisation of Estimates from Selected Studies

Vessel recordings were collated from 17 scientific studies, each reporting estimated source levels for between one and 66 small vessels (4.3 to 25 m in length), from between one and 349 recordings, for a total of 224 vessels (Figure 1, Tables 1 and 4). The initial dataset comprised 1719 estimated source levels where the mean level across all reports was 170 dB re 1 μPa m (max = 195, min = 130.5, standard deviation = 7.3 dB). These studies were conducted at various locations around the world in water depths between 4 and 200 m, at ranges of 10 to 2100 m, using a variety of equipment, with source level estimates analysed across a range of frequency bands (see Table 4, for technical details of each report). The initial test of a relationship between speed and source level across all vessel types produced a velocity coefficient (C_v) of 10.4, with an intercept of 148.9.

Some of these reports modelled the vessel as monopole sources, although most back-propagated their received levels as surface-affected dipole sources (either RNLs or ASLs). Each study reported on different vessel classes, with some including a breakdown of received one-third octave band levels or source levels for each recording, others included full source spectra and some reported only a single source level value (Table 4). Richardson et al. [53] provided information on measures from Buck and Chalfont [54], Miles et al. [55], Malme et al. [56], and Green [57], though full details of their studies were not available. Seven of the assessed reports provided estimates that included data points with associated information on variables that met the criteria for inclusion in the GAMM model (see Table 4 for summary information on each report).

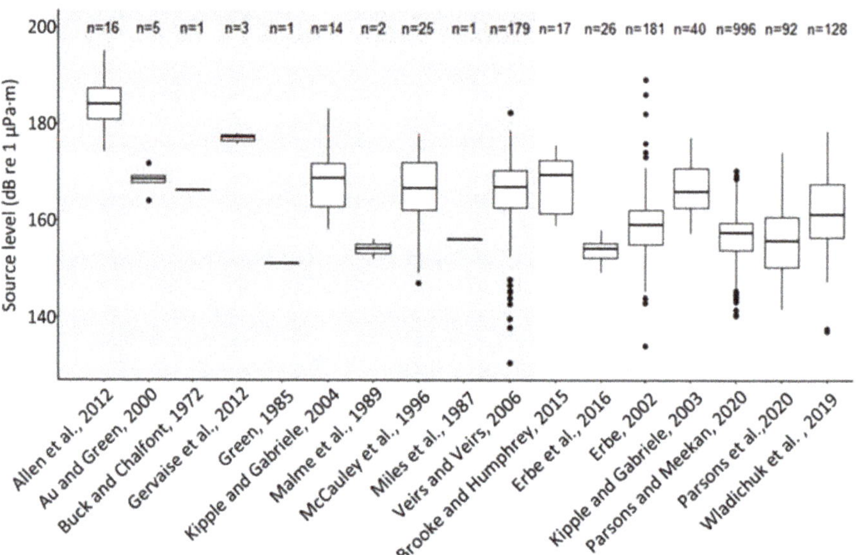

Figure 1. Box and whisker plot of reported source levels in 17 'Reference' studies showing mean, inter-quartile range, 5th/95th percentiles and outliers (vertical black line, upper/lower bounds of boxes, upper and lower extents of vertical lines, and dots, respectively). Data from studies in white area were included in the final model, while data in greyed area did not provide sufficient information or data points per vessel for statistical modelling. Number of samples (*n*) available from each reference are shown [3,31,34,36,38,51,52,54–63].

Table 4. Details of the experimental designs and data processing of examined studies.

				Protocol					
				Sample Size (n)					
				From Reference Study		In Statistical Model			
Article	Year	Location	Water Column (m)	Recordings	Vessels	Recordings	Vessels	Standard	CPA (km)
Allen et al. [58]	2011	Bar Harbour, ME, USA	30–62	16	16	0	0	x	0.2–2.1
Au and Green [59]	1999	Maui, HI, USA	15–30	5	5	0	0	x	0.091
Brooker and Humphrey [62]	2014	Tyneside, UK	100	17	1	17	1	x	0.1
Erbe [36]	1999	Haro Strait, BC, CAN and WA, USA	70	181	17	157	14	x	0.01–1.85
Erbe et al. [38]	2016	Cockburn Sound, WA, AUS	8	26	1	26	1	x	0.008–0.018
Gervaise et al. [60]	2009	Saguenay Fjord, QC, CAN	170	3	3	0	3	x	0.214
Kipple and Gabriele [61]	2000, 2002	Glacier Bay, AK, USA	55–61	39	14	36	12	x	0.5
Kipple and Gabriele [63]	2003	Glacier Bay, AK, USA.	55	14	9	0	0	x	0.5
McCauley et al. [34]	1996	Hervey Bay, QLD, AUS	20	26	12	0	0	x	Undefined
Parsons and Meekan [31]	2019	Lizard Island, QLD, AUS	10	996	3	996	3	x	0.01–0.1
Parsons et al. [52]	2018	Swan River, WA, AUS	4	34	2	0	1	x	0.015–0.05
Parsons et al. [52] *	2018	Swan River, WA, AUS	4	58	51	0	0	x	0.015
Wladichuk et al. [50,51]	2017	Haro Strait, BC, CAN and WA, USA	200	128	25	111	15	x	0.11
Veirs et al. [3]	2011	Haro Strait, BC, CAN and WA, USA	70	179	66	0	0	x	1–3
Total				1727	230	1355	49		

| | | | | Hydrophones | |
Article	Manufacturer	Hydrophone	No. Devices	Depth (m)	Sensitivity with Pre-Amp (dB re 1 V μPa^{-1})	Bandwidth Applied (kHz)
Allen et al. [58]	Cetacean Research	C54XRS	3	5, 10, 25	−20	0.001–2.5
Au and Green [59]	Int. Trans. Corp. US Navy U.S.R.D.	TC-1032 and H-52	2	7.6	Not given	0.1–6
Brooker and Humphrey [62]	Reson	TC4032	3	50	−170	0.1–10
Erbe [36]	International Transducer Corp.	TC-4123	1	5,10	−145	0.01–22
Erbe et al. [38]	High Tech, Inc.	92WB	1	8	−160	0.01–40

Table 4. *Cont.*

Protocol

Sample Size (n)

| | | | Water Column (m) | | From Reference Study | | In Statistical Model | | | |
Article	Year	Location			Recordings	Vessels	Recordings	Vessels	Standard	CPA (km)
Gervaise et al. [60]	High Tech, Inc.	96 MIN	4	30–140			−165			0.01–20
Kipple and Gabriele [61]	Not given	Not given	1	30, 50			Not given			0.01–35
Kipple and Gabriele [63]	Not given	Not given	1	30			Not given			0.01–35
McCauley et al. [34]	Massa and Clevite	1025-C and CH17	4	20			−195 and −196			0.016–12.5
Parsons and Meekan [31]	OceanInstruments	SoundTrap ST300 STD	9	8.5			−169 to −176			0.08–20
Parsons et al. [52]	OceanInstruments	SoundTrap ST300 STD	1	4			−173.6			0.1–20
Parsons et al. [52] *	OceanInstruments	SoundTrap ST300 STD	1	4			−173.6			0.1–20
Wladichuk et al. [50,51]	Geospectrum Technologies Inc.	M36	2	197			−165 ± 3			0.02–100
Veirs et al. [3]	Reson	TC4032	1	10			−164			0.012–40

Data Processing

Article	Propagation Loss *	Surface-Image Correction	Source Approximation	Source Depth	Source Level Type	Vessel Type Reported
Allen et al. [58]	HG	x	Dipole	x	RNL	7
Au and Green [59]	SG	x	Dipole	x	RNL	2,3,6
Brooker and Humphrey [62]	SG	x	Dipole	x	RNL	6
Erbe [36]	SG, RAMGeo, RAY	x✓✓	Dipole	x	RNL and MSL	2,3,6,8
Erbe et al. [38]	SG, RAMGeo	x✓	Dipole and Monopole	x	RNL and MSL	2
Gervaise et al. [60]	SG	x	Dipole	x	RNL	Undefined
Kipple and Gabriele [61]	SG	x	Dipole	x	RNL	1,3,4,5,8
Kipple and Gabriele [63]	SG	x	Dipole	x	RNL	1,3,4,6,7
McCauley et al. [34]	Linear regression	x	Dipole	x	RNL and ASL	1,3,8,6
Parsons and Meekan [31]	Linear regression	x	Dipole	x	RNL and ASL	1,3
Parsons et al. [52]	SG, RAMGeo	x✓	Dipole and Monopole	✓	RNL and MSL	3,9
Parsons et al. [52] *	SG	x	Dipole	x	RNL	1,2,3
Wladichuk et al. [50,51]	SG and RAY	x✓	Dipole and Monopole	x	RNL and MSL	1,2,3,5,6,8
Veirs et al. [3]	HG	x	Dipole	x	RNL	1,2,3,4,7,10,11

* SG: Spherical Geometry, i.e., $20 \log_{10}(r)$. HG: Hybrid Geometry, i.e., $[10.20] \log_{10}(r)$. RAMGeo: Range-dependent Acoustic Model. RAY: Ray tracing model. Linear regression: Least squares linear regression with a log function [31]. N.B. Insufficient information on the methods of Buck and Chalfont [54], Miles et al. [55], Malme et al. [56], and Greene [57] could be found for inclusion in Table 4. Upper panel: date and location where the recordings took place, height of the water column on deployment site, number of recordings taken, number of vessels recorded, number of vessels analysed in this study, whether or not observations were carried out using standard protocols, and the range of the closest points of approach (CPAs) of the vessels. Middle panel: hydrophones' technical details, number of devices used, hydrophone deployment depth, and the bandwidth of the recordings reported. Bottom panel: backpropagation methods and corresponding source approximation and vessel types reported.

Several other studies have reported the characteristics of acoustic signals of small vessels, which although informative on the drivers of vessel noise, did not meet the criteria to not be included here. Some of these provided maximum and minimum source levels without the mean estimate or information on the number of individual passes, e.g., [40]; gave received levels without sufficient information to determine RNLs, e.g., [13,64]; only published individual characteristics of the signal such as directionality or intensity at specific frequencies, e.g., [39,65]; did not provide vessel length and so could not be confirmed as being from a small vessel [66]; or may have reported on a vessel that was so quiet it was not detected [67].

Across all variables, the estimated source levels generally increased through the vessel types from the electric vessel emitting the lowest levels, through the planing and then semi-displacement hulls, to the displacement hull type vessels (Figure 2). Variability within vessel types was high (Figure 2). For example, estimated levels of fishing vessels ranged from 145 to 195 dB re 1 µPa m. In fact, all vessel types with more than 20 measures displayed source level ranges of >29 dB, except the single electric vessel, which was measured at only one speed.

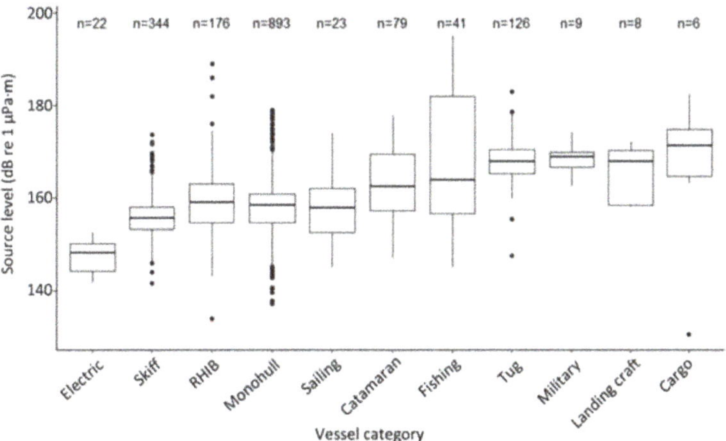

Figure 2. Reported source levels of vessels, aggregated by Vessel Type, showing mean, inter-quartile, 5th/95th percentiles and outliers (thick black line, upper/lower bounds of boxes, upper and lower extents of vertical lines, and dots, respectively). Number of samples (n) available for each vessel type before data cleaning shown.

Model

After removing all data points that did not meet the model criteria, 1355 samples from 49 vessels and four vessel types remained for statistical analysis. The final GAMM, that produced the most parsimonious model with the lowest AIC, explained approximately 22.7% of the variation encountered in this dataset (Table 5) and followed the form:

$$\mathrm{SL}_{ivr} = \beta + s(\mathrm{CPA}_{ivr}) + s_{vtype}(\log_{10}(\mathrm{Speed}_{ivr})) + \alpha_v + \alpha_r + \varepsilon_{ivr} \tag{4}$$

$$\alpha_v \sim N(0, \sigma^2_v) \tag{5}$$

$$\alpha_r \sim N(0, \sigma^2_r) \tag{6}$$

$$\varepsilon_{ivr} \sim N(0, \sigma^2) \tag{7}$$

where the response variable SL_{ivr} is the estimated source level reported for observation i, for vessel v and reported by reference r. β is the intercept, $s(\mathrm{CPA}_{ivr})$ and $s_{vtype}(\mathrm{Speed}_{ivr})$ are the smoothing functions representing CPA and speed at observation i of vessel v and published by reference r, respectively. $s_{vtype}(\text{Speed}_{ivr})$ represented the four smoothing functions describing the speed profiles of the four vessel types included in the model: Skiffs, RHIBS, Monohulls and Catamarans. The random effects α_v and α_r were assumed to be randomly distributed with means 0 and variances σ^2_v or σ^2_r, respectively. ε_{ivr} consists of the residual noise of observation i from vessel v by reference r, and was assumed to be randomly distributed with mean 0 and variance σ^2. Model validation did not indicate any issues with this final model. The final GAMM did not select Hull Type, Length, Engine Power, Water Depth, Hydrophone Depth and Propagation Model for inclusion. Potential reasons for this are described in the respective sections of the discussion.

Table 5. Results of the final GAMM. s() presents the details of the estimated smoother for each vessel type. Estimates for the intercept α and each smoothing term are given, as well as the standard error (SE), estimated degrees of freedom (edf), and approximate significance of the smooth terms as the *t*-value, F-statistics (F), and *p*-value. The variance and standard error (SE) are given for the random effects in the model.

Parametric Terms						
Fixed Effects	**Estimate**	**SE**	*t*-**Value**	**edf**	**F**	*p*-**Value**
Intercept α	157.99	2.23	70.74			$<2 \times 10^{-16}$
s(CPA)	−2.78	2.65	−1.05	6.84	13.41	$<2 \times 10^{-16}$
s(Speed)$_{\text{skiff}}$	2.76	0.79	3.47	1.96	89.4	$<2 \times 10^{-16}$
s(Speed)$_{\text{RHIB}}$	18.36	7.06	2.6	5.15	36.87	$<2 \times 10^{-16}$
s(Speed)$_{\text{monohull}}$	7.86	3.84	2.05	8.35	110.22	$<2 \times 10^{-16}$
s(Speed)$_{\text{catamaran}}$	5.58	3.35	1.67	2.36	28.57	4.58×10^{-14}
Random effects	**Variance**	**SE**				
Vessel ID	15.09	3.89				
Reference	30.32	5.51				

Speed and CPA were significant contributors and had the greatest effects on estimated source level (Table 5). The relationship and the level of influence of Speed varied between Vessel Types, with RHIBs displaying the most variation (Table 5), but all estimated speed smoothers showed an increasing effect on source level with speed (Figure 3). Skiffs displayed a linear effect of speed on source level. In comparison, RHIBs had three local maxima at $\approx$3, 7 and 11 ms^{-1}, followed by a continual increase. Monohulls increased to a local maximum at $\approx$6 ms^{-1}, plateauing until $\approx$14 ms^{-1}, before source levels and confidence intervals increased with speed (the latter due to the small number of data points). The estimated smoother for Catamarans increased to $\approx$13 ms^{-1} before plateauing until $\approx$17 ms^{-1}, after which the confidence intervals increased significantly. The estimated smoother for CPA showed a decrease to 50 m before increasing slowly until 200 m, after which the effect plateaued (Figure 4). Other factors did not provide greater explained deviance of the effect on source level and were not chosen in the final model.

The increase in the AIC value between the model using all vessels (7186) and the one only using vessels with three or more recordings (7618) supported the use of the latter in the model to remove an unbalanced random vessel effect. The AIC values also showed that the model including random structures for vessels and references, as $\alpha_v + \alpha_r$, fitted the data better than the model including a nested random structure of the form $\alpha_v + \alpha_{v/r}$ (AICs 7186 and 7212, respectively).

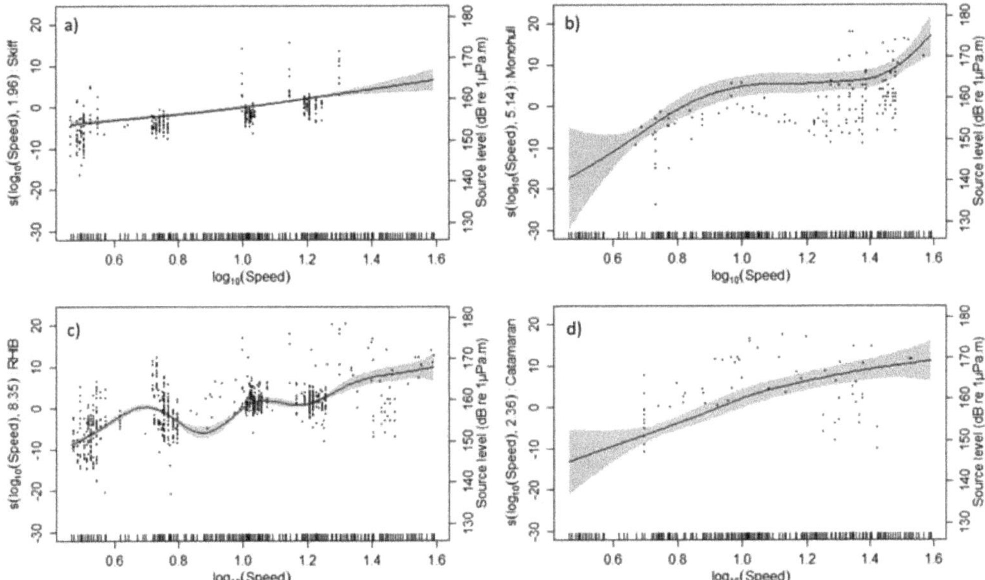

Figure 3. Estimated speed smoothing functions (solid line) with 95% confidence intervals (shaded area) illustrating the relationship between speed and source level (SL) for the four vessel types: Skiff, Monohull, RHIB, and Catamaran (**a**–**d**, respectively). The recorded data were added as points with the secondary y-axis (right). The secondary y-axis also displays the smoother with the estimated intercept added (presenting the fitted values).

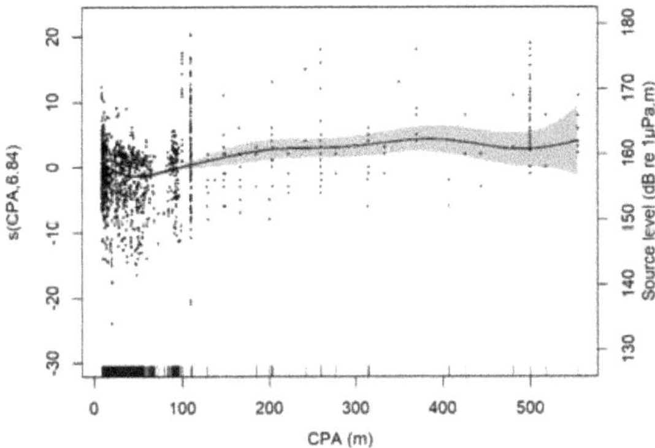

Figure 4. Estimated smoothing function (solid line) with 95% confidence intervals (shaded area) for CPA, illustrating the relationship between CPA and source level (SL). The recorded data were added as points with the secondary y-axis (right). The secondary y-axis also displays the smoother with the estimated intercept added (presenting the fitted values).

5.2. Within Study Assessments

5.2.1. Frequency Band

For the 95 examples taken from Wladichuk et al. [51], increasing the lower boundary of the frequency band used to produce MSL estimates from 10 Hz to 100 Hz led to a decrease in MSLs of up to 37 dB (Figure 5). Across all speeds and vessel types, the average decrease in MSL between measurements made with the lower boundary of 10 Hz and those using a lower boundary of 50 Hz was 9.1 dB (s.d. = 9.3, max = 27.5, min = 0), whereas the same comparison with a lower limit of 100 Hz produced a mean decrease of 11.7 dB (11.2, 37.2,

0). For individual vessel types (but across all speeds), the average difference between full broadband levels and estimates with a 100 Hz lower frequency was 0.3, 12.4, 3.5, 4.1 and 11.3 dB for Catamaran, Landing craft, Monohull, RHIB and Sailboat, respectively (n = 16, 5, 43, 20, and 11, respectively). In contrast, if the energy recorded between 20 kHz and 63 kHz is removed from the broadband estimates, this results in an average reduction of only 0.11 dB and a maximum reduction of 1 dB, i.e., energy below 100 Hz provides a significantly higher contribution to the broadband estimate than energy >20 kHz.

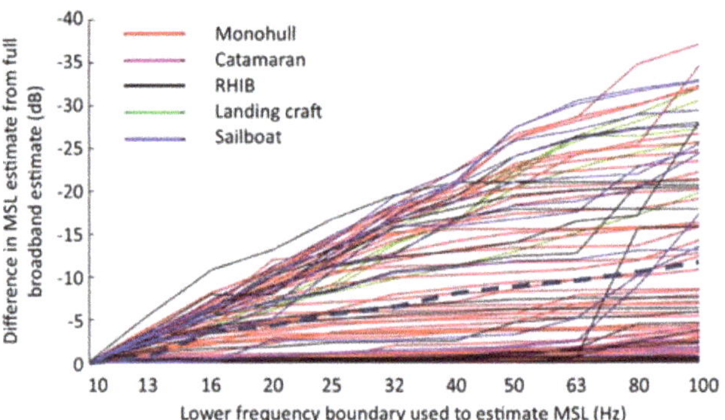

Figure 5. Reduction in MSL estimates for 95 example vessels and speeds (vessel type denoted by colours shown in the legend) when comparing the full broadband estimate (10 Hz to 63 kHz) to estimates made with a different low-frequency boundary (up to 100 Hz). Data taken from Wladichuk et al. [51]. Dashed, thick grey line denotes the mean of all vessels.

5.2.2. Propagation Model

Across 109 of the RNL and MSL estimates provided by Wladichuk et al. [51], in 70 m of water, the mean value of MSL minus RNL was 0.66 dB (max = 4.4, min = −3.2, s.d. = 1.1, mean of absolute difference = 0.96). However, at frequencies below 100 Hz, RNL and MSL one-third octave levels diverged with decreasing frequency with MSL levels reaching up to ≈20 dB higher than RNL levels at 20 Hz (see Figure 6, for levels from two example vessels).

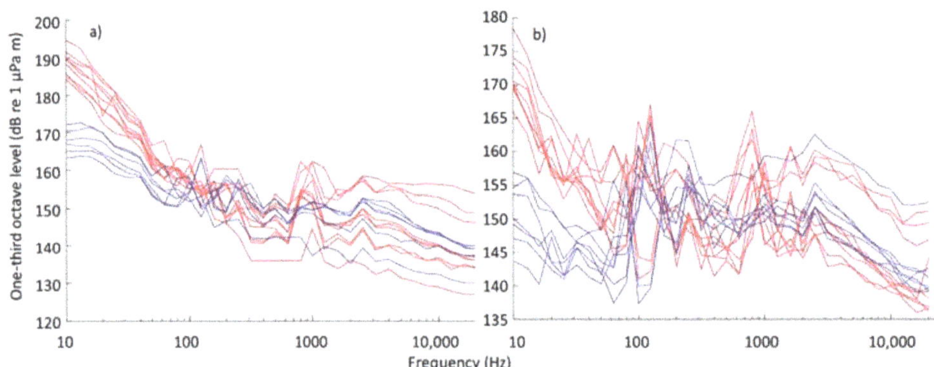

Figure 6. One-third octave band monopole source levels (red) and radiated noise levels (blue) for two example vessels (**a,b**) travelling at the same speed (data taken from Wladichuk et al. [51]).

Parsons et al. [52] reported a +1 dB difference between 50th percentile RNL and MSL estimates for an electric ferry (125–16,000 Hz band) and a conventional-powered ferry (400–16,000 Hz band), respectively. However, one-third octave band levels referenced to

1 m from the source displayed a +18 dB difference between 100 and 200 Hz and a -18 dB difference at 500 Hz between the MSL and RNL levels, for the same percentiles. Parsons et al. [52] also noted the troughs in propagation loss models at 100–200 Hz and 300–500 Hz, attributing some of this to a lack of available accurate information on the geophysical characteristics of the seabed at the study site, required to improve the propagation model. Parsons and Meekan [31] recorded multiple measures at multiple ranges, between 10 and 100 m, from three vessels conducting repeated transects at four different speeds, in 9 m of water. The mean difference between the 50th percentiles of broadband RNL and linear regression-modelled ASLs was 6.4 ($\pm$0.2) dB. In contrast, using linear regression models developed for each one-third octave band and then integrating the ASL band levels over the full frequency range had a mean difference between RNLs and ASLs of -3.5 ($\pm$2.6) dB. The factor behind the difference between the broadband and integrated one-third octave band source level estimates was that the mean linear regression-modelled one-third octave losses peaked at $40\log_{10}(range)$ at 30 Hz and had declined to 10 to $15\log_{10}(range)$ at frequencies above 1 kHz, compared to the modelled broadband loss at $16\log_{10}(range)$.

5.2.3. Hydrophone Depth

Of the measures recorded by Erbe [36] at three different water depths, the MSL estimate was 7 dB greater at 25 m depth than at 10 m (n = 1), while the RNL was 6 dB higher, for the same comparison. In comparison, the MSL estimate was 4.2 dB higher (s.d = 4.6, max = 14, min = -4) at 10 m depth than at 5 m (n = 17, shallower estimate higher twice), while the RNL was 0.5 dB higher (s.d. = 3.5, max = 6, min= -4) for the same comparison (n = 17, shallower estimate higher nine times).

6. Discussion: Variation in Small Vessel Source Levels

There are few published datasets that provide source level estimates of small vessels and even fewer that supply full spectra or breakdown of the broadband estimate into smaller bandwidths (e.g., octave or one-third octave) for individual vessel passes. The reports reviewed here span nearly 50 years of field sampling and those that included methodology and vessel data that could be used to model the factors that affect source levels were spread over the last 20 years. Consequently, the types of vessels, recordings systems and back-propagation methods applied have changed significantly across this time span and the available datasets collated in this study.

For the small vessels in the broader dataset, source level estimates varied by 65 dB, compared to the 73 dB variation observed by Chion et al. [19] in their meta-analysis of large vessels. Similar to Chion et al. [19], we found that the highest levels were reported by Allen et al. [58] and the lowest levels by Veirs and Veirs [3]. The difference between mean levels reported by these two studies was 35 dB for large vessels, compared with 20 dB for small vessels, suggesting this difference is driven by methodology. The estimates by Allen et al. [58] were not included in either model and were removed from the large vessel study as outliers and excluded from our small vessel dataset as they did not contain multiple measures of individual vessels.

6.1. Research Objective

The purpose of this study was to review and investigate factors that influence small vessel noise output, with an objective of teasing out whether there are details of vessels that can be readily accessed and commonly used by acoustic researchers and managers to assist in better modelling of marine noise in a given area. However, there is a lack of integration in the literature, between the research fields interested in understanding drivers of small-vessel source levels. Although naval architects and engineers focus on the performance with some regard to acoustic output, there is less interest in the propagation of the noise, the impact on marine fauna, or how the regulations around noise may mitigate its impact. In contrast, acousticians are primarily concerned with the propagation of the noise rather than the design specifics of how it is created, and managers are concerned with the impact of the signal mitigating it. As a result, reports of source levels authored by acousticians and propagation modellers rarely include information on ship design, such as cavitation coefficients, Admiralty numbers, or block coefficients, e.g., [3,19–22,31,34,36,38,54–56,58,59,62,63]. In contrast, reports focussed on vessel design often omit appropriate methodology regarding signal propagation or acknowledgment of the frequency-dependent impacts of the noise on marine fauna and information with which this can be assessed, e.g., [68–72]. For example, although errors in source level estimates that are caused by interference patterns, can be reduced by applying a surface reflection correction for every frequency of the spectrum measured, this is quite impractical and does not entirely remove the issue.

Collectively, all of these factors contribute to variance in the estimates of spectra and source levels from field recordings that essentially validate the models created by ship designers. However, the specifications used in design calculations are rarely reported. We therefore investigated what could be gleaned about the driving factors of small vessel noise from the information that is readily available in the literature. The application of these factors by managers and regulators do not necessarily require complicated models but may be estimated from the simplest appropriate data that can be collected on all vessels operating within the area of concern. Indeed, the ship design coefficients mostly comprise variables we have noted as required factors to better understand their impacts on noise (Table 2). Therefore, although a correlation of noise with Admiralty coefficient, propeller cavitation number, and propeller loading coefficients may be more accurate, not all of the variables required to calculate them may be accessible for all vessels. In addition, current literature on large vessels has shown that a significant proportion of the variation in noise output can be broadly estimated without knowledge of some of these specifications. We attempted to investigate their combined relationship with estimated source levels of small vessels, to better understand which factors have the biggest influence and, in doing so, found a shortfall in the information provided with typical source level reports.

While numerical methods suggest the error in RNL estimates is small [73], field data displays greater levels of variance. To reconcile this difference, a more collaborative approach is needed that includes more detailed and matched efforts between computational modelling and acquisition of validation data.

6.2. Factors That Drive Acoustic Output and Estimated Source Levels within Vessel Types

We observed increases in source level estimates across the 11 vessel types from one specifically designed to be quiet (electric-powered vessels), through those designed to skim across the water (planing hulls), to vessels that plough through the water at any speed (displacement hulls). Estimated source levels for vessels designed to have a combination of planing and displacement (semi-displacement) hulls fell roughly in between the two designs. Further, vessel type (four levels available) was chosen as a categorical predictor in the final GAMM model. Thus, while sufficient data were only available to test four Vessel Types and Hull Type was not chosen in the model (likely because it explains similar variation to vessel type) the design of the vessel is a key determinant of acoustic output.

Numerous intrinsic and extrinsic variables must be considered when estimating vessel source levels. Some of these include frequency-specific energy and thus have implications for any application that focusses on specific frequency bands. Additionally, the effect of these factors may be synergistic and a change or error in one factor may compound the effect of another increasing uncertainty in the final source level estimate.

6.2.1. Intrinsic Factors
Speed

For any vessel, if the only factor that changes is speed, then the cavitation-driving tip speed correlates with vessel speed. If other factors change, such as load in the case of tug boats, then the relationship between propeller tip speed and vessel speed changes. The small boats that were considered in our study did not experience load or other changes, and so boat speed is expected to be a good indicator of cavitation noise energy, which the statistical model confirmed. With regard to the management of boat noise (by vessel operators or government regulators), boat speed is the more readily observable parameter.

Chion et al. [19] modelled the change of source level with speed for 17 individual studies of large vessels and found that C_v, the velocity coefficient in the $C_v\log_{10}(v)$-space, varied from 11.7 to 49.9. When modelled within their generalised linear mixed model, it was estimated at 15.4 for large vessels. MacGillivray et al. [20] found that a 40% speed reduction decreased MSLs by approximately 10 dB in a study of voluntary speed reduction in the number of large vessels in the Port of Vancouver. For small vessels, Wladichuk et al. [51] observed the loss coefficient ranging between 10 and 58, whereas Parsons and Meekan [31] noted a non-linear relationship with speed, but generalised an average C_v of 9.7, for three vessels of $\approx$6 m length. In their study, the initial test of the relationship across all datapoints (all recordings of all vessel types) produced a C_v of 10.4. Although we found a non-linear relationship between speed and source level for three of the four vessel types tested in the GAMM, all four models displayed a positive relationship between speed and source level. These relationships could be generalised as C_v coefficients of 10, 16, 11 and 8 for skiffs, RHIBS, monohulls and catamarans, respectively.

Both Erbe et al. [38] and Parsons and Meekan [31] observed non-linear relationships between source levels and speed for individual vessels of <8 m length and suggested that as speed increased and vessels moved from ploughing to planing positions, the power requirement and therefore the acoustic output decreased. In fact, as the vessel moves from a displacement to planing position, the hull and propeller rise and the noise sources become shallower. Although this reduces RNLs at low frequencies the MSLs remains the same, though the source position must be accounted for in MSL estimates. In this study, although speed displayed a linear effect on skiff source level, both RHIBs and monohull vessels displayed local maxima, suggesting that one or more vessels were contributing a 'hump' speed to the effect.

In addition to broadband source level, frequency content of small vessel noise has also been observed to change with increasing speed. Veirs and Veirs [3] suggested that small vessels emitted more energy at high frequencies than large vessels, and that this difference increased with speed. Parsons and Meekan [31] observed that levels of higher frequency (>1 kHz) one-third octave bands increased significantly more than lower frequency (<1 kHz) one-third octave levels as speed of three vessels increased, as was also found in selected vessels from Wladichuk et al. [51]. Even individual vessels travelling at the same speed have produced source level estimates that vary by up to 20 dB [31,38,52]. Erbe et al. [38] observed this at high speed and suggested that vessel slap contributing additional noise to the estimate was the cause as the 6.5 m length, planing vessel bounced on the water. In contrast, the difficulty of maintaining vessel speed, position and direction at slow speed (5 kmh^{-1}, 2.7 kns) was hypothesised to cause the high variation observed by Parsons and Meekan [31], and Parsons et al. [52] attributed the variation observed in short-range (15 m) estimates of both an electric (10 m-long) and conventional (25 m-long) powered vessel to the uncertainty in position and therefore range.

As with large commercial vessels, speed is a significant driver of vessel source level and can be used as a means of mitigating the impacts of noise on taxa in shallow waters. This could be achieved through a direct reduction in the number of impacts of a single vessel travelling at a slower speed, or by facilitating the presence of a larger number of slower-moving vessels in a given area (such as a whale-watching troupe) without increasing the instantaneous exposure levels. However, the additional time spent in a given area due to slower speeds, and the resulting cumulative exposures, requires consideration. Further, the relationship between speed and source level in planing hull vessels appears more complicated than for displacement hulls and may be vessel-specific, thus a single speed limit may not be appropriate for all vessels. The findings of this study demonstrate that to fully understand the impact of speed on the source level of an individual vessel, or a vessel design type, requires a greater number of recordings of individual passes at the same speed and across a wide range of speeds than has been previously achieved.

Size and Weight

Acoustic output has been shown to have a positive relationship with length [21], beam [19] and power [31]. McCauley et al. [34] and McKenna et al. [21] observed a positive increase in source level with increasing vessel load, the latter study finding gross tonnage explained deviance in octave bands (centre frequencies) between 63 and 500 Hz. However, our dataset lacked sufficient information on beam or weight-related factors for small vessels, thus only length was analysed. Length was not included in the final GAMM in our analysis, likely due to collinearity with multiple other predictor variables. Draft is rarely reported and although this is intuitively likely to have a positive relationship with source level [34], these are generally small differences compared to large vessels (1–2 m, compared with sometimes > 10 m draft).

Propeller Specifications

The relationship between propeller specifications and emitted spectra are well documented [22] with increases in engine firing rate and propeller blade rate driving increases in the intensity and frequency of tonal components of the spectra [38]. In addition, broadband energy from propeller cavitation is one of the dominant sources of noise varying with size, shape, load (as a function of displacement, speed and power, amongst other factors) and depth [18]. However, several of these specifications are rarely reported in bioacoustics literature and therefore, could not be assessed within our model. Although these characteristics can provide vessel specific information [39,40] and increase energy over specific frequency bands, they are a function of speed. Changing the depth of the propeller (source) by a few metres, as an input parameter for propagation models, has been shown to vary MSL estimates up to 10 dB [74,75]. Small vessel propellers are typically positioned only 1–2 m below the surface (though less when planing), thus changes of a few metres depth that are possible in large vessels do not occur in small vessels. Modern vessels now have the ability to vary pitch of the propeller, however, while the ability to control trim is common in current small vessels, variable pitch control is not and there is limited publicly available data in the literature. Therefore, we have not explored the impact this may have on the noise emitted by the propeller. Propeller damage (such as chips or gauges on blades) increases cavitation and the resulting bubbles have a positive relationship with acoustic energy [22]. Indeed, in some instances, cavitation noise has been shown to be more correlated to the level of propeller damage, than cavitation number [76]. As with other characteristics, condition was not reported and could not be assessed in this study.

Engine Power and Type

The final GAMM model in this study did not select engine power or type as a significant predictor of source level. However, engine power did display an imbalance with vessel type, thus any model that had selected both variables would have required cautious consideration. This is because, in general, vessel power increases with vessel type and length, thus teasing out the exact driver is non-trivial. In previous studies of small vessels, Young and Miller [77] observed that an 18 hp engine was noisier than a 7.5-hp engine and Erbe [36] compared Evinrude 175 and 225 hp engines, also finding the larger engine to be noisier, particularly at slower speeds. Indeed, Parsons and Meekan [31] characterized the ASLs for three small vessels of similar lengths, but differing engine power (30, 90, 180 hp), at multiple speeds and found 5–6 dB difference between the 30 and 180 hp vessels, which decreased in difference with increasing speed (i.e., the relationship between estimated source level and speed differed for the three vessels). Engine type was not recorded sufficiently to be included in statistical modelling in this study and we found no reports that provided source levels from different engine types (e.g., diesel, two-stroke, four-stroke petrol) of the same power operating under the same conditions. However, engine types have been suggested to impact fauna differently, with fish behaviour varying in response to the noise of two-stroke or four-stroke engines [78].

Onboard Machinery

Whereas propeller (and subsequent cavitation) noise dominates the broadband source levels, onboard machinery vibrates at fundamental frequencies which generate narrowband spectra that are radiated through a variety of points in the hull of the vessel. Diesel service generators emit energy (dominated by the hit of the piston against the cylinder wall) at their firing rate and multiple harmonics of the firing rate, and are therefore independent of ship speed [40]. Heine and Gray [79] related the radiated power of this noise to the firing rate and engine power, with additional noise dependent on the condition of the engine. Additionally, alternating current power generators can be found on large and small vessels, typically emitting energy at 50 Hz (and harmonics) that can be distinct in the ships' spectra [40].

Directionality

Source level estimates are typically conducted using the CPA, often abeam to the vessel or over a range in azimuth about the CPA, and assume symmetry either side of the vessel [38]. This assumes that received levels are reduced by acoustic shielding from the hull towards the bow of the vessel and by the bubble cloud caused by propeller cavitation at the rear [31,43]. However, McCauley [34] observed some vessels categories displayed greater source levels fore or aft of the vessel, than abeam, whereas Malinowski et al. [65] observed increased levels at the quarter-stern to the sides of the bubble cloud. Indeed, McCauley et al. [34] hypothesised that the reason catamaran source levels were higher than those of the monohull (as observed in this meta-analysis), particularly fore and aft of the vessel, was due to the reduced acoustic shielding in the multi-hull vessels. All data assessed in the GAMM model here were from recordings taken abeam of the source vessels, however, it is noted that they may not be reflective of the maximum output for every vessel.

6.2.2. Extrinsic Factors

Closest Point of Approach

The CPA is often chosen to be at a distance where (1) the signal-to-noise ratio is sufficient that the level of acoustic energy emitted from the vessel can be separated from the background noise, (2) the difference between the actual and modelled propagation losses are minimal and (3) the receiver is positioned in the acoustic far-field and the vessel can be considered as a point source.

One wavelength from the source, theoretically the boundary of the acoustic near-field and transition ranges [80], is approximately 150 and 75 m, for 10 and 20 Hz, respectively. Although little is known about hearing sensitivity below 10 Hz for any species (indeed, few taxa are tested for infrasound (<20 Hz) hearing thresholds [2,81,82]), these are appropriate low-frequency boundaries to consider when reporting vessel source levels for biological impact assessments. Additionally, propagation of a 150 dB re 1 µPa m signal using spherical spreading would reduce to approximately ambient levels of 100 dB re 1 µPa at around 250 m range and would only have a signal-to-noise ratio of 10 dB at ≈100 m range. Add to that the Lloyd's mirror effect, and propagation loss at small/shallow slant angles is greater than the spherical $20 \log_{10}(r)$ approximation. Thus, there is a relatively fine boundary between being considered a point source at lower frequencies of interest and the received signal slipping into ambient noise.

The final GAMM chose CPA as a significant driver of source level estimates, in line with Chion et al. [19] for large vessels. Although the estimated source levels of large vessels decreased with increasing CPA, in our study, the estimated source level only decreased in the first tens of metres range, before slowly increasing. Three potential reasons for this are that:

1. The vessel types that produced higher source levels were generally recorded at a greater range than the quieter vessel types.

2. The small vessel recordings were predominantly taken at shorter ranges than in the Chion et al. [19] study. Therefore, inaccurate propagation models would have less distance to impact the source level estimates.

3. For a near-surface source, the slant angle to the receiver affects the received level (due to the Lloyd's mirror effect), and received levels decrease with decreasing hydrophone depth (i.e., when the hydrophone is closer to the sea surface and the slant angle is smaller). Gassman et al. [75] provided an example where source level estimates (0.02–1 kHz band) made using spreading laws for an angle to the hydrophone of 0.2° were 5–10 dB lower than those made at a 10° angle. This was reduced to 3–7 dB by applying a surface reflection correction, but not removed entirely [19]. The CPA and hydrophone depths applied in the small vessel studies analysed here were such that ≈85% of recordings were made at >8° and ≈77% at >10°.

Propagation Loss Model Used in Analysis

Of the types of model considered in this review, RNLs are computed by applying simple geometrical spreading laws to a dipole source [32,33], MSLs are computed with a numerical sound propagation model to estimate the range- and environment-dependent losses of a (surface reflection corrected) monopole source [80] and ASLs use empirical methods to model the losses of a dipole source [31]. The ISO (2016, 2019) criteria for RNLs require water depths that are not necessarily safe for many small vessels, to conduct opportunistic recordings. As a result, many studies of small vessels are conducted in shallower waters than recommended by ISO [32,33], for which there are no current standards. Some studies report RNLs only, e.g., [3], others report RNLs and MSLs to allow comparison, e.g., [38,51], and some report RNLs and ASLs [31,34].

Chion et al. [19] found that reports of RNLs could be underestimating source levels by up to 35 dB and suggested appropriate conditions for RNLs need to be met or estimates require correcting for the surface interactions, potentially using propagation models. Where accurate physical-chemical water column and geophysical seabed data are available, a number of range-dependent and -independent propagation models may be used to estimate the source level, each of which is appropriate for different conditions [80]. However, the effect of inaccurately modelling the propagation typically increases with range, thus a large source-receiver distance requires accurate quantification of the seabed across the propagation profile, which is not always achievable. Without these data, MSL estimates can be significantly affected even if an appropriate propagation model is applied [52].

Both Wladichuk et al. [51] and Parsons et al. [52] found around 1 dB difference between their RNL and MSL estimates, conducted in ≈70 m and ≈3 m of water depth, respectively. However, both found significant differences between RNL and MSL in low-frequency components of the spectra, with 18–20 dB difference between one-third octave band levels at some frequencies. Parsons and Meekan [31] found similar divergence between RNL and ASL one-third octave estimates in 10 m of water, as would be expected for a dipole source at low frequencies. Underestimating low-frequency energy emitted in shallow water has significant implications for assessments of impacts on site-attached marine fauna that have hearing sensitivity that overlaps with the noise (e.g., fishes) as they may be close to the source (i.e., directly beneath). There is the potential to empirically model site-specific propagation losses in controlled conditions using a tone generator as a synthetic source to identify frequency-dependent propagation in shallow water [83], although the vessel itself can provide a good proxy for this source under some conditions [31].

Frequency Band

Estimated source levels can be highly dependent on the frequency of the lower boundary of the analysed bandwidth. Although small vessels emit less low-frequency energy than large vessels [3,51], it is significant and might be enough to impact marine fauna. Omitting low-frequency energy can lead to underestimates in RNLs, MSLs, and ASLs of up to 37 dB. Even if the bandwidth for the estimates includes this low-frequency energy, it is likely underestimated if processing employs a simple spreading rule without accounting for the low-frequency energy lost due to acute source-receiver angles (i.e., shallow hydrophone depth) or energy lost due to shallow-water propagation. Even if a sound propagation model is used for MSL computation, the model might have been inappropriate at low frequencies (e.g., a RAY model) or the geophysical seafloor characteristics might not have been quantified accurately [52].

Hydrophone Depth

At locations that have a negative sound speed gradient near the surface, there is the possibility that a receiver positioned too close to the surface will be in the acoustic shadow due to the downward refraction of sound waves [41]. Additionally, interference between the direct path from source to receiver and the surface-reflected path means that recordings taken nearer the water surface receive lower energy than those deeper in the water column [41]. Data from Erbe [36] was, on average, in agreement with this phenomenon, however, individual recordings made by the shallower hydrophones were, on occasion, higher than those of the deeper hydrophone. This is believed to be caused by the highest 1 s-averaged received level (taken as reflective of the CPA) being different between the two recordings. An additional note is that the MSL estimates made using the deeper hydrophones were higher by an average of >4 dB than the MSL estimates of the shallower hydrophones, yet it would be anticipated that the propagation model would account for the receiver depth and Lloyd's mirror.

System Response

Many hydrophones are piston-calibrated at a single frequency and assumed to have a flat response across the entire frequency band. However, system frequency response (i.e., including the hydrophone and pre-amplifiers) has been shown to be reduced at low (<200 Hz) frequencies [42]. If the system response is lower than expected at a given frequency, it leads to an underestimate in the received energy at that frequency and the resulting estimated broadband level.

Environmental Conditions

Finally, the environmental conditions at the time measurements were taken (e.g., wind, currents, waves) can require minor changes in power, trim or propeller blade angle to maintain vessel course and speed [34] and therefore affect acoustic output (e.g., greater power and therefore source level when travelling upstream or upwind, compared with downstream or downwind). Additionally, rain, wind-driven waves, and swell all increase background noise, the removal of which affects confidence in source level estimates. This is important if the signal-to-noise ratio is low, as can be found when the CPA is in the hundreds of metres or the vessels being recorded have low source levels. McKenna et al. [21] found that wave height and direction, and current direction contributed minor levels of explained deviance to their overall GAM assessing drivers of large ships' source levels, specifically in the lowest octave bands.

Environmental Impacts

Species-specific impact assessments can be related to frequency-dependent auditory thresholds of the species in question and the acoustic energy emitted by the vessel at those frequencies [84–87]. In addition, source spectra are vessel- and speed-specific, and propagation losses are frequency-specific, thus the impact a chosen propagation model has on the final broadband and band level estimates differs between vessels. Addressing these two factors requires that one-third octave levels be made available for every vessel pass recorded at a minimum and where possible, full spectra be provided. Further, to assist in comparing propagation models, received band levels are preferential to source band levels, or sufficient details to replicate propagation losses are needed.

Although this review has examined the pressure-related characteristics of small vessels, fishes and invertebrates respond to particle acceleration and potentially sound-driven ground motion, rather than pressure [81,88]. Pressure and particle motion have a linear relationship in the acoustic far-field [41]. However, the water depths associated with coastal water and inland waterways are shallow enough that site-attached or low-mobility demersal and benthic species experience the greatest sound levels as the vessel passes near or directly over the top of them, often within a few metres. These ranges occur within the near-field, where the pressure-particle motion relationship is complex and ground motion from vessel noise is yet to be defined [88]. We therefore recommend that studies include particle acceleration and, where possible, ground motion measurements and report both near- and far-field measures. These factors should also be considered when conducting experimental studies to assess the impact of specific vessels. Research experiments into the impacts of noise are beginning to be designed to assess management strategies (e.g., driving at reduced speed to minimise impact on coral reef fishes [89]). As small vessel acoustic output can be highly variable, these experiments should report multiple measures of source level estimates and provide enough information about the vessel and its operation to position the exposure levels within those exhibited by a broad range of small vessels.

Finally, in addition to underwater radiated noise, airborne noise from vessels can have significant impacts on avian and terrestrial fauna, including humans [90,91], and characterizing this pollutant has been the subject of research for many years [92–95]. Similar to regulations mitigating the impact of road noise on the surrounding environment, legislation is continually being developed to limit the effects of airborne vessel noise [96,97]. Methods of mitigating noise impacts in air versus water can on occasion be in competition with each other, such as locating ship exhaust above or below the water level, which can transfer noise to either the air or water, rather than reducing the level itself.

7. Conclusions

1. There is little available published data on small (<25 length) vessels' source levels. Our dataset revealed a high level of variation among studies, vessel designs and within vessel types (up to 20 dB difference, even for the same vessel at the same speed). The lowest estimated source levels were emitted by electric vessels and these levels broadly increased as vessel designs moved from planing to displacement hulls. Our analysis revealed a significant positive relationship between small vessel speed-over-ground and estimated source level, generalised as speed coefficient C_v values of 8 to 16, among the four vessel types tested. These support the consensus that regulating speed is a viable method of reducing instantaneous noise levels and their potential impact on marine fauna, with consideration for the additional exposure time from slower travel.
2. We have shown that data acquisition and analysis protocols for opportunistic estimates of vessel source level contribute to variability between studies. This was confirmed by the inclusion of CPA as a statistically significant explanatory variable in the final GAMM, and by the demonstrated difference in estimates caused by changing the lower frequency of the analysed bandwidth and the difference in low frequency energy observed between MSL and RNL and between ASL and RNL estimates.
3. Few studies provide enough measures to assess confidence limits in the estimates, enough supplementary information in intrinsic and extrinsic factors associated with the recordings to create a dataset of a size needed to tease out the significant drivers of acoustic characteristics, or a breakdown of the vessel source spectra to assess how the signal would be perceived by different marine taxa.
4. Vessels emit significant low-frequency acoustic energy. For small vessels, this is particularly true at slow speeds. Studies that omit low-frequency energy (e.g., report a broadband level that starts at 50 or 100 Hz) may underestimate source level. Similarly, estimates of source spectra using geometrical spreading or propagation models that do not accurately reflect seabed geophysical characteristics may also underestimate

low-frequency energy. Inadequately calibrated recording systems have also been shown to underestimate low-frequency received levels. The effect these factors have on the estimated broadband source level are dependent on the vessel source spectra and therefore, errors are vessel-, speed- and system-specific and may be synergistic. Any management protocols developed using such estimated source levels would be applicable to species with good low-frequency (<100 Hz) hearing sensitivity. These are typically fish and invertebrates, many of which are the sessile, site-attached or low-mobility animals that are most likely to experience noise from vessels at short range when vessels pass near or directly above fauna in shallow coastal waters. Any regulations designed to mitigate the impact of noise on these animals therefore need accurate estimates of low-frequency source levels.

This study recommends that:

Although ISO criteria for measuring vessel noise require a minimum hydrophone and water depths of 70 m and 150 m, respectively, this is not possible in many locations as such depths may only occur offshore, beyond the safe operating range of small vessels. A set of shallow water standards to conduct such measurements should be developed. This includes increasing (1) the replication of measures at each speed, (2) the number of ranges at which measures are recorded, (3) the range of speeds at which the vessel is measured with all measures being made available to assist meta-analysis such as this study.

A collection of synergistic studies from which a direct comparison between deep- and shallow-water methodologies and equipment can be made is recommended to reduce method-driven variability so that the key influential variables driving acoustic output can be identified in greater detail.

Managers and regulators need meaningful and accessible ways to estimate anthropogenic noise and its potential ecological impacts in a given area. Therefore, the identification of factors influencing acoustic output that can be understood and collected is preferable to needing data that require specialist knowledge. To do this, there is a need for greater collaboration between naval architects/engineers, bioacousticians, propagation modellers, managers and regulators in the determination and reporting of vessel source levels, as all have interest and contribute knowledge to different aspects of the research.

Impacts of noise on marine fauna are dependent on the hearing sensitivity of the taxa of interest. Therefore, reports of vessel source levels must provide octave band levels, at a minimum, or one-third octaves/full spectra, so that the acoustic energy can be broken down into the appropriate energy bands that relate to appropriate frequency thresholds. Assessment of the impacts of vessel noise on taxa of interest can only be made if the reported frequency bandwidth includes the frequencies at which that taxa has reported sensitivity. This is of particularly note for species that are sensitive to low-frequency sound.

Future reports should include acoustic particle velocity and pressure measures from vessels and, if possible, sound-driven ground motion in the near- and far-fields, to allow assessment of the impacts on fauna that detect components of the acoustic signal other than pressure.

Author Contributions: Conceptualization, M.J.G.P., M.G.M. and C.E.; analysis M.J.G.P., C.E. and S.K.P.; investigation, M.J.G.P., C.E. and S.K.P.; data curation, M.J.G.P.; writing—M.J.G.P., C.E., M.G.M. and S.K.P.; writing—review and editing, M.J.G.P., C.E., M.G.M. and S.K.P.; project administration, M.G.M.; funding acquisition, M.G.M. All authors have read and agreed to the published version of the manuscript.

Funding: This work was undertaken for the Marine Biodiversity Hub, a collaborative partnership supported through funding from the Australian Government's National Environmental Science Program.

Institutional Review Board Statement: Not applicable.

Informed Consent Statement: Not applicable.

Conflicts of Interest: The authors declare no conflict of interest.

Nomenclature

Abbreviation	Symbol	Full Name
AIC		Akaike's information criterion
ASL	L_{ASL}	Affected source level
CPA		Closest Point of Approach
GAM		Generalised additive model
GAMM		Generalised additive mixed model
ISO		International Standards Organisation
RHIB		Rigid hull inflatable boat
RNL	L_{RN}	Radiated noise level
MSL	L_{SL}	Monopole source level
s.d.		Standard deviation
SL		Source level
VIF		Variance inflation factors
	C_v	Velocity coefficient
	d	Depth
	L_p	Sound pressure level
	p_{rms}	Root mean-square sound pressure
	N_{PL}	Propagation loss
	SL_{ivr}	Estimated source level reported for observation i, for vessel v and reported by reference r
	v	Vessel velocity
	v_r	Reference velocity
	σ^2	variance

References

1. Duarte, C.M.; Chapuis, L.; Collin, S.P.; Costa, D.P.; Devassy, R.P.; Eguiluz, V.M.; Erbe, C.; Gordon, T.A.C.; Halpern, B.S.; Harding, H.R.; et al. The Ocean Soundscape of the Anthropocene. *Science* **2021**, *371*, 581. [CrossRef]
2. Mooney, T.A.; Di Iorio, L.; Lammers, M.; Lin, T.H.; Nedelec, S.L.; Parsons, M.J.G.; Radford, C.A.; Urban, E.; Stanley, J. Listening forward: Approaching marine biodiversity assessments using acoustic methods. *R. Soc. Open Sci.* **2020**, *7*, 201287. [CrossRef] [PubMed]
3. Veirs, S.; Veirs, V. Vessel noise measurements underwater in the Haro Strait, WA. *J. Acoust. Soc. Am.* **2006**, *120*, 3382. [CrossRef]
4. Frisk, G. Noiseonomics: The relationship between ambient noise levels in the sea and global economic trends. *Sci. Rep.* **2012**, *2*, 437. [CrossRef]
5. Hatch, L.; Clark, C.W.; Merrick, R.; Van Parijs, S.M.; Ponirakis, D.; Schwehr, K.; Thompson, M.; Wiley, D. Characterizing the relative contributions of large vessels to total ocean noise fields: A case study using the Gerry E. Studds Stellwagen Bank National Marine Sanctuary. *Environ. Man.* **2008**, *42*, 735–752.
6. Stanley, J.A.; Van Parijs, S.M.; Hatch, L.T. Underwater sound from vessel traffic reduces the effective communication range in Atlantic cod and haddock. *Sci. Rep.* **2017**, *7*, 14633. [CrossRef]
7. Blane, J.; Jaakson, R. The impact of ecotourism boats on the St. Lawrence beluga whales. *Environ. Cons.* **1994**, *21*, 267–269. [CrossRef]
8. Graham, A.L.; Cooke, S.J. The effects of noise disturbance from various recreational boating activities common to inland waters on the cardiac physiology of a freshwater fish, the largemouth bass (*Micropterus salmoides*). *Aquat. Cons. Mar. Fresh. Ecosyst.* **2008**, *18*, 1315–1324. [CrossRef]
9. Williams, R.; Wright, A.J.; Ashe, E.; Blight, L.K.; Bruintjes, R.; Canessa, R.; Clark, C.W.; Cullis-Suzuki, S.; Dakin, D.T.; Erbe, C.; et al. Impacts of anthropogenic noise on marine life: Publication patterns, new discoveries, and future directions in research and management. *Ocean Coast. Man.* **2015**, *115*, 17–24. [CrossRef]
10. Erbe, C.; Reichmuth, C.; Cunningham, K.; Lucke, K.; Dooling, R. Communication masking in marine mammals: A review and research strategy. *Mar. Pollut. Bull.* **2016**, *103*, 15–38. [CrossRef]
11. Erbe, C.; Marley, S.; Schoeman, R.; Smith, J.N.; Trigg, L.; Embling, C.B. The effects of ship noise on marine mammals—A review. *Front. Mar. Sci.* **2019**, *6*, 606. [CrossRef]
12. Southall, B.L.; Finneran, J.J.; Recihmuth, C.; Nachtigall, P.E.; Ketten, D.R.; Bowles, A.E.; Ellison, W.T.; Nowacek, D.P.; Tyack, P.L. Marine Mammal noise exposure criteria: Updated scientific recommendations for residual hearing effects. *Aquat. Mam.* **2019**, *45*, 125–232. [CrossRef]
13. Nedelec, S.L.; Radford, A.N.; Pearl, L.; Nedelec, B.; McCormick, M.I.; Meekan, M.G.; Simpson, S.D. Motorboat noise impacts parental behaviour and offspring survival in a reef fish. *Proc. R. Soc. B* **2017**, *284*, 20170143. [CrossRef]
14. Nedelec, S.L.; Mills, S.C.; Lecchini, D.; Nedelec, B.; Simpson, S.D.; Radford, A.N. Repeated exposure to noise increases tolerance in a coral reef fish. *Environ. Pollut.* **2016**, *216*, 428–436. [CrossRef]

15. Nedelec, S.L.; Mills, S.C.; Radford, A.N.; Belade, R.; Simpson, S.D.; Nedelec, B.; Côté, I.M. Motorboat noise disrupts co-operative interspecific interactions. *Sci. Rep.* **2017**, *7*, 6987. [CrossRef]
16. Harding, H.R.; Gordon, T.A.C.; Havlik, M.N.; Predragovic, M.; Devassy, R.P.; Radford, A.N.; Simpson, S.D.; Duarte, C.M. A systematic literature assessment on the effects of human-altered soundscapes on marine life [Dataset]. *Zenodo* **2020**. [CrossRef]
17. Mensinger, A.F.; Putland, R.L.; Radford, C.A. The effect of motorboat sound on Australian snapper *Pagrus auratus* inside and outside a marine reserve. *Ecol. Evol.* **2018**, *8*, 6438–6448. [CrossRef] [PubMed]
18. Carlton, J.S. 10 Propeller noise. In *Marine Propellers and Propulsion*, 2nd ed.; Carlton, J.S., Ed.; Butterworth Heinemann: Oxford, UK, 1994; 584p.
19. Chion, C.; Lagrois, D.; Dupras, J. A meta-analysis to understand the variability in reported source levels of noise radiated by ships from opportunistic studies. *Front. Mar. Sci.* **2019**, *6*, 714. [CrossRef]
20. MacGillivray, A.O.; Li, Z.; Hannay, D.E.; Trounce, K.B.; Robinson, O.M. Slowing deep-sea commercial vessels reduces underwater radiated noise. *J. Acoust. Soc. Am.* **2019**, *146*, 340–351. [CrossRef]
21. McKenna, M.F.; Wiggins, S.M.; Hildebrand, J.A. Relationship between container ship underwater noise levels and ship design, operational and oceanographic conditions. *Sci. Rep.* **2013**, *3*, 1760. [CrossRef]
22. Ross, D. *Mechanics of Underwater Noise*; Pergamon Press: New York, NY, USA, 1976.
23. Simpson, S.; Radford, A.; Nedelec, S.; Ferrari, M.C.O.; Chivers, D.P.; McCormick, M.I.; Meekan, M.G. Anthropogenic noise increases fish mortality by predation. *Nat. Comm.* **2016**, *7*, 10544. [CrossRef]
24. Simpson, S.D.; Radford, A.N.; Holles, S.; Ferarri, M.C.O.; Chivers, D.P.; McCormick, M.I.; Meekan, M.G. Small-Boat Noise Impacts Natural Settlement Behavior of Coral Reef Fish Larvae. In *The Effects of Noise on Aquatic Life II. Advances in Experimental Medicine and Biology*; Popper, A., Hawkins, A., Eds.; Springer: New York, NY, USA, 2016; Volume 875, pp. 1041–1048.
25. Erbe, C. Underwater noise of small personal watercraft (jet skis). *J. Acoust. Soc. Am.* **2013**, *133*, EL326–EL330. [CrossRef]
26. Erbe, C.; Parsons, M.J.G.; Duncan, A.J.; Allen, K. Underwater acoustic signatures of recreational swimmers, divers, surfers and kayakers. *Acoust. Aust.* **2016**, *44*, 333–341. [CrossRef]
27. Erbe, C.; Parsons, M.J.G.; Duncan, A.J.; Osterrieder, S.K.; Allen, K. Aerial and underwater sound of unmanned aerial vehicles (UAV). *J. Unman. Veh. Sys.* **2017**, *5*, 92–101. [CrossRef]
28. Erbe, C.; Williams, R.; Parsons, M.J.G.; Parsons, S.K.; Hendrawan, I.G.; Dewantama, D. Underwater noise from airplanes: An overlooked source of ocean noise. *Mar. Pollut. Bull.* **2018**, *137*, 656–661. [CrossRef]
29. McLean, D.L.; Parsons, M.J.G.; Gates, A.R.; Benfield, M.C.; Bond, T.; Booth, D.J.; Bunce, M.; Fowler, A.M.; Harvey, E.S.; Macreadie, P.I.; et al. Enhancing the Scientific Value of Industry Remotely Operated Vehicles (ROVs) in Our Oceans. *Front. Mar. Sci.* **2020**, *7*, 02200. [CrossRef]
30. Marley, S.A.; Erbe, C.; Salgado-Kent, C.P.; Parsons, M.J.G.; Parnum, I.M. Spatial and Temporal Variation in the Acoustic Habitat of Bottlenose Dolphins (*Tursiops aduncus*) within a Highly Urbanized Estuary. *Front. Mar. Sci.* **2017**, *4*, 197. [CrossRef]
31. Parsons, M.J.G.; Meekan, M.G. Acoustic Characteristics of Small Research Vessels. *J. Mar. Sc. Eng.* **2020**, *8*, 970. [CrossRef]
32. International Organization for Standardization. *Underwater Acoustics—Quantities and Procedures for Description and Measurement of Underwater Sound from Ships—Part 1: Requirements for Precision Measurements in Deep Water Used for Comparison Purposes (ISO 17208-1)*; International Organization for Standardization: Geneva, Switzerland, 2016; 20p.
33. International Organization for Standardization. *Underwater Acoustics—Quantities and Procedures for Description and Measurement of Underwater Sound from Ships—Part 2: Determination of Source Levels from Deep Water Measurements (ISO 17208-2)*; International Organization for Standardization: Geneva, Switzerland, 2019; 13p.
34. McCauley, R.D.; Cato, D.H.; Jeffrey, A.F. *A Study of Impacts of the Impacts of Vessel Noise on Humpback Whales in Hervey Bay*; Report to the Queensland Department of Environment and Heritage, Maryborough Branch; Queensland Department of Environment and Heritage: Maryborough, QSL, Australia, 1996; 163p.
35. Zuur, A.F.; Ieno, E.N.; Smith, G.M. *Analysing Ecological Data*; Springer: New York, NY, USA, 2007; 672p.
36. Erbe, C. Underwater noise of whale-watching boats and potential effects on killer whales (*Orcinus orca*), based on an acoustic impact model. *Mar. Mamm. Sci.* **2002**, *18*, 394–418. [CrossRef]
37. Zuur, A.F.; Ieno, E.N.; Walker, N.J.; Saveliev, A.A.; Smith, G.M. *Mixed Effects Models and Extensions in Ecology with R*; Springer: New York, NY, USA, 2009; 574p.
38. Erbe, C.; Liong, S.; Koessler, M.W.; Duncan, A.J.; and Gourlay, T. Underwater sound of rigid-hulled inflatable boats. *J. Acoust. Soc. Am.* **2016**, *139*, EL223–EL227. [CrossRef]
39. Gloza, I. Identification methods of underwater noise sources generated by small ships. *Acoust. Biomed. Engin.* **2011**, *119*, 961–965. [CrossRef]
40. Malinowski, S.J.; Gloza, I. Underwater noise characteristics of small ships. *Acta Acust. United Acust.* **2002**, *88*, 718–721.
41. Urick, R.J. *Principles of Underwater Sound*, 3rd ed.; McGraw Hill: New York, NY, USA, 1983.
42. McCauley, R.D.; Thomas, F.; Parsons, M.J.G.; Erbe, C.; Cato, D.H.; Duncan, A.J.; Gavrilov, A.N.; Parnum, I.M.; Salgado-Kent, C.M. Developing an Underwater Sound Recorder: The Long and Short (Time) of It. *Acoust. Aust.* **2017**, *45*, 301–311. [CrossRef]
43. McKenna, M.F.; Ross, D.; Wiggins, S.M.; Hildebrand, J.A. Underwater radiated noise from modern commercial ships. *J. Acoust. Soc. Am.* **2012**, *131*, 92–103. [CrossRef]
44. Harrell, F.E., Jr. *Regression Modeling Strategies*; Springer: New York, NY, USA, 2001; 582p.

45. Wood, S.; Scheipl, F. gamm4: Generalized Additive Mixed Models Using 'mgcv' and 'lme4'. R Package Version 0.2-6. 2020. Available online: https://CRAN.R-project.org/package=gamm4 (accessed on 24 May 2021).

46. Flom, P.L.; Cassell, D.L. Stopping stepwise: Why stepwise and similar selection methods are bad, and what you should use. In Proceedings of the NorthEast SAS Users Group 20th Annual Conference, Baltimore, MD, USA, 11–14 November 2007.

47. Zuur, A.F.; Saveliev, A.A.; Ieno, E.N. *A Beginner's Guide to Additive Mixed Models with R*; Highland Statistics Ltd.: Newburgh, UK, 2014.

48. Burnham, K.P.; Anderson, D.R. *Model Selection and Multimodel Inference: A Practical Information-Theoretic Approach*; Springer: New York, NY, USA, 2002.

49. R Core Team. *R: A Language and Environment for Statistical Computing*; R Foundation for Statistical Computing: Vienna, Austria, 2020; Available online: https://www.R-project.org/ (accessed on 24 May 2021).

50. Wladichuk, J.L.; Hannay, D.E.; MacGillivray, A.O.; Li, Z.; Thornton, S.J. Systematic source level measurements of whale watching vessels and other small boats. *J. Ocean. Technol.* **2019**, *14*, 110–126.

51. Wladichuk, J.L.; Hannay, D.E.; MacGillivray, A.O.; Li, Z. *Whale Watch and Small Vessel Underwater Noise Measurement Study: Final Report*; Document 01522, V. 3.0. Technical Report by JASCO Applied Sciences for Vancouver FraserPort Authority ECHO Program; JASCO Applied Sciences (Canada) Ltd.: Victoria, BC, Canada, 2019; 110p.

52. Parsons, M.J.G.; Duncan, A.J.; Parsons, S.K.; Erbe, C. Reducing vessel noise: An example of a solar-electric passenger ferry. *J. Acoust. Soc. Am.* **2020**, *147*, 3575–3583. [CrossRef]

53. Richardson, W.J.; Green, C.R.; Malme, C.I.; Thomson, D.H. *Marine Mammals and Noise*; Academic Press: New York, NY, USA, 1995; 294p.

54. Buck, B.M.; Chalfant, D.A. *Deep Water Narrowband Radiated Noise Measurement of Merchant Ships*; Delco TR72-28. Rep. from Delco Electronics; Delco Electronics: Dayton, OH, USA, 1972.

55. Miles, P.R.; Malme, C.I.; Richardson, W.J. *Prediction of Drilling Sit Specific Interaction of Industrial Acoustic Stimuli and Endangered Whales in the Alaskan Beaufort Sea*; BBN Rep. 6509; OCS Study MMS 87-0084; BBN Labs Inc.: Cambridge, MA, USA; LGL Ltd.: King City, ON, Canada; U.S. Minerals Manage. Serv.: Anchorage, AK, USA, 1987; 341p.

56. Malme, C.I.; Miles, P.R.; Miller, G.W.; Richardson, W.J.; Roseneau, D.G.; Thomson, D.H.; Greene, C.R., Jr. *Analysis and Ranking of the Acoustic Disturbance Potential of Petroleum Industry Activities and Other Sources of Noise in the Environment of Marine Mammals in Alaska*; BBN Rep. 6945; OCS Study MMS 89-0006. Rep; BBN Systems & Technol. Corp.: Cambridge, MA, USA; U.S. Minerals Manage. Serv.: Anchorage, AK, USA, 1989.

57. Greene, C.R. Characteristics of waterborne industrial noise, 1980–84. In *Behavior, Disturbance Responses and Distribution of Bowhead Whales Balaena mysticetus in the Eastern Beaufort Sea, 1980–84*; OCS Study MMS 85-0034. Rep; Rjchardson, W.J., Ed.; LGL Ecol. Res. Assoc. Inc.: Bryan, TX, USA; U.S. Minerals Manage. Serv.: Reston, VA, USA, 1985; pp. 197–253. 306p.

58. Allen, J.K.; Peterson, M.L.; Sharrad, G.V.; Wright, D.L.; Todd, S.K. Radiated noise from commercial ships in the Gulf of Maine: Implications for whale/vessel collisions. *J. Acoust. Soc. Am.* **2012**, *132*, EL229–EL235. [CrossRef]

59. Au, W.W.L.; Green, M. Acoustics interaction of humpback whales and whale-watching boats. *Mar. Environ. Res.* **2000**, *49*, 469–481. [CrossRef]

60. Gervaise, C.; Simard, Y.; Roy, N.; Kinda, B.; Menard, N. Shipping noise in whale habitat: Characteristics, sources, budget, and impact on belugas in Saguenay–St. Lawrence Marine Park hub. *J. Acoust. Soc. Am.* **2012**, *132*, 76–89. [CrossRef]

61. Kipple, B.M.; Gabriele, C.M. *Glacier Bay Watercraft Noise–Noise Characterization for Tour, Charter, Private, and Government Vessels*; Naval Surface Warfare Center Technical Report NSWCCD-71-TR-2004/545; Naval Surface Warfare Center: Bremerton, WA, USA, 2004.

62. Brooker, A.; Humphrey, V. Measurement of radiated underwater noise from a small research vessel in shallow water. *Ocean Eng.* **2015**, *120*, 182–189. [CrossRef]

63. Kipple, B.M.; Gabriele, C.M. *Glacier Bay Watercraft Noise*; Naval Surface Warfare Center Technical Report NSWCCD-71-TR-2003/522; Naval Surface Warfare Center: Bremerton, WA, USA, 2003.

64. Buckstaff, K.C. Effects of watercraft noise on the acoustic behaviour of bottlenose dolphins *Tursiops truncates* in Sarasota Bay, Florida. *Mar. Mam. Sci.* **2004**, *20*, 709–725. [CrossRef]

65. Malinowski, S.J.; Gloza, I.; Domagalski, J. The character of underwater noise radiated by small vessels. *Hydroacoustics* **2001**, *4*, 161–164.

66. Rudd, A.B.; Richlen, M.F.; Stimpert, A.K.; Au, W.W.L. Underwater sound measurements of a high-speed jet-propelled marine craft: Implications for large whales. *Pac. Sci.* **2015**, *69*, 155–164. [CrossRef]

67. Schevill, W.E. Quiet Power Whaleboat. *J. Acoust. Soc. Am.* **1968**, *44*, 1157. [CrossRef]

68. Park, C.; Kim, G.D.; Yim, G.T.; Park, Y.; Moon, I. A validation study of the model test method for propeller cavitation. *Ocean Eng.* **2020**, *213*, 107655. [CrossRef]

69. Atlar, M.; Fitzsimmons, P.; Zoet, P.; Troll, M.; Stark, C.; Sezen, S.; Shi, W.; Aktas, B.; Sasaki, N.; Turkmen, S.; et al. Underwater Noise Measurements with a Ship retrofitted with PressurePoresTM Noise Mitigation Technology and Using HyDroneTM System. 2021. Available online: https://strathprints.strath.ac.uk/76550/ (accessed on 28 July 2021).

70. Sharma, S.D.; Mani, K.; Arakeri, V.H. Cavitation noise studies on marine propellers. *J. Sound Vib.* **1990**, *138*, 255–283. [CrossRef]

71. Kellett, P.; Turan, O.; Incecik, A. A study of numerical ship underwater noise prediction. *Ocean Eng.* **2013**, *66*, 113–120. [CrossRef]

72. Tani, G.; Viviani, M.; Hallander, J.; Johansson, T.; Rizzuto, E. Propeller underwater radiated noise: A comparison between model scale measurements in two different facilities and full scale measurements. *Appl. Ocean Res.* **2016**, *56*, 48–66. [CrossRef]
73. Wittekind, D.; Schuster, M. Propeller cavitation noise and background noise in the sea. *Ocean. Eng.* **2016**, *120*, 116–121. [CrossRef]
74. Arveson, P.T.; Vendittis, D.J. Radiated noise characteristics of a modern cargo ship. *J. Acoust. Soc. Am.* **2000**, *107*, 118–129. [CrossRef]
75. Gassmann, M.; Wiggins, S.M.; Hildebrand, J.A. Deep-water measurements of container ship radiated noise signatures and directionality. *J. Acoust. Soc. Am.* **2017**, *142*, 1563–1574. [CrossRef] [PubMed]
76. Lush, P.A.; Angell, B. Correlation of Cavitation Erosion and Sound Pressure Level. *J. Fluids Eng.* **1984**, *106*, 347–351. [CrossRef]
77. Young, R.W.; Miller, C.N. Noise data for two outboard motors in air and in water. *Noise Cont.* **1960**, *6*, 22–25. [CrossRef]
78. Holmes, L.J.; McWilliam, J.N.; Ferrari, M.C.O.; McCormick, M.I. Juvenile damselfish are affected but desensitize to small motor boat noise. *J. Exp. Mar. Bio. Eco.* **2017**, *494*, 63–68. [CrossRef]
79. Heine, J.C.; Gray, L.M. *Merchant Ship Radiated Noise Model, Bolt, Beranek, and Newman Inc. Report 3020*; Bolt, Beranek, and Newman Inc.: Cambridge, MA, USA, 1976.
80. Erbe, C. *Underwater Acoustics: Noise and the Effects on Marine Mammals, a Pocket Handbook*, 3rd ed.; JASCO Applied Sciences: Victoria, BC, Canada, 2013.
81. Popper, A.N.; Hawkins, A.; Halvorsen, M. *Anthropogenic Sound and Fishes*; Report by ICF for Washington State Department of Transportation, Research Office, Report No. WA-RD 891.1; Washington State Department of Transportation: Olympia, WA, USA, 2019; 170p.
82. Farcas, A.; Thompson, P.M.; Merchant, N.D. Underwater noise modelling for environmental impact assessment. *Environ. Impact Assess. Rev.* **2016**, *57*, 114–122. [CrossRef]
83. Thilges, K.; Potty, G.; Freeman, S.; Freeman, L.; Van Uffelen, H. Measurements and models of acoustic transmission loss on two Hawaiian coral reefs. *Proc. Meet. Acoust.* **2019**, *39*, 070005. [CrossRef]
84. Erbe, C.; MacGillivray, A.O.; Williams, R. Mapping cumulative noise from shipping to inform marine spatial planning. *J. Acoust. Soc. Am.* **2012**, *132*, EL423–EL428. [CrossRef] [PubMed]
85. Erbe, C.; Williams, R.; Sandilands, D.; Ashe, E. Identifying modeled ship noise hotspots for marine mammals of Canadas' Pacific Region. *PLoS ONE* **2014**, *9*, e89820. [CrossRef]
86. Merchant, N.D.; Pirotta, E.; Barton, T.R.; Thompson, P.M. Monitoring ship noise to assess the impact of coastal developments on marine mammals. *Mar. Pollut. Bull.* **2014**, *78*, 85–95. [CrossRef]
87. Sigray, P.; Mathias Andersson, M.; Pajala, J.; Laanearu, J.; Klauson, A.; Tegowski, J.; Boethling, M.; Fischer, J.; Tougaard, J.; Wahlberg, M.; et al. BIAS: A Regional Management of Underwater Sound in the Baltic Sea. In *The Effects of Noise on Aquatic Life II*; Advances in Experimental Medicine and, Biology; Popper, A., Hawkins, A., Eds.; Springer: New York, NY, USA, 2016; Volume 875, pp. 1015–1023. [CrossRef]
88. McCauley, R.D.; Meekan, N.G.; Parsons, M.J.G. Acoustic measurements of a 2600 cubic inch seismic airgun array source and the relationship between acoustic pressure, particle motion and induced ground motion. *J. Mar. Sci. Eng.* **2021**, *9*, 571. [CrossRef]
89. McCloskey, K.P.; Chapman, K.E.; Chapuis, L.; McCormick, M.I.; Radford, A.N.; Simpson, S.D. Assessing and mitigating impacts of motorboat noise on nesting damselfish. *Environ. Poll.* **2020**, *266*, 115376. [CrossRef] [PubMed]
90. Borelli, D. Maritime Airborne Noise: Ships and Harbours. *Int. J. Acoust. Vib.* **2019**, *24*, 631. [CrossRef]
91. Borelli, D.; Gaggero, T.; Rizzuto, E.; Schenone, C. Holistic control of ship noise emissions. *Noise Mapp.* **2016**, *3*, 1. [CrossRef]
92. Fredianelli, L.; Nastasi, M.; Bernardini, M.; Fidecaro, F.; Licitra, G. Pass-by Characterization of Noise Emitted by Different Categories of Seagoing Ships in Ports. *Sustainability* **2020**, *12*, 1740. [CrossRef]
93. Badino, A.; Borelli, D.; Gaggero, T.; Rizzuto, E.; Schenone, C. Airborne noise emissions from ships: Experimental characterization of the source and propagation over land. *App. Acoust.* **2016**, *104*, 158–171. [CrossRef]
94. Lambert, D.R. *Airborne Noise Levels on Merchant Ships. A Compilation of Data*; Technical Report ADA079356; Naval Ocean Systems Centre: San Diego, CA, USA, 1979; 35p.
95. Nastasi, M.; Fredianelli, L.; Bernardini, M.; Teti, L.; Fidecaro, F.; Licitra, G. Parameters Affecting Noise Emitted by Ships Moving in Port Areas. *Sustainability* **2020**, *12*, 8742. [CrossRef]
96. Bernardini, M.; Fredianelli, L.; Fidecaro, F.; Gagliardi, P.; Nastasi, M.; Licitra, G. Noise Assessment of Small Vessels for Action Planning in Canal Cities. *Environments* **2019**, *6*, 31. [CrossRef]
97. Badino, A.; Borelli, D.; Gaggero, T.; Rizzuto, E.; Schenone, C. Normative framework for ship noise: Present and situation and future trends. *Noise Control Eng. J.* **2012**, *60*, 740–762. [CrossRef]

Journal of
*Marine Science
and Engineering*

Article

Acoustic Characteristics of Small Research Vessels

Miles Parsons * and Mark Meekan

Australian Institute of Marine Science, Perth, WA 6009, Australia; m.meekan@aims.gov.au
* Correspondence: m.parsons@aims.gov.au; Tel.: +61-(8)-6369-4053

Received: 20 October 2020; Accepted: 19 November 2020; Published: 27 November 2020

Abstract: Vessel noise is an acute and chronic stressor of a wide variety of marine fauna. Understanding, modelling and mitigating the impacts of this pollutant requires quantification of acoustic signatures for various vessel classes for input into propagation models and at present there is a paucity of such data for small vessels (<25 m). Our study provides this information for three small vessels (<6 m length and 30, 90 and 180 hp engines). The closest point of approach was recorded at various ranges across a flat, $\approx$10 m deep sandy lagoon, for multiple passes at multiple speeds ($\approx$5, 10, 20, 30 km h^{-1}) by each vessel at Lizard Island, Great Barrier Reef, Australia. Radiated noise levels (RNLs) and environment-affected source levels (ASLs) determined by linear regression were estimated for each vessel and speed. From the slowest to fastest speeds, median RNLs ranged between 153.4 and 166.1 dB re 1 µPa m, whereas ASLs ranged from 146.7 to 160.0 dB re 1 µPa m. One-third octave band-level RNLs are provided for each vessel–speed scenario, together with their interpolated received levels with range. Our study provides data on source spectra of small vessels to assist in understanding and modelling of acoustic exposure experienced by marine fauna.

Keywords: vessel noise; radiated noise levels; monopole source levels; propagation loss

1. Introduction

Through evolutionary time, sound has become an important sensory cue for many marine taxa. The efficient transmission of sound underwater has meant that a wide variety of species have developed frequency-specific hearing sensitivity and rely on the detection of acoustic cues and subtle changes in the biophony of their local soundscape during vital life functions [1–5]. These important signals, such as the spawning calls of fishes or the sound of healthy habitat in which larvae will settle, can be masked by anthropogenic noise, disrupting natural behaviors [6–8]. Sound produced by vessels is a major element of marine anthrophony and has been recognized as a chronic stressor [9], negatively impacting communication, health and behavior of many species [4,10–15]. As human populations have increased, so too has anthrophony in oceans and inland waterways [16–21], creating what has now been termed the 'Ocean soundscape of the Anthropocene' [22].

Management strategies that aim to mitigate the impact of vessel noise on marine fauna [23–28] require information about source levels and vessel movements. Although the Automatic Information System (AIS) can be used to track passages of the majority of commercial vessels [29,30], noise is also dependent on vessel size, speed, load and power, as well as other design characteristics [24,31,32]. This requires characterization of source signatures from different types and sizes of vessels.

At present, there is little data on the noise characteristics of small (<25 m length) vessels [26,33–37]. This is important because in coastal waters, these vessels often vastly outnumber larger ships ferrying commercial cargos. The data required to accurately model the propagation of signals from small vessels are rarely reported, or typically provided as one or two measures at limited numbers of speeds [36]. For this reason, we lack data on the variability in noise among vessels of different classes (e.g., monohull, catamarans, tugs, landing craft) within this size range or even different passes of the same vessel. This is problematic for management strategies that aim to set useful guidelines to mitigate noise for boating activities [38], particularly in shallow coastal waters, inland waterways, and coral reefs, where small vessels have the potential to significantly change the local soundscape and, due to proximity, are more likely to affect fishes, invertebrates and small marine mammals [38–42].

To address these issues, our study aimed to characterize the source spectra of three small vessels under 10 m length that are commonly used in shallow coastal marine environments. We took multiple measurements at the closest point of approach (CPA) at multiple ranges and speeds, over multiple passes, in shallow water. Source characteristics of noise can be specific to a vessel and have multiple engine-, propeller- and hull-related origins [43] and their impacts on fauna are frequency-dependent. Therefore, one-third octave levels were also calculated, and their propagation across the measured ranges investigated.

2. Materials and Methods

The International Standards Organization (ISO) protocols for the measurement of vessel radiated noise levels (RNLs) and monopole source level (MSL) focus on large vessels in deep water (see ISO 17208-1 [44] and 17208-2 [45]). The ISO criteria require a minimum water depth equal to the greater of 150 m or 1.5 times the overall ship length. For the highest standard of estimates, this comprises the deployment of three hydrophones positioned vertically at depths that result in 15°, 30°, and 45° angles from the sea surface at a CPA distance of either 100 m or one overall ship length, whichever is greater. Neither are these requirements achievable, nor is the procedure applicable, in shallow water. Indeed, meeting these requirements in Australia would require vessels to travel a significant distance offshore, which may not be appropriate for all classes of small vessels. Standards for measurements of RNL in shallow water are under development. However, we had sufficient replication of measurements to accurately estimate both RNLs and affected source levels (ASLs), in lieu of any current shallow-water ISO protocols.

2.1. Study Site

Lizard Island is a granitic island located approximately 30 km off the north Queensland coastline (14°40.88′ S, 145°27.82′ E, Figure 1). The Lizard Island group comprises four late-Permian granite islands—Lizard, Palfrey, South and Bird Islands—which, together with the surrounding fringing reef, encircle an up to 10 m-deep flat, sandy-bottomed lagoon [46]. Tidal range at Lizard Island reaches a maximum of 3 m and current speeds into the lagoon can be >30 cm s^{-1} [47,48]. Measurements were collected in the 10 m deep area of the lagoon, to the south of Lizard Island (Figure 1).

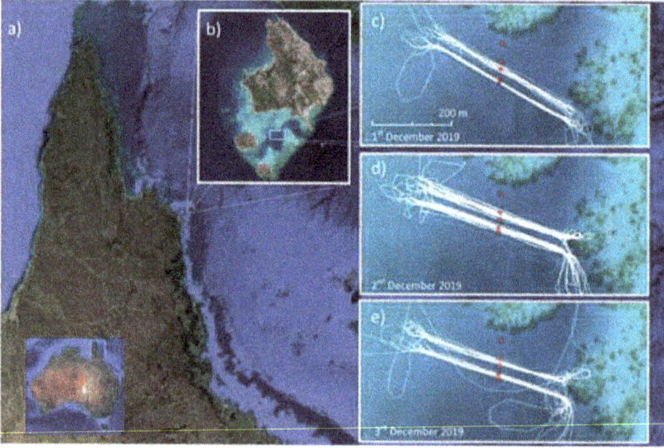

Figure 1. (**a**) Map of Australia with expansion of Cape York Peninsula, Queensland; (**b**) expansion of Lizard Island on the Great Barrier Reef; (**c–e**) expansions of the island lagoon with the vessel tracks from three consecutive survey days (**c**, **d**, and **e**, respectively) displayed in white and the positions of pairs of seafloor-mounted SoundTraps shown by the red dots.

2.2. Vessel Recordings

Vessel recordings were acquired using Ocean Instruments ST300 SoundTraps. These are piston-phone calibrated passive acoustic pressure sensors with a flat response of ±3 dB over the 20 Hz to 60 kHz system bandwidth, calibrated by the manufacturer using a 121 dB re 1 µPa source at 250 Hz. Divers deployed 10 ST300s on the seafloor of the lagoon, each orientated vertically, attached to the top of a star picket, and positioned approximately 50 cm above the sand. Two ST300s were deployed, 1 m apart, at each of five sites, forming a 100 m-long transect with relative spacing of 0, 10, 33.5, 52.1 and 100 m from the first site, running approximately north-south, at the northern end of the lagoon (Figure 1, red dots). All Sound Traps recorded 290 of every 300 s, at a sampling frequency of 48 kHz. Gaps between the recordings were kept to separate files into manageable sizes and minimize the potential for losing recordings due to buffering issues, and the 97% duty cycle minimized the likelihood of missing the CPA of a vessel pass. At each SoundTrap, a tight vertical line was run from the ST300 to a surface buoy where the exact GPS location was recorded with a Garmin 64SX.

Between the 1st and 3rd December 2019, three vessels (Figure 2, characteristics shown in Table 1) each conducted ten transects at speeds as close as possible to 5, 10, 20, and 30 km h^{-1} (1.4, 2.8, 5.6, and 8.3 m s^{-1}, respectively). For each speed, five transects were conducted across the southern end of the line of SoundTraps and five across the midway point (Figure 1, white lines), totaling 100 potential recordings of each vessel at each speed. Vessel transects were planned to be orthogonal to the line of SoundTraps, though in reality were not completely perpendicular to the SoundTrap transect (Figure 1). Each vessel conducted all transects over a three-hour period, with one vessel completed each day. Vessel positions were recorded using a handheld Garmin 64SX. Wind over the three days remained at Beaufort scale two or below.

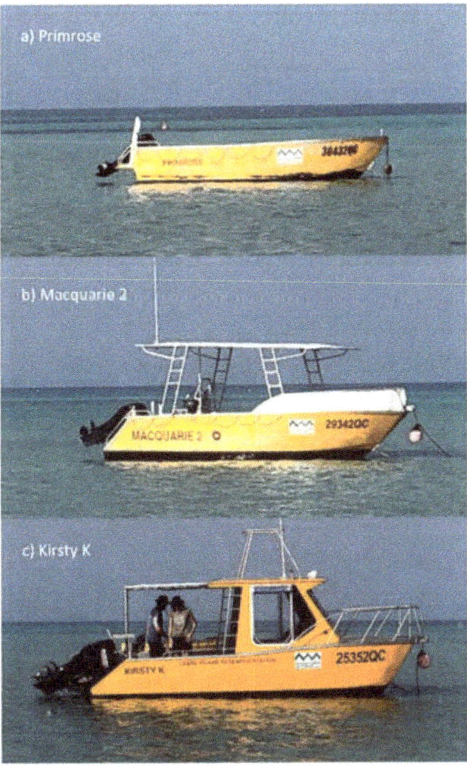

Figure 2. Photos of research vessels (**a**) *Primrose*, (**b**) *Macquarie 2* and (**c**) *Kirsty K.*

Table 1. Specifications of vessels recorded in Lizard Island lagoon.

		Primrose	Macquarie 2	Kirsty K
Vessel specifications	Length (m)	5	5.96	5.95
	Width (m)	2.1	2.4	2.6
	Draught (cm)	0.9	1.18	1.23
	Mass (kg)	360	825	1500
Propeller	No. blades	3	3	3
	Prop radius (cm)	25	32.5	32.5
	Depth below water (cm)	47	70	63
Engine	Engine	Suzuki	Suzuki	Suzuki
	Horsepower	30	90	2x90
	Fuel	Petrol (4 stroke)	Petrol (4 stroke)	Petrol (4 stroke)

2.3. Data Analysis

Data processing involved the following steps: (1) calibrating recordings, (2) selecting sections that matched the known time of each vessel pass, (3) computing mean squared pressure and power spectral density for the CPA of each transect, (4) computing RNL for each pass, and (5) least squares linear regression modelling of the recordings across all ranges to produce dipole source spectra and ASL for each vessel and speed.

The 290-s recordings were extracted, and each dataset processed with a MATLAB (version R2014, The MathWorks Inc., Nantucket, MA, USA) user interface designed for analysis of underwater passive acoustic recordings, CHORUS [49] and purpose-written functions, as well as inspected via audio and visual scrutiny of the recordings and spectrograms, respectively. Spectrograms were created in MATLAB (version R2019) using the spectrogram function with 1-s Hanning windows and 80% (=0.8) overlap. Mean squared pressure was calculated for each transect and the 1-s window that contained the maximum mean square pressure was selected to represent the received level (RL) for the vessel's CPA. Background noise levels were calculated from one-minute periods without vessel noise from each day, for each SoundTrap. RLs for each CPA were compared between pairs of SoundTraps at the same recording distance and any CPA RL with a difference greater than 5 dB were removed from analysis. CPA distance for each SoundTrap and vessel pass were determined from the known recorder positions and depth, and the GPS track of the vessel. In each case, the SoundTrap and Garmin 64SX CPA time were compared to confirm a match.

Spherical spreading losses, in the form of $20\log_{10}(\text{range})$, were back-propagated to a range of 1 m for each slant range and vessel pass to provide an estimate of RNL [50]. However, the shallow-water depth in the study area means these estimates are for comparison only as spherical spreading is unlikely reflective of the actual propagation losses (PL) at the site. Estimation of MSL is conducted by collecting measurements in the far-field and combining these with an appropriate PL model for the site, to determine the effective source level, accounting for the surface ghost, i.e., an effective RL at 1 m range [26]. However, propagation modelling, particularly in shallow water, also requires precise knowledge of the local seafloor geology, sometimes multiple meters into the substrate [37]. As this information was not available at the study site, RLs were recorded at multiple ranges from the source CPA during each transect, and broadband PLs were estimated empirically using least squares regression for a best fit curve of the RLs with range in the form of:

$$SL = C_{PL} \times \log_{10}(range) \tag{1}$$

where C_{PL} is the PL coefficient. As these RLs are a function of the propagation environment of the signal (i.e., the signal is affected by the surface reflections and interactions [51]), the value of the curve at a range of 1 m from the source may be considered an environment-affected source level (ASL).

Although hydrophones positioned directly beneath the vessel were within the acoustic near-field (for frequencies approximately $\leq$200 Hz) [50], these ranges reflected the type of exposure likely experienced on shallows reefs. We therefore modelled the PLs and resulting ASL estimates for each vessel and speed scenario using all data and with data from within 15 m slant range excluded, to minimize the effect of frequencies still in the near field. RNLs calculated in a shallow-water environment using spherical spreading are unlikely reflective of true PL. However, the regression-modelled PLs provide a useful proxy for acoustic propagation models, thus the broadband RLs were also back-propagated to 1 m using the average of all the regression-modelled PLs (i.e. mean C_{PL}) across all vessels and speeds (as calculated with and without measurements from the <15 m range), for comparison. Median and quartile estimates were then produced for each vessel and speed for all RNLs and ASLs. Thus, this study provides results on the RNL (percentiles are given for data including and excluding measurements taken at <15 m), the empirically measured and linear regression-modelled ASLs (with and without data from within the 15 m range) of three small vessels in shallow water.

As acoustic signals are vessel and speed specific [31,36,37,42,52], linear regression models were applied to the estimated RNL and ASLs, to investigate the relationship between source level and speed, in the form of:

$$\text{SL-SL}_{\text{ref}} = C_{v1} \times 10\log_{10}(v/v_{\text{ref}}), \tag{2}$$

where SL and SL_{ref} were the source levels at the tested and reference speeds, respectively, v and v_{ref} were the tested and reference vessel speeds, respectively, and C_{v1} was the velocity-related coefficient of the slope of the curve (Ross, 1976). If a single-coefficient, log-based regression model failed to produce a clear fit for the data, it was also investigated with a polynomial function with two coefficients:

$$\text{SL-SL}_{\text{ref}} = C_{v1} \times (v/v_{\text{ref}})^2 + C_{v2} \times (v/v_{\text{ref}}), \tag{3}$$

where C_{v2} was the second velocity-related coefficient.

Finally, received spectra for each CPA measurement were also integrated into one-third octave band levels. These band levels were interpolated across one-third octave frequency bands and range in logarithmic bins to compare the one-third octave band PL in the environment, again for each speed and vessel. One-third octave band RNLs were calculated and the frequency-dependent C_{PL} for each one-third octave band determined using linear regression of the respective band RLs recorded at various ranges, using the form of Equation (1).

3. Results

3.1. Measurements

A total of 120 transects were conducted across all three vessels and speeds (Table 2). One SoundTrap failed to provide calibrated recordings, leaving nine datasets. Of the remaining 1080 potential recordings of passes (360 for each vessel), 330, 336 and 344 CPAs were recorded of the *Primrose*, *Macquarie* and *Kirsty K*, respectively, at slant ranges between 9.0 m (almost directly below the vessel) and 98.8 m. Only two recordings were removed due to the RL of a CPA as noted by two SoundTraps at the same range differed by more than 5 dB. Average background noise levels across all sites and days were 106.8 dB re 1 µPa (max = 109.8, min = 102.3, s.d. 1.4 dB). Although the SoundTraps positioned at the greatest range were closer to the reef to the north and did contain more sounds of snapping shrimp than at the other recording sites, the mean noise at that location was <2 dB greater (108 compared with 106.2 dB re 1 µPa, respectively).

Table 2. Transect speeds conducted by each vessel (standard deviation, minimum and maximum speeds and the number of analyzed recordings shown in parentheses).

Target Speed (km h^{-1})	*Primrose* (km h^{-1})	*Macquarie* (km h^{-1})	*Kirsty K* (km h^{-1})
5	5.86 (0.12, 5.37, 6.55, 73)	6.12 (0.23, 5.63, 6.43, 88)	6.41 (0.61, 5.45, 7.66, 83)
10	10.32 (0.50, 9.73, 10.91, 79)	10.61 (1.64, 9.86, 11.61, 85)	10.62 (0.59, 9.73, 11.61, 92)
20	19.44 (0.39, 18.95, 20.28, 75)	20.63 (3.32, 18.95, 3.32, 86)	19.74 (0.52, 19.20, 20.87, 88)
30	30.23 (1.23, 28.80, 32.73, 81)	30.75 (3.30, 30.0, 33.49, 82)	30.40 (1.72, 27.17, 33.49, 86)

The approach of the vessel could be heard and observed in all recordings and spectrograms, respectively (Figure 3), though the exact CPA was more difficult to discern at the furthest ranges (>50 m) and slowest speed (5 km h^{-1}category). Lloyd's mirror interference pattern was evident on all spectrograms and propeller and engine/motor tones visible from 20 Hz through 24 000 Hz (the Nyquist frequency in this recording, Figure 3). Additionally, while not investigated fully in this study, at the close ranges (<50 m) energy in the 1–10 kHz band was typically higher in the final seconds of approach than the initial seconds of departure (Figure 3, rows a and b).

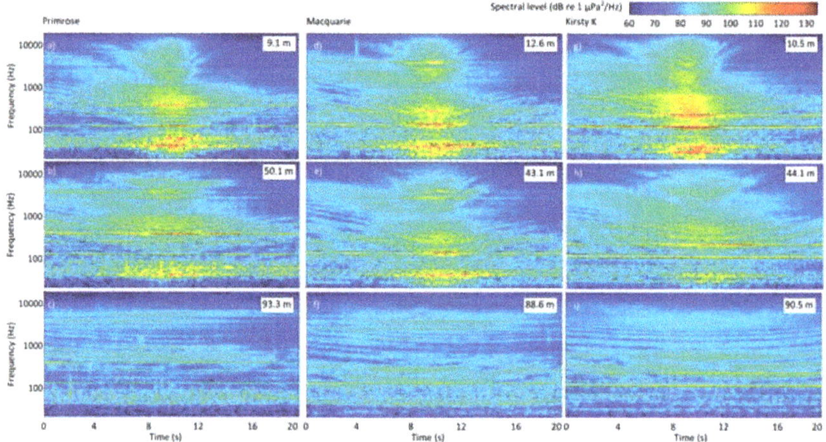

Figure 3. Spectrograms of 20 s recordings of individual passes of (**a–c**) *Primrose*, (**d–f**) *Macquarie* and (**g–i**) *Kirsty K* (all travelling at ≈30 km h^{-1}), as recorded almost directly (**a,d,g**) beneath the vessel (≈9 m range), (**b,e,h**) at the 25 m and (**c,f,i**) the ≈100 m ranges. CPA distance shown in top right of each spectrogram.

3.2. Received Levels with Range

In general, at a given range, RLs increased with speed for *Primrose* and *Macquarie*, though this was not as evident for the *Kirsty K* (Figure 4). Standard deviations in RLs at the same approximate speed were greatest closest to the source and standard deviations in RLs at the same range were greatest at the slowest speed category (Supplementary information). Regression models of PL varied between vessels and speeds and correlation with the recorded data was greater for transects conducted at 20 and 30 km h^{-1} than 5 and 10 km h^{-1} (Table 3). This variation was still evident when the recordings at the closest (<15 m) ranges were removed, though less so than when analyzing the full datasets. The average PL for individual datasets (vessels and speeds) for recordings across all ranges was 17.9log$_{10}$(*range*), becoming 16.0log$_{10}$(*range*) when data from <15 m were excluded. Standard deviation of all modelled losses decreased from 3.1 to 2.3 when removing the data taken at the <15 m range (Table 3). Percentiles of RNLs (with and without data from the <15 m range) and ASLs (using average

PL with and without data from the <15 m range) all increased with increasing engine size and speed, except the *Kirsty K* between 10 and 20 km h^{-1}and 20 and 30 km h^{-1}.

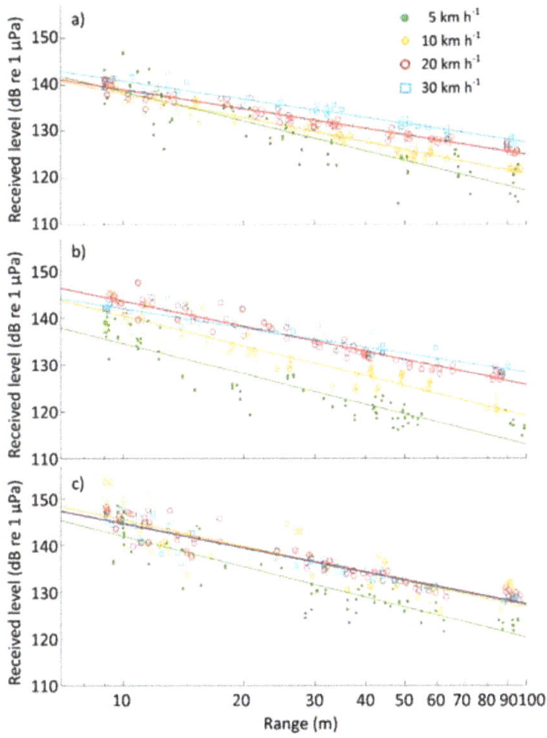

Figure 4. Received sound pressure levels from directly beneath the vessel (9 m range) to the 100 m range for (**a**) *Primrose*, (**b**) *Macquarie* and (**c**) *Kirsty K*, at speeds of 5, 10, 20 and 30 km h^{-1} (green circles, beige diamonds, red circles and blue squares, respectively).

Table 3. Propagation loss coefficients (C_{PL}) and affected source levels for individual vessels and speeds determine from Equation (1) for all ranges and for only ranges >15 m. Mean $\log_{10}(range)$ loss values (and standard deviation) found across all regression models with all data and for ranges >15 m are also included.

		All Ranges			Excluding <15 m Range		
Vessel	Speed	C_{PL}	ASL (dB re 1 µPa)	R^2	C_{PL}	ASL (dB re 1 µPa)	R^2
Primrose	5	−21.2 (−23.79, −18.64)	159.8 (156.1, 163.6)	0.79	−19.5 (−22.98, −15.96)	156.9 (151.2, 162.7)	0.72
	10	−17.3 (−18.20, −16.33)	155.4 (154.0, 156.8)	0.95	−16.2 (−17.44, −14.90)	153.5 (151.4, 155.6)	0.92
	20	−14.0 (−14.72, −13.24)	153.0 (151.9, 154.1)	0.95	−13.6 (−14.70, −12.50)	152.4 (150.6, 154.2)	0.92
	30	−13.2 (−14.02, −12.38)	154.1 (152.9, 155.3)	0.93	−14.5 (−15.65, −13.41)	156.4 (154.5, 158.2)	0.92
Macquarie	5	−21.5 (−23.50, −19.52)	156.1 (153.2, 159.0)	0.84	−12.2 (−14.95, −9.43)	140.5 (135.9, 145.0)	0.58
	10	−21.3 (−23.31, −19.31)	161.8 (158.8, 164.8)	0.85	−19.8 (−22.43, −17.10)	159.2 (155.0, 163.4)	0.78
	20	−17.9 (−18.96, −16.75)	161.5 (159.9, 163.1)	0.93	−18.3 (−19.92, −16.67)	162.2 (159.6, 164.8)	0.90
	30	−13.7 (−14.68, −12.64)	155.7 (154.2, 157.2)	0.90	−14.3 (−16.05, −12.57)	156.7 (153.9, 159.6)	0.84
Kirsty K	5	−21.6 (−23.83, −19.43)	163.6 (160.4, 166.8)	0.83	−16.7 (−20.03, −13.42)	155.3 (149.9, 160.7)	0.67
	10	−18.7 (−21.38, −16.11)	164.2 (160.3, 168.0)	0.69	−15.3 (−18.33, −12.21)	158.5 (153.7, 163.2)	0.58
	20	−17.2 (−18.41, −15.97)	161.9 (160.1, 163.7)	0.90	−15.0 (−16.62, −13.43)	158.3 (155.7, 160.8)	0.86
	30	−17.3 (−18.40, −16.13)	161.8 (160.1, 163.4)	0.92	−16.9 (−18.09, −15.71)	161.2 (159.2, 163.1)	0.94
Mean C_{PL} (s.d.)		−17.9 (3.1)			−16.0 (2.3)		

When PL has been estimated correctly, the resulting RNL or ASL should not increase if plotted against the original range at which the measurement was taken, as can be seen for *Primrose*, using the averaged PL with data from <15 m removed (i.e., using $16.0\log_{10}(range)$, Figure 5). This plot also illustrates the increased variation in RLs taken at ranges <15 m, compared with those at greater ranges at all speeds (Figure 5). The reduction in RL over the entire measured range would be 40 dB for spherical spreading RNLs and 32 dB for the average modelled losses at $16.0\log_{10}(range)$. Therefore, the median ASLs, using $16.0\log_{10}(range)$, for the *Primrose* (at 5, 10, 20 and 30 km h^{-1}, respectively) were 151.7, 153.3, 156.2, and 158.7; for the *Macquarie*, they were 146.7, 152.9, 158.2, and 159.2; and for the *Kirsty K*, they were 154.4, 159.5, 160.0, and 159.7, respectively (Table 4).

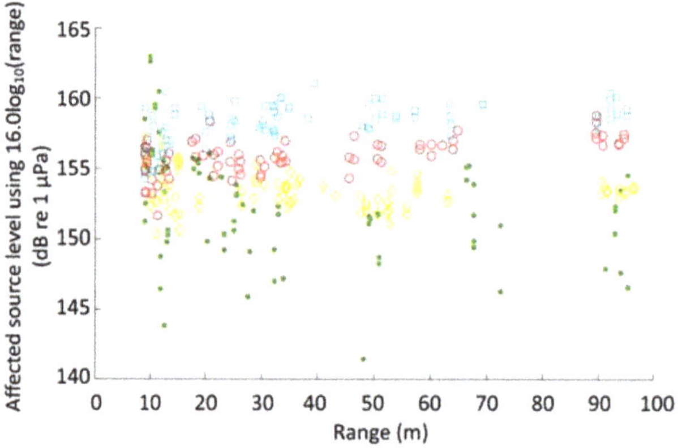

Figure 5. Affected source levels using the average linear regression-determined propagation loss across all vessels, excluding ranges below 15 m ($16.0\log_{10}\{range\}$) for *Primrose* at category speeds of 5, 10, 20 and 30 km h^{-1} (green dots, beige diamonds, red circles and blue square, respectively).

Table 4. Percentiles and maximum ranges of radiated noise levels and of affected source levels for the average $\log_{10}(range)$ losses across modelled losses for individual vessels and speeds using all ranges and using all ranges >15 m.

| | | Percentiles of Radiated Noise Level (dB re 1 µPa m) | | | | | | | | Percentiles of Affected Source Level (dB re 1 µPa m) | | | | | | | |
| | | All Ranges | | | | Excluding <15 m Range | | | | Loss = 17.9log₁₀ (*range*) | | | | Loss = 16.0log₁₀ (*range*) | | | |
Vessel	Speed	25%	50%	75%	Range	25%	50%	75%	Range	25%	50%	75%	Range	25%	50%	75%	Range
Primrose	5	155.5	**158.4**	160.2	18.8	155.6	**158.3**	159.9	14.3	152.4	**155.1**	157.4	20.2	149.3	**151.7**	153.9	14.6
	10	158.6	**159.7**	160.5	7.3	159.0	**159.9**	160.7	6.6	155.4	**156.6**	157.2	6.7	152.6	**153.3**	153.9	5.2
	20	160.3	**161.6**	163.5	10.8	161.3	**162.3**	163.9	6.9	157.6	**158.4**	159.9	8.9	155.5	**156.2**	156.9	4.7
	30	163.6	**164.8**	166.1	9.4	164.1	**165.4**	166.4	6.6	160.1	**161.3**	162.4	8.1	158.1	**158.7**	159.5	4.7
Macquarie	5	151.5	**153.4**	155.5	15.1	151.5	**153.2**	155.2	8.6	148.7	**150.5**	153.3	15.6	145.4	**146.7**	148.4	11.2
	10	158.1	**159.6**	161.2	11.4	158.0	**159.3**	161.0	10.6	154.8	**156.4**	158.0	12.2	151.4	**152.9**	154.3	10.7
	20	163.8	**164.6**	165.9	8.0	163.8	**164.5**	165.9	6.6	160.5	**161.3**	162.4	8.1	157.5	**158.2**	159.3	6.8
	30	163.9	**165.4**	166.6	12.0	164.7	**165.8**	167.1	7.9	160.0	**161.9**	163.1	10.8	158.6	**159.2**	160.4	6.4
Kirsty K	5	159.7	**161.2**	162.7	14.3	159.8	**161.0**	162.4	14.3	156.6	**158.1**	160.1	15.7	153.2	**154.4**	155.4	15.3
	10	162.9	**165.3**	169.0	16.2	163.5	**165.5**	169.7	11.9	159.7	**162.1**	166.0	17.1	157.1	**159.5**	162.7	13.3
	20	165.1	**166.0**	167.1	9.7	165.5	**166.3**	167.2	6.6	161.8	**163.0**	164.1	9.8	159.1	**160.0**	160.9	7.9
	30	165.2	**166.1**	167.1	9.6	165.7	**166.2**	167.1	4.5	162.0	**162.9**	163.6	9.5	158.9	**159.7**	160.2	3.4

3.3. One-Third Octave Band Levels

The interpolated one-third octave band levels received across the different slant ranges illustrate the interference pattern with frequency and range (Figure 6). In all three vessels, the higher frequencies increased with increasing speed, shown by the one-third octave band levels and RNLs for each vessel (Figures 6 and 7).

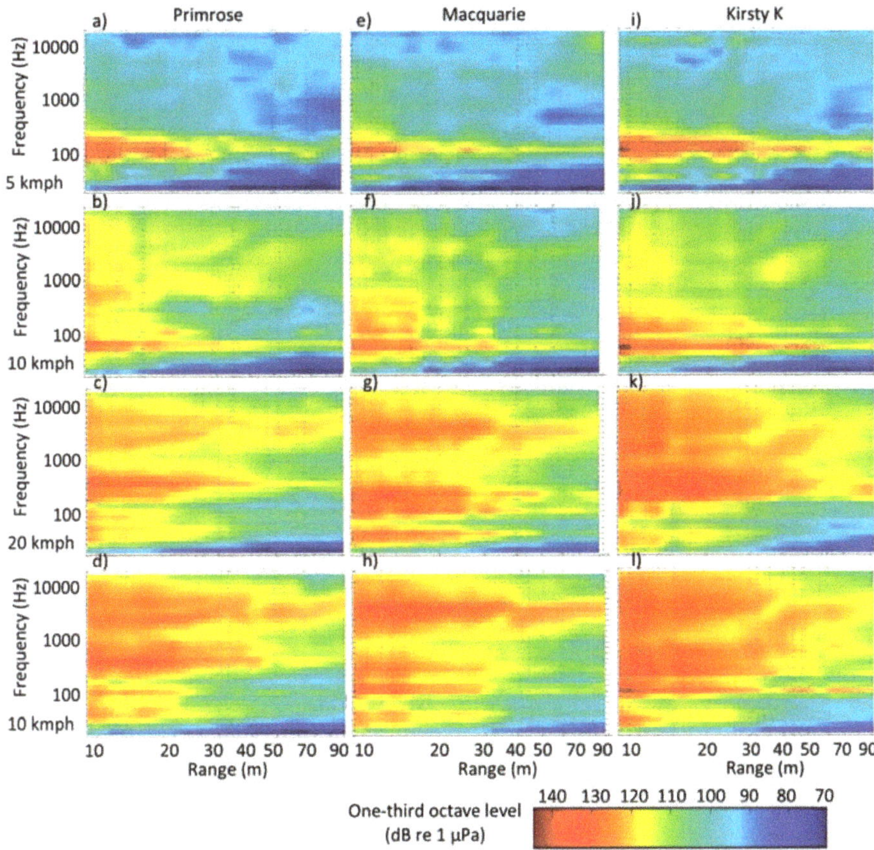

Figure 6. Interpolated (in log frequency and log range bins) one-third octave band received levels with range for each vessel (**a–d**, *Primrose*; **e–h**, *Macquarie*; **i–l**, *Kirsty K*), at each speed. (**a,e,i**, 5 km h^{-1}; **b,f,j**, 10 km h^{-1}; **c,g,k**, 20 km h^{-1}; **d,h,l**, 30 km h^{-1}). Black dots show the ranges at which measurements were taken to highlight gaps in interpolated data.

Linear regression determined one-third octave band PLs decreased with increasing frequency (Figure 8) from between 20 and 40 at frequencies below 50 Hz, to between 10 and 25 for frequencies between 50 and 2000 Hz. Above 2000 Hz, the propagation loss coefficients split with the vessels travelling at 5 km h^{-1} that exhibited little energy at these frequencies having coefficients of <10 (Figure 8, dots at >2000 Hz) and models for vessels with greater power and faster speeds displaying higher coefficients of 5 to 15 (Figure 8, red and green lines at >2000 Hz). Applying these frequency-dependent PL coefficients to the one-third octave band RLs produced ASL estimates higher than those derived using a single frequency-independent PL coefficient (Supplementary information).

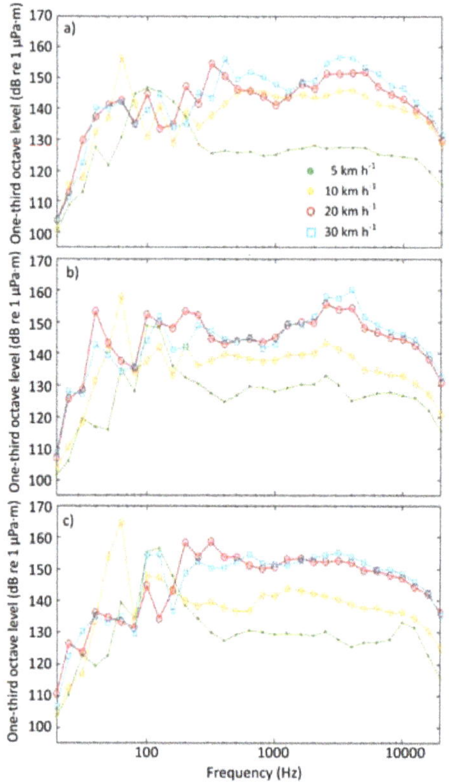

Figure 7. One-third octave radiated noise levels for (**a**) *Primrose*, (**b**) *Macquarie*, and (**c**) *Kirsty K*, at speeds of 5, 10, 20 and 30 km h^{-1} (green circles, beige diamonds, red circles and blue squares, respectively).

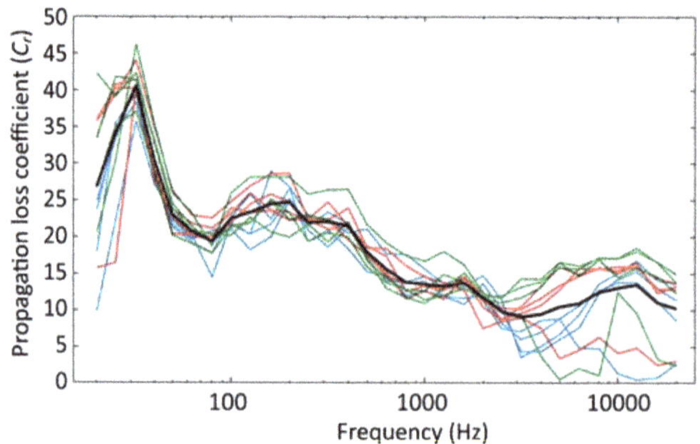

Figure 8. One-third octave band propagation loss coefficients (C_r) determined from linear regression for *Primrose*, *Macquarie* and *Kirsty K* (blue, red and green, respectively). Mean one-third octave band propagation loss coefficients across all vessels and speeds shown by the black line.

3.4. Received Levels with Speed

All three vessels displayed a general increasing trend in source level with speed, with C_{v1} values of 1.05 (s.d. = 0.09), 1.86 (0.11), and 0.68 (0.15) for the *Primrose*, *Macquarie* and *Kirsty K*, respectively (Table 5). While the *Primrose* ASLs clearly increased with each speed increase, the *Macquarie* appeared close to asymptote by 30 km h^{-1} and the *Kirsty K* appeared to reduce in source level after 20 km h^{-1} (Figure 9).

Table 5. Linear regression logistic (or polynomial for the modelled affected source levels for *Macquarie*) relationship between radiated noise levels, ASLs and averaged log$_{10}$(range) ASLs with speed, showing the theoretical source level at 0 km h^{-1} and remaining coefficients. Standard deviations shown in parentheses.

	Vessel	Level at 0 km h^{-1}	Coefficient C_{v1}	Coefficient C_{v2}	R^2	RMSE
RNL	*Primrose*	155.9 (0.7)	0.84 (0.11)	–	0.43	2.60
	Macquarie	151.5 (0.7)	1.55 (0.11)	–	0.68	2.87
	Kirsty K	161.6 (0.9)	0.53 (0.14)	–	0.14	3.33
ASL	*Primrose*	160.1 (0.7)	−0.78 (0.11)	–	0.43	2.44
	Macquarie	150.1 (1.2)	0.94 (0.12)	−0.50 (0.06)	0.50	2.62
	Kirsty K	165.1 (0.8)	−0.37 0.12)	–	0.09	2.97
Full data (17.9 loss)	*Primrose*	152.9 (0.6)	0.84 (0.1)	–	0.48	2.37
	Macquarie	148.4 (0.75)	1.56 (0.11)	–	0.69	2.82
	Kirsty K	158.6 (0.9)	0.53 (0.14)	–	0.15	3.32
Range >15 m (16.0 loss)	*Primrose*	148.8 (0.55)	1.05 (0.09)	–	0.73	1.73
	Macquarie	143.5 (0.75)	1.86 (0.12)	–	0.81	2.36
	Kirsty K	154.5 (11.0)	0.68 (0.15)	–	0.27	2.86

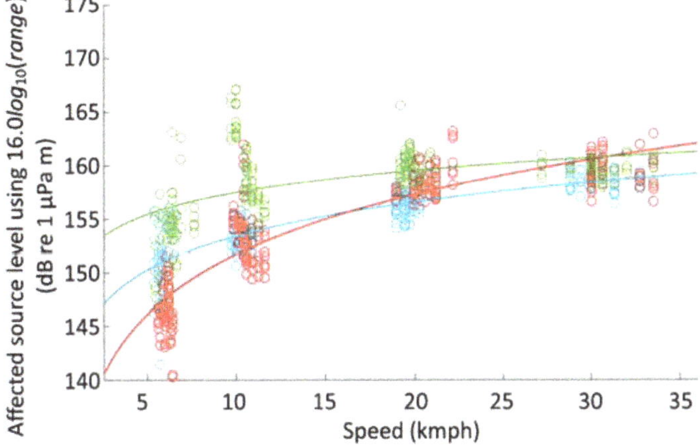

Figure 9. Relationships between affected source levels using the average linear regression-determined propagation loss across all vessels, excluding ranges below 15 m and speed for the *Primrose* (blue circles), *Macquarie* (red circles) and *Kirsty K* (green circles).

4. Discussion

Management strategies to mitigate the impacts of noise pollution by vessels in marine environments rely on an understanding of the acoustic signature of different vessel types. The variation in frequency bands in which species can detect sound means that reporting broadband RNLs or MSLs alone is insufficient information to support management decisions [36,43]. We characterized the acoustic pressure signatures for the CPA of three small (<10 m length) vessels of similar lengths, but differing engine power, at multiple speeds. We empirically measured the propagation of their noise in shallow water (10 m depth) over the flat, sandy lagoon of a tropical granitic island, from within the near field (directly beneath the vessels at the 9 m range), out to the 100 m range. RNL and empirically modelled ASL have been determined for each speed and vessel. RNLs for small vessels, particularly those under 10 m length, are rarely reported and MSLs even less so [26,36,37,53]. Thus, the estimated RNLs, ASLs, and one-third octave band levels described here provide useful information on both shallow-water propagation of these signals and the source levels and spectra that can be applied to modelling of vessel noise.

The ISO have provided criteria for quantifying RNLs. However, the highest criteria cannot be met in Australia's shallow coastal waters. Estimation of MSLs require accurate knowledge of seafloor geomorphology (and other marine geophysical characteristics) to ensure that PL is modelled appropriately, otherwise unexpected features may arise, potentially resulting in misleading estimates [42]. Our study recorded >80 measurements at a slant ranges of between 9 and 100 m for each vessel–speed scenario to ensure PL could be estimated with confidence. RLs recorded within the closest 15 m displayed greater variation than those taken at greater ranges (Supplementary information), justifying the removal of these data from PL models on the basis that they were in the acoustic near field. This still left >50 measurements for each scenario. PL coefficients in individual vessel/speed models excluding these data ranged between 13.6 and 19.8 and produced an average coefficient of 16.0 across the 12 PL models (Table 3), approximately halfway between cylindrical and spherical spreading losses, a reasonable estimate of PL in the tested conditions [54]. As the level of required information makes the accuracy of MSL estimates difficult to assess retrospectively, we argue that measurement of RLs for multiple passes at multiple ranges to model PL through linear regression of empirical measurements was an appropriate alternative to propagation modelling for similar shallow-water cases.

Not all low-frequency energy is accounted for in this method, as indicated by the low RNLs (Figure 7) and high propagation losses (Figure 8) of one-third octave bands at frequencies below 80 Hz, as would be expected with estimates made using a dipole model. However, similar errors may also often be found in the determination of MSLs in shallow water where geomorphological information has not been accurately identified. Thus, the RNL and ASL estimates provided here are representative of the 80-20,000 Hz band. Applying linear-regression to the one-third octave band RLs to determine frequency-dependent PLs can identify where some of the low-frequency energy is missed in the RNL and ASL estimates (Figure 8, Supplementary information). However, using the vessel noise as a source can be misleading as it does not necessarily contain sufficient energy across the desired bandwidth to provide accurate PLs from linear regression. For example, vessels travelling at 5 km h^{-1} emitted relatively little sound above 2 kHz, thus the PLs appeared to be low (Figure 8, lower values of C_r at frequencies above 2 kHz). This can be alleviated by using a standardized source, such as a speaker and tone generator at the desired frequencies, though these unfortunately not available during this study. The difference between estimates of RNL (excluding data at the <15 m range) and ASL (using averaged PL models and excluding data at the <15 m range) across all vessels and speeds was 6.4 dB, while the mean difference between ASLs with and without data within the 15 m range for each scenario was -3.2 dB (s.d. = 4.9 dB, Tables 3 and 4). Additionally, the average correlation coefficient (R^2) of the individual models decreased from 0.87 to 0.80 when the data <15 m were removed (Table 3). The average difference between 25th and 75th percentiles for RNLs and averaged-loss ASLs was 2.8 dB (s.d. = 1.3, Table 4), whereas the average range was 8.8 dB (s.d. = 3.3 dB). In contrast, the average standard deviation in ASL estimates from individual speed-/vessel-specific models was 6.7 dB (Table 3). These findings show that multiple measurements of an individual vessel at the same speed can be variable and greater than the changes found by moving from a spherical spreading calculation of RNL to the linear regression-modelled ASL. Removing the near-field measurements was prudent, as this reduced the PL estimate in almost every model and the resulting estimated ASLs, albeit by less than 3 dB (Table 3).

Wladichuk et al. [36] measured multiple vessel categories, of which rigid hull inflatable boats (RHIBs) were the most similar to the vessels in this study. Broadband RNLs for these vessels were 160.9 and 167.0 dB re 1 µPa m for ≈13 and 28 km h^{-1}, respectively [36], though RHIBs were larger both in length (5.2 to 8.2 m) and engine power (150 to 700 hp) than the vessels we recorded at Lizard Island. The *Kirsty K*, the closest vessel in size and power measured here to the previously measured RHIBs [36], produced ASLs of ≈160 dB re 1 µPa m and spherical spreading RNLs of 165–166 dB re 1 µPa, at both of these speeds (Table 4). Erbe [26] and Erbe et al. [37] estimated RNLs and MSLs of various RHIBs, ranging from 7 to 9.1 m length and engine powers from 90 to 450 hp. Differences in estimated levels at any one speed, across all vessels, were found to be up to >20 dB, comparable with the variation found between the *Primrose*, *Macquarie* and *Kirsty K* (Tables 3 and 4). One vessel (6.5 m length, 90 hp) in Erbe et al. [37], produced MSLs of approximately 154, 156 and 153 dB re 1 µPa m at speeds of approximately 12, 20 and 30 km h^{-1}, respectively, in 8 m of flat water. ASL estimates of the *Macquarie*, the vessel of most similar size and power measured at Lizard Island were 152.9, 158.2 and 159.2 dB re 1 µPa m, respectively, for similar speeds (Table 4). Thus, the ASLs estimated here using PLs of 16log$_{10}$(*range*) were comparable with previous measures.

Estimations of RNL and MSL typically use recordings of the CPA, or an averaged azimuth about the CPA, and assume that maximum broadband levels are received when broadside to the vessel or that there is no horizontal source directionality [37,44]. However, recorded noise levels can depend on azimuth and inclination, with some reports suggesting acoustic shielding at fore, aft or broadside, depending on the vessel design [42,55]. For example, similar to Parsons et al. [42], the recordings taken at Lizard Island displayed higher levels of acoustic energy in front of the vessel than behind, suggesting that entrained bubbles behind the vessels, generated by cavitation, have a shadowing effect after the vessel has passed. In this study, maximum levels were assumed to be broadside and directionality to be symmetrical about the centerline of the vessels, thus data from both port and starboard recordings were pooled during analysis. Any adjustment (e.g., altering power or trim) to maintain the correct speed and direction to account for wind and current can alter acoustic output of the vessel. Variations in RL recorded at Lizard Island were greatest at the slowest speed, which matched anecdotal comments from the skipper that holding the appropriate direction and line was more difficult when moving slowly. Erbe et al. [37] noted similar variability in their estimates of RNLs and MSLs of rigid-hulled inflatable boats (RHIBS) and hypothesized potential reasons for them. Acoustic output and potential shielding (e.g., by factors such as the bubble cloud and vessel hull) both contribute to the RL. Thus, changes to hull inclination and size of the cavitation vortices in the vessel wake, which occur with changes in speed or power, may be indicative of both acoustic output and dampening. Measurement of these factors could provide greater information to characterize the acoustic signature.

Ross [31] defined a relationship between vessel speed and changes in source level that Erbe [26] and then Wladichuk et al. [36] investigated for small vessels, albeit with few measures at different speeds. Broadband level increases with speed in this study comprised C_v values of 1.05, 1.86 and 0.68 for the *Primrose*, *Macquarie* and *Kirsty K*, respectively (Table 5). Wladichuk et al. [36] also observed a wide range in the same coefficient for their RHIB class (1.3 to 1.8), though they were generally higher than the three vessels studied here (Table 5). It is possible that the greater variability found at the slower speeds tested at Lizard Island (5 km h^{-1}), lack of measurements at the higher speeds (>30 km h^{-1}), yet significantly more measurements at speeds in between, can account for this difference. However, Wladichuk et al. [36] also found a much greater variation in C_v for monohulls ranging between 1.0 and 3.6, thus it is likely that the relationship between source spectra and speed is highly vessel specific.

Erbe et al. [37] noted that the decline in MSL after 20 km h^{-1} (which continued to speeds of ≈40 km h^{-1}) was likely due to reaching its 'hump' speed, when the bow-up angle was largest, producing a large trim and water resistance and therefore propeller loading. Above this speed, the vessel begins to plane, resistance and propeller loading drops, and so too does the noise level. The three vessels at Lizard Island did not display a distinct hump across the speeds measured (Figure 9). However, the asymptotic nature of the *Macquarie* and *Kirsty K* RNL and ASL curves with speed suggest that this could have occurred between 20 and 30 km h^{-1} or about to occur with increasing speed. If measures at Lizard Island had extended to the maximum speeds of each vessel, such a relationship might have been observed. At planing speeds, Erbe et al. [37] also found that variation in RLs increased with speed, attributing this to the additional sound of vessel slap on the water for some measurements, highlighting that multiple measures are required, even for an individual speed.

At the slower speeds tested at Lizard Island, almost all energy in all three vessels occurred below 200 Hz, increasing up to 10 kHz at the faster speeds (Figure 7). Energy at all frequencies was detected at the furthest ranges for the highest vessel speeds (Figure 6). Ideally, for the same speed and vessel, the interpolated measurements of one-third octave levels would represent the similar Lloyd's mirror pattern seen in the spectrograms of the individual recordings as the vessel approaches and passes the hydrophone. This pattern can be seen to varying degrees in all scenarios in Figure 6. However, these patterns are blurred by the interpolation across range and the integration of the spectra into one-third octave bands, which blurs the narrowband energy. These frequencies were all within the hearing sensitivity range of most fishes and overlapped with those of many cetaceans [22], with the majority of energy at the most sensitive frequencies and main communication band of most fishes [56–60]. Although several studies have been conducted on the impacts of noise from such vessels on the ecology of fishes and invertebrates in recent years [61–64], and specific ways to mitigate these impacts in shallow water are being investigated [38,42], there remains a knowledge gap in this area.

Together with previous studies, the data from the *Primrose*, *Macquarie* and *Kirsty K* highlight that to fully understand a vessel's acoustic signature requires taking multiple measurements at multiple speeds and that within a given vessel class, large differences can be observed between vessels. Additional measures (e.g., data on vessel inclination, wake and size or vortices behind the vessel) may help assess the noise signature and understand differences between vessels and speeds. For appropriate modelling to be conducted with a view to mitigating the impact of noise from small vessels, a meta-analysis of all reported data is required.

Supplementary Materials: The following are available online at http://www.mdpi.com/2077-1312/8/12/970/s1, Table S1: Standard deviation in RNL estimates within shown range brackets for the various vessels and speeds, Table S2: Percentiles and maximum ranges of affected source levels calculated using propagation loss coefficients derived from linear regression of one-third octave band received levels at multiple ranges

Author Contributions: Conceptualization, M.P.; methodology, M.P. and M.M.; formal analysis, M.P.; writing—original draft preparation, M.P.; writing—review and editing, M.P. and M.M.; visualization, M.P. and M.M.; funding acquisition, M.M. All authors have read and agreed to the published version of the manuscript.

Funding: This work was undertaken for the Marine Biodiversity Hub, a collaborative partnership supported through funding from the Australian Government's National Environmental Science Program.

Acknowledgments: The authors would like to thank Drs. Mark McCormick and Jen Kelley, and the Lizard Island Research Station for logistical support and assistance during data collection. The authors would also like to thank three reviewers for the time to consider the manuscript and for their valuable suggestions.

Conflicts of Interest: The authors declare no conflict of interest. The funders had no role in the design of the study; in the collection, analyses, or interpretation of data; in the writing of the manuscript, or in the decision to publish the results.

References

1. Winn, H.E. The biological significance of fish sounds. *Mar. Bioacoust.* **1964**, *2*, 213–231.
2. Au, W.W.L.; Hastings, M.C. *Principles of Marine Bioacoustics*, 2nd ed.; Springer: New York, NY, USA, 2008. [CrossRef]
3. Radford, C.A.; Stanley, J.A.; Hole, W.; Montgomery, J.C.; Jeffs, A. Localised coastal habitats have distinct underwater sound signatures. *Mar. Ecol. Prog. Ser.* **2010**, *401*, 21–29. [CrossRef]
4. Erbe, C.; Reichmuth, C.; Cunningham, K.; Lucke, K.; Dooling, R. Communication masking in marine mammals: A review and research strategy. *Mar. Pollut. Bull.* **2016**, *103*, 15–38. [CrossRef] [PubMed]
5. Simpson, S.; Meekan, M.G.; Montgomery, J.; McCauley, R.D.; Jeffs, A. Homeward Sound. *Science* **2005**, *308*, 221. [CrossRef]

6. Simpson, S.; Radford, A.; Nedelec, S.; Ferrari, M.C.O.; Chivers, D.P.; McCormick, M.I.; Meekan, M.G. Anthropogenic noise increases fish mortality by predation. *Nat. Commun.* **2016**, *7*, 10544. [CrossRef]

7. Parsons, M.J.G.; McCauley, R.D.; Mackie, M.C.; Siwabessy, P.J.; Duncan, A.J. *In situ* source levels of mulloway (*Argyrosomus japonicus*) calls. *J. Acoust. Soc. Am.* **2012**, *132*, 3559–3568. [CrossRef]

8. Stanley, J.A.; Van Parijs, S.M.; Hatch, L.T. Underwater sound from vessel traffic reduces the effective communication range in Atlantic cod and haddock. *Sci. Rep.* **2017**, *7*, 14633. [CrossRef]

9. Van der Graaf, A.J.; Ainslie, M.A.; Andre, M.; Brensing, K.; Dalen, J.; Dekeling, R.P.A.; Robinson, S.M.; Tasker, M.I.; Thomsen, F.; Werner, S. *European Marine Strategy Framework Directive-Good Environmental Status (MSFD GES): Report of the Technical Subgroup on Underwater Noise and Other Forms of Energy (JRC Scientific and Technical Report)*; TSG Noise & Milieu Ltd.: Brussels, Belgium, 2012; 75p.

10. Weilgart, L.S. The impacts of anthropogenic ocean noise on cetaceans and implications for management. *Can. J. Zool.* **2007**, *85*, 1091–1116. [CrossRef]

11. Graham, A.L.; Cooke, S.J. The effects of noise disturbance from various recreational boating activities common to inland waters on the cardiac physiology of a freshwater fish, the largemouth bass (*Micropterus salmoides*). *Aquat. Cons. Mar. Fresh. Ecosyst.* **2008**, *18*, 1315–1324. [CrossRef]

12. Williams, R.; Erbe, C.; Ashe, E.; Beerman, A.; Smith, J. Severity of killer whale behavioral responses to ship noise: A dose response study. *Mar. Poll. Bull.* **2014**, *79*, 254–260. [CrossRef]

13. Simpson, S.D.; Radford, A.N.; Holles, S.; Ferarri, M.C.O.; Chivers, D.P.; McCormick, M.I.; Meekan, M.G. Small-Boat Noise Impacts Natural Settlement Behavior of Coral Reef Fish Larvae. In *The Effects of Noise on Aquatic Life II. Advances in Experimental Medicine and Biology*; Popper, A., Hawkins, A., Eds.; Springer: New York, NY, USA, 2016; Volume 875. [CrossRef]

14. Nedelec, S.L.; Mills, S.C.; Radford, A.N.; Belade, R.; Simpson, S.D.; Nedelec, B.; Côté, I.M. Motorboat noise disrupts co-operative interspecific interactions. *Sci. Rep.* **2017**, *7*, 6987. [CrossRef] [PubMed]

15. Erbe, C.; Marley, S.; Schoeman, R.; Smith, J.N.; Trigg, L.; Embling, C.B. The effects of ship noise on marine mammals—A review. *Front. Mar. Sci.* **2019**, *6*, 606. [CrossRef]

16. Ross, D. Ship sources of ambient noise. *IEEE J. Ocean. Eng.* **2005**, *30*, 257–261. [CrossRef]

17. McDonald, M.A. Increase in deep ocean ambient noise in the Northeast Pacific west of San Nocolas Island. *J. Acoust. Soc. Am.* **2006**, *120*, 711–718. [CrossRef] [PubMed]

18. Hildebrand, J.A. Anthropogenic and natural sources of ambient noise in the ocean. *Mar. Ecol. Prog. Ser.* **2009**, *395*, 5–20. [CrossRef]

19. Frisk, G. Noiseonomics: The relationship between ambient noise levels in the sea and global economic trends. *Sci. Rep.* **2012**, *2*, 437. [CrossRef]

20. Williams, R.; Wright, A.J.; Ashe, E.; Blight, L.K.; Bruintjes, R.; Canessa, R.; Clark, C.W.; Cullis-Suzuki, S.; Dakin, D.T.; Erbe, C.; et al. Impacts of anthropogenic noise on marine life: Publication patterns, new discoveries, and future directions in research and management. *Ocean Coast. Man.* **2015**, *115*, 17–24. [CrossRef]

21. McWhinnie, L.; Smallshaw, L.; Serra-Sogas, N.; O'Hara, P.D.; Canessa, R. The grand challenges in researching marine noise pollution from vessels: A horizon scan for 2017. *Front. Mar. Sci.* **2017**, *4*, 31. [CrossRef]

22. Duarte, C.M.; Chapuis, L.; Collin, S.P.; Costa, D.P.; Devassy, R.P.; Eguiluz, V.M.; Erbe, C.; Halpern, B.S.; Harding, H.R.; Havlik, M.N.; et al. The Ocean Soundscape of the Anthropocene. *Science* **2020**. accepted.

23. Blane, J.; Jaakson, R. The impact of ecotourism boats on the St. Lawrence beluga whales. *Environ. Cons.* **1994**, *21*, 267–269. [CrossRef]

24. Erbe, C. Underwater noise of whale-watching boats and potential effects on killer whales (*Orcinus orca*), based on an acoustic impact model. *Mar. Mamm. Sci.* **2002**, *18*, 394–418. [CrossRef]

25. Erbe, C.; Williams, R.; Sandilands, D.; Ashe, E. Identifying modeled ship noise hotspots for marine mammals of Canadas' Pacific Region. *PLoS ONE* **2014**, *9*, e89820. [CrossRef] [PubMed]

26. Merchant, N.D.; Pirotta, E.; Barton, T.R.; Thompson, P.M. Monitoring ship noise to assess the impact of coastal developments on marine mammals. *Mar. Poll. Bull.* **2014**, *78*, 85–95. [CrossRef] [PubMed]

27. New, L.F.; Hall, A.J.; Harcourt, R.; Kaufman, G.; Parsons, E.C.M.; Pearson, H.C.; Cosentino, A.M.; Schick, R.S. The modelling and assessment of whale-watching impacts. *Ocean Coast. Manag.* **2015**, *115*, 10–16. [CrossRef]

28. Cominelli, S.; Devillers, R.; Yurk, H.; MacGillivray, A.O.; McWhinnie, L.; Canessa, R. Noise exposure from commercial shipping for the southern resident killer whale population. *Mar. Poll. Bull.* **2018**, *136*, 177–200. [CrossRef] [PubMed]

29. Hatch, L.; Clark, C.W.; Merrick, R.; Van Parijs, S.M.; Ponirakis, D.; Schwehr, K.; Thompson, M.; Wiley, D. Characterizing the relative contributions of large vessels to total ocean noise fields: A case study using the Gerry E. Studds Stellwagen Bank National Marine Sanctuary. *Environ. Man.* **2008**, *42*, 735–752.

30. Erbe, C.; MacGillivray, A.O.; Williams, R. Mapping cumulative noise from shipping to inform marine spatial planning. *J. Acoust. Soc. Am.* **2012**, *132*, EL423–EL428. [CrossRef]

31. Ross, D. *Mechanics of Underwater Noise*; Pergamon Press: New York, NY, USA, 1976.

32. Chion, C.; Lagrois, D.; Dupras, J. A meta-analysis to understand the variability in reported source levels of noise radiated by ships from opportunistic studies. *Front. Mar. Sci.* **2019**, *6*, 714.

33. Au, W.W.L.; Green, M. Acoustics interaction of humpback whales and whale-watching boats. *Mar. Environ. Res.* **2000**, *49*, 469–481. [CrossRef]

34. Buckstaff, K.C. Effects of watercraft noise on the acoustic behaviour of bottlenose dolphins *Tursiops truncates* in Sarasota Bay. *Fla. Mar. Mam. Sci.* **2004**, *20*, 709–725. [CrossRef]

35. Brooker, A.; Humphrey, V. Measurement of radiated underwater noise from a small research vessel in shallow water. *Ocean Eng.* **2015**, *120*, 182–189. [CrossRef]

36. Wladichuk, J.L.; Hannay, D.E.; MacGillivray, A.O.; Li, Z.; Thornton, S.J. Systematic source level measurements of whale watching vessels and other small boats. *J. Ocean. Technol.* **2019**, *14*, 110–126.

37. Erbe, C.; Liong, S.; Koessler, M.W.; Duncan, A.J.; Gourlay, T. Underwater sound of rigid-hulled inflatable boats. *J. Acoust. Soc. Am.* **2016**, *139*, EL223–EL227. [CrossRef] [PubMed]

38. McCloskey, K.P.; Chapman, K.E.; Chapuis, L.; McCormick, M.I.; Radford, A.N.; Simpson, S.D. Assessing and mitigating impacts of motorboat noise on nesting damselfish. *Environ. Poll.* **2020**, *266*, 115376. [CrossRef]

39. Marley, S.A.; Erbe, C.; Salgado-Kent, C.P.; Parsons, M.J.G.; Parnum, I.M. Spatial and Temporal Variation in the Acoustic Habitat of Bottlenose Dolphins (*Tursiops aduncus*) within a Highly Urbanized Estuary. *Front. Mar. Sci.* **2017**, *4*, 197. [CrossRef]

40. Parsons, M.J.G.; McCauley, R.D.; Mackie, M.C. Characterisation of mulloway (*Argyrosomus japonicus*) advertisement sounds. *Acoust. Aust.* **2013**, *196*, 196–201.

41. Nedelec, S.L.; Radford, A.N.; Pearl, L.; Nedelec, B.; McCormick, M.I.; Meekan, M.G.; Simpson, S.D. Motorboat noise impacts parental behaviour and offspring survival in a reef fish. *Proc. R. Soc. B.* **2017**, *284*, 20170143. [CrossRef]

42. Parsons, M.J.G.; Duncan, A.J.; Parsons, S.K.; Erbe, C. Reducing vessel noise: An example of a solar-electric passenger ferry. *J. Acoust. Soc. Am.* **2020**, *147*, 3575–3583. [CrossRef]

43. Malinowski, S.J.; Gloza, I. Underwater noise characteristics of small ships. *Acta Acust. United Acust.* **2002**, *88*, 718–721.

44. International Organization for Standardization. *Underwater Acoustics—Quantities and Procedures for Description and Measurement of Underwater Sound from Ships—Part 1: Requirements for Precision Measurements in Deep Water Used for Comparison Purposes (ISO 17208-1)*; International Organization for Standardization: Geneva, Switzerland, 2016; 20p.

45. International Organization for Standardization. *Underwater Acoustics—Quantities and Procedures for Description and Measurement of Underwater Sound from Ships—Part 2: Determination of Source Levels from Deep Water Measurements (ISO 17208-2)*; International Organization for Standardization: Geneva, Switzerland, 2019; 13p.

46. Hamylton, S.M.; Leon, J.X.; Saunders, M.I.; Woodroffe, C.D. Simulating reef response to sea-level rise at Lizard Island: A geospatial approach. *Geomorphology* **2014**, *222*, 151–161. [CrossRef]

47. Daly, M. Wave Energy and Shoreline Response on a Fringing Reef Complex, Lizard Island, Qld, Australia. Bachelor's Thesis, Env. Sci., University of New South Wales, Sydney, Australia, 2005; 105p.

48. Frith, C.; Leis, J.; Goldmand, B. Currents in the Lizard Island region of the Great Barrier Reef Lagoon and their relevance to potential movements of larvae. *Coral Reefs* **1986**, *5*, 81–92. [CrossRef]

49. Gavrilov, A.G.; Parsons, M.J.G. A Matlab toolbox for the Characterisation of Recorded Underwater Sound (CHORUS). *Acoust. Aust.* **2014**, *42*, 191–196.

50. Urick, R.J. *Principles of Underwater Sound*, 3rd ed.; McGraw Hill: New York, NY, USA, 1983.

51. De Jong, C.A.F. Characterization of ships as sources of underwater noise. In Proceedings of the NAG-DAGA 2009, Rotterdam, The Netherlands, 23–26 March 2009; pp. 271–274.

52. Veirs, S.; Veirs, V. Vessel noise measurements underwater in the Haro Strait, WA. *J. Acoust. Soc. Am.* **2006**, *120*, 3382. [CrossRef]

53. Kipple, B.; Gabriele, C. Underwater noise from skiffs to ships. In Proceedings of the Fourth Glacier Bay Science Symposium, Junea, AK, USA, 26–28 October 2004; Piatt, J.F., Gende, S.M., Eds.; U.S. Geological Survey Scientific Investigations Report 2007-5047; 2007 U.S. Geological Survey. pp. 172–175.

54. Cato, D.H. Simple methods of estimating source levels and locations of marine animal sounds. *J. Acoust. Soc. Am.* **1998**, *104*, 1667–1678. [CrossRef] [PubMed]

55. McCauley, R.D.; Cato, D.H.; Jeffrey, A.F. *A Study of Impacts of the Impacts of Vessel Noise on Humpback Whales in Hervey Bay*; Report to the Queensland Department of Environment and Heritage; Maryborough Branch: Queensland, Australia, 1996; 163p.

56. Ladich, F.; Fay, R.R. Auditory evoked potential audiometry in fish. *Rev. Fish. Biol. Fish.* **2013**, *23*, 317–364. [CrossRef] [PubMed]

57. Kasumyan, A.O. Acoustic signaling in fish. *J. Ichthyol.* **2009**, *49*, 963–1020. [CrossRef]

58. Parsons, M.J.G.; Longbottom, S.; Lewis, P.; McCauley, R.D.; Fairclough, D.V. Sound production by the West Australian dhufish (*Glaucosoma hebraicum*). *J. Acoust. Soc. Am.* **2013**, *134*, 2701–2709. [CrossRef]

59. McWilliam, J.N.; McCauley, R.D.; Erbe, C.; Parsons, M.J.G. Patterns of biophonic periodicity on coral reefs in the Great Barrier Reef. *Sci. Rep.* **2017**, *7*, 17459. [CrossRef]

60. Montgomery, J.C.; Jeffs, A.; Simpson, S.D.; Meekan, M.G.; Tindle, C. Sound as an Orientation Cue for the Pelagic Larvae of Reef Fishes and Decapod Crustaceans. *Adv. Mar. Biol.* **2006**, *51*, 143–196.

61. Radford, A.N.; Kerridge, E.; Simpson, S.D. Acoustic communication in a noisy world: Can fish compete with anthropogenic noise? *Behav. Ecol.* **2014**, *25*, 1022–1030. [CrossRef]

62. Nedelec, S.L.; Mills, S.C.; Lecchini, D.; Nedelec, B.; Simpson, S.D.; Radford, A.N. Repeated exposure to noise increases tolerance in a coral reef fish. *Environ. Pollut.* **2016**, *216*, 428–436. [CrossRef] [PubMed]

63. Putland, R.L.; Merchant, N.D.; Farcas, A.; Radford, C.A. Vessel noise cuts down communication space for vocalizing fish and marine mammals. *Glob. Chang. Biol.* **2018**, *24*, 1708–1721. [CrossRef] [PubMed]

64. Mensinger, A.F.; Putland, R.L.; Radford, C.A. The effect of motorboat sound on Australian snapper *Pagrus auratus* inside and outside a marine reserve. *Ecol. Evol.* **2018**, *8*, 6438–6448. [CrossRef] [PubMed]

Publisher's Note: MDPI stays neutral with regard to jurisdictional claims in published maps and institutional affiliations.

Journal of
Marine Science and Engineering

Article

A Reference Spectrum Model for Estimating Source Levels of Marine Shipping Based on Automated Identification System Data

Alexander MacGillivray [1],* and **Christ de Jong [2]**

[1] JASCO Applied Sciences, Victoria, BC V8Z 7X8, Canada
[2] Netherlands Organisation for Applied Scientific Research (TNO), 2597 AK The Hague, The Netherlands; christ.dejong@tno.nl
* Correspondence: alex.macgillivray@jasco.com; Tel.: +1-250-483-3300

Abstract: Underwater sound mapping is increasingly being used as a tool for monitoring and managing noise pollution from shipping in the marine environment. Sound maps typically rely on tracking data from the Automated Information System (AIS), but information available from AIS is limited and not easily related to vessel noise emissions. Thus, robust sound mapping tools not only require accurate models for estimating source levels for large numbers of marine vessels, but also an objective assessment of their uncertainties. As part of the Joint Monitoring Programme for Ambient Noise in the North Sea (JOMOPANS) project, a widely used reference spectrum model (RANDI 3.1) was validated against statistics of monopole ship source level measurements from the Vancouver Fraser Port Authority-led Enhancing Cetacean Habitat and Observation (ECHO) Program. These validation comparisons resulted in a new reference spectrum model (the JOMOPANS-ECHO source level model) that retains the power-law dependence on speed and length but incorporates class-specific reference speeds and new spectrum coefficients. The new reference spectrum model calculates the ship source level spectrum, in decidecade bands, as a function of frequency, speed, length, and AIS ship type. The statistical uncertainty (standard deviation of the deviation between model and measurement) in the predicted source level spectra of the new model is estimated to be 6 dB.

Keywords: source levels; underwater noise; marine shipping; automated identification system; sound mapping

Citation: MacGillivray, A.; de Jong, C. A Reference Spectrum Model for Estimating Source Levels of Marine Shipping Based on Automated Identification System Data. *J. Mar. Sci. Eng.* **2021**, *9*, 369. https://doi.org/10.3390/jmse9040369

Academic Editor: Alessandro Ridolfi

Received: 17 February 2021
Accepted: 25 March 2021
Published: 30 March 2021

Publisher's Note: MDPI stays neutral with regard to jurisdictional claims in published maps and institutional affiliations.

1. Introduction

Underwater sound mapping is becoming an important tool in support of marine spatial planning of human activities at sea while protecting the marine environment [1]. Though the relationship between the environmental pressure caused by ambient noise and the state of the ecosystem is not yet fully understood, the European Union (EU) advises its Member States using underwater sound maps, combined with measurements, to quantify levels and trends of ambient noise for the implementation of its Marine Strategy Framework Directive [2]. European North Sea countries are jointly developing a framework for monitoring ambient noise in the North Sea in the Interreg Joint Monitoring Programme for Ambient Noise North Sea (JOMOPANS; https://northsearegion.eu/jomopans/; accessed on 26 March 2021). A key task in the project is to develop and demonstrate verified and validated modelling methods applicable for generating maps of ambient noise in the North Sea, with a focus on ships and wind as the dominant sources of sound. Noise from shipping is a dominant contributor to the global marine soundscape, and can adversely impact aquatic life via several effects pathways, including behavioral disturbance, stress, and masking [3,4]. The issue of underwater noise has also been recognized by the International Maritime Organization (IMO), which has published guidelines aimed at reducing vessel noise emissions from commercial shipping [5].

Various previous efforts have demonstrated that the approach for modelling shipping noise based on density and distribution of ship traffic is feasible [6–10]. However, large uncertainties remain. The diversity of ship characteristics and the various noise source mechanisms at different operational conditions make it impossible to include an exact prediction of the underwater radiated noise of individual vessels in the calculation of shipping sound maps. Moreover, information available from the vessel tracking data from Automated Identification System (AIS) is limited and not easily related to vessel noise emissions. In an international workshop [1], it was concluded that the speed variance remains a fundamental uncertainty in estimating source levels from AIS information: "AIS can provide information about the presence of ships (GT > 300) or shipping densities, but the Wales and Heitmeyer [11] model that is often used to estimate source levels for this traffic data does not include ship speed dependence as earlier models (e.g., the RANDI model [12]) had. For regulation purposes and for noise mapping, a new model that includes speed dependence and associated uncertainty is required. This requires coherent empirical measurements to inform model development". The need for more coherent underwater radiated noise measurements on commercial vessels, to support the development of statistical ship source level models, was also concluded in two large European research programs, SONIC [13] and AQUO [8].

One such data set, consisting of a large collection of systematic source level measurements for a wide variety of vessels, was collected during the Vancouver Fraser Port Authority-led Enhancing Cetacean Habitat and Observation (ECHO) Program's 2017 vessel slowdown trial in Haro Strait [14]. During 2017, the ECHO Program carried out their first voluntarily slow down trial to reduce underwater noise from marine shipping within the critical habitat of the endangered southern resident killer whale population [15]. To support underwater noise studies associated with the trial, JASCO Applied Sciences collected a total of 1862 monopole source level measurements, over a four-month period, on three different hydrophone systems [16]. This data set was used to establish speed scaling relationships for source levels of several different categories of vessels [17]. This data set is unique, not only because it provides a large collection of source levels for many different types of vessels, but also because the voluntary slow down protocol provided a strong experimental control for determining the effects of vessel speed on noise emissions. As such, the ECHO data set provided an ideal validation data set for testing speed dependence in statistical source level models.

In this study, we have applied the ECHO data set to test the validity of previously published statistical models for estimating ship source level from AIS data. In this comparison we observed systematic differences between the model predictions for different vessel classes. Therefore, we propose an updated reference spectrum model that incorporates ship type as well as speed and length. The parameters of this model are fitted to the ECHO data. In this paper, we describe the various source level models and the comparison of the model predictions with measured source levels from the ECHO data set.

2. Methods

2.1. Source Level Data and Validation

A large collection of 1862 source level measurements from ships of opportunity, collected near Vancouver (Canada), were used for validating source level models in JO-MOPANS. This data set was collected by JASCO Applied Sciences, in partnership with the ECHO Program, and included source level measurements collected in Haro Strait and Strait of Georgia during the 2017 ECHO Program voluntary slowdown trial [16,17]. Measurements were collected on calibrated hydrophones (Geospectrum M36, -165 ± 3 dB re 1 V/µPa sensitivity), using JASCO AMAR-G3 (Autonomous Multichannel Acoustic Recorder, Generation 3) recorders, at three locations, situated adjacent to the international shipping lanes in the Salish Sea. Hydrophones were deployed near the seabed in water depths ranging from 173 to 250 m. The mean closest point of approach (CPA) distance of vessels in the data set was 361 m and the maximum CPA distance was 1 km. The acoustic data were collected with 24-bit resolution at a minimum sampling frequency of 64 kHz. The location, speed, draft, and identifying details of vessels near the hydrophones were recorded using an automated identification system (AIS) receiver. Vessel classification and design details were obtained from the MarineTraffic.com web service [18], based on the maritime mobile service identity (MMSI) number. Additional details regarding the data set employed in the source level validation are provided in [17].

The acoustic data were analyzed using JASCO's ShipSound system, which analyzes hydrophone data and AIS broadcasts from passing vessels to calculate vessel source levels in terms of monopole source level (SL). For time periods when a passing vessel was detected on AIS, the system processes hydrophone data to obtain standard decidecade (i.e., 1/3-octave) band sound pressure level (SPL) inside a data window encompassing $\pm30°$ of the vessel's CPA to the hydrophone, according to the methods specified in the ANSI ship noise measurement standard (S12.64, 2009). SL was calculated in 36 decidecade bands (with nominal centre frequencies 10 Hz to 31.5 kHz), using a frequency-dependent propagation loss (PL) model, based on numerical solution of the acoustic wave equation, which accounts for the effect of the environment on sound transmission. Following a similar methodology to [11], the source depth in the PL model was represented as normally distributed random variable, where the mean source depth d_s was assumed to 50% of the static vessel draft as broadcast over AIS at the time of measurement and the distribution parameter was taken to be $\sigma = d_s/3.4$. Additional details regarding the source level analysis methodology employed in ShipSound are given in [17,19]. Source level measurements were anonymized and assigned to one of twelve different categories (discussed below).

To correct the comparison of the ECHO data set to the RANDI 3.1 source level model (Section 2.2.1), an adjustment of the ECHO source levels to a standard depth of 6 m has been made, according to the procedure described in Appendix A. For the JOMOPANS noise mapping, the selection of a fixed 6 m reference source depth for all vessels has the advantage that it offers the possibility to decrease the complexity of the propagation loss model calculations.

Ship source level models were validated against the ECHO data set by comparing model predictions to measured SL values, in each of the 36 decidecade frequency bands. Residual differences, in decibels, between predicted and observed source levels were calculated as:

$$e_k(f) = \widehat{L_{S_f}}(f, l_k, V_k, C_k) - L_{S_f k}(f), \tag{1}$$

where e_k is the residual difference between predicted and observed source level spectral density (L_{S_f}) for measurement k at frequency f and, depending on the model, the predicted spectrum level could be a function of vessel length, l, speed, V, and class, C. The observed source level spectral density (L_{S_f}) was calculated from the measured SL (L_S) in decidecade bands by subtracting the bandwidth (in dB re Hz):

$$L_{S_f}(f_i) = L_S(f_i) - 10\log_{10}\left(0.231 \times \frac{f_i}{1\,\text{Hz}}\right) \text{dB}, \tag{2}$$

where the centre frequencies of the standard decidecade bands were calculated according to the IEC 61260-1 formula:

$$f_i = 10^{i/10} \times 1000\,\text{Hz}, \tag{3}$$

and the band number, i, is an integer in the range –20 to 15.

2.2. Source Level Models

2.2.1. RANDI

The Research Ambient Noise Directionality noise model (RANDI 3.1) is a naval ambient noise model, designed to support prediction of the performance of low- to mid-frequency sonar receivers [12]. RANDI 3.1 includes a semi-empirical ship SL model consisting of a baseline spectrum of an "average ship" (with reference speed $V_0 = 12$ kn and length

$l_0 = 300$ ft), modified by a dependence on ship length and speed. The source level spectral density L_{S_f} as a function of frequency f (Hz), speed V (kn) and length l (ft) is:

$$L_{S_f,\,RANDI}(f,V,l) = L_{S_f,0}(f) + 60\log_{10}(V/V_0)\,\text{dB} + 20\log_{10}(l/l_0)\,\text{dB} + df \times dl + 3.0\,\text{dB}, \tag{4}$$

with reference spectrum:

$$L_{S_f,0}(f) = \begin{cases} -10\log_{10}\left(10^{-1.06\log_{10}(f/f_{\text{ref}})-14.34} + 10^{3.32\log_{10}(f/f_{\text{ref}})-21.425}\right)\,\text{dB} & \text{for } f < 500\,\text{Hz} \\ 173.2\,\text{dB} - 18\log_{10}(f/f_{\text{ref}})\,\text{dB} & \text{for } f \geq 500\,\text{Hz} \end{cases} \tag{5}$$

where $f_{\text{ref}} = 1\,\text{Hz}$ and $dl = (l/l_{\text{ref}})^{1.15}/3643.0$, with $l_{\text{ref}} = 1\,\text{ft}$, and

$$df = \begin{cases} 8.1\,\text{dB} & \text{for } f \leq 28.4\,\text{Hz} \\ 22.3\,\text{dB} - 9.77\log_{10}(f/f_{\text{ref}})\,\text{dB} & \text{for } 28.4\,\text{Hz} < f \leq 191.6\,\text{Hz} \end{cases} \tag{6}$$

The dependence of the RANDI 3.1 SL model on ship speed and length is based on empirical relations derived from World War II acoustic data and theoretical considerations related with propeller cavitation noise [20]. The original measurements on which this model is based are no longer available. Most of these measurements were made in shallow-water sheltered environments. The propagation loss was determined from the slant range assuming spherical spreading. Hence, the spectra were not corrected for the actual propagation loss in the environment in which the measurements were taken, and therefore represent a radiated noise level (as defined in ISO 17208-1) and not a source level in the sense of ISO 18405. The source depth associated with the SL model was not reported in [12], but the RANDI 3.1 user's guide suggests a source depth of 6 m ("the average propeller depth of a merchant ship"). The statistical uncertainty of the RANDI 3.1 model predictions is not reported.

2.2.2. Wales and Heitmeyer (WH02)

Wales and Heitmeyer [11] concluded from an analysis of 54 merchant ship source spectra, measured by sonobuoys in the Mediterranean Sea and the Eastern Atlantic Ocean, that the correlation between the source level and the ship speed and length is negligible. Due to the relatively small sample of ships, probably all sailing at their optimum transit speed, it is unclear to what extent this conclusion can be generalized. They proposed an alternative ensemble source spectra model for merchant ship-radiated noise, with source spectral density level (in the frequency range between 30 Hz and 1200 Hz):

$$L_{S_f,\text{WH02}}(f) = 230.0\,\text{dB} - 10\log_{10}\left(\left(\frac{f}{1\,\text{Hz}}\right)^{3.594}\right)\,\text{dB} + 10\log_{10}\left[\left(1 + \left(\frac{f}{340\,\text{Hz}}\right)^2\right)^{0.917}\right]\,\text{dB}, \tag{7}$$

Wales and Heitmeyer applied a Gaussian source distribution across the upper quadrant of the region swept out by the propeller to determine the ship source level from the measurements, but did not report the range of source depths for which the proposed model is valid.

The ship source levels observed by Wales and Heitmeyer are normally distributed around the source spectra model, with an associated standard deviation [21]:

$$\sigma_{S_f,\text{WH02}}(f) \approx \begin{cases} 5.3\,\text{dB} & f < 150\,\text{Hz} \\ 5.3\,\text{dB} - 0.0088\,\text{dB}\left(\frac{f-150\,\text{Hz}}{1\,\text{Hz}}\right) & 150\,\text{Hz} \leq f < 400\,\text{Hz} \\ 3.1\,\text{dB} & f \geq 400\,\text{Hz} \end{cases} \tag{8}$$

2.2.3. Updated Reference Spectrum Model (JOMOPANS-ECHO)

The observed discrepancy between the predictions by the RANDI and WH02 models with the measured source levels from the ECHO data set (Section 3.1) triggered the development of an updated reference spectrum model. As the data [17] confirm that there is a correlation between the source level and the ship speed and length, and because these two parameters are readily available from ship traffic systems such as AIS, we decided to maintain the speed and length dependencies from the RANDI 3.1 model:

$$L_{S_f,\text{J–E}}(f,V,l,C) = L_{S_f,0}(f,C) + 60\log_{10}(V/V_C)\,\text{dB} + 20\log_{10}(l/l_0)\,\text{dB} \tag{9}$$

To reduce the ship class dependent deviation between measured and modelled source levels, the RANDI 3.1 model was adapted, by replacing the generic reference speed $V_0 = 12$ kn by a reference speed per vessel class (C). This new reference speed V_C was obtained from minimizing the mean model-data residuals $e_k(f,C)$ (Equation (1)) for broadband source level per vessel class. Next, an updated baseline spectrum per vessel class was developed, from minimizing the model-data residuals for the mean spectrum in each category per decidecade band:

$$L_{S_f,0}(\hat{f},C) = K - 20\log_{10}(\hat{f}_1)\,\text{dB} - 10\log_{10}\left(\left(1 - \frac{\hat{f}}{\hat{f}_1}\right)^2 + D^2\right)\,\text{dB} \tag{10}$$

with $\hat{f} = \frac{f}{f_{\text{ref}}}$, $\hat{f}_1 = 480\,\text{Hz} \times \left(\frac{V_{\text{ref}}}{V_C}\right)$, $f_{\text{ref}} = 1$ Hz and $V_{\text{ref}} = 1$ kn, and $K = 191$ dB, $D = 3$ for all classes, except $D_{\text{cruise vessel}} = 4$.

For the cargo vessels (container ships, vehicle carriers, bulkers, tankers) the updated model includes an additional peak in the baseline spectrum below 100 Hz:

$$L_{S_f,0}(\hat{f} < 100, \text{Cargo}) = K^{LF} - 40\log_{10}(\hat{f}_1^{LF})\,\text{dB} + 10\log_{10}(\hat{f})\,\text{dB} - 10\log_{10}\left(\left(1 - \left(\frac{\hat{f}}{\hat{f}_1^{LF}}\right)^2\right)^2 + \left(D^{LF}\right)^2\right)\,\text{dB}, \tag{11}$$

with $K^{LF} = 208$ dB and $\hat{f}_1^{LF} = 600\text{Hz}(V_{\text{ref}}/V_c)$ and $D^{LF} = 0.8$ for container ships and bulkers) or $D^{LF} = 1.0$ for vehicle carriers and tankers. The above model expressions are for source spectral density level. In the final modelling these are converted to source level in decidecade frequency bands by adding $10\log_{10}(0.231\,\hat{f})$ dB.

A source level model for dredgers was added, based on data from measurements by TNO in a project during the construction of Maasvlakte 2 (Rotterdam port extension; [22]). While dredging, the source level is much higher than would be predicted based on the speed, due to the propeller loading associated with the dredging. Based on the measurement results, the JOMOPANS-ECHO model prediction of the dredger source level at a sailing speed of 14 knots is applied as an estimation of the source level when the dredger is dredging (independent of the actual dredging speed). The AIS data available to JOMOPANS do not provide an indication when dredger is dredging. Based on the speeds observed during dredging for Maasvlakte 2 [22], it is tentatively assumed that a dredger is dredging when its speed is lower than 3 knots.

The measured vessel source level spectra in the ECHO data set are approximately normally distributed around the source spectra model, with an associated standard deviation of about 6 dB (Section 3.2).

3. Results

3.1. Source Level Validation

Statistics of the residual differences between the RANDI and WH02 model predictions and the ECHO data set were calculated for decidecade bands ranging from 10 Hz to 31.5 kHz (Figure 1). For this comparison we have extended the WH02 model beyond the frequency range for which it was developed, by applying Equation (7) between 1.2 kHz and 20 kHz and by assuming a constant SL spectral density at 30 Hz and below [23]. This analysis showed that both models overestimated observed source levels below 250 Hz and underestimated observed source levels at 250 Hz and above. While the mean residual differences for the two models were very similar, the RANDI model had significantly greater standard deviation of the residual differences than the WH02 model, particularly at higher frequencies. While the prediction error of the RANDI model was undoubtedly larger, scatter plots of the data nonetheless showed that the speed and length terms in the RANDI model more accurately reproduced trends evident in the ECHO source level measurements (Figure 2). Furthermore, it was evident that source levels for different types of vessels exhibited systematic offsets when compared to the RANDI model. The results of the validation therefore suggest that an improved reference spectrum model could be devised via straightforward modifications to the basic RANDI formulae.

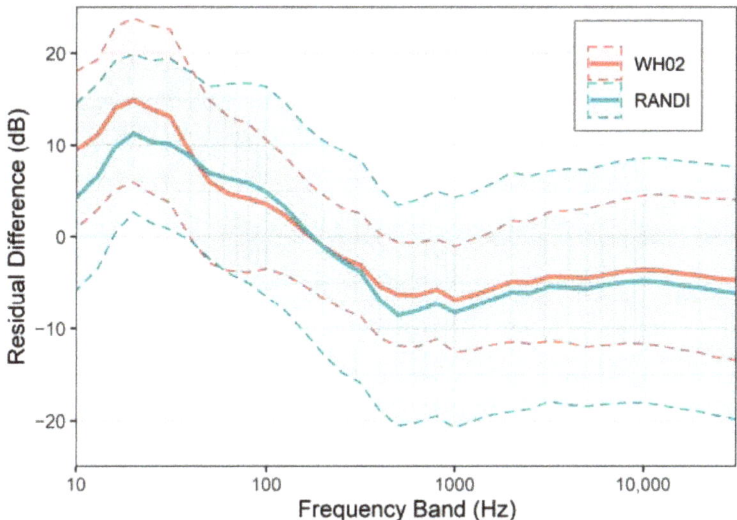

Figure 1. Mean residual differences (solid lines RANDI and WH02) versus frequency between the source level models and the ECHO data set. The dashed lines show the standard deviation of the residual differences about the mean.

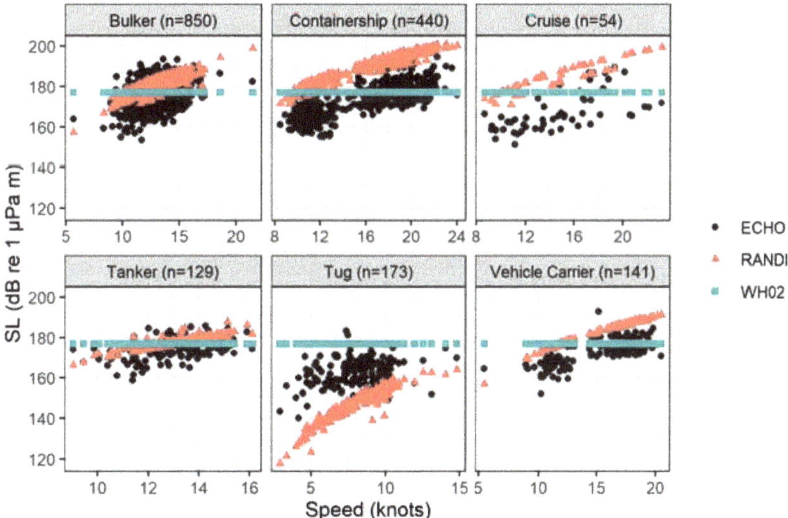

Figure 2. Scatter plots of predicted and observed (ECHO) source levels versus speed, in the 63 Hz decidecade band (one of the two MSFD indicator bands [2]), for the RANDI and WH02 models, for six vessel types. The n value in each panel indicates the number of measurements for the specified vessel type. Predictions of the WH02 model do not vary with vessel speed and length, whereas predictions of the RANDI model vary with vessel speed and length according to power law functions, see Equation (4).

3.2. JOMOPANS-ECHO Model

As described in Section 2.2.3, a modified source level model was devised that preserves the speed and length dependencies of the RANDI model but introduces a modified reference spectrum and category-specific reference speeds to achieve a better fit to the validation data set. Source level measurements were available for 12 different classes of vessels, representing a wide range of vessel sizes, speeds, and roles. Table 1 provides the applied conversion between the ship type identification (ID) in the AIS data set and vessel class. The AIS types 'Passenger' (ID = 60–69) and 'Cargo' (I = 70–79) do not provide a clear identification of larger and faster vessels such as containerships and cruise vessels. Without access to additional ship information, these vessel classes are tentatively identified by ship length, observed mean speed and AIS hazard class (see Table 1). Coefficients of a modified reference spectrum (Equations (10) and (11)) were chosen to match as closely as possible the mean source level versus frequency data for the different vessel types. Each vessel class was assigned an appropriate reference speed (V_C), so chosen to minimize the residual differences between the new model and the validation data set, in decidecade bands ($f_i \geq 20$ Hz) (Figure 3). Each vessel class was furthermore associated with specific AIS ship type ID codes, with additional speed and length criteria to disambiguate between sub-types of vessels that could not be identified based purely on the AIS ID (Table 1). A reference implementation of the JOMOPANS-ECHO source level formulae is provided in an Excel spreadsheet (File S1).

Table 1. The vessel class (*C*) is obtained from the AIS 'ship type' parameter, according to the following table, which also presents the reference speed (V_C) per vessel class, the number of unique vessels, the number of measurements (n), and the mean length ($\bar{l}$) per vessel type of the measured ships in the ECHO data set and the dredgers from [22] used for the present model development. Many vessels in the data set were measured more than once.

Vessel Class (*C*)	AIS SHIPTYPE ID	V_C (kn)	Unique Vessels	n	$\bar{l}$ (m)
Fishing	30	6.4	10	21	32
Tug	31, 32, 52	3.7	67	173	28
Naval	35	11.1	9	19	79
Recreational	36, 37	10.6	7	15	45
Government/Research	51, 53, 55	8.0	2	2	58
Cruise	60–69 (length $l > 100$ m)	17.1	23	54	268
Passenger	60–69 (length $l \leq 100$ m)	9.7	2	6	52
Bulker	70, 75–79 (speed $V \leq 16$ kn)	13.9	360	850	211
Containership	71–74 (all speeds) 70, 75–79 (speed $V > 16$ kn)	18.0	195	440	294
Vehicle Carrier	n/a	15.8	65	141	194
Tanker	80–89	12.4	53	129	186
Other	All other type IDs	7.4	6	12	81
Dredger	33	9.5	7	52	128

Over the 20 Hz to 20 kHz decidecade bands, the maximum absolute value of the mean residual difference (in decidecade bands, averaged over all vessels) between the new model and the ECHO data is 2 dB, and the mean standard deviation of the residual differences per decidecade band is 6 dB (Figure 4). The larger deviations observed in the decidecade bands below 20 Hz is likely due to tonal noise at propeller blade rates and harmonics which are not represented in the statistical source level models. The speed and length trends in the original RANDI model appear to follow the trends observed the ECHO data set and so these coefficients were not modified in the new model (Figures 5 and 6). The new model was found to provide a significantly better match to the ECHO data, when compared with the RANDI and WH02 models (compare Figure 4 with Figures 1 and 5 with Figure 2).

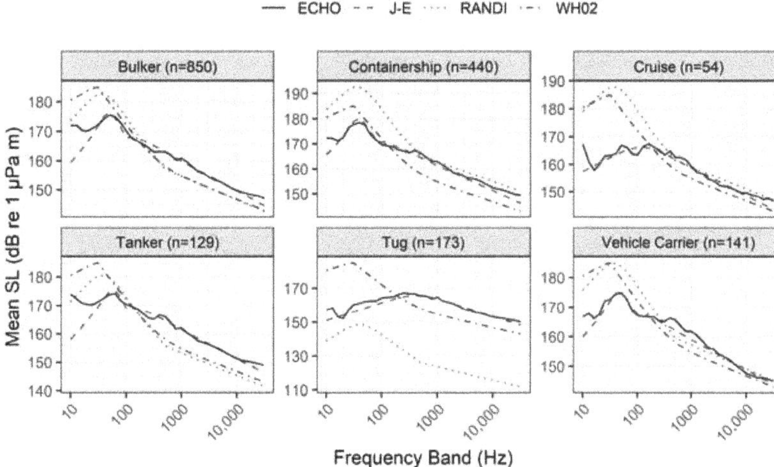

Figure 3. Comparison of the mean source level versus frequency predictions of the WH02, RANDI and JOMOPANS-ECHO (J-E) models with mean measured source levels from ECHO data set (black). The n value in each panel indicates the number of measurements that were averaged for the specified vessel type.

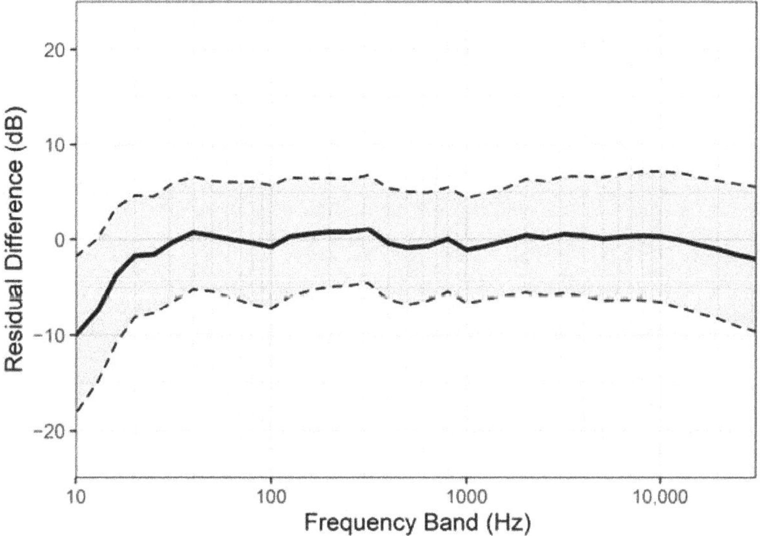

Figure 4. Mean residual differences (solid line) versus frequency between JOMOPANS-ECHO model and the ECHO data set. The dashed lines show one standard deviation of the residual differences about the mean. The mean standard deviation of the residual differences is 6 dB over the frequency range 20 Hz to 20 kHz. Over the same frequency range, the mean absolute value and interquartile range of the residual differences are 5 dB and 8 dB, respectively.

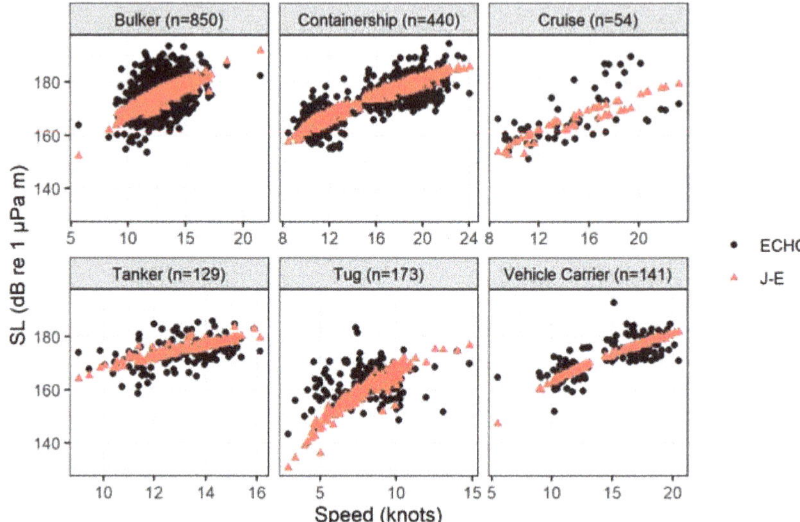

Figure 5. Scatter plots of predicted and observed (ECHO) source levels versus speed, in the 63 Hz decidecade band, for the JOMOPANS-ECHO (J-E) model, for six vessel types. The n value in each panel indicates the number of measurements for the specified vessel type.

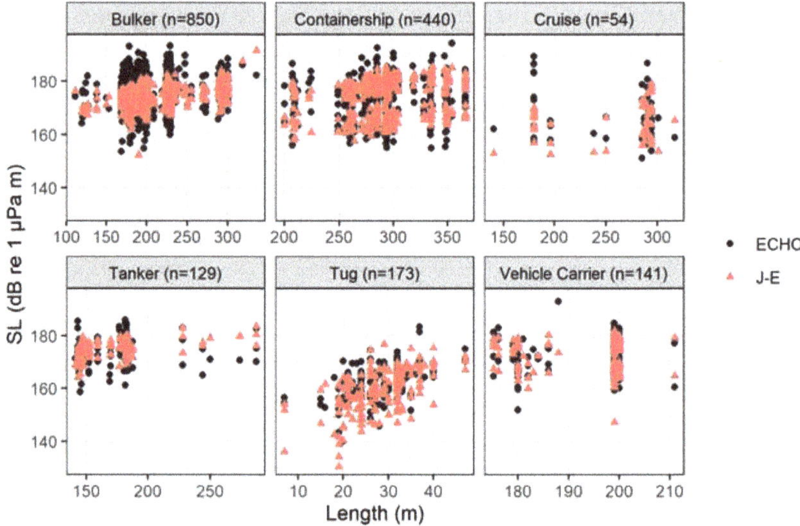

Figure 6. Scatter plots of predicted and observed (ECHO) source levels versus length, in the 63 Hz decidecade band, for the JOMOPANS-ECHO (J-E) model, for six vessel types. The n value in each panel indicates the number of measurements for the specified vessel type.

4. Discussion and Conclusions

4.1. Comparison with Other SL Models

Since the RANDI and WH02 models were originally published, the widespread availability of high-quality AIS data and advanced digital hydrophone recorders have allowed for the collection of newer, more extensive source level data sets for ships of opportunity [24–29]. Using these newer data sets, other recent studies have sought to develop improved reference spectrum models for marine vessels, many of which are summarized in a recent review article [30]. While it is not possible to provide an exhaustive review here, several notable examples are discussed below, and their details are compared to the present work.

Chion et al. [31] used source levels from 11 merchant vessels, measured in the St. Lawrence Estuary, to develop a modified version of the RANDI model that included an updated reference spectrum and modified speed dependence. Chion et al. reported that the original RANDI model overestimated their measurements, so they introduced a speed-dependent correction parameter (K_0) that minimized the residuals between their data and the corrected model. They retained the common 12-knot reference speed for all vessels. The modified Chion et al. model has a much weaker speed dependence than the original RANDI model, predicting an SL increase of only 1.7 dB when speed is doubled from 10 to 20 knots (Equation (4) predicts an increase of 18 dB over the same speed range). Chion et al. did not report whether they assumed a specific source depth in their modified RANDI model, but their use of geometrical propagation loss suggests that their model employed a dipole source representation (i.e., neglecting the influence of the sea surface on predicted source levels).

Simard et al. [26] tested several source-spectrum models, including RANDI and WH02, using source levels from 255 merchant vessels measured in the St. Lawrence Estuary. Their validation data employed a monopole source representation, with propagation loss computed using the wavenumber integration method. The assumed source depth was not reported. Simard et al. reported that both the RANDI and WH02 models generally underestimated their measurements, so they fit several polynomial models, involving length, breadth, draft, and speed to their data, discarding insignificant terms. Simard et al. presented three possible models that were consistent with their data, although the length and speed dependencies in these models (if present) were weaker than the $20 \times \log_{10}(l)$ and $60 \times \log_{10}(V)$ trends in RANDI. Their simplest model, denoted as AS^4, which depends only on frequency in a similar manner to WH02, explained 72% of their data variance. For comparison, the JOMOPANS-ECHO model explains 64% of the ECHO data variance. However, based on the statistics reported in their paper, the spread of their measured source levels around the AS^4 model predictions appears to be approximately the same as the spread observed in the JOMOPANS-ECHO model-data comparison (interquartile range of the residuals for both models is 8 dB).

Most recently, Jiang et al. [27] used source levels from 27 merchant vessels, measured near the port of Qingdao in the Yellow Sea, to develop a modified version of the RANDI model that included updated reference spectra and modified speed and length dependencies. Jiang et al. employed a monopole source representation, having calculated propagation loss using a wavenumber integration model, with a Gaussian source distribution, as in [11], and the source in the upper quadrant of the propeller, as in [32]. Jiang et al. reported that the original RANDI model overestimated their data for vessels over 200 m (category I, mostly container ships), but gave a good fit to their data for vessels under 200 m (category II, mostly bulkers and tankers). Thus, they developed two new sets of formulae, that provided a better fit to their data for these two categories of vessels. Jiang et al. reported a strong correlation between vessel speed and source level, with best-fit speed coefficients similar to the original RANDI value ($43.5 \times \log_{10}(V)$ and $65 \times \log_{10}(V)$ for categories I and II, respectively), whereas they reported a weak correlation between length and source level—particularly for category I—with best-fit length coefficients less than the original RANDI value ($14 \times \log_{10}(l)$ and $-2.7 \times \log_{10}(l)$ for categories I and II, respectively). In comparing their modified model with their data, Jiang et al. reported a mean absolute error of 4 dB, which was slightly less than the mean absolute error of the JOMOPANS-ECHO model, which was 5 dB.

Figure 7 illustrates that the difference between the source level spectra predicted by these models exceeds the statistical uncertainty reported for the models. This comparison is limited to merchant vessels over 100 m in length because that is, to our understanding, the range of ships for which the Chion et al., Simard et al. and Jiang et al. models are applicable.

The four models compared in Figure 7 are all semi-empirical, with fitted parameters based on source level data for ships of opportunity. These data sets were collected at different locations, for different ship populations, using different measurement and analysis procedures. Differences between source level data sets, due to a combination of these factors, are thus most likely responsible for differences between the model predictions. For example, obtaining accurate source levels is more difficult in shallow water than in deep water, particularly at frequencies below 100 Hz, due to the influence of the seabed. Furthermore, calculated source levels are sensitive to the estimated propagation loss, and so differences in the methods used to estimate propagation loss can introduce systematic differences into the source level estimates. The Chion et al. and Simard et al. studies both measured source levels in relatively deep water, similar to the ECHO Program's study (~200–300 m), whereas the Jiang et al. study measured source levels in much shallower water (~30 m). On the other hand, the propagation loss method employed by the Jiang et al. study was most similar to that of the ECHO Program's study, whereas the propagation loss methods employed by Chion et al. and Simard et al. studies were quite different. Karasalo et al. [28] estimated the source level spectra of over 900 ships from measurements of 2088 ship passages along a hydrophone deployed near a major shipping lane in the Baltic Sea. Propagation loss for the shallow water environment (water depth ~41 m) was calculated using acoustic seabed parameters obtained from geo-acoustic inversion of data from a propagation loss trial. In the frequency range between 100 Hz and 1000 Hz, to which the source level estimations in [28] are limited, the JOMOPANS-ECHO model predicts the median source levels of cargo vessels and tankers in [28] within the model uncertainty (~6 dB). For tugs and passenger vessels, which are less well represented in the data sets, the differences are greater.

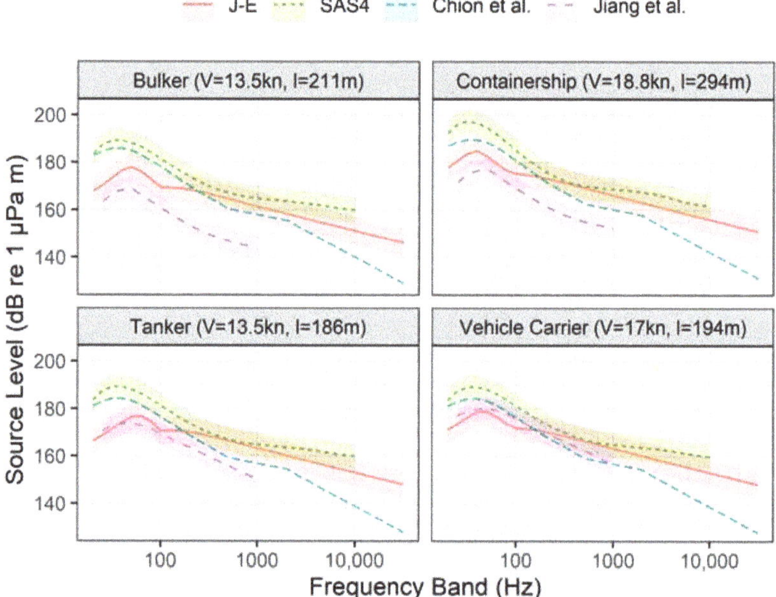

Figure 7. Decidecade band source levels predicted by the JOMOPANS-ECHO model (J-E) compared with three other recent source level models. Predictions are shown for four representative merchant vessels, with mean length (l) and speed (V) from the ECHO data set. Lines correspond to different models as follows: J-E is the model from this work, SAS4 is the simple AS^4 model from ref [26], Chion et al. is the model from ref [31], and Jiang et al. is the model from ref [27]. Color-shaded areas indicate the model uncertainties, where reported (± one estimated standard deviation, assuming normal distribution of errors).

Uncertainties about certain assumptions involved in the models (e.g., Chion et al. and Simard et al. do not report the assumed source depths) also make comparisons more difficult. A recent meta-analysis found differences in source level of as much as 30 dB between published data sets for similar vessels operating under similar conditions [30]. We rely upon the ECHO data set for this work because it includes a significantly larger number of measurements, and we have a detailed understanding of the underlying calculation methods. It is hoped that ongoing efforts toward standardization of source level estimation procedures (e.g., by the ISO subcommittee on underwater acoustics, ISO/TC 43/SC3) will improve the consistency of different data sets in future [33].

The new JOMOPANS-ECHO source level model addresses several key requirements for marine sound mapping that were identified during the JOMOPANS project and which were not fulfilled by pre-existing models:

1. The new model provides an explicit relationship between AIS ship type and source level spectrum (i.e., as described in Table 1), to facilitate making reproducible sound maps. The incorporation of class-dependent reference speeds and reference spectra in the model is consistent

with prior studies which have established that ship class is a strong predictor of vessel source levels [24,25].

2. The new model retains an explicit dependence of source level on vessel speed, which allows for the ability to consider, e.g., slow down mitigation scenarios in marine spatial planning. The speed dependence follows an established power-law trend for cavitation noise [34] which is strongly supported by experimental evidence [17].

3. The new model employs a monopole source representation with a consistently specified reference source depth (z_s = 6 m). Ambiguities surrounding monopole and dipole source level representations have been identified as a key source of inconsistencies between prior vessel noise studies [30].

4. The new model includes an estimate of statistical uncertainty ($\pm$6 dB rms error in the range 0.02–20 kHz), which has been calculated from a large validation data set. This permits a robust analysis of statistical uncertainties associated with the inherent variability in source levels of marine vessels.

4.2. Sources of Uncertainty

There are several sources of uncertainty that need to be considered, when applying the proposed model for estimating sound levels from marine shipping. Information gleaned from AIS data provides an imperfect means of identifying vessel class and estimating source levels. For example, using AIS alone, container ships and cruise ships cannot be identified with certainty, and vehicle carriers seemingly cannot be distinguished from other types of cargo vessels (see Table 1). AIS broadcasts often contain errors in vessel length, ship type ID, and speed, and many smaller vessels (typically under 300 gross tonnes) do not broadcast on AIS at all [35,36].

Furthermore, most of the vessel design details that truly relate to noise emissions are entirely absent from AIS data. Models that calculate individual contributions of propeller and machinery noise to the source level, as proposed by Wittekind [37], by the AQUO project [8] and by Jalkanen et al. [38] have the potential to reduce part of the uncertainty in the source level predictions, but require more information than available from AIS and cannot address the fundamental uncertainty associated with the influence of details such as vessel maintenance condition and the effect of environmental conditions on propeller cavitation. Moreover, complex detailed source level models based on many parameters are impractical for large scale sound mapping.

Other sources of uncertainty in the proposed model relate to the simplifying assumptions in the predicted source level model itself. The proposed model assumes isotropic sound radiation, but real vessels can exhibit strong directivity (both fore-aft and port-starboard) in their noise emissions [39,40]. The proposed model also assumes a point-like sound source, but noise sources on real vessels (e.g., the engine room and propeller) originate from different positions along the hull and at different depths [41]. Hence, the underwater radiated noise in the vicinity of a real vessel deviates from the assumed radiation from a monopole source at a depth of 6 m below the water surface. Various data sets indicate that there is a significant, seemingly random, component to vessel source level measurements that ultimately limit the precision with which noise emissions from any particular vessel can be estimated.

Comparisons between empirical vessel noise models (see Figure 7) suggest the need for independent validation against different source level data sets. Such validation efforts are hampered, at present, by the lack of an agreed-upon standard for measurement of source levels for ships of opportunity. Existing standards (e.g., ISO 17208) were developed for acoustic ranging of co-operating vessels under controlled test conditions. As such, they can only be approximately adhered to for ships of opportunity under the best of circumstances. As a result, past studies that collected large numbers of measurements for ships of opportunity employed widely varying methodologies, often yielding very different results [30]. The present work has attempted to address such uncertainties by employing a large, statistically robust source level data set, collected according to a well-defined measurement protocol. Nonetheless, it is clear that more work is needed to address sources of error between vessel noise data sets and source level models. This is especially important for marine sound mapping applications because source level models are a key source of uncertainty in ambient noise prediction.

Supplementary Materials: The following are available online at https://www.mdpi.com/article/10.3390/jmse9040369/s1, File S1: JOMOPANS-ECHO vessel source level calculator (Excel format).

Author Contributions: A.M. and C.d.J. contributed equally to the preparation of this paper. A.M. was responsible for the data analysis and optimization of model-data residuals. C.d.J. was responsible for formulation and implementation of the reference spectrum model for JOMOPANS. Both authors have read and agreed to the published version of the manuscript.

Funding: This research was funded under the Joint Monitoring Programme for Ambient Noise in the North Sea (JOMOPANS) by the North Sea Programme of the European Regional Development Fund (InterReg) of the European Union and several national funding bodies. See https://northsearegion. eu/jomopans/ (Accessed on 26 March 2021).

Institutional Review Board Statement: Not applicable.

Informed Consent Statement: Not applicable.

Data Availability Statement: Restrictions apply to the availability of these data. Data was obtained from the Enhancing Cetacean Habitat and Observation (ECHO) Program and are available from the authors with permission from the ECHO Program.

Acknowledgments: The anonymized vessel noise data set used in this research was collected by the Vancouver Fraser Port Authority-led Enhancing Cetacean Habitat and Observation (ECHO) Program.

Conflicts of Interest: The authors declare no conflict of interest.

Appendix A. Source Depth Conversion

International standard ISO DIS 17208-2 summarizes the theory for conversion between deep water Radiated Noise Level (L_{RN}) and Source Level (L_s):

$$L_s = L_{RN} + \Delta_L \tag{A1}$$

A simple approximation, valid when $kd_s \ll 1$ is:

$$\Delta_L \approx -20\log_{10}(2\sin(kd_s\sin)) \text{ dB} = -10log_{10}\left(4\sin^2(kd_s\sin)\right) \text{ dB} \tag{A2}$$

This may be approximated by:

$$\Delta_L \approx \begin{cases} -10\log_{10}\left(4\sin^2(kd_s\sin)\right)dB & kd_s\sin \leq 3\pi/4 \\ -10\log_{10}(2)dB & kd_s\sin > 3\pi/4 \end{cases} \tag{A3}$$

Here, $k = 2\pi/c$ is the acoustic wavenumber, d_s the source depth and α the vertical observation angle.

The ECHO data set contains source levels for an assumed Gaussian source depth distribution, with mean value d_s and standard deviation $\sigma_s = d_s/0.85/4$, limited to interval [1 m, 24 m]. For the purpose of the present modelling, the source levels have been adjusted to be applicable for a single fixed source depth of 6 m. Conversion of the SL reported for one (mean) source depth $d_{s,1}$ to another source depth $d_{s,2}$ can be done by applying the correction

$$L_s(d_{s,2}) \approx L_s(d_{s,1}) + \Delta_L(d_{s,2}) - \Delta_L(d_{s,1}) \tag{A4}$$

For a small vessel, e.g., a tug with an assumed mean source depth $d_{s,1} = 2.27$ m, this leads to a correction $\Delta_L(d_{s,2}) - \Delta_L(d_{s,1})$ that is approximately equal to 8 dB in the lowest frequency bands, where $kd_s \ll 1$.

References

1. International Whaling Commission (IWC); IQOE; US NOAA & ONRG; Netherlands TNO & Ministry of Infrastructure and The Environment. Joint Workshop Report: Predicting Sound Fields—Global Soundscape Modelling to Inform Management of Cetaceans and Anthropogenic Noise. In *Joint Workshop Sponsored By the IWC, IQOE, US NOAA & ONRG, and Netherlands TNO & Ministry of Infrastructure and The Environment*; NOAA Fisheries: Leiden, The Netherlands, 2014.
2. Dekeling, R.; Tasker, M.; Van Der Graaf, S.; Ainslie, M.; Andersson, M.; André, M.; Borsani, J.F.; Brensing, K.; Castellote, M.; Cronin, D.; et al. *Monitoring Guidance for Underwater Noise in European Seas, Part II: Monitoring Guidance Specifications*; JRC Scientific and Policy Report EUR 26557 EN; Publications Office of the European Union: Luxembourg, 2014.
3. Erbe, C.; Marley, S.A.; Schoeman, R.P.; Smith, J.N.; Trigg, L.E.; Embling, C.B. The Effects of Ship Noise on Marine Mammals—A Review. *Front. Mar. Sci.* **2019**, *6*, 606. [CrossRef]

4. Richardson, W.J.; Greene, C.R.; Malme, C.I.; Thomson, D.H. *Marine Mammals and Noise*; Academic Press: San Diego, CA, USA, 1995; p. 576.

5. International Maritime Organization (IMO). *MEPC.1/Circ.833. Guidelines for the Reduction of Underwater Noise from Commercial Shipping to Address Adverse Impacts on Marine Life*; IMO: London, UK, 2014; 8p.

6. Erbe, C.; MacGillivray, A.O.; Williams, R. Mapping Cumulative Noise from Shipping to Inform Marine Spatial Planning. *J. Acoust. Soc. Am.* **2012**, *132*, EL423–EL428. [CrossRef]

7. Aulanier, F.; Simard, Y.; Roy, N.; Gervaise, C.; Bandet, M. Effects of shipping on marine acoustic habitats in Canadian Arctic estimated via probabilistic modeling and mapping. *Mar. Pollut. Bull.* **2017**, *125*, 115–131. [CrossRef] [PubMed]

8. Audoly, C.; Rizzuto, E. *Ship Underwater Radiated Noise Patterns*; Technical Report AQUO European Collaborative Project Deliverable D2.1; AQUO Project Consortium: Val-de-Reuil, France, 2015.

9. Joy, R.; Tollit, D.; Wood, J.; MacGillivray, A.; Li, Z.; Trounce, K.; Robinson, O. Potential Benefits of Vessel Slowdowns on Endangered Southern Resident Killer Whales. *Front. Mar. Sci.* **2019**, *6*, 34. [CrossRef]

10. Farcas, A.; Powell, C.F.; Brookes, K.L.; Merchant, N.D. Validated shipping noise maps of the Northeast Atlantic. *Sci. Total Environ.* **2020**, *735*, 139509. [CrossRef] [PubMed]

11. Wales, S.C.; Heitmeyer, R.M. An ensemble source spectra model for merchant ship-radiated noise. *J. Acoust. Soc. Am.* **2002**, *111*, 1211–1231. [CrossRef]

12. Breeding, J.E.; Pelug, L.A. *Research Ambient Noise Directionality (RANDI) 3.1 Physics Description*; NRL/FS/7176–95-9628; Naval Research Laboratory: Stennis Space Center, MS, USA, 1996.

13. Brooker, A.G.; Humphrey, V.F. *Noise Model for Radiated Noise/Source Level (Intermediate)*; Document Number FP7-314394-SONIC, Deliverable 2.3. Tech. Rep.; SONIC, Suppression of Underwater Noise Induced by Cavitation Consortium: The Amsterdam, The Netherlands, 2015.

14. Siceloff, L.; Howell, W.H. Fine-scale temporal and spatial distributions of Atlantic cod (*Gadus morhua*) on a western Gulf of Maine spawning ground. *Fish. Res.* **2013**, *141*, 31–43. [CrossRef]

15. Trounce, K.; Robinson, O.; MacGillivray, A.; Hannay, D.; Wood, J.; Tollit, D.; Joy, R. The Effects of Vessel Slowdowns on Foraging Habitat of the Southern Resident Killer Whales. In Proceedings of the 2019 International Congress on Ultrasonics, Bruges, Belgium, 3–6 September 2019; Volume 37.

16. MacGillivray, A.O.; Li, Z. *Appendix A. Vessel Noise Measurements from the ECHO Slowdown Trial, in Voluntary Vessel Slowdown Trial Summary Findings*; Trouce, K., Ed.; Vancouver Fraser Port Authority: Vancouver, BC, Canada, 2018.

17. MacGillivray, A.O.; Li, Z.; Hannay, D.E.; Trounce, K.B.; Robinson, O.M. Slowing deep-sea commercial vessels reduces underwater radiated noise. *J. Acoust. Soc. Am.* **2019**, *146*, 340–351. [CrossRef]

18. Marine Traffic. Marine Traffic: Global Ship Tracking Intelligence. Available online: https://www.marinetraffic.com/ (accessed on 19 February 2020).

19. Hannay, D.E.; Mouy, X.; Li, Z. An Automated Real-Time Vessel Sound Measurement System for Calculating Monopole Source Levels Using a Modified Version of ANSI/ASA S12.64-2009. *Can. Acoust.* **2016**, *44*, 166–167.

20. Ross, D. *Mechanics of Underwater Noise*; Pergamon Press: New York, NY, USA, 1987.

21. Ainslie, M.A.; de Jong, C.A.F.; MacGillivray, A.O. *Quantitative Source Descriptions for Soundscape Modeling*; Technical Memorandum by TNO and JASCO Applied Sciences, Version 1.0; Atlantic Deepwater Ecosystem Observatory Network, University of New Hampshire: Hampshire, UK, 2018; 29p.

22. De Jong, C.A.F.; Ainslie, M.; Dreschler, J.; Jansen, E.; Heemskerk, E.; Groen, W. *Underwater Noise of Trailing Suction Hopper Dredgers at Maasvlakte 2: Analysis of Source Levels and Background Noise*; Document Number TNO-DV 2010 C335; TNO: The Hague, The Netherlands, 2010.

23. Colin, M.E.G.D.; Ainslie, M.A.; Binnerts, B.; De Jong, C.A.; Karasalo, I.; Östberg, M.; Sertlek, H.O.; Folegot, T.; Clorennec, D. *Definition and Results of Test Cases for Shipping Sound Maps, in OCEANS 2015—Genova*; IEEE: Genoa, Italy, 2015; pp. 1–9.

24. McKenna, M.F.; Ross, D.; Wiggins, S.M.; Hildebrand, J.A. Underwater Radiated Noise from Modern Commercial Ships. *J. Acoust. Soc. Am.* **2012**, *131*, 92–103. [CrossRef]

25. Veirs, S.; Veirs, V.; Wood, J.D. Ship noise extends to frequencies used for echolocation by endangered killer whales. *PeerJ* **2016**, *4*, e1657. [CrossRef]

26. Simard, Y.; Roy, N.; Gervaise, C.; Giard, S. Analysis and modeling of 255 source levels of merchant ships from an acoustic observatory along St. Lawrence Seaway. *J. Acoust. Soc. Am.* **2016**, *140*, 2002–2018. [CrossRef] [PubMed]

27. Jiang, P.; Lin, J.; Sun, J.; Yi, X.; Shan, Y. Source spectrum model for merchant ship radiated noise in the Yellow Sea of China. *Ocean Eng.* **2020**, *216*, 107607. [CrossRef]

28. Karasalo, I.; Östberg, M.; Sigray, P.; Jalkanen, J.-P.; Johansson, L.; Liefvendahl, M.; Bensow, R. Estimates of Source Spectra of Ships from Long Term Recordings in the Baltic Sea. *Front. Mar. Sci.* **2017**, *4*, 164. [CrossRef]

29. Jansen, E.; de Jong, C.A.F. Experimental Assessment of Underwater Acoustic Source Levels of Different Ship Types. *IEEE J. Ocean. Eng.* **2017**, *42*, 439–448. [CrossRef]

30. Chion, C.; Lagrois, D.; Dupras, J. A Meta-Analysis to Understand the Variability in Reported Source Levels of Noise Radiated by Ships from Opportunistic Studies. *Front. Mar. Sci.* **2019**, *6*, 714. [CrossRef]

31. Chion, C.; Lagrois, D.; Dupras, J.; Turgeon, S.; McQuinn, I.H.; Michaud, R.; Ménard, N.; Parrott, L. Underwater acoustic impacts of shipping management measures: Results from a social-ecological model of boat and whale movements in the St. Lawrence River Estuary (Canada). *Ecol. Model.* **2017**, *354*, 72–87. [CrossRef]
32. Gray, L.M.; Greeley, D.S. Source level model for propeller blade rate radiation for the world's merchant fleet. *J. Acoust. Soc. Am.* **1980**, *67*, 516–522. [CrossRef]
33. Audoly, C. Standardization in Underwater Acoustics with Focus on Ship Radiated Measurement—An Update. In Proceedings of the WESPAC, New Delhi, India, 1–18 November 2018.
34. Ross, D.; Alvarez, F.F. Radiated underwater noise of surface ships. *U.S. Navy J. Underw. Acoust.* **1964**, *14*, 331.
35. Sotirov, S.; Alexandrov, C. Improving AIS Data Reliability. In Proceedings of the International Association of Maritime Universities Conference (IAMUC), Varna, Bulgaria, 11–15 October 2017.
36. Iphar, C.; Napoli, A.; Ray, C. Detection of False AIS Messages for the Improvement of Maritime Situational Awareness. In Proceedings of the Oceans 2015-MTS/IEEE, Washington, DC, USA, 19–22 October 2015.
37. Wittekind, D.K. A simple model for the underwater noise source level of ships. *J. Ship Prod. Des.* **2014**, *30*, 7–14. [CrossRef]
38. Jalkanen, J.-P.; Johansson, L.; Liefvendahl, M.; Bensow, R.; Sigray, P.; Östberg, M.; Karasalo, I.; Andersson, M.; Peltonen, H.; Pajala, J. Modelling of ships as a source of underwater noise. *Ocean Sci.* **2018**, *14*, 1373–1383. [CrossRef]
39. Arveson, P.T.; Vendittis, D.J. Radiated noise characteristics of a modern cargo ship. *J. Acoust. Soc. Am.* **2000**, *107*, 118–129. [CrossRef]
40. Gassmann, M.; Wiggins, S.M.; Hildebrand, J.A. Deep-water measurements of container ship radiated noise signatures and directionality. *J. Acoust. Soc. Am.* **2017**, *142*, 1563–1574. [CrossRef] [PubMed]
41. Tollefsen, D.; Dosso, S.E. Ship source level estimation and uncertainty quantification in shallow water via Bayesian marginalization. *J. Acoust. Soc. Am.* **2020**, *147*, EL339–EL344. [CrossRef] [PubMed]

Journal of
*Marine Science
and Engineering*

Article

Acoustic Pressure, Particle Motion, and Induced Ground Motion Signals from a Commercial Seismic Survey Array and Potential Implications for Environmental Monitoring

Robert D. McCauley [1], Mark G. Meekan [2] and Miles J. G. Parsons [2,*]

[1] Centre for Marine Science and Technology, Curtin University, Bentley, WA 6102, Australia;
     r.mccauley@cmst.curtin.edu.au
[2] Australian Institute of Marine Science, Perth, WA 6009, Australia; m.meekan@aims.gov.au
*   Correspondence: m.parsons@aims.gov.au; Tel.: +61-(8)-6369-4053

Abstract: An experimental marine seismic source survey off the northwest Australian coast operated a 2600 cubic inch (41.6 l) airgun array, every 5.88 s, along six lines at a northern site and eight lines at a southern site. The airgun array was discharged 27,770 times with 128,313 pressure signals, 38,907 three-axis particle motion signals, and 17,832 ground motion signals recorded. Pressure and ground motion were accurately measured at horizontal ranges from 12 m. Particle motion signals saturated out to 1500 m horizontal range (50% of signals saturated at 230 and 590 m at the northern and southern sites, respectively). For unsaturated signals, sound exposure levels (SEL) correlated with measures of sound pressure level and water particle acceleration (r^2 = 0.88 to 0.95 at northern site and 0.97 at southern) and ground acceleration (r^2 = 0.60 and 0.87, northern and southern sites, respectively). The effective array source level was modelled at 247 dB re 1μPa m peak-to-peak, 231 dB re 1 μPa^2 m mean-square, and 228 dB re 1 μPa^2·m^2 s SEL at 15° below the horizontal. Propagation loss ranged from -29 to $-30\log_{10}$ (*range*) at the northern site and -29 to $-38\log_{10}$(*range*) at the southern site, for pressure measures. These high propagation losses are due to near-surface limestone in the seabed of the North West Shelf.

Keywords: seismic airgun source; particle motion; ground motion; propagation loss

Citation: McCauley, R.D.; Meekan, M.G.; Parsons, M.J.G. Acoustic Pressure, Particle Motion, and Induced Ground Motion Signals from a Commercial Seismic Survey Array and Potential Implications for Environmental Monitoring. *J. Mar. Sci. Eng.* **2021**, *9*, 571. https://doi.org/10.3390/jmse9060571

Academic Editors: Michel André and Christine Erbe

Received: 13 April 2021
Accepted: 4 May 2021
Published: 25 May 2021

Publisher's Note: MDPI stays neutral with regard to jurisdictional claims in published maps and institutional affiliations.

1. Introduction

Geophysical compressed-air (seismic) sources generate high-energy, low-frequency acoustic signals (most energy in band 10–100 Hz) with short rise times. The signals are produced by multiple airguns grouped in arrays, designed to direct maximum energy downward into the seabed. Travel time and character of signals reflected from density discontinuities in the seabed provide information on the layering of strata and potential hydrocarbon traps [1,2]. The frequencies produced by seismic sources fall within the hearing sensitivity of fishes [3,4], many invertebrates [5], reptiles [6,7], and marine mammals [8]. The combination of frequency spectra, intensity, and the extended duration of seismic survey operations (often weeks to months) can result in varying degrees of acute and chronic impacts on marine taxa [5,8–16]. These are primary considerations for regulators and industry in the approval and environmental management of exploration permits using seismic surveys.

Acoustic characteristics of an airgun array signal are dependent on the number, size, pressure, relative position, depth, and design of the airguns. Airgun arrays are typically spaced over 15–20 m along the array tow line and 10–20 m across it, containing multiple strings and a total of 12–40 individual airguns. Seismic survey array volumes range considerably, from < 1000 cubic inch (16.4 l) for shallow (<20 m) water operations, to 2500–5000 cui (41-81.9 l) for a 'typical' petroleum survey, up to in excess of 5000–6000 cui for deep geological imaging surveys in water depths of many thousands of metres [17].

For operational purposes and to estimate ranges of biological impact, airgun arrays are modelled to generate estimates of signal directionality, frequency content, and source level. The propagation of the signal from the array can be derived using the modelled source level combined with propagation loss to give an estimated received level with range, depth, and azimuth about the source. The estimated received level is used to identify significant exposure levels that may be experienced by fauna at given locations around the survey area [18]. Operational procedures and limitations can be put in place to mitigate the impact of the seismic signals on those fauna. However, propagation of sound energy in the ocean depends on the water depth, bathymetry profile along the propagation path, the geological layering of the seabed and the associated geo-acoustic properties, and the sound speed profile of the water column in vertical and, to a lesser extent, horizontal axes. Given the number of unknowns, significant measures are required at various distances and azimuths from the noise source to validate modelled exposure levels, particularly if the signals have to travel through environments of differing propagation losses [2,19–21]. Although the modelling step is usually implemented for mitigation of biological impacts these models are rarely validated.

Mitigation of the impact of an individual airgun signal or the accumulated exposure to multiple discharges on stationary or mobile fauna requires an understanding of: (1) what component of the acoustic signal animals respond to; (2) the threshold that elicits a response; and (3) how the intensity of this component varies with distance from the source. This is complicated by the fact that different marine taxa detect different components of the acoustic signal. For example, marine mammals [8] and some species of fish are sensitive to acoustic pressure, whereas all fishes are sensitive to sound-driven particle motion e.g., [3,22,23]. Marine invertebrates are predominantly sensitive to waterborne particle acceleration for animals that live above the seabed e.g., [24–27] or waterborne particle acceleration and ground acceleration for animals that live on or within it e.g., [15,16,28–30].

For a plane wave in the acoustic 'far-field', sound pressure and waterborne particle motion are related, and one can be estimated from the other. However, in the 'near-field', individual signals from a group of time-synchronized point sources (e.g., an array of airguns) will have travelled different distances and arrive at a receiver with different phases, leading to constructive and destructive interference. Far from the sources, the differences in distance travelled between signals originating from each source are minimal, waves arrive in phase, and intensity decreases with range (r) in the form of spherical spreading (i.e., $1/r^2$), assuming a free space [31]. The ranges at which this near- and far-field transition occurs are also dependent on interactions with boundaries, such as the sea surface, seafloor, or, to a lesser extent, mid-water boundary layers originating from sea water layering of density discontinuities. Thus, the distance at which a signal can be considered to be in the far-field is dependent on source geometry, frequency, water depth, substrate geo-acoustic parameters and changes in the physical water properties through the water column. Accurately quantifying these characteristics across a survey area prior to operation, is non-trivial.

Management of exposure levels generated by proposed seismic surveys would benefit from increased knowledge of how airgun signal metrics are related in the field and whether a single or handful of metrics can be used to evaluate the effects on multiple receptors. Sound exposure level (SEL) is one common metric used to quantify the energy levels, to assess their potential impact on fauna [18,32]. However, whether this provides an exposure measure appropriate for all taxa or types of biological impact has not been thoroughly explored. The measurement of pressure, particle motion, and ground motion energy levels from a seismic survey source are logistically and technically complex, and therefore uncommon. As a result, there is yet to be an empirically validated dataset that encompasses near- and far-fields to compare the three acoustic components and the numerous metrics that can be derived from them for a seismic source signal of commercial size. To explore these relationships, this study used calibrated acoustic pressure, particle motion, and ground motion measurements (including several co-located sensors), recorded during a seismic source exposure experiment that was operated under as-near-as-possible real-world (commercial survey) conditions off northwest Australia. In addition, we describe the technique used to derive measures of sound exposure for selected locations that were sampled for various biological impacts from the seismic operations.

2. Materials and Methods

Acoustic data were collected during a seismic exposure experiment to determine the responses of tropical demersal fishes and pearl oysters (*Pinctada maxima*) to a commercial-size geophysical compressed-air survey at two sites off the North West Shelf of Western Australia: a northern-shallow ($\approx$15 m water depth lowest astronomical tide, LAT) and a southern-deep ($\approx$55 m water depth LAT) one. For this paper, a line of seismic signals has been termed a 'sail line' and an individual airgun signal is the result of the airguns being 'discharged'.

2.1. Study Sites

The northern site where the pearl oyster experiment occurred was located $\approx$ 40 km south southwest of Broome, in waters ranging between 10 and 25 m depth (Figure 1). A seafloor mapping study highlighted that the shoals to the northeast of the sail lines were 10–15 m deep at the time of the seismic survey, and covered by a fine layer of coarse sand overlying rock (limestone), with no sand at all in some areas (Supplementary Material, Figure S1). A total of six 23 km-long seismic sail lines were conducted in a roughly west to east direction (109°), starting 17 km west of the target location and finishing 6 km east. Water depths exceeded 40 m at the beginning of the sail line, in the west, and were 20–25 m deep at the opposite end of the sail line, in the east. At the western end, a 1–2 m deep layer of sand covered the limestone base, whereas at the eastern end this layer was thinner and over shoals the limestone was exposed.

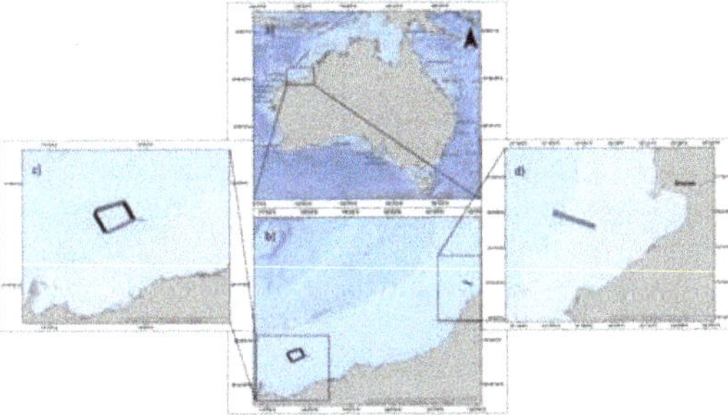

Figure 1. Map of Australia (**a**) with expansion of northwest Australia (**b**). Additional expansions of the southern ((**c**); off Point Samson) and northern ((**d**); off Broome) sites and positions of seismic sail lines.

At the demersal fish experiment site in the south, the centres of the seismic sail lines were located $\approx$ 93 km north northeast of Cape Lambert. Sail lines were oriented on a 150° heading (Figure 1). The seafloor mapping study characterised the area as predominantly flat with a gentle slope from water depths of 55 m at the southern side of the site to depths of 80 m at the northern side, over a distance of 30 km (Supplementary Material, Figure S1). A combination of historical data from the region, towed video, and sediment grabs showed the seafloor was composed of a thin layer of coarse sand over a limestone base of relatively uniform hardness. Sessile biota (sponges, corals, sea whips) were present where the sand layer was at its thinnest (centimetres) or absent.

Pilot studies conducted with a single airgun provided provisional estimates of propagation losses greater than spherical spreading, with losses greater at the northern than the southern site (Supplementary Material, Figure S2).

2.2. Seismic Source and Operation

Two 2600 cui (41.61 l) airgun arrays were discharged alternately. Each array comprised two 12.5 m strings of guns (tow direction), each string comprised ten Sercel G Gun II airguns with each string spaced 5 m from the array central point (across tow direction, Supplementary Material, Figure S3, Tables S1 and S2). The arrays were towed 102 m astern of the vessel *BGP Explorer*, operated at 2000 psi (13.8 MPa), towed at 5 m depth and spaced 20 m to port and starboard of the vessel's tow line. The arrays operated asynchronously at a mean 5.88 ± 0.004 s (95% CI) signal spacing (median 6 s) and median along-track distance of 12.5 m between signals. The modelled array beam pattern supplied by the contractor displayed little horizontal directionality (Supplementary Material, Figures S4 and S5).

At the northern site, one 23-km control sail line (airguns not operated) and six 23 km-long active sail lines were conducted with the first two active lines separated by 24 h and the following four active lines by 12 h. At the southern site, the seismic vessel operated eight control (eastern lines on Figure 1) and eight active (western) sail lines, every 12–13 h. The first two active sail lines were 25 km long and the last six, 20 km long. Control sail lines were 20 km (first two) or 15 km (last six) in length. All sail lines were operated with 500 m offsets between sequential lines, and all ran west to east.

The *BGP Explorer* provided: (1) airgun navigation data (*.p190 files for airgun signals with centre of source location only and UTC time to nearest second); (2) ships navigation data from a prescribed aerial; (3) ship specifications; (4) layout of ship, aerials, and source configuration; (4) seismic source details with modelled outputs in standard industry formats; and (5) daily logs.

2.3. Passive Acoustic Measurements

Three types of passive acoustic sensors were repeatedly deployed at different distances from the source to measure in-water acoustic pressure (underwater sound recorders, USR $\times$ 6) [33], particle motion (GeoSpectrum M20, GS-M20 $\times$ 2), and ground motion (three-axis geophones, GM $\times$ 2), during operation of the seismic sail lines (Table 1). Particle and ground motion measures were converted to provide measures of acceleration. Instruments were placed at different ranges around each site (Figure 2) for separate sail lines to quantify directionality in the beam pattern of the airgun arrays, account for sound propagation anomalies occurring due to variation in composition of the seabed, and to get a variety of closest-point-of approach (CPA) ranges. All USR and GM sensors were placed on the seafloor, whereas the GS-M20 sensor package was located 47 centimetres above the seabed, suspended from a tripod frame.

Table 1. Instrument configurations, input signal tolerances, and expected operational ranges, including pre-amplifier gain (dB), the secondary gain (dB) applied in the digitizing electronics, and for the CMST-DSTO instruments: calculated maximum peak pressure (kPa and dB re 1 μPa), the maximum voltage at the hydrophone output.

Instrument	Pre-Amplifier Gain (dB)	System Gain (dB)	Maximum Peak Pressure (kPa)	Maximum Peak Pressure Level (dB re 1 μPa)	Maximum Voltage (V)	Expected Working Range (m)
USR	−40	0	1451.2	243.2	230	<200
USR	−20	0	145.1	223.2	23	100–400
USR	0	0	14.5	203.2	2.3	200–1000
USR	0	20	4.6	193.2	0.727	500–1500
USR	20	0	1.5	183.2	0.23	>1000
USR	20	20	0.1	163.2	0.023	>4000
GS-M20 pressure	20	0–18				Unknown
GS-M20 particle motion	20	0–18				Unknown

Multiple gain settings were used on instruments to measure airgun signals in the near-field without saturation and in the far-field without noise levels dropping into system electronic noise (see below). As the relationships between acoustic pressure, particle acceleration, and ground acceleration are complex in the near-field, deployments of the particle motion and ground motion sensors focussed on ranges of less than 2 km, whereas deployments of pressure sensors extended to ranges of > 30 km (Figure 2). All instruments used for measuring airgun signals collected 10-min samples running as frequently as possible (usually a 10 s gap between samples) at 4 kHz (GM, geophone and pressure), 8 kHz (USR), or 16 kHz (GS-M20, all channels) sample rates.

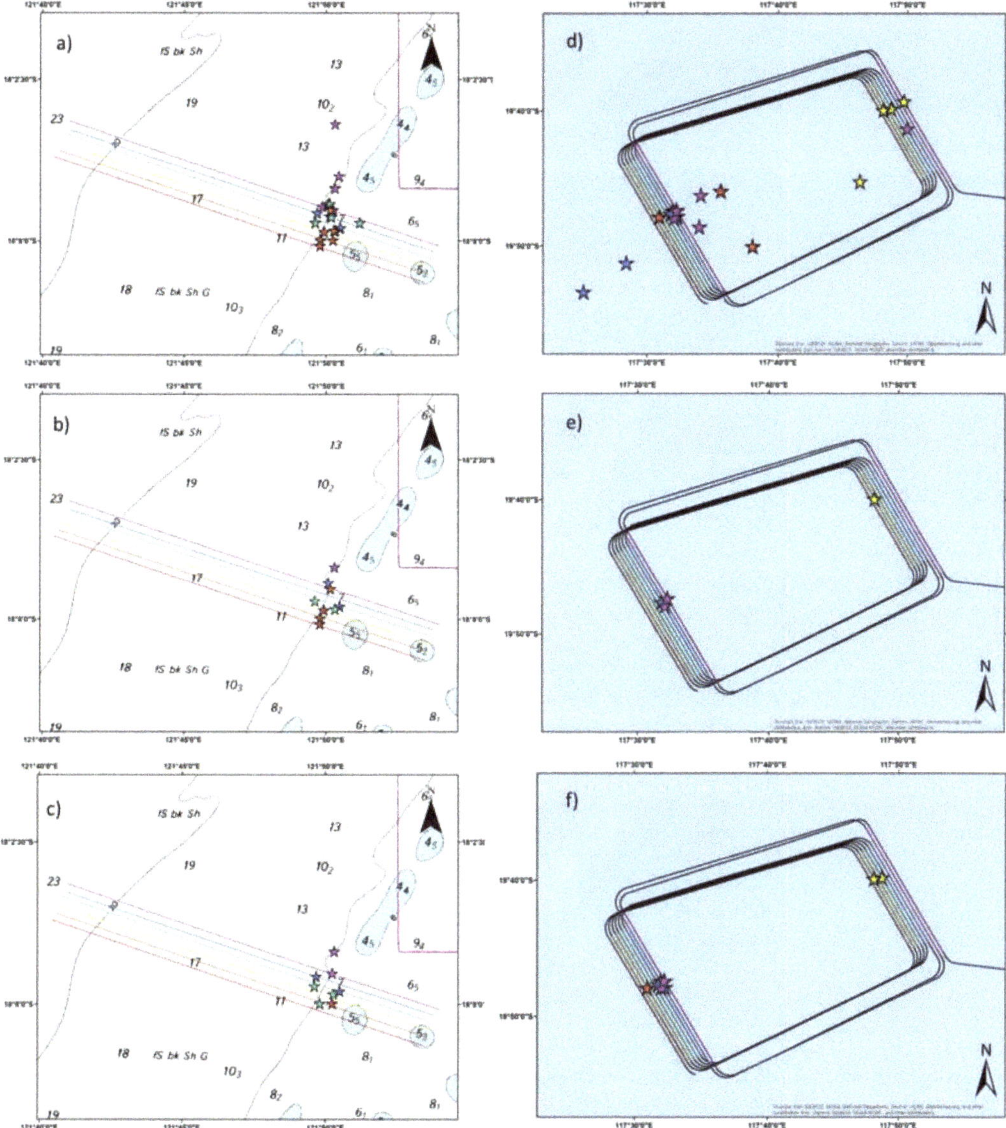

Figure 2. Map of all acoustic sensor deployments for (**a,d**) pressure, (**b,e**) ground motion and (**c,f**) particle motion for the northern (left images) and southern (right images) experimental sites. Sensors notations (stars) are colour coded with the respective seismic sail lines they recorded. Seismic survey at the southern site was conducted as a racetrack design where the airgun array was only operated during sail lines at the western end of the track (coloured lines running north northwest to south southeast to the left of the images).

USRs were individually calibrated across all sampled frequencies using white noise of known level injected in series with the hydrophone [33]. Hydrophones used in the GM and USR instruments were Massa TR1025C or High Tech HTI U-90 (without in-built pre-amplifier), all individually calibrated with nominal sensitivity −198 to −196 dB re V/μPa. The white noise injection technique spans all frequencies and is particularly important for measuring airgun signals that have high intensities in low frequencies (<400 Hz typically) where system impedance mismatches between hydrophone, pre-amplifier, and recording electronics will cause loss of sensitivity. The saved white noise was analysed to retrieve a frequency with gain curve (see Supplementary Material Figure S6 for an example calibration curve) which, when combined with hydrophone sensitivity, enabled calibration in the time or frequency domain, as required. USR clocks were synchronised with GPS transmitted time before each field trip and clock drift read at the field trip end using GPS transmitted time. Corrected clock times were then interpolated for the time in question to assist correct assignment of airgun signals between the *BGP Explorer* navigation logs and USR recording times.

Geophone measurements were taken using two USRs modified to include three-axis manufacturer-calibrated geophones, with an ION Geophysical, SM-6/U-B 10 Hz geophone in the vertical axis and two SM-6/H-B 10 Hz geophones aligned 90° apart in the horizontal plane. The instrument containing sensors was deployed flat with sensors coupled to the seabed. The frequency-dependent calibration response combined with the system gain provided calibration values for the system settings. The geophone sensors exhibited a noise floor of −15 dB re 1 ms^{-2}, which was above ambient levels of ground acceleration. The GM instrument also included a pressure sensor. The GM pressure sensor pre-amplifier gain was fixed at 0 dB and the secondary gain set to either 0 or 20 dB dependent on the closest point of approach (CPA). The geophone channel gains were fixed.

Acoustic particle velocity was measured using two manufacturer-calibrated (generic sensor-type calibration curves, i.e., assumed to apply to all instruments of that type) GeoSpectrum GS-M20, three-axis particle velocity sensors connected to a JASCO AMAR logger. These sensors (x and y horizontal and z vertical) were mounted in a tripod frame set on the seabed with the sensor hanging from the apex of the frame ≈ 47 cm above the seafloor (inbuilt tilt sensors confirmed the mooring had been deployed in an upright position). The x, y, and z channel phase responses of the GS-M20 velocity sensor were included in the calibration process. The GS-M20 sensor package had a 20 dB fixed gain that could not be modified and a secondary gain at the AMAR recorder of 0–18 dB, set depending on expected CPA. The calibration specifications (particle velocity sensitivity and phase or pressure sensor sensitivity, respectively) were combined with the system gain and the AMAR analogue to digital electronics rail (5 V) to the *.wav file format (±1), to return calibrated waveforms (ms^{-1} or Pa) in the time domain (method defined below). The GeoSpectrum-AMAR units allowed for clocks to be synchronised to laptop times (UTC) and the drift read post deployments. The GS-M20 data included roll, pitch, yaw, and magnetic declination that when corrected to the horizontal and vertical plane gave the sensor tilt in three-axis and calibrated compass headings.

Airgun signals have high peak amplitudes close to the source (<1 km) that can result in the sensor voltage saturating (or overloading) the recording system electronics if the input voltage at the pre-amplifier output is too high. Saturated signals can simply be clipped (i.e., the top of the signal cut off above some voltage threshold) or the electronics can produced artefacts that may persist for much longer than the input signal. Diode protection for the pre-amplifier and digitizing electronics are present in some systems to limit this effect. The diodes begin to reduce the input signal at a lower voltage than the maximum voltage they allow through, thus biasing signals close to the protection threshold. For saturated signals the correct received airgun level cannot be calculated.

We used eight instrument configurations to measure airgun signals. Each configuration had an optimal range bracket at which it could be deployed from the airgun array, based on the system dynamic range (lowest to highest peak signal levels) and noise floor (level the signal will be detected above background or system self-noise). For each configuration, we calculated the maximum signal peak pressures that could be tolerated (Table 1), assuming a hydrophone sensitivity of -196 dB re V/µPa and a voltage threshold before diode protection started at levels above or below $2.3/-2.3$ V, respectively, for the USR instruments. These details were not calculated for the GS-M20, which only had secondary gain, as precise details of the system electronics, particularly any voltage protection applied were not available.

2.4. Signal Metrics

There are numerous metrics or signal characteristics that may be important in driving responses from marine fauna to a 'noise' signal [18,27,28,32,34]. The response thresholds for these metrics are likely species-specific and possibly vary with life function or condition [35]. Although the three acoustic pressure components and indeed many of the metrics are correlated, their relationships are not necessarily consistent or linear as the signal propagates away from the source [36]. It is probable that different metrics are applicable to different forms of an animal's response, such as propensity for physiological impact where the mechanical response and so forces driving an organ are important or behavioural responses where the neurological interpretation of signals are important. Sound exposure level (SEL) is the most common metric for quantifying impulsive airgun signals with practical techniques to determine this and other metrics described by McCauley et al. [18], Madsen [32], or defined in ISO standard 18405-2017 [37]. Multiple signal parameters have been derived here from pressure, particle acceleration and ground acceleration. Metrics include SEL, mean-square sound pressure level (SPL), peak-to-peak sound pressure level (P-P), peak values of an airgun signal's horizontal and vertical vectors of particle acceleration (differentiated particle velocity), maximum magnitude of the airgun signal's particle acceleration vector (the 3-axis vectors combined into a single magnitude with two angles per time point) and the same peak ground acceleration vector and magnitude components (Table 2). Note, as per ISO 18405-2017 the acceleration values used here are "field" quantities which do not involve any averaging or root mean squared values being calculated. On page 4, Section 3.1.2.11, Note 3, ISO 18405:2017 defines "sound particle acceleration" with units specified as ms^{-2} [37]. These units have been used throughout the paper. The conversion of acceleration units to decibels is listed in ISO 18405:2017 as $20log_{10}\left(\frac{a}{a_0}\right)$ where a is the value and a_0 the reference value used, which is stated as 1 μms^{-2} for acceleration.

In addition to the above metrics, McCauley and Duncan [34] speculated that it is the high positive peak value immediately followed by a high negative peak value received over a short time-period (often referred to as 'jerk') that causes physical trauma in some taxa. This jerk is observed in the pressure and three-axis motion component of the signal. The aspects which are important in this measure are the positive and negative peak values and the time between these peaks. To test this, we created a unit from the airgun signal pressure waveform, which divides the sum of absolute values of maximum and minimum pressures within the airgun signal waveform, by the time between these peaks (dB re 1 $\mu Pa \cdot s^{-1}$, Table 2). For simplicity in calculation and future replication, we used maximum and minimum values experienced across the defined time window of the airgun signal. Here, this metric has been called peak pressure gradient (PPG). At low received airgun signal levels, the PPG values are low and random in distribution as the received signals have no clear peak values and the time between peaks is random. Once a clear positive peak immediately followed by a high negative peak value appears in the waveform the PPG measure stabilizes and rapidly increases as the time between peaks drops. An increase in positive or negative peak values experienced increases the PPG value, as does a shorter time between the two peaks. In contrast, differentiating the pressure waveform as suggested by some can provide the slope or rate of change of the pressure signal, but it does not correctly account for the time between maximum positive and minimum negative peaks.

Table 2. Description of acoustic metrics quantified for individual discharges of the 2600 cui airgun array.

Metric (Abbreviation)	Description	Units	Derivation
Mean-square sound pressure level (SPL)	Ten times the logarithm (base 10) of the ratio of the mean-square sound pressure, $\overline{p^2}$, to the specified reference value, $p_0{}^2$, in decibels, here taken over the time for 90% of total signal energy to pass ($T_{90\%}$).	dB re 1 μPa^2	$L_p = 10 \times \log_{10}\left(\frac{1}{T_{90\%}p_0^2} \int_{T_5}^{T_{95}} p_{s+n}^2(t)dt \right)$, where L_p is mean-square sound pressure level, $T_{90\%}$ is the signal length, p_{s+n}^2 is signal plus noise, and T_5 and T_{95} are the times where 5% and 95% of the cumulative p^2·s (signal plus noise) has passed and p_0 is 1 μPa.
Peak-to-peak pressure level (P-P)	Twenty times the logarithm (base 10) of the ratio of pressure difference between the compressional and rarefactional pressure within the signal (not necessarily consecutive peaks) using the appropriate reference value	dB re 1 μPa	$L_{p_{pk-pk}} = 20 \times \log_{10}\left(\frac{(\max(p(t))+|\min(p(t))|)}{p_0} \right)$, where $L_{p_{pk-pk}}$ is the peak-to-peak level, p is sound pressure, t is time, p_{pk-pk} is peak-to-peak pressure, and p_0 is 1 μPa.
Sound exposure level (SEL)	Ten times the logarithm (base 10) of the ratio of time-integrated (over the 90% energy duration) squared pressure, less ambient noise contribution, to the reference exposure.	dB re 1 μPa^2·s	$E_p(t) = \int_{T5}^{T95} p_{s+n}^2(t)dt - \int_{T_n}^{T_n+T} p_n^2(t)dt$, where E_p is the time-integrated pressure, p is pressure, s and n denote signal and noise, respectively, T_5 and T_{95} are time points bracketing the 90% energy duration, and T_n and $T_n + T$ denote a period away from the signal where the noise is stationary. $T = T_{95} - T_5$. $L_{E_p} = 10 \times \log_{10}\left(\frac{E_p}{E_{p,0}} \right)$, where L_{E_p} is the SEL and $E_{p,0}$ is the reference level using $p_0^2 = 1$ μPa^2·s.
Peak pressure gradient (PPG)	Twenty times the logarithm (base 10) of the peak-to-peak pressure, divided by the time taken for this pressure difference to occur, using the appropriate reference value.	dB re 1 μPa·s^{-1}	$L_{PPG} = 20 \times log_{10}\left(\frac{P_{pk-pk}}{T_p p_0} \right)$, where L_{PPG} is the *peakpressuregradient* level, p_{pk-pk} is the peak-to-peak value, T_p is time between maximum positive and minimum negative peaks and p_0 is 1 μPa.
Max. hor. particle acc. (MxHPA)	Twenty times the logarithm (base 10) of the maximum magnitude of each of the horizontal particle acceleration vectors	dB re 1 μms^{-2}	$A_{mh} = 20 \times \log_{10}\left(\frac{|a_h|}{a_0} \right)$, where A_{mh} is maximum horizontal particle acceleration level, a_h isairgun signal's maximum horizontal particle acceleration value with a reference a_0 of 1 μms^{-2}
Max. vert. particle acc. (MxVPA)	Twenty times the logarithm (base 10) of the maximum vertical particle acceleration	dB re 1 μms^{-2}	$A_{mv} = 20 \times \log_{10}\left(\frac{|a_v|}{a_0} \right)$, where A_{mv} is maximum vertical particle acceleration level, a_v is airgun signal's maximum vertical particle acceleration with a reference a_0 of 1 μms^{-2}
Max. particle acc. 3-axis (MxMPA)	Twenty times the logarithm (base 10) of the maximum magnitude of the vector sum of the X, Y, and Z particle acceleration values (where X, Y, and Z are north, east, and vertical vectors).	dB re 1 μms^{-2}	$A_{m3} = 20 \times \log_{10}\left(\frac{|a_3|}{a_0} \right)$, where A_{m3} is maximum of 3-axis particle acceleration magnitude, a_3 is maximum magnitude of airgun signal's 3-axis particle acceleration with a reference a_0 of 1 μms^{-2}
Max. hor. ground acc. (MxHGA)	Twenty times the logarithm (base 10) of the maximum magnitude of the vector sum of the two horizontal ground acceleration vectors	dB re 1 μms^{-2}	$G_{mh} = 20 \times \log_{10}\left(\frac{|g_h|}{a_0} \right)$, where G_{mh} is maximum horizontal ground acceleration level, g_h is airgun signal's maximum horizontal ground acceleration value with a reference a_0 of 1 μms^{-2}
Max. vert. ground acc. (MxVGA)	Twenty times the logarithm (base 10) of the maximum vertical ground acceleration	dB re 1 μms^{-2}	$G_{mv} = 20 \times \log_{10}\left(\frac{|g_v|}{a_0} \right)$, where G_{mv} is maximum vertical ground acceleration level, g_v is airgun signal's maximum vertical ground acceleration value with a reference a_0 of 1 μms^{-2}
Max. ground acc. 3-axis. (MxMGA)	Twenty times the logarithm (base 10) of the maximum magnitude of the vector sum of the X, Y, and Z ground acceleration values (where X, Y, and Z are north, east, and vertical vectors)	dB re 1 μms^{-2}	$G_{m3} = 20 \times \log_{10}\left(\frac{|g_3|}{a_0} \right)$, where G_{m3} is maximum of 3-axis ground acceleration magnitude, g_3 is maximum magnitude of airgun signal's 3-axis ground acceleration with a reference a_0 of 1 μms^{-2}

2.5. Airgun Signal Processing

Airgun pressure signals were extracted and analysed for each metric by the following steps:

1. Identify samples in a data set with airgun signals by aligning seismic navigation logs with sea-noise sample start and end times;
2. Load a sample, and down-sample to 4 kHz to match the geophone pressure channel and high-pass filter at 2 Hz;
3. Display the full sample waveform (10 min, volts using a 5–100 Hz band-pass filtered signal) and spectrogram (10–500 Hz displayed) for each recording that included airgun array signals;
4. Identify the 'leading edge' of each airgun signal (band-pass filtered) by applying a voltage threshold (specific to a sample, which is dependent on the range of the recording site to the airgun source), combined with a minimum time separation (5 s, based on *BGP Explorer* navigational log data) between consecutive airgun signals;
5. Delete any identified airgun signals which had overlapping 'noise' sources;
6. Remove each identified signal from the high-pass only filtered sample (volts at this stage) by extracting the signal from 4 s before to 4 s after the identified time of leading edge;
7. Calibrate the signal by obtaining the fast Fourier transform (FFT) of the airgun signal voltage waveform (frequency resolution of $\approx$ 0.12 Hz, or 32,768 points using 4 kHz sampling frequency), multiplying the real FFT part by the amplitude correction for that frequency, then converting back to a calibrated signal with an inverse FFT (see McCauley et al. [19], for further detail);
8. Calculate the level of each metric including those in Table 2, for each signal;
9. Calculate power spectra of each extracted signal;
10. Save extracted airgun signals, level metrics, start and end time of airgun signal (as given by times for 5% and 95% of signal energy to pass), a flag for if the signal had saturated or not and the power spectra.

The airgun signal times were used to extract and analyse the three-axis geophone and particle velocity (GS-M20) values using the appropriate data set. Each particle velocity channel for an airgun signal was checked for overload and flagged for the channel if the saved *.wav volts exceeded ±0.98 V (to allow for some diode protection). The geophone channels did not overload. The geophone and particle velocity signals were extracted similarly to the pressure signals and as for the pressure signals, calibrated in the time domain using their respective gain with frequency curves (independent sensor types) and for the GS-M20 data, including the appropriate sensor phase calibrations (correcting the real and imaginary part of the FFT for amplitude and phase response, respectively, before the inverse FFT was calculated).

For all airgun signals, the arrival time was aligned with the seismic navigation data and the airgun signal identified in the recordings, after assuming a sound speed and accounting for range between discharged and received signal. Once the airgun signal was defined the source to receiver geometry was established (horizontal range and angle from array tow direction to receiver).

All saturated airgun signals were removed in all analysis. Saturated signals were defined by using the flag from the saved data or identifying ranges below which a particular data set had attained a plateau and removing all values below this range.

To compare different metrics, co-located signals from various sensors were found. Linear regression models were used to assess correlation between selected metrics and investigate how one metric such as SEL, predicted other metrics at the individual airgun signal level (i.e., same airgun signal measured by different sensors) using unsaturated signals.

To estimate received levels for each airgun signal at biological sampling sites:

1. The airgun strings were symmetrical along the centre of their towlines. Therefore, any potential beam pattern was considered similar on the port and starboard sides, and port and starboard measurements were collated;
2. Measurements for each metric were gridded into a 2D space of horizontal range and azimuth (angle of receiver from tow direction) using linear interpolation and a constant sized grid spacing. No smoothing was applied in this step, data were linearly interpolated;
3. The resulting 2D grid was interpolated for missing values within the data matrixes (this was only required for ≈ 30 points, for some metrics);
4. The edges of the gridded matrix were populated for ranges greater or less than the maximum or minimum measured range, respectively, for any particular azimuth, or for azimuths less than or greater than as measured at a particular range, using a variety of techniques, specific to each metric;
5. All airgun signal received levels at ranges greater than ≈ 30 km (depending on the metric) were set to the ambient noise level as it was not possible to analyse these received airgun signal levels. This was because at this range signals were within 2 dB of ambient noise levels and had smeared in time so that no recognisable peak occurred.

Linear regression was applied to the measured data using the equations:

$$RL = \left(a \log_{10} R\right) + b \tag{1}$$

$$RL = \left(a \log_{10} R\right) + (bR) + c \tag{2}$$

where RL is received level in the appropriate metric, a, b, and c are constants derived from the fit, and R is horizontal range of source to receiver (m). Correlation coefficients (r^2) were calculated for each fit. An alternative technique for defining trends of received level with range was to calculate statistics of dB values for the appropriate metric in logarithmic range bins, with bin centres and widths defined by one-third octave bands:

$$bc = 1000 * 2^{\frac{N}{3} - 10} \tag{3}$$

$$bl = \frac{bc}{2^{\frac{1}{6}}} \tag{4}$$

$$bu = 2^{\frac{1}{6}} * bc \tag{5}$$

where bc is centre of bin (m), N is an increasing integer value, bl is the lower range limit for that bin-centre, and bu is the upper range limit (m). The value N is iterated to include the maximum range to be encountered.

Horizontal range (great circle path between source and receiver), rather than slant range (the direct path between the source near the surface and the receiver at the seabed), was used to determine propagation loss across the experimental sites. The distance at which slant vs. horizontal range would be significant at each site was estimated by applying spherical spreading from the known array source level at a given elevation to calculate horizontal received level and slant range received level, accounting for water depth, array depth and maximum tidal excursion over the experimental period. The two estimated received level values for that range were then compared. The maximum tidal excursion during the survey period at the northern site was 5.3 m. When combined with the water depth and the mean lowest astronomical tide (LAT) in the area, less 5 m for the source depth, a 1 dB difference between slant and horizontal range occurred at 70 m horizontal range. At ranges >70 m, this error declined rapidly. The equivalent range at the southern site, for a 1 dB difference, was $\approx$120 m horizontal range, beyond which the difference again dropped rapidly. For signals recorded at shorter ranges, measurement plots of logarithmic range with received level would appear more consistent if slant range was used. However: (1) inter-discharge variability in SEL has been shown to range between 1 and 3 dB [38] and the different received level estimates for slant and horizontal ranges fell below 3 dB at horizontal ranges of approximately 10 and 30 m for the northern and southern sites, respectively; and (2) all levels at biological sites were calculated using horizontal range (provided a consistent range type was used this made no difference to exposure estimates).

3. Results

3.1. Measurements

Over the experimental exposure period, 39 instruments were deployed 108 times on 58 moorings. From the airgun source navigation data, 27,770 active airgun signals were logged, 14,227 at the northern site and 13,543 at the southern site. The mean discharge interval was 5.9 s (median 6 s, 95%CI $\pm$ 0.0044 s, n = 27,769). Of these airgun signals, 128,313 unsaturated pressure signals were measured from different instruments, from horizontal ranges of 12 m (almost directly below the array at northern site) to 20.3 km. At the northern site, 58,402 pressure signals were analysed and 69,911 at the southern site. Although all measurements included the pressure signal, 17,832 included three-axis ground-borne geophone measures (8688 and 9144 at the northern and southern sites, respectively) and 38,907 included three-axis particle velocity measurements (19,160 and 19,747 at the northern and southern sites, respectively). Matched geophone and particle velocity measurements were available for 877 airgun signals (505 and 372 at the northern and southern sites, respectively), with the geophone and particle velocity sensors within 30–50 m of each other and on the seabed (geophone) or 47 cm above the seabed (particle motion). Histograms of ambient noise levels for each metric at the time of the seismic survey can be found in Supplementary Material, Figure S7 and received SELs for an example sail line at each site are shown in Figures S8 and S9.

Using the source model of Duncan [2] the 2600 cui airgun array was estimated to have effective source levels of: (1) 252 dB re 1 μPa m peak-to-peak pressure, 234 dB re 1 μPa^2 m^2 mean-square pressure, and 228 dB re 1 μPa$^2\cdot$m^{2}s SEL, directly below the array; and (2) 247 dB re 1 μPa m peak-to-peak pressure, 231 dB re 1 μPa^2 m mean-square pressure, and 228 dB re 1 μPa$^2\cdot$m^{2}s SEL at 15° below the horizontal (more relevant to longer range horizontal source level than the higher levels directly below the array). The maximum measured levels and the horizontal ranges at which they were measured, can be found in Table 3. Within a few hundred m of the array most energy emitted occurred with uniform spectral density between 2 and 100 Hz with almost all energy <1000 Hz (Figure 3). The passage of the *BGP Explorer* past a USR, during a seismic sail line, illustrated the frequency-dependent propagation of the signal showing that with range, almost all energy remained under ≈100 Hz, at the northern site, while at the southern site, the highest energy was under 100 Hz with peaks around 10 Hz and 50 Hz (Figure 4). The 'waisting' with frequency evident at each site on Figure 4) is due to the limestone seabed preferentially attenuating certain frequencies [38]. Energy > 100 Hz propagated only short distances (Figure 4). A Lloyd's mirror interference pattern, typical of a point source varying with increasing range was more clearly visible at the southern site where energy propagated further (compare Figure 4a with 4b).

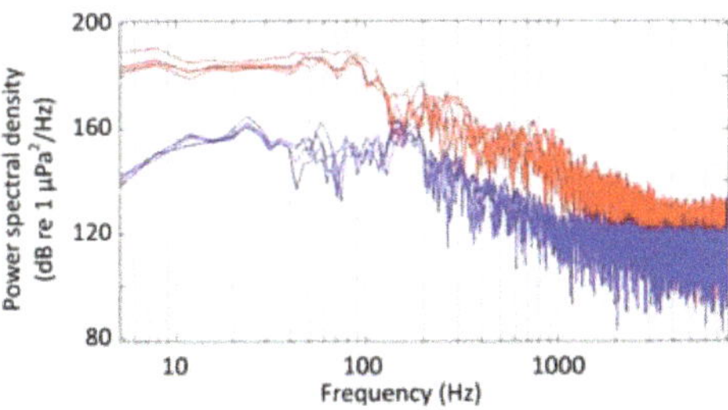

Figure 3. Pressure power spectra measured almost directly underneath airgun array (red) and at 190 m (blue).

Table 3. Maximum levels measured at the northern and southern sites, and the horizontal range at which they occurred for peak-to-peak pressure level (P-P), sound exposure level (SEL), mean-square sound pressure level (SPL), pressure peak-gradient (PPG), seabed ground acceleration (maximum horizontal and vertical vectors and magnitude), and particle acceleration (maximum horizontal and vertical vectors and magnitude) levels, as defined in Table 2.

Measure and Unit	Northern Site		Southern Site	
	Level	Horizontal Range (m)	Level	Horizontal Range (m)
P-P (dB re 1 µPa)	230	12	209	56
SEL (dB re 1 µPa^2·s)	217	14	187	56
SPL (dB re 1 µPa^2)	218	14	191	5
PPG (dB re 1 µPa·s^{-1})	314	12	275	231
MxVGA (dB re 1 µms^{-2})	147	60	142	68
MxHGA (dB re 1 µms^{-2})	148	60	148	80
MxMGA (dB re 1 µms^{-2})	149	60	149	80
MxVPA (dB re 1 µms^{-2})	138	65	141	93
MxHPA (dB re 1 µms^{-2})	141	65	141	99
MxMPA (dB re 1 µms^{-2})	142	39	142	99

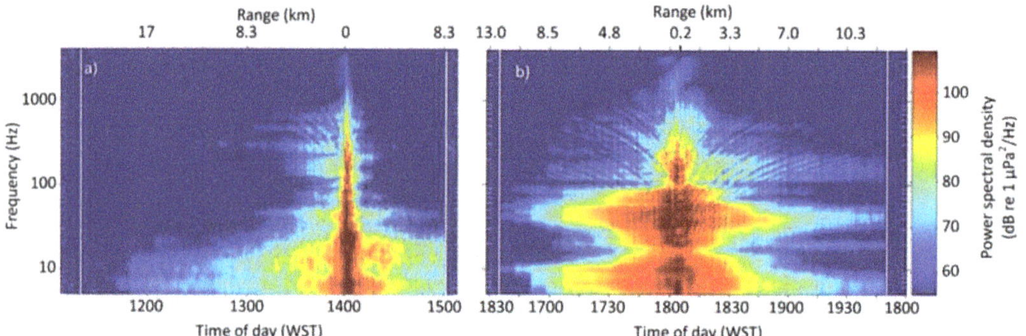

Figure 4. Power spectral density of pressure signals from the airgun array as recorded by a USR throughout an individual seismic sail line operated at the northern ((**a**), closest point of approach, ≈12 m horizontal range, 18 September) and southern ((**b**), closest point of approach, ≈190 m horizontal range, 21 September) sites. The white lines mark the start and end of airgun operations along the seismic line.

3.2. Saturated Signals

The proportion of saturated signals to total airgun array signals measured illustrated how the different instruments and gain settings performed with increasing range from the airgun array (Figure 5b). The pressure channel of the GS-M20 using the lowest secondary gain setting of 0 dB saturated at ranges between 300 and 700 m. Particle motion signals recorded by the GS-M20 began saturating at 1.2 and 1.5 km from the source at the northern and southern sites, respectively (Figure 5a), and increased in the proportion of saturated signals to the closest ranges at which recordings were taken. At ranges of 230 and 590 m (northern and southern sites, respectively), 50% of the particle motion measures were saturated. The ground motion accelerometers' signals were not saturated at any range. However, at long ranges (or under low level ambient conditions) the sensor response fell into the system electronic noise floor and the sensors did not provide ambient level ground acceleration.

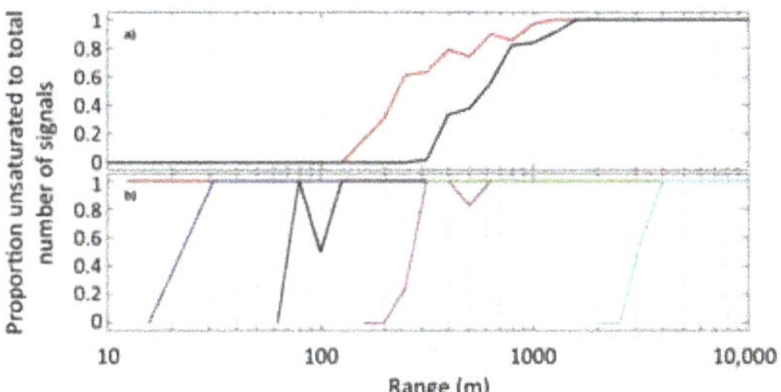

Figure 5. Proportion of unsaturated to total signals within logarithmic horizontal range brackets for: (**a**) particle motion at the northern (red) and southern (black) sites; and (**b**) pressure only for USR instruments with pre-amp/secondary gain (dB) settings of −40/0 (red), −20/0 (blue), 0/0 (black), 0/20 (magenta), 20/0 (green) and 20/20 (cyan). The expected working ranges for each sensor can be found in Table 1.

3.3. Received Levels with Range

At the northern and southern sites received levels of all metrics decreased with range at rates greater than spherical spreading and did so at a greater rate at the northern (shallower), compared to southern (deeper) site (Figure 6). The trends shown for each metric averaged in logarithmic range bins (Figure 7, bin ranges defined by Equations (3)–(5)) clearly highlighted the poorer sound propagation conditions at the northern site. Mean levels for five metrics at 250, 500, and 1000 m displayed up to 17 dB difference between the sites (Table 4). The confidence limits in averaging in the logarithmic range bins were low at each site, with the mean error < 0.7 dB at either site. At short-range (low hundreds of metres), the trends with range shown on Figure 7 converged for all metrics.

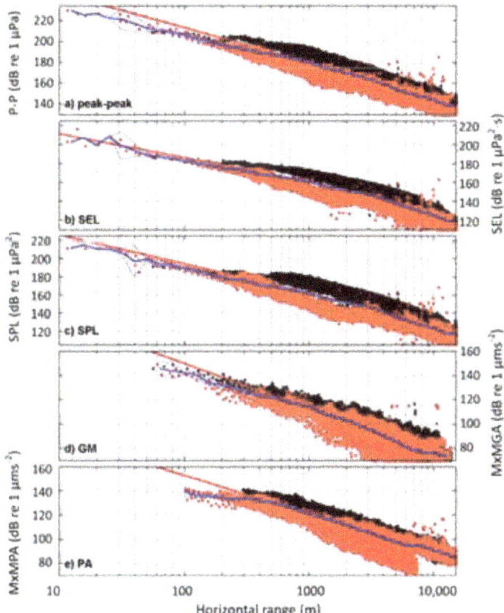

Figure 6. Received levels of the airgun array signal with range for the southern (black) overlaid with northern (red) sites for: (**a**) peak-to-peak pressure level; (**b**) sound exposure levels; (**c**) mean-square sound pressure levels; (**d**) maximum magnitude ground acceleration; (**e**) and maximum magnitude particle acceleration. Mean values at each range (all data) interpolated along $\log_{10}(range)$ (continuous blue line), together with the 95% confidence intervals (dotted blue lines) are shown along with the fitted curve (Equation (2)) using all data as the red curve.

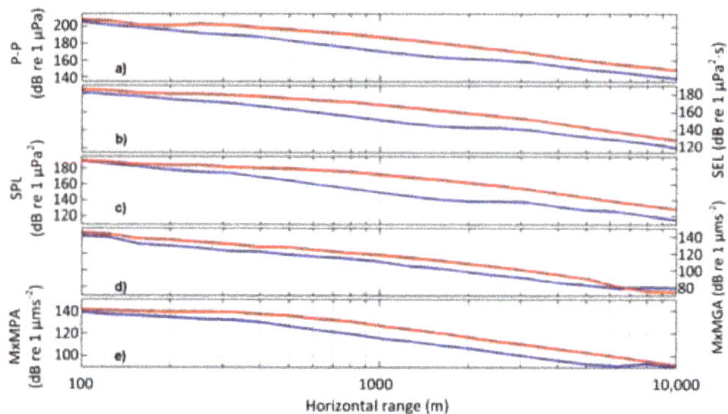

Figure 7. Received levels of the airgun array signal averaged in logarithmic range bins with horizontal range for the northern (blue) and southern (red) sites for peak-to-peak pressure level (**a**), sound exposure level (**b**), mean-square sound pressure level (**c**), maximum magnitude ground acceleration (**d**), and maximum magnitude particle acceleration (**e**) levels.

Table 4. Received levels with 95% confidence limits ($\pm$) and sample size (n), for peak-to-peak pressure level (P-P), sound exposure level (SEL), mean-square sound pressure level (SPL), maximum magnitude ground acceleration (MxMGA), and maximum magnitude particle acceleration (MxMPA), with logarithmic range bin centred at 250, 500, and 1000 (m), for the two sites.

	250 m		500 m		1000 m	
	Northern	**Southern**	**Northern**	**Southern**	**Northern**	**Southern**
P-P	193 ± 0.5 (99)	204 ± 0.6 (18)	184 ± 0.3 (501)	197 ± 0.2 (244)	171 ± 0.3 (899)	188 ± 0.2 (1131)
SEL	173 ± 0.4 (99)	181 ± 0.4 (18)	164 ± 0.3 (501)	176 ± 0.1 (244)	152 ± 0.2 (899)	168 ± 0.1 (1131)
SPL	176 ± 0.7 (99)	185 ± 0.4 (18)	165 ± 0.4 (501)	180 ± 0.2 (244)	150 ± 0.2 (899)	172 ± 0.2 (1131)
MxMGA	127 ± 1.5 (24)	134 ± 1.1 (18)	119 ± 0.6 (128)	127 ± 0.4 (116)	111 ± 0.7 (210)	118 ± 0.4 (213)
MxMPA	133 ± 0.8 (57)	na	126 ± 0.6 (186)	136 ± 0.2 (143)	116 ± 0.3 (433)	127 ± 0.20 (426)

Coefficients for the linear regression curves formed for each of the five metrics using Equation (2), showed the propagation loss coefficient ranged between −17 and −31 at the northern site and −16 and −38 across all metrics at the southern site (Table 5). In comparison, when Equation (1) was applied at short ranges and Equation (2) was applied at long ranges (minimum and maximum ranges for each metric defined in Tables 6 and 7), short-range propagation loss decreased significantly at the southern site (compare column 'a' values for the southern site between Tables 6 and 7).

Table 5. Parameters for fits over the full sampling range using Equation (2) at the northern and southern sites using Figure 6 and the appropriate metric against horizontal range with correlation coefficient (r^2). Abbreviations defined in Table 2.

	Northern Site				**Southern Site**			
	a	**b**	**c**	r^2	**a**	**b**	**c**	r^2
P-P	−30.73	−0.00034	265.4	0.901	−38.37	−0.00014	303.4	0.945
SEL	−24.03	−0.00094	226.5	0.928	−28.79	−0.00120	256.3	0.943
SPL	−29.84	−0.00061	242.1	0.917	−36.64	−0.00076	283.7	0.921
MxMGA	−41.30	0.0011	231.3	0.704	−35.48	−0.00108	225.9	0.855
MxMPA	−34.52	0.0008	218.3	0.716	−32.04	−0.00028	223.2	0.952

Table 6. Parameters for fits to metrics for northern and southern sites at short-range using horizontal-range limited curve fitting (Equation (1)), with correlation coefficient (r^2) for each fit and the maximum range used in the curve fitting. Abbreviations defined in Table 2.

Metric	Northern Site				Southern Site			
	Max Range (m)	a	b	r^2	Max Range (m)	a	b	r^2
P-P	300	−30.24	265.81	0.905	200	−19.69	245.90	0.819
SEL	300	−25.96	235.17	0.904	200	−14.06	212.78	0.893
SPL	300	−30.69	250.05	0.824	200	−15.82	220.14	0.888
PPG	300	−51.61	364.92	0.270	200	0.38	233.72	0.000
MxHPA	1000	−31.55	209.08	0.617	1000	−24.37	199.65	0.427
MxVPA	1000	−28.14	195.91	0.438	1000	−29.54	208.38	0.239
MxMPA	1000	−28.91	203.67	0.680	1000	−27.06	208.92	0.632
MxHGA	500	−27.62	193.53	0.671	500	−23.80	190.65	0.812
MxVGA	500	−22.01	172.24	0.464	500	−23.57	184.60	0.807
MxMGA	500	−27.83	194.10	0.672	500	−23.93	191.27	0.823

Table 7. Parameters for fits to metrics for northern and southern sites at long horizontal range using Equation (2), with correlation coefficient (r^2) for each fit and the minimum range used in the curve fitting. Abbreviations defined in Table 2.

Metric (Levels)	Northern Site						Southern Site					
	Min Range (m)	Type	a	b	c	r^2	Min Range (m)	Type	a	b	c	r^2
P-P	200	1	−30.55	−0.00035	264.8	0.897	200	1	−38.73	−0.00012	304.6	0.946
SEL	200	1	−23.48	−0.00098	224.8	0.926	200	1	−29.04	−0.00118	257.1	0.943
SPL	200	1	−29.14	−0.00067	239.9	0.914	200	1	−37.03	−0.00074	285.0	0.921
PPG	200	1	−33.54	328.4	0.00	0.185	200	2	−22.5	292.9	0	0.109
MxHPA	400	2	−25.87	189.7	0.00	0.597	1000	1	−34.10	−0.00004	227.2	0.886
MxVPA	400	2	−27.44	191.2	0.00	0.623	1000	1	−33.00	−0.00022	220.0	0.772
MxMPA	400	2	−26.56	194.0	0.00	0.632	1000	1	−32.86	−0.00021	225.8	0.932
MxHGA	50	2	−16.48	159.9	0.00	0.709	50	2	−22.25	183.3	0.00	0.893
MxVGA	50	2	−10.23	134.9	0.00	0.482	50	2	−16.56	160.0	0.00	0.823
MxMGA	50	2	−32.98	206.9	0.00	0.690	50	2	−44.78	253.7	0.00	0.848

There was no major evidence of horizontal beam pattern of the airgun array or localised sound propagation variation within an experimental area in any of the measured metrics (Figures 8–10). All metrics displayed comparatively uniform propagation loss away from the source across all azimuths (0° to 180°), with the exception of PPG (compare received levels with range from the source for Figures 8–10 with that of Figure 8d). Although PPG displayed a general decline in received levels with increasing range, there were patches of sudden changes in received levels at all ranges and azimuths. The spatial propagation plots suggested a weak beam pattern was present (semi-circular 'step' declines in level with range when plotted as azimuth against range in a cartesian, rather than polar plot), however, these were in fact unavoidable artefacts of the sampling design, created by the multiple straight line passes of the seismic vessel past stationary recording sensors, (areas of high-density sampling; Figures 8–10).

3.4. Correlation of Airgun Signal Metrics

SEL was strongly correlated with P-P and SPL, with the correlation coefficients for SEL/P-P and SEL/SPL, 0.88 and 0.95, respectively, at the northern site, both 0.97 at the southern site, and 0.94 and 0.95, respectively, when using all data combined (Table 8). When comparing SEL with maximum magnitude of ground acceleration and particle acceleration, these correlations dropped to 0.60 and 0.64, respectively, at the northern site, 0.87 and 0.93, respectively, at the southern site, and 0.74 and 0.77, respectively, for all data combined (Table 8). Particle acceleration and ground acceleration showed strong correlation, particularly at the southern site (0.87), though also at the northern site (0.64), and overall (0.80, all data). In the units used, dB re 1 μms^{-2}, ground acceleration was given by particle acceleration 47 cm above the seabed, minus $\approx$ 23 dB (Figure 11). In general, correlations were stronger at the southern, deeper site, than the northern site (Figure 11, Table 8). SEL did a poor job of predicting PPG.

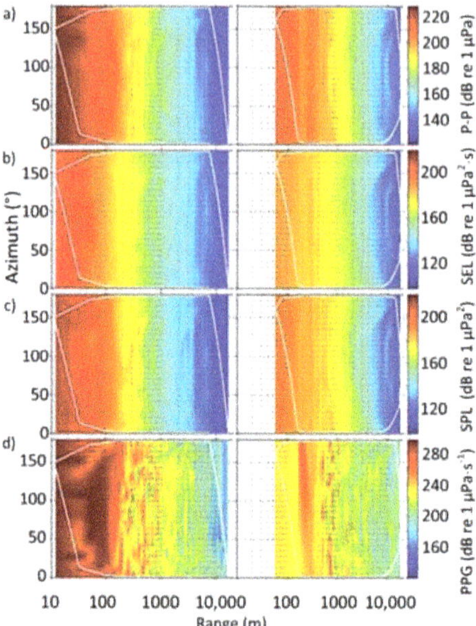

Figure 8. Peak-to-peak pressure levels (**a**), sound ex-posure levels (**b**), mean-square sound pressure levels (**c**) and pressure peak gradient (**d**) for the northern (N = left) and southern (S = right) sites, shown as a function of logarithmic horizontal range and azimuth (y-axis, o) from tow direction (assumed array symmetrical, 0° is ahead). The white lines encapsulate measured data bounds. Gridding used linear interpolation.

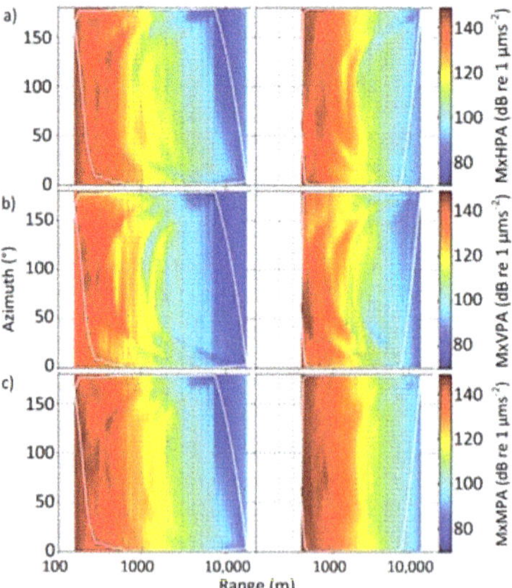

Figure 9. Maximum values of horizontal (**a**), vertical (**b**), and magnitude (**c**) particle acceleration levels for the northern (N = left) and southern (S = right) sites, shown as a function of logarithmic horizontal range and azimuth (y-axis, °) from tow direction (assumed array symmetrical, 0° is ahead). The white lines encapsulate measured data bounds. The colour axes are common across panels. Gridding used linear interpolation.

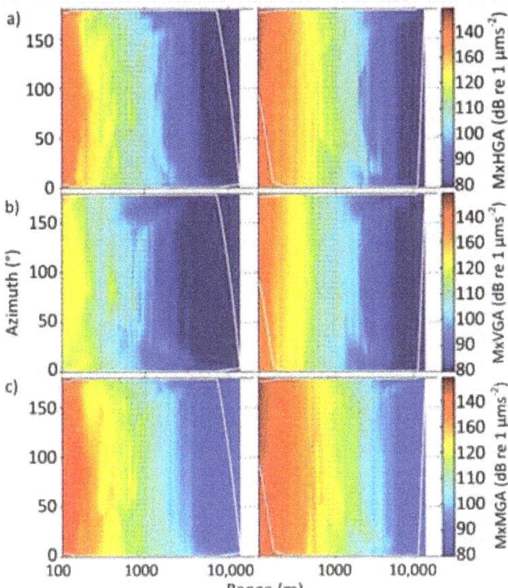

Figure 10. Maximum values of horizontal (**a**), vertical (**b**), and magnitude (**c**), ground acceleration levels for the northern (N = left) and southern (S = right) sites, shown as a function of logarithmic horizontal range and azimuth (y-axis, °) from tow direction (assumed array symmetrical, 0° is ahead). The white lines encapsulate measured data bounds. The colour axes are common across panels. Gridding used linear interpolation.

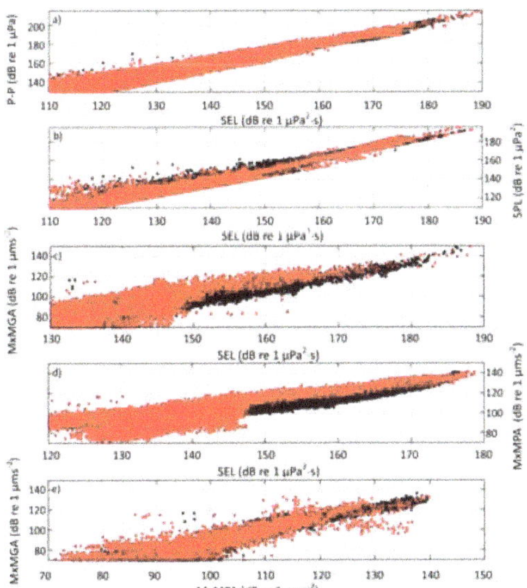

Figure 11. Relationship between pairs of metrics, for the southern site (black dots) overlaid with the northern site (red dots) for sound exposure level against peak-to-peak pressure level (**a**), sound exposure level against mean-square sound pressure level (**b**), SEL against magnitude of ground acceleration (**c**), sound exposure level against magnitude of particle acceleration (**d**), and magnitude of particle acceleration against magnitude of ground acceleration (**e**).

Table 8. Statistics of linear fits for the correlations shown including first and second coefficients of the linear regression fit (a and b), correlation coefficient, standard error (SE) and the 95% confidence error of the first coefficient of the linear fit. Abbreviations are defined in Table 2.

	Northern Site					Southern Site					All Data				
Pair	a	b	r^2	SE-O	95% CI-a	a	b	r^2	SE-O	95% CI-a	a	b	r^2	SE-O	95% CI-a
SEL/P-P	0.98	19.9	0.88	4.29	0.0030	0.93	28.70	0.97	2.43	0.0014	0.96	23.4	0.94	3.52	0.0015
SEL/SPL	1.06	−12.1	0.95	2.93	0.0021	1.06	−7.99	0.97	2.62	0.0015	1.12	−17.1	0.95	3.45	0.0014
SEL/MxMGA	0.96	−43.1	0.60	9.31	0.0017	1.06	−54.45	0.87	6.00	0.0080	0.97	−46.6	0.74	8.01	0.0084
SEL/PPG	0.92	83.1	0.16	24.47	0.0173	0.53	134.17	0.12	21.09	0.0121	0.70	110.9	0.13	23.28	0.0118
SEL/MxMPA	0.74	−1.01	0.64	7.32	0.0082	0.87	−23.09	0.93	3.13	0.0037	0.74	−1.8	0.77	6.06	0.0043
MxMPA/MxMGA	1.07	−18.01	0.64	6.99	0.0329	1.13	−25.25	0.87	4.15	0.0124	1.16	−23.0	0.80	5.25	0.0132

4. Discussion

Received levels recorded in this study showed that the exposure levels experienced from the 2600 cui source operated at both sites were similar to those likely to be experienced by fauna around typical commercial surveys under similar conditions [19,39]. This comprised energy at frequencies that would be detected by many marine taxa [40,41]. Propagation loss through the experimental areas were typical of environments with comparable water depths (50–70 m) uniform bathymetry and a similar seafloor composition (a thin layer of sand over limestone pavement [2]). The relatively high propagation losses at our study sites were due to the interaction with the underlying or exposed limestone seabed compared with thicker layers of sand elsewhere [2]. This limestone seabed is typical of continental shelf waters across southern and western Australia into the far northwest [19]. The lack of directionality in the beam pattern combined with comparatively uniform depth and seabed substrate allowed robust estimates of exposure levels at biological sampling sites to be predicted for the two experimental areas with high levels of confidence, based on the correlation coefficients found for different metrics.

Sound exposure level was strongly correlated with mean-square sound pressure level ($r^2 = 0.95$) and peak-to-peak pressure levels ($r^2 = 0.94$). SEL was also correlated (though to a lesser degree) with the maximum magnitudes of particle acceleration ($r^2 = 0.77$) and maximum magnitude of ground acceleration ($r^2 = 0.74$) when using all data. The correlation of SEL and ground or particle acceleration improved at the more uniform southern site ($r^2 > 0.87$). The relationships between pressure and ground acceleration metrics were valid across all ranges as unsaturated recordings were collected even at ranges closest to the source. However, although some measures of particle acceleration in the near-field did not overload and remain correlated with SEL, relationships involving particle acceleration in the near-field < 500 m could not be fully assessed due to the saturation of >50% of measured signals. The correlation between SEL and ground acceleration was weaker at the shallower northern site ($r^2 = 0.60$) than at the deeper southern site ($r^2 = 0.87$). The reasons for this were not explored though are likely due to the northern site having a more diverse range of seabed types(cm to m of sand over limestone or no sand and the instrument lying on limestone, compared with tens cm to m of sand at the southern site).

In this situation, harder seafloors will dampen the ground acceleration. The results highlighted a strong correlation between particle acceleration near the seafloor and seabed ground acceleration ($r^2 = 0.80$ when using all data).

These results imply that SEL is a good proxy for other conventional pressure metrics, particle acceleration and ground acceleration, when assessing the impact of noise on fauna at horizontal ranges of more than ≈250 and ≈600 m (northern and southern sites, respectively) with a similar source for sites with similar depths and geophysical characteristics (e.g., sound speed profile and seafloor types).

The metric PPG correlated relatively poorly with other metrics and was not predicted well by SEL. This was believed to be due to the time between peaks at low level (SEL) signals. At low received levels, the peaks arrived randomly within the signal causing large natural variability in PPG values. Even at a short range, with low to modest received airgun signal levels the time between peaks was highly variable, causing differences in the PPG measure. This metric was introduced as a simple measure of airgun signal 'jerk', which McCauley and Duncan [34] speculated generates physiological damage in invertebrates. To reduce variation, the measure may need to be more complex in its derivation, for example, by constraining the measures to consecutive peaks instead of the time between maximum positive and negative peaks within the signal. The application of PPG cannot be assessed until measured biological impacts have been correlated and compared to other more standard metrics (i.e., SEL). It may be that the variation observed in PPG around an airgun array correlates with physiological trauma in some marine fauna.

The empirically measured correlations between metrics in this study show that where environmental management is assessing the impact of impulsive acoustic signals, such as those from seismic surveys, at ranges of 100 s of m and greater (i.e., acoustic transition- and far-fields) then SEL may be used as a proxy for other metrics. This minimizes the need for acquiring particle or ground acceleration data unless there are specific concerns regarding benthic, sessile, or species of low-mobility.

Supplementary Materials: The following are available online at https://www.mdpi.com/article/10.3390/jmse9060571/s1, Figure S1: Map of sonar backscatter from a multibeam survey of the southern-deep site (left images) and the northern-shallow site (right images). Still images captured from towed video at points indicated on the map show examples of seabed corresponding to different levels of backscatter and benthic organisms (mostly sponges and gorgonians) within zones., Figure S2: Sound exposure levels with range of a 150 cui airgun towed and discharged every 60 s along one transect at the northern site (red dots) and two transects conducted at the southern site (blue–offshore side and black dots–inshore side), Figure S3: Airgun (black squares) positions and sizes for the 2600 cui arrays towed by *BGP Explorer* where size of marker has been scaled to reflect the relative volume of the airgun and arrow denotes direction of travel, Figure S4: Modelled (PGS Nucleus model) source signal waveform (a) and relative power spectral density (b) for the far-field signature of the 2600 cui airgun array signal (5 m source depth, 41.8 bar m primary pressure). Red line in (b) marks the -6 dB limit. Images supplied by Exploiter PTE. LTD, Figure S5: Modelled (PGS Nucleus) airgun array signal directivity patterns for (a) horizontal plane directivity at 60 Hz, (b) vertical plane along-track directivity at 50 Hz, and (c) vertical plane across-track directivity at 50 Hz. Images supplied by Exploiter PTE. LTD, Figure S6: Example of system gains with frequency response for USRs from the white noise injection calibration, Figure S7: Distribution of ambient noise levels for peak-to-peak pressure level (a), sound exposure level (b), mean-square sound pressure level (c), PPG (d), maximum horizontal particle acceleration (e), maximum vertical particle acceleration (f) and maximum magnitude of particle acceleration (g) at the northern (left columns) and southern (right columns) sites, Figure S8: Sound exposure levels with range for a seismic sail line operated at the northern site on 18th September 2018 ((a) and (b)) by USRs positioned at various distances from the sail line. Coloured dots relate to the USR datasets from the recording positions shown in (c) using the respective colours from a) and b), Figure S9: Sound exposure levels with range for a seismic sail line operated at the southern site on 23rd September 2018 ((a) and (b)) by USRs positioned at various distances from the sail line. Coloured dots relate to the USR datasets from the recording positions shown in (c) using the respective colours from a) and b). Periods where an increasing number of guns were operated prior to the start of the seismic sail line (i.e., the 'ramp-up') are highlighted, Table S1: Configuration of each 2600 cui array, where X is the across-track axis (negative to port and positive to starboard direction, referenced to the centreline of the vessel) and Y is the along-track axis (positive to forward, referenced to the forward guns), Table S2: Characteristics of seismic vessel operations at each experimental site, Table S3: Details of the source vessel, *MV BGP Explorer*. Ancilliary data [42].

Author Contributions: Conceptualization, M.J.G.P., R.D.M. and M.G.M.; methodology, formal analysis, writing—original draft preparation, visualization, R.D.M. and M.J.G.P.; writing—review and editing, M.J.G.P., R.D.M. and M.G.M.; funding acquisition, M.G.M. All authors have read and agreed to the published version of the manuscript.

Funding: This work was conducted as part of the North West Shoals to Shore Program, which is proudly supported by Santos as part of the company's commitment to better understand WA's marine environment.

Institutional Review Board Statement: The experiment during which this data was collected was conducted according to the guidelines of the Declaration of Helsinki, and approved by the Ethics Committee of the University of Western Australia (RA/3/100/159, 5/6/2018).

Informed Consent Statement: Not applicable.

Data Availability Statement: The acoustic data supporting this study is available on request from the Australian Institute of Marine Science.

Acknowledgments: The authors would like to acknowledge the crew of the RV Solander who facilitated the deployment of all acoustic sensors. The authors would also like to thank the anonymous reviewers for their time to consider the manuscript and useful comments to improve the content.

Conflicts of Interest: The authors declare no conflict of interest. The funders participated in a stakeholder workshop during the development of the experimental design of the overall program. This workshop was attended by collaborators and stakeholders including: AIMS, Curtin University, University of Tasmania, Western Australia's Department of Primary Industry and Regional Development, WA Fishing Industry Council, commercial fishers, members of the WA pearling industry, geophysical contractors, members of the petroleum industry, as outlined in the Supplemental Material. The funders had no role in the collection, analyses, or interpretation of data; in the writing of the manuscript, or in the decision to publish the results.

References

1. Gisiner, R.C. Sound and marine seismic surveys. *Acoust. Today* **2016**, *12*, 10–18.
2. Duncan, A.J. Airgun Arrays for Marine Seismic Surveys-Physics and Directional Characteristics. In Proceedings of the Acoustics 2017, Perth, Australia, 19–22 November 2017; p. 10.
3. Popper, A.N.; Hawkins, A.D.; Fay, R.R.; Mann, D.A.; Bartol, S.; Carlson, T.J.; Coombs, S.; Ellison, W.T.; Gentry, R.L.; Halvorsen, M.B. *ASA S3/SC1. 4 TR-2014 Sound Exposure Guidelines for Fishes and Sea Turtles: A Technical Report Prepared by ANSI-Accredited Standards Committee S3/SC1 and Registered with ANSI*; Springer: Berlin/Heidelberg, Germany, 2014.
4. Ladich, F.; Fay, R.R. Auditory evoked potential audiometry in fish. *Rev. Fish Biol. Fish.* **2013**, *23*, 317–364. [CrossRef]
5. Carroll, A.; Przeslawski, R.; Duncan, A.J.; Gunning, M.; Bruce, B. A critical review of the potential impacts of marine seismic surveys on fish & invertebrates. *Mar. Pollut. Bull.* **2017**, *114*, 9–24.
6. Martin, K.J.; Alessi, S.C.; Gaspard, J.C.; Tucker, A.D.; Bauer, G.B.; Mann, D.A. Underwater hearing in the loggerhead turtle (Caretta caretta): A comparison of behavioral and auditory evoked potential audiograms. *J. Exp. Biol.* **2012**, *215*, 3001–3009. [CrossRef] [PubMed]
7. Chapuis, L.; Kerr, C.C.; Collin; Hart, N.S.; Sanders, K.L. Underwater hearing in sea snakes (Hydrophiinae): First evidence of auditory evoked potential thresholds. *J. Exp. Biol.* **2019**, *222*. [CrossRef] [PubMed]
8. Southall, B.L.; Finneran, J.J.; Reichmuth, C.; Nachtigall, P.E.; Ketten, D.R.; Bowles, A.E.; Ellison, W.T.; Nowacek, D.P.; Tyack, P.L. Marine mammal noise exposure criteria: Updated scientific recommendations for residual hearing effects. *Aquat. Mam.* **2019**, *45*, 125–232. [CrossRef]
9. Slabbekoorn, H.; Dalen, J.; de Haan, D.; Winter, H.V.; Radford, C.; Ainslie, M.A.; Heaney, K.D.; van Kooten, T.; Thomas, L.; Harwood, J. Population-level consequences of seismic surveys on fishes: An interdisciplinary challenge. *Fish Fish.* **2019**, *20*, 653–685. [CrossRef]
10. Paxton, A.B.; Taylor, J.C.; Nowacek, D.P.; Dale, J.; Cole, E.; Voss, C.M.; Peterson, C.H. Seismic survey noise disrupted fish use of a temperate reef. *Mar. Pol.* **2017**, *78*, 68–73.
11. Bruce, B.; Bradford, R.; Foster, S.; Lee, K.; Lansdell, M.; Cooper, S.; Przeslawski, R. Quantifying fish behaviour and commercial catch rates in relation to a marine seismic survey. *Mar. Environ. Res.* **2018**, *140*, 18–30. [CrossRef]
12. Davidsen, J.G.; Dong, H.G.; Linné, M.; Andersson, M.H.; Piper, A.; Prystay, T.S.; Hvam, E.B.; Thorstad, E.B.; Whoriskey, F.; Cooke, S.J.; et al. Effects of sound exposure from a seismic airgun on heart rate, acceleration and depth use in free-swimming Atlantic cod and saithe. *Cons. Physiol.* **2019**, *7*. [CrossRef]
13. Kavanagh, A.S.; Nykänen, M.; Hunt, W.; Richardson, N.; Jessopp, M.J. Seismic surveys reduce cetacean sightings across a large marine ecosystem. *Sci. Rep.* **2019**, *9*, 19164. [CrossRef]
14. Hubert, J.; Neo, Y.Y.; Winter, H.V.; Slabbekoorn, H. The role of ambient sound levels, signal-to-noise ratio, and stimulus pulse rate on behavioural disturbance of seabass in a net pen. *Behav. Proc.* **2020**, *170*, 103992. [CrossRef] [PubMed]
15. Day, R.D.; McCauley, R.D.; Fitzgibbon, Q.P.; Hartmann, K.; Semmens, J.M. Seismic air guns damage rock lobster mechanosensory organs and impair righting reflex. *Proc. R. Soc. B.* **2019**, *286*, 20191424. [CrossRef] [PubMed]
16. Day, R.D.; Fitzgibbon, Q.P.; McCauley, R.D.; Hartmann, K.; Semmens, J.M. Lobsters with pre-existing damage to their mechanosensory statocyst organs do not incur further damage from exposure to seismic air gun signals. *Envir. Poll.* **2020**, *267*, 115478. [CrossRef]
17. Gisiner, R.; Harper, S.; Livingston, E.; Simmen, J. Effects of Sound on the Marine Environment (ESME): An underwater noise risk model. *IEEE J. Ocean. Eng.* **2006**, *31*, 4–7. [CrossRef]

18. McCauley, R.D.; Fewtrell, J.; Duncan, A.J.; Jenner, C.; Jenner, M.-N.; Penrose, J.D.; Prince, R.I.T.; Adhitya, A.; Murdoch, J.; McCabe, K. Marine seismic surveys: Analysis and propagation of airgun signals; and effects of exposure on humpback whales, sea turtles, fishes and squid. In *(Anon) Environmental Implications of Offshore Oil and Gas Development in Australia: Further Research*; Australian Petroleum Production Exploration Association: Canberra, Australia, 2003; pp. 364–521. Available online: www.cmst.curtin.edu.au$\backslash$publications (accessed on 11 May 2021).

19. McCauley, R.D.; Duncan, A.J.; Gavrilov, A.N.; Cato, D.H. Transmission of marine seismic survey, air gun array signals in Australian waters. In Proceedings of the Acoustics 2016, Brisbane, Australia, 9–11 November 2016.

20. Dunlop, R.A.; Noad, M.J.; McCauley, R.D.; Kniest, E.; Slade, R.; Paton, D.; Cato, D.H. The behavioural response of migrating humpback whales to a full seismic airgun array. *Proc. Roy. Soc. B Biol. Sci.* **2017**, *284*, 20171901. [CrossRef] [PubMed]

21. Morris, C.; Cote, D.; Martin, B.; Kehler, D. Effects of 2D seismic on the snow crab fishery. *Fish. Res.* **2018**, *197*, 67–77. [CrossRef]

22. Popper, A.N.; Hastings, M. The effects of anthropogenic sources of sound on fishes. *J. Fish Biol.* **2009**, *75*, 455–489. [CrossRef]

23. Popper, A.; Hawkins, A.; Halvorsen, M. Anthropogenic sound and fishes. In *Report by ICF for Washington State Department of Transportation*; Research Office: Washington, DC, USA, 2019; Available online: https://www.wsdot.wa.gov/research/reports/fullreports/891-1.pdf (accessed on 22 May 2021).

24. Meyer-Rochow, V.B.; Penrose, J.D. Sound production by the western rock lobster *Panulirus longipes* (Milne Edwards). *J. Exp. Mar. Biol. Ecol.* **1976**, *23*, 191–209. [CrossRef]

25. André, M.; Kaifu, K.; Sole, M.; van der Schaar, M.; Akamatsu, T.; Balastegui, A.; Sanchez, A.M.; Castell, J.V. Contribution to the Understanding of Particle Motion Perception in Marine Invertebrates. In *The Effects of Noise on Aquatic Life II. Advances in Experimental Medicine and Biology*; Popper, A., Hawkins, A., Eds.; Springer: New York, NY, USA, 2016; Volume 875. [CrossRef]

26. Montgomery, J.C.; Radford, C.A. Marine bioacoustics. *Curr. Biol.* **2017**, *27*, R502–R507. [CrossRef]

27. Popper, A.N.; Hawkins, A.D. The importance of particle motion to fishes and invertebrates. *J. Acoust. Soc. Amer.* **2018**, *143*, 470–488. [CrossRef] [PubMed]

28. Day, R.D.; McCauley, R.D.; Fitzgibbon, Q.P.; Semmens, J.M. *Assessing the Impact of Marine Seismic Surveys on Southeast Australian Scallop and Lobster Fisheries*; FRDC Report, Project No 2012/008; University of Tasmania: Hobart, Australia, 2016; p. 167. Available online: https://frdc.com.au/project/2012-008 (accessed on 22 May 2021).

29. Day, R.D.; McCauley, R.D.; Fitzgibbon, Q.P.; Hartmann, K.; Semmens, J.M. Scallops shaken following seismic surveys: Exposure to air gun signals affects mortality, physiology and behaviour. *Glo. Chan. Biol.* **2017**. [CrossRef]

30. Day, R.D.; McCauley, R.D.; Fitzgibbon, Q.P.; Hartmann, K.; Semmens, J.M. Exposure to seismic air gun signals causes physiological harm and alters behavior in the scallop *Pecten Fumatus*. *Proc. Nat. Acad. Sci. USA* **2017**, *114*. [CrossRef]

31. Erbe, C. *Underwater Acoustics: Noise and the Effects on Marine Mammals a Pocket Handbook*, 3rd ed.; JASCO Applied Sciences: Brisbane, Australia, 2013.

32. Madsen, P. Marine mammals and noise: Problems with root mean square sound pressure levels for transients. *J. Acoust. Soc. Amer.* **2005**, *117*, 3952–3957. [CrossRef]

33. McCauley, R.D.; Thomas, F.; Parsons, M.J.G.; Erbe, C.; Cato, D.H.; Duncan, A.J.; Gavrilov, A.N.; Parnum, I.M.; Salgado-Kent, C.P. Developing an Underwater Sound Recorder: The Long and Short (Time) of It . . . . *Acoust. Aus.* **2017**, *45*, 301–311. [CrossRef]

34. McCauley, R.D.; Duncan, A.J. How do Impulsive Marine Seismic Surveys Impact Marine Fauna and How Can We Reduce Such Impacts? In Proceedings of the Acoustics 2017, Perth, Australia, 20 November 2017.

35. Harding, H.R.; Gordon, T.A.C.; Wong, K.; McCormick, M.I.; Simpson, S.D. Condition-dependent responses of fish to motorboats. *Biol. Lett.* **2020**, *16*, 20200401.

36. Urick, R.J. *Principles of Underwater Sound*, 3rd ed.; Peninsula Publishing Los Atlos: California, CA, USA, 1983.

37. International Organization for Standardization. *Underwater Acoustics—Terminology*; IOS: Geneva, Switzerland, 2017.

38. Duncan, A.J.; Gavrilov, A.N.; McCauley, R.D.; Parnum, I.M.; Collis, J.M. Characteristics of sound propagation in shallow water over an elastic seabed with a thin cap-rock layer. *J. Acoust. Soc. Am.* **2013**, *134*, 207–215. [CrossRef] [PubMed]

39. Greene, C.R., Jr. Characteristics of marine seismic survey sounds in the Beaufort Sea. *J. Acoust. Soc. Am.* **1988**, *83*, 2246. [CrossRef]

40. Duarte, C.M.; Chapuis, L.; Collin, S.P.; Costa, D.P.; Devassy, R.P.; Eguiluz, V.M.; Erbe, C.; Gordon, T.A.C.; Halpern, B.S.; Harding, H.R.; et al. The Ocean Soundscape of the Anthropocene. *Science* **2021**, *371*, 581. [CrossRef]

41. Mooney, T.A.; Di Iorio, L.; Lammers, M.; Lin, T.H.; Nedelec, S.L.; Parsons, M.J.G.; Radford, C.A.; Urban, E.; Stanley, J. Listening forward: Approaching marine biodiversity assessments using acoustic methods. *R. Soc. Open Sci.* **2020**, *7*, 201287. [CrossRef]

42. Whiteway, T.G. Australian Bathymetry and Topography Grid, June 2009. In *Geoscience Australia Record 2009/21*; Geoscience Australia: Canberra, Australia, 2009; p. 46.

Journal of
*Marine Science
and Engineering*

Article

Above and below: Military Aircraft Noise in Air and under Water at Whidbey Island, Washington

Lauren M. Kuehne [1,2,*], **Christine Erbe** [3], **Erin Ashe** [4], **Laura T. Bogaard** [4], **Marena Salerno Collins** [4] **and Rob Williams** [4]

[1] School of Aquatic and Fishery Sciences, University of Washington, 1122 NE Boat Street, Seattle, WA 98105, USA
[2] Omfishient Consulting, 2333 Seringa Avenue, Bremerton, WA 98310, USA
[3] Centre for Marine Science and Technology, Curtin University, GPO Box U1987, Perth, WA 6845, Australia; C.Erbe@curtin.edu.au
[4] Oceans Initiative, 117 E Louisa St #135, Seattle, WA 98102, USA; erin@oceansinitiative.org (E.A.); laura@oceansinitiative.org (L.T.B.); marena@oceansinitiative.org (M.S.C.); rob@oceansinitiative.org (R.W.)
* Correspondence: lkuehne@uw.edu

Received: 26 September 2020; Accepted: 4 November 2020; Published: 16 November 2020

Abstract: Military operations may result in noise impacts on surrounding communities and wildlife. A recent transition to more powerful military aircraft and a national consolidation of training operations to Whidbey Island, WA, USA, provided a unique opportunity to measure and assess both in-air and underwater noise associated with military aircraft. In-air noise levels (110 ± 4 dB re 20 µPa rms and 107 ± 5 dBA) exceeded known thresholds of behavioral and physiological impacts for humans, as well as terrestrial birds and mammals. Importantly, we demonstrate that the number and cumulative duration of daily overflights exceed those in a majority of studies that have evaluated impacts of noise from military aircraft worldwide. Using a hydrophone deployed near one runway, we also detected sound signatures of aircraft at a depth of 30 m below the sea surface, with noise levels (134 ± 3 dB re 1 µPa rms) exceeding thresholds known to trigger behavioral changes in fish, seabirds, and marine mammals, including Endangered Southern Resident killer whales. Our study highlights challenges and problems in evaluating the implications of increased noise pollution from military operations, and knowledge gaps that should be prioritized with respect to understanding impacts on people and sensitive wildlife.

Keywords: military aircraft; noise pollution; ocean noise; Endangered species; human health; animal behavior

1. Introduction

Military aircraft activity in the Salish Sea, Washington State, has been increasing over the past decade due to changes in operations and training for personnel out of the Naval Air Station Whidbey Island (NASWI). Although naval flights have been operating in the area for decades, the recent transition from Northrop Grumman EA-6B Prowler to the more powerful Boeing EA-18G Growler aircraft for electronic warfare has led to increases in the number of complaints about noise, including concern for area wildlife. Consolidation of nationwide training for these aircraft to NASWI increased the fleet size by 44% (from 82 to 118 aircraft) in 2019, with corresponding increases in air carrier practice, electronic warfare training, and overall base operations [1]. The changes at NASWI are reflective of a broader national trend in military base closures and consolidation, which are likely to intensify community noise and air pollution in some areas [2]. The implications of a concurrent change to more powerful aircraft and increased operations for noise pollution have not been measured, leaving knowledge gaps in the ability to assess vulnerability of both people and wildlife, including Endangered, Threatened or sensitive species.

Worldwide, military transportation and activities are among the least studied sources of noise pollution [3,4]. This is due to a combination of differing regulations in areas surrounding military bases and airfields [2,5], the complexity of conducting research when operation schedules are not publicly accessible, and analytical challenges in measuring and characterizing noise from periodic and intermittent activity [3,6]. For example, the NASWI and other airfields in the United States are not regulated by the Federal Aviation Administration as civilian airports are, resulting in more limited legal bases on which to contest aircraft noise that is disruptive. Military air traffic schedules may be classified or largely unobtainable, making it difficult to conduct monitoring or validate modeled noise data. Lastly, high-intensity but intermittent activities require alternatives to standard community noise metrics, which are geared toward more continuous sources of noise [7]. For these reasons, although studies and reviews exist on the impacts of community noise from civilian airports and highways, independent studies related to military activity are relatively rare [8] and likely to be opportunistic [3,9]. This creates crucial information gaps when the public or agencies try to evaluate the impact of proposed changes in activities [10].

While often considered an acoustic barrier, the air-water interface may effectively transmit sound in certain situations (e.g., in calm conditions and for vertical incidence; [11,12]). This opens up the need to more critically examine underwater impacts of civil and military aviation noise, which have typically been considered negligible [13]. Of paramount concern in the Salish Sea are the Southern Resident killer whales (*Orcinus orca*, SRKW), which were listed as Endangered under Canada's Species at Risk Act in 2001 and the U.S. Endangered Species Act (ESA) in 2005. Endangered by chemical pollution, food shortages, and vessel traffic, additional anthropogenic stressors should be examined for the potential to put recovery of SRKW out of reach. Protecting foraging areas is important because SRKWs are food-limited, and because they are more vulnerable to disturbance while feeding than during any other activity [14]. Another species of particular concern in this region is Threatened marbled murrelet *(Brachyramphus marmoratus)*, a non-migratory seabird that makes use of protected and shallow coastal areas for foraging. In short, we see a number of compelling and timely reasons to measure in-air sound levels from Growlers to assess impacts on humans and terrestrial wildlife, and explore whether Growler noise is audible under water in areas used by SRKWs, marbled murrelet, and other wildlife.

In this study, we evaluated the potential bioacoustic impacts of noise from Growlers and implications for the Puget Sound and Salish Sea region. Noise pollution is usually studied as a public health issue for people [3], or, less commonly, as an anthropogenic impact on wildlife [8]. Terrestrial and aquatic impacts and species are usually considered and studied separately. Though understandable, this compartmentalizing does not acknowledge the cascading changes that can occur in ecosystems as a result of new noise sources, or that anthropogenic noise impacts species and taxonomic groups broadly [15,16]. In this study, we therefore adopted an integrative approach by measuring both in-air and underwater noise, and then interpreting those levels against established impacts for humans as well as sensitive terrestrial and aquatic wildlife. Our study seeks to answer two questions: (1) Does noise from military aircraft have the potential to impact aquatic as well as terrestrial habitats? and (2) How do measured levels in air and under water compare with thresholds known from previous studies to impact humans, terrestrial, and marine wildlife? We then use these results to critically examine the processes by which noise impacts are assessed and mitigated, helping to bridge the gap between monitoring and management of noise pollution.

2. Methods

2.1. Study Area

Whidbey Island, where NASWI is located, is near the border of Canada and the United States of America, and forms the northern border of Puget Sound. The island is approximately 88 km long, and 2.4–19 km wide; it is the largest island in Washington State. Whidbey Island was historically inhabited by people of multiple Native American tribes that maintain reservations in the surrounding area today, including the Lower Skagit, Swinomish, Suquamish, and Snohomish. The island is currently home to 70,000 residents living in multiple medium- and small-sized communities. The majority of the island's economic activity is directly or indirectly related to the Navy's presence, but other economically important activities include farming, fishing, tourism, and real estate/vacation home purchases. Public concerns about impacts from NASWI are not limited to Whidbey Island, but are present throughout Island County, which relies on a reputation for remote and peaceful tourist opportunities (e.g., the San Juan Islands).

The NASWI is the largest single employer on Whidbey Island, with a base population of approximately 10,000 soldiers, civilians, and contractors. NASWI was first commissioned and constructed in the early 1940s and has undergone various eras of expansion and contraction. Currently, NASWI consists of two airfields (Ault Field and OLF Coupeville) with four runways (Figure 1a). Aircraft are housed at Ault Field, but both airfields are used for field carrier landing practice (FCLP) by Growlers. FCLPs are intended to replicate conditions for carrier-based takeoffs and landings and feature repeated "touch-and-go" flights; a certain number of these must be conducted at night to adequately prepare pilots. Although FCLP is the dominant type of aircraft training at NASWI, other base aircraft activities include electronic warfare and air-to-air combat training in nearby military operations areas [10], submarine detection, and cargo aircraft training [17].

Schedules for base activities are not published, with the exception of a courtesy (i.e., non-official) notification of the airfield and approximate time frame for FCLPs for that week [18]. Information on base operations, aircraft activity, and corresponding noise impacts is otherwise available only in the form of general estimates (e.g., annual operations, modeled maximum loudness in selected areas) conducted as part of the Environmental Impact Statement (EIS) process [17] and corresponding Biological Opinions for ESA-listed species [19]. The U.S. Department of Defense policy is to model rather than monitor noise from military operations, and no noise monitoring has been done by the U.S. Navy to date.

2.2. In-Air Acoustic Data Collection

Growlers were recorded in air at Moran County Beach Park (48.3693, −122.6662), the nearest public location from the underwater recording site (see below) on September 13 and 16, 2019, and located under FCLP flight track 14 for Ault Field (Figure 1b). On both days, FCLPs were scheduled from "Morning to Late Afternoon", and were done on track 14, with jets circling south to north; as a result, the recorder was capturing sound associated with landings. An observer logged the type and number of all visible and (in the case of takeoffs) audible aircraft events and noted the direction of travel and flight activity as landing, pass, or takeoff. A Songmeter SM4 autonomous recorder (Wildlife Acoustics, Maynard, MA, USA) collected audio data from aircraft landings and flyover events. Sound was sampled at 48 kHz and with zero gain added. The Songmeter was deployed between 0930 and 1530 on September 13 and between 1100 and 1500 on September 16. A sound level data logger (Extech 407760; Nashua, NH, USA) was deployed at the same time recording A-weighted sound levels (dBA) at 1-s intervals; however, the data logger failed to record on the 13th, so simultaneous sound pressure levels were collected with audio data on September 16 only.

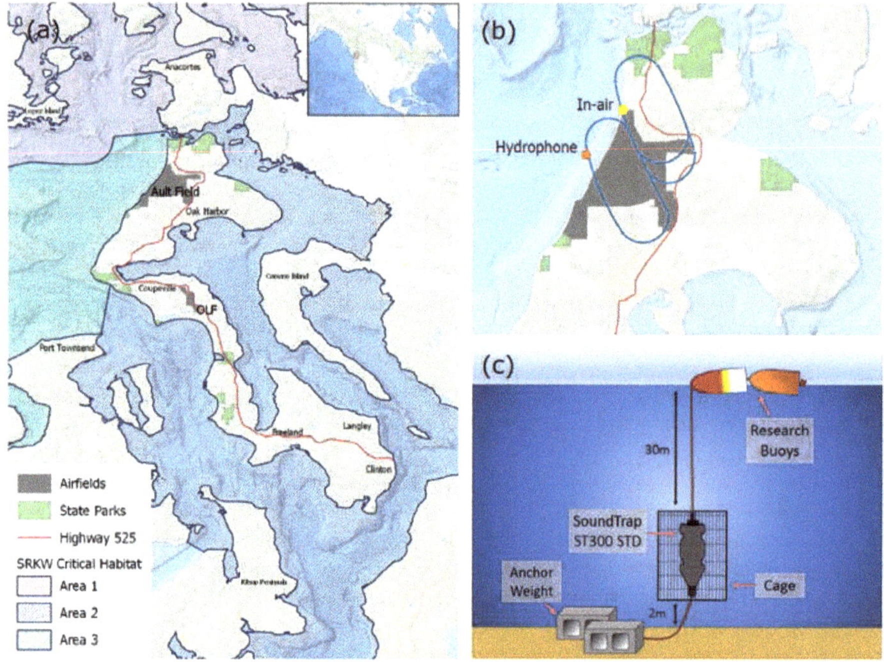

Figure 1. Map of the study area, location of underwater and in-air sound recordings, and schematic of hydrophone mooring. (**a**) Whidbey Island and surrounding areas, showing the location of Naval Air Station Whidbey Island's two airfields (Ault Field and OLF), largest cities and towns, and state parks. Critical habitat areas (Areas 1–3) for Endangered Southern Resident killer whales are shown. (**b**) Location of SoundTrap (orange circle) and in-air recordings (yellow circle) relative to Ault Field and field carrier landing practice (FCLP) flight tracks 32 and 14 (blue lines). Flight tracks were georeferenced from the Environmental Impact Statement [17]; two other flight tracks around Ault Field (tracks 7 and 25, oriented east-west) are not shown. (**c**) Schematic of the SoundTrap mooring, which was deployed from 15 August–11 September 2019. The SoundTrap was housed in a coated wire cage attached to a concrete block anchor with a 2 m line; a 30 m line attached to buoys kept the SoundTrap suspended above the sea floor.

The number of aircraft events during the FCLPs on September 13 and 16 was summarized from the visual observations for the period of 1100–1500 (when FCLP activity occurred) on both days. To provide a visual representation of the timing and duration of FCLPs, long-term spectral averages (LTSAs) were generated for the same periods each day in MATLAB, using 1-s and 1-Hz resolutions.

2.3. Underwater Acoustic Data Collection

Growlers were recorded under water with a SoundTrap 300 STD autonomous recorder (Ocean Instruments, Auckland, New Zealand) that was factory-calibrated and programmed to record continuously at 96 kHz sampling frequency (fs) prior to deployment. The SoundTrap was deployed off the northwest coast of Whidbey Island, approximately 1400 m from the end of the east-west runway and 1000 m from the shoreline (Figure 1b). This location is below the path of aircraft taking off to the west, and FCLP flight tracks 7 and 32.

The SoundTrap was suspended in a metal cage 2 m above the sandy mud sea floor, and was moored using a system of concrete blocks, sinking line, and two floats (Figure 1c). The SoundTrap was deployed twice for two weeks, totaling 28 days of data collection. In between deployments, the SoundTrap was retrieved for charging and downloading data. It was first deployed on 15 August

2019, at 13:54 PDT at 48.3492, −122.6917, at a depth of 33.2 m, and then on August 29, 2019, at 12:12 PDT, in a similar location (48.3494, −122.6907), at a depth of 34.7 m. The weather throughout this month was variable, consisting of rain, wind, sun, and clouds. Growlers taking off to the west flew over the SoundTrap at an altitude of 120–190 m above sea level.

2.4. Data Analysis

Visual observations confirmed the occurrence of 23 single Growler flights over the Songmeter on September 16, during an FCLP session with only one aircraft operating. Opportunistic observations on four dates (August 15, 28, 29, and September 12) visually confirmed the occurrence of ten Growler flights over the SoundTrap. These overflights were manually identified in the recordings using Audacity© (Version 2.3.2; retrieved 20 September 2019 from https://audacityteam.org/) and a 15-s audio file was saved for each overflight. The audio files were analyzed using a custom script in MATLAB (version 2018b; The MathWorks Inc., Natick, MA, USA). Each 15-s file was calibrated and then Fourier-transformed in 1-s Hann windows (i.e., the number of Fast Fourier Transform components NFFT equaled the number of samples per second) with 67% overlap. A time series of band levels was computed between 20 Hz and 20 kHz, corresponding to the frequency band occupied by Growler overflights. The peak band level was assumed to correspond to the time when the plane was directly overhead. The mean-square sound pressure spectral density (in short, power spectral density, PSD; [20]) from the corresponding 1-s window was saved. A 140-min sample of underwater ambient noise (i.e., sound received at this location from all sources but the signal of interest: airplane noise) was collated from before, in between, and after the overflights, and also calibrated and Fourier-transformed in 1-s Hann windows with 67% overlap.

Over all overflights, median and quartile PSD levels, one-third octave band levels, and weighted levels were computed. One-third octave band levels were obtained by integrating the PSD into bands that are 1/3 of an octave wide, then applying $10 \times \log_{10}$ [20]. One-third octave band levels were compared to published audiograms to estimate which parts of the in-air and underwater noise spectra might be audible to the two ESA-listed species (SRKW and marbled murrelet), and at what levels. Audiogram data were extracted from publications using the software program WebPlotDigitizer (Version 4.2; A. Rohatgi, Pacifica, CA, USA) if data tables were not published. The killer whale underwater audiogram followed the model proposed by Branstetter et al. (2017) [21]. In the absence of killer whale critical ratio data across the frequency band of Growler noise, one-third octave bands were used as a conservative estimate (see Figure 4A in Erbe et al. 2016) [22]. There is no audiogram available for marbled murrelet, so audiograms of other seabirds were used as surrogates. In-air and underwater audiograms for cormorant (*Phalacrocorax carbo sinensis*) were measured by Johansen et al. (2016) [23], the in-air audiogram for the lesser scaup (*Aythya affinis*) duck by Crowell et al. (2016) [24], and the in-air audiograms for common murre (*Uria aalge*) and Atlantic puffin (*Fratercula arctica*) by Mooney et al. (2019) [25]. One-third octave bands were also used for the birds in air and under water [24,26]. We report A-weighted levels for humans as well as audiogram-weighted levels for the animals in air and under water. Audiogram-weighting involved filtering the sound spectrum by the animal audiogram prior to integration over frequency. In praxis, the audiogram was interpolated to 1-Hz resolution for comparison to the noise spectrum, also in 1-Hz resolution. Over the range of frequencies where the noise PSD levels exceeded the audiogram levels, the audiogram levels were subtracted from the noise PSD levels at each frequency, yielding differences in dB at each frequency. Differences were converted to linear quantities (by applying $10^{\wedge}(\text{level}/10)$), which were then integrated over frequency, and the result was converted to a level-quantity (by taking $10 \times \log_{10}$), yielding the audiogram-weighted level in dBth.

To evaluate the scope of potential impact at the ecosystem level, we compared the distribution of recorded (i.e., received) levels in air and under water with thresholds of behavioral and physiological stress responses for humans and a suite of representative terrestrial and marine species. Selection of representative species, responses, and thresholds from the literature was guided by two criteria: if the species occurred in or was a reasonable surrogate for species in the Salish Sea area, and if the study used noise stimuli that was a sensible proxy (i.e., low-mid frequency, broadband) for aircraft noise. Whenever possible, studies that established or modeled a noise–dose relationship were used; in the case of modeled probability, the 50% likelihood of response was used as the threshold. Despite recognition that human-weighting of sound pressure levels is understood to be potentially unsuitable for wildlife [16,27], we found that most terrestrial studies nonetheless evaluated responses to A-weighted sound pressure levels. In addition to thresholds for people [28,29], the final suite of terrestrial species (or genus) contrasted against in-air received levels were: marbled murrelet [30], owls [31–33], harlequin duck (*Histrionicus histriónicas*) [34], and caribou (*Rangifer tarandus*) [35]. Marine species selected for contrast with underwater received levels were: killer whales [36], common murre [37], harbor porpoise (*Phocoena phocoena*) [38], herring (*Clupea harengus*) [39], and California sea lion (*Zalophus californianus*) [40].

Although our study design did not allow for comparison of underwater sound from Growlers with other surface-confirmed anthropogenic sources (in this area, primarily boats), we used LTSAs to visually represent and contrast underwater sound from Growlers and vessels. We used the weekly notifications of FCLPs (Table S2) to focus on dates and time periods (e.g., "Midmorning", "Late Afternoon") when training was scheduled for Ault Field, and created LTSAs for these periods (1-s and 1-Hz resolutions) to identify periods when both Growler noise and vessel noise were present. Three 1-h LTSAs were generated to visualize the underwater soundscape under varying flight and vessel activity.

2.5. Comparison of Sound Levels and Flight Activity with Prior Studies

To place sound levels and flight activity at NASWI in the context of those documented in other studies, we conducted a literature review to identify studies of impacts of military low-altitude flights (MLAF) on people and wildlife. We restricted our search to these studies because the noise strength, onset rate, and intermittent nature of MLAF are distinct from commercial or general aviation aircraft [7,41]. In particular, comparable environmental noise levels (>100 dBA) are encountered only rarely in other contexts [3]. Our initial search resulted in 26 primary research articles that evaluated impacts of MLAF on people or communities (i.e., annoyance, hearing damage or loss, and effects on mental and physical health), and 34 articles that examined impacts on wildlife (Data S1). A subset was removed before extracting noise data; reasons for exclusion included inability to obtain full articles, reporting of events only (vs. noise), or non-relevant context (e.g., air shows) (Data S1); some studies also had multiple publications related to the same dataset (Table S1). The final number of studies from which noise metrics were extracted was 12 (people) and 18 (wildlife) (Table S1). The number of studies that have measured or modeled impacts of underwater noise from aircraft was too low for meaningful analysis, and included studies therefore only reflect in-air conditions.

From each study, three metrics were extracted or estimated: (1) maximum received sound level, (2) typical or average number of daily events > 100 dBA, and (3) total daily duration in seconds > 100 dBA (Table S1). If a typical number of daily MLAF events was not reported, we calculated the average number of daily events as the total reported events divided by the number of days when recording took place; since military activity usually occurs almost exclusively on weekdays, weekend days were excluded from this formulation. A threshold of 100 dBA was used because it was relevant to the current study and is frequently used as a reporting threshold, facilitating the extraction of metrics across disparate studies that could include events both below and above that threshold.

The region of study, year in which the study was conducted, and (for wildlife) the focal taxonomic group were also extracted. If a study included multiple noise treatments or examined geographic areas with different received levels, metrics were extracted for each treatment or geographic area. If the duration of individual flight events was not reported, a conservative estimate of 4 s per overflight event (based on mean reported event duration across all field studies) was used (Table S1). The same metrics were then calculated for the current study using the sound pressure level data and observed flight events from September 13 and 16. These two dates do not represent maximum daily periods with FCLPs, which is up to 8 time periods per day, but typical and moderate activity on training days (Table S2). The relative positions of different studies with respect to the three metrics were contrasted separately for people and wildlife.

3. Results

3.1. Sound Levels, Audibility, and Response Thresholds

The waveform and spectrum of an example overflight recorded in air are shown in Figure 2a. Broadband levels (20 Hz–20 kHz) exceeded 117 dB re 20 µPa for about 1 s in this example.

Averaged over all 23 overflights, the received level was 110 ± 4 dB re 20 µPa rms and 107 ± 5 dBA; maximum received levels were 119 dB re 20 µPa and 118 dBA. In-air noise covered a frequency band from 20 Hz to greater than 10 kHz, peaking between 50 Hz and 1 kHz (Figure 3a). Comparing $^1/_3$ octave band levels with audiograms indicated that in-air noise from Growlers would be audible to all species within the limits of the audiogram measurements available, which ranged from a minimum of 250 Hz for cormorants to a maximum of 8 kHz for ducks (Figure 3b). Audiogram-weighted levels suggested that murre might experience less disturbance (18–28 dBth) from Growlers compared with puffins (60–65 dBth), cormorants (65–71 dBth), and ducks (81–88 dBth; Table 1). A-weighted noise levels experienced by people ranged from 104 to 109 dBA.

The waveform and spectrum of an example overflight recorded under water are shown in Figure 2b. Broadband levels exceeded 131 dB re 1 µPa for about 1 s in this example. Averaged over the 10 overflights, the received level in the strongest 1-s window was 134 ± 3 dB re 1 µPa rms. The underwater noise recorded during the 10 overflights covered a frequency band from 20 Hz to 30 kHz, peaking between 200 Hz and 1 kHz (Figure 4a). Based on intersection with audiograms, Growler noise penetrating the water was expected to be audible to killer whales between 200 Hz and 40 kHz, and to cormorants between 1 kHz and 4 kHz (Figure 4b). Audiogram-weighted levels indicated Growler flights would result in 48–56 dBth of noise for killer whales, and 40–44 dBth for cormorants (Table 1).

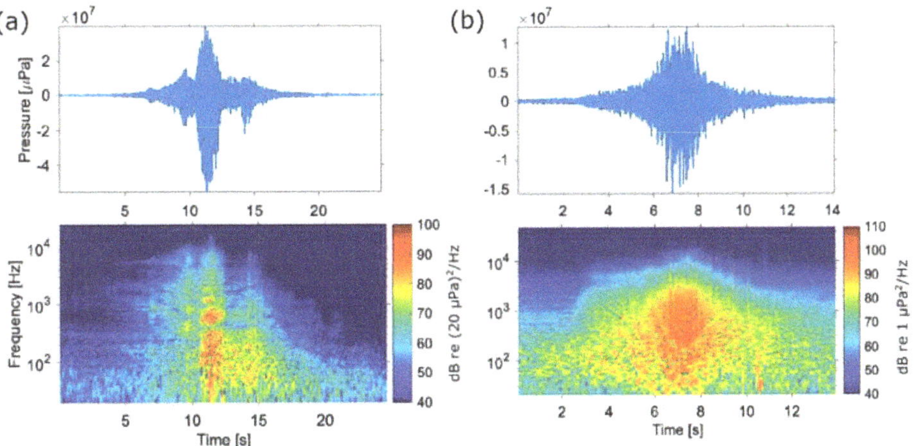

Figure 2. Waveform (top) and spectrogram (bottom) of (**a**) a Growler overflight recorded in air (fs = 48 kHz, NFFT = 12,000, 50% overlap) and (**b**) a Growler overflight recorded under water (fs = 96 kHz, NFFT = 24,000, 50% overlap).

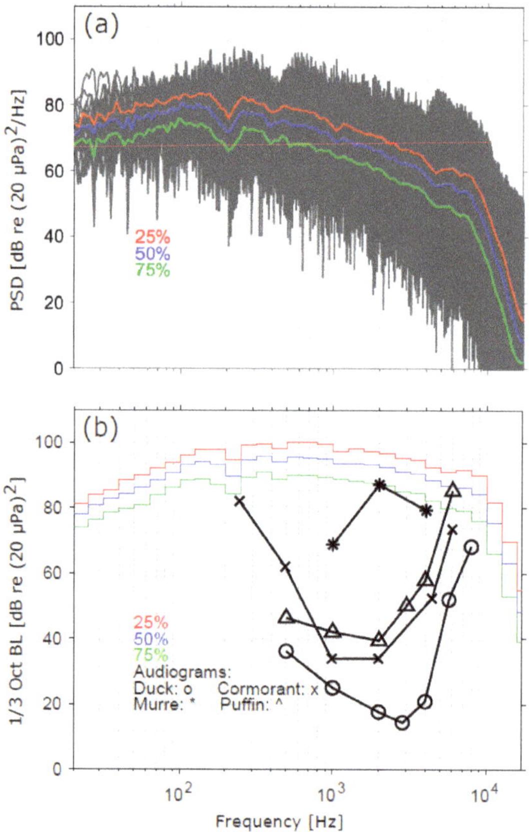

Figure 3. In-air (**a**) received power spectral density (PSD) from 23 overflights (grey), and median (blue) and quartile (red and green) levels. (**b**) One-third octave band levels (median and quartiles; blue, red, and green, respectively) are compared to the in-air audiograms of cormorants, ducks (i.e., lesser scaup), murres, and puffins. Noise above the audiogram lines is expected to be audible.

Table 1. Median and quartile audiogram-weighted levels (dBth) for killer whales and seabirds, and A-weighted levels for humans.

	Under Water			In Air			
	Orca whale (dBth)	Cormorant (dBth)	Duck (dBth)	Cormorant (dBth)	Murre (dBth)	Puffin (dBth)	Human (dBA)
25%	56	44	88	71	28	65	109
50%	54	42	84	69	25	63	107
75%	48	40	81	65	18	60	104

When compared with thresholds of behavioral and physiological stress responses in humans and a suite of terrestrial wildlife (i.e., terrestrial birds and mammals), we found that in-air received levels exceeded all identified thresholds (Figure 5a). Underwater received levels exceeded thresholds of startle response for common murre and avoidance by killer whales. The strongest received levels exceeded the threshold of startle response for herring and harbor porpoise, but were below those associated with avoidance in California sea lions (Figure 5b).

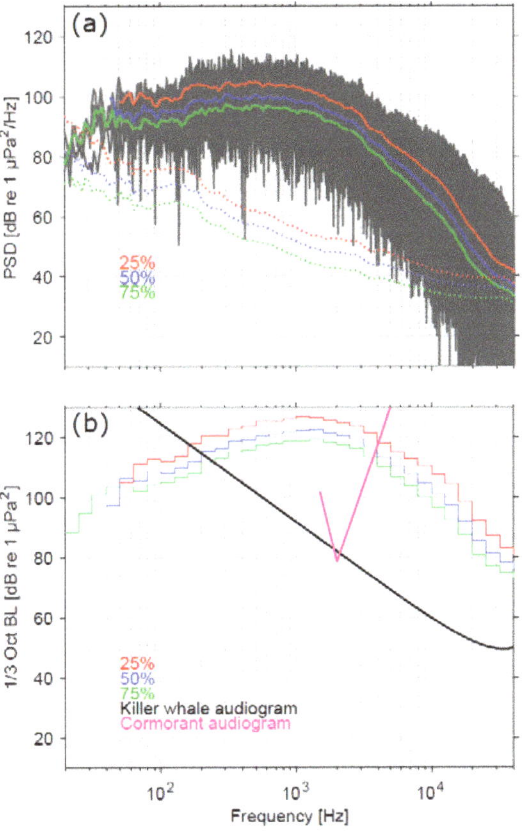

Figure 4. Underwater (**a**) received power spectral density (PSD) from 10 Growler overflights (grey), with median (blue) and quartile (red and green) levels. Ambient noise percentiles at the time of recording are shown in dotted curves. (**b**) One-third octave band levels (median and quartiles; blue, red, and green, respectively) shown together with the killer whale (black) and cormorant (pink) underwater audiograms. Noise above the audiogram lines is expected to be audible.

LTSAs of in-air recordings show the pattern of FCLPs as 30–60 min periods of rapid consecutive flights interspersed with shorter intervals of reduced or no flights (Figure S1). Underwater noise was detected on multiple dates and time periods when FCLPs were scheduled, with the same pattern of clustered activity (Figure 6a,b). Visual contrasts of underwater noise from FCLPs and routine takeoffs show the unique characteristics of sound from Growlers compared to vessels, and that received levels from Growlers are likely to exceed those associated with a range of typical vessel noise (Figure 6a–c).

3.2. Comparison of Sound Levels and Flight Activity with Prior Studies

On September 13, 185 landings and overhead passes of aircraft at the north end of Ault Field occurred between 1100 and 1500. Of these, all but two (1 Boeing 737 and 1 DC-9) were EA-18G Growlers engaged in FCLPs. The majority of overhead passes were a single aircraft, but passes with up to three aircraft simultaneously were observed. Seventeen events of Growlers taking off to the south were audible but not visible. On September 16, 83 passes or landings were observed during the same time period; of these, three were Boeing 737 and 10 were P-3s. The remaining 70 events were Growlers, with a maximum of two aircraft observed at any one time; 13 events of Growlers taking off were also audible but not visible.

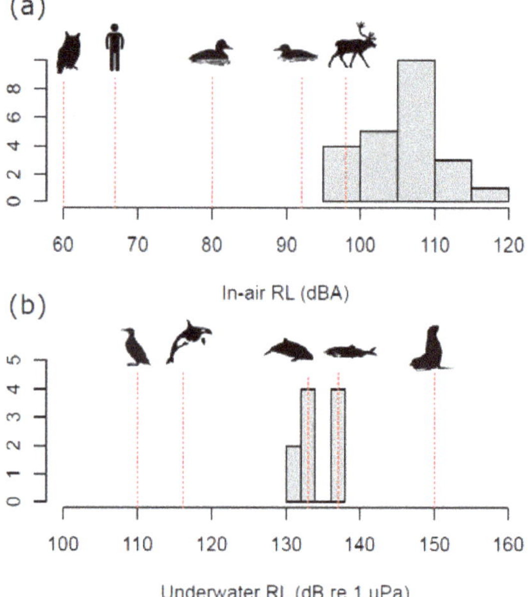

Figure 5. The distribution of received levels (RL) for (**a**) 23 in-air overflight events and (**b**) 10 flight events recorded under water relative to thresholds known to cause behavioral and physiological responses in humans and representative suites of terrestrial and marine wildlife. In-air: owls, 60 dBA = physiological stress responses [31], 50% chance of nest flushing [32], and 50% reduction in the probability of prey detection and hunting strikes [33]; humans, 67 dBA = 50% probability of awakening at night [29] and increases in nighttime blood pressure [28]; harlequin duck, 80 dBA = reduced courtship and increased vigilance and agonism [34]; marbled murrelet, 92 dBA = risk of disturbance in nesting marbled murrelets [30]; caribou, 98 ASEL (A-weighted sound exposure level) = interrupted resting bouts and increased activity [35]. (Note: The threshold for caribou was reported in ASEL which likely overestimates RL (dBA).) Underwater: common murre, 110 dB re 1 µPa = startle response and interrupted feeding [37]; killer whales, 116 dB re 1 µPa = evading noise from small boats [36]; harbor porpoise, 133 dB re 1 µPa = 50% probability of startle response to low- and mid-frequency up/downsweeps [38]; herring, 137 dB re 1 µPa = startle response to recorded boat noise [39]; California sea lions, 150 dB re 1 µPa = 50% probability of avoidance of area with a simulated mid-frequency tactical sonar signal [40].

When the three metrics of maximum received level, daily number of events, and daily duration > 100 dBA were contrasted with those in previous studies that assessed impacts of MLAF, the combined sound levels and flight activity associated with FCLPs exceeded those in most other studies (Figure 7). In studies related to people, some documented louder maximum received levels, but with fewer events and cumulative daily durations (Figure 7a). Similarly, cumulative daily duration was substantially exceeded in only one previous study; however, the received levels were lower (110 vs. 118 dBA). Overall, the sound levels and flight activity we describe in this study bear the strongest similarity to the most extreme areas around airfields on Okinawa, which were measured opportunistically between 1968 and 1972, and then systematically in 1998 (Table S1a). Contrasts with studies for wildlife show that when all three metrics are considered, sound levels and flight activity at NASWI are largely incomparable to most prior studies (Figure 7b). The taxonomic groups that have been evaluated for impacts of MLAF include ungulates (caribou, sheep, deer, and horse), one species of raptor, four species of ducks, two rodents, and one reptile (Table S1b).

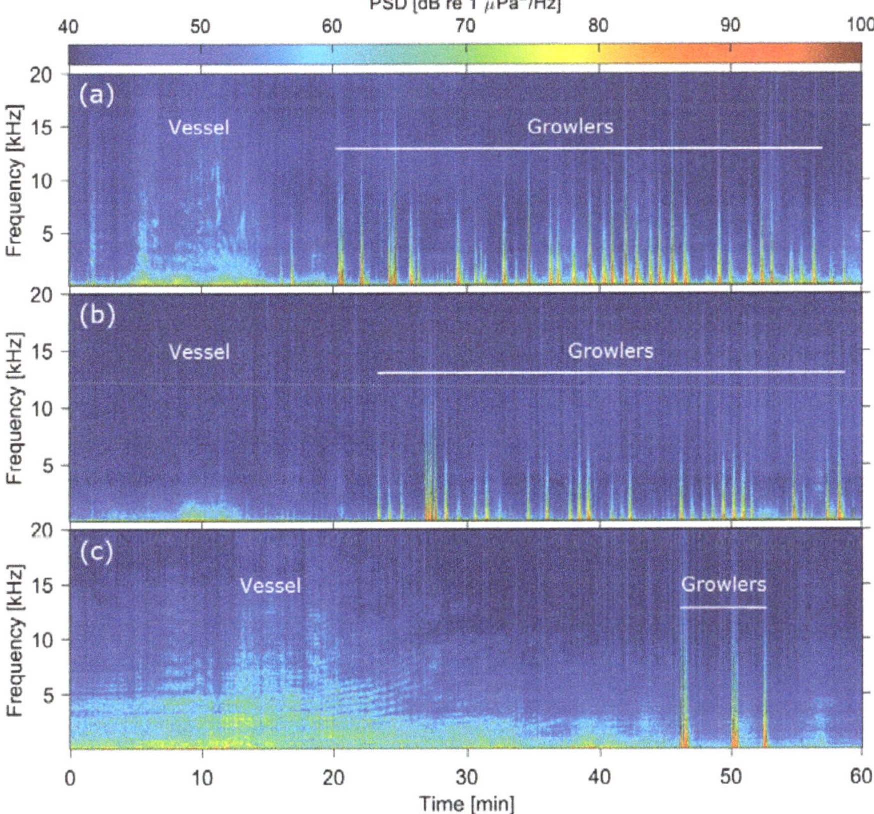

Figure 6. One-hour spectrograms (fs = 96 kHz, NFFT = 96,000, 0% overlap) contrasting underwater sound from Growlers juxtaposed with examples of vessel noise: (**a**) FCLP sessions on 20 August (0940–1040), (**b**) FCLP sessions on August 27 (1930–2030) and (**c**) typical clusters of consecutive takeoffs (2–3 Growlers per cluster) on September 4 (0915–1015).

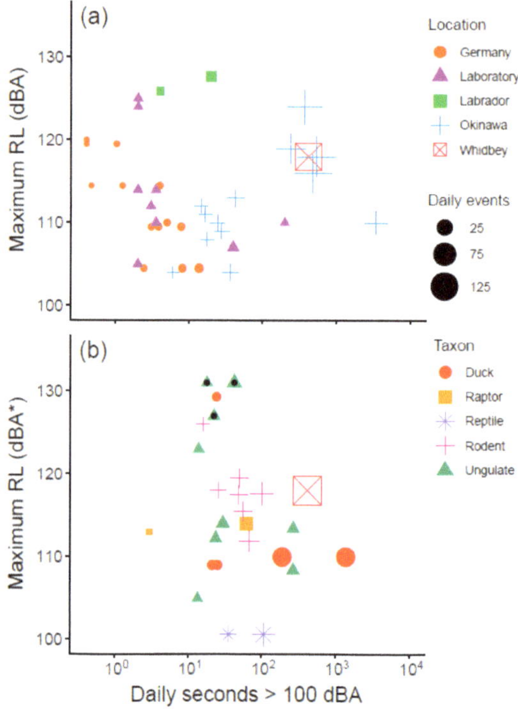

Figure 7. Maximum received level (RL), cumulative daily duration in seconds > 100 dBA, and the daily number of events in the current study (Whidbey) and previous studies of impacts of military low-altitude flights (Table S1) related to (**a**) people and communities and (**b**) wildlife. Colored symbols reflect (**a**) geographic region and (**b**) broad taxonomic group, with the daily number of events represented by the size of the symbol. Maximum received levels in studies were typically reported as A-weighted sound pressure levels (dBA) or dBA could be calculated, (*) with the exception of 3 data points from wildlife studies (black dot inside symbol) that exclusively reported either C-weighted sound pressure levels or A-weighted sound exposure level (Table S1); the exceptions were included as these metrics are expected to overestimate RL (dBA).

4. Discussion

In this study, we measured noise from an infrequently studied source of MLAF, operating in close proximity to residential sites, recreational areas, and habitat for multiple sensitive marine species. Our goal was to evaluate potential impacts on people and wildlife, using thresholds of response that have been established in previous studies. We measured sound both in air and under water and compared received levels with species-specific audiograms to demonstrate the extent to which noise from Growlers is perceived by sensitive wildlife. We also place the measured noise (i.e., received levels, total daily duration, and number of events) in the context of studies that have assessed impacts of MLAF on people and wildlife. By adopting this integrated approach, our study is uniquely positioned to illustrate knowledge gaps that can undermine assessment of noise impacts.

When we considered noise as a totality of received level, frequency of events, and total daily duration, the sound levels and flight events exceeded those in most previous studies. This finding is critical because it indicates that assessments of impact (e.g., the EIS) are, by definition, based largely on studies that have evaluated responses of people and wildlife to fewer and quieter MLAF events. To find where similar sound levels are experienced by people, we would have to turn to industrial and occupational noise studies, including those for military personnel (e.g., [42]). However, extrapolating from these studies to a community noise context is largely inappropriate given differences in the type and duration of exposure as well as occupational regulations such as time exposure limits, use of hearing protection, and testing [43].

In our review of MLAF studies for people, we found that comparable community or environmental noise has been studied in only one other region of the world, on Okinawa Island, Japan. From World War II until 1998, Okinawa Island had 39 U.S. military facilities (today there are 28), including two major bases of Kadena Air Base and Futenma Air Station. Noise from aircraft was measured opportunistically around these bases in 1968 and 1972 (during the Vietnam War), but were not measured systematically until 1998, when a multi-year study evaluated consequences for health and well-being. The study was launched because at the time it was estimated that 38% of Okinawa's population were living in conditions that exceeded the national standards for exposure to aircraft noise [6]. It is notable that the sound levels and flight activity we document around Whidbey Island are similar to Okinawan measurements during the Vietnam War, prior to passage and adoption of national noise regulations by Japan's Environment Agency in 1973 and the Defense Facilities Administration Agency in 1980.

The same trend was apparent when we compared the maximum received level, number of events, and total daily duration with studies related to wildlife. Only one wildlife study, conducted in a laboratory, exceeded the total daily duration that we measured. Although some studies evaluated exposure to stronger maximum received levels, cumulative daily duration was less. For example, one of the most comprehensive assessments conducted by Goudie and Jones (2004) examined behavioral responses of harlequin ducks to MLAF, finding reduced courtship and increased agonism at a threshold of 80 dBA, with recovery requiring about two hours [34,44]. Although Goudie and Jones (2004) recorded a higher maximum sound pressure level, the typical number of daily events was just 3% of the number we document in the current study. This example illustrates the difficulty in assessing impacts of increased Growler training on area wildlife. We face not only general research limitations in how noise impacts communication, behavior, foraging, and ultimately fitness of wildlife [8,16] but an added burden of extrapolating to a number of events, received levels, and cumulative daily duration that is largely unstudied.

We considered carefully whether sampling decisions or assumptions made during analysis could have inflated the number of events, received levels, or cumulative daily duration. Recordings of in-air noise were done over two days, which could be considered non-representative (e.g., if an extreme number of events were recorded). However, when we compiled the FCLP schedule for the past 4 years, we found that our sampling days, with two published active time frames, represented typical and moderate training activity for a single day. The number of flight events and received levels we recorded were also consistent with previous monitoring of Growlers on Whidbey Island. Between 2013 and 2019, noise and events from FCLPs have been measured periodically at 12 locations around Coupeville OLF. The daily number of flight events associated with FCLPs ranged from 69 to 239, easily encompassing our calculated average of 127 events per training day [45–48]. Similarly, maximum sound levels in previous sampling ranged from 97 to 121 dBA (depending on distance from the flight track in use), a range that also encompasses our maximum of 118 dBA.

Knowledge of the consequences for health and well-being of people experiencing these numbers of flight events and cumulative daily duration of noise exposure is necessarily limited, given that similar conditions have been rarely studied. The investigations by the Okinawa Prefecture in the mid-1990s offer the best available information on implications for public health. One of these associated studies found that noise from aircraft around the Kadena airbase was hazardous and sufficient to cause hearing loss among the population as a whole [49], while an epidemiological study identified individuals with noise-induced hearing loss that was likely due to living in proximity to the base [6,50]. In a different region of the world, Finnish Air Force investigations found that two claims of hearing loss from MLAF events were plausible based on measured exposures [51]. A unique laboratory study examining temporary threshold shift and stapedius reflex period concluded that 114 dBA (below our measured maximum) was a critical threshold, where repeated exposure to military aircraft noise above this was likely to result in noise-induced hearing loss [52].

Studies have also documented consequences for cardiovascular health within the noise exposures that we measured. Clear dose-response relationships existed between blood pressure and aircraft noise surrounding Kadena and Futenma airbases, with noise-exposed groups exhibiting a 30% increase over control groups [6]; risk of hypertension due to noise exposure was highest for older age groups [53]. Laboratory studies have demonstrated short-term increases in blood pressure following exposures to military aircraft noise, with suggested response thresholds ranging from 90 [54] to 106 dBA [55]. Other aspects of human health and well-being including annoyance, sleep disturbance, resident dissatisfaction, and even low birth weights have been studied and associated with high-intensity exposure specifically from military aircraft [6,41,56–58] as well as at lower levels of community noise from civilian aircraft and other sources [4,59]. As a result of reviewing these and other studies, in 2017 the Washington State Department of Health recommended that the U.S. Navy conduct a health impact assessment as part of the EIS process [60]. Our results confirm the need for such an assessment to occur.

The strength of the noise from flight events resulted in another critical finding from this study, where we document sound from Growlers 30 m below the sea surface, and at levels known to trigger behavioral changes for aquatic wildlife. Sound levels between the hydrophone and the surface may have been stronger than those we measured (though complex noise fields arise, particularly in shallow water (see, e.g., Figure 2i–l in Erbe et al. 2017) [61]. For Endangered SRKW, received levels were above those associated with changes in call amplitude [62,63] and avoidance or changes in behavior [36]. Other Salish Sea marine mammals that have been shown to react with avoidance or startle responses to low-mid frequency sound in this range include harbor porpoise [38,64,65], harbor seals (*Phoca vitulina*) [66], and gray whales (*Eschrichtius robustus*) [67]. At lower levels, communication masking has been demonstrated for bottlenose dolphins (*Tursiops aduncus*) [68], and is likely for other species such as humpback whales (*Megaptera novaeangliae*) [69]. Although the number of studies that have looked at behavioral responses of fish to noise is small, our received levels overlapped or exceeded thresholds for herring and some other marine fish including sea bass (*Dicentrarchus labrax*) [39,70]. Studies of how noise may be perceived by and impact seabirds underwater are almost nonexistent. In 2011, an expert panel was convened to establish underwater thresholds for injury to Threatened marbled murrelet from pile driving noise, but injury was defined to include only permanent loss of cochlear hair cells or barotrauma [71]. However, two very recent studies have demonstrated startle, avoidance, and changes in foraging of common murre and Gentoo penguins (*Pygoscelis papua*) at 105–115 dB re 1 µPa, suggesting that marbled murrelet and other seabirds around Whidbey Island may be impacted by underwater (and in-air) noise from jet aircraft [37,72].

Our results indicate underwater impacts that have been unstudied, underestimated, or otherwise dismissed in the two relevant EIS [17,73] and corresponding Biological Opinion(s) for ESA-listed species [19,30]. Of chief concern in this region is SRKW. In the EIS, underwater noise from aircraft were deemed unlikely to adversely affect SRKW (and humpback whales), and to have no effect on critical habitat. The rationale for this conclusion included assertions that whales would have to be at the surface of the water and directly underneath low-altitude aircraft (<300 m), and that whales were already exposed to boat and ship noise that could "drown out or lessen" any noise from aircraft. Our results indicate instead that noise from Growlers is measurable at least 30 m under water, with sound levels known to impact whales. Furthermore, these sound levels are comparable to those documented by studies of noise that is experienced by SRKW from small and large vessels [74,75]. The reason that no effects on SRKW critical habitat are assumed is not due to evaluation of noise impacts, but rather to an exemption of waters within the boundaries of military installations from critical habitat designation (see Appendix C, Sections 4.1 and 3.2.5 in [17]). Lastly, the rationale for not considering impacts of aircraft flying higher than 300 m is based on an agreement with the National Marine Fisheries Service in 2015 to assume that underwater noise from any event where aircraft exceed this altitude will cause no reaction in marine mammals (see Appendix C and Section 4.1 in [17]). This is despite the fact that modeled underwater noise for aircraft at altitudes of 300–3000 m is 128–152 dB re 1 µPa (see Table 3.0–4 in [73]), exceeding known thresholds for behavioral reactions and adverse impacts on marine mammals, including SRKW (Figure 5). A recent synthesis of underwater noise and vessel disturbance on SRKW by the Washington State Academy of Sciences recommended "defining every interaction with an SRKW as an opportunity to disturb a whale", due to the "fragile condition" of the population [76]. Collectively, we believe that our results create a case for revisiting these impact assessments as well as future inclusion of military aircraft noise in cumulative effects models for SRKW [77].

Evaluating the EIS process against our results further illuminates a problem wherein risk from noise effects is calculated based on the likelihood that individuals (e.g., an individual SRKW or marbled murrelet group) will be exposed to sound levels that result in physical damage (i.e., hearing damage or barotrauma) or direct changes in behavior such as foraging, breeding, or nesting. This does not account for the problem that the habitat itself is being impacted by noise, and becomes less hospitable [78], nor that noise may be added to other stressors [79,80]. In this scenario, the very rarity of the species becomes a factor in assuming impacts are discountable, negligible, or insignificant (see Appendix C and Section 4.1 in [17]). Our purpose in outlining these inconsistencies in environmental impact assessment is not to point fingers at federal oversight agencies or the U.S. military, but to exemplify how knowledge gaps [81], exemptions [5], and use of high noise thresholds for harm intersect to discount or underestimate noise impacts on wildlife, which are increasingly understood to include indirect effects on habitat, abundance and fitness of populations [16,82].

The above challenges are part of an evolving understanding of how to evaluate and mitigate growing noise pollution worldwide. However, our study reveals added challenges specific to noise from MLAF, which is a scarcity of studies resulting in large knowledge gaps with respect to impact. There are substantial logistical and bureaucratic hurdles in monitoring military operations; these have been pointed out in reviews [3,9] as well as experienced by the authors of this study. In particular, the fact that schedules are usually not available in advance and access to operational areas may be restricted increases time and costs of doing these studies. Despite the challenges, there is a strong need to close knowledge gaps, as increased noise from MLAF is predicted to become more common in the future due to base consolidation [2]. Other countries (i.e., Finland and Australia) have recently adopted the Growler platform and may find similar issues in locating training facilities. The problem is not limited to Growler aircraft; the new F-35 is also causing similar concerns and discomfort in areas around airfields [83,84]. And while the trend within the U.S. is toward consolidation, the building of new bases and the expansion of military aircraft activity continues worldwide [41,58,85].

In summary, our study suggests the need for underwater noise from Growlers to be included in cumulative effects models [77] and Biological Opinions for ESA-listed species [19,30], as well as more broadly evaluated outside of the immediate vicinity of Whidbey Island. Furthermore, our results show that sound levels and flight operations around NASWI are largely beyond those that have been previously evaluated, supporting calls for a comprehensive health assessment to evaluate consequences for human health and well-being [60]. Finally, we hope that this study stimulates consideration of how to evaluate impacts of intense noise exposure not only for the benefit of this region, but other areas that may face similar challenges now and in the future.

Supplementary Materials: The following are available online at http://www.mdpi.com/2077-1312/8/11/923/s1, Figure S1: LTSAs of in-air recordings, Table S1: Noise metrics and attributes extracted from MLAF studies, Table S2: Summary of FCLP days and time periods, 2015–2019, Data S1: Results of MLAF literature review.

Author Contributions: R.W., E.A. and L.M.K. conceived and designed the study and data collection; R.W., L.T.B., M.S.C. and L.M.K. collected the data; L.M.K. conducted the literature review; C.E., R.W., L.M.K. and M.S.C. analyzed the data; R.W., E.A., L.M.K. and C.E. contributed equipment and analytical tools and software; L.M.K., C.E. and R.W. wrote the paper; all authors contributed to editing and final reviews of the paper. All authors have read and agreed to the published version of the manuscript.

Funding: Multiple small grants and sources of funding made this work possible. The National Parks Conservation Association contributed funds and led a campaign to allow individual donors to contribute to the project. A project grant was also awarded from The Suquamish Foundation (Appendix X Award 2018Q226). The funders had no role in study design, data collection and analysis, decision to publish, or preparation of the manuscript.

Acknowledgments: We are grateful for the assistance of several volunteers on this project and manuscript. Heather McCauliffe loaned two Extech sound pressure data loggers. Kimberly Nielsen assisted with the creation of Figure 1, and Toby Hall assisted with deployment and retrieval of the SoundTrap. The manuscript was substantially improved by comments from two anonymous reviewers. Lastly, Rob Williams thanks and acknowledges the Pew Fellows Program in Marine Conservation for support.

Conflicts of Interest: The authors declare no conflict of interest.

References

1. US Department of the Navy. *Record of Decision for the Final Environmental Impact Statement for EA-18G "Growler" Airfield Operations at Naval Air Station Whidbey Island Complex, Island County, WA*; United States Department of the Navy: Washington, DC, USA, 2019.
2. Waitz, I.A.; Lukachko, S.P.; Lee, J.J. Military aviation and the environment: Historical trends and comparison to civil aviation. *J. Aircr.* **2005**, *42*, 329–339. [CrossRef]
3. Pepper, C.B.; Nascarella, M.A.; Kendall, R.J. A review of the effects of aircraft noise on wildlife and humans, current control mechanisms, and the need for further study. *Environ. Manag.* **2003**, *32*, 418–432. [CrossRef]
4. Basner, M.; Clark, C.; Hansell, A.; Hileman, J.I.; Janssen, S.; Shepherd, K.; Sparrow, V. Aviation noise impacts: State of the science. *Noise Health* **2017**, *19*, 41–50.
5. Truban, E. Military exemptions from environmental regulations: Unwarranted special treatment or necessary relief. *Villanova Environ. Law J.* **2004**, *15*, 139–171.
6. Okinawa Prefectural Government. *A Report on the Aircraft Noise as a Public Health Problem in Okinawa*; Department of Culture and Environmental Affairs: Okinawa, Japan, 1999.
7. Kerry, G.; Weeler, P.D.; Hempstock, T.I.; James, D.J. Impulse noise metrics and their application to noise from low flying military jet aircraft. *J. Acoust. Soc. Am.* **1998**, *103*, 2800. [CrossRef]
8. Shannon, G.; McKenna, M.F.; Angeloni, L.M.; Crooks, K.R.; Fristrup, K.M.; Brown, E.; Warner, K.A.; Nelson, M.D.; White, C.; Briggs, J.; et al. A synthesis of two decades of research documenting the effects of noise on wildlife: Effects of anthropogenic noise on wildlife. *Biol. Rev.* **2016**, *91*, 982–1005. [CrossRef] [PubMed]
9. Efroymson, R.A.; Suter, G.W., II. Ecological risk assessment framework for low-altitude aircraft overflights: II. Estimating effects on wildlife. *Risk Anal.* **2001**, *21*, 263–274. [CrossRef] [PubMed]
10. Kuehne, L.; Olden, J. Military flights threaten the wilderness soundscapes of the Olympic Peninsula, Washington. *Northwest Sci.* **2020**, *94*, 188–202.
11. Erbe, C.; Williams, R.; Parsons, M.; Parsons, S.K.; Hendrawan, I.G.; Dewantama, I.M.I. Underwater noise from airplanes: An overlooked source of ocean noise. *Mar. Pollut. Bull.* **2018**, *137*, 656–661. [CrossRef]
12. Williams, R.; Erbe, C.; Dewantama, I.M.I.; Hendrawan, I.G. Effect on ocean noise: Nyepi, a Balinese day of silence. *Oceanography* **2018**, *31*, 16–18. [CrossRef]
13. Eller, A.I.; Cavanagh, R.C. *Subsonic Aircraft Noise at and Beneath the Ocean Surface: Estimation of Risk for Effects on Marine Mammals*; Science Applications International Corporation: McLean, VA, USA, 2000.
14. Ashe, E.; Noren, D.P.; Williams, R. Animal behaviour and marine protected areas: Incorporating behavioural data into the selection of marine protected areas for an endangered killer whale population. *Anim. Conserv.* **2010**, *13*, 196–203. [CrossRef]
15. Kunc, H.P.; Schmidt, R. The effects of anthropogenic noise on animals: A meta-analysis. *Biol. Lett.* **2019**, *15*, 20190649. [CrossRef] [PubMed]
16. Francis, C.D.; Barber, J.R. A framework for understanding noise impacts on wildlife: An urgent conservation priority. *Front. Ecol. Environ.* **2013**, *11*, 305–313. [CrossRef]
17. US Department of the Navy. *Final Environmental Impact Statement for EA-18G Growler Airfield Operations at Naval Air Station Whidbey Island Complex*; United States Department of the Navy: Washington, DC, USA, 2018.
18. US Department of the Navy. Flight Operations Notification for NAS Whidbey Island Complex. Available online: https://www.cnic.navy.mil/regions/cnrnw/installations/nas_whidbey_island/news/news_releases/field-carrier-landing-practice-at-nas-whidbey-island-complex-for.html (accessed on 20 June 2020).
19. US Fish and Wildlife Service. *Biological Opinion: Naval Air Station Whidbey Island Complex EA-18G "Growler" Airfield Operations Project*; United States Fish and Wildlife Service: Washington, DC, USA, 2018.

20. International Organization for Standardization. *International ISO Standard 18405, Underwater Acoustics—Terminology*, 1st ed.; International Organization for Standardization: Geneva, Switzerland, 2017.

21. Branstetter, B.K.; Leger, J.S.; Acton, D.; Stewart, J.; Houser, D.; Finneran, J.J.; Jenkins, K. Killer whale (*Orcinus orca*) behavioral audiograms. *J. Acoust. Soc. Am.* **2017**, *141*, 2387–2398. [CrossRef]

22. Erbe, C.; Reichmuth, C.; Cunningham, K.; Lucke, K.; Dooling, R. Communication masking in marine mammals: A review and research strategy. *Mar. Pollut. Bull.* **2016**, *103*, 15–38. [CrossRef] [PubMed]

23. Johansen, S.; Larsen, O.N.; Christensen-Dalsgaard, J.; Seidelin, L.; Huulvej, T.; Jensen, K.; Lunneryd, S.-G.; Boström, M.; Wahlberg, M. In-air and underwater hearing in the great cormorant (*Phalacrocorax carbo* sinensis). In *Effects of Noise on Aquatic Life II*; Springer: Berlin/Heidelberg, Germany, 2016; pp. 505–512.

24. Crowell, S.E.; Wells-Berlin, A.M.; Therrien, R.E.; Yannuzzi, S.E.; Carr, C.E. In-air hearing of a diving duck: A comparison of psychoacoustic and auditory brainstem response thresholds. *J. Acoust. Soc. Am.* **2016**, *139*, 3001–3008. [CrossRef]

25. Mooney, T.A.; Smith, A.; Larsen, O.N.; Hansen, K.A.; Wahlberg, M.; Rasmussen, M.H. Field-based hearing measurements of two seabird species. *J. Exp. Biol.* **2019**, *222*. [CrossRef]

26. Okanoya, K.; Dooling, R.J. Hearing in passerine and psittacine birds: A comparative study of absolute and masked auditory thresholds. *J. Comp. Psychol.* **1987**, *101*, 7. [CrossRef]

27. McKenna, M.F.; Shannon, G.; Fristrup, K. Characterizing anthropogenic noise to improve understanding and management of impacts to wildlife. *Endanger Species Res.* **2016**, *31*, 279–291. [CrossRef]

28. Haralabidis, A.S.; Dimakopoulou, K.; Vigna-Taglianti, F.; Giampaolo, M.; Borgini, A.; Dudley, M.-L.; Pershagen, G.; Bluhm, G.; Houthuijs, D.; Babisch, W. Acute effects of night-time noise exposure on blood pressure in populations living near airports. *Eur. Heart J.* **2008**, *29*, 658–664. [CrossRef]

29. Basner, M.; Buess, H.; Mueller, U.; Plath, G.; Samel, A. Aircraft noise effects on sleep: Final results of DLR laboratory and field studies of 2240 polysomnographically recorded subject nights. In Proceedings of the 33rd International Congress and Exposition on Noise Control Engineering, Prague, Czech Republic, 22–25 August 2004; pp. 22–25.

30. US Fish and Wildlife Service. *Biological Opinion: Navy's Northwest Training and Testing Activities*; United States Fish and Wildlife Service: Washington, DC, USA, 2016.

31. Hayward, L.S.; Bowles, A.E.; Ha, J.C.; Wasser, S.K. Impacts of acute and long-term vehicle exposure on physiology and reproductive success of the northern spotted owl. *Ecosphere* **2011**, *2*, 1–20. [CrossRef]

32. Delaney, D.K.; Grubb, T.G.; Beier, P.; Pater, L.L.; Reiser, M.H. Effects of helicopter noise on Mexican spotted owls. *J. Wildl. Manag.* **1999**, *63*, 60–76. [CrossRef]

33. Mason, J.T.; McClure, C.J.; Barber, J.R. Anthropogenic noise impairs owl hunting behavior. *Biol. Conserv.* **2016**, *199*, 29–32. [CrossRef]

34. Goudie, R.I.; Jones, I.L. Dose-response relationships of harlequin duck behaviour to noise from low level military jet over-flights in central Labrador. *Biol. Conserv.* **2004**, *31*, 289–298. [CrossRef]

35. Maier, J.A.; Murphy, S.M.; White, R.G.; Smith, M.D. Responses of caribou to overflights by low-altitude jet aircraft. *J. Wildl. Manag.* **1998**, *62*, 752–766. [CrossRef]

36. Williams, R.; Bain, D.E.; Ford, J.K.; Trites, A.W. Behavioural responses of male killer whales to a 'leapfrogging' vessel. *J. Cetacean Res. Manag.* **2002**, *4*, 305–310.

37. Hansen, K.A.; Hernandez, A.; Mooney, T.A.; Rasmussen, M.H.; Sørensen, K.; Wahlberg, M. The common murre (*Uria aalge*), an auk seabird, reacts to underwater sound. *J. Acoust. Soc. Am.* **2020**, *147*, 4069–4074. [CrossRef]

38. Kastelein, R.A.; Steen, N.; Gransier, R.; Wensveen, P.J.; de Jong, C.A.F. Threshold received sound pressure levels of single 1–2 kHz and 6–7 kHz up-sweeps and down-sweeps causing startle responses in a harbor porpoise (*Phocoena phocoena*). *J. Acoust. Soc. Am.* **2012**, *131*, 2325–2333. [CrossRef] [PubMed]

39. Doksæter, L.; Handegard, N.O.; Godø, O.R.; Kvadsheim, P.H.; Nordlund, N. Behavior of captive herring exposed to naval sonar transmissions (1.0–1.6 kHz) throughout a yearly cycle. *J. Acoust. Soc. Am.* **2012**, *131*, 1632–1642. [CrossRef]

40. Houser, D.S.; Martin, S.W.; Finneran, J.J. Behavioral responses of California sea lions to mid-frequency (3250–3450 Hz) sonar signals. *Mar. Environ. Res.* **2013**, *92*, 268–278. [CrossRef]

41. Gelderblom, F.B.; Gjestland, T.T.; Granoien, I.L.; Taraldsen, G. The impact of civil versus military aircraft noise on noise annoyance. In *INTER-NOISE and NOISE-CON Congress and Conference Proceedings*; Institute of Noise Control Engineering: Reston, VA, USA, 2014; Volume 249, pp. 786–795.

42. Wu, Y.-X.; Liu, X.L.; Wang, B.-G.; Wang, X.Y. Aircraft noise-induced temporary threshold shift. *Aviat. Space Environ. Med.* **1989**, *60*, 268–270. [PubMed]

43. Yong, J.S.; Wang, D.-Y. Impact of noise on hearing in the military. *Mil. Med. Res.* **2015**, *2*, 1–6. [CrossRef] [PubMed]

44. Goudie, R.I. Multivariate behavioural response of harlequin ducks to aircraft disturbance in Labrador. *Environ. Conserv.* **2006**, *33*, 28–35. [CrossRef]

45. Lilly, J. *Whidbey Island Military Jet Noise Measurements*; JGL Acoustics, Inc.: Issaquah, WA, USA, 2013.

46. Lilly, J. *Whidbey Island Military Jet Noise Measurements*; JGL Acoustics, Inc.: Issaquah, WA, USA, 2016.

47. Lilly, J. *Military Jet Noise Measurements OLF Coupeville Whidbey Island, WA*; JGL Acoustics, Inc.: Issaquah, WA, USA, 2020.

48. Pipkin, A. *Ebey's Landing National Historical Reserve Acoustical Monitoring Report*; National Park Service: Fort Collins, CO, USA, 2016.

49. Hiramatsu, K.; Matsui, T.; Ito, A.; Miyakita, T.; Osada, Y.; Yamamoto, T. The Okinawa study: An estimation of noise-induced hearing loss on the basis of the records of aircraft noise exposure around Kadena Air Base. *J. Sound Vib.* **2004**, *277*, 617–625. [CrossRef]

50. Miyakita, T.; Yoza, T.; Matsui, T.; Ito, A.; Hiramatsu, K.; Osada, Y.; Yamamoto, T. An epidemiological study regarding the hearing acuity of residents in the area with high level of aircraft noise. *Jpn. J. Hyg.* **2001**, *56*, 577–587. [CrossRef] [PubMed]

51. Kuronen, P.; Pääkkönen, R.; Savolainen, S. Low-altitude overflights of fighters and the risk of hearing loss. *Aviat. Space Environ. Med.* **1999**, *70*, 650–655.

52. Ising, H.; Joachims, Z.; Babisch, W.; Rebentisch, E. Effects of military low-altitude flight noise Part I: Temporary threshold shift in humans. *Z. Fur Audiol.* **1999**, *38*, 118–127.

53. Matsui, T.; Uehara, T.; Miyakita, T.; Hiramatsu, K.; Yamamoto, T. Dose-response relationship between hypertension and aircraft noise exposure around Kadena airfield in Okinawa. In Proceedings of the 9th International Congress on Noise as a Public Health Problem, Foxwoods, CT, USA, 21–25 July 2008.

54. Miyazaki, M. Circulatory effect of jet noise, with special reference to cerebral circulation. *Jpn. Circ. J.* **1978**, *42*, 1019–1024. [CrossRef]

55. Ising, H.; Michalak, R. Effects of noise from military low-level flights on humans (part II. Noise as a Public Health Problem). *New Adv. Noise Res.* **1990**, *1*, 21–25.

56. Ising, H.; Rebentisch, E.; Poustka, F.; Curio, I. Annoyance and health risk caused by military low-altitude flight noise. *Int. Arch. Occup. Environ. Health* **1990**, *62*, 357–363. [CrossRef]

57. Lukas, J.S.; Dobbs, M.E.; Kryter, K.D. *Disturbance of Human Sleep by Subsonic Jet Aircraft Noise and Simulated Sonic Booms*; National Aeronautics and Space Administration: Washington, DC, USA, 1971.

58. Tokuda, Y.; Barnett, P.B. Constructing a new US Military Base: A health threat to Okinawan people. *Environ. Justice* **2017**, *10*, 23–25. [CrossRef]

59. Basner, M.; Babisch, W.; Davis, A.; Brink, M.; Clark, C.; Janssen, S.; Stansfeld, S. Auditory and non-auditory effects of noise on health. *Lancet* **2014**, *383*, 1325–1332. [CrossRef]

60. Washington State Department of Health. *Comments on the Environmental Impact Statement for EA-18G Growler Airfield Operations at Naval Air Station Whidbey Island*; Washington State Department of Health: Tumwater, VA, USA, 2017.

61. Erbe, C.; Parsons, M.; Duncan, A.; Osterrieder, S.K.; Allen, K. Aerial and underwater sound of unmanned aerial vehicles (UAV). *J. Unmanned Veh. Syst.* **2017**, *5*, 92–101. [CrossRef]

62. Holt, M.M.; Noren, D.P.; Emmons, C.K. Effects of noise levels and call types on the source levels of killer whale calls. *J. Acoust. Soc. Am.* **2011**, *130*, 3100–3106. [CrossRef] [PubMed]

63. Holt, M.M.; Noren, D.P.; Veirs, V.; Emmons, C.K.; Veirs, S. Speaking up: Killer whales (*Orcinus orca*) increase their call amplitude in response to vessel noise. *J. Acoust. Soc. Am.* **2009**, *125*, EL27–EL32. [CrossRef] [PubMed]

64. Kastelein, R. Brief behavioral response threshold levels of a harbor porpoise (*Phocoena phocoena*) to five helicopter dipping sonar signals (1.33 to 1.43 kHz). *Aquat. Mamm.* **2013**, *39*, 162–173. [CrossRef]

65. Kastelein, R.A.; van Heerden, D.; Gransier, R.; Hoek, L. Behavioral responses of a harbor porpoise (*Phocoena phocoena*) to playbacks of broadband pile driving sounds. *Mar. Environ. Res.* **2013**, *92*, 206–214. [CrossRef]

66. Kastelein, R.A.; van der Heul, S.; Verboom, W.C.; Triesscheijn, R.J.; Jennings, N.V. The influence of underwater data transmission sounds on the displacement behaviour of captive harbour seals (*Phoca vitulina*). *Mar. Environ. Res.* **2006**, *61*, 19–39. [CrossRef]

67. Malme, C.I.; Miles, P.R.; Clark, C.W.; Tyack, P.; Bird, J.E. *Investigations of the Potential Effects of Underwater Noise from Petroleum Industry Activities on Migrating Gray Whale Behavior*; Bolt Beranek and Newman Inc.: Cambridge, MA, USA, 1983.

68. Lemon, M.; Lynch, T.P.; Cato, D.H.; Harcourt, R.G. Response of travelling bottlenose dolphins (*Tursiops aduncus*) to experimental approaches by a powerboat in Jervis Bay, New South Wales, Australia. *Biol. Conserv.* **2006**, *127*, 363–372. [CrossRef]

69. Risch, D.; Corkeron, P.J.; Ellison, W.T.; Van Parijs, S.M. Changes in humpback whale song occurrence in response to an acoustic source 200 km away. *PLoS ONE* **2012**, *7*, e29741. [CrossRef]

70. Kastelein, R.A.; Van Der Heul, S.; Verboom, W.C.; Jennings, N.; Van Der Veen, J.; de Haan, D. Startle response of captive North Sea fish species to underwater tones between 0.1 and 64 kHz. *Mar. Environ. Res.* **2008**, *65*, 369–377. [CrossRef] [PubMed]

71. Science Applications International Corporation. *Environmental Science Panel for Marbled Murrelet Underwater Noise Injury Threshold*; Science Applications International Corporation: Bothell, WA, USA, 2011.

72. Sørensen, K.; Neumann, C.; Dähne, M.; Hansen, K.A.; Wahlberg, M. Gentoo penguins (*Pygoscelis papua*) react to underwater sounds. *R. Soc. Open Sci.* **2020**, *7*, 191988. [CrossRef] [PubMed]

73. US Department of the Navy. *Northwest Training and Testing Final Supplemental Environmental Impact Statement/Overseas Environmental Impact Statement (FEIS/OEIS)*; United States Department of the Navy: Washington, DC, USA, 2020.

74. Houghton, J.; Holt, M.M.; Giles, D.A.; Hanson, M.B.; Emmons, C.K.; Hogan, J.T.; Branch, T.A.; VanBlaricom, G.R. The relationship between vessel traffic and noise levels received by killer whales (*Orcinus orca*). *PLoS ONE* **2015**, *10*, e0140119. [CrossRef] [PubMed]

75. Veirs, S.; Veirs, V.; Wood, J.D. Ship noise extends to frequencies used for echolocation by endangered killer whales. *PeerJ* **2016**, *4*, e1657. [CrossRef] [PubMed]

76. Washington State Academy of Sciences. *Summary of Key Research Findings about Underwater Noise and Vessel Disturbance*; Washington State Academy of Sciences: Seattle, WA, USA, 2020.

77. Lacy, R.C.; Williams, R.; Ashe, E.; Balcomb, K.C., III; Brent, L.J.; Clark, C.W.; Croft, D.P.; Giles, D.A.; MacDuffee, M.; Paquet, P.C. Evaluating anthropogenic threats to endangered killer whales to inform effective recovery plans. *Sci. Rep.* **2017**, *7*, 1–12. [CrossRef] [PubMed]

78. Williams, R.; Clark, C.W.; Ponirakis, D.; Ashe, E. Acoustic quality of critical habitats for three threatened whale populations. *Anim. Conserv.* **2014**, *17*, 174–185. [CrossRef]

79. Maxwell, S.M.; Hazen, E.L.; Bograd, S.J.; Halpern, B.S.; Breed, G.A.; Nickel, B.; Teutschel, N.M.; Crowder, L.B.; Benson, S.; Dutton, P.H. Cumulative human impacts on marine predators. *Nat. Commun.* **2013**, *4*, 1–9. [CrossRef]

80. Williams, R.; Thomas, L.; Ashe, E.; Clark, C.W.; Hammond, P.S. Gauging allowable harm limits to cumulative, sub-lethal effects of human activities on wildlife: A case-study approach using two whale populations. *Mar. Policy* **2016**, *70*, 58–64. [CrossRef]

81. Southall, B.L.; Finneran, J.J.; Reichmuth, C.; Nachtigall, P.E.; Ketten, D.R.; Bowles, A.E.; Ellison, W.T.; Nowacek, D.P.; Tyack, P.L. Marine mammal noise exposure criteria: Updated scientific recommendations for residual hearing effects. *Aquat. Mamm.* **2019**, *45*, 125–232. [CrossRef]

82. Williams, R.; Wright, A.J.; Ashe, E.; Blight, L.K.; Bruintjes, R.; Canessa, R.; Clark, C.W.; Cullis-Suzuki, S.; Dakin, D.T.; Erbe, C. Impacts of anthropogenic noise on marine life: Publication patterns, new discoveries, and future directions in research and management. *Ocean Coast. Manag.* **2015**, *115*, 17–24. [CrossRef]

83. Mizokami, K. The F-35 Could Make Some Neighborhoods in the U.S. Unliveable. Available online: https://www.popularmechanics.com/military/aviation/a28625774/f-35-too-loud/ (accessed on 20 September 2020).

84. Verburg, S. Residents Weary of Jet Noise Worry about F-35. Available online: https://madison.com/wsj/news/local/environment/residents-weary-of-jet-noise-worry-about-f-35/article_d1a3c79a-34bb-5f65-8093-5c3c98cecae3.html (accessed on 20 September 2020).
85. Cabestan, J.-P. China's military base in Djibouti: A microcosm of China's growing competition with the United States and new bipolarity. *J. Contemp. China* **2020**, *29*, 731–747. [CrossRef]

Publisher's Note: MDPI stays neutral with regard to jurisdictional claims in published maps and institutional affiliations.

Journal of
Marine Science and Engineering

Article

Source Levels of 20 Hz Fin Whale Notes Measured as Sound Pressure and Particle Velocity from Ocean-Bottom Seismometers in the North Atlantic

Andreia Pereira [1,*], Miriam Romagosa [2], Carlos Corela [1], Mónica A. Silva [2] and Luis Matias [1]

[1] FCUL—Campo Grande Edifício C1, Instituto Dom Luiz (IDL), University of Lisbon, Piso 1, 1749-016 Lisboa, Portugal; ccorela@fc.ul.pt (C.C.); lmmatias@fc.ul.pt (L.M.)
[2] OKEANOS—R&D Centre, Institute of Marine Research (IMAR), University of the Azores, 9900-138 Horta, Portugal; m.romagosa4@gmail.com (M.R.); monica.silva.imar@gmail.com (M.A.S.)
* Correspondence: afpereira@fc.ul.pt

Abstract: Source level is one factor that determines the effectiveness of animal signal transmissions and their acoustic communication active space. Ocean-bottom seismometers (OBS) are platforms of opportunity to monitor marine species because they record data as pressure fluctuations in the water using a hydrophone and/or as particle velocity of the seabed using a seismometer. This study estimates source levels of 20 Hz fin whale notes recorded simultaneously in these two OBS channels and in two areas of the North Atlantic (Azores and southwest Portugal). It also discusses factors contributing to the variability of the estimates, namely geographical (deployment areas), instrumental (recording channels and sample size), and temporal factors (month of detected notes, inter-note interval, and diving duration). The average source level was 196.9 dB re 1 μPa m for the seismometer (derived from particle velocity measurements) and 186.7 dB re 1 μPa m for the hydrophone. Variability was associated with sample size, instrumental characteristics, acoustic propagation, and month of recordings. Source level estimates were very consistent throughout sequences, and there was no indication of geographical differences. Understanding what causes variation in animal sound source levels provides insights into the function of sounds and helps to assess the potential effects of increasing anthropogenic noise.

Keywords: geophysical instruments; bioacoustics of marine mammals; underwater acoustic propagation; animal communication

Citation: Pereira, A.; Romagosa, M.; Corela, C.; Silva, M.A.; Matias, L. Source Levels of 20 Hz Fin Whale Notes Measured as Sound Pressure and Particle Velocity from Ocean-Bottom Seismometers in the North Atlantic. *J. Mar. Sci. Eng.* **2021**, *9*, 646. https://doi.org/10.3390/jmse9060646

Academic Editors: Michel André and Christine Erbe

Received: 7 May 2021
Accepted: 8 June 2021
Published: 10 June 2021

Publisher's Note: MDPI stays neutral with regard to jurisdictional claims in published maps and institutional affiliations.

1. Introduction

Baleen whales produce low-frequency and high-intensity sounds that can be classified as calls or songs [1–3]. Calls are normally produced irregularly and have been associated with foraging activities [4], group cohesion [5], and mother–calf bonding [6]. Songs, on the other side, are regularly repeated sequences, with a widely accepted reproductive function, either to attract females, compete with other males, or both [7,8], and are especially suited for long-range communication [9]. However, the precise significance of most baleen whale sounds is still unknown because it is difficult to correlate a sound with a behavioral response and to determine its information content [2]. The distance over which an animal sound can be detected by individuals and by recording instruments is influenced by its source level [10,11], among other factors like source depth and sound propagation speed. Source level is one major factor to determine the transmission effectiveness of a sound and the acoustic communication active space among animals, which is defined as the area over which animals can exchange information [12]. Over the past few decades, ambient noise levels in the ocean, especially below 500 Hz, have increased in some regions, which seems to be linked with global economic growth [13,14]. These sounds can potentially have a more extensive impact on animals because they can travel farther distances due to their lower attenuation [15]. Their impact can be even more extensive in low-frequency biological sounds that can also travel great distances, and therefore, they can affect the propagation of these signals. Even though long-term measurements of ambient noise levels are not available for the Atlantic Ocean, several studies suggest sound levels below 100 Hz are increasing [16,17]. Since baleen whales seem to rely so strongly on sounds for their daily activities, changes in their acoustic environment may cause masking of their sounds and affect their behavior and ultimately their survival [18,19]. A noisy environment may cause animals to produce more complex sounds and increase source levels to compensate for the potential decrease in communication active space [18]. Under high noise conditions, baleen whales have been shown to produce calls with a higher average fundamental frequency [20,21], amplitude [22], and duration [21], and with higher source levels [23,24]. Disruptions in baleen whale call production have also been reported, but the relationship with noise is not clear due to opposing observed tendencies [20,21,24–26].

Fin whales produce long sequences of sounds with patterned intervals, classified as songs, lasting for numerous hours [3,27]. The most common sound included in these songs is the 20 Hz note, a 1 s pulse that downsweeps in frequency from 30 to 15 Hz [28,29]. Since signaling in long sequences, which is characteristic of this species, is considered a singing behavior, the sounds included in songs are termed "notes". Source level estimates of the 20 Hz fin whale note have been reported throughout the world's oceans and range from 159 to 220 dB re:1 1 µPa at 1 m [10,30–33]. The combined high source levels and small propagation loss of the 20 Hz note result in communication active spaces of hundreds of km for fin whales under favorable conditions [9,32,34]. Although there is evidence of regional differences in fin whale songs that may be linked to population structure [34–36], there is no indication of differences in source levels of the 20 Hz note between populations, nor have there been any significant changes in the estimates over the past 50 years [31]. However, the variability in source level estimates in published studies can be high, up to approximately 40 dB [31]. In addition, within a fin whale song, 20 Hz notes can also show highly variable source levels [28,31,37].

To make any inferences about the variability caused by biotic factors, it is necessary to undertake an assessment of the instrumental and methodological factors that can cause source levels to vary. Miksis-Odls et al. (2019) [31] described several factors that can contribute to the variability in estimates of source levels of fin whale notes: hardware configuration, signal detection methods, sample size, location, recorded time, and acoustic propagation modeling. Four Component Ocean-Bottom Seismometers (4C-OBS) can provide two types of received sound levels, pressure fluctuations in the water recorded by their hydrophones and seabed particle velocity recorded by the directional components of the seismometer. Although recent OBS usually record acoustic data as sound pressure levels and seabed particle velocity, older seismic instruments only included a seismometer component to record seismic data. Many seismic recordings were obtained throughout the years from several areas of the world's oceans that have not yet been analysed in terms of their baleen whale acoustic data, and they could provide legacy data that could have crucial information to assess long-term temporal changes in source levels of fin whale notes. Estimated source levels of fin whale notes recorded on OBS have been calculated only with the seismometer component in the Pacific Ocean [33] and only with the hydrophone in the North Atlantic Ocean [31].

This study presents estimates of source levels of 20 Hz fin whale notes recorded concurrently in the two types of recording channels of the OBS, and discusses several types of factors contributing to the variability of the estimates, namely, geographical (deployment areas), instrumental (recording channels and sample size), and temporal factors (time of the detected notes, inter-note interval and dive duration). In addition, this study shows that the directional components of the seismometer of OBS can also provide reliable source level estimates of fin whale notes.

2. Materials and Methods

2.1. Study Area and Data Collection

The recordings used in this study were derived from a set of four 4C-OBS deployed off the channel Pico–Faial Islands and from one 4C-OBS of an array deployed in the seas to the southwest of Portugal (Figure 1).

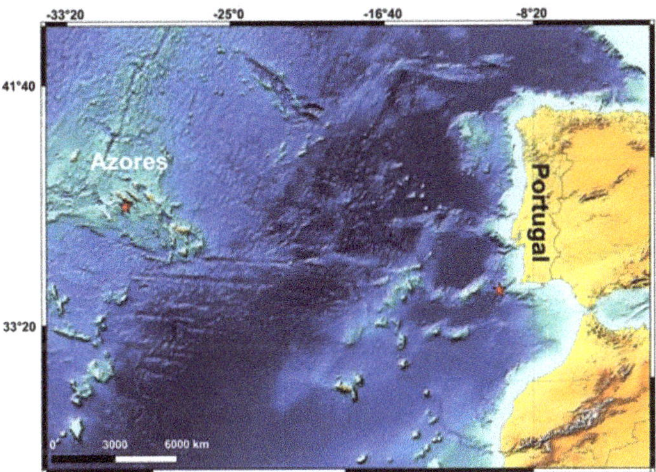

Figure 1. Locations of the OBS (red stars) that recorded fin whale notes used in this study.

The waters around the Azores archipelago are part of a fin whale migratory corridor towards northern latitudes [38]. This species is present year-round in the Azores, but sightings are rare between autumn and winter [39]. Between spring and early summer, fin whales suspend their journeys northwards and stay around the Azores to feed [38], which is correlated with the spring bloom primary productivity [40] and the consequent increase in prey availability [41]. The deployment area in the Azores waters was chosen according to the sighting records of fin whales and the proximity to the coast to facilitate simultaneous visual observations. In the seas to the southwest of Portugal, whaling records of the 20th century suggest the presence of a local nonmigratory subpopulation of fin whales that was heavily exploited [42]. There is no current evidence of the suggested resident subpopulation, and relative abundance of fin whales has not yet recovered to past numbers [42]. Recent visual and acoustics data show two fin whale groups in the seas to the southwest of Portugal, at least between autumn and spring: fin whales from the northeast North Atlantic Ocean and fin whales from the Mediterranean waters [34,35,43].

The array deployed in the seas to the southwest of Portugal was part of the seismic monitoring project NEAREST (Integrated Observation from Near Shore Sources of Tsunamis: Towards an Early Warning System) [44,45]. A full description of the project and the OBS can be found in [46] and in [47]. For this study, only OBS04 was used because it was the most superficial instrument, and its deployment depth was closest to the Azores OBS.

Each 4C-OBS was composed of three directional channels in the seismometer, allowing the recording of the 3-component seabed particle velocity (two channels for the horizontal components, X and Y, and one channel for the vertical component, Z) and the sound pressure in the water (H-channel). The OBS had a 200 Hz sampling rate in the Azores area and 100 Hz in the southwest Portugal area (Table 1). The Azores OBS included a short period three-component seismometer SM6-4.5Hz and a hydrophone HTI-01-PCA/ULF. All OBS were laid on the seabed, and the seismometers were in direct contact with the seabed. The hydrophones were placed in the water above the seabed, tied to the OBS frame. Additional technical specifications used for this study can be found in Table 1.

Table 1. Technical specifications of each OBS used in this study.

OBS	Depth (m)	Sensitivity-H (dB re 1 V/μPa)	Sensitivity-Z (V/m/s)	Gain-H (dB)	Gain-Z (dB)	Conversion Factor-H (count/Pa)	Conversion Factor-Z (count/m/s)
PO2	830	−168.5	27.6	21.32	11.9	1140.43	19,898,719.47
PO3	760	−168.3	27.6	21.32	11.9	1166.99	19,898,719.47
PO4	776	−168.1	27.6	26	11.9	2046.72	19,898,719.47
PO5	790	−167.9	27.6	21.32	11.9	1221.99	19,898,719.47
OBS04	1993	−194.7	1918	4	1	1797.21	1,434,804,623

The OBS were deployed at different depths, ranging from 760 m (PO3) to 1993 m (OBS04) (Table 1), and recorded during different periods. The instruments in the Azores waters recorded between March and September 2019, while the OBS deployed off southwest Portugal recorded between September 2007 and June 2008. The hydrophone (H-channel) of two OBS of the Azores area, PO2 and PO4, was not working correctly and received levels of these two instruments were only recorded for the vertical Z-channel of their seismometer.

2.2. Signal Detection and Localization

Acoustic data were processed to identify 20 Hz fin whale notes with high signal-to-noise ratio (SNR) and high amplitude to estimate source levels. Recordings of the southwest Portugal OBS04 Z-channel were manually inspected by an analyst that listed the start time of all 20 Hz fin whale notes. The Azores OBS recordings in the H-channel were also inspected manually, prior to the use of an automatic detection algorithm, to identify the days with the highest number of SNR 20 Hz notes. Those days were then fed to an automatic detector based on a matched filter which is included in the software program Ishmael version 3.0. A sample of automatic detections (10%) for each day and each Azores OBS was manually checked to obtain daily false positive rates. During the manual checking of the recordings and the false positives of the Azores dataset, it was possible to identify the highest-quality 20 Hz notes that could potentially be used in the estimates of source level.

Fin whale notes used in the following analysis were recorded in April 2019 (n = 1459) in the Azores, and between November 2007 and April 2008 (n = 14,733) in the southwest Portugal. Regarding the Azores dataset, within the same sequence of 20 Hz notes, sometimes notes could not be detected in the spectrograms because of the presence of varying levels of noise. Therefore, the number of recorded notes was not equal across recording channels and instruments in the Azores OBS. The two channels in the OBS of southwest Portugal recorded the same number of fin whale notes.

The range (i.e., the horizontal distance) between the source of each 20 Hz note and the OBS was estimated according to two methods. For the southwest Portugal dataset, range was estimated using the direct signal to a single OBS [47,48]. This method has been used to locate fin whale notes, but it only works well in an area that is defined by a critical range, which is a distance at which parameters used to estimate ranges are reliable [47,48]. A coherency threshold between the horizontal and vertical seismometer channels was set to 0.1 to identify the 20 Hz notes inside this critical range. A more detailed description of the coherency factor can be found in [48]. After the coherency filtering, there were 46 days with 20 Hz notes detected inside the critical range of OBS04 (n = 4866). The ranges of the 20 Hz fin whale notes of the Azores dataset were obtained by triangulation, which meant that notes had to be detected in at least three instruments. After the signal detection process, there were only 3 days with 20 Hz fin whale notes detected in at least three OBS.

2.3. Sound Source Levels Measurements

Underwater sound source level (SL) is defined as the sound pressure level at a reference distance of 1 m from a given source (expressed in dB re: 1 µPa at 1 m). Source levels of 20 Hz fin whale notes were estimated using the passive form of the Sonar Equation (1):

$$RL = SL - TL, \tag{1}$$

where RL is the received level of the sound and TL is the transmission loss. Received levels of the 20 Hz note were measured as pressure fluctuations in the water in the hydrophone and as particle velocity of the seabed in the seismometer.

Received levels of the 20 Hz notes recorded by the OBS H-channel and Z-channel were automatically measured as the amplitude root-mean-square (RMS) of each note waveform based on a matched filter code that used the library routines of the seismological software package SEISAN [49]. The preferred measurement of received levels was RMS because it is the most used in the literature. Usually, the matched filter is used in the automatic detection process of acoustic signals, but in this study, it was used to obtain several parameters that characterize the 20 Hz notes, including RMS. During the matched filter run, the OBS recordings were cross-correlated with a signal template. Measurements of each 20 Hz note were calculated whenever the matched filter function was maximized 0.5 s around the provided manual start time of each note. This buffer was used to accommodate potential errors made by the researcher in the start time of each note. The end of the measurement window was defined by the duration of a template of high-amplitude and high-SNR 20 Hz note. For each 20 Hz note processed automatically, the in-house code calculated the amplitudes of the H-channel and the Z-channel, as well as the correlation value between the recordings and the template, SNR, relative azimuth, and incidence angles, following the methods described in [47] and [48]. Relative azimuth and incidence angles calculated in SEISAN were only reliable for the southwest Portugal dataset because fin whale notes were inside the critical range. In the case of the Azores dataset, most estimated ranges of fin whale notes were outside the critical range of each OBS, and several angles had to be calculated according to the methodology described below.

Measured RMS amplitudes of each 20 Hz note recorded in the OBS Z-channel were transformed in substrate particle velocity (m/s) and then in dB received levels using the procedure described in [33]. In addition to the angles of the incoming and transmitted signals, the procedure needs information about the water column and seabed properties, namely the density and the velocities of the signals in the two mediums (Table 2).

Table 2. Properties of the water column and the seabed of the Azores and southwest Portugal deployment areas.

Parameter	Azores	Southwest Portugal
Water column sound (P-wave) velocity (m/s)	1500	1500
Water column density (kg/m^3)	1000	1000
Seabed S-wave velocity (m/s)	800	300
Seabed P-wave velocity (m/s)	1800	1700
Seabed density (kg/m^3)	1300	1400

The amplitudes of the transmitted P- (T_{PP}) and S- (T_{PS}) waves at a fluid–solid interface relative to an incident pressure wave of unit amplitude are determined by the Zoeppriiz equations [50]:

$$T_{PP} = \left(\frac{V_{P1}}{V_{P2}}\right) \frac{2B\rho_1 V_{P2}\cos(\theta_i)}{A_1\rho_2 V_{P2}\cos(\theta_i) + A_2\cos(\theta_i)\cos(\theta_t) + \rho_1 V_{P1}\cos(\theta_t)} \tag{2}$$

$$T_{PS} = \left(\frac{V_{P1}}{V_{P2}}\right) \frac{2C\cos(\theta_i)\cos(\theta_t)}{A_1\rho_2 V_{P2}\cos(\theta_i) + A_2\cos(\theta_i)\cos(\theta_t) + \rho_1 V_{P1}\cos(\theta_t)} \tag{3}$$

where,

$$A_1 = B^2 = cos^2(2\phi_t) \tag{4}$$

$$A_2 = 4\rho_2 V_{S2} sin^2(\phi_t)\cos(\phi_t) \tag{5}$$

$$C = 2\rho_1 V_{S2}\sin(\phi_t) \tag{6}$$

and V_{P1} is the incident P-wave velocity, V_{P2} is the transmitted P-wave velocity, V_{S2} is the transmitted S-wave velocity, θ_i is the incidence angle of the acoustic wave and the normal to the interface, θ_t is the transmitted P wave angle, ϕ_t is the transmitted S wave angle, and ρ_1 and ρ_2 are the densities in the fluid and solid layers, respectively [40]. In the southwest Portugal dataset, the incidence angle θ_i was obtained with the range estimates of the 20 Hz notes and the adjustments in [48]. In the Azores dataset, θ_i was calculated assuming a homogenous medium:

$$\theta_i = \tan^{-1}\left(\frac{range}{OBS\ depth}\right) \tag{7}$$

The angles of the transmitted P- (θ_t) and S- (ϕ_t) waves were calculated using Snell's law.

The amplitude RMS of each fin whale note was transformed in measured vertical particle velocity u_v based on the seismometer sensitivity, system gain, and information about the digital conversion (Table 1). Following [33], u_v was then scaled by the vertical projection of the Zoeppritz equations to obtain particle velocity in the direction of the incoming wave u:

$$u = u_v\left(\frac{1}{T_{PP}\cos\theta_t + T_{PS}\sin\phi_t}\right) \tag{8}$$

The received sound pressure level p_m, expressed in decibels relative to 1 µPa, of each fin whale note recorded in the Z-channel of the OBS was calculated as:

$$p_m = 20 \times \log_{10}\left(\frac{u\rho_1 V_{P1}}{10^{-6}}\right) \tag{9}$$

Properties of the seabed in the southwest Portugal area were selected according to work shown in [48]. This set of values, combined with the application of some adjustments, resulted in the most accurate ranging estimates for a dataset of airgun shots, for which the location was known. Properties of the Azores area seabed were based on a top strata composed of poorly consolidated water-saturated sediments [51]. Weirathmueller et al. (2013) [33] recognized that the Zoeppritz correction to convert vertical ground velocity to the velocity of the incoming acoustic wave was too sensitive at incidence angles close to and larger than the critical angle ($i_c = \sin^{-1}(V_{P1}/V_{P2})$ which could result in bias of the source level estimates. They limited source level estimates to incidence angles that resulted in a steady increase of the Zoeppritz correction and discarded the data with a rapid increase of the Zoeppritz correction. In this study, the southwest Portugal dataset was already limited by the critical angle ($i_c = 61.9°$), and therefore no additional filtering was needed. Most of the fin whale notes in the Azores dataset showed incidence angles greater than the critical angle ($i_c = 56.4°$). In this case, and for source level estimates recorded in the Z-channel, the Azores dataset was filtered to retain only notes with incidence angles showing a steady increase of the Zoeppritz correction (<47.0°). For the H-channel of the Azores dataset, all fin whale notes were used to estimate source levels. Measured RMS amplitudes of the OBS hydrophone were converted into dB received levels using a conversion factor obtained from the hydrophone sensitivity, system gain, and information about the digital conversion (Table 1).

Transmission loss was calculated using the ray-trace package BELLHOP [52]. The signal source depth was assumed to be 15 m, based on fin whale tag data from [53]. The signal frequency was assumed to be 22 Hz, based on median frequency measurements of a sample of 20 Hz notes from the OBS dataset (n = 2952) and according to the values used in [32]. The sound speed profile used to model the transmission loss for the Azores dataset was calculated using the salinity and temperature profiles from the World Ocean Atlas climatological data for a point at 38° N 29° W in April [54]. The sound speed profile for the southwest Portugal dataset was calculated from conductivity, temperature, and depth (CTD) data collected in situ on 24 August 2007. The models were developed with the multiple OBS receiver depths (Table 1) and maximum estimated ranges for each OBS: 20 km for PO2, 9 km for PO3, 13 km for PO4, 8 km PO5, and 4 km for OBS04.

2.4. Statistical Analysis

The variability of source level estimates for the 20 Hz note was assessed based on the deployment area, type of recording channel, sample size, month, dive duration, and inter-note interval. Variability in source level estimates for each recording channel and deployment area was assessed using the coefficient of variation (CV), which is a measure of spread of the data and allows comparison between datasets with different distributions [55]. The effect of sample size on source estimates was assessed by filtering the southwest Portugal dataset based on note quality (cross-correlation value $\geq$0.7 and SNR $\geq$ 3.1), following [31]. Differences in source level estimates calculated for each month of the southwest Portugal dataset were evaluated using the non-parametric Kruskal–Wallis test followed by a Dunn's multiple comparison test with a Bonferroni adjustment method. The null hypothesis of this test assumes that the samples (groups) are from identical populations. The significance level used to decide the acceptance of the null hypothesis was $p = 0.05$. Preliminary data exploration showed that the relationship between source level estimates of fin whale notes and elapsed time within a sequence was not linear. Therefore, the relationship between source level estimates and elapsed time within a sequence was evaluated using Generalized Additive Models (GAMs) with a Gaussian distribution and an identity link function [56]. A sequence was defined as several 20 Hz notes with inter-note intervals shorter than 45 s [28,36]. Intervals longer than 45 s were associated with potential surfacing episodes, and associated sequences were evaluated separately. Statistical analysis were undertaken with the *nlme* package in R software version 5.474 [57].

A summary of the data acquisition and processing stages can be found in Figure 2.

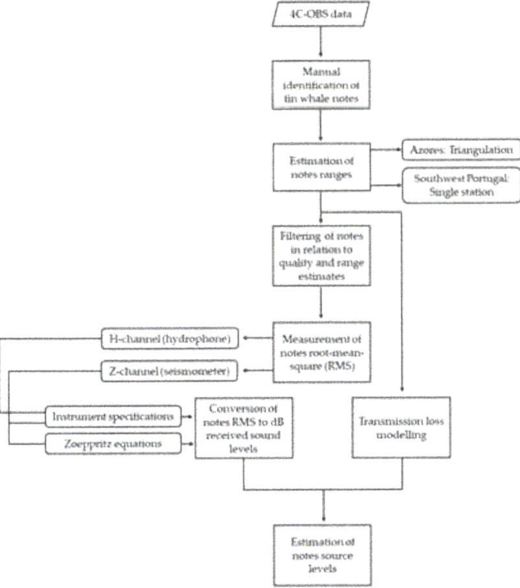

Figure 2. Flowchart of data acquisition and processing stages to estimate source levels of 20 Hz fin whale notes.

3. Results

3.1. Geographical and Instrumental Differences in Source Level Estimates

The most significant difference in average source level estimates of fin whale notes between deployment areas (34 dB) was observed for the Z-channel, mainly associated with one instrument of the Azores deployment, PO5 (Table 3). The remaining two OBS from the Azores showed relatively similar source level estimates compared with the southwest Portugal OBS. However, the two distributions of estimated source levels for the Z-channel were considerably different (Figure 3), which was related to the small sample size of the Azores dataset (n = 70, except PO4) and the higher coefficient of variation. The observed difference of estimated source levels for the H-channel between deployment areas (~15 dB) was not as clear as the one observed for the Z-channel.

Table 3. Estimated fin whale mean source levels (SL) and the corresponding coefficient of variation as a function of the deployment location, detection period, and sample size. The sample size of the Z-channel of the Azores dataset corresponds to notes that were accepted after the incidence angle filtering.

Location	Detection Period	OBS	Z-Channel			H-Channel		
			Sample (n)	SL (dB re: 1 μPa at 1 m)	CV (%)	Sample (n)	SL (dB re: 1 μPa at 1 m)	CV (%)
Southwest PT	11/07–04/08	OBS04	4866	196.9	1.8	4866	186.7	1.7
		OBS04	1443	197.3	1.6	1443	187.2	1.5
Azores		PO2	38	194.6	5.0	0	-	-
	04/19	PO3	0	-	-	260	172.2	3.2
		PO4	2	201.8	0.2	0	-	-
		PO5	32	163.4	1.8	281	171.3	3.7

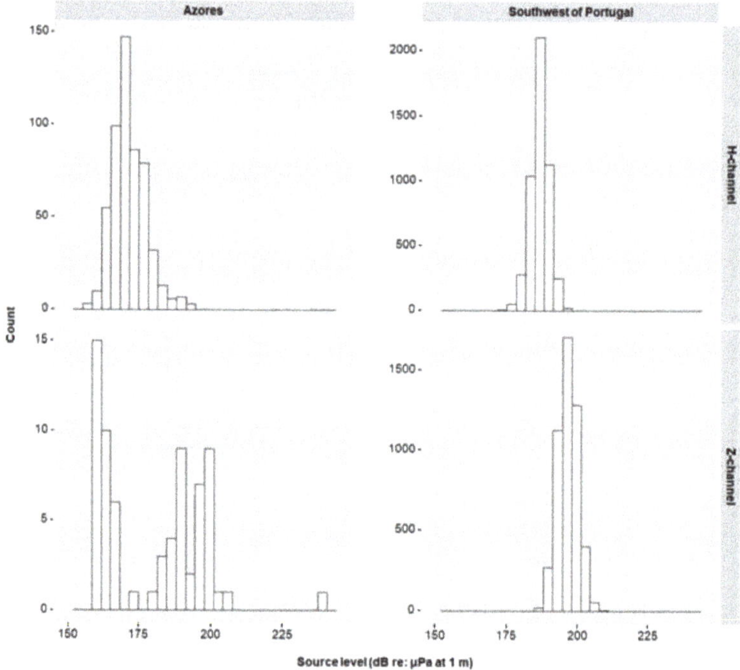

Figure 3. Distribution of estimated source levels from 20 Hz fin whale notes manually detected in the H-channel (**top**) and the Z-channel (**bottom**) of three OBS in the Azores (**left**) and one OBS in southwest Portugal (**right**). Estimated source levels for the Z-channel of PO4 were not included because of the small sample size.

Within each OBS, differences between recording channels were small: 8 dB in the Azores and 10 dB in southwest Portugal. The reduction of the sample size of the southwest Portugal dataset to retain only high-quality fin whale notes did not result in a significant change of estimated source levels. The differences were 0.4 dB for the Z-channel and 0.5 dB for the H-channel (Table 3). The source level estimates did not change, as observed in the Azores dataset, because the sample was still large (n = 1443).

3.2. Temporal Differences in Source Levels

Estimated source levels from the southwest Portugal dataset showed significant statistical differences between months in both OBS channels (Z-channel: Chi-square = 978.54, df = 5, $p < 2.2 \times 10^{-16}$; H-channel: Chi-square = 255.32, df = 5, $p < 2.2 \times 10^{-16}$). In the Z-channel, pairwise comparisons using a post hoc Dunn's test indicated that January and February had significantly higher estimated source levels than all other months, while March and April were not different from each other and to November and December (Table S1). The H-channel also showed significantly higher estimated source levels in January compared to other months and in February compared only to April and December. April showed lower source levels than all other months and March showed no differences to November or December (Table S1; Figure 4).

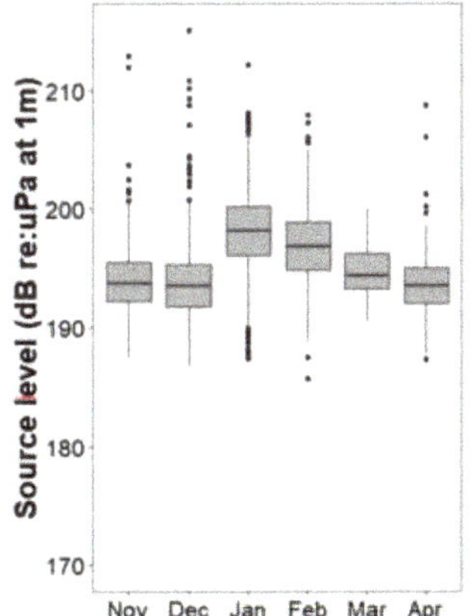
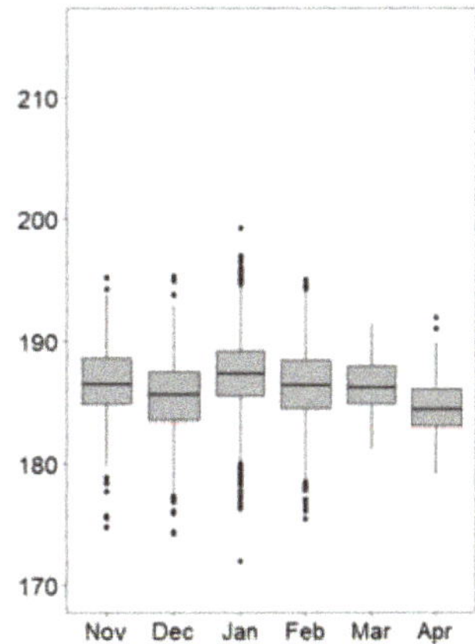

Figure 4. Estimated source levels of fin whale notes by month recorded in the Z-channel (**left**) and the H-channel (**right**) of the OBS off southwest Portugal. The upper and lower whiskers represent the maximum and minimum value of the data within 1.5 times the interquartile range over the 75th percentile and under the 25th percentile, respectively.

Results from GAMs showed how estimated source levels slightly decreased at greater elapsed times within a sequence in both channels (GAM smoother for elapsed time in H channel: edf = 7.67, F = 20.91, $p < 0.001$; and Z channel: edf = 2.7, F = 26.39, $p < 0.001$) (Figure 5). However, the lower deviance explained by these models (H-channel: 3.62%; Z-channel: 1.55%) indicated that other variables apart from elapsed time within a sequence affected the variability of source levels. Estimated source levels only started decreasing with elapsed times greater than 1400 s in both channels (Figure 5). A linear regression for elapsed times above 1400s showed a clear decreasing trend in both channels (H-channel: $p < 0.001$, Adj. $R^2 = 0.6$; Z-channel: $p < 0.01$, Adj. $R^2 = 0.3$) (Figure 6).

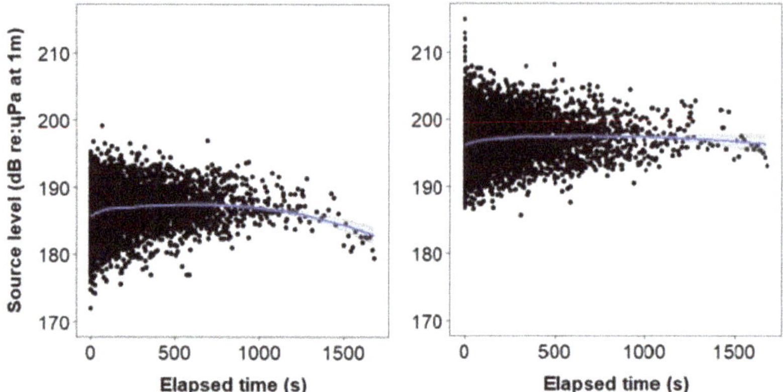

Figure 5. Estimated source levels for 20 Hz fin whale notes detected in the H-channel (**left**) and the Z-channel (**right**) of the OBS off southwest Portugal plotted against the elapsed time within a sequence. A fitted smoothed line (blue) is fitted to the data with its associated 95% confidence intervals in shaded grey. Black dots represent estimated source levels.

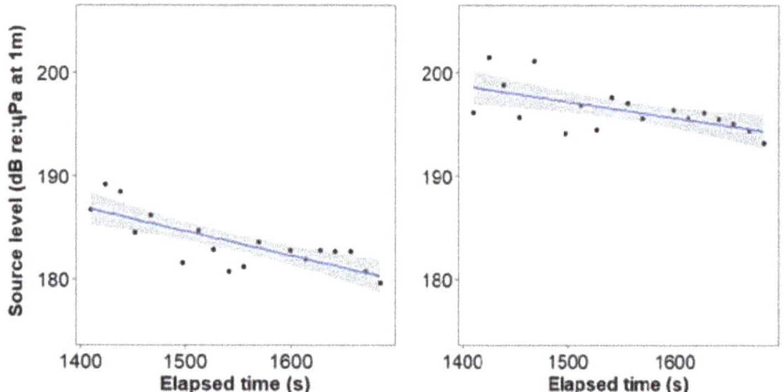

Figure 6. Estimated source levels from 20 Hz fin whale notes manually detected in the H-channel (**left**) and the Z-channel (**right**) from the OBS off southwest Portugal against elapsed time within a sequence longer than 1400 s. A linear regression line (blue) is fitted to the data with its associated 95% confidence intervals (shaded grey). Black dots represent estimated source levels.

Since there was a decrease of estimated source levels with longer sequences, four of the longest sequences (longer than 1000 s, which is equivalent to ~16 min) were further explored individually. The estimates were very consistent throughout the four sequences, and on some occasions, they followed a wave pattern (Figure 7). The average differences between consecutive source levels were 2.2 dB in the H-channel and 2.3 dB in the Z-channel. Estimated source levels did not vary significantly with increasing inter-note intervals in neither of the channels (H-channel: $p = 0.03$, Adj. $R^2 = 0.0007$; $p = 0.4$, $R^2 = -0.0001$) (Figure 8).

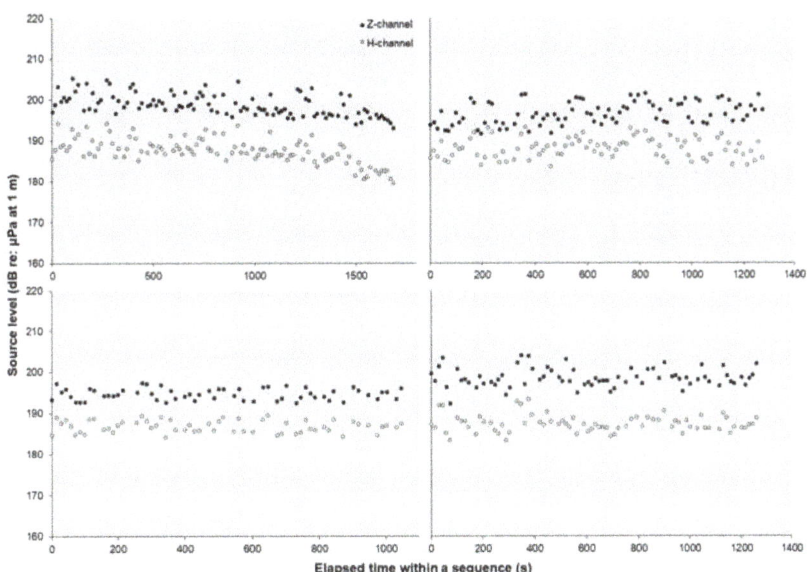

Figure 7. Examples of estimated source levels over elapsed time within sequences recorded on 11 January 2008 (**two top**), 26 December 2007 (**bottom-left**), and 1 February 2008 (**bottom-right**) in the southwest Portugal.

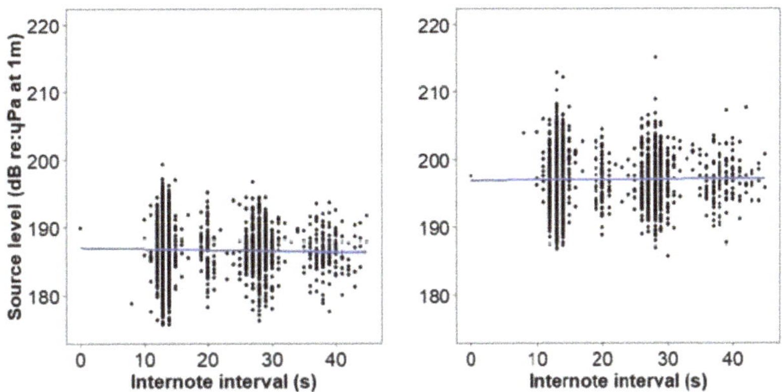

Figure 8. Estimated source levels from 20 Hz fin whale notes manually detected in the H-channel (**left**) and the Z-channel (**right**) from the OBS off southwest Portugal against inter-note interval. The blue line represents a linear regression fitted to the data (black dots).

4. Discussion

The average source level estimates of fin whale notes calculated for two types of recording channels in five ocean-bottom seismometers (OBS) were in line with the values reported for other deployment areas and recording instruments [28,30–33,58]. The highest variability was found for the Z-channel source level estimates with a 35 dB difference between instruments and deployment areas. However, this seemed to be associated with an unaccounted technical issue of one seismometer of the Azores deployment. Within all OBS of the Azores dataset, PO5 measured the lowest received levels of fin whale notes. If we discard the instruments with technical issues and small sample sizes (n < 10), the difference between deployment areas was only 2 dB, with an average source level for the

Z-channel of 196 dB re: 1 μPa at 1 m. There was a 7 dB difference between the Z-channel source level estimates in this study and the values described by [33]. Considering the potential effects that could not be accounted for, like differences in the amplitudes of notes, bias in ranging methods, and the parameters included in the acoustic propagation settings, differences between both studies can be considered small. Therefore, estimated source levels obtained from the OBS Z-channel were generally consistent throughout deployment areas, signal processing techniques, and acoustic propagation modeling.

The distribution and average (187 dB re: 1 μPa at 1 m) H-channel source levels estimates for the southwest Portugal area were equivalent to those found for the same channel and area but for a different OBS (189 dB re: 1 μPa at 1 m) [31], indicating minor differences caused by the use of different acoustic propagation models and signal detection and processing. In addition, H-channel source levels for the Azores dataset were similar between OBS. This means that within the same deployment area, instrumental variability for the hydrophone was small. Since datasets were processed using the same methodology, we can speculate that the observed 15 dB difference for the H-channel between deployment areas could be mainly related to signal propagation effects.

In the Azores dataset, almost all 20 Hz notes showed changes in their waveform, which were caused by multipaths and acoustic interference. On the other hand, the effects of acoustic interference on the notes of the southwest of Portugal dataset was very small in relation to the large sample size of unaltered fin whale notes. The acoustic propagation settings of the seabed in southwest Portugal resulted in very low amplitudes of the multipaths. Furthermore, estimated ranges for the southwest Portugal dataset were shorter than the critical range (<3700 m) and were inside an area where direct path arrivals dominated the recordings [41]. The associated ranges of fin whale notes of the Azores dataset were estimated to be inside an area where multipaths dominated, resulting in changes of the received signal. Other studies have shown that acoustic interference can also affect the received levels of fin whale notes and therefore cause bias in the source level estimates [10]. Therefore, local effects of signal propagation seem to be the major factor contributing to the differences between source level estimates of the 20 Hz fin whale notes recorded in the H-channel of the two deployment areas. Propagation effects and transmission loss differences can be associated with the 9 dB difference of average source levels between the recording channels since fin whale notes were measured in different mediums. However, the differences in source level estimates found here seem small, given that variability within the same study can sometimes exceed 40 dB [31].

The large sample size in the southwest Portugal OBS resulted in robust source level estimates and allowed the exploration of temporal factors related to fin whale behavior: month, duration of the note sequence, and inter-note interval. The Z-channel source level estimates varied throughout the recording period, peaking in January and February and gradually decreasing during spring. In the northern hemisphere, fin whales show a seasonal pattern in the production of 20 Hz notes, with notes being recorded more frequently between autumn and spring, with a maximum in winter [28,35,55,59]. If fin whale song functions as a reproductive display, as suggested by several studies [7,28], then male fin whales could produce louder notes during the time of highest vocal activity. Assuming that loud signals require a higher energetic cost, the loudness could be an acoustic trait that indicates the condition or quality of the sender and therefore be a proxy of male quality [60–62]. Male fin whales could also produce loud notes to maximize the signal transmission distance and potentially increase the number of receivers. If the variability of source levels is associated with the seasonal pattern of note production, then both recording channels of the OBS should follow these changes. The H-channel of the OBS also showed a similar peak in source level estimates in winter months, but the variation was not as pronounced as the one observed in the Z-channel. Seasonal changes in the seabed properties were not expected to contribute significantly to the estimates, which suggests that further analyses are needed to clarify the intra-annual variability of source levels recorded in OBS.

Estimated source levels were relatively constant throughout four long sequences, which [33] also found for recorded fin whales in the Pacific Ocean. During the longest sequence, after 21 min, source levels seemed to decrease with elapsed time. If the number of produced notes and associated source levels are limited by the individual capacity in retaining and using a volume of air to vocalize during a dive, then this tendency should be observed more often. Weirathmueller et al. (2013) [33] identified both positive and negative trends in source level estimates over time in equal proportions. While the small number of long sequences recorded in this study limited this analysis, future research is needed to determine the proportion of individual variability in source level estimates.

Results from this study add valuable information on the variation of fin whale 20 Hz note source levels due to biotic and abiotic factors. Understanding what causes variation in animal call source levels provides insights into the function of calls and helps in assessing the potential effects of increasing anthropogenic noise.

Supplementary Materials: The following are available online at https://www.mdpi.com/article/10.3390/jmse9060646/s1, Table S1: Results of the post hoc pairwise comparison Dunn's test of estimated source levels per month and for the OBS H- and Z-channel deployed off southwest Portugal.

Author Contributions: Conceptualization, A.P., M.R., M.A.S. and L.M.; methodology, A.P., M.R., C.C. and L.M.; formal analysis, A.P. and M.R.; investigation, A.P., C.C., M.R. and L.M.; writing—original draft preparation, A.P.; writing—review and editing, A.P., M.R., C.C., M.A.S. and L.M.; visualization, A.P., M.R., M.A.S. and L.M.; project administration, L.M. and M.A.S.; funding acquisition, L.M. and M.A.S. All authors have read and agreed to the published version of the manuscript.

Funding: This research was funded by FCT–Fundação para a Ciência e a Tecnologia, through project AWARENESS "PTDC/BIA-BMA/30514/201" and "UIDB/50019/2020–IDL". The acoustic data of southwest Portugal were collected for the NEAREST project, on behalf of the EU Specific Programme "Integrating and Strengthening the European Research Area," Sub-Priority 1.1.6.3, "Global Change and Ecosystems," Contract No. 037110. A.P. was supported by project AWARENESS "PTDC/BIA-BMA/30514/201". M.R. was supported by a DRCT doctoral grant "M3.1.a/F/028/2015". M.A.S. was funded by FCT (IF/00943/2013) and by PO Açores 2020, DRCT/Açores, and PRO-SCIENTIA through the project "Investigadores Mar AZ".

Institutional Review Board Statement: Not applicable.

Informed Consent Statement: Not applicable.

Data Availability Statement: The data presented in this study are available on request from the corresponding author. The data are not publicly available because the funding project is not finished yet. Data will be available when the project is finished, by the end of 2021.

Acknowledgments: We thank José Luís Duarte and all of the skippers and crew members that collaborated in the deployment and retrieval of the OBS and EAR in the Azores study area. We also thank Irma Cascão and Cláudia Oliveira for reviewing initial versions of the article and providing critical insights.

Conflicts of Interest: The authors declare no conflict of interest.

References

1. Au, W.W.L.; Hastings, M.C. *Principles of Marine Bioacoustics*, 1st ed.; Springer: New York, NY, USA, 2008; p. 695. [CrossRef]
2. Edds-Walton, P.L. Acoustic communication signals of mysticete whales. *Bioacoustics* **1997**, *8*, 47–60. [CrossRef]
3. Clark, C.W. Acoustic behavior of mysticete whales. In *Sensory Abilities of Cetaceans*, 1st ed.; Thomas, J., Kastelein, R., Eds.; Plenum: New York, NY, USA, 1990; pp. 571–583. [CrossRef]
4. Parks, S.E.; Cusano, D.A.; Stimpert, A.K.; Weinrich, M.T.; Friedlaender, A.S.; Wiley, D.N. Evidence for acoustic communication among bottom foraging humpback whales. *Sci. Rep.* **2014**, *4*, 7508. [CrossRef]
5. Tyack, P.L. Differential response of humpback whales, *Megaptera novaeangliae*, to playback of song or social sounds. *Behav. Ecol. Sociobiol.* **1983**, *13*, 49–55. [CrossRef]
6. Zoidis, A.M.; Smultea, M.A.; Frankel, A.S.; Hopkins, J.L.; Day, A.J.; McFarland, A.S.; Whitt, A.D.; Fertl, D. Vocalizations produced by humpback whale (*Megaptera novaeangliae*) calves recorded in Hawaii. *J. Acoust. Soc. Am.* **2008**, *123*, 1737–1746. [CrossRef]
7. Croll, D.; Clark, C.W.; Acevedo, A.; Tershy, B.; Flores, S.; Gedamke, J.; Urban, J. Only male fin whales sing loud songs. *Nature* **2002**, *417*, 809. [CrossRef]

8. Smith, J.N.; Goldizen, A.W.; Dunlop, R.A.; Noad, M.J. Songs of humpback whales, *Megaptera novaeangliae*, are involved in intersexual interactions. *Anim. Behav.* **2008**, *76*, 467–477. [CrossRef]
9. Payne, R.; Webb, D. Orientation by means of long range acoustic signaling in baleen whales. *Ann. N. Y. Acad. Sci.* **1970**, *188*, 110–141. [CrossRef] [PubMed]
10. Charif, R.A.; Mellinger, D.K.; Dunsmore, K.J.; Fristrup, K.M.; Clark, C.W. Estimated sound source levels of fin whale (*Balaenoptera physalus*) vocalizations: Adjustments for surface interference. *Mar. Mamm. Sci.* **2002**, *18*, 81–98. [CrossRef]
11. Cato, D.H. Simple methods of estimating source levels and locations of marine animal sounds. *J. Acoust. Soc. Am.* **1998**, *3*, 1667–1678. [CrossRef] [PubMed]
12. Hatch, L.T.; Clark, C.W.; Parijs, S.M.V.; Frankel, A.S.; Ponirakis, D.W. Quantifying loss of acoustic communication space for right whales in and around a U.S. National Marine Sanctuary. *Conserv. Biol.* **2012**, *6*, 983–994. [CrossRef] [PubMed]
13. Duarte, C.M.; Chapuis, L.; Collin, S.P.; Costa, D.P.; Devassy, R.P.; Eguiluz, V.M.; Erbe, C.; Gordon, T.A.C.; Halpern, B.S.; Harding, H.R.; et al. The soundscape of the Anthropocene ocean. *Science* **2021**, *371*, eaba4658. [CrossRef]
14. Frisk, G.V. Noiseonomics: The relationship between ambient noise levels in the sea and global economic trends. *Sci. Rep.* **2012**, *2*, 437. [CrossRef] [PubMed]
15. Hildebrand, J.A. Anthropogenic and natural sources of ambient noise in the ocean. *Mar. Ecol. Prog. Ser.* **2009**, *395*, 5–20. [CrossRef]
16. Ross, D. Ship sources of ambient noise. *IEEE J. Ocean. Eng.* **2005**, *30*, 257–261. [CrossRef]
17. Širović, A.; Hildebrand, J.; McDonald, M. Ocean ambient sound south of Bermuda and Panama Canal traffic. *J. Acoust. Soc. Am.* **2016**, *139*, 2417–2423. [CrossRef]
18. Erbe, C.; Reichmuth, C.; Cunningham, K.; Lucke, K.; Dooling, R. Communication masking in marine mammals: A review and research strategy. *Mar. Pollut. Bull.* **2015**, *103*, 15–38. [CrossRef] [PubMed]
19. Kunc, H.P.; McLaughlin, K.E.; Schmidt, R. Aquatic noise pollution: Implications for individuals, populations, and ecosystems. *Proc. R. Soc. B* **2016**, *283*, 20160839. [CrossRef]
20. Parks, S.E.; Clark, C.W.; Tyack, P.L. Short- and long-term changes in right whale calling behavior: The potential effects of noise on acoustic communication. *J. Acoust. Soc. Am.* **2007**, *122*, 3725–3731. [CrossRef] [PubMed]
21. Castellote, M.; Clark, C.W.; Lammers, M.O. Acoustic and behavioural changes by fin whales (*Balaenoptera physalus*) in response to shipping and airgun noise. *Biol. Conserv.* **2012**, *147*, 115–122. [CrossRef]
22. McKenna, M.F. Blue Whale Response to Underwater Noise from Commercial Ships. Ph.D. Thesis, University of California, San Diego, CA, USA, 2011.
23. Parks, S.E.; Johnson, M.; Nowacek, D.; Tyack, P.L. Individual right whales call louder in increased environmental noise. *Biol. Lett.* **2010**, *7*, 33–35. [CrossRef]
24. Melcón, M.L.; Cummins, A.J.; Kerosky, S.M.; Roche, L.K.; Wiggins, S.M.; Hildebrand, J.A. Blue whales respond to anthropogenic noise. *PLoS ONE* **2012**, *7*, 681. [CrossRef]
25. Parks, S.E.; Urazghildiiev, I.; Clark, C.W. Variability in ambient noise levels and call parameters of North Atlantic right whales in three habitat areas. *J. Acoust. Soc. Am.* **2009**, *125*, 1230–1239. [CrossRef]
26. Di Ioro, L.; Clark, C.W. Exposure to seismic survey alters blue acoustic communication. *Biol. Lett.* **2010**, *5*, 51–54. [CrossRef] [PubMed]
27. Širović, A.; Oleson, E.M.; Buccowich, J.; Rice, A.; Bayless, A.R. Fin whale song variability in southern California and the Gulf of California. *Sci. Rep.* **2017**, *7*, 10126. [CrossRef] [PubMed]
28. Watkins, W.A.; Tyack, P.; Moore, K.E. The 20-Hz signals of finback whales (*Balaenoptera physalus*). *J. Acoust. Soc. Am.* **1987**, *82*, 1901–1912. [CrossRef]
29. Thompson, P.O.; Findley, L.T.; Vidal, O. 20-Hz pulses and other vocalizations of fin whales, *Balaenoptera physalus*, in the Gulf of California, Mexico. *J. Acoust. Soc. Am.* **1992**, *92*, 3051–3057. [CrossRef] [PubMed]
30. Wang, D.; Huang, W.; Garcia, H.; Ratilal, P. Vocalization source level distributions and pulse compression gains of diverse baleen whale species in the Gulf of Maine. *Remote Sens.* **2016**, *8*, 881. [CrossRef]
31. Miksis-Olds, J.L.; Harris, D.V.; Heaney, K.D. Comparison of estimated 20-Hz pulse fin whale source levels from the tropical Pacific and Eastern North Atlantic Oceans to other recorded populations. *J. Acoust. Soc. Am.* **2019**, *146*, 2373–2384. [CrossRef]
32. Širović, A.; Hildebrand, J.A.; Wiggins, S.M. Blue and fin whale call source levels and propagation range in the Southern Ocean. *J. Acoust. Soc. Am.* **2007**, *122*, 1208–1215. [CrossRef] [PubMed]
33. Weirathmueller, M.J.; Wilcock, W.S.; Soule, D.C. Source levels of fin whale 20 Hz pulses measured in the Northeast Pacific Ocean. *J. Acoust. Soc. Am.* **2013**, *133*, 741–749. [CrossRef]
34. Hatch, L.T.; Clark, C.W. Acoustic differentiation between fin whales in both the North Atlantic and North Pacific Oceans, and integration with genetic estimates of divergence. Presented at IWC Scientific Committee, Sorrento, Italy, 19–22 July 2004; Paper No. SC/56/SD6; pp. 1–37.
35. Pereira, A.; Harris, D.; Tyack, P.; Matias, L. Fin whale acoustic presence and song characteristics in seas to the southwest of Portugal. *J. Acoust. Soc. Am.* **2020**, *147*, 2235–2249. [CrossRef]
36. Delarue, J.; Todd, S.K.; VanParijs, S.M.; DiIorio, L. Geographic variation in Northwest Atlantic fin whale (*Balaenoptera physalus*) song: Implications for stock structure assessment. *J. Acoust. Soc. Am.* **2009**, *125*, 1774–1782. [CrossRef] [PubMed]
37. Watkins, W.A. Activities and underwater sounds of fin whales. *Sci. Rep. Whales Res. Inst.* **1981**, *33*, 83–117.

38. Silva, M.A.; Prieto, R.; Jonsen, I.; Baumgartner, M.F.; Santos, R.S. North Atlantic blue and fin whales suspend their spring migration to forage in middle latitudes: Building up energy reserves for the journey? *PLoS ONE* **2013**, *8*, e76507. [CrossRef]
39. Silva, M.A.; Prieto, R.; Cascão, I.; Seabra, M.I.; Machete, M.; Baumgartner, M.F.; Santos, R.S. Spatial and temporal distribution of cetaceans in the mid-Atlantic waters around the Azores. *Mar. Biol. Res.* **2014**, *10*, 123–137. [CrossRef]
40. Prieto, R.; Tobeña, M.; Silva, M.A. Habitat preferences of baleen whales in a mid-latitude habitat. *Deep Sea Res. Part II Top. Stud. Oceanogr.* **2017**, *141*, 155–167. [CrossRef]
41. Pérez-Jorge, S.; Tobeña, M.; Prieto, R.; Vandeperre, F.; Calmettes, B.; Lehodey, P.; Silva, M.A. Environmental drivers of large-scale movements of baleen whales in the mid-North Atlantic Ocean. *Divers. Distrib.* **2020**, *26*, 683–698. [CrossRef]
42. Clapham, P.J.; Aguilar, A.; Hatch, L.T. Determining spatial and temporal scales for management: Lessons from whaling. *Mar. Mammal Sci.* **2008**, *24*, 183–201. [CrossRef]
43. Panigada, S.; Notarbartolo di Sciara, G. Balaenoptera physalus in IUCN 2017, IUCN Red List of Threatened Species, Version 2017.3. Available online: http://www.iucnredlist.org (accessed on 27 April 2021).
44. Geissler, W.H.; Matias, L.; Stich, D.; Carrilho, F.; Jokat, W.; Monna, S.; IbenBrahim, A.; Mancilla, F.; Gutscher, M.-A.; Sallarès, V.; et al. Focal mechanisms for subcrustal earthquakes in the Gulf of Cadiz from a dense OBS deployment. *Geophys. Res. Lett.* **2010**, *37*, L18309. [CrossRef]
45. Silva, S.D.M.M.F. Strain Partitioning and the Seismicity Distribution Within a Transpressive Plate Boundary: SW Iberia-NW Nubia. Ph.D. Thesis, University of Lisbon, Lisbon, Portugal, 2017.
46. Schmidt-Aursch, C.M.; Haberland, C. DEPAS (Deutscher Geräte-Pool für amphibische Seismologie): German Instrument Pool for Amphibian Seismology. *J. Largescale Res. Facilit.* **2017**, *3*, A122. [CrossRef]
47. Harris, D.; Matias, L.; Thomas, L.; Harwood, J.; Geissler, W.H. Applying distance sampling to fin whale calls recorded by single seismic instruments in the Northeast Atlantic. *J. Acoust. Soc. Am.* **2013**, *134*, 3522–3535. [CrossRef]
48. Matias, L.; Harris, D. A single-station method for the detection, classification and location of fin whale calls using ocean-bottom seismic stations. *J. Acoust. Soc. Am.* **2015**, *138*, 504–520. [CrossRef] [PubMed]
49. Ottemöller, L.; Voss, P.; Havskov, J. SEISAN: Earthquake analysis software for Windows, Solaris, Linux and Mac OSX, 2011. Available online: http://seisan.info (accessed on 9 June 2021).
50. Ikelle, L.; Amundsen, L. *Introduction to Petroleum Seismology*, 2nd ed.; Society of Exploration Geophysicists: Tulsa, OK, USA, 2005; pp. 92–95. [CrossRef]
51. Ludwig, W.J.; Nafe, J.E.; Drake, C.L. Seismic refraction. In *The Sea*, 1st ed.; Maxwell, A.E., Ed.; Wiley-Interscience: New York, NY, USA, 1970; Volume 4, pp. 53–84.
52. Porter, M. The BELLHOP manual and user's guide: Preliminary draft. 2011. Available online: http://oalib.hlsresearch.com/Rays/HLS-2010-1.pdf (accessed on 27 April 2021).
53. Stimpert, A.K.; DeRuiter, S.L.; Falcone, E.A.; Joseph, J.; Douglas, A.B.; Moretti, D.J.; Friedlaender, A.S.; Calambokidis, J.; Gailey, G.; Tyack, P.L.; et al. Sound production and associated behavior of tagged fin whales (*Balaenoptera physalus*) in the Southern California Bight. *Anim. Biotelem.* **2015**, *3*, 23. [CrossRef]
54. Boyer, T.P.; Garcia, H.E.; Locarnini, R.A.; Zweng, M.M.; Mishonov, A.V.; Reagan, J.R.; Weathers, K.A.; Baranova, O.K.; Seidov, D.; Smolyar, I.V. World Ocean Atlas 2018. NOAA National Centers for Environmental Information. Dataset. Available online: https://accession.nodc.noaa.gov/NCEI-WOA18 (accessed on 27 April 2021).
55. Romagosa, M.; Cascão, I.; Merchant, N.D.; Lammers, M.O.; Giacomello, E.; Marques, T.A.; Silva, M.A. Underwater ambient noise in a baleen whale migratory habitat off the Azores. *Front. Mar. Sci.* **2017**, *4*, 109. [CrossRef]
56. Wood, S.N. *Generalised Additive Models: An Introduction with R*, 2nd ed.; Chapman & Hall: Boca Raton, FL, USA, 2017. [CrossRef]
57. RStudio Team. RStudio: Integrated Development for R. RStudio, PBC, Boston, MA, USA 2020. Available online: http://www.rstudio.com/ (accessed on 27 April 2021).
58. Northrop, J.; Cummings, W.; Thompson, P. 20-Hz signals observed in the central Pacific. *J. Acoust. Soc. Am.* **1968**, *43*, 383–384. [CrossRef]
59. Stafford, K.M.; Mellinger, D.K.; Moore, S.E.; Fox, C.G. Seasonal variability and detection range modeling of baleen whale calls in the Gulf of Alaska, 1999–2002. *J. Acoust. Soc. Am.* **2007**, *122*, 3378–3390. [CrossRef] [PubMed]
60. Zollinger, S.A.; Brumm, H. Why birds sing loud songs and why they sometimes don't. *Anim. Behav.* **2015**, *105*, 289–295. [CrossRef]
61. Searcy, W.A. Sound-pressure levels and song preferences in female red-winged blackbirds (*Agelaius phoeniceus*) (Aves, Emberizidae). *Ethology* **1996**, *102*, 187–196. [CrossRef]
62. Benítez, M.E.; Roux, A.; Fischer, J.; Beehner, J.C.; Bergman, T.J. Acoustic and Temporal Variation in Gelada (*Theropithecus gelada*) Loud Calls Advertise Male Quality. *Int. J. Primatol.* **2016**, *37*, 568–585. [CrossRef]

Journal of
*Marine Science
and Engineering*

Article

Sea Lice Are Sensitive to Low Frequency Sounds

**Marta Solé [1,*], Marc Lenoir [2], José-Manuel Fortuño [3], Steffen De Vreese [1], Mike van der Schaar [1]
and Michel André [1,*]**

[1] Laboratory of Applied Bioacoustics, Technical University of Catalonia-Barcelona Tech, 08800 Vilanova i la
   Geltrú, Spain; steffen.de.vreese@upc.edu (S.D.V.); mike.vanderschaar@upc.edu (M.v.d.S.)
[2] Institute of Neurosciences of Montpellier, INSERM, 34091 Montpellier, France; m.lenoir118@laposte.net
[3] Institute of Marine Sciences, Spanish National Research Council, 08003 Barcelona, Spain; jmanuel@icm.csic.es
* Correspondence: marta.sole@upc.edu (M.S.); michel.andre@upc.edu (M.A.)

Abstract: The salmon louse *Lepeophtheirus salmonis* is a major disease problem in salmonids farming
and there are indications that it also plays a role in the decline of wild salmon stocks. This study
shows the first ultrastructural images of pathological changes in the sensory setae of the first antenna
and in inner tissues in different stages of *L. salmonis* development after sound exposure in laboratory
and sea conditions. Given the current ineffectiveness of traditional methods to eradicate this plague,
and the strong impact on the environment these treatments often provoke, the described response to
sounds and the associated injuries in the lice sensory organs could represent an interesting basis for
developing a bioacoustics method to prevent lice infection and to treat affected salmons.

Keywords: sea lice; *Lepeophtheirus salmonis*; acoustic trauma; transmission electron microscopy;
scanning electron microscopy

Citation: Solé, M.; Lenoir, M.;
Fortuño, J.-M.; De Vreese, S.; van der
Schaar, M.; André, M. Sea Lice Are
Sensitive to Low Frequency Sounds. *J.
Mar. Sci. Eng.* **2021**, *9*, 765. https://
doi.org/10.3390/jmse9070765

Academic Editor: Milva Pepi

Received: 12 June 2021
Accepted: 6 July 2021
Published: 12 July 2021

Publisher's Note: MDPI stays neutral
with regard to jurisdictional claims in
published maps and institutional affiliations.

1. Introduction

Ectoparasitic sea lice represent the most important parasite problem to date for the
salmon farming industry. The salmon louse *Lepeophtheirus salmonis* (Figure 1) is a Caligid
copepod that infests both wild and farmed salmonid fish in the northern and southern
hemispheres. Salmon lice are a major disease problem in the farming of Atlantic salmon,
Salmo salar, and it has been suggested that salmon lice also play a role in the decline of wild
stocks [1,2].

Severe infestations produce pathological lesions on the host that are caused by attachment and feeding of sea lice in both the adult and juvenile stages. *L. salmonis* induces
stress-related responses in the host skin and gills and modulates the immune system [3].
L. salmonis is exceptional among parasite species in infecting adult wild Atlantic salmon
(*Salmo salar*) with 100% prevalence. The infective planktonic larva is extraordinarily effective at locating and infecting wild Atlantic salmon. For this reason, *L. salmonis* has the
potential to become a pest disease to salmonid fish [2].

Understanding the nature of the interactions between *L. salmonis* and its host is crucial
for identifying possible ways to resolve the negative impacts of this infection [4]. *L. salmonis*
is able to detect different stimuli (e.g., pressure/moving water, chemicals and light) in its
habitat. However, the response thresholds to these stimuli and the role that they play in
the context of host location are still unknown. Sea lice use physical (light and salinity)
and chemical (kairomones) cues to locate and recognize their host. Another fundamental
sensory modality to fish location is mechanoreception through sensory organs, which
allows sea lice to detect and land on their host [5].

The life cycle of *L. salmonis* includes ten stages, three of which are pelagic [1]. The
third, the copepodid, represents the infective stage of the salmon louse. It carries both
chemosensory aesthetes and mechanosensory setae on its antennules, indicating that both
mechanical and chemical signals may be important in host-finding [6]. Initial attachment

for the copepodid normally occurs on the fish fins where it hooks into the host tissue. After some time (depending on the temperature) the copepodid undergoes a moult to the first sessile stage in the life cycle, the chalimus, which attaches itself to the fish with a penetrative thread referred to as the frontal filament [7]. Adult *L. salmonis* present different types of sensitive setae on their antennae. Heuch and Karlsen [8,9] found that salmon louse copepodids are sensitive to low frequency water accelerations such as those produced by a swimming fish.

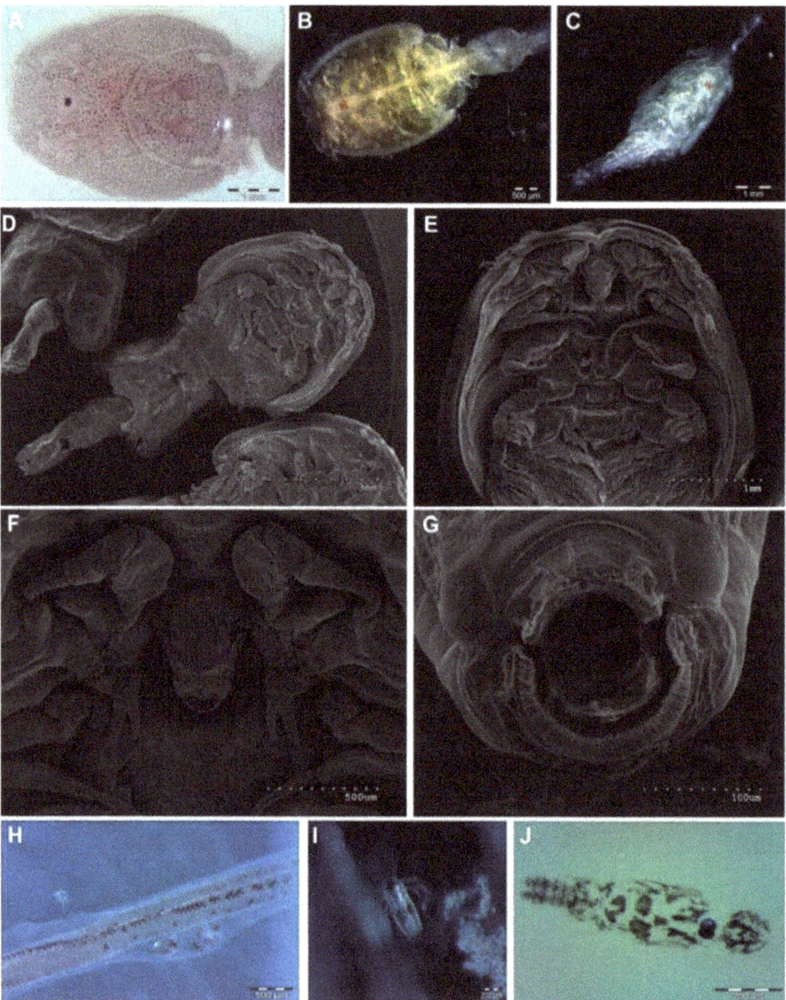

Figure 1. Light microscopy (LM) (**A–C,H–J**) and Scanning Electron Microscopy (SEM) (**C–G**). External morphology of *L. salmonis*. (**A**) Dorsal view of the head of an adult sea lice. (**B**) Dorsal view of pre-adult. (**C**) Dorsal view of chalimus. (**D**) Ventral view of the whole body of an adult *L. salmonis*. (**E**) Ventral view of a sea lice head. (**F**) Ventral view of an adult *L. salmonis* showing the mouth, maxilas and abdominal arms. (**G**) Detail of the ventral cavity showing the mouth and the three maxila. (**H**) String of sea lice eggs. In the low part of the string, two larvae are being extruded from the eggs. (**I**) Nauplius just hatched from the egg. (**J**) Copepodid of sea lice. Scale bar: (**A,C,E**) = 1 mm; (**D**) = 3 mm; (**B,F,H**) = 500 µm; (**I,J**) = 200 µm, G = 100 µm.

Recent findings on cephalopods [10], gastropods [11], crustaceans [12], bivalves [13] and cnidarians [14] have shown that exposure to anthropogenic noise: (i) had a direct consequence on the functionality and physiology of the statocysts, which are sensory organs responsible for these species equilibrium and movements (linear and angular accelerations) in the water column; and (ii) was challenging to the exposed individuals' survival. These experiments demonstrated the sensitivity of invertebrates to sound and described the associated pathological effects. Electron microscopy revealed injuries in the statocyst sensory epithelium after exposure to sound. The lesions present on the exposed animals were consistent with the manifestation of a massive acoustic trauma observed in other species.

In contrast to other invertebrate species that require statocysts to sense the water column in order to maintain balance, zooplankters (copepods, protists) use external mechanosensors for sensing spatial velocity gradients generated by preys or predators [15]. It is understood that the absence of gravity receptors (i.e., statocysts) in planktonic animals has to do with the specific gravity of the zooplankton body, which is the same or slightly higher than water. In *L. salmonis*, these external mechanoreceptors are located on the first antenna [16].

The present study aimed at addressing the problem of sea lice infestation on salmon by using acoustic and bioacoustics techniques and by evaluating the potential effects of these techniques on the parasite's sensory organs.

2. Methods

2.1. Laboratory Experiments

We initially looked at the physiological response of lice after exposure to sound in a controlled laboratory environment.

2.1.1. Sound Exposure

Frequency

We tested a series of discrete frequencies, ranging from 100 Hz to 1 kHz. After determining which frequencies would trigger a stronger reaction to the sound exposure, we exposed with combinations of the optimal frequencies and compared the results with the discrete frequency experiments (Tables 1 and 2).

Table 1. Number of lice per frequency.

Discrete Frequencies	Number of Copepodids	
	Control	Exposed
100 Hz	500	500
150 Hz	500	500
200 Hz	500	500
250 Hz	500	500
300 Hz	500	500
350 Hz	500	500
400 Hz	500	500
450 Hz	500	500
500 Hz	500	500
550 Hz	500	500
600 Hz	500	500
650 Hz	500	500
700 Hz	500	500

Table 1. *Cont.*

Discrete Frequencies	Number of Copepodids	
	Control	Exposed
750 Hz	500	500
800 Hz	500	500
850 Hz	500	500
900 Hz	500	500
950 Hz	500	500
1000 Hz	500	500

Table 2. Combinations of frequencies and corresponding number of lice exposed and used as control.

Frequencies Combination	Number of Copepodids	
	Control	Exposed
350 Hz–450 Hz	500	500
350 Hz–550 Hz	500	500
450 Hz–550 Hz	500	500
300 Hz–400 Hz	500	500
400 Hz–500 Hz	500	500

Sea Lice Specimens

Fifty (50) sets of five hundred (n = 500) copepodids from *L. salmonis* were shipped to our laboratory facilities immediately after they had moulted. In the laboratory they were held in a closed system of natural seawater (at 7–10 °C, salinity 35‰) consisting of plastic tanks with a capacity of 20L until required for the experiments. An air pump facilitated the copepodid movements in the water column.

Individuals were maintained in the tank system until exposure. Several specimens (see below) were used as controls and were kept in the same conditions as the experimental animals until they were exposed to noise.

Sound Exposure Protocol

Sequential Controlled Exposure Experiments (CEE) were conducted on copepodids (25 × 500) of *L. salmonis*. A same number of copepodids (25 × 500) was used as a control and were kept in the same conditions as the exposed ones (Table 1 andTable 2).

The sound was produced and amplified through an in-air loudspeaker while the received levels were measured by a calibrated B&K 8106 hydrophone. The sound production was tuned such that each constant tone from 100 Hz to 1000 Hz was measured at 150 dB re $1~\mu Pa^2/Hz$ at a fixed point in the tank.

The copepodids were exposed to sound for 4 h. The sample collection and the fixation of the lice were performed immediately after the end of the sound exposure session. The controls remained for the same time as those exposed in the isolated exposure tank (4 h) without being exposed to playback. The sacrificing process after exposure was identical for both the controls and exposed animals.

Amplitude

In this experiment, we tested the level of lice trauma after exposing them to different combinations of exposure duration, frequencies, and levels of exposure in order to define the required SEL that would trigger potential lesions and, accordingly, determine the combination that would best induce damage to the lice (Table 3).

Table 3. Frequencies and combination of frequencies and amplitudes used.

	Copepodids
500 Hz-2 h-194 SEL	500
500 Hz-4 h-193 SEL	500
500 Hz-4 h-189 SEL	500
350 Hz-2 h-191 SEL	500
350 Hz-2 h-167 SEL	500
350 Hz-2 h-500 Hz-2 h-195 SEL	500
350 Hz-2 h-500 Hz-2 h-191 SEL	500
350 Hz-3 h-192 SEL	500
350 Hz-2 h-500 Hz-1 h-194 SEL	500
Control-152 SEL	500

Sea Lice Specimens

Ten (10) sets of five hundred (n = 500) copepodids from *L. salmonis* were kept (until required for the experiments) in a closed system of natural seawater (at 7–10 °C, salinity 35‰) consisting of plastic tanks with a capacity of 20 L. An air pump facilitated the copepodid movements in the water column.

Individuals were maintained in the tank system until exposure. Several specimens (see below) were used as controls and were kept in the same conditions as the experimental animals until they were exposed to noise.

Sound Exposure Protocol

The results of the previous experiments allowed us to determine the best response from the lice and to choose the corresponding frequencies or combinations of frequencies to be used for further experimentation. Sequential Controlled Exposure Experiments (CEE) were then conducted on copepodids (5 × 500) of *L. salmonis*. Five additional sets of copepodids (5 × 500) were used as control (Table 3). The same sound exposure set up and sacrifice protocol used as in Section Frequency was followed.

2.1.2. Imaging Techniques

The same imaging techniques were used as with previous experiments with cephalopods [17]. Individuals were processed according to routine Scanning (SEM) and Transmission (TEM) electron microscopy procedures.

Light Microscopy (LM)

Previous to preparing the samples for analysis by SEM and TEM procedures, some light microscopy images of live individuals were taken in order to clarify the morphology and location of the sensory setae of the first antenna.

Scanning Electron Microscopy (SEM)

All sets of *L. salmonis* copepodids (control and treatments of exposed sea lice) were used to analyse the lesions after sound exposure.

Fixation was performed in glutaraldehyde 2.5% for 24–48 h at 4 °C. Samples were dehydrated in graded ethanol solutions and critical-point dried with liquid carbon dioxide in a Bal-Tec CPD030 unit (Leica Mycrosystems, Austria). The dried specimens were mounted on specimen stubs with double-sided tape. The mounted samples were gold coated with a Quorum Q150R S sputter coated unit (Quorum Technologies, Laughton, East Sussex, UK) and viewed with a variable pressure Hitachi S-3500N scanning electron microscope (Hitachi High-Technologies Co., Tokio, Japan) at an accelerating voltage of 5 kV in the Institute of Marine Sciences of the Spanish Research Council (CSIC) facilities.

SEM images were used to determine the number of fused setae and to calculate the rate (%) of irregular branching tips of the first antenna that were fused after each treatment.

Transmission Electron Microscopy (TEM)

Ten (10) exposed and ten control *L. salmonis* copepodids were used for this study. Fixation was performed in 2.5% glutaraldehyde-2% paraformaldehyde for 24 h at 4 °C. Subsequently, the samples were osmicated in 1% osmium tetroxide, dehydrated in acetone, and embedded in Spurr. To orient the specimens properly, semithin sections (1 mm) were cut transversally or tangentially with a glass knife, stained with methylene blue, covered with Durcupan, and observed on an Olympus CX41. Ultrathin (around 100 nm) sections of the samples were then obtained by using a diamond knife (Diatome) with an Ultracut Ultramicrotome from Reichert-Jung. Sections were double-stained with uranyl acetate and lead citrate and viewed with a Jeol JEM 1010 at 80 kV. Images were obtained with a Bioscan camera model 792 (Gatan) at the University of Barcelona technical services.

2.2. Sea Trials

Sea trials were performed at an experimental fish farm located in Averøy (Norway). Two isolated test cages were exposed to sound and four cages at different distances from the test cages were used as a control (Figure 2).

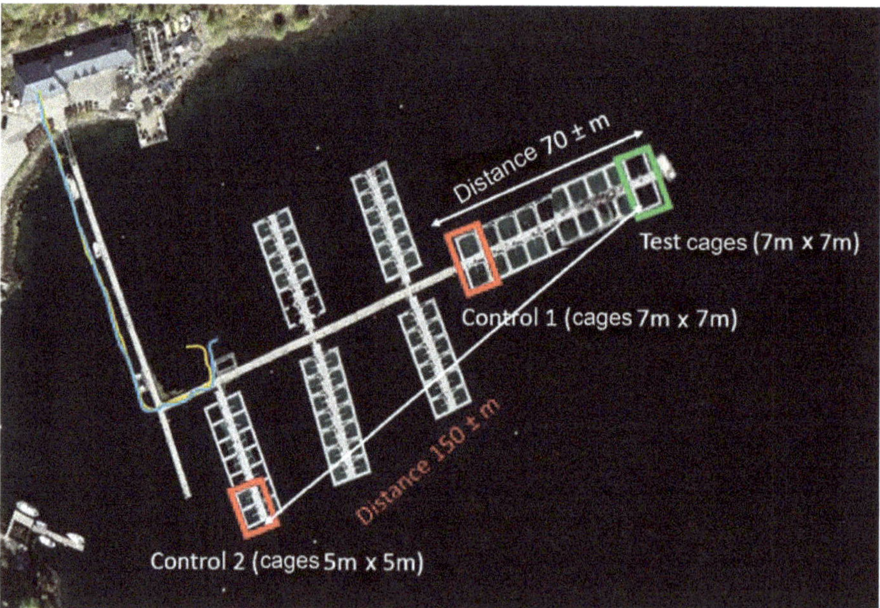

Figure 2. Control and test cages distribution.

2.2.1. Sound Exposure
Sound Exposure Level (SEL)

The protocol refers to the term "Sound Exposure Level (SEL)" as the total cumulative squared sound pressure that an organism is exposed to, expressed in decibels. In other words, it is equivalent to exposing the target organisms to a certain dose of sound. Here, the duration of the exposure represents the cumulative time interval that is necessary to induce lesions.

Sound Pressure Level (SPL)

To expose the target organisms to the necessary SEL, the protocol also refers to the term "Sound Pressure Level (SPL)" as the average sound pressure in a 1 s time interval expressed in decibel, which is produced by an underwater speaker, called a transducer.

Acoustic and Time Parameters for the Sea Trials

The protocol was based on a method and system (Figures 3 and 4) where lice are exposed to continuous acoustic signals over time until a target Sound Exposure Level (SEL) is achieved for the organism: the SEL is chosen at a level that induces sufficient lesions in the sensory organs of the lice to disrupt vital functions necessary for survival, and particularly to detect (and attach) salmon.

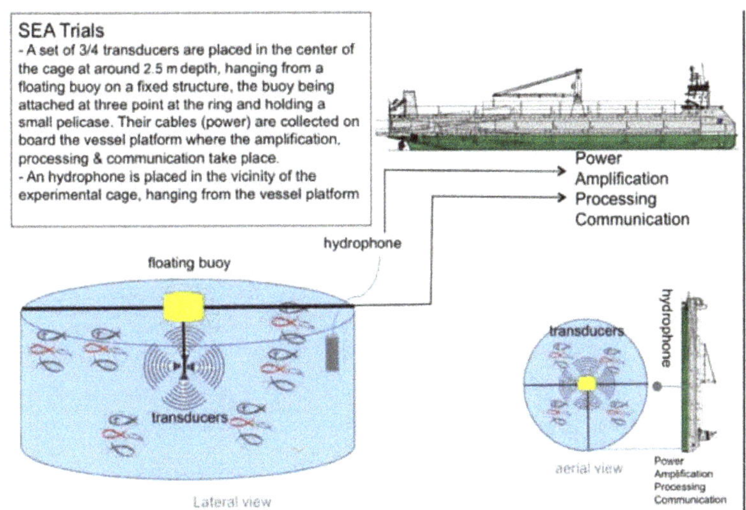

Figure 3. Sound exposure system.

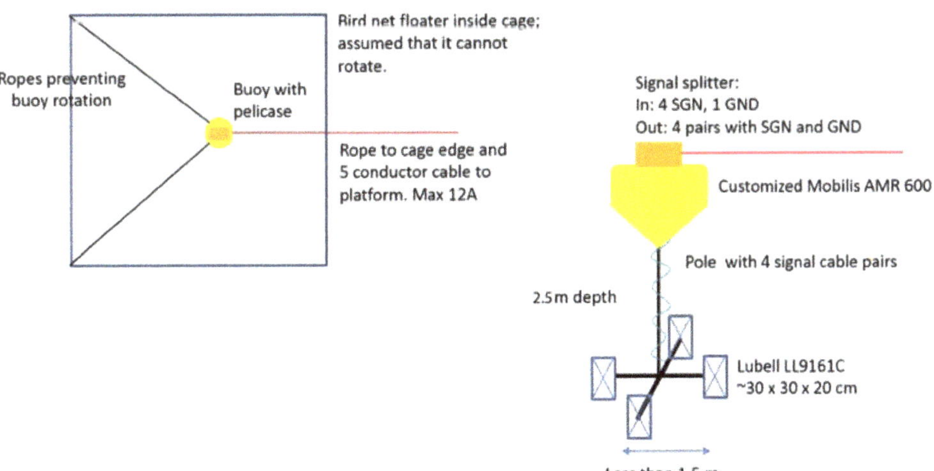

Figure 4. Drawing of the experimental setup. Note that the depth of the structure that holds the loud speakers was modified along the duration of the experiments. M9 loud speakers were lowered to −5 m.

Based on the output of the previous laboratory experiments (see Section 3, Results) where the lice showed sensitivity to a rather broad range of frequencies (and particularly to continuous exposure to individual 350 Hz and 500 Hz signals) during, respectively, a cumulative cycle of 2 h and 1 h, this combination was initially played back every 4 h (Figure 5). Due to the presence of continuous free-swimming lice from external sources, it was decided to increase the exposure time as well as the SPL at one cage so that the required SEL could be reached faster (see below).

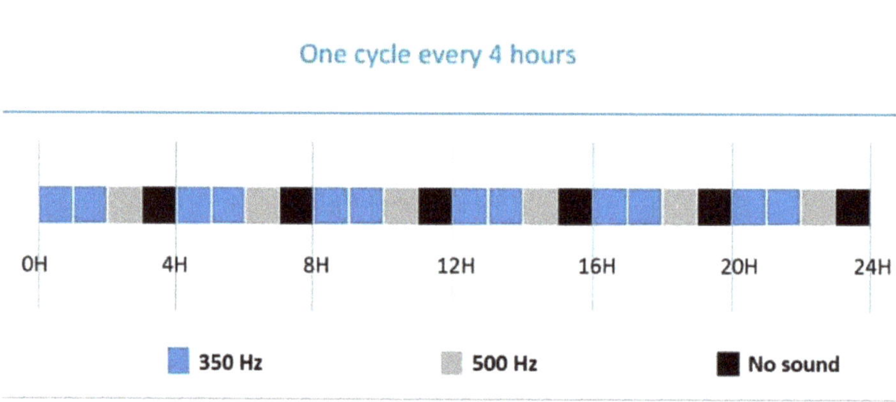

Figure 5. Sound exposure protocol. Note that this cycle was modified along the duration of the experiments favouring the exposure to 500 Hz to produce higher SPL.

Hardware

Under the sea trial protocol, the system and method included producing the sounds using calibrated transducers capable of reproducing sound covering the essential part of the sensitivity range for the lice, particularly from 300 Hz to 600 Hz. The transducers had a source level of at least up to 160 dB re 1 μPa^2 at 1 m for individual frequencies and 180 dB re 1 μPa^2 at 1 m for each selected third octave band. The transducers were driven by amplifiers that could reach the required voltages to reach these levels. Typical peak voltage levels were below 100 V. The sound production system was calibrated as a whole and for each individual frequency.

Calibrated hydrophones recorded the acoustic pressure in a given frequency range with maximum sound pressure levels at least up to 180 dB re 1μPa^2 without saturation. The hydrophone system was arranged to provide digitized data to a sound exposure control system.

2.2.2. Scanning Electron Microscopy (SEM)

Ten (10) exposed and ten control *L. salmonis* for each pre-adult and adult stage were used for this study. The fish in sea conditions were naturally exposed to lice infestation. The lice specimens were collected after two weeks from the start of the sound exposure experiments from the fishes in the cages and processed immediately for the analysis. Individuals were processed according to routine SEM procedures (See Section Scanning Electron Microscopy (SEM)).

2.2.3. Transmission Electron Microscopy (TEM)

Ten (10) exposed and ten control *L. salmonis* for each chalimus, pre-adult and adult stage were used for this study. The specimens were collected after three and six weeks (chalimus), after six weeks (adults and pre-adults) from the start of the sound exposure experiments, and from the fishes in the cages and were processed immediately for the analysis. Individuals were processed according to routine TEM procedures (See Section Transmission Electron Microscopy (TEM)).

3. Results

3.1. Laboratory Experiments

3.1.1. *L. salmonis* Copepodids Sensory Setae Morphology

Copepodids (length 0.7 ± 0.01; width 0.2 ± 0.01) (Figure 6) present 10 pairs of setules arranged symmetrically about the medial longitudinal axis (6 pairs of simple setules, 4 pairs bifurcates) on the dorsal shield of the cephalothorax and 2 simple setules near the base of the rostrum.

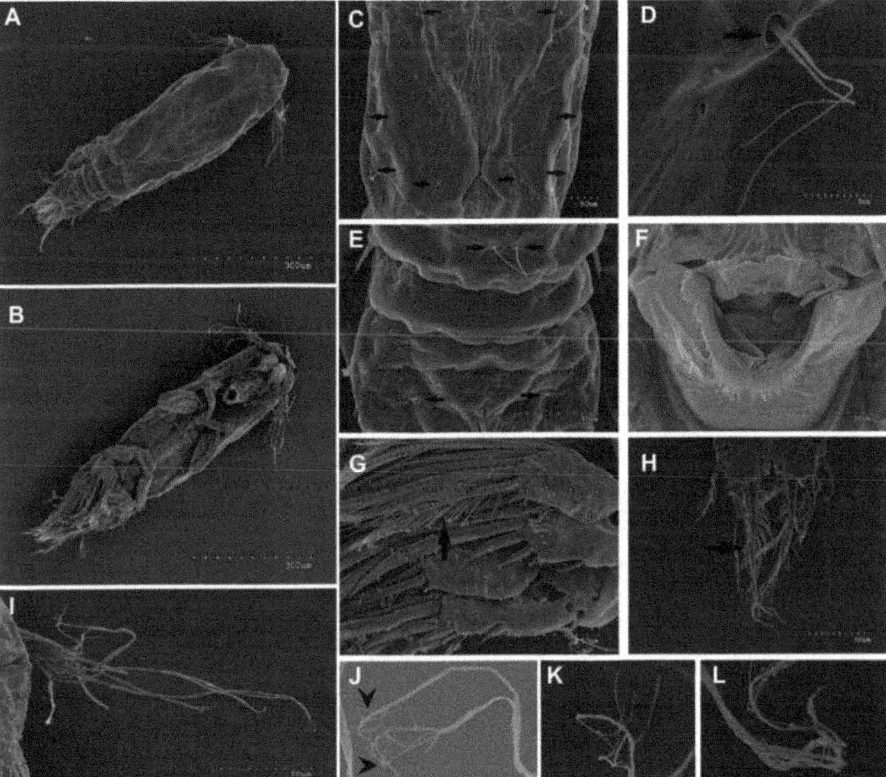

Figure 6. SEM. Copepodid setae morphology. Control animals: (**A**) Dorsal and (**B**) Ventral view of a *L. salmonis* copepodid. (**C**) Cephalothorax dorsal view showing some paired setae distributed along the body (arrows). (**D**) Detail from C shows the structure of a birrame setae (arrow). (**E**) Dorsal view of the abdomen showing some paired setae (arrows). (**F**) Mouth of the copepodid. (**G**) Ventral arms showing pinnate setae (arrow). (**H**) Caudal ramus showing the distal setae (arrow). (**I**) First antenna. The irregular branching tips are visible (arrowheads). (**J–L**) Detail of the first antenna setae showing their irregularly branching tips. Scale bar: (**A,B**) = 300 μm; (**C,G,H,I**) = 50 μm; (**E**) = 30 μm; (**F**) = 20 μm; (**J,K**) = 10 μm; (**D,L**) = 5 μm.

The first antenna (Figure 6) presents a proximal segment with 3 unramed setae and a distal segment with 5 setae with irregularly branching tips, 7 unramed setae, and 1 aesthete. The second antenna exhibits 3 segments with a spiniform process.

The 3 thoracic legs (Figure 6) present plumose setae, semipinnate setae, pinnate setae, spines, spiniform process, and fine setules. Caudal ramus (Figure 6) shows both short and long pinnate setae and aesthete.

3.1.2. Ultrastructural Analysis of Copepodids Setae after Noise Exposure

Ultrastructural changes took place on *L. salmonis* copepodid setae following acoustic exposure. In the control animals, the first antenna presented completely free setae with irregularly branching tips on the distal segment (Figure 7A–D). All the exposed copepodids presented different degrees of fusion of the irregularly branching tips of the setae on the distal segment of the first antenna (Figure 7E–T).

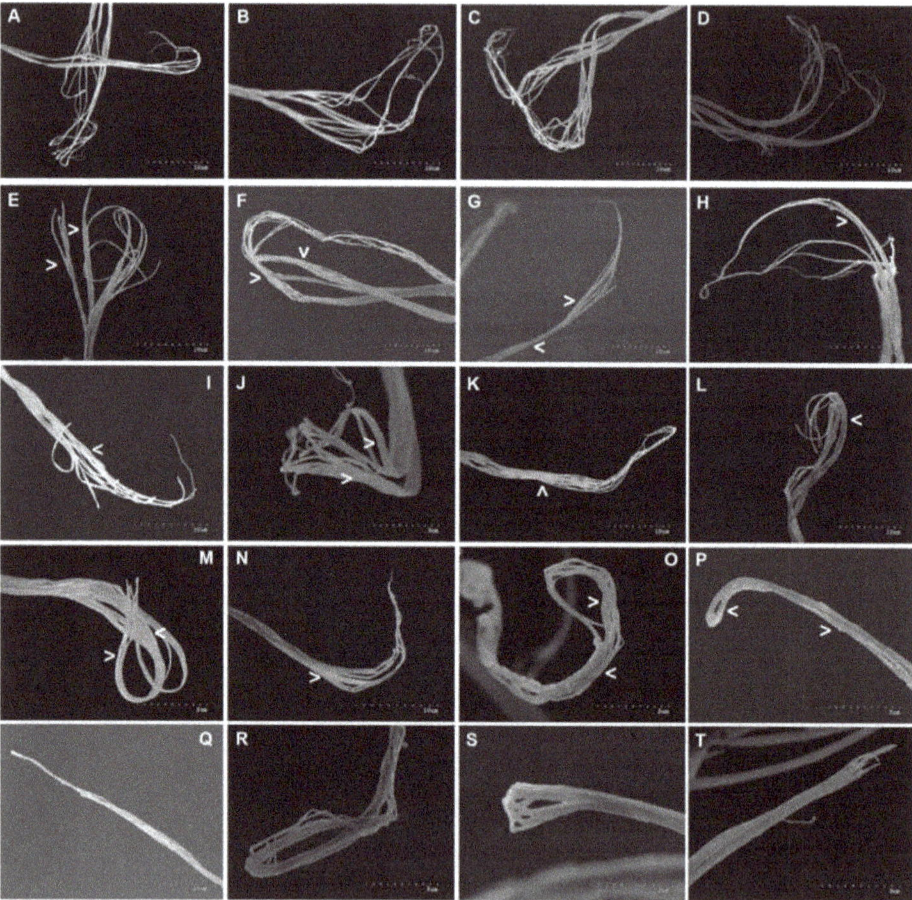

Figure 7. SEM: (**A**) *L. salmonis* copepodids. Setae on distal segment of first antenna; (**A–D**) Normal setae on control animal. The tips on the setae distal segments are entirely free (not fused); (**E–H**) Different views of exposed animals showing fusion (arrowheads) on the basal segment of the setae on the distal segment of the first antenna; (**I–N**) Different views of exposed animals showing the almost entirely fused (arrowheads) distal segment of the first antenna; (**O–T**) Different views of exposed animals showing completely fused distal segment of the first antenna. Scale bar: (**A–I,K,L,N,Q**) = 10 μm; (**M,O,P,R**) = 5 μm; (**J,S,T**) = 3 μm.

3.1.3. Exposure Parameters vs. Lesions

Frequency

After sound exposure and the analysis of the first antenna setae of the sea lice, we found maximum fusion at 350 Hz (95.5%). However, with frequencies between 300 Hz and 550 Hz, we achieved a percentage of setae fusion higher than 90% (Figure 8).

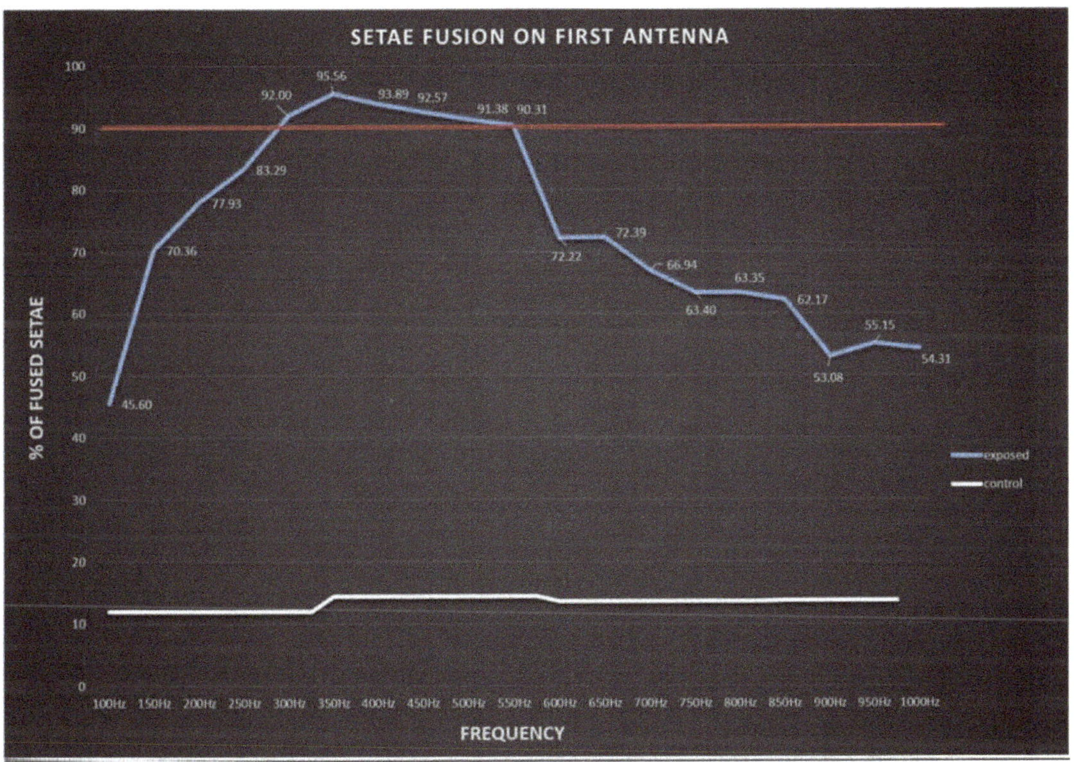

Figure 8. Setae fusion on sea lice first antenna (%) in function of frequency. 350 Hz achieved the maximum percentage of setae fusion. Between 350 Hz and 550 Hz the fusion percentage was higher than 90% (red bar).

After exposure to combinations of frequencies that previously had achieved the maximum fusion and the analysis of the first antenna setae of the sea lice, we found that 350 Hz–450 Hz and 350 Hz–550 Hz were the two combinations that achieved the maximum percentage of setae fusion (95.2%) (Figure 9).

We appointed that there was not a significant increase in the level of fusion with the combination of two frequencies.

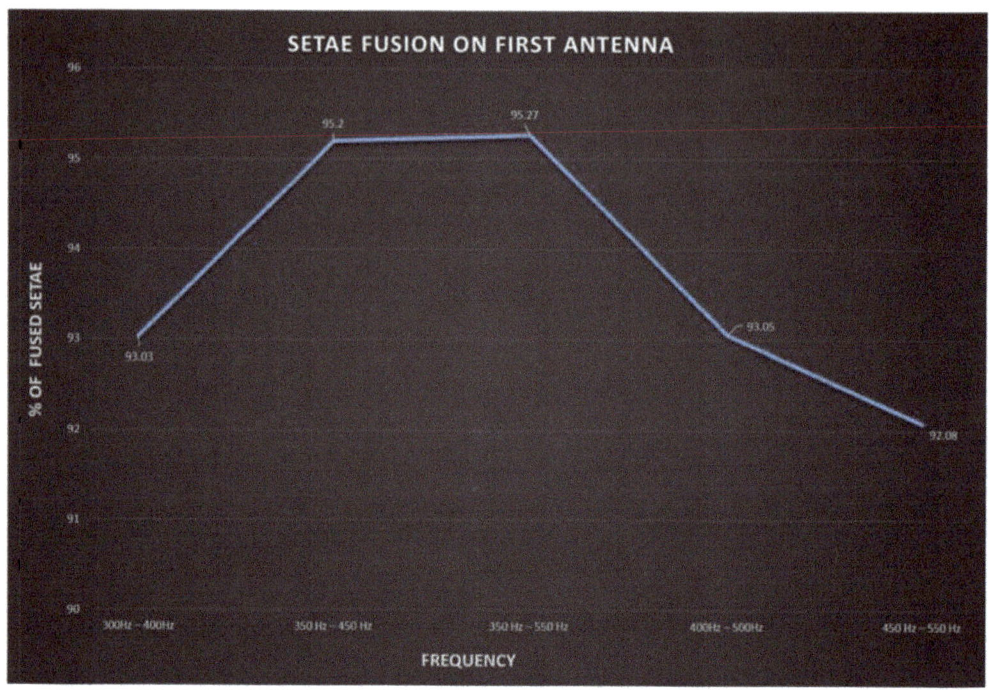

Figure 9. Setae fusion on sea lice first antenna (%) in function of frequency combinations of 350 Hz–450 Hz and 350 Hz. 550 Hz are the combination that achieve the maximum percentage of setae fusion (95.2%).

Amplitude

We found maximum setae fusion with the combination of 350 Hz-2 h-65 V and 500 Hz-2 h-65 V (93.02%). Other combinations (e.g., 350 Hz-2 h-65 V and 500 Hz-1 h-65 V, 350 Hz-3 h-65 V) achieved a percentage of setae fusion higher than 90% (Table 4).

Table 4. Setae fusion on sea lice first antenna (%) in function of frequency, time, and level of exposure in our tank conditions.

	% Fused Antenna
350 Hz-2 h-191 SEL	89.68
350 Hz-2 h-167 SEL	70.17
500 Hz-2 h-194 SEL	84.75
500 Hz-4 h-193 SEL	87.91
500 Hz-4 h-189 SEL	77.58
350 Hz-2 h-500 Hz-2 h-195 SEL	93.02
350 Hz-2 h-500 Hz-1 h-194 SEL	90.11
350 Hz-3 h-192 SEL	92.13
350 Hz-2 h-500 Hz-2 h-191 SEL	83.32
Control-152 SEL	12.89

3.1.4. Determination of Ultrastructural Lesions in Inner Tissues of Sea Lice Copepodids

Sound exposure affected the nervous system and A/B cells, which are responsible for the precursor secretions of the frontal filament. The lesions present in all the samples of exposed sea lice copepodids (but not in any of the control animals) were characterized by an exuberant accumulation of dark material and by cell cytoplasm vacuolization (Figures 10 and 11). These pathological features suggest the involvement of massive autophagy processes.

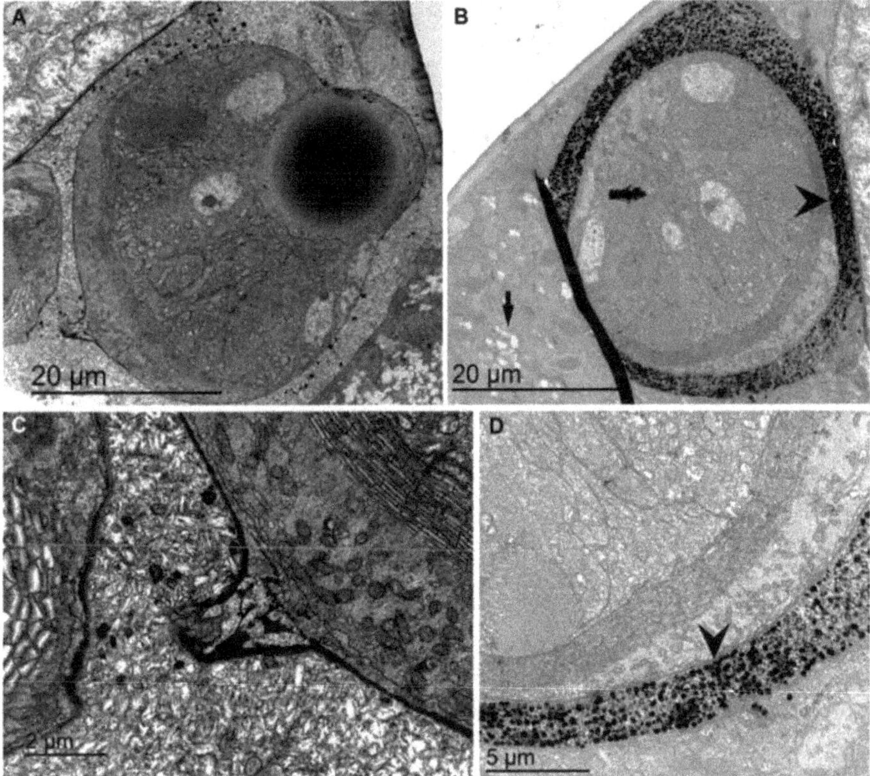

Figure 10. TEM. Sagittal section of the copepodid anterior cephalotorax showing the copepodid eye: (**A,C**) Control copepodid; (**B,D**); Exposed copepodid. (**A**) In control animals the dark inclusions around the eye are scarce. (**B**) In exposed copepdids a large amount of dark inclusions are visible in the axons of the central nervous system surrounding the eye (arrowhead). Vacuolization is visible on the tissue surrounding the eye (arrow); (**C**) Detail of A showing the optic nerve. Note the low quantity of dark inclusions around the eye; (**D**) Detail from B. Arrowhead point to the large amount of dark inclusions. Scale bar: (**A,B**) = 20 μm; (**D**) = 5 μm; and (**C**) = 2 μm.

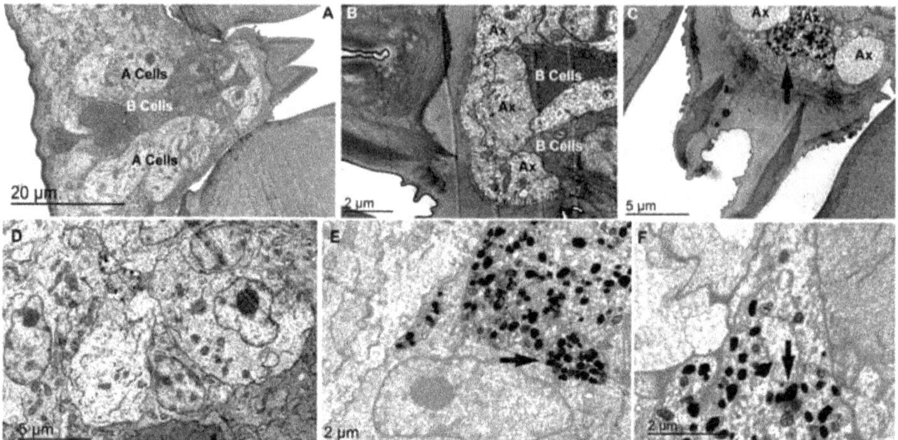

Figure 11. TEM. Frontal medial section of the copepodid anterior cephalotorax showing A and B Cells involved in Frontal Filament production: (**A,B,D**) Control; (**C,E,F**) Exposed: In control animals A and B cells do not show inner dark inclusions. (**B**) Axons sited next to A and B cells present normal aspect. (**C**) In exposed copepodids dark inclusions are visible in the axons of the nervous system (arrow). (**D**) Normal aspect of cells without dark inclusions. (**E,F**) Dark inclusions in the cells of exposed copepodids (arrows). Scale bar: (**A**) = 20 μm; (**C,D**) = 5 μm; and (**B,E,F**) = 2 μm.

3.2. Sea Trials

3.2.1. *L. salmonis* Adult and Pre-Adult Sensory Setae Morphology

In adult and pre-adult specimens (length 9.9 ± 1; width 4.5 ± 0.2), the first antenna (Figure 12) presents a proximal segment with 27 setae (25 pinnate on ventral and 2 unramed on dorsal surface) and a distal segment bearing 1 seta on posterior margin and 11 setae and 2 aesthetes at apex. Some setae are visible on maxilas too. The five pairs of legs present unramed setae, pinnate setae, spines and rows of simple setules. Caudal ramus shows distal setae with distinctly inflated bases that are relatively short.

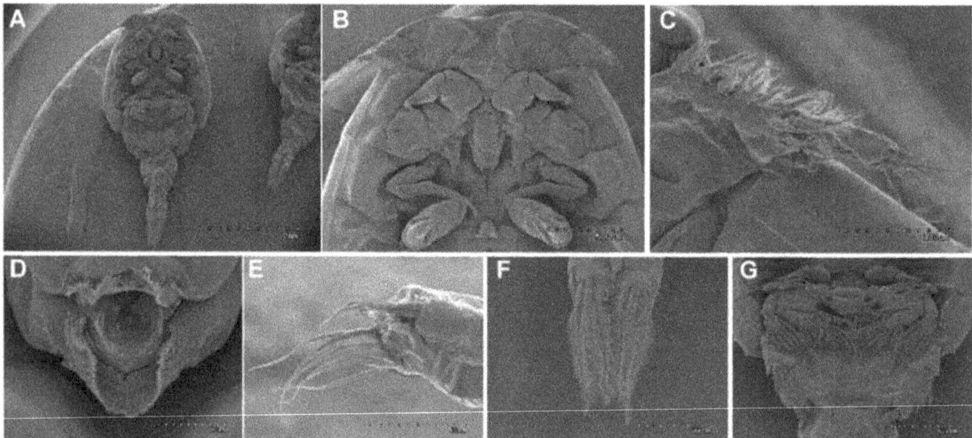

Figure 12. SEM. Adult and pre-adult *L. salmonis* morphology. Control animals: (**A**) Ventral view of the whole body of a pre-adult; (**B**) Ventral anterior view of an adult; (**C**) First antenna of an adult *L. salmonis*; (**D**) Mouth of a pre-adult; (**E**) Distal segment of the first antenna of *L. salmonis* adult; (**F**) Caudal ramus showing the distal setae; (**G**) Ventral arm showing the distribution of the pinnate setae. Scale bar: (**A**) = 2 mm; (**B,D,G**) = 500 μm; (**F**) = 300 μm; (**C**) = 100 μm; and (**E**) = 50 μm.

3.2.2. Ultrastructural Analysis of Pre-Adult and Adult First Antenna Setae after Noise Exposure

Ultrastructural changes took place on setae in adults and pre-adults of *L. salmonis* following acoustic exposure. Damaged setae were either extruded, completely missing, or presented flaccid, fused, or missing sensory hairs.

After two weeks of sound exposure (Figure 13B,D,E,G and Figure 14B–E) in comparison with the same tissues from control animals (Figures 13A and 14A), damage was observed on the first antenna setae. The setae of the proximal segment of the first antenna presented sensory hairs flaccid, fused or showed blebs. Some setae had lost part of the sensory hairs. Sometimes the setae had lost their rigidity and appeared unstructured and almost empty. Some animals presented setae partially or completely ejected above the antenna surface (Figures 13G and 14E).

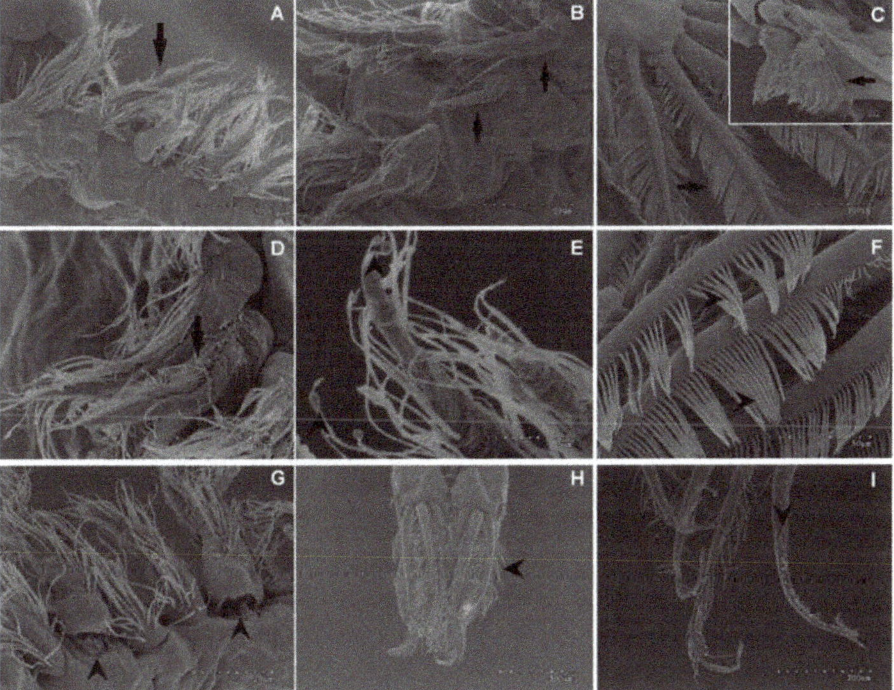

Figure 13. SEM. Pre-adult *L-salmonis* setae on proximal segment of first antenna: (**A,C,H**) Control animals; (**B,D–G,I**) animals after 2 weeks of sound exposure on sea trials. (**A**) Image of healthy setae bearing organized sensory hairs (arrow). (**B**) Setae showing flaccid or fused sensory hairs. Some of them have almost entirely lost the sensory hairs (arrows). (**C**) Pinnate setae on ventral arms presenting normal aspect (arrow). Insert in (**C**), lateral fraction of ventral arms presenting sensory hairs with normal aspect (arrow). (**D**) Section of proximal segment of the first antenna showing all the setae bearing bend and flaccid sensory hairs (arrow). (**E**) Sensory hairs showing blebs (arrowheads). (**F**) Pinnate setae on ventral arms are fused. (**G**) Setae partially ejected (arrowheads) above the antenna surface. (**H**) Normal aspect of the sensory hairs in the distal setae on caudal ramus. (**I**) Caudal ramus has partially lost the sensory hairs. Scale bar: (**H**) = 300 µm; (**C,I**); Insert in (**C**) = 100 µm; (**F**) = 50 µm; (**A**) = 30 µm; (**B,D,G**) = 20 µm; and (**E**) = 5 µm.

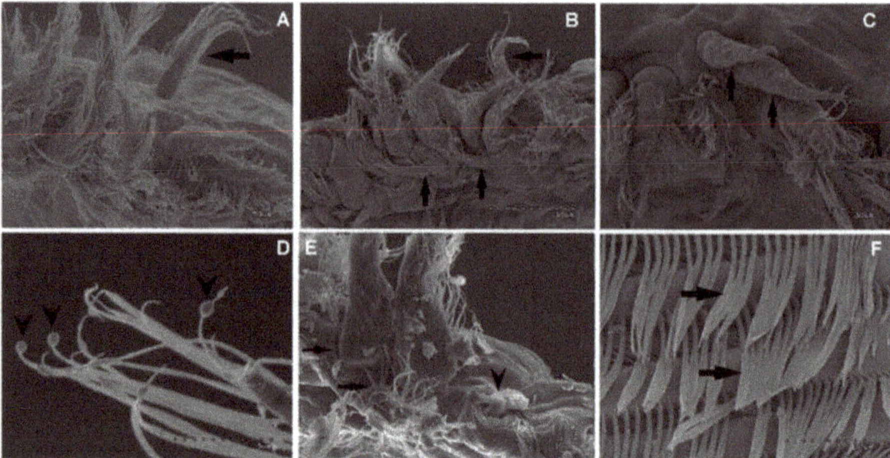

Figure 14. SEM. Adult *L. salmonis* setae on proximal segment of first antenna: (**A**) Control animals; (**B–F**): animals after two weeks of sound exposure on sea trials. (**A**) Image of healthy setae bearing organized sensory hairs (arrow). (**B**) Setae showing flaccid or fused sensory hairs. Some of them have almost lost totally the sensory hairs (arrows). (**C**) Some of the setae have almost lost totally the sensory hairs (arrows). (**D**) Sensory hairs showing blebs (arrowheads). (**E**) Two setae are partially (arrows) or totally ejected (arrowhead) above the antenna surface. (**F**) Pinnate setae on ventral arms are fused (arrows). Scale bar: (**A**) = 500 μm; (**B,C,E,F**) = 30 μm; and (**D**) = 5 μm.

In some specimens, in addition to the lesions on the first antenna, we found some effects on setae located in other positions. The arms showed sections of different lengths of fused pinnate setae (Figures 13F and 14F). Some specimens presented loss of sensory hairs or broken bases of the distal setae on caudal ramus (Figure 13I).

3.2.3. Ultrastructural Analysis of Chalimus, Pre-Adult and Adult *L. salmonis* Inner Tissues after Sound Exposure

Sound effects were observed in chalimus, pre-adult, and adult stages in comparison with the control animals. All noise-exposed individuals presented a similar variety of ultrastructural changes in cells of the inner tissues surrounding the eyes (in chalimus, specifically in cells involved in frontal filament production, namely A and B cells) and in axons of neurons around the eyes. In cell cytoplasm, ultrastructural changes were essentially a massive accumulation of dark inclusions (including ribosomes), the presence of numerous large double-membrane bounded autophagic vacuoles and of numerous lysosomes, and vacuolization of the cell cytoplasm (Figures 15–20). In addition, some B cells showed a deformed cell nucleus and chromatin compaction into the nucleus (Figure 15).

In the axons of neurons around the eyes, prominent ultrastructural changes were the presence of both abundant double-membrane-bounded autophagic vacuoles and myelin-like formations (Figures 15–18 and 21), and the accumulation of lysosomes at different stages of evolution (Figures 19 and 21). Large areas of empty cytoplasm (Figures 16–18, 20 and 21) and the accumulation of dark inclusions including ribosomes were also a common feature in these nervous cells (Figures 15, 17 and 18).

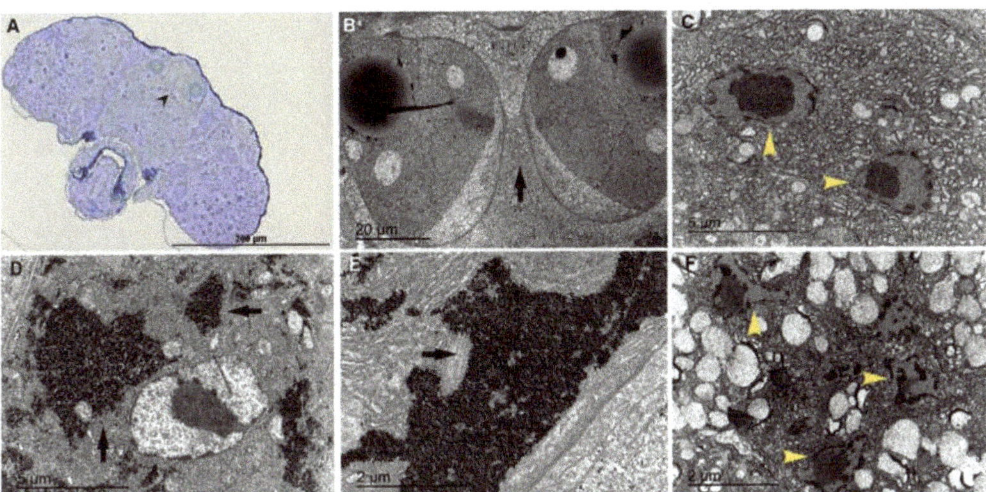

Figure 15. (**A**) Light microscopy; (**B–F**) TEM. Frontal medial section of a chalimus stage of sea lice; (**A–C**) Control specimens; (**D–F**) 3 weeks exposed specimens. (**A**) In light microscopy images of control animals there are not specially stained areas (corresponding to dark inclusions in TEM). Arrowhead shows the eyes. (**B**) Eyes of a control animal. The axons of the central nervous system surrounding the eye did not show dark inclusions (arrow). (**C**) Normal B cells nucleus (yellow arrowheads) in a control animal. (**D**) Dark inclusions in A cell cytoplasm are visible (arrows). (**E**) A large section of the cytoplasm of A cells are filled with ribosomes (dark inclusions, arrow). (**F**) By comparison to C, the B cell nuclei on exposed animals presented irregular shapes and chromatin compaction (yellow arrowheads). Scale bar: (**A**) = 200 μm; (**B**) = 20 μm; (**C,D**) = 5 μm; (**E,F**) = 2 μm.

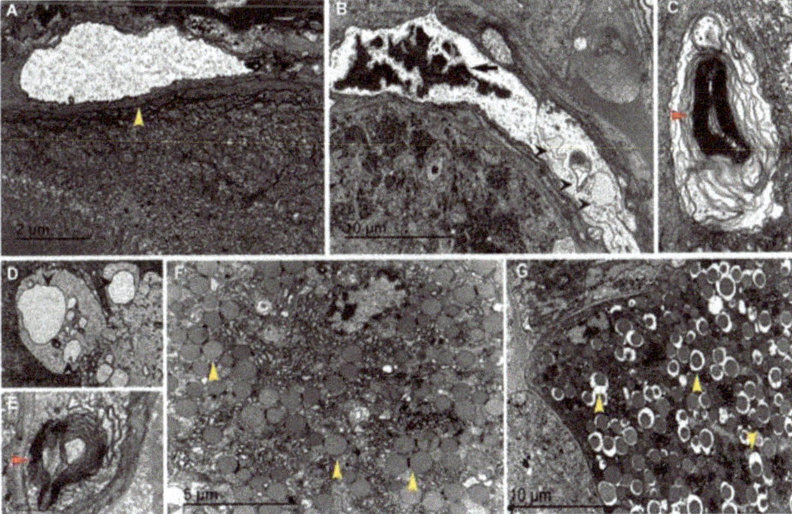

Figure 16. TEM. Frontal medial section of a chalimus anterior cephalothorax: (**A,F**) Control specimens; (**B–E,G**) 3 weeks exposed specimens. (**A**) There are no dark inclusions visible in the central nervous system axons (yellow arrowhead) neighbours to the eye. (**B**) One axon of the central nervous system shows ribosome accumulation (arrow) and presence of double-membrane-bounded autophagic vacuoles (arrowheads). (**C,E**) "Myelin-like formations" (red arrowhead). (**D**) Double-membrane-bounded autophagic vacuoles (arrowheads). (**F**) Normal aspect of the tissue located next to the B cells. (**G**) the tissue shows a process of vacuolization (yellow arrowheads). Scale bar: (**B,G**) = 10 μm; (**F**) = 5 μm; (**A,C,D**) = 2 μm; (**E**) = 1 μm.

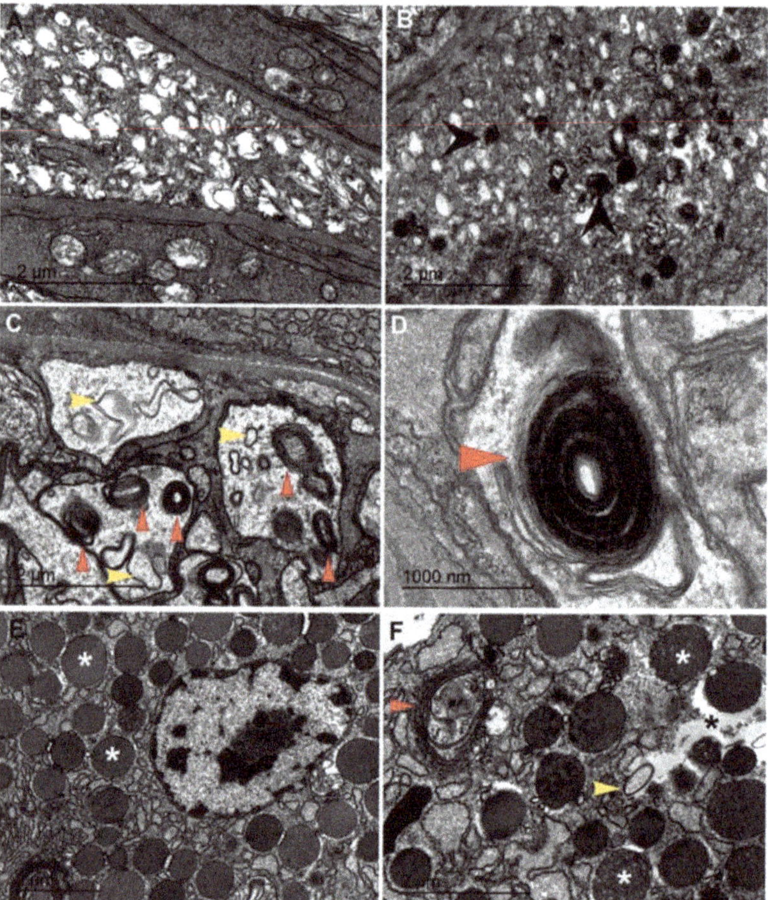

Figure 17. TEM. Frontal medial section of a chalimus anterior cephalothorax: (**A,E**) Control specimens; (**B,C,D,F**) Six weeks exposed specimens. (**A**) Normal aspect of the central nervous system between the two eyes. (**B**) By comparison to A, large dark inclusions (black arrowheads) are visible in the central nervous system between the two eyes. (**C**) "Myelin-like formations" (red arrowheads) and double-membrane-bounded autophagic vacuoles (yellow arrowheads) are present in axons. (**D**) Detail of "myelin-like formations" in an axon (red arrowhead). (**E**) Normal lysosomes (white asterisks) next to the B cell nuclei. (**F**) Some degraded lysosomes (white asterisks) are visible in a B cell of an exposed animal. "Myelin-like formations" (red arrowhead) and autophagic vacuoles (yellow arrowhead) are present. Note the empty cytoplasm in some areas of the tissue (black asterisks). Scale bar: (**A–C,E,F**) = 2 μm; (**D**) = 1000 nm.

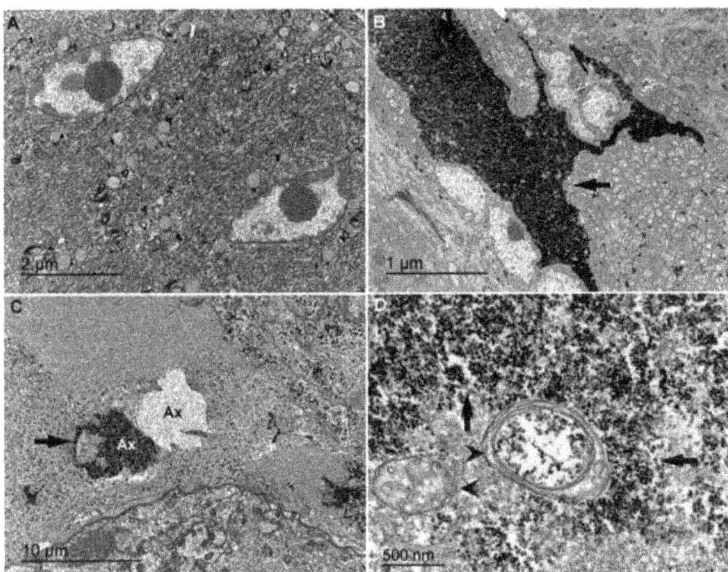

Figure 18. TEM. Frontal medial section of a pre-adult anterior cephalothorax showing tissues located around the eyes: (**A**) Control specimens; (**B,C,D**) Exposed specimens. (**A**) Normal aspect of cells in tissue surrounding the eyes. No dark inclusions are visible. (**B**) A large section of the cytoplasm is filled with ribosomes (dark inclusions, arrow). (**C**) One axon filled with dark inclusions (arrow) is adjacent located to one normal axon. (**D**) In the cytoplasm of an axon note the accumulation of ribosomes (dark inclusions, arrows) and the presence of autophagic vacuoles (arrowheads). Scale bar: (**C**) = 10 μm; (**A**) = 2 μm; (**B**) = 1 μm; (**D**) = 500 nm.

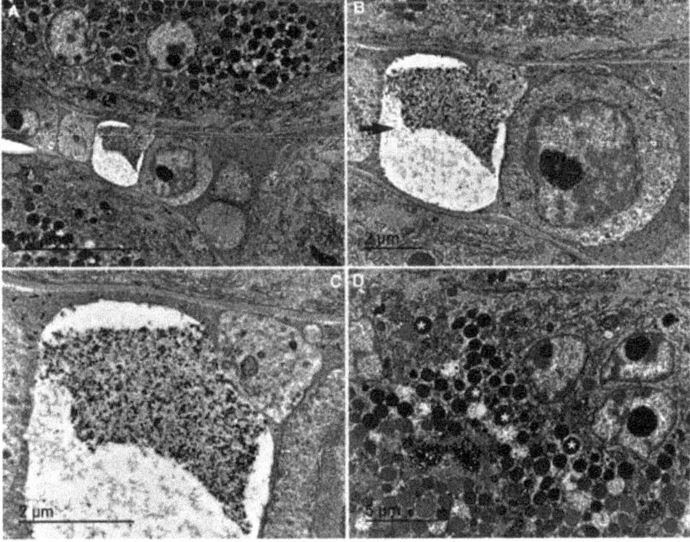

Figure 19. TEM. Frontal medial section of an exposed pre-adult anterior cephalothorax. Exposed animal: (**A**) General view of a section of tissue showing different features on sound exposed cells (Details in **B–D**); (**B**) Two adjacent cells are visible. On the right a normal cell shows its nucleus with inner nucleolus. On the left (arrow) a sound affected cell shows organelles destroyed by an enzymatic process; (**C**) Detail from (**B,D**), presence of very dark lysosomes (asterisks). Scale bar: (**A**) = 10 μm; (**D**) = 5 μm; and (**B,C**) = 2 μm.

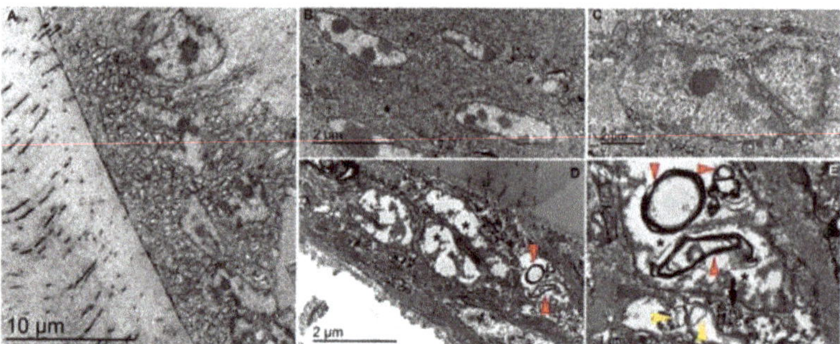

Figure 20. TEM. Sagittal section of adult sea lice anterior cephalothorax: (**A–C**) control animals; (**D,E**) Exposed animals; (**A–C**) No abnormal features are visible. (**D**) Cells present large empty areas in the cytoplasm (asterisks). Red triangles point to "myelin-like formations". (**E**) Detail from D. Note the double-membrane-bounded autophagic vacuoles (yellow arrowheads), the empty areas of cytoplasm (asterisk), the "myelin-like formations" (red arrowheads) and ribosome accumulation (arrow). Scale bar: (**A**) = 10 μm; (**B–D**) = 2 μm; (**E**) = 1 μm.

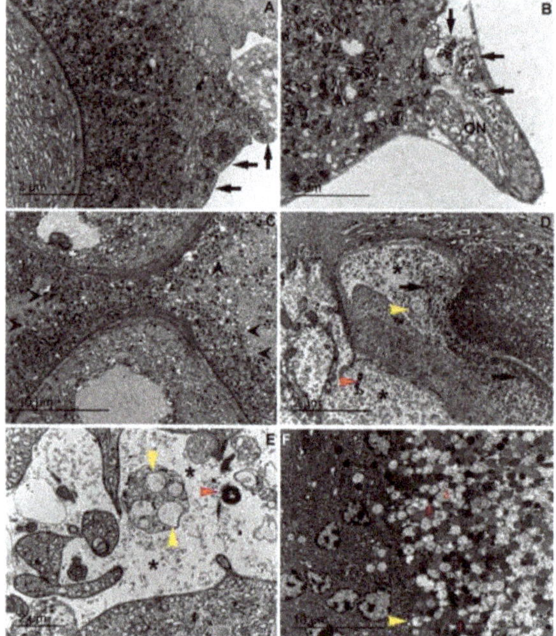

Figure 21. TEM. Frontal medial section of exposed adult anterior cephalothorax showing tissues located around the eye. (**A**) Arrows point to cells of the central nervous tissue filled with dark inclusions (ribosomes). (**B**) Detail from the optic nerves (ON) showing ribosome accumulation (arrows). (**C**) The axons of the central nervous system between the two eyes present some large dark inclusions (arrowheads). (**D,E**) Cells of the central nervous system shows autophagic vacuoles (yellow arrowheads), empty areas of cytoplasm (asterisks), "myelin-like formations" (red triangles) and ribosome accumulation (arrows). (**F**) The large amount of lysosomes type 3 suggest the evolution sequence of lysosomes going from the darkest (1) to lightest (3) appearance probably linked to sustained autophagy. Yellow arrowhead points to a type 3 lysosome releasing their inner content to a next autophagosoma. Scale bar: (**C,F**) = 10 μm; (**A,B,D,E**) = 2 μm.

4. Discussion

The sea louse *L. salmonis* causes millions of dollars in commercial losses to the salmon aquaculture industry globally. It reduces the productivity at fish farms through either low feed efficiency or growth reduction of the fish. In addition to such an industrial problem, it was recently shown that lice from salmon farms can have an adverse impact on wild migratory salmonids by increasing the abundance of this parasite in bays and estuaries adjacent to the farms [1].

Different methods have been used in the fight against the sea lice infestation. In-feed treatments and usage of skirts (sheets hung around the salmon cages to prevent sea lice from entering) are very expensive methods. Skirts have low impact on the salmon welfare and the environment but reduce oxygen flow, which may cause a detrimental effect on fish respiratory functions [18].

Other methods such as cleaner fish, fresh water, physical removal measures, and veterinary medicines (sea lice have built up resistance to most of the chemicals that are used in medicines [19]) have environmental, health, and welfare impacts [19]. Less cost-effective methods include the use of hydrogen peroxide baths that, in addition, have effects on fish welfare and, consequently, on environmental and human health. From this perspective, the complexity of sea lice control requires a global holistic approach [18].

In this context, sound exposure methods can constitute an effective, innovative, and promising technology to address sea lice infestation. Our results showed the first ultrastructural images that characterize pathological changes in copepodids, chalimus, adult and pre-adults *L. salmonis* sensory first antenna setae after sound exposure. Essentially, the lesions were partial or complete fusions of the setae irregular branching tips of the first antenna. Fusion of fine sensory structures is typically the result of mechanical constraints due loud sound vibrations. Fusion of stereocilia has been shown to occur in statocysts sensory epithelium and lateral line systems of cephalopods [17,20] and in statocyst of cnidarians [14] after underwater noise exposure. Moreover, stereocilia fusion on auditory hair cells is a morphological characteristic of acoustic trauma in terrestrial animals [21,22]. Such pathological changes that directly affect the main *L. salmonis* sensory organ could make difficult finding of a host for a copepodid [8,9]. Additionally, exposed sea lice showed lesions on some distal pinnate setae on the caudal ramus and ventral arms. These abnormalities could provoke difficulties for the sea lice to move around the fish, which could also contribute to the decrease in the number of attachments to the salmons.

Moreover, to the best of our knowledge, this study shows the first published ultrastructural images of sea lice inner tissues affected by sound. The exposure affected the central nervous system of all analysed stages, likely altering their normal behaviour and challenging their survival as has been shown in other invertebrates [10,14]. In the copepodids and chalimus stages the A/B cells, which are responsible for the secretion of the precursor of frontal filament [7], were affected, thus probably challenging a correct anchoring to the fish. Similar lesions were shown on the tissues located in the same regions in both adults and pre-adults. In all of the stages, those lesions appear to be produced by autophagic and apoptotic processes [23–25].

Typical features of autophagic processes include the presence of numerous lysosomes and double-membrane-bounded vacuoles, "myelin-like formations" resulting from cell membrane destruction, large aggregates of dark material, and massive accumulation of ribosomes in the cell cytoplasm. Interestingly, in our exposed tissues we could follow the normal evolution sequence from primary to mature secondary lysosome with decreasing activity, which posteriorly released its content to an adjacent autophagosome [26,27]. The presence of a large amount of secondary lysosomes and residual bodies in the exposed animals is a clear sign of the extreme functioning of the cytological mechanisms caused by the stress situation originated after sound exposure. Moreover, the frequent presence of large areas of empty cytoplasm was the hallmark of advanced stages of cell degeneration through autophagy. Beside autophagy, pathological features such as deformed cell nuclei, chromatin compaction into the nuclei, and cytoplasm condensation strongly suggest the occurrence of apoptotic processes.

The mechanisms by which sound induced massive autophagy and apoptosis in the present specimens, have yet to be precisely determined. One possible hypothesis is that acoustic exposure primarily induced an oxidative stress that is known to regulate the expression of both autophagy and apoptosis [28]. In support of the oxidative stress hypothesis, the accumulation of dark inclusions around the eye could correspond to mitochondrial autophagic profiles. Mitochondria in fact play a key role in oxidative stress [29–31], and autophagy-damaged mitochondria has been described as the effect of sound exposure in the optic nerve [22]. Another possibility is that part of the dark inclusions was due to hyperpigmentation. Hyperpigmentation actually may occur as a consequence of an oxidative stress following acoustic trauma in several tissues, including the eyes [32].

Altogether, the present study indicates that the central nervous system in all stages and the A/B cells (responsible for the secretion of the precursor of frontal filament) [7] in copepodids and chalimus stages were affected by sound exposure. These encouraging findings therefore indicate that sound exposure can lead to severe consequences on the capacity of the sea lice to infest its host. The present results were completed by an exhaustive health status analysis of the exposed salmons [33] that showed that the use of these frequency combinations did not affect the fish. In this assessment through gross pathology and histopathological analysis, salmons didn't show any lesion that could be related to sound exposure. In addition, the analysis of the otolith's organs did not show any effects on the auditory organs of the fish. Although some consequences of the sound-induced lesions found in lice remain to be further studied, this method constitutes a promising approach to address lice plagues while also reducing the need for chemical treatments of the fish.

5. Patents

André M., Solé M., Van der Schaar, De Vreese S (International Patent WO 2018/167003 A1). 20-09-2018. A method for inducing lethal lesions in sensory organs of undesirable aquatic organisms by use of sound. Licensed to SEASEL SOLUTIONS AS [NO/NO]; P.O. BOX 93 N-6282 BRATTVÅG (NO).

André M., Solé M., Van der Schaar (Norwegian patent WO/2020/048945). 03-09-2019. System and method for reducing sea lice exposure in marine fish farming. Licensed to SEASEL SOLUTIONS AS [NO/NO]; P.O. BOX 93 N-6282 BRATTVÅG (NO).

Author Contributions: M.S. and M.A. planned the research and designed the study; M.S., M.A., M.v.d.S. and S.D.V. conducted experimental/lab work; M.S. and M.A. conducted sea trial work; M.S. and J.-M.F. performed SEM analysis; M.S. performed TEM analysis; M.S., M.L. and M.A. analysed the data; M.S. and M.A. prepared the figures; M.S. and M.A. wrote the article. All authors reviewed the manuscript. All authors have read and agreed to the published version of the manuscript.

Funding: Funding for this project was provided by SEASEL SOLUTIONS AS. Project: *An acoustic and Bioacoustic solution to sea lice infestation on salmon* P.O. BOX 93 N-6282 BRATTVÅG. Norway.

Institutional Review Board Statement: Although there are no legal requirements for studies involving crustaceans in Spain, the experimental protocol strictly followed the necessary precautions to comply with current ethical and welfare considerations when dealing with animals in scientific experimentation (Royal Decree 1386/2018, of 19 November). This process was also carefully analysed and approved by the Ethical Committee for Scientific Research of the Technical University of Catalonia, BarcelonaTech (UPC) (approval code B9900085).

Informed Consent Statement: Not applicable.

Acknowledgments: We would like to thank the staff of Mowi Fish Feed A/S Avd Averøy for their assistance and helpful cooperation during the experiments at sea. Special thanks to Josep M. Rebled, Eva Prats, Adriana Martínez, and Rosa Rivera (Unitat de microscòpia electrònica, Hospital Clínic, Universitat de Barcelona) for assistance and advice in obtaining TEM images.

Conflicts of Interest: The authors declare no conflict of interest.

References

1. Whelan, K.A. Review of the Impacts of the Salmon Louse, *Lepeophtheirus salmonis* (*Krøyer, 1837*) on Wild Salmonids. Atl. Salmon Trust **2010**. Available online: https://atlanticsalmontrust.org/wp-content/uploads/2016/12/ast-sea-lice-impacts-review1.pdf (accessed on 12 May 2021).
2. Revie, C.; Dill, L.; Finstad, B.; Todd, C. *Sea Lice Working Group Report*; NINA Special Report 39; World Wildlife Fund: Norway. Available online: http://www.nina.no/archive/nina/pppbasepdf/temahefte/039.pdf (accessed on 12 May 2021).
3. Thorstad, E.; Todd, C.; Uglem, I.; Alamaru, A.; Bronstein, O. Effects of salmon lice *Lepeophtheirus salmonis* on wild sea trout, Salmo trutta—A literature review. *Aquac. Environ. Interact.* **2015**, *7*, 91–113. [CrossRef]
4. Wagner, G.; Fast, M.; Johnson, S. Physiology and immunology of *Lepeophtheirus salmonis* infections of salmonids. *Rev. Cell.* **2008**, *24*, 176–183. [CrossRef]
5. Mordue, A.; Birkett, M. A review of host finding behaviour in the parasitic sea louse, *Lepeophtheirus salmonis (Caligidae: Copepoda)*. *J. Fish Dis.* **2009**, *32*, 3–13. [CrossRef] [PubMed]
6. Strickler, J.; Bal, A. Setae of the first of the copepod *Cyclops scutifer* (Sars): Their structure and importance. *Proc. Natl. Acad. Sci. USA* **1973**, *70*, 2656–2659. [CrossRef] [PubMed]
7. Gonzalez-Alanis, P.; Wright, G.M.; Johnson, S.C.; Burka, J.F. Frontal filament morphogenesis in the salmon louse *Lepeophtheirus salmonis*. *J. Parasitol.* **2001**, *87*, 561–574. [CrossRef]
8. Heuch, P.A.; Karlsen, E. Detection of infrasonic water oscillations by copepodids of *Lepeophtheirus salmonis* (Copepoda 10.1093/plankt/19.6.735, Caligida). *J. Plankton Res.* **1997**, *19*, 735–747. [CrossRef]
9. Heuch, P.A.; Doall, M.H.; Yen, J. Water flow around a fish mimic attracts a parasitic and deters a planktonic copepod. *J. Plankton Res.* **2007**, *29*, i3–i16. [CrossRef]
10. Solé, M.; Sigray, P.; Lenoir, M.; van der Schaar, M.; Lalander, E.; André, M. Offshore exposure experiments on cuttlefish indicate received sound pressure and particle motion levels associated with acoustic trauma. *Sci. Rep.* **2017**, *7*, 45899. [CrossRef]
11. Solé, M.; Fortuño, J.-M.; van der Schaar, M.; André, M. An acoustic treatment to mitigate the effects of the apple snail on agriculture and natural ecosystems. *J. Mar. Sci. Eng.* **2021**, *9*.
12. Day, R.D.; McCauley, R.D.; Fitzgibbon, Q.P.; Hartmann, K.; Semmens, J.M. Seismic air guns damage rock lobster mechanosensory organs and impair righting reflex. *Proc. R. Soc. B Biol. Sci.* **2019**, *286*, 20191424. [CrossRef] [PubMed]
13. Day, R.D.; McCauley, R.D.; Fitzgibbon, Q.P.; Hartmann, K.; Semmens, J.M. Exposure to seismic air gun signals causes physiological harm and alters behavior in the scallop Pecten fumatus. *Proc. Natl. Acad. Sci. USA* **2017**, *114*, E8537–E8546. [CrossRef]
14. Solé, M.; Lenoir, M.; Fontuño, J.M.; Durfort, M.; van der Schaar, M.; André, M. Evidence of Cnidarians sensitivity to sound after exposure to low frequency noise underwater sources. *Sci. Rep.* **2016**, *6*, 37979. [CrossRef] [PubMed]
15. Fuchs, H.L.; Christman, A.J.; Gerbi, G.P.; Hunter, E.J.; Diez, F.J. Directional flow sensing by passively stable larvae. *J. Exp. Biol.* **2015**, *218*, 2782–2792. [CrossRef]
16. Johnson, S.; Albright, L. The developmental stages of *Lepeophtheirus salmonis* (Kroyer, 1837) (Copepoda:Caligidae). *Can. J. Zool.* **1991**, *69*, 929–950. [CrossRef]
17. Solé, M.; Lenoir, M.; Durfort, M.; López-Bejar, M.; Lombarte, A.; van der Schaar, M.; André, M. Does exposure to noise from human activities compromise sensory information from cephalopod statocysts? *Deep. Res. Part II Top. Stud. Oceanogr.* **2013**, *95*, 160–181. [CrossRef]
18. Barrett, L.T.; Oppedal, F.; Robinson, N.; Dempster, T. Prevention not cure: A review of methods to avoid sea lice infestations in salmon aquaculture. *Rev. Aquac.* **2020**, *12*, 2527–2543. [CrossRef]
19. Hannisdal, R.; Nøstbakken, O.J.; Hove, H.; Madsen, L.; Horsberg, T.E.; Lunestad, B.T. Anti-sea lice agents in Norwegian aquaculture; surveillance, treatment trends and possible implications for food safety. *Aquaculture* **2020**, *521*, 735044. [CrossRef]
20. Solé, M.; Lenoir, M.; Fortuño, J.-M.; van der Schaar, M.; André, M. A critical period of susceptibility to sound in the sensory cells of cephalopod hatchlings. *Biol. Open* **2018**, *7*, bio033860. [CrossRef]
21. Engström, B. Stereocilia of sensory cells in normal and hearing impaired ears. A morphological, physiological and behavioural study. *Scand. Audiol. Suppl.* **1983**, *19*, 1–34. [PubMed]
22. Liberman, M. Chronic ultrastructural changes in acoustic trauma: Serial-section reconstruction of stereocilia and cuticular plates. *Hear. Res.* **1987**, *26*, 65–88. [CrossRef]
23. D'Arcy, M.S. Cell death: A review of the major forms of apoptosis, necrosis and autophagy. *Cell Biol. Int.* **2019**, *43*, 582–592. [CrossRef] [PubMed]
24. Glick, D.; Barth, S.; MacLeod, K.F. Autophagy: Cellular and molecular mechanisms. *J. Pathol.* **2010**, *221*, 3–12. [CrossRef]
25. Elmore, S. Apoptosis: A review of programmed cell death. T.P. 2007 J.-516. Apoptosis: A review of programmed cell death. *Toxicol. Pathol.* **2007**, *35*, 495–516. [CrossRef]
26. Liberge, M.; Gros, O.; Frenkiel, L. Lysosomes and sulfide-oxidizing bodies in the bacteriocytes of Lucina pectinata, a cytochemical and microanalysis approach. *Mar. Biol.* **2001**, *139*, 401–409. [CrossRef]
27. Trivedi, P.C.; Bartlett, J.J.; Pulinilkunnil, T. Lysosomal Biology and Function: Modern View of Cellular Debris Bin. *Cells* **2020**, *9*, 1131. [CrossRef] [PubMed]

28. Kaminskyy, V.O.; Zhivotovsky, B. Free radicals in cross talk between autophagy and apoptosis. *Antioxid. Redox Signal.* **2014**, *21*, 86–102. [CrossRef] [PubMed]
29. Peoples, J.N.; Saraf, A.; Ghazal, N.; Pham, T.T.; Kwong, J.Q. Mitochondrial dysfunction and oxidative stress in heart disease. *Exp. Mol. Med.* **2019**, *51*, 1–13. [CrossRef]
30. Lejri, I.; Agapouda, A.; Grimm, A.; Eckert, A. Mitochondria- and Oxidative Stress-Targeting Substances in Cognitive Decline-Related Disorders: From Molecular Mechanisms to Clinical Evidence. *Oxid. Med. Cell Longev.* **2019**, *2019*, 1–26. [CrossRef]
31. Subramaniam, S.R.C.M. Mitochondrial dysfunction and oxidative stress in Parkinson's disease. *Prog. Neurobiol.* **2013**, *106–107*, 17–32. [CrossRef]
32. Taubitz, T.; Tschulakow, A.V.; Tikhonovich, M.; Illing, B.; Fang, Y.; Biesemeier, A.; Julien-Schraermeyer, S.; Schraermeyer, U. Ultrastructural alterations in the retinal pigment epithelium and photoreceptors of a Stargardt patient and three Stargardt mouse models: Indication for the central role of RPE melanin in oxidative stress. *PeerJ* **2018**, *2018*, e5215. [CrossRef]
33. Solé, M.; Constenla, M.; Padrós, F.; Lombarte, A.; Fortuño, J.M.; André, M. Farmed salmons show no pathological alterations when exposed to acoustic sea lice infestation treatment. *J. Mar. Sci. Eng.* **2021**, *9*.

Journal of
*Marine Science
and Engineering*

Article

Farmed Salmon Show No Pathological Alterations When Exposed to Acoustic Treatment for Sea Lice Infestation

Marta Solé [1,*], Maria Constenla [2], Francesc Padrós [2], Antoni Lombarte [3], José-Manuel Fortuño [3], Mike van der Schaar [1] and Michel André [1,*]

[1] Laboratory of Applied Bioacoustics, Technical University of Catalonia-Barcelona TECH, 08800 Barcelona, Spain; mike.vanderschaar@upc.edu
[2] Fish Diseases Diagnostic Service, Veterinary School, Universitat Autònoma de Barcelona, 08193 Barcelona, Spain; Maria.Constenla@uab.cat (M.C.); francesc.padros@uab.cat (F.P.)
[3] Institut de Ciències del Mar, Spanish National Research Council (CSIC), 08003 Barcelona, Spain; toni@icm.csic.es (A.L.); jmanuel@icm.csic.es (J.-M.F.)
* Correspondence: marta.sole@upc.edu (M.S.); michel.andre@upc.edu (M.A.)

Abstract: The use of bioacoustic methods to address sea lice infestation in salmonid farming is a promising innovative method but implies an exposure to sound that could affect the fish. An assessment of the effects of these techniques related to the salmon's welfare is presented here. The fish were repeatedly exposed to 350 Hz and 500 Hz tones in three- to four-hour exposure sessions, reaching received sound pressure levels of 140 to 150 dB re 1 μPa^2, with the goal of reaching total sound exposure levels above 190 dB re 1 μPa^2 s. Gross pathology and histopathological analysis performed on exposed salmons' organs did not reveal any lesions that could be associated to sound exposure. The analysis of their otoliths through electron microscopy imaging confirmed that the sound dose that was used to impair the lice had no effects on the fish auditory organs.

Keywords: salmon; *Salmo salar*; acoustic trauma; scanning electron microcopy; otolith organ; lateral line; histopathology; vaterite; neuromast

Citation: Solé, M.; Constenla, M.; Padrós, F.; Lombarte, A.; Fortuño, J.-M.; van der Schaar, M.; André, M. Farmed Salmon Show No Pathological Alterations When Exposed to Acoustic Treatment for Sea Lice Infestation. *J. Mar. Sci. Eng.* **2021**, *9*, 1114. https://doi.org/10.3390/jmse9101114

Academic Editor: Francesco Tiralongo

Received: 21 September 2021
Accepted: 11 October 2021
Published: 14 October 2021

Publisher's Note: MDPI stays neutral with regard to jurisdictional claims in published maps and institutional affiliations.

1. Introduction

Sea lice infection is still one of the most devastating diseases in the salmon industry [1] and many different methodologies and strategies to prevent or reduce the impact of the disease have been developed [2]. The efficiency of these strategies in the field has been shown to be variable due to different environmental and husbandry factors but, together with these treatments, salmon welfare in response to the use of certain preventive and delousing methods has become more relevant. This is the reason why nowadays the implementation of these strategies must be based on good efficacy scores but also well-balanced safety procedures. Recent findings on the use of acoustic treatments have demonstrated the potential of this approach to address lice infestation [3] but, together with these very encouraging results, it was also necessary to conduct studies to ensure that the risk to the acoustically treated salmon was negligible.

Fishes are indeed able to detect and respond to a wide range of sounds. Experimental studies can investigate the range of frequencies that a fish could detect, and then determine the lowest levels of the detected sound at each frequency (the 'threshold', or lowest signal that an animal will detect in some statistically determined percentage of signal presentations—most often 50%). However, for most commercial species there is no audiogram available in the literature to assess their sensitivity to noise. *Salmo salar* audiograms based on mean (±SE) minimum received levels (dB re 1 μPa) that elicited a characteristic auditory evoked potential (electrical response that is produced anytime a sound is perceived and that can be recorded out of the brain from electrodes) revealed that salmon are most sensitive to 200 Hz frequency sounds [4].

As most fish can hear sounds, a series of published experiments and studies have looked at the effects of intense sound sources on the otolith organ that is responsible for sound reception. Nine months of exposure to broadband noise in an aquaculture facility did not induce hearing loss in two species of fish (Nile tilapia, *Oreochromis niloticus* and bluegill sunfish, *Lepomis macrochirus*) [5]. Other studies showed that salmonids did not trigger temporary threshold shifts (equivalent to a temporary acoustic trauma), whereas northern pike (*Esox lucius*) and sandbar shark (*Carcharhinus plumbeus*) suffered from hearing loss that recovered within 24 h after exposure [6,7]. Several animals examined after a post-exposure period up to 52 days showed 2–7% sensory cell loss after sound exposure [8]. Other studies analysed the relationship between auditory hair cell damage and hearing loss [9,10].

Experimental studies have examined the effects of pile driving and airguns on fish. Although the associated source levels did not lead to mortality in any of the exposed animals, and despite no effects being found on external and internal anatomy or damage on sensory hair cells of the otolith organs [11,12], more recent studies suggest that damage to the sensory hairs of fish inner ear tissues are likely to occur at levels considerably higher than those inducing other physiological effects, such as swim bladder ruptures [13].

Teleost fishes present an inner ear that contains three calcareous structures (otoliths), overlaying the sensory epithelia that enable their capacity for hearing and balance. The sagitta, the largest otolith, is usually composed of calcium carbonate crystals in the form of aragonite. A deformity, extremely common in farmed fish, where the aragonite is replaced by vaterite (a clearer crystallised form of the calcium carbonate), heavily affects the farmed salmon [14].

Some studies looked at the effects of anthropogenic sounds on the fish behaviour, sowing minor behavioural and startle responses of fish maintained in cages at the start of the air-gun exposure, responses which appeared to decline at subsequent air-gun emissions, but this sound level did not appear to elicit a decline in catch [15–18]. In addition there were no permanent changes in the behaviour of the fish or invertebrates throughout the course of the studies, and animals did not appear to leave the exposed area [16]. Contrary to these findings, more recent studies reported alarm responses, flight reactions, aggressive behaviours, changes in antipredator defence behaviour and reproduction related behaviour (courtship vocal activity, spawning), alterations on schooling behaviour [19] and, in some cases, effects in larval development in addition to these behavioural changes [20].

Fish physiological responses to noise exposure, like stress, were also measured. Complementary to other health indicators, corticosteroid levels are considered a measure of stress. Several studies that looked at stress levels in different fish species concluded they were not influenced by noise exposure [5,21]. Nevertheless, it is fair to mention that these experimental approaches were undertaken in cages where fish could not avoid the exposed areas, making it difficult to definitively conclude on the short- and long-term physiological impact of noise exposure [22].

In the context of a project that aimed at addressing the problem of sea lice *Lepeophtheirus salmonis* infestation on salmon (*Salmo salar*) by using acoustic and bioacoustic techniques (SEASEL SOLUTIONS AS. Project: An acoustic and bioacoustic solution to sea lice infestation on salmon- P.O.BOX 93 N-6282 BRATTVÅG. Norway, [3]) and given the inconclusive results of previous studies, an evaluation of the possible effects on salmon after sound exposure was necessary. Here, we proceeded with a series of controlled exposure experiments to determine the salmon sensitivity to the sounds that would be used in our method against lice infestation, under laboratory and field conditions. Given the industrial nature of this project, this assessment was absolutely necessary in order to ensure that the risk for the commercial caged salmon was negligible and would not result in an economic burden for the companies using the method.

2. Methods

2.1. Laboratory Experiments

After determining the effective combination of sound parameters (frequency, time of exposure and amplitude) on the sea lice *L. salmonis* through controlled exposure experiments (CEE) [3], we repeated these experiments on salmon in the same configuration and assessed the potential alterations and lesions that the method could induce with respect to behavioural levels, the vestibular system, the lateral line and also in the fish organs after sound exposure.

2.1.1. Salmon Specimens Maintenance and Health Assessment

A set of salmon (*S. salar*) (*n* = 50; weight 204.6 ± 25.5 g; total length 26.6 ± 2.5 cm), was received and kept in continuous observation at the LAB infrastructure in a closed system of recirculating natural seawater (at 7–10 °C, salinity 35‰ and natural oxygenation) consisting of two mechanically filtered (physicochemical self-filtration system with activated carbon and sand, driven by a circulation pump and refrigeration system) fiberglass-reinforced plastic tanks with a capacity of 2000 L each and connected to each other (Figure S1, Supplemental Material). The fish stock was then progressively acclimatised to the test conditions for two weeks. Fish were regularly fed ad libitum with commercial food and feeding rates were also monitored. After these two weeks, 10 fish were sampled in order to proceed with a complete histopathological examination to guarantee an adequate health status for the test. These fish were taken as controls and sampled and processed in the same way as fish exposed to sound (Figure 1).

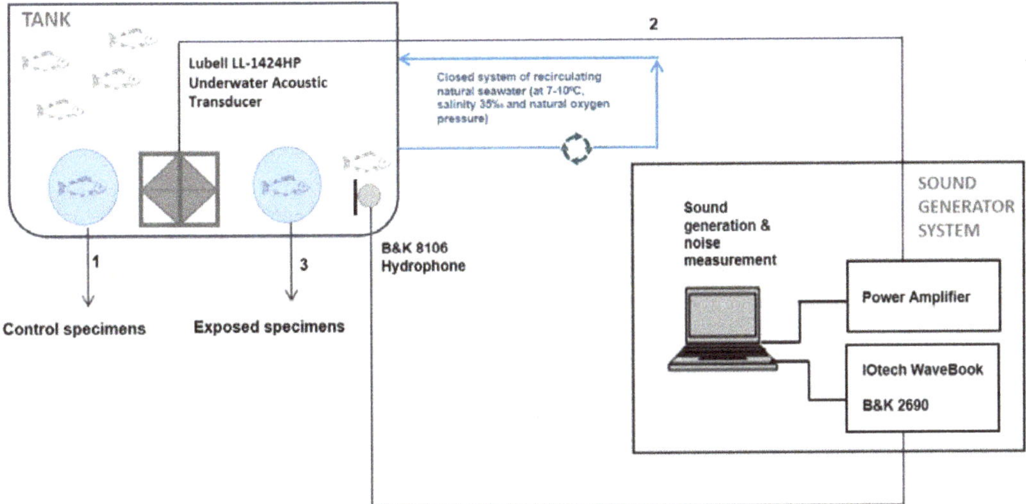

Figure 1. Sound exposure protocol, sampling collection, and analysis. (1) After 2 weeks of acclimation, 10 salmon were taken for sample analysis as controls previous to the sound exposure. (2) Forty salmon were exposed to sound (Weeks 1, 2, 3:2 cycles [350 Hz (65 V-2 h) and 500 Hz (65 V-2 h)], daily with 2 h rest in between two cycles. Week 4:3 cycles [350 Hz (65 V-2 h) and 500 Hz (65 V-2 h)], daily with 2 h rest in between two cycles). (3) Samples of exposed salmon were sequentially taken for analysis after exposure (week 1 to 4).

2.1.2. Sound Controlled Exposure Experiment (CEE) Protocol

A set of 40 salmon were exposed daily to sounds during 4 weeks (Figure 1). The CEE protocol consisted of a cycle of 350 Hz (65 V-2 h) and 500 Hz (65 V-2 h) exposure, twice daily with a 2 h period of rest in between the two cycles, during the first three weeks (see [3] for the description of the selection of these sound exposure levels in concurrence with the sea lice exposure experiments). In the last week of the experiments we added another cycle of exposure (i.e., three times 4 h exposure a day) to increase the sound dose received by the fish.

The transducer used was the Lubell LL-1424HP with the capacity to reach levels of at least 180 dB re 1 µPa at the frequencies of interest, although it was driven at 65 vrms, well below its maximum rating. The transducer was driven by a Monacor PA-12040. The sound production system was calibrated as a whole and for each individual frequency. A calibrated hydrophone (B&K 8106 with Nexus signal conditioner and IOtech WaveBook/516 ADC) was used to make spot measurements in the exposure and control tanks to verify the levels and then further used to monitor the exposure experiments. The hydrophone system was arranged to provide its digitized data to a sound exposure control system that was driving the transducer.

The received sound pressure levels were estimated to be 152 dB re 1 μPa^2 at 350 Hz and 155 dB re 1 μPa^2 at 500 Hz. With the animals moving around the tank the levels would vary. The exposure target was to reach a sound exposure level dose of at least 190 dB re 1 μPa^2 s. The estimated SEL for the frequency and duration combination used above was 195 dB re 1 μPa^2 s.

2.1.3. Sample Collection

Ten exposed salmon were euthanised by bath immersion in 2-phenoxyethanol (2-PE) each week during 4 weeks. All salmon (both control and exposed individuals) were equally treated: samples from otoliths, inner ear and internal organs were processed for histopathological and SEM analysis.

2.1.4. Analysis of Salmon Otolith Organs by Scanning Electron Microscopy

Otolith organs epithelia from individual fishes were inspected with SEM imaging techniques to detect any possible alteration of the sensory epithelia. The samples were processed with routine SEM procedures.

Fixation was performed in glutaraldehyde 2.5% for 24–48 h at 4 °C. Samples were dehydrated in graded ethanol solutions and critical-point dried with liquid carbon dioxide in a Bal-Tec CPD030 unit (Leica Mycrosystems, Vienna, Austria). The dried specimens were mounted on specimen stubs with double-sided tape. The mounted samples were gold coated with a Quorum Q150R S sputter coated unit (Quorum Technologies, Laughton, East Sussex, UK) and viewed with a variable pressure Hitachi S-3500N scanning electron microscope (Hitachi High-Technologies Co., Tokyo, Japan) at an accelerating voltage of 5 kV in the Institute of Marine Sciences of the Spanish Research Council (CSIC) facilities.

2.1.5. Quantification and Data Analysis

For quantification of lesions, the region comprising the whole sensory area of the saccule, utricle and lagena was considered. The length of the sensory epithelium areas comprising hair cells was determined for each sample, and 2500 μm^2 (50 µm × 50 µm) sampling squares were placed along the centre length of the area at 5%, 25%, 50%, 75% and 95% of the length axe of the macula statica princeps (Figure 2). Numbers of hair cell bundles were counted in sampling squares of both saccules, utricles and lagenas of each fish. In order to identify whether there were lesions due to the acoustic exposure the samples were treated as follows:

1. All controls taken at different times were grouped together into a single control group.
2. The two samples that were taken for each animal were combined by summing both the intact hair cells and the extruded or missing hair cells over the samples.
3. The hair cells were summed over all the regions, obtaining a single intact hair cells count and single extruded/missing count per animal.
4. The extruded/missing count was divided by the intact hair cell count.

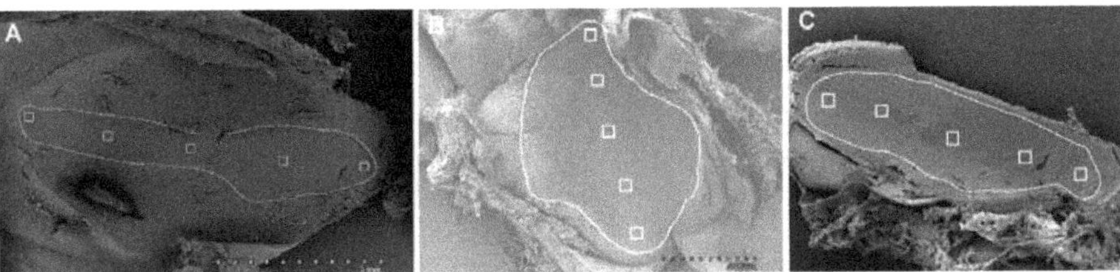

Figure 2. SEM. *S. salar* macula of saccule (**A**), utricle (**B**) and lagena (**C**). Hair cell bundle count locations on macula. Hair cell counts were sampled at five predetermined locations: 5, 25, 50, 75 and 95% of the total macular length. A 2500 μm^2 box was placed at each sampling area and hair cells were counted within each box. Scale bar (**A**) = 1 mm. (**B**,**C**) = 500 μm.

For each experiment this resulted in a series or damaged hair cell ratios for the control group (10) and each sampled group over time (after weeks 1, 2 and 3 in the LAB with 10 samples each), and for each organ that was analysed (lagena, saccule, utricle). A Kruskal–Wallis test was performed for each organ and each experiment to identify if there was a difference in median hair cell damage between the control group and any of the time-sampled groups. All calculations were performed with Matlab R2019a.

2.1.6. Salmon Gross Pathology and Histopathological Analysis

Salmon were anaesthetised and sacrificed with an overdose of 2-phenoxyethanol and spinal severance and were subjected to a gross pathology examination after a standardised necropsy procedure in order to identify potential external lesions or alterations. Particular attention was paid to identify lesions such as haemorrhages or other vascular disturbances, and mainly in the swim bladder, as these are the most frequently described lesions in previously papers.

Immediately after examination, samples of different tissues (liver, digestive system, swim bladder, spleen, kidney, gonads and skeletal musculature) were fixed in 10% buffered formalin. Fixed samples were processed for routine histological studies by progressive dehydration, clearing, embedding in paraffin, block sectioning, staining with haematoxylin and eosin (H/E) and examined under the microscope.

2.1.7. Behavioural Observations

Salmon behaviour was monitored before, during and after the sound exposure, for a period of 10 min each time, in order to determine behavioural alterations (expected behavioural reactions were jumps, rolls and twitches). Jumps were defined as fast accelerations in swimming speeds that ended in a jump, rolls involved turning 90° in the horizontal or vertical plane, and twitches were defined as rapid spasmodic contractions of the body of the salmon.

2.2. Sea Trial Experiments

2.2.1. Sound Exposure

Hardware. Under the sea trial protocol (Figures 3 and 4; see complete description in [3]), the system and method included producing the sounds using calibrated transducers capable of reproducing sound covering from 300 Hz to 600 Hz. The transducers used were Lubell LL916C projectors, installed at the centre of each cage. They were driven by a Monacor PA-12040. A control system consisting of an HTI-99-HF hydrophone connected to an MCCDAQ USB-1608G which was connected to a Raspberry Pi monitored correct functioning of the system, the exposure time periods, and the accumulated sound exposure levels.

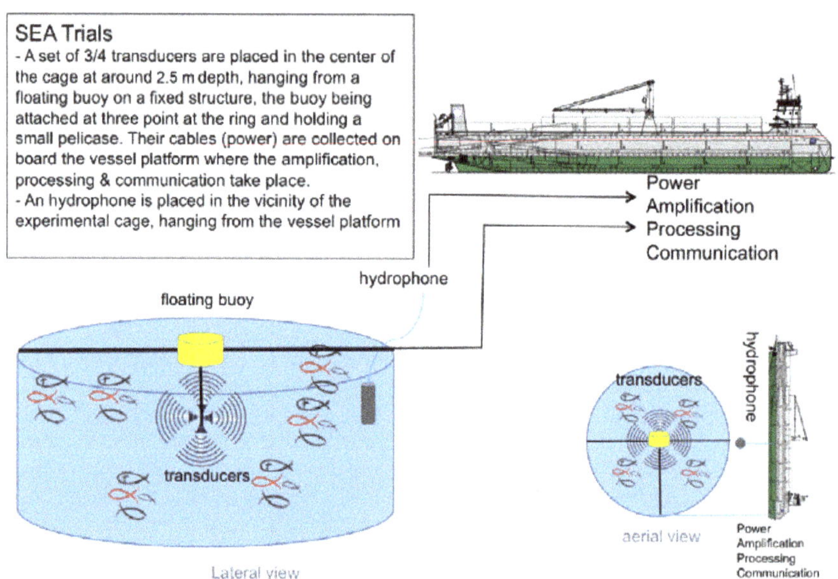

Figure 3. Sound exposure system [3].

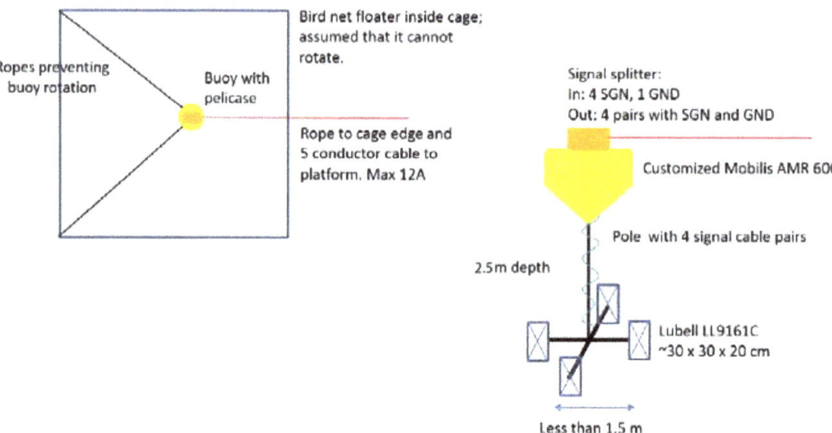

Figure 4. Drawing of the experimental setup. Note that the depth of the structure that holds the loud speakers was modified along the duration of the experiments. M9 loud speakers were lowered to -5 m [3].

Acoustic and time parameters: salmon were exposed to continuous exposure (Figures 3 and 4; see complete description in [3]), to individual 350 Hz and 500 Hz signals during, respectively, a cumulative cycle of 1 h and 2 h, and this combination was initially played back every 4 h. The received sound pressure levels were estimated conservatively (as the measurement point was fixed while the cages provided a lot of space for the fish to move around) to be 139 dB re 1 μPa^2 at 350 Hz and 142 dB re 1 μPa^2 at 500 Hz. The received levels at the monitoring hydrophone could vary by 5 dB between exposure windows. The estimated SEL for the frequency and duration combination used above and one 3-h exposure session was 179 dB re 1 μPa^2 s. In order to be able to more rapidly reach a target SEL similar to those obtained during the laboratory experiments (195 dB re 1 μPa^2 s) it was decided to increase the received level by 10 dB for both frequencies at one of the cages, playing both frequencies for 2 h each continuously.

2.2.2. Sample Collection

Every 3 weeks (in weeks 3, 6, 9 and 12 after sound exposure) samples from salmon were collected:

1. Week 3 and week 6: three salmon from each cage (taken from the two exposed cages and from the four control cages). Total: six exposed individuals and 12 control individuals per week.
2. Week 9 and week 12: three salmon (from the two exposed cages and from the two control cages). Total: six exposed individuals and six control individuals per week.

All collected salmon were used to assess salmon health status. External surfaces, mouth and internal organs of each individual were checked macroscopically for gross pathologies. Then, an anterior and a posterior body wedge, which included skeletal muscle (including white and red muscle), head and trunk kidney, swimbladder, stomach, intestine, liver, perivisceral fat (with pancreatic tissue) and gonads were extracted and processed for histological analysis at the Pathological Diagnostic Service in Fish (SDPP) of the Autonomous University of Barcelona (UAB). A portion of the body containing lateral lines as well as the whole head of the salmon were taken and processed to assess possible lesions in the inner ear structures (otolith organ) and lateral lines at the LAB (Laboratory of Applied Bioacoustics).

2.2.3. Analysis of Salmon Otolith Organs by Scanning Electron Microscopy

Otolith organs epithelia from individual fishes were observed by SEM imaging techniques to detect any possible alteration of the sensory epithelia. The samples were processed by routine SEM procedures.

Fixation was performed in glutaraldehyde 2.5% for 24–48 h at 4 °C. Samples were dehydrated in graded ethanol solutions and critical-point dried with liquid carbon dioxide in a Bal-Tec CPD030 unit (Leica Mycrosystems, Vienna, Austria). The dried specimens were mounted on specimen stubs with double-sided tape. The mounted samples were gold coated with a Quorum Q150R S sputter coated unit (Quorum Technologies, Laughton, East Sussex, UK) and viewed with a variable pressure Hitachi S-3500N scanning electron microscope (Hitachi High-Technologies Co., Tokyo, Japan) at an accelerating voltage of 5 kV in the Institute of Marine Sciences of the Spanish Research Council (CSIC) facilities.

2.2.4. Quantification and Data Analysis

We considered for the quantification the region comprising the whole sensory area of the saccule, utricle and lagena. The length of the sensory epithelium areas comprising hair cells was determined for each sample, and 2500 μm^2 (50 × 50 μm) sampling squares were placed along the centre length of the area at 5, 25, 50, 75 and 95% of the length axe of the macula statica princeps (Figure 2). Numbers of hair cell bundles were counted in sampling squares of both saccules, utricles and lagenas of each fish. In order to identify whether there were lesions due to the acoustic exposure, the samples were treated as follows:

1. All controls taken at different times were grouped together into a single control group.
2. The two samples that were taken for each animal were combined by summing both the intact hair cells and the extruded or missing hair cells over the samples.
3. The hair cells were summed over all the regions, obtaining a single intact hair cells count and single extruded/missing count per animal.
4. The extruded/missing count was divided by the intact hair cell count.

For each experiment this resulted in a series of damaged hair cell ratios for the control group (12) and each sampled group over time (after weeks 3, 6, and 12 with six samples each), and for each organ that was analysed (lagena, saccule, utricle). A Kruskal–Wallis test was performed for each organ and each experiment to identify if there was a difference in median hair cell damage between the control group and any of the time-sampled groups. All calculations were performed with Matlab R2019a.

2.2.5. Analysis of Presence of Vaterite in Salmon Otolith

Sagittal otoliths are primary hearing structures in the inner ear of all teleost (bony) fishes and are normally composed of aragonite, though abnormal vaterite replacement is sometimes seen. Additional to the analysis of salmon otolith epithelia by scanning electron microscopy, the evaluation and quantification of the presence of vaterite in otoliths were performed by optic microscopy. The proportion of otoliths presenting vaterite was quantified.

2.2.6. Analysis of Superficial Neuromasts of the Salmon Lateral Line by Scanning Electron Microscopy

A portion of the body containing lateral lines of all the fishes (48, control and exposed) were collected during the three weeks of sampling. Superficial neuromasts of the lateral line from individual fishes were observed by SEM imaging techniques to detect any possible alteration of the sensory epithelia. The samples were processed by routine SEM procedures.

Fixation was performed in glutaraldehyde 2.5% for 24–48 h at 4 °C. Samples were dehydrated in graded ethanol solutions and critical-point dried with liquid carbon dioxide in a Bal-Tec CPD030 unit (Leica Mycrosystems, Vienna, Austria). The dried specimens were mounted on specimen stubs with double-sided tape. The mounted samples were gold coated with a Quorum Q150R S sputter coated unit (Quorum Technologies, Laughton, East Sussex, UK) and viewed with a variable pressure Hitachi S-3500N scanning electron microscope (Hitachi High-Technologies Co., Tokio, Japan) at an accelerating voltage of 5 kV in the Institute of Marine Sciences of the Spanish Research Council (CSIC) facilities.

2.2.7. Salmon Gross Pathology and Histopathological Analysis

The salmon were subjected to a gross pathology and histological analysis to assess possible lesions in internal organs. The same procedure as in LAB experiments was followed to collect and analyse of the samples.

Salmon were anaesthetised and sacrificed with an overdose of 2-phenoxyethanol and spinal severance and were subjected to a gross pathology examination after a standardised necropsy procedure in order to identify potential external lesions or alterations. Particular attention was paid to identify lesions such as haemorrhages or other vascular disturbances and mainly in the swim bladder as these were the most frequently described lesions in previous papers.

Immediately after examination, samples of different tissues (liver, digestive system, swim bladder, spleen, kidney, gonads and skeletal musculature) were fixed in 10% buffered formalin. Fixed samples were processed for routine histological studies by progressive dehydration, clearing, embedding in paraffin, block sectioning, staining with haematoxylin and eosin (H/E) and examined under the microscope.

3. Results

3.1. Laboratory Experiments

3.1.1. Analysis of Salmon Otolith Organs by Scanning Electron Microscopy

The fish basic inner ear structure consists of three semicircular canals and their sensory epithelia, the cristae and three otolithic end organs (utricle, saccule, lagena) including their maculae with respective sensory epithelia, the macula saccule, the macula utricle and the macula lagena (Figure 5).

Otolith organs epithelia from individual fishes were observed through SEM imaging techniques to detect any possible alteration of the sensory epithelia. No effects were detected in any of the exposed animals nor in control animals. Figures 6–8 show a healthy aspect of hair cells on the three epithelia from the different times of sample collection.

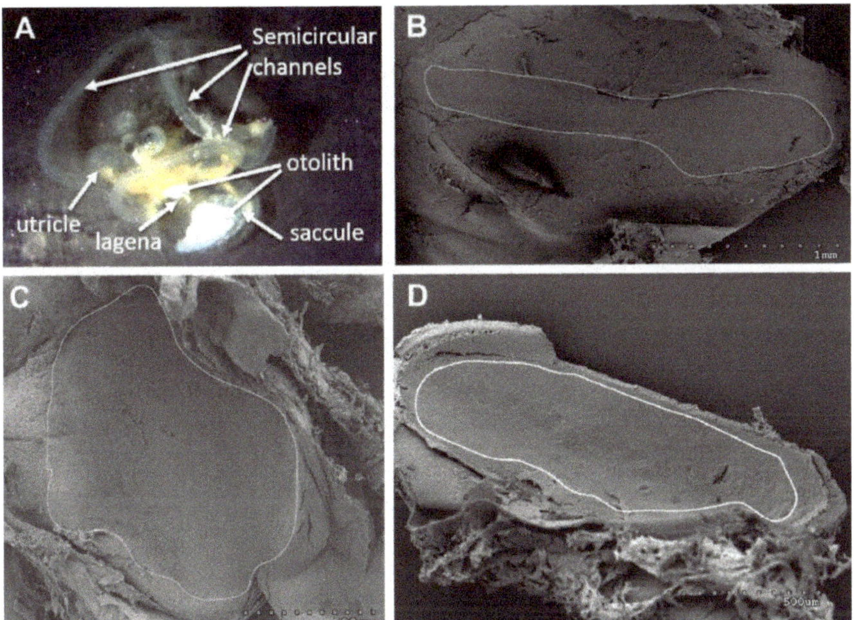

Figure 5. (**A**) LM (light microscopy). (**B–D**): SEM (scanning electron microscopy). Salmon inner ear. (**A**) The location of the three sensory epithelia and the three semicircular channels in the salmon inner ear is visible. (**B**) White line encloses the saccule sensory epithelium. (**C**) White line encloses the utricle sensory epithelium. (**D**) White line encloses the lagena sensory epithelium. Scale bar: (**B**) = 1 mm. (**C**,**D**) = 500 μm.

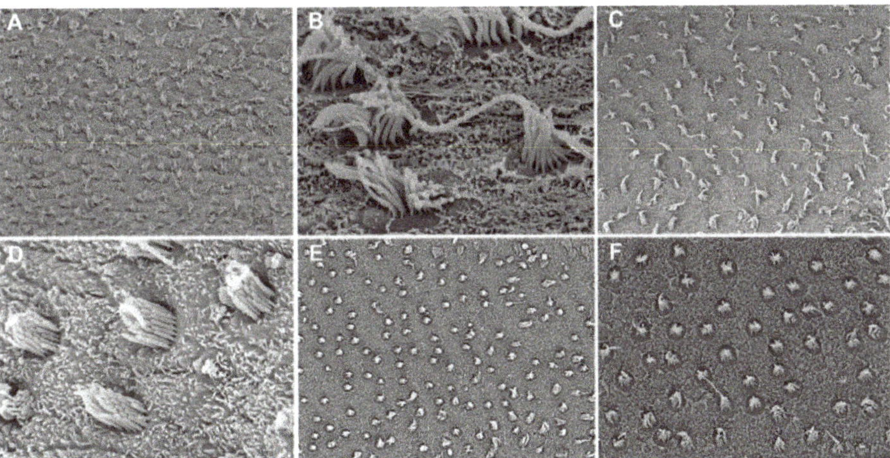

Figure 6. SEM (scanning electron microscopy). Saccule sensory epithelium. (**A**,**B**) Exposed animals. Control. (**C–F**) By comparison with control animals, the images of the saccule epithelium show healthy sensory hair cells in all cases of exposed animals (1 week (**C**), 2 weeks (**D**), 3 weeks (**E**) and 4 weeks (**F**) of sound exposure). Scale bar: E = 30 μm. (**A**,**C**) = 20 μm. (**F**) = 10 μm. (**D**) = 5 μm. (**B**) = 3 μm.

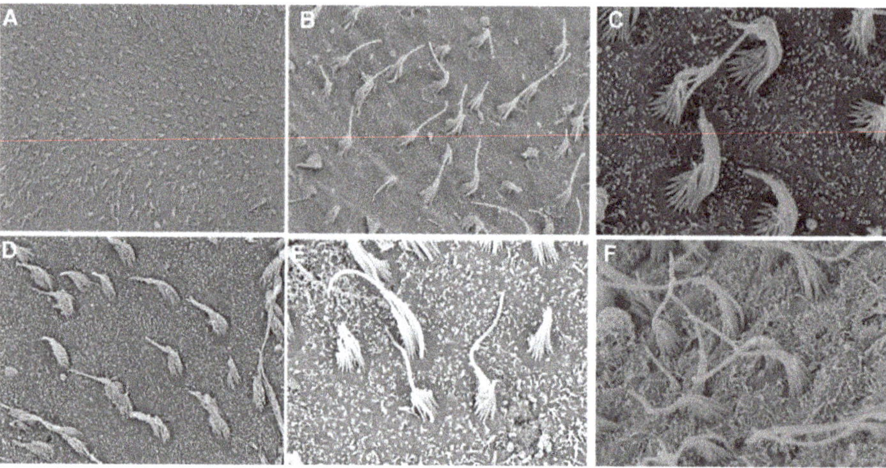

Figure 7. SEM (scanning electron microscopy). Utricle saccule sensory epithelium. (**A,B**) Exposed animals. Control. (**C–F**) By comparison with control animals, the images of the saccule epithelium show healthy sensory hair cells in all cases of exposed animals (1 week (**C**), 2 weeks (**D**), 3 weeks (**E**) and 4 weeks (**F**) of sound exposure). Scale bar: (**A**) = 30 μm. (**B,D**) = 10 μm. (**C,E,F**) = 5 μm.

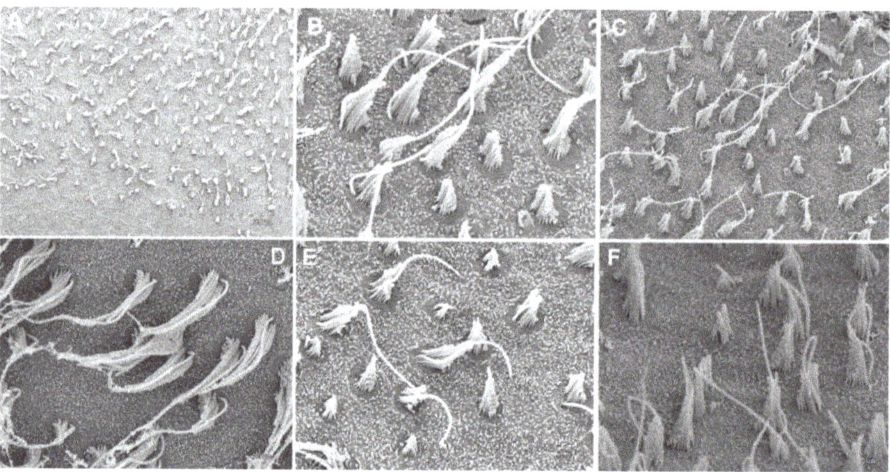

Figure 8. SEM (scanning electron microscopy). Lagena sensory epithelium. (**A,B**) Exposed animals. Control. (**C–F**) By comparison with control animals, the images of the saccule epithelium show healthy sensory hair cells in all cases of exposed animals (1 week (**C**), 2 weeks (**D**), 3 weeks (**E**) and 4 weeks (**F**) of sound exposure). Scale bar: (**A**) = 30 μm. (**C,D**) = 10 μm. (**B,E,F**) = 5 μm.

3.1.2. Quantification and Data Analysis

No significant differences were found between hair cell assessment in control and exposed animals during LAB experiments (Kruskal-Wallis test) (Figure 9).

3.1.3. Gross Pathology and Histological Analysis

The gross morphological (Figure 10) and histopathological analysis (Figures 11–14) on sampled salmon did not show any lesions that could be associated to sound exposure. All dissected salmon were presenting a completely normal aspect.

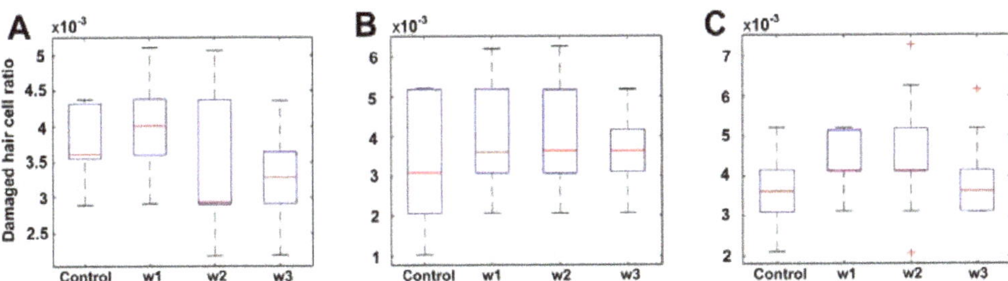

Figure 9. Overview of the damaged hair cell ratio from data collected at the LAB experiments. (**A**) lagena, (**B**) saccule and (**C**) utricle. The red line is the median with the boxes defined by the 25 and 75 percentiles. The whiskers are defined by 1.5 times the interquartile distance and outliers (+) beyond that range. No significant differences were found between controls and exposed salmons (*p*-values: (**A**) 0.21, (**B**) 0.92, (**C**) 0.54).

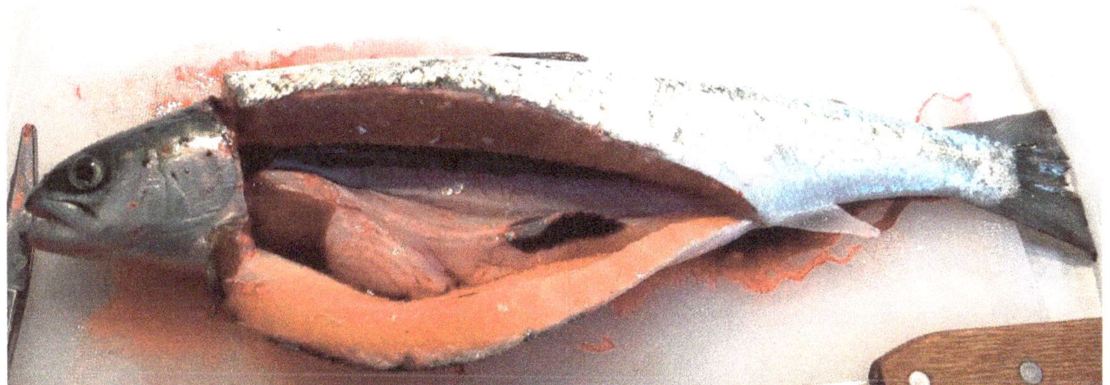

Figure 10. Dissected salmon showing the inner organs with normal aspect. No internal or external haemorrhages or lesions, nor alterations in the swim bladder were detected.

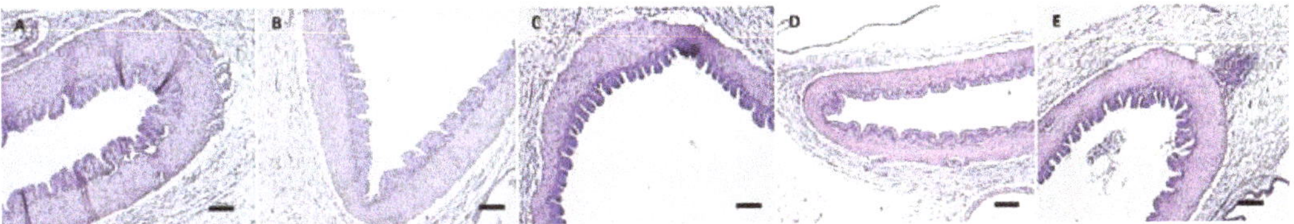

Figure 11. Swim bladder with normal aspect of salmon at different samplings times. (**A**) Control. (**B**) 1 week after exposure. (**C**) 2 weeks after exposure. (**D**) 3 weeks after exposure. (**E**) 4 weeks after exposure. Scale bar: 100 μm.

3.1.4. Behavioural Observations

No behavioural reaction was observed before, during and after the sound exposure. No jumps, rolls and twitches were observed during the 10 min of observation on each period of analysis. A major proportion of salmon acquired a perpendicular position near to the transducer when it was placed in the tank and maintained this position during the entire experiment. The rest of the fishes placed themselves counter current in the tank while the sound exposure was performed. All salmon were observed eating consistently over the whole experimental period (Figure 15).

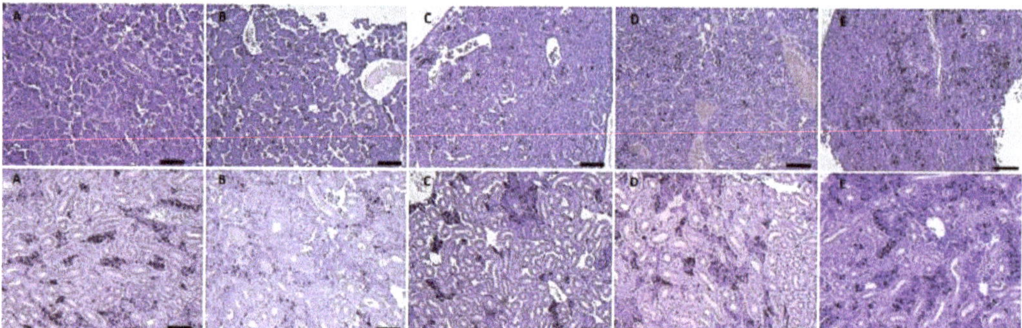

Figure 12. Head kidney (**first line**) and posterior kidney (**second line**) of salmon at different sampling times. (**A**) Control. (**B**) 1 week after exposure. (**C**) 2 weeks after exposure. (**D**) 3 weeks after exposure. (**E**) 4 weeks after exposure. Scale bar: 100 µm.

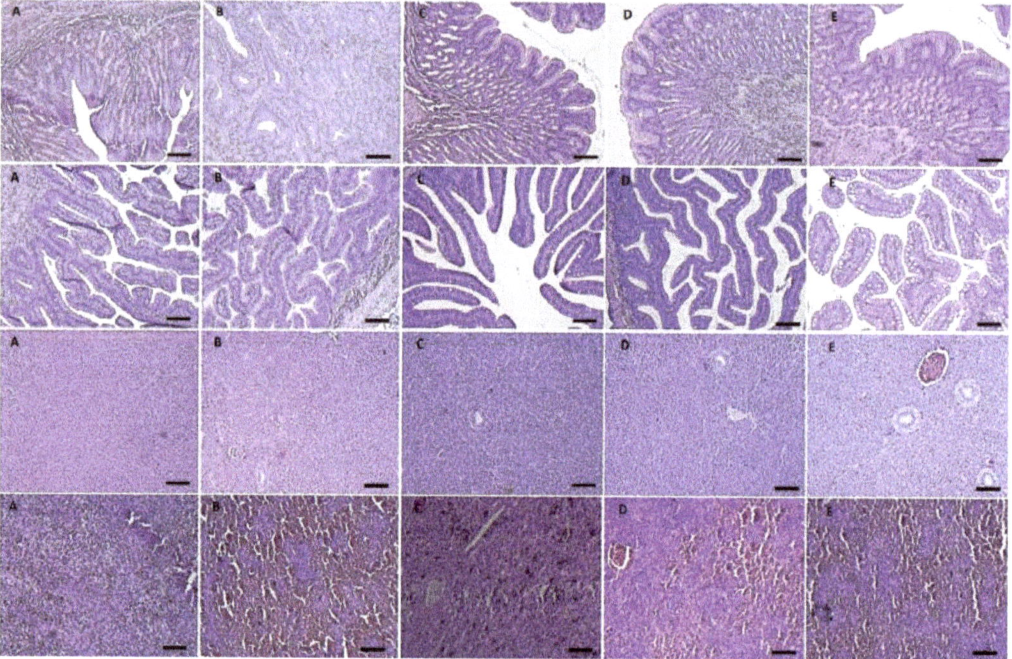

Figure 13. Stomach (**first line**), intestine (**second line**), liver (**third line**) and spleen (**fourth line**) of salmon at different sampling times. (**A**) Control. (**B**) 1 week after exposure. (**C**) 2 weeks after exposure. (**D**) 3 weeks after exposure. (**E**) 4 weeks after exposure. Scale bar: 100 µm.

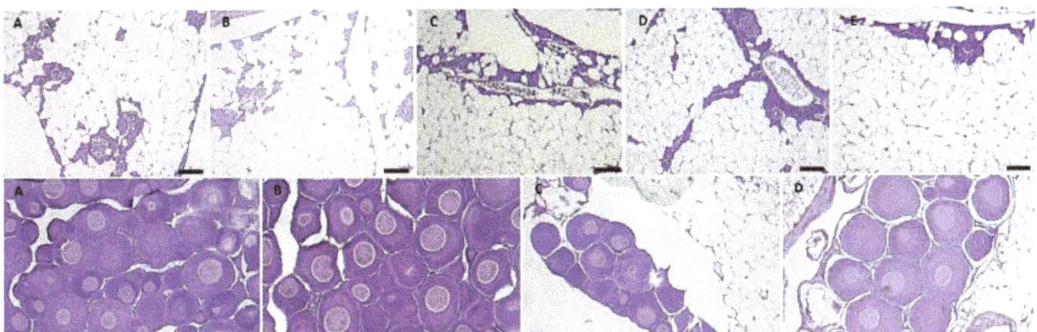

Figure 14. Pancreas and adipose tissue (visceral fat) (**first line**) and gonad (**second line**) of salmon at different sampling times. (**A**) Control. (**B**) 1 week after exposure. (**C**) 2 weeks after exposure. (**D**) 3 weeks after exposure. (**E**) 4 weeks after exposure. Scale bar: 100 µm.

Figure 15. Salmon swimming in the tanks and showing normal behaviour.

3.2. Sea Trial Experiments

3.2.1. Analysis of Salmon Otolith Organs by Scanning Electron Microscopy

Otolith organs observed by SEM imaging techniques did not show any alteration on the sensory epithelia. Figures 16–18 show a healthy aspect of hair cells on the three epithelia at different times of sample collection at sea trials.

3.2.2. Quantification and Data Analysis

No significant differences were found between hair cell assessment during control and exposed animals at sea trial experiments (Kruskal–Wallis test). (Figure 19).

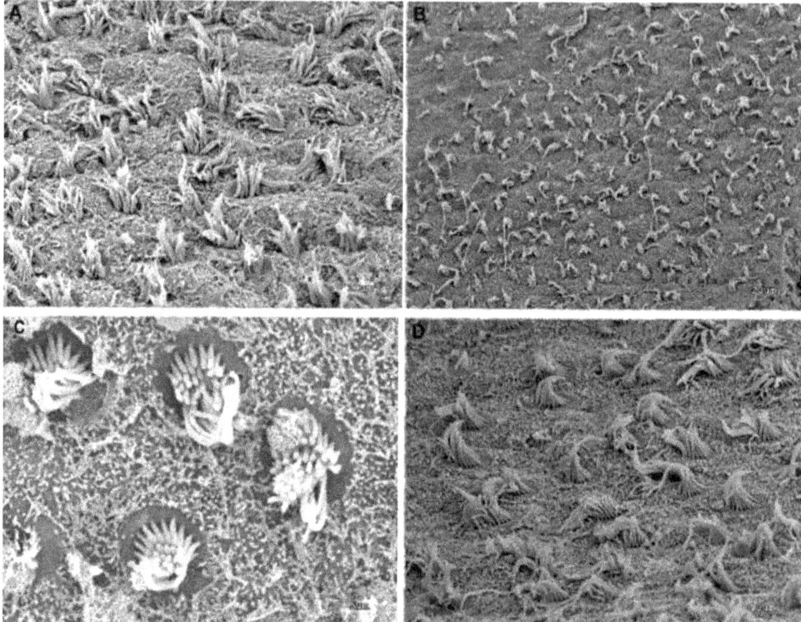

Figure 16. SEM. Salmon inner ear. Saccule sensory epithelium. (**A**) Control. (**B**) 3 weeks after sound exposure. (**C**) 6 weeks after sound exposure. (**D**) 12 weeks after sound exposure. By comparison with control animals, the images of the saccule epithelium show healthy sensory hair cells in all cases of exposed animals. Scale bar: (**B**) = 20 μm. (**A,D**) = 10 μm. (**C**) = 5 μm.

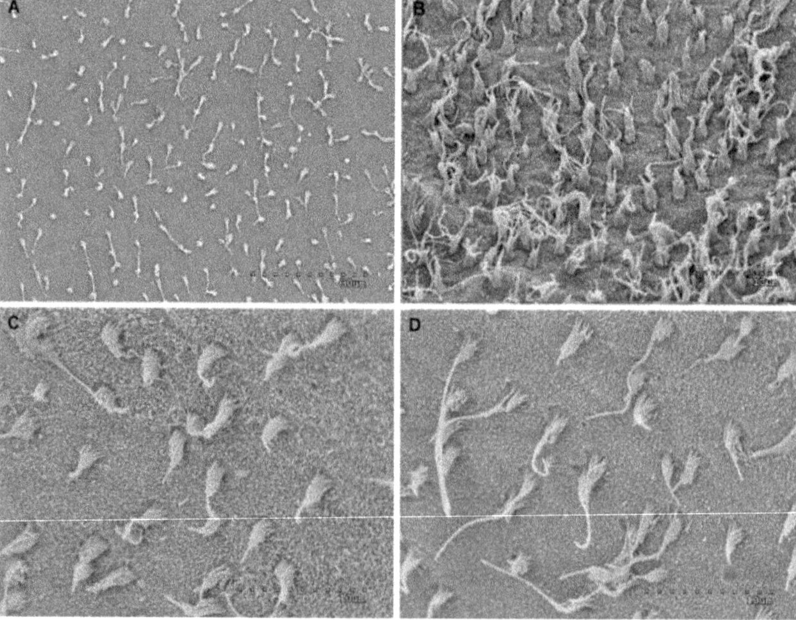

Figure 17. SEM. Salmon inner ear. Utricle sensory epithelium. (**A**) Control. (**B**) 3 weeks after sound exposure. (**C**) 6 weeks after sound exposure. (**D**) 12 weeks after sound exposure. By comparison with control animals, the images of the saccule epithelium show healthy sensory hair cells in all cases of exposed animals. Scale bar: (**A**) = 30 μm. (**B–D**) = 10 μm.

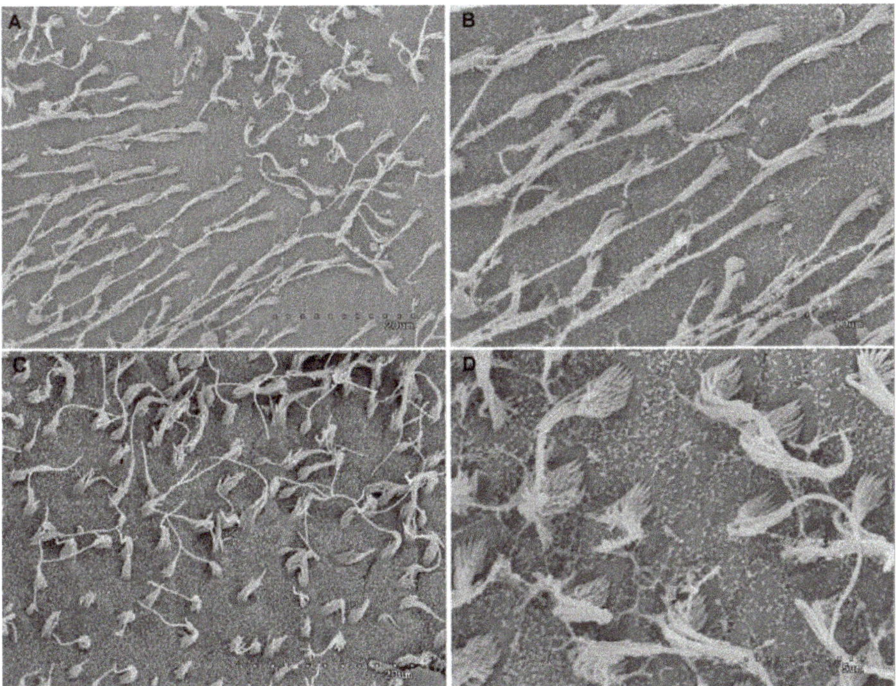

Figure 18. SEM. Salmon inner ear. Lagena sensory epithelium. (**A**) Control. (**B**) 3 weeks after sound exposure. (**C**) 6 weeks after sound exposure. (**D**) 12 weeks after sound exposure. By comparison with control animals (**A**), the images of the saccule epithelium show healthy sensory hair cells in all cases of exposed animals (**B–D**). Scale bar: (**C**) = 30 μm. (**A**) = 20 μm. (**B**) = 10 μm. (**D**) = 5 μm.

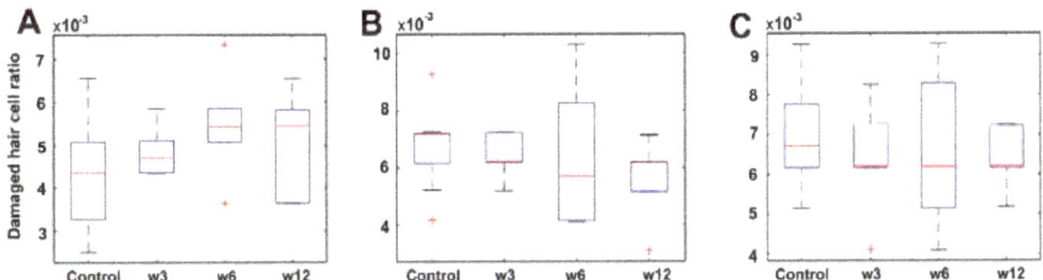

Figure 19. Overview of the damaged hair cell ratio from data collected at sea experiments. (**A**) Lagena, (**B**) saccule, (**C**) utricle. The red line is the median with the boxes defined by the 25th and 75th percentiles. The whiskers are defined by 1.5 times the interquartile distance and outliers (+) beyond that range. No significant differences were found between controls and exposed salmons (*p*-values (**A**) 0.29, (**B**) 0.38, (**C**) 0.78).

3.2.3. Analysis of Salmon Otolith

Vaterite incidence on salmon otoliths (Figure 20) did not present a linear pattern over time (Table 1). The percentage of fish affected increased in week 3 (75%) and week 6 (83, 3%), and decreased in the week 12 (16, 6%) (Figure S2A,B, Supplemental Material). Considering the total incidence of vaterite in the sampled fishes, there were more otoliths affected (65, 6%) than otoliths with no vaterite (34, 3%) (Figure S2C). Right otoliths were more likely to be vateritic than left otoliths, with 52% and 48% of otoliths showing vaterite formation. Out of all fish sampled, 9, 4% had two vateritic otoliths (Figure S2D,E). Considering the total incidence of vaterite on the sampled fishes, the proportion otolith affected by vaterite in control cages (75%) was higher than in exposed cages (50%) (Figure S2F). The analysis of the vaterite presence in otoliths was not related to sound exposure, but to a deficiency in nutrition associated to captivity (Table 1).

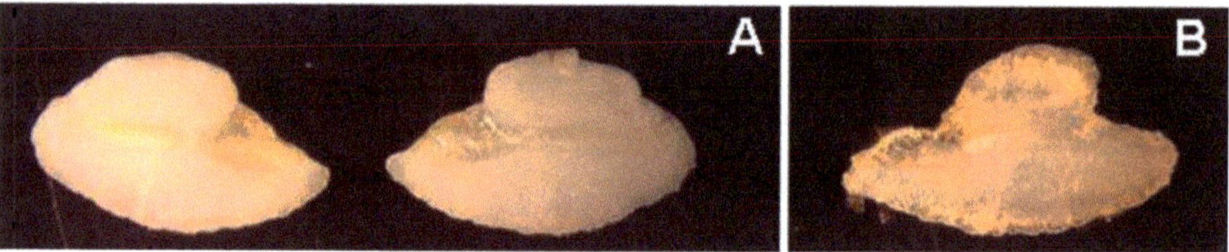

Figure 20. LM. (**A**) Left and right otolith composed entirely of aragonite. (**B**) Right otolith composed basically of vaterite.

Table 1. Vaterite incidence rate on the salmon otoliths (%).

Time	Vaterite Incidence Rate (%)
Week 3	75
Week 6	83.3
Week 12	16.6
Total Incidence	-
Vateritic otolith	65.6
No vateritic otolith	34.3
Location	-
Right otolith	52
Left otolith	48
Cages	-
Control	75
Exposed	50

3.2.4. Analysis of Superficial Neuromast of the Salmon Lateral Line by Scanning Electron Microscopy

Neuromasts are either contained in canals or are located on the epithelium of the head, trunk and tail. Each lateral line scale has a single canal neuromast and three groups of superficial neuromasts (Figure 21). In comparison with control animals the lateral line superficial neuromasts observed by SEM imaging techniques did not show any alteration of the sensory epithelia (Figure 22) in exposed salmon.

3.2.5. Gross Pathology and Histological Analysis

No macroscopic pathology nor histopathological alteration were observed in any of the tissue analysed that could be associated to sound exposure, exactly in the same way to those observed in LAB experiments. See Figure S3 (Supplementary Material) as an example of the normal aspect of the swim bladder of salmon (the most sensitive organ to sound exposure) throughout the samplings of the sea trial.

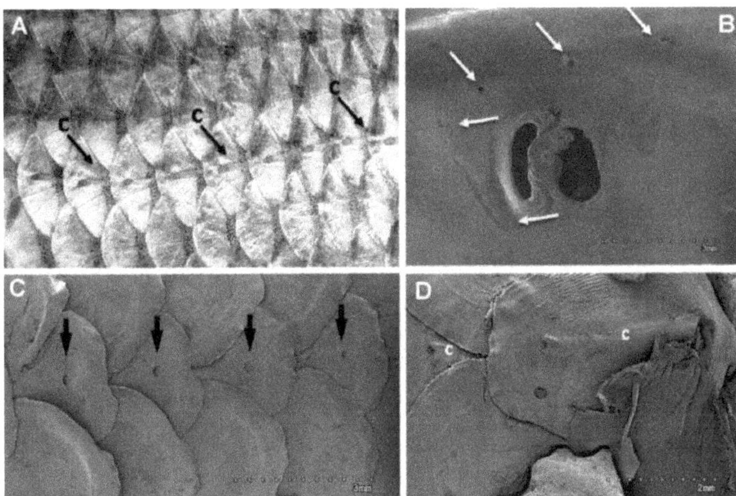

Figure 21. SEM. (**A**) Light microscopy (LM). (**B–D**): SEM. (**A**) Lateral line system located on the salmon scales of trunk (c signs the lateral line channel). (**B**) Lateral system located on the epithelium of the head (arrows). (**C**) Holes (arrows) of the channel of the lateral line system on the trunk. (**D**) Upper channel view on the scales (c). Scale bar: (**B,C**) = 3 mm. (**D**) = 2 mm.

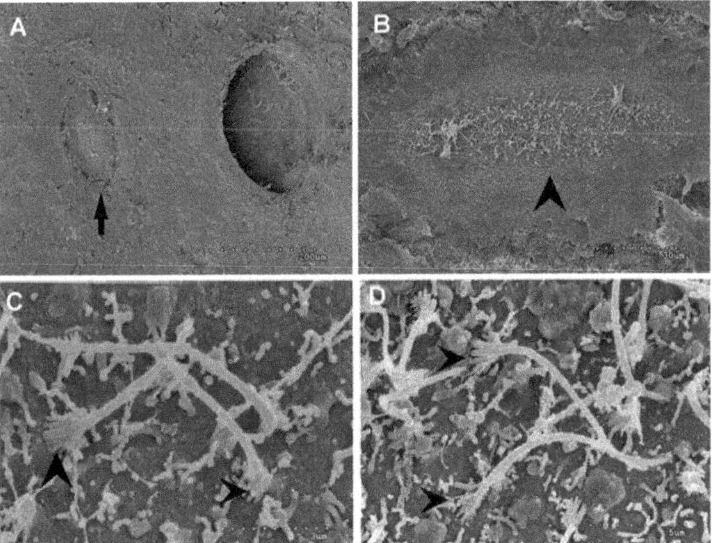

Figure 22. SEM. Superficial neuromast. (**A**) Upper view of a superficial neuromast. (**B**) Normal aspect of hair cell bundles on superficial neuromast of a control salmon. (**C,D**) Hair cell bundles on superficial neuromast of a 12-week exposed salmon. No lesions are visible. Scale bar: (**A**) = 200 μm. (**B**) = 30 μm. (**C**) = 5 μm. (**D**) = 3 μm.

4. Discussion

The use of acoustic methods as an effective new approach to address sea lice infestation (*L. salmonis*) should first ensure that these methods do not negatively affect the fish. In this context, the present results showed that salmon exposed to sound with the same characteristics as previous lice exposure experiments [3] did not present any lesion in internal organs, nor in the otolith and lateral line sensory epithelia and they also did not induce behavioural alterations. We did not find an increase in mortality due to sound exposure, which is consistent with some previous experiments of sound exposure where much higher sound levels (e.g., pile driving and airgun) [11,12,15–17] did not show an effect on fish survival.

Although some authors reported sensory epithelia cell loss in the otolith organ after sound exposure [8,10,23], our analysis did not reveal any damage in auditory tissues. We proceeded to analyse the vaterite proportion of the otoliths. Sagittal otoliths are essential components of the sensory organs that are composed of calcium carbonate. In abnormal otoliths, aragonite (the normal crystal form) is replaced with vaterite that decreases hearing sensitivity, reducing growth rates [14]. In some Chinook salmon studies vateritic sagittae were bigger and less dense than the aragonitic form, and vaterite presence was associated with moderately altered saccular epithelia and a significant decrease in auditory sensitivity [24]. The functional cause of the degradation remains speculative and a variety of physiological factors may be involved. Among other hypotheses the most likely explanation is that hearing loss is caused by the lower density of the vaterite otoliths, which could induce a decrease in the differential movement between the saccular epithelium and its otolith. It would require a higher force to stimulate the sensory epithelium and trigger a neural response and, therefore, auditory sensitivity would be reduced [24].

After assessment (in our samples the otolith proportion that was affected by vaterite in control cages was higher than in exposed cages) we concluded that differences of the vaterite presence in otoliths had no relation with sound exposure, but was probably explained by a deficiency in nutrition associated to captivity as has been shown in previous studies [14].

In terms of behaviour, despite minor and non-permanent behavioural responses were reported (small level of startle response with habituation at subsequent exposures) to very high sound levels produced by air-guns (196 dB re 1 µPa at 1 m from the source) [16,17] we did not find any change in the behaviour of the salmon.

In addition, the lateral line superficial neuromasts did not show any alteration in the sensory epithelia. Acoustic trauma was, however, observed in larval zebrafish lateral line hair cells using underwater cavitation to stimulate a response [25].

Gross pathology and histopathological analysis on salmon did not show any lesion that could be associated to sound exposure. Some works have described different pathologies associated to sound exposure. Various levels of barotrauma after sound exposure—mild injuries (eye haemorrhage, fin haematoma), moderate injuries (liver haemorrhage, bruised swim bladder), and mortal injuries (intestinal haemorrhage and kidney haemorrhage)—were observed after exposing Chinook salmon (*Oncorhynchus tshawytscha*) to impulsive pile-driving sounds using a high-intensity controlled impedance fluid filled wave tube [26]. However, the sound exposure level received during this experiment greatly exceeded the levels we reached in our study. None of these consequences were observed during our experiments. Parasitic sea lice (*L. salmonis*) infestations are one of the most important concerns in salmonid farming, reducing productivity and causing economic losses to the aquaculture industry (around 0.39 € per kg of salmon produced) [27,28]. Currently the most common treatment against salmon lice are chemicals. The pharmaceuticals currently used for the control of sea lice (cypermethrin, deltamethrin, azamethiphos, hydrogen peroxide) are applied through in situ immersion treatments [28]. Although these treatments have been effective in managing sea lice outbreaks, they have negative effects on fish welfare, reducing appetite and growth. Furthermore, over time salmon lice have built up a resistance to the three major classes of chemicals being used [29]. In addition, these pharmaceuticals are released into the surrounding environment, exposing non-target species to lethal and sub-lethal doses, and are harmful to the human health [28]. Other methods (e.g., in-feed treatments and use of skirts) are very expensive. Skirts have low impact on salmon welfare and the environment but reduce oxygen flow, affecting the respiratory functions of fish. From this perspective, this exhaustive assessment of the effects of acoustic treatment against lice infestation on salmon confirmed that the chosen sound dose did not affect the exposed fish and, therefore, confirmed it as a potentially safe and sustainable protocol to address the problem.

Supplementary Materials: The following are available online at https://www.mdpi.com/article/10.3390/jmse9101114/s1, Figure S1: Salmon tanks and refrigeration system; Figure S2: Incidence of vaterite on the salmon otoliths. (A,B) Incidence of vaterite versus time. (C) Incidence of vaterite on total sampled fishes. (D) Incidence of vaterite in right and left otolith on total sampled fishes. (E) Incidence of vaterite on right and left otolith versus time. (F) Incidence of vaterite on total sampled fishes versus control/exposed cages; Figure S3: Swimbladder with normal aspect of salmons from sea trial experiment. (A) Control. (B) 3 weeks after exposure. (C) 6 weeks after exposure. (D) 9 weeks after exposure. (E) 12 weeks after exposure. Scale bar: 100μm.

Author Contributions: M.S. and M.A. planned the research and designed the study. M.S., M.A. and M.v.d.S. conducted experimental/lab work. M.S. and M.A. conducted sea trial work. A.L. and M.S. performed the otolith organ dissection. M.S. and J.-M.F. performed SEM analysis. M.C. and F.P. performed the histopathological examination. M.S. and M.v.d.S. analysed the data. M.S. and M.v.d.S. prepared the figures. M.S. wrote the article and all authors reviewed the manuscript. All authors have read and agreed to the published version of the manuscript.

Funding: Funding for this project was provided by SEASEL SOLUTIONS AS. Project: An acoustic and Bioacoustic solution to sea lice infestation on salmon- P.O.BOX 93 N-6282 BRATTVÅG. Norway.

Institutional Review Board Statement: The experimental protocol strictly followed the necessary precautions to comply with current ethical and welfare considerations when dealing with animals in scientific experimentation (Royal Decree 1386/2018, of 19 November). This process was also carefully analysed and approved by the Ethical Committee for Scientific Research of the Technical University of Catalonia, BarcelonaTech (UPC) (approval code B9900085).

Informed Consent Statement: Not applicable.

Acknowledgments: We would like to thank the staff of Mowi Fish Feed A/S Avd Averøy for their assistance and helpful cooperation on Sea trials experiments.

Conflicts of Interest: The authors declare no conflict of interest.

References

1. Iversen, A.; Hermansen, Ø.; Andreassen, O.; Brandvik, R.K.; Marthinussen, A.; Nystøyl, R. Cost Drivers in Salmon Farming; Trondheim. 2015. Available online: https://nofima.no/en/pub/1281065/ (accessed on 13 October 2021).
2. Barrett, L.T.; Oppedal, F.; Robinson, N.; Dempster, T. Prevention not cure: A review of methods to avoid sea lice infestations in salmon aquaculture. *Rev. Aquac.* **2020**, *12*, 2527–2543. [CrossRef]
3. Solé, M.; Lenoir, M.; Fortuño, J.M.; De Vreese, S.; van der Schaar, M.; André, M. Sea Lice are sensitive to low frequency sounds. *J. Mar. Sci. Eng.* **2021**, *9*, 765. [CrossRef]
4. Harding, H.; Bruintjes, R.; Radford, A.N.; Simpson, S.D. *Measurement of Hearing in the Atlantic Salmon (Salmo salar) Using Auditory Evoked Potentials, and Effects of Pile Driving Playback on Salmon Behaviour and Physiology*; Marine Scotland Science: Aberdeen, UK, 2016; Volume 7, 47p. [CrossRef]
5. Smith, M.E.; Kane, A.S.; Popper, A.N. Noise-induced stress response and hearing loss in goldfish (*Carassius auratus*). *J. Exp. Biol.* **2004**, *207*, 427–435. [CrossRef] [PubMed]
6. Popper, A.N.; Smith, M.E.; Cott, P.A.; Hanna, B.W.; MacGillivray, A.O.; Austin, M.E.; Mann, D.A. Effects of exposure to seismic airgun use on hearing of three fish species. *J. Acoust. Soc. Am.* **2005**, *117*, 3958–3971. [CrossRef] [PubMed]
7. Hastings, M.C. Coming to terms with the effects of ocean noise on marine animals. *Acoust. Today* **2008**, *4*, 22–34. [CrossRef]
8. McCauley, R.D.; Fewtrell, J.; Popper, A.N. High intensity anthropogenic sound damages fish ears. *J. Acoust. Soc. Am.* **2003**, *113*, 638–642. [CrossRef] [PubMed]
9. Uribe, P.M.; Sun, H.; Wang, K.; Asuncion, J.D.; Wang, Q.; Chen, C.W.; Steyger, P.S.; Smith, M.E.; Matsui, J.I. Aminoglycoside-induced hair cell death of inner ear organs causes functional deficits in adult zebrafish (*Danio rerio*). *PLoS ONE* **2013**, *8*, e58755. [CrossRef]
10. Smith, M.E.; Schuck, J.B.; Gilley, R.R.; Rogers, B.D. Structural and functional effects of acoustic exposure in goldfish: Evidence for tonotopy in the teleost saccule. *BMC Neurosci.* **2011**, *12*, 19. [CrossRef] [PubMed]
11. Nedwell, J.R.; Turnpenny, A.W.H.; Lovell, J.M.; Edwards, B. An investigation into the effects of underwater piling noise on salmonids. *J. Acoust. Soc. Am.* **2006**, *120*, 2550–2554. [CrossRef]

12. Ruggerone, G.T.; Goodman, S.E.; Miner, R. *Behavioral Response and Survival of Juvenile Coho Salmon to Pile Driving Sounds*; Natural Resources Consultants, Inc.: Seattle, DC, USA, 2008; Available online: http://home.comcast.net/$\sim$ruggerone/FishTerminal%0APileDriveStudy.pdf (accessed on 13 October 2021).
13. Casper, B.M.; Smith, M.E.; Halvorsen, M.B.; Sun, H.; Carlson, T.J.; Popper, A.N. Effects of exposure to pile driving sounds on fish inner ear tissues. *Comp. Biochem. Physiol. A Mol. Integr. Physiol.* **2013**, *166*, 352–360. [CrossRef]
14. Reimer, T.; Dempster, T.; Wargelius, A.; Fjelldal, P.G.; Hansen, T.; Glover, K.A.; Solberg, M.F.; Swearer, S.E. Rapid growth causes abnormal vaterite formation in farmed fish otoliths. *J. Exp. Biol.* **2017**, *220*, 2965–2969. [CrossRef]
15. Skalski, J.R.; Pearson, W.H.; Malme, C.I. Effects of sounds from a geophysical survey device on catch-per-unit-effort in a hook-and-line fishery for rockfish (*Sebastes* spp.). *Can. J. Fish. Aquat. Sci.* **1992**, *49*, 1357–1365. [CrossRef]
16. Wardle, C.S.; Carter, T.J.; Urquhart, G.G.; Johnstone, A.D.F.; Ziolkowski, A.M.; Hampson, G.; Mackie, D. Effects of seismic air guns on marine fish. *Cont. Shelf Res.* **2001**, *21*, 1005–1027. [CrossRef]
17. Boeger, W.A.; Pie, M.R.; Ostrensky, A.; Cardoso, M.F. The effect of exposure to seismic prospecting on coral reef fishes. *Brazilian J. Oceanogr.* **2006**, *54*, 235–239. [CrossRef]
18. Herbert-Read, J.E.; Kremer, L.; Bruintjes, R.; Radford, A.N.; Ioannou, C.C. Anthropogenic noise pollution from pile-driving disrupts the structure and dynamics of fish shoals. *Proc. R. Soc. B Biol. Sci.* **2017**, *284*. [CrossRef] [PubMed]
19. Weilgart, L.S. *The Impact of Ocean Noise Pollution on Fish and Invertebrate*; Ocean Care and Dalhousie University. 2018, p. 34. Available online: https://www.oceancare.org/wp-content/uploads/2017/10/OceanNoise_FishInvertebrates_May2018.pdf (accessed on 13 October 2021).
20. Nedelec, S. Impacts of Anthropogenic Noise on Behaviour, Development and Fitness of Fishes and Invertebrates. Ecosystems. École Pratique Des Hautes Études-EPHE PARIS. University of Bristol. 2015. Available online: https://tel.archives-ouvertes.fr/tel-02099631/document (accessed on 13 October 2021).
21. Wysocki, L.E.; Davidson, J.W.I.; Smith, M.E.; Frankel, A.S.; Ellison, W.T.; Mazik, P.M.; Popper, A.N.; Bebak, J. Effects of aquaculture production noise on hearing, Growth, and disease resistance of rainbow trout. *Oncorhynchus Mykiss. Aquac.* **2007**, *272*, 687–697. [CrossRef]
22. Popper, A.N.; Hawkins, A.D.; Halvorsen, M.B. *Anthropogenic Sound and Fishes*; 2019. Available online: https://www.wsdot.wa.gov/research/reports/fullreports/891-1.pdf (accessed on 13 October 2021).
23. Angelica, M.D.; Fong, Y. Cell proliferation follows acoustically-induced hair cell bundle loss in the zebrafish saccule. *Hear. Res.* **2009**, *253*, 67–76. [CrossRef]
24. Oxman, D.S.; Barnett-Johnson, R.; Smith, M.E.; Coffin, A.; Miller, D.L.; Josephson, R.; Popper, A.N. The effect of vaterite deposition on sound reception, otolith morphology, and inner ear sensory epithelia in hatchery-reared Chinook salmon (*Oncorhynchus tshawytscha*). *Can. J. Fish. Aquat. Sci.* **2007**, *64*, 1469–1478. [CrossRef]
25. Uribe, P.M.; Villapando, B.K.; Lawton, K.J.; Fang, Z.; Gritsenko, D.; Bhandiwad, A.; Sisneros, J.A.; Xu, J.; Coffin, A.B. Larval zebrafish lateral line as a model for acoustic trauma. *eNeuro* **2018**, *5*. [CrossRef] [PubMed]
26. Halvorsen, M.B.; Casper, B.M.; Woodley, C.M.; Carlson, T.J.; Popper, A.N. Threshold for onset of injury in chinook salmon from exposure to impulsive pile driving sounds. *PLoS ONE* **2012**, *7*, e38968. [CrossRef] [PubMed]
27. Whelan, K. A review of the impacts of the salmon louse, *Lepeophtheirus salmonis* (Krøyer, 1837) on wild salmonids. *Atl. Salmon Trust* **2010**. Available online: https://atlanticsalmontrust.org/wp-content/uploads/2016/12/ast-sea-lice-impacts-review1.pdf (accessed on 13 October 2021).
28. Urbina, M.A.; Cumillaf, J.P.; Paschke, K.; Gebauer, P. Effects of pharmaceuticals used to treat salmon lice on non-target species: Evidence from a systematic review. *Sci. Total Environ.* **2018**, *649*, 1124–1136. [CrossRef] [PubMed]
29. Hannisdal, R.; Nøstbakken, O.J.; Hove, H.; Madsen, L.; Horsberg, T.E.; Lunestad, B.T. Anti-sea lice agents in Norwegian aquaculture; surveillance, treatment trends and possible implications for food safety. *Aquaculture* **2020**, *521*, 735044. [CrossRef]

Journal of
Marine Science and Engineering

Article

Severity Scoring of Behavioral Responses of Sperm Whales (*Physeter macrocephalus*) to Novel Continuous versus Conventional Pulsed Active Sonar

Charlotte Curé [1,*], Saana Isojunno [2], Marije L. Siemensma [3], Paul J. Wensveen [2,4], Célia Buisson [1], Lise D. Sivle [5], Benjamin Benti [1,2], Rune Roland [6], Petter H. Kvadsheim [7], Frans-Peter A. Lam [8] and Patrick J. O. Miller [2]

1 Acoustics Group, Cerema-University Gustave Eiffel, UMRAE, F-67035 Strasbourg, France; celia.buisson@gmail.com (C.B.); benjamin.benti@protonmail.com (B.B.)
2 Sea Mammal Research Unit, University of St. Andrews, St. Andrews, Fife KY16 8LB, UK; si66@st-andrews.ac.uk (S.I.); pjw@hi.is (P.J.W.); pm29@st-andrews.ac.uk (P.J.O.M.)
3 Marine Science & Communication, Bosstraat 123, NL-3971 XC Driebergen-Rijsenburg, The Netherlands; m.siemensma@msandc.nl
4 Falculty of Life and Environmental Sciences, University of Iceland, 101 Reykjavik, Iceland
5 Department of Ecosystem Acoustics, Institute of Marine Research, P.O. Box 1870 Nordnes, NO-5817 Bergen, Norway; lise.doksaeter.sivle@hi.no
6 Department of Biosciences; University of Oslo, Marine Biological Station, Biologveien 2, NO-1440 Drøbak, Norway; r.r.hansen@ibv.uio.no
7 Sensor and Surveillance Systems, Norwegian Defense Research Establishment FFI, NO-3191 Horten, Norway; petter-helgevold.kvadsheim@ffi.no
8 Acoustics & Sonar Research Group, Netherlands Institute for Applied Scientific Research TNO, Oude Waalsdorperweg 63, NL-2597 AK The Hague, The Netherlands; frans-peter.lam@tno.nl
* Correspondence: charlotte.cure@cerema.fr

Citation: Curé, C.; Isojunno, S.; Siemensma, M.L.; Wensveen, P.J.; Buisson, C.; Sivle, L.D.; Benti, B.; Roland, R.; Kvadsheim, P.H.; Lam, F.-P.A.; et al. Severity Scoring of Behavioral Responses of Sperm Whales (*Physeter macrocephalus*) to Novel Continuous versus Conventional Pulsed Active Sonar. *J. Mar. Sci. Eng.* **2021**, *9*, 444. https://doi.org/10.3390/jmse9040444

Academic Editors: Michel André and Christine Erbe

Received: 30 March 2021
Accepted: 15 April 2021
Published: 19 April 2021

Publisher's Note: MDPI stays neutral with regard to jurisdictional claims in published maps and institutional affiliations.

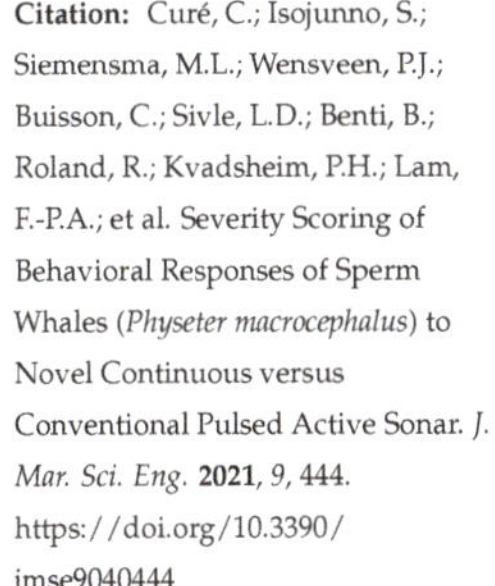

Abstract: Controlled exposure experiments (CEEs) have demonstrated that naval pulsed active sonar (PAS) can induce costly behavioral responses in cetaceans similar to antipredator responses. New generation continuous active sonars (CAS) emit lower amplitude levels but more continuous signals. We conducted CEEs with PAS, CAS and no-sonar control on free-ranging sperm whales in Norway. Two panels blind to experimental conditions concurrently inspected acoustic-and-movement-tag data and visual observations of tagged whales and used an established severity scale (0–9) to assign scores to putative responses. Only half of the exposures elicited a response, indicating overall low responsiveness in sperm whales. Responding whales (10 of 12) showed more, and more severe responses to sonar compared to no-sonar. Moreover, the probability of response increased when whales were previously exposed to presence of predatory and/or competing killer or long-finned pilot whales. Various behavioral change types occurred over a broad range of severities (1–6) during CAS and PAS. When combining all behavioral types, the proportion of responses to CAS was significantly higher than no-sonar but not different from PAS. Responses potentially impacting vital rates i.e., with severity ≥ 4, were initiated at received cumulative sound exposure levels (dB re 1 μPa^2 s) of 137–177 during CAS and 143–181 during PAS.

Keywords: behavioral response studies; severity scoring of responses; controlled exposure experiments; cetaceans; *Physeter macrocephalus*; continuous naval sonar

1. Introduction

All marine mammals and particularly cetacean species use sound as a primary sensory modality to perform vital functions such as finding prey, communicating with their congeners (e.g., for mating or maintaining group cohesion) and detecting predators [1,2]. Facing the urgent need to quantify the impacts of anthropogenic noise on cetacean species, the last decades have seen a growing number of controlled exposure experiment (CEE) studies in which animals are exposed to an acoustic stimulus to assess their behavioral and/or physiological responses. Information collected during CEEs in the field is used to calculate dose–response functions [3] and in modeling frameworks that have the ultimate goal to determine population-level impacts of the noise source [4,5]. Among the anthropogenic sound sources potentially impacting cetaceans, there has been a particular concern with long-range anti-submarine sonars since their use has been spatiotemporally correlated to various cetacean stranding events [6–8]. These naval sonars are very powerful sources, generating sound within 1–10 kHz and thus overlapping with the hearing sensitivity and sounds of most cetacean species [9]. Beside the risk of direct physical injuries (e.g., hearing impairment) [10], behavioral responses (e.g., avoidance responses) may contribute to the chain of events leading to lethal strandings [5].

CEEs have been a key experimental approach to get a better understanding of the impacts of anthropogenic noise sources such as naval sonar on free-ranging cetacean behavior. Their goal is to record short-term individual or group behavioral changes, specify the dose (e.g., received sound pressure level and received sound exposure level) and source proximity at which responses occurred, and to extrapolate the effects judged to have a relevant impact on individual fitness to long-term population effects [11]. The basic procedure involves monitoring individual behavior, before, during and after sonar exposure, in order to identify potential behavioral changes and quantify response duration to sonar. The development of animal-borne archival tags carrying various sensors provided a key tool for CEEs to track animals and to directly measure their behavior through the dive cycle (movement and sound recordings tags [12]). CEEs have been carried out to characterize behavioral responses to sonar and to investigate the factors driving those responses such as the particular ecological and/or social context of the exposure [13], the received level thresholds of response onset (by conducting CEE with a controlled escalating RL dose [14]), the sonar signal characteristics (e.g., the frequency range: CEE with 1–2 kHz or 6–7 kHz naval sonar [14]), or the animal–source distance [15].

Previous work focused on assessing behavioral responses of various cetacean species to conventional pulsed active sonar (PAS) systems that transmit short pulses separated by relatively long pauses for listening to returning echoes. PAS exposures in the 1–2 kHz frequency band induced costly behavioral responses in sperm whales [14,16]. Indeed, behavioral responses such as cessation of feeding indicated a potential for impacting individual vital rates if sonar exposures were sufficiently common and if animals continued to respond to the exposures. Comparing behavioral responses to sonar with how animals of the same species react when they detect a known natural high-level threat such as increased predation risk (simulated by predator sound exposures) provided a useful approach to interpret the biological significance of responses to sonar [17]. In sperm whales in particular, responses to 1–2 kHz PAS sonar were highly concordant with the observed antipredator behavioral template, including horizontal avoidance, interruption of foraging and increase of social sound production, providing evidence that sonar is perceived as a high-level threat [16,17].

Since then, new generation Continuous Active Sonar (CAS) systems transmitting almost continuously have been developed by navies as an alternative to traditional PAS in order to improve the opportunities of target detection. CAS sonar can be operated at lower source levels than PAS, potentially leading to less environmental impact compared to PAS. However, the higher duty cycle of CAS might increase the disturbance and risk of masking with smaller temporal gaps without sonar transmission. Therefore, the potential future use of CAS by navies raised further concerns on potential impacts of such new types of sonar on animals and how different those effects are compared to PAS sonar. To address this question, we conducted CEEs using both 1–2 kHz CAS and 1–2 kHz PAS exposures, and no-sonar (NS) controls on sperm whales (*Physeter macrocephalus*) in Northern Norway [18]. Using this dataset, Isojunno and colleagues [19] focused on quantifying the effects of CAS and PAS on sperm whale foraging behavior and movement effort. The authors found that responses to CAS were similar to responses to PAS as long as the energy levels of the transmissions were similar, even though the peak pressure levels of PAS were much higher.

This highlights the cumulative sound energy (received sound exposure level) rather than the received sound pressure level as a main driver of behavioral responses to sonar in feeding sperm whales.

In the present work, using established procedures to score putative responses [4,14,20], we aimed to identify the nature and severity of responses to CAS versus PAS, and to test whether CAS can lead to significant effects on the behavior repertoire of sperm whales in a different way than PAS. In particular, we investigated potential avoidance responses, cessation of feeding or resting behaviors, and exhibition of social responses. These responses have the potential to reduce individual fitness if expressed for a biologically relevant duration, and ultimately may have negative impacts at the population level. The basic principle of severity scoring is expert examination of multivariate timeseries of behavioral observations to assess potential changes across a range of predefined behavioral response categories (e.g., changes in diving behavior, changes in vocal behavior etc.). Such severity scores are attributed by experts using an existing qualitative severity scale ranging from no effect (0), to effects potentially impacting (4–6) or likely to impact vital rates (7–9) [4]. Previously, this severity scoring method was used consistently with a range of cetacean species that were subjected to sonar CEEs, (e.g., long-finned pilot whales, sperm whales and killer whales [14]; humpback whales, bottlenose whales and minke whales [15,20], and blue whales [21]). The present work adds to this list, presenting a unique dataset on sperm whales which increases comparison perspective across different sound exposures and species.

2. Materials and Methods

2.1. Animal Welfare Considerations

All animal research activities were licensed under permit provided by the Norwegian Animal Research Authority (Permit n° 2015–223 222) and were approved by the Animal Welfare Ethics Committee of the University of St Andrews (UK). Our experimental protocol followed a safety plan, designed to protect the welfare of our research subjects and to reduce risk to any other animals present in the studied area [22,23]. Expert marine mammal observers visually scanned continuously for research subjects and other cetaceans throughout the experimental exposures, with a detailed plan in place to stop sonar transmission if potentially hazardous responses occurred i.e., response which might bring the animal in danger of direct harm (e.g., animals showing signs of panic and fast swimming towards the shore or into confined areas) or if any animal came too close to the sonar source. The stand-off range between source and animals during full-power transmission was 100 m. If any animals were to approach this safety zone, an emergency shut-down of sonar transmission would be ordered. Transmission would cease immediately if any animals showed any signs of pathological effects, disorientation, severe behavioral reactions, or if any animals swam too close to the shore or entered confined areas that might limit escape routes. Moreover, other aspects of our protocol design also reduced the risk of harm to experimental subjects, such as the limited duration of the sound exposure periods, the limited number of tested whales, and the change of whales and area between the experiments.

2.2. Study Species and General Protocol

This study was conducted in two boreal summers (3–17 May 2016 and 22 June to 14 July 2017) on free-ranging sperm whales encountered on their feeding grounds in Northern Norway between 69–70.5° northern latitude and 12.5–19.5° eastern longitude. There, sperm whales are mostly solitary males typically spending ~80% of their time foraging and ~20% of their time resting and engaging in other activities [24–26].

The experiments were designed and conducted by the 3S (Sea mammals, Sonar, Safety) research consortium and detailed protocols used in the experiments can be found in two dedicated cruise reports [22,23]. Fieldwork was carried out from the 55 m FFI research vessel R/V H.U. Sverdrup II (hereafter "research vessel"), hosting scientists and crew members. The general protocol consisted of the following phases: (1) searching for the target species using both visual observers posted on the flying bridge of the research vessel and acoustic monitoring by the mean of a towed hydrophone array (DELPHINUS) developed by TNO [27]; (2) tagging operations from a dedicated 8 m workboat launched from the research vessel followed by a post-tagging period of at least 30 min to allow the animal to recover from any potential effects of the tagging procedure; (3) baseline data collection of the tagged animals for about 4h; (4) controlled exposure experiments; (5) end of tracking once the tag released, and tag recovery. Once the tag was recovered, the vessel transited at least 20 nautical miles away from the exposed area before tagging another whale and conducting the next set of experimental sessions.

The tag was a noninvasive multisensor suction-cup tag (DTAG) [12] attached to the whales using a cantilever pole or a pneumatic airgun [28] and was set to release after approximately 15 h. In all but two cases, only one whale was tagged, becoming the "focal animal" for which visual observations were collected throughout the following period of the tag deployment using the radio tracking of the VHF-beacon on the tag when the whale surfaced. In two cases, a second whale was tagged but not tracked visually (hereafter "non-focal animal").

2.3. Experimental Exposures

Our aim was to expose each whale to a no-sonar control (abbreviated 'NS'), followed by three sonar exposures conducted in an alternated order. The three sonar exposures consisted of the repetition of a 1–2 kHz hyperbolic upsweep signal. CAS had a 95% duty cycle (repetition of a 19-s signal + 1-s silence, Figure S1) and was generated with a maximum sound pressure level (SPL) of 201 dB re. 1 µPa m, and a maximum sound exposure level (SEL$_{19s}$) of 214 dB re. 1 µPa^2 m^2 s. PAS had a duty cycle of 5% (1-s signal + 19-s silence, Figure S1) and was transmitted either at a medium source level (signal MPAS) matching the SPL of CAS, (201 dB re. 1 µPa·m), or at a higher source level (signal HPAS) matching the SEL of CAS (214 dB re. 1 µPa·m).

Before each exposure session, the sonar source (SOCRATES) was deployed and towed by the research vessel at an average depth of 55 m (range 35–100 m). At the start of exposure, the vessel was positioned at 4 nautical miles (7.4 km) from the focal whale to approach from the front at an angle of about 45° relative to the animal's estimated course of travel. Each exposure session lasted 40min and consisted of the vessel approaching and then sometimes passing the whale, at a constant speed of 8 knots while either transmitting sonar or not transmitting sonar (NS). Sonar transmissions always started with a 20min ramp-up procedure, consisting of a gradual increase of the source level starting at a level of 60 dB below the maximum level and increasing in 1 dB steps per pulse. For the remaining 20 min of exposure, the sonar transmitted at maximum level. The vessel approach and sonar transmissions scheme aimed to achieve dose escalation, with a gradual increase of the sound levels received by the focal whales. This protocol was specifically designed to determine the received levels (RL) thresholds associated with response onsets. The NS sessions allowed separating the effects of the sonar from possible effects of the approaching vessel. When present, the NS session was always conducted as the first session for each set of experimental exposures, in order to test the effect of the vessel approach before any potential sensitization to the vessel if the whales had been previously exposed to the vessel towing a transmitting sonar.

The successive exposure sessions were separated by a minimum of 1 h 20 min allowing the animal to return to normal behavior following any behavioral response, and to plan the geometry of the next exposure session by relocating the research vessel relative to the expected course of the focal whale.

The goal was to expose each focal whale to all four exposure types (NS, CAS, MPAS, HPAS), however, due to logistical issues (e.g., tag released prematurely or whale track lost), some individuals were not exposed to all four experimental sessions. Two whales (sw16_126a and sw17_179a) were not exposed to NS. In two cases (NS session of sw16_134a and HPAS session of sw17_182b), the tag came off prematurely during the exposure, leading to an interruption of the behavioral response data collection during exposure. In one case (sw16_134b exposed to MPAS), visual contact to the whale was temporarily lost, reducing the number of whale geographical positions and leading to a particularly low quality track that prevented assessment of movement response.

2.4. Data Recording and Processing

Whale subjects were tagged with movement and sound recording DTAGs (version 3). These tags carry a suite of sensors, enabling the monitoring of the behavior of whales throughout their dive cycle [12]. All tags were equipped with a pressure sensor, temperature sensor, and three-axis accelerometer and magnetometer sensors sampling at 50 Hz. Moreover, they contained hydrophones that recorded stereo sound with 16-bit resolution at 96kHz sampling rate. A VHF beacon on the tags was used to identify, localize, and visually track the focal whale when it surfaced. In addition to the DTAG in its standard housing, the 'mixed-DTAG' was used which contained the DTAG version-3 core unit (i.e., all sensors of DTAG version-3) and VHF beacon, a GPS sensor (Fastloc2, Sirtrack, New Zealand) and an Argos transmitter (SPOT, Wildlife Computers, Redmond, WA).

Depth, heading, and pitch were calculated using established techniques [12]. The swim speed during dives of each tagged animal was calculated by regressing the acoustic flow noise in the 22.4–28.2 Hz frequency band to kinematic speed estimates during ascent and descent periods ($|\,pitch\,| > 60°$) [29]. The horizontal turning angle was calculated as a centered moving circular average of heading with a +/- 1 min window size.

Horizontal tracks of the tagged whales were reconstructed to 1 s resolution based on (1) the tag-derived movement data and visual and GPS position fixes using a state-space model implemented in a Bayesian framework [29], or (2) linear interpolations between visual and GPS position fixes when tag-derived heading data were not available due to failure of the magnetometer (lower-resolution method).

The acoustic recordings from the DTAGs were aurally and visually inspected via spectrograms using Adobe Audition software (Blackman-Harris window, FFT length: 4096) to identify sounds produced by the tagged sperm whale, sounds produced by conspecifics or other species present in the area, and sonar sounds received by the tagged whale. Typical sperm whale vocalizations were identified and included regular echolocation clicks and buzzes associated with foraging behavior [24], and other types of sounds associated with social behavior (slow echolocation clicks, codas, clangs and trumpet sounds) [30,31].

Moreover, incidental anthropogenic sonar, as well as sounds produced by other whale species in the research area, i.e., typically killer whales or long-finned-pilot whales (hereafter grouped as "blackfish" species), were annotated. Killer and pilot whales are considered as potential threatening stimuli for sperm whales as they represent a potential food competitor and/or a predator species [32,33].

The acoustic dose of the experimental sonar received by the tagged whales was quantified from the tag recordings of those sonar signals (see detailed method [14,20]). For each sonar pulse, the received maximum sound pressure levels (SPL_{max}) was determined using a sliding window of 200 ms, and the received cumulative sound exposure level (SEL_{cum}) was measured since the start of the sonar exposure session. Both received level metrics were analyzed in the 890–2240 Hz frequency band as it included the fundamental frequencies of the transmitted signal (the contribution of the harmonic frequencies on the broadband levels was determined to be negligible to the sound metrics we quantified, [34]).

Simultaneously with visual recording of the tagged whale positions at the surface, the best estimate of group size, defined as the number of individuals within 200 m of the focal animal during the surfacing period [35], was recorded. Visual data collection including the geographic position of whale resightings and group size was recorded using the software Logger. Moreover, sightings of blackfish species present in the area were reported (time and geographic position recorded).

2.5. Scoring Severity of Expert-Identified Behavioral Responses

Expert scoring of putative responses was used to assess the severity of identified behavioral responses on a numeric scale [4] ranging from no effect (0), effects not likely to influence vital rates (scores 1–3), effects that could impact vital rates (scores 4–6), to effects that are likely to impact vital rates (scores 7–9). The severity score of an identified response depends on the type of response and its duration relative to the duration of the exposure. The scale provided by Southall et al. [4] was further modified slightly to add some behavioral changes that were not covered in the original scale (Table 1). Each of the experimental exposures was visualized using a series of standardized data plots (available in [18]) where the exposure period was indicated but the experimental condition i.e., whale ID, type and order of exposure, was hidden (see one example of data plots on Figure S2). Data plots included a geographic track of the tagged whale and the research vessel, and time-series data plots of group size, swim speed, heading and turning angle, pitch, depth and whale sounds. Based on examination of those data plots, behavioral changes likely to be responses to the experimental exposures were identified. This was based upon a clearly visible behavioral change not observed during baseline periods. The severity was scored based upon the type of response and its duration by two independent groups of experts in accordance with the severity scale (Table 1). All scorers were part of the field team or familiar with the process of data collection, and seven out of eight coauthors had participated in previous similar scoring work [14,15,17,20]. One group consisted of authors C.C., L.S., P.W., B.B., and the second of authors M.S., R.R., P.K. and F-P.L. Each group conducted separate scoring, blind to the other team's scoring. Thereafter, the two groups met and assimilated their results in the presence of an adjudicator (author P.J.O.M.) to reach a consensus scoring.

One methodological improvement in the present study compared to the previous ones was that scorers were blind to the experimental conditions (NS, CAS, MPAS or HPAS). This blind procedure was applied to ensure that unconscious biases of panel members would not result in differences in scoring across different exposure types. Data plots were presented to the scorers as shown on one example presented in Figure S2, for the baseline period (i.e., period between end of post-tagging period until start of the 60 min of pre-exposure of the first experimental exposure) and the period covering 60 min of pre-exposure until 60 min of post-exposure, but excluding the full time track and full time series per tag deployment. Given the minimum period of 1 h 20 min between two successive exposure sessions, the remaining 20-min period between the end of a 60-min post-exposure and the start of the next 60-min pre-exposure was not shown on the plots, making the time series disconnected for the scorers. A random exposure number (RE#) was attributed to each experimental exposure session. For each RE's set of plots, the scorers evaluated whether the behavior exhibited during the exposure was different compared to the 60 min immediately preceding the exposure period and to the baseline period.

The behavioral response assessment was conditional to the studied species and context, i.e., solitary male sperm whales in their feeding grounds. Therefore, and similarly to the scoring methodology previously applied on sperm whales studied in this area, the scores of severity of the following eight behavioral response categories were systematically recorded for each experimental exposure (Figure 1): avoidance (vertical and/or horizontal), change in orientation other than avoidance (based on horizontal turns, pitch and vertical movements such as wiggles), change in locomotion (speed and directivity) not related to avoidance, change in dive behavior (based on dive profile), cessation of feeding (based on cessation of buzzing), cessation of resting (based on previous observations that sperm whales rest with a sharp pitch, i.e., with head up or down), modification of vocal behavior (including production of foraging and social sounds) and change in group distribution (group size). Other potential behavioral response categories existing in the severity scale, e.g., associated to reproduction, mother–calf association or aggressive behavior, were not assessed because they are not relevant to the behavioral context of the tested population's subjects (Table 1).

Table 1. Severity scale used for scoring behavioral responses. The original scale provided by Southall et al. 2007 [4] was slightly modified with some added behavioral responses (in bold) by Miller et al. 2012 [14] and Sivle et al. 2015 [20], and in the present work (in bold underlined). Given the exposure scheme of 40 min, a "Brief" response was defined as to be significantly shorter than the exposures (0–5 min), a "Minor" response was shorter than the exposure but longer than Brief (5–30 min) and stopped during the exposure, a "Moderate" response lasted roughly the duration of the exposure (30–60 min) and ceased soon after the end of exposure, and a "Prolonged" response was significantly longer than the exposure (> 60 min).

Score	Behavioral Responses
0	No observable response
1	Brief orientation response
2	Moderate or multiple orientation responses Brief or minor changes in respiration rates Brief cessation/modification of vocal behavior **Brief change in dive profile**
3	Prolonged orientation behavior Minor change in locomotion (speed/direction) and or dive profile but no avoidance of sound source Minor cessation/modification of vocal behavior Individual alert behavior Moderate change of respiration rate **Brief shift in group distribution**
4	Moderate change in locomotion (speed/direction) and or dive profile but no avoidance of sound source **Brief avoidance of sound source** Minor shift in group distribution Moderate cessation/modification of vocal behavior **Brief cessation of feeding/resting**
5	Extended change in locomotion (speed/direction) and or dive profile but no avoidance of sound source **Minor avoidance of sound source** Moderate shift in group distribution Change in inter-animal distance and/or group size Prolonged cessation or modification of vocal behavior **Minor cessation of feeding/resting**
6	Moderate avoidance of sound source Extended cessation or modification of vocal behavior Visible startle response **Moderate cessation of feeding** Prolonged shift in group distribution Brief or minor separation of female and dependent offspring Aggressive behavior related to noise exposure Brief cessation of reproductive behavior Moderate cessation of resting behavior
7	**Prolonged cessation of feeding** Moderate separation of female and dependent offspring Severe and or sustained avoidance of sound source Extensive or prolonged aggressive behavior Clear antipredator response Moderate cessation of reproductive behavior **Prolonged avoidance**
8	Obvious aversion and/or progressive sensitization Long-term avoidance of area Prolonged or significant separation of female and dependent offspring with disruption of acoustic reunion mechanisms Prolonged cessation of reproductive behavior
9	Outright panic, flight, stampede, or attack Avoidance related to predator detection

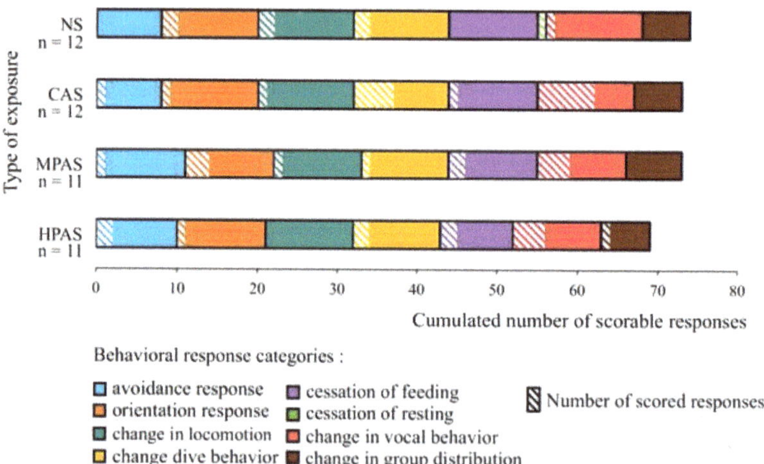

Figure 1. Scored responses (hatched) over scorable sessions across the different potential behavioral response categories and exposure types. The eight behavioral response categories (avoidance response, orientation response, change in locomotion, change in dive profile, cessation of feeding, cessation of resting, change in locomotion, change in dive profile, change in vocal behavior and change in group distribution) are represented by a color code (e.g., blue for avoidance response). For each of the four exposure types (NS, CAS, MPAS, HPAS), the total number of scorable sessions per behavioral response category is represented by the width of the box, of which the number of scored responses (i.e., non-zero value scores) is indicated as hatched. For example, for HPAS, there were a total of 69 scorable potential responses, all behavioral response categories combined. The avoidance behavioral response category was scorable in 10 out of the 11 HPAS exposure sessions, and scored responses were attributed in two of these 10 exposures. By contrast, for NS, there were no avoidance responses scored among the eight scorable sessions of the 12 conducted NS sessions. n: number of exposure sessions; CAS: Continuous Active Sonar; MPAS: Medium-level Pulsed Active Sonar; HPAS: High-level Pulsed Active Sonar; NS: No-Sonar control.

Typical feeding and resting behaviors were clearly identified based on characteristics of the dive profile and vocal behavior (e.g., presence of buzzes indicating feeding activity) [24,25]. Scorers also assessed whether the behavior of the whale during the 60-min pre-exposure period and the quality of the collected data allowed for a proper assessment of all behavioral response categories. For instance, the assessment of cessation of resting or feeding was conditional to whether the whale was resting or feeding at the time the exposure started. Moreover, the whale's geographic positions (visual or GPS fixes) and group size data could be collected only when the whale was at surface (between dive cycles). Therefore, the ability to assess potential 'avoidance' responses depended on the resolution of the whale track, and, the evaluation of potential changes in group distribution could be achieved only if group size data was recorded over the pre- and during-exposure phases. Non-focal whales were not tracked visually, preventing data collection on group size.

To account for cases of inability to assess potential behavioral changes, we differentiated a score "zero" (i.e., scorable, but no identified behavioral change judged to be response to the exposure) from the absence of a score (i.e., impossible to assess because data were missing or the behavioral context of the animal did not allow for it).

Once the two teams had reached a consensus on the scored putative responses, the experimental conditions of all RE were revealed and the received levels of the onset times of responses were identified. All unblinded data plots are published in Kvadsheim et al. 2021 [18].

2.6. Behavioral Response Analyses

A descriptive analysis was conducted in order to assess whether the scorability of the experiments, i.e., the ability to assess potential responses, was homogeneous across the panel of behavioral response categories and exposure types, and to reveal the distribution of the different behavioral response categories and magnitude (severity of scored responses) across the exposure types.

For each exposure session, we calculated two variables. The first was the proportion of scored behavioral responses (%), expressed as the total number of behavioral response categories for which a non-zero score was attributed (i.e., scored responses), normalized to the maximum number of potential scored responses (i.e., number of scorable behavioral response categories for which potential scores could be assessed). The second was the maximum score of severity among the different scored behavioral response categories.

Statistical analyses were carried out to model the proportion of scored responses (of all behavioral response categories combined) and the maximum score of severity per exposure session, in order to test the null hypothesis that the response variable was randomly assigned with respect to the exposure types. Since the whales were exposed to several exposure sessions, we used Generalized Estimating Equation (GEE) models that accounted for repeated measures in R v.3.0.2 for binomial response variables (geepack [36] in R Development Core Team 2013) and SAS 9.4 for categorical response variables (genmod procedure in SAS Institute 2011). GEE models also tested for potential influence of the first sonar exposure presentation compared to the subsequent sonar exposures (covariate Order) as well as for the previous presence of blackfish (covariate Blackfish) on the response variables.

Given that the protocol of exposures (range and direction of the approaching vessel relative to the animal) was specifically designed in relation to the focal whale position, the two non-focal whales (Table 2, sw17_182a and sw17_186a) received lower exposure levels at greater distances than the focal whales. Consequently, the behavioral response scoring data of the two non-focal whales were excluded from the statistical analyses.

2.6.1. Quantitative Analysis of the Proportion of Scored Responses and the Maximum Score Per Session Variables

For both severity scoring response variables (i.e., Proportion of scored responses and Maximum score per session) the full model tested whether the three covariates Signal, Order and Blackfish had an effect on the response variable. The covariate Signal was assigned with two factor levels, no-sonar and sonar, aiming at testing effect of sonar, or with four factor levels, CAS, HPAS, MPAS and NS, that aimed at testing potential effect of the sonar exposure type on the response variables. The covariate Order had two factor levels: one noted « 1st » including the NS and first sonar exposure sessions, and one noted « diff_1st » for which all subsequent sonar exposure sessions following the first sonar session were assigned. The Blackfish covariate was encoded as a variable that linearly decreases with time after a detected blackfish event (visually sighted and/or acoustically detected from the tag recordings). Specifically, it corresponds to the duration needed to recover at the start of exposure since the last identified blackfish event. We assumed that a full recovery from a detected blackfish event lasted a maximum 15 h because this corresponded approximately to the duration of data collection (i.e., programmed release time of the tag) [19]. Therefore, the more recently the whale was exposed to blackfish, the higher the value of the Blackfish covariate. If at start of exposure the whale had not been exposed to blackfish for more than 15 h, then the Blackfish covariate was given a zero value (blackfish event was considered as "absent"). If a blackfish event was detected within the 15 h preceding the start of exposure, the Blackfish covariate was 15 minus the number of hours since the event (blackfish event defined as "present"). If the blackfish event was detected during the exposure, the Blackfish was applied the maximum value of 15.

Table 2. Overview of collected data. The whale ID code includes information regarding the species ("sw" for sperm whale), the year (e.g., "16" for 2016), the Julian date (e.g., 126) and a letter (e.g., "a") identifying the tag deployed. Sixteen tagged whales were subjected to the exposure experiments, from which 11 whales were equipped with a regular DTAG (version 3) and five had a mixed-DTAG (including a GPS logger, indicated with a *). In two occasions, two whales were simultaneously tagged, one focal whale and one non-focal whale (in italic), and both were exposed to the same exposures. For one of the non-focal whales (sw17_186a), the tag came off before the last exposure session started (MPAS). For each whale ID, the exposure sessions are listed by order of exposure. Exposure sessions indicated in bold are those for which a blackfish event has been detected within the 15-h period preceding the start of exposure or during the exposure. Blackfish events were defined as acoustic and/or visual detections of blackfish presence (i.e., killer whales and/or long-finned pilot whales) based on the visual sightings and inspection of the sound recordings on the tags. The focal whale dataset includes a total of 34 sonar exposures (12 CAS, 11 MPAS, 11 HPAS) and 12 NS. See abbreviations defined in Figure 1.

Whale ID.	Exposure Sessions (by Order of Exposure)
Sw16_126a	CAS, MPAS, HPAS
Sw16_130a	NS, MPAS, CAS
Sw16_131a	NS
Sw16_134a	NS
Sw16_134b	NS, CAS, HPAS, MPAS
Sw16_135a	NS, **HPAS, CAS, MPAS**
Sw16_136a	NS, CAS, MPAS, HPAS
Sw17_179a *	**MPAS, HPAS, CAS**
Sw17_180a *	**NS, HPAS, MPAS, CAS**
Sw17_182a *	*NS, MPAS, CAS, HPAS*
Sw17_182b	**NS, MPAS, CAS, HPAS**
Sw17_184a *	NS, CAS, **HPAS**
Sw17_186a *	*NS, HPAS, CAS*
Sw17_186b	**NS, HPAS, CAS, MPAS**
Sw17_188a	NS, MPAS, HPAS, CAS, **CAS**
Sw17_191a	**NS, HPAS, MPAS**

For the multiple GEE models (i.e., with two or four factor levels for the covariate Signal) applied on the two severity scoring variables, the full model with all three candidate explanatory variables was first run. Hypothesis-based model selection using *p*-values given by ANOVA (sequential Wald test) and backwards selection was conducted. After fitting each model, an ANOVA was conducted and the covariate with the highest *p*-value was removed and the GEE model refitted (for detailed method, see [37]). This was repeated until all terms retained in the ANOVA were significant at 5% level. The best fitted GEE models were then fitted to obtain results.

2.6.2. Dose–Response Function Analysis

To obtain probabilistic relationships of received level and response onset, we generated dose–response functions by fitting marginal stratified Cox proportional hazards models to the severity scoring data (see [38], for full details of this approach). This form of recurrent event survival analysis allowed us to combine the results from individual CEE to estimate the likelihood of response as a function of the acoustic dose while accounting for contextual covariates. Models were stratified by severity level (i.e., low for severity 1–3, moderate for severity 4–6, high for severity > 6) to produce the dose–response functions for different severity levels from the same fitted model.

The input data for the stratified Cox models was the received SELcum of the sonar at the first occurrence of each response level within each sonar exposure session. In the case of a severity score of 0 (no response), we allocated the received SELcum of the entire exposure session and labeled the data as right-censored. Statistical analyses were carried out to model the dose–response function in order to test the null hypothesis that the response variable (low or medium severity level) was randomly assigned with respect to the received level and exposure types. The full candidate model consisted of the covariate Stimulus with three factor levels (CAS, MPAS, HPAS) in addition to covariates Order and Blackfish (as defined previously). All possible model combinations including the null model were fitted and AIC-based model selection was used. The standard errors of the model estimates were corrected for the correlations within individuals using a grouped jack-knife procedure [39]. For the selected model, we verified the assumptions of proportional hazards, no influential outliers, and no interaction between covariates and strata [38,40]. Analyses were conducted in R version 3.6.1 (R Core Team 2019) and the dose–response functions were generated as survival curves using the survfit function package [41].

3. Results

In total we recorded behavioral data of 14 focal whales and two non-focal whales. All whales were exposed from one to four exposure sessions except for whale sw17_188a which was exposed to a 5th exposure (repeat CAS, see Table 2). The focal whale dataset consisted of 46 exposure sessions including 12 NS control, 12 CAS, 11 MPAS and 11 HPAS exposures, and the non-focal whale dataset included 2 NS, 2 CAS, 1 MPAS and 2 HPAS exposures (Table 2).

3.1. Scorability of the Data Across the Behavioral Response Categories and Exposure Types

In order to investigate potential differences in the 'proportion of scored responses' and 'maximum score per session' variables across the exposure types, we first inspected whether the scorability of experimental sessions, i.e., our ability to assess potential responses, were similar for all behavioral response categories and across the four exposure types (Figure 1).

Most behavioral response categories (orientation response, change in locomotion, change in dive profile, and change in vocal behavior) were scorable in all sessions, i.e., we were able to assess whether a potential behavioral change occurred (non-zero score) or not (zero score). In the focal whale dataset, for the four exposure types, the change in group distribution was scorable for half of the exposure sessions and the avoidance response category was scorable for two-thirds of the exposure sessions (Table S1). As predicted by the location of the studied population (on their feeding grounds), most whales were foraging immediately before an exposure session, making a potential cessation of feeding scorable for most exposure sessions. By contrast, it happened twice that a whale was in a resting mode at the start of an experimental session, making a potential cessation of resting scorable only for these two cases, one in the focal whale dataset (sw16_134a exposed to NS) and one in a non-focal whale dataset (sw17_182a exposed to MPAS). Therefore, the cessation of resting category of behavioral response could not be investigated across exposure types and was excluded from the statistical analyses.

The distribution of scorable sessions among all behavioral response categories except cessation of resting, was homogeneous across the four experimental conditions (NS, CAS, MPAS, HPAS), making the comparison of scored responses across the exposure types suitable for the remaining seven behavioral response categories.

3.2. Overview of the Scoring of Behavioral Responses
3.2.1. Expert Scoring Process

The scores of severity of responses made by the two teams of expert scorers were mostly the same and the adjudicator was needed only in two cases to achieve consensus (Table S1). For four identified behavioral scored responses of the focal whale dataset (sw16_126a exposed to MPAS: foraging and vocal behavior, sw17_191a exposed to NS: dive profile and sw16_130a exposed to MPAS: locomotion) and one of the non-focal whale dataset (sw17_182a exposed to HPAS: vocal behavior,), it was difficult to determine whether the change in behavior was a response to the exposure or a coincidental change. For instance, one focal whale (sw16_126a) was attributed a scored 'cessation of feeding' in response to MPAS although the scorers reported a low confidence in assigning this scored response given that an interruption of buzzing had occurred also during the baseline period. For those potential false positive cases, the scores of severity were noted as being of "low confidence" (Table S1), with provided justification. For two exposures, sw17_182b exposed to HPAS and sw16_134a subjected to NS, the tag came off prematurely about halfway into the exposure, precluding assessment of the duration of identified responses and of other behavioral changes that might have occurred over the remaining duration, thus leading to minimum estimates of severity.

3.2.2. Summary of Scored Behavioral Responses

Only two out of 14 focal whales and one out of two non-focal whales were judged not to have responded to any of the exposure types (i.e., no scored behavioral changes): sw16_131 (exposed only to NS), sw17_184 (exposed to NS, CAS and HPAS) and non-focal sw17_186a (exposed to NS, HPAS and CAS) (Table 2). Half of the 46 experimental sessions in the focal whale dataset elicited at least one scored behavioral response, i.e., a score different than zero attributed at least for one of the behavioral response categories assessed. Specifically, scored behavioral responses were obtained in response to 5 out of 12 NS trials, 7 out of 12 CAS trials, 4 out of 11 HPAS trials and 7 out of 11 MPAS trials. A total of 48 putative scored responses were assigned, of which 31 had a severity 1–3 (24 in response to sonar and 7 to NS) corresponding to responses considered not likely to influence vital rate, and 17 had severity 4–6 (16 in response to sonar, and 1 in response to NS), thus considered to have the potential to impact vital rates (Table 1). No behavioral response of severity higher than 6 was identified in this data set.

3.3. Description and Distribution of the Types of Behavioral Responses within Exposure Type

There was a high diversity of scored behavioral response categories assigned to the experimental exposures, with all behavioral response categories represented at least once in the total set of scored responses (Figure 1). The two scorable 'cessation of resting' cases were scored: one focal whale interrupted resting at least briefly in response to a NS exposure session (severity $\geq$ 4), and one non-focal whale ceased resting in response to MPAS for about the duration of the exposure (severity 6) (Table S1). The distribution of the other scored behavioral response categories varied across the experimental conditions (Figure 2). Scored behavioral responses to NS were mainly represented by orientation and locomotion responses of low severity and changes in the dive profile (together representing 86% of the scored responses), and to a lesser extent, by modifications in vocal behavior. Putative responses were also observed during sonar exposure sessions (Figure 2). Most of them had severities ranging from 1 to 3 (i.e., not likely to impact vital rates), and only two reached a severity 4 (i.e., that could impact vital rate) during CAS sessions (moderate change in the dive profile or moderate change in the vocal behavior).

Overall, changes in locomotion were mostly represented by a heading turn towards the source, orientation responses included primarily vertical wiggles, and changes in the dive profile corresponded mainly to switching to shallower and shorter dives (Table S1). Changes in vocal behavior represented a large part of the scored responses to the sonar exposures (30% for HPAS, 33% for MPAS, and 42% for CAS) compared to the NS (only 14%). They involved modifications in the production of foraging sounds and/or social sounds, and were scored more often in response to CAS (in 7 out of 12 CAS trials) compared to other exposure types (4 out of 11 for both HPAS and MPAS trials, and only 1 out of 12 NS trials). Modifications of social sound production included unusual occurrences of slow clicks, codas or other types of social sounds (trumpet sounds, clangs).

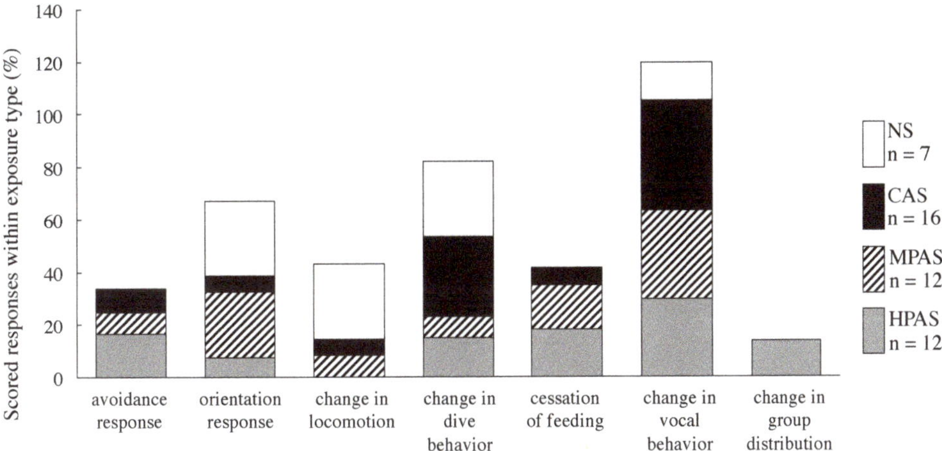

Figure 2. Distribution of the proportion of scored responses across the behavioral response categories, summing to 100% within each exposure type. The figure includes the dataset of the focal whales tested in the present study with the four exposure types NS, CAS, HPAS, and MPAS. A potential 'cessation of resting' could be assessed only once, for a NS see Figure 1. because the whale was rarely resting at the start of exposure. This category of response could not be evaluated (unscorable) for any of the other exposures and was thus excluded from this figure. n: number of scored responses per exposure type. See Figure 1 for definition of abbreviations CAS, MPAS, HPAS, NS.

The scores of severity associated with changes in the production of social sounds had a maximum severity 4, obtained once during a CAS exposure session, whereas they were always $\leq$ severity 3 during HPAS, MPAS or NS sessions (Table S1). The production of codas occurred in response to all three sonar exposure types but never during NS. Moreover, for CAS sessions, 30% of the scored behavioral responses corresponded to changes in the dive profile (Figure 2), during 5 out of 12 CAS sessions (Figure 1, Table S1). Most of these changes in the dive profile were not accompanied with a scored cessation of feeding (i.e., the animal was still buzzing while diving). By contrast, changes in the dive profile contributed to less than 15% of the total scored behavioral responses for MPAS (only 1 out of 11 MPAS trials) and for HPAS (2/11 HPAS) and were always associated with concomitant cessation of feeding, i.e., interruption of buzzing (Figures 1 and 2).

In addition to these types of behavioral changes, the sonar exposures were scored to have triggered changes in group distribution, and minor to moderate avoidance and cessation of feeding responses. These represented 45% of the scored responses to HPAS, 25% for MPAS and 19% of responses to CAS (Figure 2). These categories of behavioral response were never assigned a score in response to NS, indicating that these types of behavioral changes were specifically induced in response to sonar transmissions and not the vessel approach.

Avoidance and cessation of feeding were scored in response to all three sonar exposure types and represented the highest levels of severity of responses with a maximum score up to 5 for minor responses to CAS and up to 6 for moderate responses to both MPAS and HPAS (Figure S3). Change in group distribution was scored only in response to one sonar exposure session, a HPAS exposure (Figures 1 and 2), during which a solitary whale (sw16_135a) grouped with another whale for at least the duration of one surfacing phase during the exposure (severity $\geq$ 3, Table S1). The portion of the dive immediately preceding the start of that exposure session coincided with sightings of killer whales (at about 5 min before the exposure started).

3.4. Quantitative Analysis of the Severity Scoring Variables in Relation to Exposure Type, Order and Recent Exposure to Blackfish

Each of the three sonar exposure types (CAS, HPAS, and MPAS) was presented as the first sonar exposure session four times, providing a well-balanced dataset for testing the potential influence of the first sonar session compared to the subsequent sonar exposure sessions (Table 2). Moreover, for about half of the total exposures of the focal whale dataset (four NS, and six of each of the three sonar exposure types), the whales had been subjected to a blackfish event within the 15 h preceding the start of exposure or during the exposure, allowing for investigation of the potential influence of blackfish exposure on the response variables.

The quantitative analysis was conducted on the focal dataset including the low confidence scores and responses of potentially under-estimated severity (e.g., when the tag had come off before the end of exposure, Table S1). Twelve out of the 14 focal whales were judged to have changed behavior in response to at least one of the exposure types (Table 2). Among the 10 responding focal whales exposed to both NS and sonar exposures, four were attributed scored behavioral changes in response to both NS and at least one of the sonar exposure sessions. In those four cases, the maximum of severity of scored responses ranged from 2 to 6 in response to sonar, versus 1 to 3 in response to NS (Figure S3). The six other whales were judged not to have responded to NS (severity 0 for all behavioral response categories) but were identified as having responded to one or more of the sonar exposure types, with a maximum severity score of up to 5.

The ANOVA conducted on the GEE models for the maximum score per session did not support any of the factors Signal, Order and Blackfish at 5% significance level, indicating that none of them substantially explained the variation in the dataset, precluding any further quantitative analysis on this response variable.

For the proportion of scored responses, the ANOVA applied to the GEE models retained the factors Signal (with Signal having two or four factor levels) and Blackfish in the best fitted model ($p < 0.05$ for each factor, Table S2). However, the ANOVA did not support Order in any of the models, indicating that there was no main effect of the first exposures compared to the subsequent sonar exposures on the proportion of scored responses (Table S2). GEE model with factor Signal represented by the two factor levels NS and Sonar (i.e., including all sonar types) showed that the proportion of scored responses was significantly higher in response to sonar exposures ($18 \pm 23\%$) compared to NS ($7 \pm 3\%$) (Table S3), meaning that the whales were more likely to respond to the approaching vessel towing a transmitting sonar than to the approaching vessel without sonar transmission. The results of GEE models with Signal represented by four factor levels (NS, CAS, MPAS, HPAS) showed that CAS led to a significantly greater proportion of scored responses ($21 \pm 7\%$) compared to NS ($7 \pm 3\%$). The mean proportion of scored responses were similar between MPAS ($17 \pm 5\%$) and HPAS ($16 \pm 9\%$), intermediate values compared to the lowest and highest proportion of scored responses represented by NS and CAS, respectively. However, MPAS and HPAS were not significantly different from CAS or NS ($p > 0.2$ for the pairwise comparisons HPAS vs. NS, HPAS vs. CAS, MPAS vs. NS and MPAS vs. CAS, Figure 3, Table S3).

For these GEE models (factor Signal with two or four levels), the Blackfish covariate had a significant main effect on the response, showing that the proportion of scored responses was more likely to be increased if whales had been recently exposed to blackfish, independent of the exposure type (Figure 3).

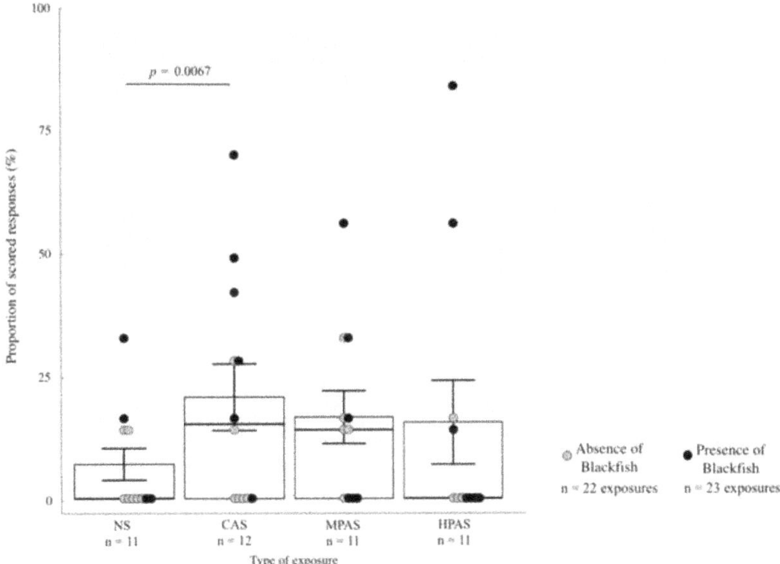

Figure 3. Proportion of scored responses per exposure type for focal whales (all behavioral response categories combined), Figure 2. GEE results (detailed in Table S3) of paired comparisons between exposure types are given for 5% significance level. For each exposure type, one dot represents the value of one exposure session being labeled in black or grey in relation to, respectively, the presence or absence of detected blackfish event within the 15 h preceding the start of exposure up to the end of exposure. GEE results showed that the proportion of scored responses for CAS was significantly higher than for NS ($p < 0.0067$). Moreover, the covariate Blackfish had a positive effect on response occurrence ($p < 0.005$, Table S3), indicating that the more recently the whale had been exposed to blackfish, the highest the proportion of scored responses. n: number of exposure sessions; see Figure 1 for definition of abbreviations CAS, MPAS, HPAS, NS.

3.5. Severity of Scored Response in Relation to Received sound Pressure Level

Changes in behavior were scored with response thresholds over a large range of received sound pressure levels, from 86 to 175 dB re 1 µPa (SPL_{max}) and from 82 to 189 dB re 1 µPa^2 s (SEL_{cum}) (Figure S4). The most severe scored responses (severity $\geq$ 4) were initiated by the tagged animal at received levels of 119–159 dB re 1 µPa (SPL_{max}) and 137–177 dB re 1 µPa^2 s (SEL_{cum}) during CAS, and of 138–175 dB re 1 µPa (SPL_{max}) and 143–181 dB re 1 µPa^2 s during PAS (HPAS and MPAS combined).

Stratified Cox models were fitted to the SEL_{cum} for responses of low (score 1–3) and moderate severity (score 4–6), as responses of high severity (score 7–9) were not observed. The selected model retained none of the covariates (Stimulus, Order or Blackfish). This null model had a slightly higher AIC ($\Delta AIC = 0.03$) than the candidate model with only Stimulus, but two fewer degrees of freedom. All four candidate models with Blackfish violated the proportional hazards assumption (global p-value from χ^2 test < 0.05), indicating a significant relationship between the Blackfish estimate and SEL_{cum}. They were thus excluded on that basis during model selection, although those models had the lowest AICs. Dose–response functions generated from the null model never reached a response probability of 0.5 for responses of moderate severity (Figure 4). For responses of low severity, a response probability of 0.5 was predicted to occur at a received SEL_{cum} of 173 dB re 1 µPa^2 s (Figure 4).

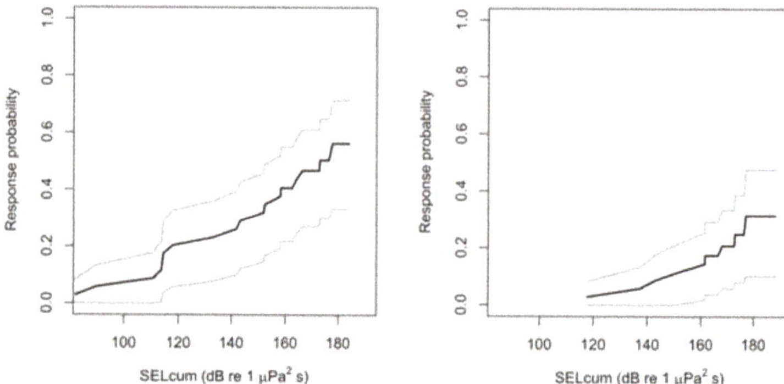

Figure 4. Dose–response probability functions for the focal sperm whales' dataset generated from the selected stratified Cox proportional hazards model for low severity responses (score 1–3) on the left, and moderate severity responses (score 4–6) on the right, with their 95% confidence limits in grey.

4. Discussion

The primary goals of the present study were to assess whether CAS leads to significant effects on free-ranging sperm whale behavior, and to investigate whether these effects differ in type and severity from previously reported effects of PAS (e.g., [14]). Using visual observations and acoustic-movement tag data, we identified and described behavioral responses of sperm whales to experimental exposure to CAS and PAS as well as to NS, and scored their level of severity (depending on their type and duration) by the mean of an objective scale [4].

Sonar exposures induced a higher diversity of scored responses across behavioral categories, more scored responses, and greater severities of scored responses (up to severity 6) compared to NS (maximum severity 4 assigned in only one case). Avoidance and cessation of feeding, typically associated with moderately higher severity scores (5–6) than the other behavioral response categories, were only induced in response to sonar and not to NS. Most other scored categories of behavioral response were common between CAS and PAS but with a distribution that did differ across the sonar types. The proportion of exposure sessions with scored responses was significantly higher during CAS compared to NS—indicating that CAS transmissions led to a significant change in sperm whale behavior beyond any effect of the approaching vessel. This higher proportion of scored responses to CAS was not significantly different from the proportion of scored responses to PAS.

4.1. Responses to NS

Behavioral responses to NS allow separating the components of the responses specifically exhibited in response to sonar from those that could occur in response to the approaching vessel alone. Previous CEEs carried out on several cetacean species including sperm whales, and using the same basic experimental design as the one conducted in the present study but for which NS and sonar exposures order were randomized, showed that animals hardly changed behavior to NS, whether conduced as first or following previous sonar exposures [14,16,17,26]. Results of these previous studies showed more severe behavioral responses to sonar than to NS, and even more severe responses to predatory sounds exposures compared to sonar (humpback whales [19], sperm whale [16,17]). Responses to the predatory killer whale sounds playbacks conducted in 2008–2009 [14] and 2010 [17] were used as a positive control for characterizing sperm whales' responses to a natural threatening stimulus, thus representing a yardstick of aversive reactions.

In the present study, NS was always conducted first allowing characterization of the response to NS excluding the effect of any potential sensitization to the vessel if the whales had been previously exposed to the vessel towing a transmitting sonar. When responding to NS, the focal whales made slight changes in horizontal and vertical movements, scored to be low severity responses. Such responses, ranging from severity 1 to 3, are unlikely to impact vital rate, and thus unlikely to lead to significant population effect. The fact that NS triggered some responses indicates that either the approaching vessel was perceived as a type of disturbance by the whales, and/or tendency of human observers to interpret and attribute behavior changes in relation to the exposure, when in fact these changes could be elicited by factors other than the experimental protocol.

In one single case of NS, a focal whale was attributed a scored response of moderate severity (score '≥ 4'), for an identified interruption of resting with potential underestimated duration because the tag came off prematurely, ceasing data collection. Cessation of resting is, similarly to cessation of feeding, considered to potentially impact fitness if repeated with sufficient duration. The probability of whales to be in a resting state at start of an exposure was lower than to be in a feeding mode, given their time budget predicted about 80% of time spent in foraging mode versus 20% in resting and other activities [26]. There was a second case of scored cessation of resting obtained in a non-focal whale exposed to MPAS (severity 6). Proximity to the sea surface during shallow resting dives [25] could be a factor driving responsiveness to an approaching vessel, and further data would be needed to compare the effects (probability and duration of response) of sonar and NS on this behavioral response category.

4.2. Responses to Sonar and Influence of Exposure Type

Response durations to sonar were identified as brief (<5min) to moderate (i.e., 30–60 min), with scored response severities ranging from 1–3 (not likely to impact vital rates) to 4–6 (considered to have potential to impact vital rates). No high-severity responses (7–9, likely to impact vital rates) were identified. A high diversity in behavioral response categories was found in response to all types of sonar: changes in the dive profile, changes in vocal behavior, orientation and/or locomotion responses, as well as avoidance and cessation of feeding. The two latter behavioral response categories were specific to sonar (they never occurred in response to NS) and had the highest severity scores (5–6) in the whole dataset. These two behavioral response categories carry a potential to impact fitness even for relatively short-term responses (severity 4 for brief duration in the scale) and they are typically part of an antipredator strategy. This highlights their biological relevance and indicates that they represent higher level disturbance similar to immediate predation risk.

The distribution of the other behavioral change categories varied with sonar exposure type. In particular, changes in vocal behavior including changes in the production of social sounds were more frequently observed in response to CAS than to PAS signals. Given the high duty cycle of CAS, it is possible that animals need more adjustment in their vocal behavior than when they are exposed to PAS, which leaves more time without sound masking between the sonar pulses. Further studies could focus on vocal responses to CAS to investigate masking effect and potential effects on the efficiency of foraging (regular clicks and buzzes) and communicating with congeners (social sounds). Moreover, changes in the dive profile in response to CAS were scored more often than for PAS, and those changes were not always associated with a cessation of feeding response scored for PAS sessions.

Four HPAS exposures were previously conducted in 2008 and 2009 (named as 'LFAS' exposures in [14]) using a similar experimental protocol as the HPAS exposures of the present study except that the ramp-up period lasted 10 min followed by 30 min full-power transmission and the vessel approach could turn towards the whale during the exposure until it was 1km away from the whale (instead of 20 min followed by 20 min full power in the present study, with the source vessel driven in a straight course throughout). The suite of behavioral change types identified in response to HPAS conducted in the present study was comparable to the one described to previously conducted HPAS exposures [14], except for one HPAS session in our dataset that obtained a scored change in group distribution. The grouping behavioral response had previously only been observed in response to predatory sound exposures, not to sonar, thus had been interpreted as a specific response component of the antipredator strategy [17]. It could be that some individuals perceive HPAS as a particularly high level of threat, leading to potential behavioral response components similar to antipredator behavioral strategy, such as the grouping behavior response. The presence of killer whales during the pre-exposure period of this HPAS session, might have sensitized those whales' responsiveness to sonar, resulting in social cohesion.

Overall, 10 out of 12 focal whales responded during at least one exposure session, however, half of the total exposure sessions obtained no scored responses i.e., they were judged not to have induced any behavioral changes. This observation indicates the relatively low probability of responses in the sample of tested subjects which concurs with the observation made by Isojunno et al. [19] who pointed out that the current study's (2016–2017) sperm whale subjects appear less responsive than the subjects tested in the same area 7–8 years ago with controlled PAS exposures (2008–2009, [14]). Moreover, the maximum scores per session in the present study had broad ranges for all sonar types: from 2 to 6 in response to HPAS, 1–6 in response to MPAS and from 2 to 5 in response to CAS). This finding indicates interindividual or other interdeployment contextual variability within exposure type, which contrasted to the consistent maximum score of 6 represented by moderate avoidance or cessation of feeding responses obtained in the four subjects that responded to the HPAS exposures conducted in 2008–2009.

In addition to the changes in exposure protocols detailed above, a possible explanation of this apparent reduced sensitivity to sonar in the 2016–2017 individuals dataset could be that these animals have been more exposed and habituated to the lower frequencies used by modern naval sonars. Naval sonar in the 5–10 kHz band has been commonly used in this area since the 1960s, but low frequency sonars in the 1–2 kHz band are a fairly recent technological development only used frequently the last 10 years. The 1–2 kHz sonar exposures conducted in 2008–2009 might have been perceived as a more novel stimulus for the whale populations, than in the current dataset. Such apparent tolerance to sonar could lead to underestimation of responses in a population living in a more pristine habitat devoid of naval sonar exercises. Novelty has been indeed suggested to potentially influence the probability of responses in marine mammals [42], including cetacean species (e.g., long-finned pilot whales [43] and bottlenose whale [15]).

There were no clear trends of individuals responding with a higher probability to one or the other sonar exposure types. The statistical analysis of the proportion of scored responses confirmed this observation: the proportion of scored responses was not significantly different between CAS and PAS. However, the proportion of behavioral scored responses to CAS was significantly higher compared to NS showing that CAS has a significant effect on sperm whale behavior. The proportion of scored response was not found to have been significantly different between PAS and NS. While our analysis accounted for Blackfish presence, other contextual variability across exposure sessions is likely to explain this result.

The motivation to include three sonar types (CAS, MPAS and HPAS) in the experimental design was to disentangle the effects of duty cycle, sound energy (SEL) and sound amplitude (SPL) as the main driver of behavioral responses. Using the same dataset, Isojunno et al. [19] used quantitative state switching analysis focusing on the effects of CAS vs. PAS on foraging effort. The authors found significant and similar reduction of foraging effort in response to both CAS and HPAS, but not to MPAS. Since CAS and HPAS exposures will have the same received sound energy level, but HPAS will have much higher peak pressure level, this result led to the conclusion that received sound energy levels might be an important driver of the response [19]. Here we did not find clear evidence of whether duty cycle, sound energy or amplitude best predicted probability of response given there was no significant differences of probability of scored responses between the three sonar signals and that all of them could lead to a broad range of max scores ranging from low (1–3) to moderate severities (4–6). The main difference between Isojunno et al. 2020 [19], and this study is that Isojunno et al. 2020 focused on feeding behavior, whereas our analysis combines different behavioral response types into one metric of response.

Given only CAS led to significantly increased proportion of scored responses compared to NS, one could conclude that the duty cycle might be a main predictor of response. However, the proportion of scored responses was not significantly different between CAS and PAS, refuting this hypothesis. The nonsignificant results to PAS vs. NS are likely due to a combined low sample size and high interindividual variability. Comparing the behavioral types between responses to MPAS and HPAS, there was no significant differences in their proportion of scored responses, and the description of max scores per session showed comparable ranges between the two (max score 1–6 in response to MPAS, and 2–6 in response to HPAS). Since HPAS had higher SPL and SEL levels than MPAS, this result indicates that received levels of sound are not a particularly reliable metric to predict the severity of responses. Further data is needed to confirm this outcome, given the high variability between individuals of the present individual dataset. Moreover, the MPAS scores dataset contained some uncertainties (particularly with the low confidence scores) which might have overestimated the proportion of scored responses to this sonar type.

In their study, using the same dataset, Isojunno et al. [19] found more cases where individuals switched from foraging to nonforaging states compared to the number of 'cessation of feeding' events scored in the present study. Moreover, they found a significantly higher reduced foraging effort in response to CAS and HPAS compared to MPAS whereas we did not find any evidence of differences in the proportion of scored responses (combining all response categories) across the three sonar types. In the present study, the distribution of behavioral response categories across exposure type did not show indication of more cessation of feeding in response to CAS and HPAS compared to MPAS. The reason for such apparent differences between studies can be explained by different analytical approaches. Here we looked at various potential behavioral response categories among which a 'cessation of feeding', that was defined as an interruption of buzzing, whereas Isojunno et al. [19] quantified the alteration of foraging effort (activity time budget) and two proxies (i.e., fluke stroke rate and buzz presence, given an activity state, indicating locomotion costs and foraging success, respectively). Movement behavior and the concurrent presence of echolocation (irrespective of type—regular or buzz clicks) were used to inform about the activity state of the whales, whereas here, buzzes were used as the primary indicator for cessation of feeding; indeed, Isojunno et al. [19] did not find differences in buzz rates between CAS vs. PAS exposure either. Moreover, in the present study, we tested for the proportion of scored responses combining all types of behavioral response categories, not only the cessation of feeding. With this analysis, we expect that the greater an individual will be impacted, the higher the number of exhibited behavioral change types, so the higher the proportion of scored responses (independently of their nature).

Previous meta-analysis study showed the importance and complementarity of different analytical approaches to fully describe behavioral responses of whales to sonar or other stimuli [17]. Similarly, the present study brings complementary information to the one of Isojunno et al. [19].

4.3. Severity of Response to Sonar Related to RL Thresholds

It is useful to correlate responses to sonar with the dose of acoustic level received by the whale, as this information can be used by navies and other anthropogenic noise producers to predict the behavioral disturbance impact of their activities. Severity of scored responses to sonar ranged from unlikely (1–3) to potentially (4–6) affecting vital rates, with response onsets occurring over a broad range of received levels for all sonar types (Figure S4). Overall, thresholds of RLs were higher for responses of moderate severity (4–6) than for responses of low severity (1–3). Response thresholds in terms of SEL_{cum} ranged from 114 to 181 dB in response to HPAS, from 82 to 166 dB in response to MPAS, and from 114 to 189 dB in response to CAS. By comparison, previous studies with HPAS also showed a broad range of response threshold SEL_{cum} ranging from 120 to 168 dB re 1 μPa^2 s [14].

The dose–response functions derived from our data predicted a low severity response probability of 0.5 at a received SEL_{cum} of 173 dB re 1 μPa^2 s whereas there was no prediction to reach a probability of 0.5 for moderate severity responses. Using the same type of analysis, Harris et al. 2015 [38] found 0.5 probability of moderate severity responses around 140 dB SEL_{cum} in feeding sperm whales in the same area. The fact that sperm whales seem less sensitive in our study could be explained by a habituation over time to low frequency sonar or the combination of an insufficient sample size for the responses of moderate severity (the dose–response analysis only considered the first response onset of the session irrespective of category) and interindividual variability of responses that together prevented to reach the probability of 0.5. Our data suggested that the severity of responses cannot be accurately predicted by the RL alone, because of large variability in response threshold for most severity scores (Figure S4), which was also observed in previous CEE studies conducted on sperm whales and other cetacean species [14,20].

Beside the potential of a higher tolerance of the sample of subjects to sonar compared to the population tested in 2008–2009, that might partly explain interindividual variation in the responses to sonar, other key questions remain regarding potential other influencing factors.

A large suite of other factors might have influenced responses to sonar: individual factors such as age, body condition, experience (e.g., habituation), environmental variation such as bathymetry, resource quality or distribution [44]. In addition, even when in the feeding grounds, sperm whales could be in a different behavioral state and at different stages of those behavioral states (at start or end of a resting phase for instance), which might lead to different perceived trade-off between the costs and benefits of interrupting their activity.

4.4. Responses Related to Order of Exposures

Previous studies pointed out potential short-term habituation or sensitization to successive exposures (i.e., respectively, an attenuation or amplification of a response over repeated exposures). For instance, Sivle et al. [20] showed that some humpback whales responded less to a second sonar session compared to the first one. However, statistical analysis did not support any order effect in previous sonar exposures conducted in sperm whales and several other cetacean species [14]. That said, a small reduction in buzz rates was found during and post repeated exposures when the first exposure was received at high sound exposure levels [19].

Similarly, we show here that the order of exposure was unlikely to influence neither the proportion nor severity of scored behavioral responses. Indeed, responses of moderate severity (≥ 4) occurred during first or subsequent exposures. One individual (sw17_188a) exposed to two successive CAS exposure sessions obtained in both cases scored changes in the dive profile and in the vocal behavior showing no apparent habituation even in this case when the same stimulus was repeated. Moreover, statistical analyses showed that the proportion of scored responses was not significantly different between first and subsequent exposures. However, further data are needed to test potential effect of the interaction between the order and other covariates such as the exposure type, on the probability of responses or other response variables.

4.5. Responses Related to Blackfish Presence

Effect of anthropogenic disturbances on prey behavior by increasing their level of vigilance to predation is reviewed in [45]. However, the corollary, i.e., that increased predation risk can influence the degree of alert/reaction to other stressors such as anthropogenic disturbances or other potential predators, has been much less considered (e.g., in ungulate species: [46,47]). In the marine environment, herring (*Clupea harengus*) exhibit stronger antipredator responses to visual predator cues when previously exposed to predator sounds [48]. Heterospecific context could have potential influence on cetacean responsiveness to anthropogenic disturbance such as sonar. The detection of potential predatory killer whales triggers consistent and clear behavioral responses in many cetacean species including sperm whales [37,43,49]. Given blackfish species might represent potential threatening stimulus for sperm whales, we thus suspected that their presence could influence the whales' behavior and responses to sonar.

Our results clearly showed that whales were more likely to respond to sonar when they had been recently exposed to blackfish events. These results are in line with previous analyses using the same dataset that showed that the whales exposed to blackfish species were more likely to switch from foraging to nonforaging active states during subsequent sonar exposures [19]. Finding the same potentiating effect of blackfish species on the response to sonar using two different analytical approaches and a range different response metrics substantially increased our confidence in this result.

Moreover, the only case of grouping behavior in the present study was during a HPAS exposure (sw16_135a) for which blackfish were present. Blackfish were acoustically detected during the pre-exposure dive immediately preceding the surfacing during HPAS exposure where the tagged whale was sighted together with other whales. It is possible that this rarely observed grouping behavior, thought to be a specific component of the antipredator behavioral response [49], was induced by the presence of killer whales rather than in response to sonar. Other types of behavioral changes composing the antipredator behavioral template such as the occurrence of codas production might have been also potentiated by blackfish presence.

Killer whales present in the area are mostly fish-eating killer whales which might not represent a predator threat for sperm whales. Sperm whales may be able to discriminate acoustically between mammal-eating and fish-eating killer whales as has been shown in other marine mammal species (pinnipeds: [50]; cetaceans: [43,51]). If not perceived as a risk of predation, the presence of blackfish species might still represent a type of threat associated to the competing interest with sympatric species to exploit of the same habitat [43]. Moreover, it could be that both blackfish species, i.e., pilot whales and killer whales, that we were not always able to distinguish, lead to different behavioral responses and that they influenced the response to sonar differently. Pilot whales are not predators of other cetacean species and detecting their presence might be perceived as less threatening than detecting killer whales, so we could expect less costly responses in sperm whales when exposed to pilot whales. Further studies are needed to characterize their reaction to the detection of pilot whale or fish-eating killer whale presence and to disentangle the potential different effects of the two blackfish species on sperm whale behavior.

Our results indicate that cumulating sonar exposure with natural stressors such as threatening blackfish species present in the area can potentiate the probability of response to sonar. An increased vigilance priming by predator presence or other interspecific threats might further exaggerate such effects in a synergistic, rather than additive, cumulative impact.

4.6. Remarks on the Scoring Method

Another aspect that might have introduced variability in our data, and thus uncertainty in the estimated sonar–response relationships, was some low confidence scores which could have actually been no-responses, and the potentially underestimated scores (noted as $\geq$). All these cases of uncertain scores were obtained for PAS sessions (mainly for MPAS) and to a less extent to NS but never for CAS sessions. Moreover, some behavioral changes could be caused by other factors rather than by our exposures and led also to low confidence scores.

Some heterogeneity in the quality of the collected data could increase the difficulty to identify behavioral changes. In particular, on a few occasions, the low quality of the track and limited group size data prevented the possibility to assess avoidance and group distribution behavioral response categories whereas the other behavioral response categories were almost always scorable. In order to minimize the introduction of such bias in the scoring data, the scoring protocol was supplemented by a judgment of scorable versus unscorable behavioral response categories [17] and we decided to exclude for some exposures, the behavioral response categories judged to be impossible to assess.

Changes in horizontal movement and social responses are two main important aspects of behavioral responses to sound stimuli in many cetacean species including sperm whales (e.g., avoidance and grouping behavior). This highlights the importance of a sufficient quality data on the related parameters i.e., the whale track and group size.

An improvement of the scoring protocol presented here compared to previous ones [14,20] is the implementation of a blind procedure where scorers were blind to the experimental condition. One caveat we think relevant to advise though is related to the fact that NS was always the first experimental session, resulting in the pre-exposure immediately following the baseline period only for the NS sessions and never for the sonar sessions. To deal with this constraint and in order to prevent the identification of the sonar versus NS session types, we had to disconnect the pre-exposure from baseline. Despite those adjustments, the feasibility of a blinding procedure was proven in the present study, and we encourage its use in future severity scoring studies.

5. Conclusions

This study provides rich descriptive material of the behavioral responses of free-ranging sperm whales to short-term CAS and PAS naval sonar experimental exposures and no-sonar controls.

The overall low probability of the animals to respond to sonar, high variability of responses between individuals within exposure type, and the potentiating influence of blackfish presence reduced the power of statistical analysis (e.g., for 'max score per session' variable), weakening our ability to accurately disentangle substantial differences in responses to the different sonar types.

However, the descriptive analyses clearly show that various behavioral change types, including avoidance and cessation of feeding that are considered having some potential to impact vital rate (if the exposure is of sufficient duration or repeated), occurred in response to all sonar types, and that the distribution of the behavioral response categories could vary with sonar type. Responding animals exhibited more responses, and more severe responses, to sonar compared to no-sonar controls. Moreover, the statistical analysis showed that the proportion of scored responses (all behavioral response categories combined) was significantly greater in response to CAS compared to NS, but responses to CAS were not statistically different from responses to PAS. Further data are needed to better characterize differences in responses to CAS and PAS within each behavioral response category and to understand the underlying mechanisms of such responses.

Supplementary Materials: The following are available online at https://www.mdpi.com/article/10.3390/jmse9040444/s1, Figure S1: Amplitude (arbitrary unit) and frequency (Hz) over time (s), at top and bottom, respectively, for both PAS (left) and CAS (right) signals, Figure S2: Example of standardized data plots of the tagged sperm whale sw17_182b during the time period from 60 min prior to a CAS exposure (random exposure RE #15) to 60 min after the exposure, Figure S3: Maximum score of severity per session across the four exposure types (NS, CAS, HPAS and MPAS) for the total number of exposure sessions (N = 46), Figure S4: Severity of the 40 scored behavioral responses of the focal whale dataset versus received level thresholds of the corresponding response onsets across the three types of sonar exposures CAS (diamond), MPAS (circle) and HPAS (cross), Table S1: Severity scoring panel results, Table S2: Results of the ANOVA (sequential Wald test) showing the contribution of each factor to the final fitted GEE models applied to the 'Proportion of scored responses', Table S3: Results of the GEE models fitted to the 'Proportion of scored responses' variable.

Author Contributions: Conceptualization, C.C., S.I., M.L.S., P.J.W., P.H.K., F.-P.A.L., P.J.O.M.; methodology, C.C., S.I., M.L.S., P.J.W., P.H.K., F.-P.A.L., P.J.O.M.; formal analysis, C.C., S.I., C.B., P.J.W.; investigation, C.C., S.I., M.L.S., P.J.W., L.D.S., B.B., R.R., P.H.K., F.-P.A.L., P.J.O.M.; writing—original draft preparation, C.C.; writing—review and editing, C.C., S.I., M.L.S., P.J.W., C.B., L.D.S., B.B., R.R., P.H.K., F.-P.A.L., P.J.O.M.; visualization, C.C., S.I., P.J.W., C.B.; resources, C.C., P.H.K., F.-P.A.L., P.J.O.M.; data curation, S.I., P.J.W.; supervision, C.C., P.H.K., F.-P.A.L., P.J.O.M.; project administration, C.C., P.H.K., F.-P.A.L., P.J.O.M.; funding acquisition, C.C., P.H.K., F.-P.A.L., P.J.O.M. All authors have read and agreed to the published version of the manuscript.

Funding: This research was funded by four naval organizations: the US Navy Living Marine Resources program (LMR), the Netherlands Ministry of Defence, the UK Ministry of Defense (Dstl) and the French Ministry of Defense (DGA-TN).

Institutional Review Board Statement: The study was conducted according to the guidelines of the Declaration of Helsinki and approved by the Animal Welfare Ethics Committee of the University of St Andrews (UK). All animal research activities were licensed under permit provided by the Norwegian Animal Research Authority (Permit n° 2015–223 222).

Informed Consent Statement: Not applicable.

Data Availability Statement: All data needed to reproduce these results are available in a published data report (see [18]).

Acknowledgments: We thank all 3S (Sea mammals, Sonar, Safety) team members and captain and crew members of the FFI R/V H.U. Sverdrup II vessel for their collaborating efforts in participating to field works. We thank Sander van IJsselmuide for his contribution to Figure S1.

Conflicts of Interest: The authors declare no conflict of interest.

References

1. Richardson, W.J.; Greene, C.R.; Malme, C.I.; Thomson, D.H. *Marine Mammals and Noise*; Academic Press: San Diego, CA, USA, 1995; ISBN 978-0-12-588441-9.
2. Tyack, P.L. Convergence of Calls as Animals Form Social Bonds, Active Compensation for Noisy Communication Channels, and the Evolution of Vocal Learning in Mammals. *J. Comp. Psychol.* **2008**, *122*, 319–331. [CrossRef]
3. Miller, P.J.O.; Antunes, R.N.; Wensveen, P.J.; Samarra, F.I.P.; Catarina Alves, A.; Tyack, P.L.; Kvadsheim, P.H.; Kleivane, L.; Lam, F.-P.A.; Ainslie, M.A.; et al. Dose-Response Relationships for the Onset of Avoidance of Sonar by Free-Ranging Killer Whales. *J. Acoust. Soc. Am.* **2014**, *135*, 975–993. [CrossRef] [PubMed]
4. Southall, B.L.; Bowles, A.E.; Ellison, W.T.; Finneran, J.J.; Gentry, R.L.; Greene, C.R.; Kastak, D.; Ketten, D.R.; Miller, J.H.; Nachtigall, P.E.; et al. Marine mammal noise exposure criteria: Initial scientific recommendations. *Aquat. Mamm.* **2007**, *33*, 411–414. [CrossRef]
5. Southall, B.L.; Nowacek, D.; Miller, P.J.O.; Tyack, P.L. Experimental Field Studies to Measure Behavioral Responses of Cetaceans to Sonar. *Endang. Species Res.* **2016**, *31*, 293–315. [CrossRef]
6. Frantzis, A. Does Acoustic Testing Strand Whales? *Nature* **1998**, *392*, 29. [CrossRef] [PubMed]
7. Balcomb III, K.C.; Claridge, D.E. A mass stranding of cetaceans caused by naval sonar in the Bahamas. *Bahamas J. Sci.* **2001**, *8*, 2–12.
8. D'Amico, A.; Gisiner, R.C.; Ketten, D.R.; Hammock, J.A.; Johnson, C.; Tyack, P.L.; Mead, J. Beaked Whale Strandings and Naval Exercises. *Aquat. Mamm.* **2009**, *35*, 452–472. [CrossRef]
9. Popper, A.N.; Ketten, D.R. Underwater Hearing. In *The Senses: A Comprehensive Reference*; Elsevier: Amsterdam, The Netherlands, 2008; pp. 225–236, ISBN 978-0-12-370880-9.
10. Erbe, C.; Dunlop, R.; Dolman, S. Effects of Noise on Marine Mammals. In *Effects of Anthropogenic Noise on Animals*; Slabbekoorn, H., Dooling, R., Popper, A., Fay, R., Eds.; Springer Handbook of Auditory Research; Springer: New York, NY, USA, 2018; Volume 66, pp. 227–309. [CrossRef]
11. New, L.; Clark, J.; Costa, D.; Fleishman, E.; Hindell, M.; Klanjšček, T.; Lusseau, D.; Kraus, S.; McMahon, C.; Robinson, P.; et al. Using Short-Term Measures of Behaviour to Estimate Long-Term Fitness of Southern Elephant Seals. *Mar. Ecol. Prog. Ser.* **2014**, *496*, 99–108. [CrossRef]
12. Johnson, M.P.; Tyack, P.L. A Digital Acoustic Recording Tag for Measuring the Response of Wild Marine Mammals to Sound. *IEEE J. Ocean. Eng.* **2003**, *28*, 3–12. [CrossRef]
13. Isojunno, S.; Sadykova, D.; DeRuiter, S.L.; Curé, C.; Visser, F.; Thomas, L.; Miller, P.J.O.; Harris, C.M. Individual, Ecological, and Anthropogenic Influences on Activity Budgets of Long-Finned Pilot Whales. *Ecosphere* **2017**, *8*, e02044. [CrossRef]
14. Miller, P.J.O.; Kvadsheim, P.H.; Lam, F.-P.A.; Wensveen, P.J.; Antunes, R.; Alves, A.C.; Visser, F.; Kleivane, L.; Tyack, P.L.; Sivle, L.D. The Severity of Behavioral Changes Observed During Experimental Exposures of Killer (*Orcinus Orca*), Long-Finned Pilot (*Globicephala Melas*), and Sperm (*Physeter Macrocephalus*) Whales to Naval Sonar. *Aquat. Mamm.* **2012**, *38*, 362–401. [CrossRef]
15. Wensveen, P.J.; Isojunno, S.; Hansen, R.R.; von Benda-Beckmann, A.M.; Kleivane, L.; van IJsselmuide, S.; Lam, F.-P.A.; Kvadsheim, P.H.; DeRuiter, S.L.; Curé, C.; et al. Northern Bottlenose Whales in a Pristine Environment Respond Strongly to Close and Distant Navy Sonar Signals. *Proc. R. Soc. B.* **2019**, *286*, 20182592. [CrossRef]
16. Isojunno, S.; Curé, C.; Kvadsheim, P.H.; Lam, F.-P.A.; Tyack, P.L.; Wensveen, P.J.; Miller, P.J.O. Sperm Whales Reduce Foraging Effort during Exposure to 1-2 KHz Sonar and Killer Whale Sounds. *Ecol. Appl.* **2016**, *26*, 77–93. [CrossRef] [PubMed]
17. Curé, C.; Isojunno, S.; Visser, F.; Wensveen, P.J.; Sivle, L.D.; Kvadsheim, P.H.; Lam, F.; Miller, P.J.O. Biological Significance of Sperm Whale Responses to Sonar: Comparison with Anti-Predator Responses. *Endang. Species. Res.* **2016**, *31*, 89–102. [CrossRef]

18. Kvadsheim, P.H.; Isojunno, S.; Curé, C.; Siemensma, M.; Wensveen, P.; Lam, F.-P.A.; Hansen, R.R.; Benti, B.; Sivle, L.D.; Burslem, A.; et al. The 3S3 experiment data report–using operational naval sonars to study the effects of continuous active sonar, and source proximity, on sperm whales. *FFI Report* **2021**, in press.

19. Isojunno, S.; Wensveen, P.J.; Lam, F.-P.A.; Kvadsheim, P.H.; von Benda-Beckmann, A.M.; Martín López, L.M.; Kleivane, L.; Siegal, E.M.; Miller, P.J.O. When the Noise Goes on: Received Sound Energy Predicts Sperm Whale Responses to Both Intermittent and Continuous Navy Sonar. *J. Exp. Biol.* **2020**, *223*, jeb219741. [CrossRef]

20. Sivle, L.D.; Kvadsheim, P.H.; Curé, C.; Isojunno, S.; Wensveen, P.J.; Lam, F.-P.A.; Visser, F.; Kleivane, L.; Tyack, P.L.; Harris, C.M.; et al. Severity of Expert-Identified Behavioural Responses of Humpback Whale, Mike Whale, and Northern Bottlenose Whale to Naval Sonar. *Aquat. Mamm.* **2015**, *41*, 469–502. [CrossRef]

21. Southall, B.L.; DeRuiter, S.L.; Friedlaender, A.; Stimpert, A.K.; Goldbogen, J.A.; Hazen, E.; Casey, C.; Fregosi, S.; Cade, D.E.; Allen, A.N.; et al. Behavioral Responses of Individual Blue Whales (*Balaenoptera Musculus*) to Mid-Frequency Military Sonar. *J. Exp. Biol.* **2019**, *222*, jeb190637. [CrossRef]

22. Lam, F.-P.A.; Kvadsheim, P.H.; Isojunno, S.; Wensveen, P.J.; van IJsselmuide, S.; Siemensma, M.L.; Dekeling, R.; Miller, P.J.O. Behavioural response study on the effects of continuous sonar on sperm whales in Norwegian waters-The 3S-2016-CAS cruise report. 2018. Available online: http://publications.tno.nl/publication/34627070/Q3bPWP/TNO-2018-R10802.pdf (accessed on 16 April 2021).

23. Lam, F.-P.A.; Kvadsheim, P.H.; Isojunno, S.; Wensveen, P.J.; van IJsselmuide, S.; Siemensma, M.L.; Dekeling, R.; Miller, P.J.O. Behavioural response study on the effects of continuous sonar on sperm whales in Norwegian waters-The 3S-2017-CAS cruise report. 2018. Available online: http://publications.tno.nl/publication/34627071/pohdo8/TNO-2018-R10958.pdf (accessed on 16 April 2021).

24. Miller, P.J.O.; Johnson, M.P.; Tyack, P.L. Sperm Whale Behaviour Indicates the Use of Echolocation Click Buzzes 'Creaks' in Prey Capture. *Proc. R. Soc. Lond. B* **2004**, *271*, 2239–2247. [CrossRef]

25. Miller, P.J.O.; Aoki, K.; Rendell, L.E.; Amano, M. Stereotypical Resting Behavior of the Sperm Whale. *Curr. Biol.* **2008**, *18*, R21–R23. [CrossRef]

26. Sivle, L.D.; Kvadsheim, P.H.; Fahlman, A.; Lam, F.-P.A.; Tyack, P.L.; Miller, P.J.O. Changes in Dive Behavior during Naval Sonar Exposure in Killer Whales, Long-Finned Pilot Whales, and Sperm Whales. *Front. Physio.* **2012**, *3*, 400. [CrossRef] [PubMed]

27. von Benda-Beckmann, A.M.; Lam, F.-P.A.; Moretti, D.J.; Fulkerson, K.; Ainslie, M.A.; van IJsselmuide, S.P.; Theriault, J.; Beerens, S.P. Detection of Blainville's Beaked Whales with Towed Arrays. *Appl. Acoust.* **2010**, *71*, 1027–1035. [CrossRef]

28. Miller, P.J.O.; Antunes, R.; Alves, A.C.; Wensveen, P.J.; Kvadsheim, P.H.; Nordlund, N.; Lam, F.P.; Ijsselmuide, S.; Visser, F.; Tyack, P.; et al. *The 3S experiments: Studying the Behavioral Effects of Sonar on Killerwhales (Orcinus orca), Sperm Whales (Physeter macrocephalus), and Long-Finned Pilot Whales (Globicephala melas) in Norwegian Waters*; IQOE: Bremerhaven, Germany, 2011; p. 290.

29. Wensveen, P.J.; Thomas, L.; Miller, P.J.O. A Path Reconstruction Method Integrating Dead-Reckoning and Position Fixes Applied to Humpback Whales. *Mov. Ecol.* **2015**, *3*, 1–16. [CrossRef]

30. Frantzis, A.; Alexiadou, P. Male Sperm Whale (Physeter Macrocephalus) Coda Production and Coda-Type Usage Depend on the Presence of Conspecifics and the Behavioural Context. *Can. J. Zool.* **2008**, *86*, 62–75. [CrossRef]

31. Oliveira, C.; Wahlberg, M.; Johnson, M.; Miller, P.J.O.; Madsen, P.T. The Function of Male Sperm Whale Slow Clicks in a High Latitude Habitat: Communication, Echolocation, or Prey Debilitation? *J. Acoust. Soc. Am.* **2013**, *133*, 3135–3144. [CrossRef]

32. Weller, D.W.; Würsig, B.; Whitehead, H.; Norris, J.C.; Lynn, S.K.; Davis, R.W.; Clauss, N.; Brown, P. Observations of an interaction between sperm whales and short-finned pilot whales in the Gulf of Mexico. *Mar. Mammal Sci.* **1996**, *12*, 588–594. [CrossRef]

33. Reeves, R.R.; Berger, J.; Clapham, P.J. Killer Whales as Predators of Large Baleen Whales and Sperm Whales. In *Whales, Whaling, and Ocean Ecosystems*; Estes, J., Ed.; University of California Press: Berkeley, CA, USA, 2007; pp. 174–187, ISBN 978-0-520-24884-7.

34. von Benda-Beckmann, A.M.; Thomas, L.; Tyack, P.L.; Ainslie, M.A. Modelling the Broadband Propagation of Marine Mammal Echolocation Clicks for Click-Based Population Density Estimates. *J. Acoust. Soc. Am.* **2018**, *143*, 954–967. [CrossRef]

35. Visser, F.; Miller, P.J.O.; Antunes, R.N.; Oudejans, M.G.; Mackenzie, M.L.; Aoki, K.; Lam, F.-P.A.; Kvadsheim, P.H.; Huisman, J.; Tyack, P.L. The Social Context of Individual Foraging Behaviour in Long-Finned Pilot Whales (*Globicephala Melas*). *Behaviour* **2014**, *151*, 1453–1477. [CrossRef]

36. Højsgaard, S.; Halekoh, U.; Yan, J. The R Package geepack for Generalized Estimating Equations. *Journal of Statistical Software*. **2006**, *15*, 1–11. [CrossRef]

37. Curé, C.; Sivle, L.D.; Visser, F.; Wensveen, P.J.; Isojunno, S.; Harris, C.M.; Kvadsheim, P.H.; Lam, F.-P.A.; Miller, P.J.O. Predator Sound Playbacks Reveal Strong Avoidance Responses in a Fight Strategist Baleen Whale. *Mar. Ecol. Prog. Ser.* **2015**, *526*, 267–282. [CrossRef]

38. Harris, C.M.; Sadykova, D.; DeRuiter, S.L.; Tyack, P.L.; Miller, P.J.O.; Kvadsheim, P.H.; Lam, F.-P.A.; Thomas, L. Dose Response Severity Functions for Acoustic Disturbance in Cetaceans Using Recurrent Event Survival Analysis. *Ecosphere* **2015**, *6*, 1–14. [CrossRef]

39. Therneau, T.M.; Grambsch, P.M. Multiple events per subject. In *Modeling Survival Data: Extending the Cox Model*; Springer: New York, NY, USA, 2000; pp. 169–230. [CrossRef]

40. Kleinbaum, D.G.; Klein, M. Competing risks survival analysis. In *Survival Analysis: A Self-Learning Text*; Springer: New York, NY, USA, 2010; pp. 425–495. [CrossRef]

41. Therneau, T. A package for survival analysis in SR package, 2014, Version 2.37-7. Available online: http://CRAN.R-project.org/package=survival (accessed on 16 March 2021).
42. Ellison, W.T.; Southall, B.L.; Clark, C.W.; Frankel, A.S.A. New Context-Based Approach to Assess Marine Mammal Behavioral Responses to Anthropogenic Sounds. *Conserv. Biol.* **2012**, *26*, 21–28. [CrossRef]
43. Curé, C.; Isojunno, S.; Vester, H.I.; Visser, F.; Oudejans, M.; Biassoni, N.; Massenet, M.; Barluet de Beauchesne, L.; Wensveen, P.J.; Sivle, L.D.; et al. Evidence for Discrimination between Feeding Sounds of Familiar Fish and Unfamiliar Mammal-Eating Killer Whale Ecotypes by Long-Finned Pilot Whales. *Anim. Cogn.* **2019**, *22*, 863–882. [CrossRef]
44. Harris, C.M.; Thomas, L.; Falcone, E.A.; Hildebrand, J.; Houser, D.; Kvadsheim, P.H.; Lam, F.-P.A.; Miller, P.J.O.; Moretti, D.J.; Read, A.J.; et al. Marine Mammals and Sonar: Dose-Response Studies, the Risk-Disturbance Hypothesis and the Role of Exposure Context. *J. Appl. Ecol.* **2017**, *55*, 396–404. [CrossRef]
45. Shannon, G.; McKenna, M.F.; Angeloni, L.M.; Crooks, K.R.; Fristrup, K.M.; Brown, E.; Warner, K.A.; Nelson, M.D.; White, C.; Briggs, J.; et al. A Synthesis of Two Decades of Research Documenting the Effects of Noise on Wildlife: Effects of Anthropogenic Noise on Wildlife. *Biol. Rev.* **2016**, *91*, 982–1005. [CrossRef]
46. Bonnot, N.C.; Couriot, O.; Berger, A.; Cagnacci, F.; Ciuti, S.; De Groeve, J.E.; Gehr, B.; Heurich, M.; Kjellander, P.; Kröschel, M.; et al. Fear of the Dark? Contrasting Impacts of Humans versus Lynx on Diel Activity of Roe Deer across Europe. *J. Anim. Ecol.* **2020**, *89*, 132–145. [CrossRef] [PubMed]
47. Proudman, N.J.; Churski, M.; Bubnicki, J.W.; Nilsson, J.-Å.; Kuijper, D.P.J. Red Deer Allocate Vigilance Differently in Response to Spatio-Temporal Patterns of Risk from Human Hunters and Wolves. *Wildl. Res.* **2020**. [CrossRef]
48. Rieucau, G.; Sivle, L.D.; Handegard, N.O. Herring perform stronger collective evasive reactions when previously exposed to killer whales calls. *Behav. Ecol.* **2016**, *27*, 538–544. [CrossRef]
49. Curé, C.; Antunes, R.; Alves, A.C.; Visser, F.; Kvadsheim, P.H.; Miller, P.J.O. Responses of Male Sperm Whales (Physeter Macrocephalus) to Killer Whale Sounds: Implications for Anti-Predator Strategies. *Sci. Rep.* **2013**, *3*, 1–7. [CrossRef]
50. Deecke, V.B.; Slater, P.J.B.; Ford, J.K.B. Selective Habituation Shapes Acoustic Predator Recognition in Harbour Seals. *Nature* **2002**, *420*, 171–173. [CrossRef]
51. Benti, B.; Miller, P.J.O.; Biuw, M.; Curé, C. Indication that the behavioural responses of humpback whales to killer whale sounds are influenced by trophic relationships. *Mar. Ecol. Prog. Ser.* **2021**, *660*, 217–232. [CrossRef]

Journal of
*Marine Science
and Engineering*

Article

Effects of Sound from Seismic Surveys on Fish Reproduction, the Management Case from Norway

Lise Doksæter Sivle [1,*], Emilie Hernes Vereide [1], Karen de Jong [1], Tonje Nesse Forland [1], John Dalen [2] and Henning Wehde [1]

[1] Ecosystem Acoustics Department, Institute of Marine Research, Postboks 1870 Nordnes, NO-5817 Bergen, Norway; emilie.hernes.vereide@hi.no (E.H.V.); karen.de.jong@hi.no (K.d.J.); tonje.nesse.forland@hi.no (T.N.F.); henningw@hi.no (H.W.)
[2] SoundMare, Helleveien 243, NO-5039 Bergen, Norway; johndal@broadpark.no
* Correspondence: lisedo@hi.no

Abstract: Anthropogenic noise has been recognized as a source of concern since the beginning of the 1940s and is receiving increasingly more attention. While international focus has been on the effects of noise on marine mammals, Norway has managed seismic surveys based on the potential impact on fish stocks and fisheries since the late 1980s. Norway is, therefore, one of very few countries that took fish into account at this early stage. Until 1996, spawning grounds and spawning migration, as well as areas with drifting eggs and larvae were recommended as closed for seismic surveys. Later results showed that the effects of seismic surveys on early fish development stages were negligible at the population level, resulting in the opening of areas with drifting eggs and larvae for seismic surveys. Spawning grounds, as well as concentrated migration towards these, are still closed to seismic surveys, but the refinement of areas and periods have improved over the years. Since 2018, marine mammals have been included in the advice to management. The Norwegian case provides a clear example of evidence-based management. Here, we examine how scientific advancements informed the development of Norwegian management and how management questions were incorporated into new research projects in Norway.

Keywords: management; fish; anthropogenic sound; seismic surveys; electromagnetic surveys

Citation: Sivle, L.D.; Vereide, E.H.; de Jong, K.; Forland, T.N.; Dalen, J.; Wehde, H. Effects of Sound from Seismic Surveys on Fish Reproduction, the Management Case from Norway. *J. Mar. Sci. Eng.* **2021**, *9*, 436. https://doi.org/10.3390/jmse9040436

Academic Editor: Giuseppa Buscaino

Received: 26 March 2021
Accepted: 14 April 2021
Published: 17 April 2021

Publisher's Note: MDPI stays neutral with regard to jurisdictional claims in published maps and institutional affiliations.

1. Introduction

Anthropogenic noise pollution is considered an important pollutant of terrestrial and aquatic ecosystems [1–3]. There are few records of systematic underwater anthropogenic noise measurements prior to 1990, but they show that ambient noise levels have increased by as much as 12 dB in 30 years in some parts of the ocean [4–6]. Impulsive anthropogenic sound is currently the subject of monitoring within the frame of the regional agreements such as OSPAR. Maximum sound exposure levels have been proposed for marine mammals and fish based on physical damage [7]. However, the masking of acoustic information from the environment may affect animals at a much lower sound level, and, thus, further away from the source. Most of the energy of anthropogenic caused sound lies in the lower frequency ranges [8]. This may affect a wide range of animals. For example, all fish can hear low-frequency sounds (<500 Hz) and can, consequently, be disturbed by man-made sound activities [7,9].

Noise disturbance can affect the physical integrity (at very high levels), the physiology, and the behavior of aquatic animals. This may affect individual fitness and could, ultimately, lead to population and ecosystem-level consequences [10–12]. The effects of noise on aquatic life have been reviewed extensively (e.g., [7,8,11–19]). These reviews highlight the absence of observational evidence of population-level impacts. Experimental data often show short-term damage or behavioral changes in individual animals only, while numerical models are needed to provide information on whether such changes can lead to population-level effects.

In Norway, the first research on the effects of sound generated by seismic sources has been developed in response to concerns from fishermen that fish may be scared away from fishing grounds. Fishing, a key industry in Norway, has been influenced by oil and gas exploration activities as the two industries operate in much of the same areas. Thus, there have been demands for balanced coexistence between fisheries and the oil industry since the start of the oil era in the 1970s, to ensure the acceptable development of both industries [20]. Therefore, protecting core habitats and fishing grounds of commercially important fish stocks with reproductive success from exposure to seismic surveys is an important element within this balance and is of major importance for the management of seismic surveys in Norway.

Reproduction is vital to population sustainability but can be very sensitive to stress and changes in environmental conditions [21]. Even if yearly variations in spawning stocks are not necessarily correlated to recruitment, long-term reduction in egg production is expected to lead to a mean decrease in the population [22]. In addition, spawning is the most clearly quantifiable investment in a specific mating and, as such, directly related to fitness. Moreover, for many fish species, the spawning period may be highly sensitive to impacts from noise, if individuals gather in dense, localized spawning aggregations [23]. Disturbances in the spawning period may thus affect a larger fraction of the population than disturbances during other periods. Additionally, fish may be vulnerable to external stressors during spawning [24], because fish are often in their poorest body condition in this period [25,26].

In this study, we review how management advice has been given through a period of about 30 years and how improved scientific knowledge has transferred into scientific advice. There was an additional focus on how external drivers have induced scientific questions and how weakly documented knowledge-based advice has exposed the need for more research to obtain better and more scientifically based advice. This process is shown in Figure 1.

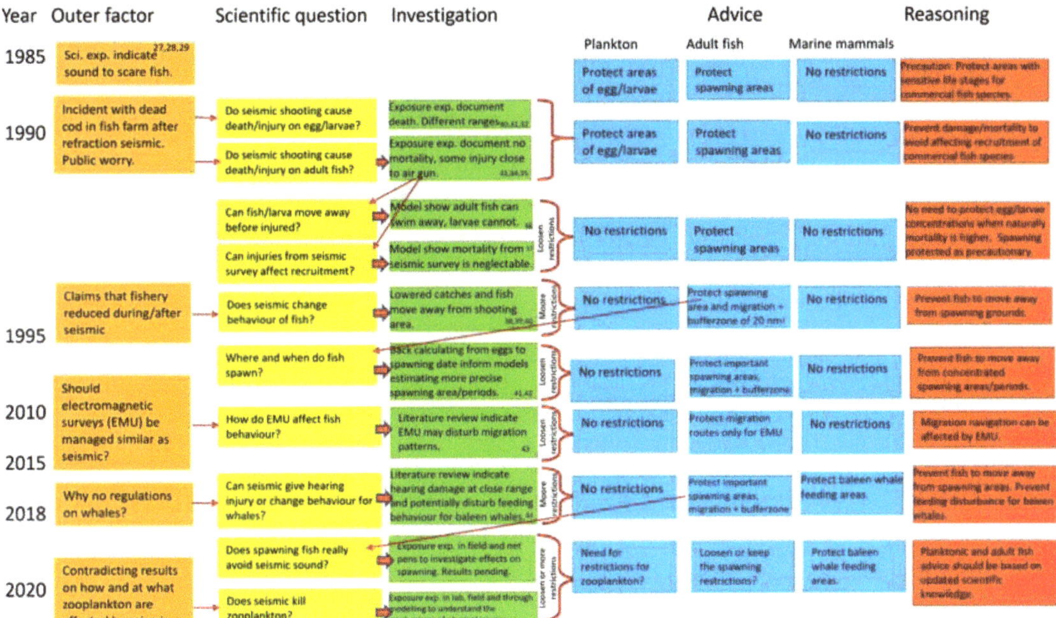

Figure 1. Overview of how management advice has changed since the beginning of the advisory practice in the late 1980s. The "Outer factor" column covers drivers from outside the scientific community that may induce new scientific questions and lead to research activity. The "Scientific question" column lists the research questions being raised to improve management advice, which may arise from outer factors or from existing research that raises new questions. The "Scientific investigation" column briefly describes the main findings from research projects. The "Advice" column summarizes the management advice from IMR to the Norwegian Petroleum Directorate and is divided into different advice given for three specific groups of animals; plankton (small organisms with very little or no self-movement, including egg, fry, larvae, and zooplankton), adult fish, and marine mammals. The "Reasoning" column describes the rationale for the particular advice given. Timeline is indicative and not to scale. The numbers in the bottom corner refer to publications given in the reference list; [27–46].

2. The Norwegian Story—A Journey through the History of Management Advice

Since the beginning of oil exploration in the 1960s, seismic surveys applying different sound sources have been carried out within the Norwegian Exclusive Economic Zone (NEZ) to locate and estimate oil and gas resources. To reach an optimal basis for development it has been a goal for the Norwegian Government to have a good coexistence between the traditionally existing fishing industry and the newly established oil industry. The Norwegian Petroleum Directorate (NPD) issues licenses for seismic surveys in Norwegian waters, but several stakeholders within governmental organizations have been asked for advice since 1983. The Institute of Marine Research (IMR) is asked for advice on the potential impact on biology and ecosystems, while the Directorate of Fisheries is asked for advice on the likely impact and potential conflicts with ongoing fishery activities. In the beginning, the advice from IMR was mainly based on a precautionary approach; preventing potential impact on presumed sensitive habitats. Therefore, spawning areas and areas with spawning migration, as well as areas with drifting eggs and larvae in the periods of the respective migration, spawning, and drifting, were recommended to be closed for seismic surveys to avoid impact on these volatile ecosystem compounds.

Recommendations from IMR on the regulation of seismic activity have always been given in the form of geographical and temporal restrictions to avoid seismic exposure of sensitive habitats, i.e., specifically to protect fish engaged in susceptible activities such as spawning and concentrated spawning migrations. The NPD and the seismic operator are not obliged to follow the advice given, and, until the late 1990s, the NPD oversaw these recommendations to some extent, but due to frequent contact and communication over the years, the advice given today is almost always followed.

In July 1989, one particular incident brought attention to how sound from seismic surveys could potentially affect fish; in a fish farm in Northern Norway, high mortality of cod was observed after explosives for refractional seismic investigations were detonated nearby [47]. This raised the question of how seismic exposures could potentially cause damage in wild fish. In order to clarify these issues, several research projects were initiated in the early 1990s. The main concern was that seismic blasting could cause injury and even death in fish.

2.1. The Early 1990s: Physical Injuries and Death

Knutsen and Dalen [30] previously analyzed mortality and damage to fish eggs, larvae, and small juveniles of cod (*Gadus morhua*) after exposure to seismic airguns, describing that some of the larger larvae developed problems with their balance, but returned to normal swimming after few minutes and, overall, there were no significant differences in injuries and death between the control and exposed groups. A larger project was initiated that investigated the effects of airguns on fish eggs, larvae, and juveniles of cod, saithe (*Pollachius virens*), turbot (*Scophthalmus maximus*), plaice (*Pleuronectes platessa*), and herring (*Clupea harengus*) [32]. Despite some differences between species, results showed significantly increased mortality rates in exposed groups, but only rather close to the air guns; up to 1.35 m for eggs, 0.9–3 m yolk sac larval stage, 2–5 m for larval stage, up to 1.5 m for post-larval stages, and up to 1.3 m for the fry stages. Different sublethal effects, e.g., injuries to neuromats and swimbladder, and changes in behavior due to buoyancy were observed for some species and life stages. The studies concluded that the highest mortality rate was observed about 1.4 m from the airgun, with potential minor damages up to 5 m away from the airgun [32]. These Norwegian studies were in line with similar international studies at the time, documenting lethal and sublethal effects at distances equal or closer than 3 m from the airguns on fish egg and larvae [47–49].

These results thus showed that fish at early life stages could experience both indirect and direct mortality, but only at rather close range within a few meters of the air guns. To propose realistic scenarios for impacts from a seismic survey, results from these experiments, together with fish biology and physiology knowledge, the vertical distributions of larvae, and the sound intensity output from the seismic source were included in a modeling study. The results from this study demonstrated that adult fish would be able to swim away from the spatial zone of potential injuries, while the smallest larvae and fry would not, as they would suffer from total exhaustion, and, therefore, would not be able to escape from the zone of injury [36,50].

In summary, these studies show that some injuries, including lethal ones, may occur, but at ranges less than 5 m from the air gun. However, in a management context, the most interesting issue is whether these effects can translate into the negative development of the stock or stock recruitment. Sætre and Ona [37], therefore, used these results to assess the potential total mortality rate on fish larvae from a regular 3D seismic survey. Assuming a lethal radius of 2 m from the air guns, the mortality rate for cod larvae was 0.45%—in a worst-case scenario, and 0.3‰ in a more realistic scenario, compared to a natural daily mortality rate of 5–15%. Therefore, they concluded that the mortality due to 3D seismic surveying is negligible compared to the natural mortality in the larval stage.

The conclusion was, therefore, that mortality can occur at the earliest life stages of fish, but only at very close range, and that the risk that such mortality negatively affects recruitment to the fishable stock is close to non-existent. Therefore, the recommendations were updated to allow seismic surveys in areas and periods with drifting eggs and larvae [47]. This meant that larger areas of the NEZ became available for seismic surveys for a larger part of the year. The restrictions for spawning areas and areas with spawning migration, however, were kept as before.

2.2. The Later 1990s: Reduced Catches and Behavioral Response

Based on the above documentation, it seemed clear that physical injuries occurred only in the nearest few meters of the air gun and mostly affected early-stage larvae. However, fishermen claimed reduced catches at much further distances from operating seismic vessels than could be explained by the injured fish close to the air guns. This could only be explained if the fish heard and responded to the sound of the seismic shooting.

Fish hearing was intensively studied in the 1960s, showing that fish hear well, with the highest sensitivity below 1 kHz (e.g., [51–53]). Fish were also shown to be able to discriminate the direction of sounds (e.g., [54–57]). Fish can, therefore, hear seismic noise and determine the direction of its source [28,58,59]. In 1969, Chapman and Hawkins reported that shoals of whiting (*Merlangius merlangus*) dive deeper and form more compact schools in response to seismic air gun shots. Later in the 1980s, Dalen and Raknes (1985) found fish distribution, mainly saithe (*Gadus virens* L.), cod, and haddock (*Melanorammus aeglefinus* L.) recorded by echosounder and echo integrator to be reduced by 36%, and for blue whiting (*Micromesistius poutassou*) by 54% after the previously compared seismic blasting. Similar results were demonstrated for rockfish [58].

Reduced catches in areas of seismic investigations were, therefore, assumed to be associated with the fish either avoiding the shooting area, or descending closer to the bottom and thus becoming less catchable. Several research projects were initiated in the early 1990s, with the purpose of documenting whether catches were actually reduced, as well as trying to understand the underlying mechanisms. Holand et al. [38] conducted a controlled experiment with cod swimming freely in a bay, and cod enclosed in a net pen. They observed startle responses at the onset of the air gun, as well as that the largest fish stopped feeding during exposures. While free-swimming fish did not react with an increased heartbeat frequency, the enclosed fish increased their heartbeat after repeated exposure. Løkkeborg and Soldal [39] analyzed catch records from logbooks of longliners and trawlers operating in areas of ongoing seismic surveys, documenting that the lowest catch rates were closest to the seismic survey area and then that the catch rates increased with increasing distance from the seismic survey. These findings were, together with [31], used to design a full-scale fishery experiment in the Barents Sea, using trawl, longline, and acoustic quantity determination within and outside set distances from a seismic shooting area of 3 × 10 nautical miles. Both trawl and longline catches of cod and haddock were considerably reduced up to at least 18 nautical miles from the seismic blasting area [40]. The reduction was largest in the center of the area, with gradually decreased impact towards the outer edges of the area. Acoustic quantity determination showed that the decline in catch rates was caused by a reduction in spatial fish density in the area. These studies all point in the same direction; seismic exposure appears to disturb the fish, and responses may be in the form of avoidance of the exposed area and/or cessation of foraging. Repeated exposure of enclosed fish, which are thus unavailable to avoid exposure, can cause increased heartbeat frequency, indicative of an increase in stress level.

These results have had great importance for management advice. Before this, spawning areas of commercial fish were recommended to be closed to seismic surveys mainly on a precautionary basis. Now, scientific results support this advice. Many of the offshore fish populations are distributed over a large area most of the year, but gather within specific, smaller defined areas during spawning. These areas are not random, but may have specific characteristics, such as bottom type for the bottom spawners (e.g., herring and capelin (*Mallotus villosus*)), and are localized so that the spawned eggs will drift with current to favorable areas with the available food supply of specific zooplankton. The same holds for the temporal component; the spawning period usually occurs so that the fish eggs will hatch during the zooplankton spring bloom, ensuring food abundance. The scientific studies as described above show that seismic may cause fish to swim away from an area of seismic exposure, and that such avoidance may be in the order of more than 18 nmi [40]. If similar avoidance occurs when fish are at the spawning grounds, they may move too far away from these optimal geographical and oceanographical conditions, or if they delay or even stop their spawning, the spawning may be less successful with regards to time and physical conditions. From 1996, therefore, scientifically based advice was given that seismic surveys should avoid spawning areas during the spawning period. Additionally, based on the result that several fish species moved away for a distance of at least 18 nmi from the blasting area, an additional 20 nmi buffer zone around spawning grounds was recommended to be closed for 3D seismic surveys [40,60,61].

2.3. Into a New Millennium with New Studies Drawing a More Complex Picture

Into the new millennium, behavioral responses continued to be the main topic of interest and scientific focus. Until now, scientific results had shown a clear trend that fish avoided areas of seismic exposure. However, as we shall exemplify here, more research does not always draw a clearer picture.

Sometimes seismic surveys are conducted in areas and periods of potentially large ecosystem consequences. This was the case for a survey in the Norwegian Sea in April 1999 overlapping with an area of a high density of migrating pelagic fish, mainly post-spawned herring migrating out from the spawning areas at the coast. Therefore, the advice from IMR was to postpone the survey. When this was not possible, IMR agreed to monitor the fish density in the shooting area to make sure the density did not exceed a predetermined limit, otherwise, the seismic survey had to be stopped. This monitoring created the opportunity to study the abundance and vertical movement of pelagic fish before, during and after a seismic survey. Results showed that schools of blue whiting (*Micromesistius poutassou*) move deeper during exposure but found no horizontal or vertical response of herring [62]. Furthermore, in 2009, NPD planned a 3D seismic exploration in Vesterålen, an area normally closed to commercial seismic activity due to its status as a highly important ecosystem. In order to evaluate the potential negative impact on fish and fisheries, several studies were initiated. Løkkeborg et al. [63] summarize the findings from acoustic mapping with echo sounders and gillnet and longline catches before, during, and after the seismic survey, documenting that most fish species did not leave the area, but rather changed their onsite behavior; increased catches in gillnets are likely caused by increased swimming behavior, while reduced longline catches indicate less feeding motivation. Another study showed that schools of young herring in the area did not respond to the seismic blasting with changes in swimming direction, speed, or vertical position in the water column [64].

Conflicts between seismic activity and fishing have occurred now and then and have at times become quite harsh. This has also occasionally initiated research to clarify whether there is any scientific basis for such claims. In Norway, a sandeel (*Ammodytes marinus*) fishery in the North Sea, close to several oil and gas fields, is such an example. Fishermen claimed to have reduced catches and explained this with the sound from seismic surveys causing sandeel to migrate away, or to bury themselves in the sand during seismic exposure. Video of caged sandeel during seismic exposure did reveal some alarm responses to the sound, but no burrowing into the sand [65]. Similar claims have also been posed for mackerel (*Scomber scombrus*), initiating a study where mackerel were kept in net pens and exposed to a small, approaching air gun, with gradually increasing sound exposure. Neither of these studies showed any particular reaction in terms of diving, startling, or increased swimming speed in mackerel [66]. Similar conflicts have been reported elsewhere in the world; scientific studies on seismic exposure in a redfish (*Sebasteds* sp.) fishery in California revealed that fish elicit alarm responses [59] and reduced longline catches [67]. However, it should be noted that studies were conducted with caged fish, which may influence the observed behavior, as well as inhibit larger-scale movements such as flight or avoidance, although this depends on the size and design of the enclosure [18,68].

These studies, as well as other studies conducted in other countries (e.g., [69–71]), show that a behavioral response is not always present, and, by nature, will vary in characteristic and strength. Some studies showed that the fish did not move away, while other studies showed that the fish left the area of exposure. Hence, responses may depend both on the species and the context. Throughout the first 10–15 years of the millennium, new knowledge was evaluated as it emerged to improve management recommendations. Despite the variable nature of how fish respond behaviorally during exposure to sound from seismic surveys, there was still a core knowledge-based conviction that fish could abandon their spawning sites if exposed to seismic blasting, so the advice to protect these areas and periods was not changed.

2.4. The 2010s: Towards Better Basis for Exclusion Zones (Refining Spawning Maps)

To effectively protect spawning areas, good knowledge of the actual spawning areas and periods of those species is crucial. An extensive report describing the spawning habitats and periods with drifting eggs and larvae from historic and recent data acquisition was published in 1991 [72]. These data were used to pinpoint areas and periods where seismic surveys should be avoided. Spawning areas are, however, dynamic habitats and change over time with changes in environmental variables, such as temperature (e.g., [73]). To better ensure that the recommendations reflect the actual spawning habitats, as well as to give more precise estimates of the relative importance of different spawning areas and periods, two projects were initiated to improve existing information on spawning areas and periods for the Norwegian and Barents Sea [41] and the North Sea [42]. These reports also took into account both historic and new knowledge from spawning surveys, as well as data from scientific surveys. Furthermore, data from fisheries on sampled egg and fish larvae were back-calculated to their spawning position using drift models. The results of these studies produced updated spawning maps, and, importantly, pinpointed those areas where the most significant and concentrated spawning occurred in time and space. For many species, restriction areas could be narrowed down to those most concentrated areas, without jeopardizing the links to recruitment. In addition to the species of greatest commercial importance that until now had been included in the advice, these projects also provided data to map spawning areas of other fish species. New questions, therefore, arose concerning which species to include in the advice in addition to those already included. Based on the evaluation of ecological importance, stock condition, and data basis, several species of less commercial importance, such as Greenland halibut (*Reinhardtius hippoglossoides*) and golden redfish (*Sebastes norvegicus*) were also included in the recommendations given from around 2015.

2.5. 2015 Onwards; New Technical Achievements Require New Advice

Another emerging trend since the 2000s was the use of electromagnetic (EM) surveys to more precisely locate and verify oil and gas deposits in the seabed. With this technique, electric and magnetic fields are generated within the water column. Several species of marine animals use electric and/or magnetic fields for orientation, migration, and prey or predator detection [74–76] thus with the potential of disturbing that behavior. During the first years of EM surveys, the recommendations from IMR for such surveys were the same as for a seismic survey, and the same areas were restricted, only without a buffer zone. This was, however, a highly questionable practice, as was pointed out by both the industry and the scientific community. A literature study was initiated and the results indicated that the main EM disturbances to fish species are likely to occur during their navigation during migration [43]. In accordance with the general goal of the recommendations: to prevent recruitment failure, the migration towards spawning grounds was considered to be most important to protect. From 2019, recommendations for EM surveys were to avoid known spawning migration routes during the migration periods.

Further studies on how the EM field induced by these surveys affect orientation and behavior of early-stage fish are currently (spring 2021) ongoing and the results thereof will be implemented within the advice as they emerge.

2.6. The Late 2010s: Inclusion of Marine Mammals

While Norway may have been very early to include fish in management advice, recommendations with respect to marine mammals have largely been lacking, and in contrast to most other countries, no restrictions were made to protect this group from potential effects from seismic blasting. The question of whether to include marine mammals in management advice has been regularly raised, e.g., by environmental organizations, the scientific community, and the general public. A challenge with including marine mammals in management advice is that Norwegian waters have many species, which are distributed over large areas, and that data on distribution, and, in particular, data on the relative importance of different habitats, is largely lacking. In response to the increasing amount of seismic activity in the Barents Sea, the demand for better management of seismic activity in these important mammal habitats increased [77]. From 2018, a new regulation made ramp-up procedures prior to seismic blasting mandatory by law to protect marine mammals from hearing injuries. In this respect, IMR also saw the need for marine mammals to be included in their recommendations. In the absence of specific studies on the impact of seismic exposure on mammals in Norway, the evaluation of seismic exposure studies elsewhere, as well as an extensive amount of scientific publications on exposure experiments of low-frequency naval sonar, another high intensive sound source, in Norwegian waters with different mammal species were considered as the basis for giving advice. Exposure to such low-frequency sonar has been documented to reduce foraging in several common species in Norwegian waters for humpback whales (*Megaptera novaeangliae*) [78], blue whales (*Balaenoptera musculus*) [79], bottlenose whales (*Hyperoodon ampullatus*) [80], sperm whales (*Physeter catodon*) [81], and killer whales (*Orcinus orca*) [82]. In particular, species that feed intensively within a season and depend on dense prey concentrations can experience severe consequences [79,83]. Baleen whales migrate to the Barents Sea to feed intensively during summer and early autumn to feed on the large concentrations of zooplankton and small fish [84]. Based on this knowledge, from 2019, areas and periods with intensive feeding of baleen whales were included in the recommendations of where to restrict seismic activity.

2.7. Into the 2020s: Increasing Scientific Effort

Since the first advice was given, spawning habitats for fish have been recommended to be avoided. As described above, the rationale for this recommendation is that if the sound from seismic airguns causes fish to avoid these habitats, this can lead to failure of reproduction and stock recruitment. However, the question of whether spawning behavior is actually hampered by the sound from seismic surveys has been raised repeatedly. Some argue that the "drive" to reproduce is so strong that other behaviors, such as avoidance, are depressed, and that they thus may likely ignore the seismic disturbance. If so, the need for strong protection of spawning habitats may not be the most efficient advice to prevent potential negative impacts on fish stocks. Therefore, a large project investigating the effects of seismic on spawning behavior and spawning performance as well as avoidance was initiated in 2018; "Effects of sound on spawning behavior and reproductive success of cod" (Spawn-Seis) (https://prosjektbanken.forskningsradet.no/project/FORISS/280367?Kilde=FORISS &distribution=Ar&chart=bar&calcType=funding&Sprak=no&sortBy=score&sortOrder=de sc&resultCount=30&offset=0&Fritekst=SpawnSeis, accessed on 17 April 2021) From this project, there are indications that continuous noise may be more hazardous for fish reproduction than intermittent blasts [18]. However, loud impulsive noise has been shown to produce stress, which could affect the spawning output. While the general literature on fish reproduction shows that stress should be avoided during spawning, the buffer zones, particularly, could be adjusted if the sound from seismic air guns does not seem to affect spawning behavior at the levels of exposure in SpawnSeis. Results from three years of experimental work in net pens and in the field are currently being analyzed and will be used to update advice.

Areas with drifting planktonic organisms have been excluded from the protection zones since the mid-1990s. Despite some documented effects on increased mortality and damage to zooplankton and other invertebrates, a large number of studies demonstrate the lack of effects unless in the very vicinity of the air gun [30–33]. However, in 2017, McCauley et al. [45] presented some noticeably contradictory results from Australian waters. Here, the abundance of zooplankton exposed to experimental airgun signals decreased by more than 50% in comparison with the control groups, and effects were observed up to 1.2 km from the airgun source. They concluded that seismic surveys have a highly negative impact on zooplankton, particularly, small copepods [45,85]. Thus, attention was created, as these organisms constitute the basis of the food web and support many of the most important fish stocks worldwide [86]. In contrast, a Norwegian study published in 2019 [46] found such effects to occur only at distances of five meters or closer to the air gun on larger copepods, and the increase in mortality did not exceed more than 30% at any distances from the airgun. In addition, no effects on escape response nor important changes in genes were detected. Together with previous similar results, these contradictions lead to the initiation of a research project on understanding the mechanisms of potential impact from the pressure and particle motion associated with seismic shooting (ZoopSeis). The project started in 2020 and will continue until 2023 and the results will inform recommendations as they emerge.

3. Advisory Tools

As described above, the recommendations from IMR on the regulation of seismic activity have always been provided in the form of geographical and seasonal restrictions to avoid sound exposure of sensitive habitats and periods. Such management rules are easily expressed in a map, and maps of sensitive habitats have always been used as tools.

In parallel to the scientific investigations described in the previous sections, advisory tools have been improved from relatively static maps in paper format to digital maps, so-called restriction maps, showing the exact area to avoid in two-week periods throughout the year (Figure 2). These maps include all the recommendations (spawning grounds, spawning migration areas, and feeding areas for marine mammals as well as buffer zones) in one map, and the operator can choose the type of survey (2D/3D seismic surveys, site survey, or EM survey) and obtain a full overview of when and where advice not to conduct seismic activity will be given. This approach simplifies planning for commercial seismic companies, and may have increased the acceptance of the recommendations. The areas and periods that are included in the restriction maps are evaluated once a year by experts on the different species of fish and marine mammals. The latest ongoing development is that these restriction maps are included in the online application process for seismic surveys. If the planned surveys overlap with a restriction area, an automatic warning will be given when either the area or the time period must be changed for the survey to be approved. Since 2018, IMR has published an annual report describing the restriction maps for that year and the scientific background behind the recommendations. The newest report can be found at https://www.hi.no/hi/nettrapporter/rapport-fra-havforskningen-2021-4 (accessed on 17 April 2021).

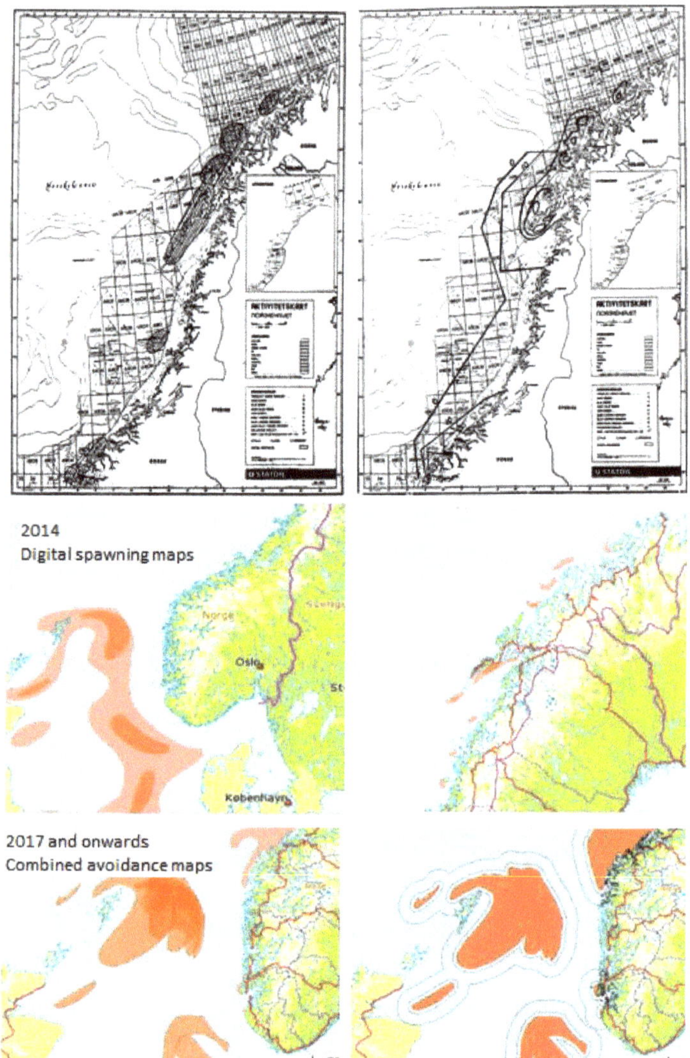

Figure 2. Upper panel: Examples of maps of spawning areas of and of areas of drifting eggs and larvae, as used to give advice in the period 1991 to 2014 [72]. These maps only existed in paper format. **Left**: Spawning areas for cod, with restriction periods between 15 March and 15 May. **Right**: Drifting cod eggs. Restriction periods between 1 April and 30 April. The restriction on drifting eggs and larvae ended in 1996, while the restrictions on the spawning areas still hold, but the exact area and period have been updated. **Middle panel**: Digital spawning maps available from 2014. Areas divided into important and less important spawning areas. This example is for cod, with darker areas being the most important ones. The important restrictions are applied in(dark) areas. **Right**: North Sea cod. **Left**: Northeast Arctic cod. **Lower panel**: Spawning maps for each 2 week period of the year merged into maps with restriction areas. **Left**: Important spawning maps for different North Sea fish species (cod, Norway pout, saithe, herring). **Right**: Avoidance map for period 1 to 15 March. These are made by combining all the maps for important spawning areas for those fish that spawn in this period (which are those shown on the map to the left), as well as two buffer zones (dotted lines) of 5 and 20 nmi, and the advised restrictions for site surveys and regular seismic surveys, respectively.

In parallel to advice from IMR on biological implications, the Norwegian Directorate of Fisheries gives advice to all seismic surveys on potential conflicts with ongoing fishery activity. In areas with traditional seasonal fishing grounds, such as those for herring and mackerel, it is advised that seismic surveys are conducted either before or after the fishing season. Additionally, their advice includes a requirement to always have a fishery liaison officer on board the seismic vessel, to handle all communication between the seismic vessel and fishing vessels in the area to ensure cooperation. The fishery liaison officer will inform the captain on the seismic vessel, e.g., about the specifics of the fishing tools used in the area and how to best avoid them, as well as talking to the fishers and telling them about the seismic production and how they can best conduct their activity without being in the way of the seismic vessel.

4. The Way Forward—The Science Needed to Make Good Management Decisions

The above example from Norway shows that scientific input can be routinely used in management decisions by including scientific institutions in the management process. To ensure that scientific research in the field is applicable to management, Prewlaski et al. [87] highlight some important issues, including identifying useful metrics and species, and the ability to generalize results to a certain degree among species and regions. Further, to ensure that regulations are applicable, recommendations should be balanced between highly restrictive regulations and the loss of resource benefits. Effective research-based management, therefore, requires close collaboration between scientists, industry, and regulators to frame scientific results into applicable regulations. For example, in theory, there may always be one or more species spawning, mating, or feeding in an area that can be argued as a reason for avoiding disturbance, and, hence, closing the area year-round. However, such a strict regulation will never be applied by managers. Therefore, instead of the manager taking a potentially arbitrary decision on where and when to allow seismic survey, the scientist should help identify those areas and time periods that are most important to protect.

During the past 20 years, several guidelines have proposed certain sound threshold levels that should not be exceeded both for marine mammals [88,89] and fish [7,90]. Such criteria are useful and relatively easy to apply. Such thresholds are effective to prevent physical injury, as these are likely to arise when the animal is exposed to sound levels exceeding a certain level. Behavioral responses, however, are far more complex and a response may also depend on factors such as time of day [91], season [92,93], context [94,95], and previous exposure (e.g., [96,97]). Hawkins et al. [98] highlight the need for research on how fish respond to sounds at different levels and changes during the course of sound presentation while the sound characteristics (pressure and particle motion) are carefully measured. Further, Duarte et al. [19] emphasized that a new, globally binding agreement on the regulation of anthropogenic sound in the sea is needed, e.g., by inclusion into the UN Law of the Sea.

Some issues of high importance for better management decisions that remain unsolved include the extent and duration of displacement, as well as the thresholds of received sound levels or distances from the source that lead to avoidance of essential habitats, such as spawning, mating, or foraging sites for various species and animal groups. Additionally, studies should preferably enable an evaluation of how the measured effects could disturb the population, stock, or habitat as a whole, as this is usually the main unit that is managed.

In Norway, management has focused on commercially important fish stocks. This was related to a focus on sustainable management of fish stocks at IMR and a focus on coexistence between oil exploration and fisheries at a government level. Currently, an ecosystem-based approach is called for, as a more productive approach for the management of sustainable harvest [99–101]. To reach this goal, a wider range of species should be included in future management advice, including key species for the ecosystem and threatened species. Because data availability is a limiting factor for many such species, this requires a continued effort to collect data on the reproductive behavior of such species in relation to noise.

Furthermore, noise is not the only stressor that affects reproduction, and multi-stressor approaches could provide more insight into anthropogenic effects on underwater life. Thus, future management should also focus more on the overall effects of human impact on the ecosystem, by integrating different types of pressures instead of managing them one by one.

Author Contributions: Conceptualization, L.D.S., K.d.J.; Original draft preparation, L.D.S., E.H.V., K.d.J., T.N.F.; Writing—Review & Editing, L.D.S., E.H.V., K.d.J., T.N.F., J.D.; Visualization, L.D.S.; Project administration, L.D.S., H.W.; Funding acquisition, L.D.S., H.W. All authors have read and agreed to the published version of the manuscript.

Funding: This research was funded by internal funding from the Insitute of Marine Research for authors L.D.S., K.d.J., T.N.F. and H.W. Funding for E.H.V. was given by the Norwegian Research Council, grant number 302675. Author J.D. was funded by Soundmare.

Institutional Review Board Statement: Not applicable.

Informed Consent Statement: Not applicable.

Data Availability Statement: Data sharing not applicable. No new data were created or analyzed in this study. Data sharing is not applicable to this article.

Acknowledgments: Egil Ona and Svein Løkkeborg are acknowledged for their background information on the research conducted in the early 1990s.

Conflicts of Interest: The authors declare no conflict of interest.

References

1. Francis, C.D.; Barber, J.R. A framework for understanding noise impacts on wildlife: An urgent conservation priority. *Front. Ecol. Environ.* **2013**, *11*. [CrossRef]
2. Shannon, G.; McKenna, M.F.; Angeloni, L.A.; Crooks, K.R.; Fristrup, K.M.; Brown, E.; Warner, K.A.; Nelson, M.D.; White, C.; Briggs, J.; et al. A synthesis of two decades of research documenting the effects of noise on wildlife. *Biol. Rev.* **2016**, *91*, 982–1005. [CrossRef]
3. Slabbekoorn, H.; Dooling, J.D.; Popper, A.N.; Fay, R.R. *Effects of Antropogenic Noise on Animals*; Springer: New York, NY, USA, 2018. [CrossRef]
4. Andrew, R.K.; Howe, B.M.; Mercer, J.A.; Dzieciuch, M.A. Ocean ambient sound: Comparing the 1960s with the 1990s for a receiver off the California coast. *Acoust. Res. Lett. Online* **2002**, *3*, 65. [CrossRef]
5. McDonald, M.A.; Hildebrand, J.A.; Wiggins, S.A. Increases in deep ocean ambient noise in the Northeast Pacific west of San Nicolas Island, California. *J. Acoust. Soc. Am.* **2006**, *120*, 711. [CrossRef]
6. Hildebrand, J.A. Anthropogenic and natural sources of ambient noise in the ocean. *Mar. Ecol. Prog. Ser.* **2009**, *5*, 20. [CrossRef]
7. Popper, A.N.; Hawkins, A.D.; Fay, R.R.; Mann, D.A.; Bartol, S.; Carlson, T.J.; Coombs, S.; Ellison, W.T.; Gentry, R.L.; Halvorsen, M.B.; et al. *Sound Exposure Guidelines for Fishes and Sea Turtles: A Technical Report Prepared by ANSI-Accredited Standards Committee S3/SC1 and Registered with ANSI*; ASA S3/SC1.4 TR-2014; Springer: New York, NY, USA, 2014.
8. Slabbekoorn, H.; Bouton, N.; van Opzeeland, I.; Coers, A.; ten Cate, C.; Popper, A.N. A noisy spring: The impact of globally rising underwater sound levels on fish. *Trends Ecol. Evol.* **2010**, *25*, 419–427. [CrossRef] [PubMed]
9. Popper, A.N.; Fewtrell, J.; Smith, M.E.; McCauley, R.D. Anthropogenic sound: Effects on the behaviour and physiology of fishes. *Mar. Technol. Sci. J.* **2004**, *37*, 35–40. [CrossRef]
10. New, L.F.; Clark, J.S.; Costa, D.P.; Fleishman, E.; Hindell, M.A.; Klanjšček, T.; Lusseau, D.; Kraus, S.; McMahon, C.R.; Robinson, P.W. Using short-term measures of behaviour to estimate long-term fitness of southern elephant seals. *Mar. Ecol. Prog. Ser.* **2014**, *496*, 99–108. [CrossRef]
11. Kunc, H.P.; McLaughlin, K.E.; Schmidt, R. Aquatic noise pollution: Implications for individuals, populations, and ecosystems. *Proc. R. Soc. B* **2016**, *283*, 20160839. [CrossRef] [PubMed]
12. Slabbekoorn, H. Noise Pollution. *Curr. Biol.* **2019**, *29*, R942–R995. [CrossRef]
13. Radford, A.N.; Kerridge, E.; Simpson, S.D. Acoustic communication in a noise world: Can fish compete with antropogenic noise? *Behav. Ecol.* **2014**, *25*, 1022–1030. [CrossRef]
14. Williams, R.; Wright, A.J.; Ashe, E.; Blight, L.K.; Bruintjes, R.; Canessa, R.; Clark, C.W.; Cullis-Suzuki, S.; Dakin, D.T.; Erbe, C. Impacts of anthropogenic noise on marine life: Publication patterns, new discoveries, and future directions in research and management. *Ocean. Coast. Manag.* **2015**, *115*, 17–24. [CrossRef]

15. Carroll, A.G.; Przeslawski, R.; Duncan, A.; Gunning, M.; Bruce, B. A critical review of the potential impacts of marine seismic surveys on fish & invertebrates. *Mar. Pollut. Bull.* **2017**, *114*, 9–24.
16. Hawkins, A.D.; Popper, A.N. A sound approach to assessing the impact of underwater noise on marine fishes and invertebrates. *ICES J. Mar. Sci.* **2017**, *74*, 635–651. [CrossRef]
17. Cox, K.; Brennan, L.P.; Gerwing, T.G.; Dudas, S.E.; Juanes, F. Sound the alarm: A meta-analysis on the effect of aquatic noise on fish behavior and physiology. *Glob. Chang. Biol.* **2018**, *24*, 3105–3116. [CrossRef]
18. de Jong, K.; Forland, T.N.; Amorim, M.C.P.; Rieucau, G.; Slabbekoorn, H.; Sivle, L.D. Predicting the effects of anthropogenic noise on fish reproduction. *Rev. Fish. Biol. Fish.* **2020**, *30*, 245–268. [CrossRef]
19. Duarte, C.M.; Chapuis, L.; Collin, S.P.; Costa, D.P.; Devassy, R.P.; Eguiluz, V.M.; Erbe, C.; Gordon, T.A.C.; Halpern, B.S.; Harding, H.R.; et al. The soundscape of the Anthropocene ocean. *Science* **2021**, *371*, eaba4658. [CrossRef] [PubMed]
20. Anonymous. *Veileder: Gjennomføring av Seismiske Undersøkelser på Norsk Kontinentalsokkel*; (Guide: Implementation of seismic investigations on the Norwegian Continental Shelf. Norwegian Ministry of Fisheries and Coastal Affairs and the Norwegian Ministry of Oil and Energy, Oslo, Norway); Fiskeri- og Kystdepartement og Olje- og Energidepartement: Oslo, Norway, 2013; 28p. (In Norwegian)
21. Bonga, S.E.W. The stress response in fish. *Physiol. Rev.* **1997**, *77*, 591–625. [CrossRef] [PubMed]
22. Szuwalski, C.S.; Vert-Pre, K.A.; Punt, A.E.; Branch, T.A.; Hilborn, R. Examining common assuptions about recruitement: A meta-analysis of recruitement dynamics of worldwide marine fisheries. *Fish Fish.* **2014**, *16*, 633–648. [CrossRef]
23. Colin, P.L.; Sadovy, Y.J.; Domeier, M.L. *Manual for the Study and Conservation of Reef Fish Aggregations*; Society for the Conservation of Reef Fish Aggregations: Fallbrook, CA, USA, 2003.
24. Portner, H.O.; Farrel, A.P. Physiologu and climate change. *Science* **2008**, *322*, 690–692. [CrossRef] [PubMed]
25. Holst, J.C. The herring. In *The Norwegian Sea Ecosystem*; Skjoldal, H., Ed.; Fagbokforlaget: Bergen, Norway, 2004.
26. Rose, G.A.; Bradbury, I.R.; de Young, B. Rebuilding Atlantic cod: Lessons from a spawning ground in coastal Newfoundland. In *Resiliency of Gadid Stocks to Fishing and Climate Change*; Kruse, G.H., Drinkwater, K., Ianelli, J.N., Link, J.S., Stram, D.L., Wespestad, V., Woodby, D., Eds.; Alaska Sea Grant College Program: Fairbanks, AK, USA, 2008; pp. 197–218.
27. Dalen, J.; Raknes, A. *Skremmeeffekter på Fisk frå 3-Dimensjonale Seismiske Undersøkjingar. (Scaring Effects in Fish from 3-Dimensional Seismic Explorations)*; Report No FO 8504; Institute of Marine Research: Bergen, Norway, 1985; 22p. (In Norwegian with English Abstract)
28. Dalen, J. *Stimulering av Sildestimer. Forsøk i Hopavågen og Imsterfjorden-Verrafjorden 1973. (Stimulating Herring Shoals. Experiments in Hopavågen and Imsterfjorden-Verrafjorden)*; Report to the Royal Norwegian Council for Scientific and Industrial Research (NTNF); No. 73-143-T; The Norwegian Institute of Technology (NTH): Trondheim, Norway, 1973; 36p. (In Norwegian)
29. Chapman, C.J.; Hawkins, A.J. *The Importance of Sound in Fish Behavior in Relation to Capture by Trawls*; FAO Fisheries Reports; FAO: Rome, Italy, 1969; Volume 621, pp. 717–729.
30. Knutsen, G.M.; Dalen, J. *Skadeeffekter på egg, Larver og Yngel Fra Seismiske Undersøkelser (Harmful Effects on Eggs, Larvae and Fry from Seismic Explorations)*; Report No FO 8505; Institute of Marine Research: Bergen, Norway, 1985; 26p. (In Norwegian with English Abstract)
31. Dalen, J.; Knutsen, G.M. Scaring effects in fish and harmful effects on eggs, larvae and fry by offshore seismic explorations. In *Progress in Underwater Acoustics*; Merklinger, H.M., Ed.; Plenum Publishing Corporation: New York, NY, USA, 1987; pp. 93–102.
32. Booman, C.; Dalen, J.; Leivestad, H.; Levsen, A.; van der Meeren, T.; Toklum, K. *Effekter av Luftkanonskyting på Egg, Larver og Yngel. Undersøkelser ved Havforskningsinstituttet og Zoologisk Laboratorium, UiB. (Effects from Air Gun Shooting on Eggs, Larvae, and Fry. Experiment at the Institute of Marine Research and Zoological Laboratorium, Univ. of Bergen)*; Fisken og Havet, No 3-1996; Institute of Marine Research: Bergen, Norway, 1996; 83p. (In Norwegian with English Summary, Figure and Table Legends)
33. Kosheleva, V. The impact of air guns used in marine seismic explorations on organisms living in the Barents Sea. In Proceedings of the Control Petro PISCIS II '92 Conference F-5, Bergen, Norway, 6–8 April 1992.
34. Soldal, A.V.; Engås, A.; Løkkeborg, S. *Refraksjonsseismiske Spregninger i Øygarden. Effekten på Vill- og Oppdrettsfisk. (Refraction Seismic Blasting in Øygarden. The Effect in Wild and Farmed Fish*; Report. Research Number 06-90; Institute of Fishery Tehcnology: Bergen, Norway, 1990; 20p. (In Norwegian)
35. Engås, A.; Olsen, S.; Soldal, A.V. *Undersøkelser av Effekten på Torsk i Mær av Refraksjonsseismiske Spregninger i Øygarden. (Investigations in Penned Cod from Refraction Seismic Blasting in Øygarden)*; Report; Institute of Fishery Tehcnology: Bergen, Norway, 1989; 21p. (In Norwegian)
36. Holmstrøm, S. *Effekter av Luftkanonseismikk på Larver og Yngel—Modellering og Simulering (Effects from Airgun Seismic in Larvae and Fry—Modelling and Simulating)*; SINTEF Report No STF48 A93007; SINTEF: Trondheim, Norway, 1993; 70p. (In Norwegian)
37. Sætre, R.; Ona, E. *Seismiske Undersøkelser og Skader på Fiskeegg og -Larver; en Vurdering av Mulige Effekter på Bestandsnivå. (Seismic Investigations and Injuries in Fish Eggs and Larvae; an Assessment of Potential Effects on Stock Level)*; Fisken og Havet, No 8-1996; Institute of Marine Research: Bergen, Norway, 1996; 25p. (In Norwegian with English Summary)
38. Holand, B.; Walsø, Ø.; Berg, T. *Seismiske Eksperimenter i Våg (Seismic Experiments in a Bay)*; SINTEF Report No STF23 A93005; SINTEF: Trondheim, Norway, 1993; 45p. (In Norwegian)
39. Løkkeborg, S.; Soldal, A.V. The influence of seismic exploration with air guns on cod (Gadus morhua) behaviour and catch rates. In Proceedings of the ICES Marine Science Symposium 201, Bergen, Norway, 21–23 June 1993; Volume 196, pp. 62–67.

40. Engås, A.; Løkkeborg, S.; Ona, E.; Soldal, A.V. Effects of seismic shooting on local abundance and catch rates of cod (Gadus morhua) and haddock (Melanogrammus aeglefinus). *Can. J. Fish. Aquat. Sci.* **1996**, *53*, 2238–2249. [CrossRef]
41. Sundby, S.; Fossum, P.; Sandvik, A.; Vikebø, F.B.; Aglen, A.; Buhl-Mortensen, L.; Folkvord, A.; Bakkeplass, K.; Buhl-Mortensen, P.; Johannessen, M.; et al. *Kunnskapsinnhenting Barentshavet—Lofoten—Vesterålen. KILO. Rapport fra KILO-Prosjektet (Knowledge Acquisition the Barents Sea—Lofoten—Vesterålen. KILO. Report of the KILO Project)*; Fisken og Havet No 3-2013; The Institute of Marine Research: Bergen, Norway, 2013; 186p. (In Norwegian)
42. Sundby, S.; Kristiansen, T.; Nash, R.; Johannessen, T.; Bakkeplass, K.; Höffle, H.; Opstad, I. *Dynamic Mapping of North Sea Spawning. Report of the KINO Project*; Fisken og Havet 2-2017; The Institute of Marine Research: Bergen, Norway, 2017; 195p, ISSN 0071-5638.
43. Nyqvist, D.; Durif, C.; Johnsen, M.G.; de Jong, K.; Forland, T.N.; Sivle, L.D. Electric and magnetic senses in marine animals, and potential behavioral effects of electromagnetic surveys. *Mar. Environ. Res.* **2020**, *155*, 104888. [CrossRef]
44. Sivle, L.D.; Forland, T.N.; de Jong, K.; Nyqvist, D.; Grimsbø, E. *Havforskningsinstituttets Rådgivning for Menneskeskapt lyd i Havet: Seismikk, Elektromagnetiske Undersøkelser og Undersjøiske Sprengninger—Kunnskapsgrunnlag, Vurderinger og Råd*; (Advice from the Institute of Marine Research of anthropocentric noise in the sea: Seismic, electromagnetic surveys and underwater explosions); No 2019-10; Institute of Marine Research: Bergen, Norway, 2019; 75p, ISSN 1893-4536. (In Norwegian with English Summary)
45. McCauley, R.D.; Day, R.D.; Swadling, K.M.; Fitzgibbon, Q.; Watson, R.A.; Semmens, J.M. Widely used marine seismic survey air gun operations negatively impact zooplankton. *Nat. Ecol. Evol.* **2017**, *1*. [CrossRef] [PubMed]
46. Fields, D.M.; Handegard, N.O.; Dalen, J.; Eichner, C.; Malde, K.; Karlsen, Ø.; Skiftesvik, A.B.; Durif, C.M.F.; Browman, H.I. Airgun blasts used in marine seismic surveys have limited effects on mortality, and no sublethal effects on behaviour or gene expression, in the copepod Calanus finmarchicus. *ICES J. Mar. Sci.* **2019**, *76*, 2033–2044. [CrossRef]
47. Dalen, J.; Ona, E.; Vold, A.S.; Sætre, R. *Seismiske Undersøkelser til Havs: En Vurdering av Konsekvenser for Fisk og Fiskerier. (Offshore Seismic Investigations: An Evaluation of Consequences for Fish and Fisheries)*; Fisken og havet, No 9-1996; Institute of Marine Research: Bergen, Norway, 1996; 26p. (In Norwegian with English Summary)
48. Holliday, D.V.; Pieper, R.E.; Clarke, M.E.; Greenlaw, C.F. *Effects of Airgun Energy Releases on the Northern Anchovy*; API Publ. No 4453; American Petroleum Institute, Health and Environmental Sciences Department: Washington, DC, USA, 1987; 108p.
49. Matishov, G.G. The reaction of bottom-fish larvae to airgun pulses in the context of the vulnerable Barents Sea ecosystem. In Proceedings of the Control Petro PISCIS II '92 F-5, Bergen, Norway, 6–8 April 1992.
50. Dalen, J. *Effektar av Luftkanonseismikk på Larvar og Yngel til Havs. NFFR Sluttrapport 1701-701.354. (Effects from Airgun Seismic on Offshore Larvae and Fry. NFFR Final Report 1701-701.354)*; Report No 9-1993; Institute of Marine Research: Bergen, Norway, 1993; 18p, ISSN 0804-2128. (In Norwegian)
51. Enger, P.S.; Anderson, R. An electrophysical field study of hearing in fish. *Comp. Biochem. Physiol.* **1967**, *22*, 517–525. [CrossRef]
52. Chapman, C.J.; Johnstone, A.D.F. Some auditory discrimination experiments on marine fish. *J. Exp. Biol.* **1974**, *61*, 521–528.
53. Chapman, C.J.; Hawkins, A.J. A field study of hearing in the cod. *J. Comp. Physiol.* **1973**, *85*, 147–167. [CrossRef]
54. Olsen, K. *Directional Responses in Herring to Sound and Noise Stimuli*; Conference and meeting documents; ICES Annual Science Conference 1969; ICES: Burnaby, BC, Canada, 1969; 8p.
55. Olsen, K. *Directional hearing in cod (Gadus morhua)*; International Working Group for Fishing Technology: Rome, Italy, 1969; pp. 77–88.
56. Hawkins, A.J.; Sand, O. Directional hearing in the median vertical plane by the cod. *J. Comp. Biol. Physiol. A* **1977**, *122*, 1–8. [CrossRef]
57. Schuijf, A. Directional hearing of cod (Gadus morhua) under approximate free field conditions. *J. Comp. Physiol.* **1975**, *98*, 307–332. [CrossRef]
58. Pearson, W.H.; Skalski, J.R.; Malme, C.I. *Effects of Sounds from a Geophysical Survey Device on Fishing Success*; OCS Study MMS-86-0032; BBN Laboratories Inc.: Cambridge, MA, USA; Battelle, Marine Research Laboratory: Washington, DC, USA, 1987; 293p.
59. Pearson, W.H.; Skalski, J.R.; Malme, C.I. Effects of sounds from a geophysical survey device on behavior of captive rockfish (Sebastes spp). *Can. J. Fish. Aquat. Sci.* **1992**, *49*, 1343–1356. [CrossRef]
60. Kramer, F.S.; Peterson, R.A.; Walter, W.C. Seismic Energy Sources 1968. Handbook. Prep: Staff Members of United Geophysical Corporation. In Proceedings of the 38th Annual Meeting of the SEG, Denver, CO, USA, 18–19 October 1968.
61. Malme, C.I.; Smith, P.W., Jr.; Miles, P.R. *Characterization of Geophysical Acoustic Survey Sounds*; OCS Study MMS-86-0032; BBN Laboratories Inc.: Cambridge, MA, USA, 1986.
62. Slotte, A.; Hansen, K.; Dalen, J.; Ona, E. Acoustic mapping of pelagic fish distribution and abundance in relation to a seismic shooting area off the Norwegian west coast. *Fish. Res.* **2004**, *67*, 143–150. [CrossRef]
63. Løkkeborg, S.; Ona, E.; Vold, A.; Salthaug, A. Sounds from seismic air guns: Gear- and species specific effects on catch rates and fish distribution. *Can. J. Fish. Aquat. Sci.* **2012**, *69*, 1278–1291. [CrossRef]
64. Peña, H.; Handegard, N.O.; Ona, E. Feeding herring schools do not react to seismic air gun surveys. *ICES J. Mar. Sci.* **2013**, *70*, 1174–1180. [CrossRef]
65. Hassel, A.; Knutsen, T.; Dalen, J.; Skaar, K.; Løkkeborg, S.; Misund, O.A.; Østensen, Ø.; Fonn, M.; Haugland, E.K. Influence of seismic shooting on the lesser sandeel (Ammodytes marinus). *ICES J. Mar. Sci.* **2004**, *61*, 1165–1173. [CrossRef]

66. Sivle, L.D.; Forland, T.N.; Hansen, R.R.; Andersson, M.; Grimsbø, E.; Linne, M.; Karlsen, H.E. *Behavioural Effects of Seismic Dose Escalation Exposure on Captive Mackerel (Scomber scombrus)*. *Rapport fra Havforskningen*; Report No 34-2017; Institute of Marine Research: Bergen, Norway, 2017; ISSN 1893-453.

67. Skalski, J.R.; Pearson, W.H.; Malme, C.I. Effects of sounds from a geophysical survey device on catch-per-unit-effort in a hook-and-line fishery for rockfish (Sebastes spp.). *Can. J. Fish. Aquat. Sci.* **1992**, *49*, 1357–1365. [CrossRef]

68. Slabbekoorn, H. Aiming for progress in understanding underwater noise impact on fish: Complementary need for indoor and outdoor studies. In *The Effects of Noise on Aquatic Life II, Advances in Experimental Medicine and Biology 875*; Popper, A.N., Hawkins, A., Eds.; Springer: Berlin, Germany, 2016; pp. 1057–1065.

69. Hirst, A.G.; Rodhouse, P.G. Impacts of geophysical seismic surveying on fishing success. *Rev. Fish. Biol. Fish.* **2000**, *10*, 113–118. [CrossRef]

70. Wardle, C.S.; Carter, T.J.; Urquhart, G.G.; Johnstone, A.D.F.; Ziolkowski, A.M.; Hampson, D.; Mackie, D. Effects of seismic air guns on marine Fish. *Cont. Shelf Res.* **2001**, *21*, 1005–1027. [CrossRef]

71. Fewtrell, J.L.; McCauley, R.D. Impact of air gun noise on the behaviour of marine fish and squid. *Mar. Pollut. Bull.* **2012**, *64*, 984–993. [CrossRef]

72. Bjørke, H.; Dalen, J.; Bakkeplass, K.; Hansen, K.; Rey, L. *Tilgjengelighet av Seismiske Aktiviteter i Forhold til Sårbare Fiskeressurser. (Seismic Explorations' Accessibility in Relation to Vulnerable Fish Resources)*; HELP-Report No 38; Institute of Marine Research: Bergen, Norway, 1991; 119p. (In Norwegian with English Summary and Figure and Table Legends).

73. Sundby, S.; Nakken, O. Spatial shifts in spawning habitats of Arcto-Norwegian cod related to multidecadal climate oscillations and climate change. *ICES J. Mar. Sci.* **2008**, *65*, 953–962. [CrossRef]

74. Collin, S.; Whitehead, D. The functional roles of passive electroreception in non-electric fishes. *Anim. Biol.* **2004**, *54*, 1–25.

75. Kalmijn, A.J. Electric and magnetic detection un elasmobranch fishes. *Science* **1982**, *218*, 916–918. [CrossRef] [PubMed]

76. Kullnick, U. Influences of Electric and Magnetic Fields on Aquatic Ecosystems. In Proceedings of the International Seminar on Effects of Electromagnetic Fields on the Living Environment, Ismaning, Germany, 4–5 October 1999; pp. 113–132.

77. Kvadsheim, P.; Sivle, L.D.; Hansen, R.R.; Karlsen, H.E. *Effekter av Menneskeskapt Støy på Havmiljø. Rapport til Miljødirektoratet om Kunnskapsstatus (Effects from Man-Made Noise on Ocean Environment. Report to the Norwegian Environment Agency of Status of Knowledge)*; FFI Report No 17/00075; Norwegian Defence Research Establishment (FFI): Horten, Norway, 2017; 75p. (In Norwegian with English Summary)

78. Sivle, L.D.; Wensveen, P.J.; Kvadsheim, P.H.; Lam, F.-P.A.; Visser, F.; Curé, C.; Harris, C.M.; Tyack, P.L.; Miller, P.J.O. Naval sonar disrupts foraging in humpback whales. *Mar. Ecol. Prog. Ser.* **2016**, *562*, 211–220. [CrossRef]

79. Goldbogen, J.A.; Southall, B.L.; DeRuiter, S.L.; Calambokidis, J.; Friedlaender, A.S.; Hazen, E.L.; Falcone, E.A.; Schorr, G.S.; Douglas, A.; Moretti, D.J. Blue whales respond to simulated mid-frequency military sonar. *Proc. R. Soc. B Biol. Sci.* **2013**, *280*, 20130657. [CrossRef] [PubMed]

80. Miller, P.J.O.; Kvadsheim, P.H.; Lam, F.P.A.; Tyack, P.L.; Cure, C.; DeRuiter, S.L.; Hooker, S.K. First indications that northern bottlenose whales are sensitive to behavioural disturbance from anthropogenic noise. *R. Soc. Open Sci.* **2015**, *2*. [CrossRef]

81. Isojunno, S.; Curé, C.; Kvadsheim, P.H.; Lam, F.P.; Tyack, P.L.; Wensveen, P.; Miller, P.J.O. Sperm whales reduce foraging effort during exposure to 1–2 kHz sonar and killer whale sounds. *Ecol. Appl.* **2016**, *26*, 77–93. [CrossRef]

82. Miller, P.J.O.; Kvadsheim, P.H.; Lam, F.P.A.; Wensveen, P.J.; Antunes, R.; Alves, A.C.; Visser, F.; Kleivane, L.; Tyack, P.L.; Sivle, L.D. The Severity of Behavioral Changes Observed During Experimental Exposures of Killer (*Orcinus orca*), Long-Finned Pilot (*Globicephala melas*), and Sperm (*Physeter macrocephalus*) Whales to Naval Sonar. *Aquat. Mamm.* **2012**, *38*, 362–401. [CrossRef]

83. Farmer, N.A.; Baker, K.; Zeddies, D.G.; Denes, S.L.; Noren, D.P.; Garrison, L.P.; Zykov, M. Population consequences of disturbance by offshore oil and gas activity for endangered sperm whales (Physeter macrocephalus). *Biol. Conserv.* **2018**, *227*, 189–204. [CrossRef]

84. Mauritzen, M.; Johannesen, E.; Bjørge, A.; Øien, N. Baleen whale distributions and prey associations in the Barents Sea. *Mar. Ecol. Prog. Ser.* **2011**, *426*, 289–301. [CrossRef]

85. Richardson, A.J.; Matear, R.J.; Lenton, A. *Potential Impacts on Zooplankton of Seismic Surveys*; CSIRO: Canberra, Australia, 2017; 34p.

86. Kaiser, M.J.; Attrill, M.J.; Jennings, S.; Thomas, D.N.; Barnes, D.K.A.; Brierley, A.S.; Graham, N.A.J.; Hiddink, J.G.; Howell, K.L.; Kaartokallio, H. *Marine Ecology: Processes, Systems, and Impacts*, 3rd ed.; Oxford University Press: New York, NY, USA, 2020.

87. Przeslawski, R.; Brooke, B.; Carroll, A.G.; Fellows, M. An integrated approach to assessing marine seismic impacts: Lessons learnt from the Gippsland Marine Environmental Monitoring project. *Ocean. Coast. Manag.* **2019**, *160*, 117–123. [CrossRef]

88. Southall, B.; Bowles, A.; Ellison, W.; Finneran, J.; Gentry, R.; Greene, C., Jr.; Kastak, D.; Ketten, D.; Miller, J.; Nachtigall, P.; et al. Marine mammal noise exposure criteria: Initial scientific recommendations. *Aquat. Mamm.* **2007**, *33*, 411–521. [CrossRef]

89. Southall, B.L.; Finneran, J.J.; Reichmuth, C.; Nachtigall, P.E.; Ketten, D.R.; Bowles, A.E.; Ellison, W.T.; Nowacek, D.P.; Tyack, P.L. Marine mammal noise exposure criteria: Updated scientific recommendations for residual hearing effects. *Aquat. Mamm.* **2019**, *45*, 125–232. [CrossRef]

90. Andersson, M.H.; Andersson, S.; Ahlsen, J.; Andersoson, B.L.; Hammar, J.; Persson, L.K.; Pihl, J.; Sigray, P.; Wisstrom, A. *A Framework for Regulating Underwater Noise during Pile Driving*; A Technical Vindal Report; Environmental Protection Agency: Stockholm, Sweden, 2017.

91. Hawkins, A.D.; Roberts, L.; Cheesman, S. Responses of freeliving coastal pelagic fish to impulsive sounds. *J. Acoust. Soc. Am.* **2014**, *135*, 3101–3116. [CrossRef]

92. Skaret, G.; Slotte, A.; Handegard, N.O.; Axelsen, B.E.; Jørgensen, R. Pre-spawning herring in a protected area showed only moderate reaction to a surveying vessel. *Fish. Res.* **2006**, *78*, 359–367. [CrossRef]
93. Vabø, R.; Olsen, K.; Huse, I. The effect of vessel avoidance of wintering Norwegian spring spawning herring. *Fish. Res.* **2002**, *58*, 59–77. [CrossRef]
94. Ellison, W.T.; Southall, B.L.; Clark, C.W.; Frankel, A.S. A New context-based approach to assess marine mammals behavioral responses to anthropogenic sounds. *Conserv. Biol.* **2011**, *26*, 21–28. [CrossRef] [PubMed]
95. Harris, C.M.; Thomas, L.; Falcone, E.A.; Hildebrand, J.; Houser, D.; Kvadsheim, P.H.; Lam, F.-P.A.; Miller, P.J.O.; Moretti, D.J.; Read, A.J.; et al. Marine mammals and sonar: Dose-response studies, the risk-disturbance hypothesis and the role of exposure context. *J. Appl. Ecol.* **2017**, 1–9. [CrossRef]
96. Neo, Y.Y.; Hubert, J.; Bolle, L.J.; Winter, H.V.; Slabbekoorn, H. European seabass respond more strongly to noise exposure at night and habituate over repeated trials of sound exposure. *Environ. Pollut.* **2018**, *239*, 367–374. [CrossRef]
97. Goetz, T.; Janik, V.M. Aversiveness of sounds in phocid seals: Psycho-physiological factors, learning processes and motivation. *J. Exp. Biol.* **2010**, *213*, 1536–1548. [CrossRef]
98. Hawkins, A.D.; Johnson, C.; Popper, A.N. How to set sound exposure criteria for fishes. *J. Acoust. Soc. Am.* **2020**, *147*. [CrossRef]
99. FAO Fisheries Departement. *The Ecosystem Approach to Fisheries*; FAO Technical Guidelines for Responsible Fisheries 4, (Suppl. 2); FAO: Rome, Italy, 2008; 112p.
100. Garcia, S.M.; Cochrane, K.L. Ecosystem approach to fisheries: A review of implementation guidelines. *ICES J. Mar. Sci.* **2005**, *62*, 311–318. [CrossRef]
101. Link, J.S.; Browman, H.I. Integrating what? Levels of marine ecosystem-based assessment and management. *ICES J. Mar. Sci.* **2005**, *71*, 1170–1173. [CrossRef]

Journal of
Marine Science and Engineering

Article

It Is Not Just a Matter of Noise: *Sciaena umbra* Vocalizes More in the Busiest Areas of the Venice Tidal Inlets

Marta Picciulin [1,*], Chiara Facca [1], Riccardo Fiorin [2], Federico Riccato [2], Matteo Zucchetta [3] and Stefano Malavasi [1]

[1] Department of Environmental Sciences, Informatics and Statistics, Ca' Foscari, University of Venice, via Torino 155, Mestre, 30172 Venice, Italy; facca@unive.it (C.F.); mala@unive.it (S.M.)
[2] Laguna Project S.N.C., Castello 6411, 30122 Venice, Italy; riccardo.fiorin@lagunaproject.it (R.F.); federico.riccato@lagunaproject.it (F.R.)
[3] Institute of Polar Sciences, ISP-CNR, via Torino 155, Mestre, 30172 Venice, Italy; matteo.zucchetta@cnr.it
* Correspondence: marta.picciulin@unive.it

Abstract: Boat noise is known to have a detrimental effect on a vulnerable Mediterranean sciaenid, the brown meagre *Sciaena umbra*. During summer 2019, two acoustic surveys were conducted at 40 listening points distributed within the inlet areas of Venice (northern Adriatic Sea). Two five-minute recordings were collected per each point during both the boat traffic hours and the peak of the species' vocal activity with the aims of (1) characterizing the local noise levels and (2) evaluating the fish spatial distribution by means of its sounds. High underwater broadband noise levels were found (sound pressure levels $(SPLs)_{50-20kHz}$ 107–137 dB re 1 μPa). Interestingly, a significantly higher background noise within the species' hearing sensibility (100–3150 Hz) was highlighted in the afternoon (113 ± 5 dB re 1 μPa) compared to the night (103 ± 7 dB re 1 μPa) recordings due to a high vessel traffic. A cluster analysis based on *Sciaena umbra* vocalizations separated the listening points in three groups: highly vocal groups experienced higher vessel presence and higher afternoon noise levels compared to the lower ones. Since the species' sounds are a proxy of spawning events, this suggests that the reproductive activity was placed in the noisier part of the inlets.

Keywords: coastal areas; fish; anthropogenic noise; passive acoustic monitoring; protected species; reproduction

Citation: Picciulin, M.; Facca, C.; Fiorin, R.; Riccato, F.; Zucchetta, M.; Malavasi, S. It Is Not Just a Matter of Noise: *Sciaena umbra* Vocalizes More in the Busiest Areas of the Venice Tidal Inlets. *J. Mar. Sci. Eng.* **2021**, *9*, 237. https://doi.org/10.3390/jmse9020237

Academic Editors: Michel André, Christine Erbe and Philippe Blondel

Received: 26 December 2020
Accepted: 17 February 2021
Published: 23 February 2021

Publisher's Note: MDPI stays neutral with regard to jurisdictional claims in published maps and institutional affiliations.

1. Introduction

Many human activities generate sounds in the aquatic environment that are very different from those arising from natural sources both at the intensity and frequency levels; as a result, man-made noise has changed the acoustic underwater landscape of many areas, and it has become a pollutant of international concern, given its potential to harm marine fish [1].

Living in a very noisy environmental condition represents a constraint for aquatic species. It has been widely recognized that anthropogenic noise can threaten animals at both the physiological and behavioral levels, increasing the hearing thresholds and stress hormones and impacting their foraging and anti-predatory ability and reproductive success, with potential consequences in terms of survival and fitness (reviewed in [2–4]). In marine ecosystems, commercial shipping and recreational boating are common sources of anthropogenic noise, and noise from vessel traffic along coastal areas is a widespread stressor [5] to which animals have to cope. Vessel noise was demonstrated to affect fish [6] by inducing changes in fish swimming, brooding, and anti-predator behaviors in both laboratory and field environments, as well as impacting fish social communication interfering with the receiver's ability to hear the signal's original content.

Continuous and chronic disturbance from boat noise is typically associated with marinas, boat channels, and harbor entrances. This is also expected to be the case of the inlets that allow for shipping traffic in and out of the Venice Lagoon; Venice is one of the principal ports of the northern Adriatic Sea, with a number of about 3500 port calls for commercial vessels and cruise ships. The large number of fishing boats and motor- and speed-boats operating along the inlets also gives a large contribution to the local anthropogenic noise levels, particularly during the summer period [7].

On the other side, the Venice inlets, which are constituted by piers made by artificial 3-D structures, represent a standardized homogeneous habitat that has the potential to attract and aggregate the local pelagic and benthic fauna in accordance with other artificial structures in coastal areas [8]. In more detail, since they are constituted by rocky reefs with holes and shelters close to the soft substrates that act as feeding grounds, the inlets resemble the typical reproductive habitat of a small-sized sciaenid occurring along most of the Mediterranean coast, the brown meagre *Sciaena umbra* [9,10]. A preliminary survey confirmed the local presence of this species, with a variable number of heterogeneously distributed individuals [11].

The brown meagre is a slow-growing species that can live for up to approximately 30 years and exceed 60 cm in total length [12,13]. In the Mediterranean, it is frequently targeted by spear-fishers and caught by the coastal commercial fleet, leading to a change in the abundance and composition of its populations [14,15]. As a result, it is listed in the Annex III (Protected Fauna Species) of the Barcelona Conventions and classified as a vulnerable species by the International Union for Conservation of Nature (IUCN), although a slow recovery has been recently reported [16].

Sciaena umbra is sedentary with a limited capability for adult dispersal, particularly during the autumn and winter months [14,17]. During the summer season, daytime site fidelity is corroborated by underwater observations [18,19], whereas there is a general lack of data on its nocturnal behavior; it is known that *S. umbra* feeds actively on crustaceans during the night [10,14], but the spatial extent of its nocturnal movements is unknown. Feeding has been proved to be mainly focused during the spring gonad maturation [17,20].

Sciaena umbra reproduces from late spring to autumn [20,21]. It emits drumming sounds as part of its reproductive process [22–24], whose acoustic features are consistent in space and time [25]. As a consequence, the species can be acoustically identified at sea by mean of its vocalizations. *S. umbra* vocalizations consist of low-frequency pulsed sounds with main energy below 1 kHz (mean dominant frequency of 200–300 Hz); they are made of 4–7 pulses, with a pulse period of approximately 70–145 ms and a pulse duration of approximately 16–27 ms [9,11,21,23,25,26]. Recently, a two-year continuous acoustic monitoring at a study site inside the no-take Réserve de Couronne (near Marseille, France) highlighted a strong consistency in the sound production along the reproduction period [21], thus further confirming the site-fidelity for breeding, vocalizing individuals.

Being sedentary, *S. umbra* benefits from protection measures inside marine protected areas (MPAs), where it is usually present at high densities [18,19,27,28]. Chorusing activity produced by spawning aggregations has been recorded within fully protected zones of old MPAs in the northwestern Mediterranean Sea, where *S. umbra* was present in approximately 70–80% of the monitored stations, though a generally lower probability of detecting *S. umbra* calls was found in younger MPAs (approximately 30% of the stations; [21]). Sound production has also been reported in anthropized coastal areas [9,11,21], where the species could be nevertheless affected by the underwater noise produced by vessel traffic. Thus far, *S. umbra* behavior and vocal activity have been proved to be influenced by boat passages by two different studies run inside MPAs [19,29]. MPAs represent an ideal situation to evaluate the impact of potential stressors on relatively pristine animal populations. However, it remains unclear whether fish avoid high-noise areas. To explore this question, the distribution of the protected fish species *Sciaena umbra* was analyzed here in a potentially highly noisy area: the Venice inlets. This research aimed at (i) evaluating the vessel traffic and the received sound pressure level of noise along the three Venice inlets while considering the hearing thresholds of the target species; (ii) monitoring the spatial distribution of *S. umbra* in the inlets by recording its acoustical activity, following a previous established methodology [9,11,21,24]; and (iii) observing and interpreting the fish distribution in relation to the local underwater anthropogenic noise pressure.

2. Materials and Methods

2.1. Data Collection

During summer 2019, two acoustic surveys were conducted at 40 listening points distributed along the three inlets that connect the Venice lagoon to the sea (Figure 1). The total number of listening points (n = 40) were allocated into the three different sea inlets (n = 13 Lido, north-eastern inlet; n = 15 Malamocco, the central inlet; and n = 12 Chioggia south-western inlet; see Figure 1). They were distributed along both the internal and external sides of the inlets, with each at about 300 m apart; this distance was based on the sound source levels reported by [22], assuming a cylindrical spreading loss.

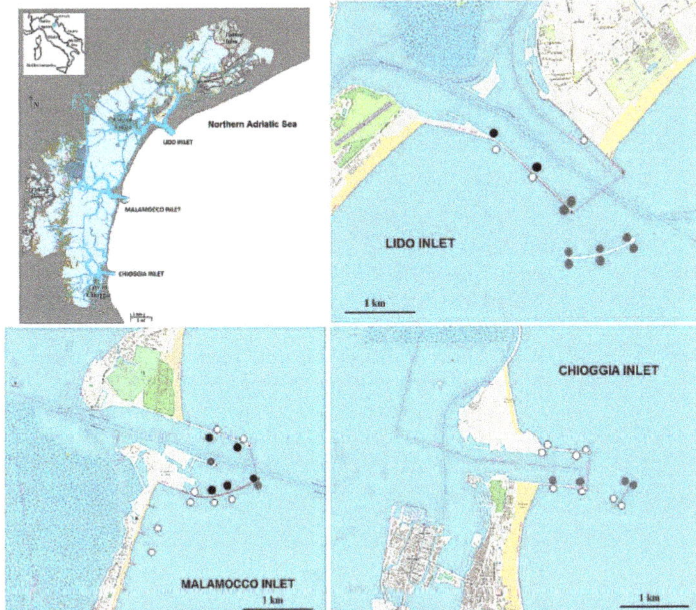

Figure 1. Map showing the 40 listening points in the three inlets (Lido, Malamocco, and Chioggia) explored during the acoustic monitoring. The map also visualizes the stations characterized by higher *Sciaena umbra* vocalization rates with different colors (black dots for group 1 and dark grey dots for group 2), whereas white dots indicate stations characterized by lower vocalization rates (group 3), in accordance with the results presented in this study.

Each listening point was replicated twice, one at the end of July and the other at the end of August (5 and 29 August for the Lido inlet; 29 July and 28 August for the Malamocco inlet; and 1 August and 27 August 2019 for the Chioggia inlet). Within the above-indicated monitoring days, each listening point was also replicated twice per station: one 5-min acoustic sample was collected in the late afternoon, corresponding to one of the local peaks of boat traffic in the summer period (17–19); a second 5-min acoustic sample was collected a few hours later, corresponding to the peak of the species' vocal activity (19.30–23) [23,25]. The earlier acoustic samples were mainly meant to instantaneously evaluate the man-made noise pressure per listening point, whereas the sunset-nocturnal ones locally monitored the presence of the target species by means of its reproductive sounds. *Sciaena umbra* vocalizations consist of pulsed sounds with main energy below 1 kHz and a mean dominant frequency of 200–300 Hz; they are made of 4–7 pulses, with a pulse period of approximately 70–145 ms and a pulse duration of approximately 16–27 ms [23,25].

Recordings were obtained using a pre-amplified Colmar GP1280 hydrophone (sensitivity of −170 dB re 1V/μPa and frequency range of 5–90 kHz) connected to a Tascam Handy Recorder (Tascam Corporation, Santa Fe Springs CA, USA; sampling rate 44.1 kHz, 16 bit) generating Waveform Audio File Format (WAV). Prior to each survey, the signal was calibrated using a generator of pure waves of known voltage. The hydrophone was lowered from a 7.5 m open boat to an average depth of 4 m (range of 2–6 m in depth). Sampling was carried out only in a sea state of less than two on the Douglas scale and a wind speed of less than 10 km/h. Surface water temperature was measured prior to each recording by using a digital thermometer (HANNA Checktemp® 1 HI98509 ± 0.1 °C), resulting in an average of 27.4 °C (range: 26.6–28.5 °C) for the acoustic samples containing *S. umbra* sounds.

A quantification of the traffic in the area was obtained by keeping a log of the vessels visible by eye per each acoustic recording; this was further confirmed by scoring the number of vessel signals that were visually and aurally identifiable for their unique signature in the acoustic files. The same vessel was never included in two different acoustic files because of a minimum of 10 min needing to pass from the end of one recording and the start of the next one.

2.2. Data Analysis

A total of 160 5-min recordings were collected and analyzed minute by minute using the Adobe Audition software by the aural and visual assessment of the spectrograms (sampling rate—FS 44.1 kHz, 16 bit, resampled at 6 kHz, Fast Fourier Transform—FFT = 512, and 50% overlap). This allowed us (i) to discriminate geophony (waves and currents), biophony (i.e., *S. umbra* sounds), and anthropophony (vessels and cargo), as well as (ii) to score their presence/absence per acoustic sample. The Adobe Audition software was also used to quantify both the vessel passages and the *S. umbra* calls.

S. umbra produces low-frequency pulsed sounds with a main energy below 1 kHz that are clearly detectable even in the presence of vessel noise (Figure 2). The pulses were identified and scored by an aural and visual assessment of the recorded file, following [11,26]; the number of pulses present per minute was defined as the pulse rate (PR). The PR was further scaled on a quantitative scale (pulse code: PC) ranging from 0 (no sound) to 5 (maximum pulse rate): 0 = no sound production, 1 = very few sounds (less than 50 pulses min^{-1}), 2 = some sounds (30–50 pulses min^{-1}), 3 = semi-continuous sound production (>50 pulses min^{-1}), 4 = continuous sound production (>100 pulses min^{-1}), and 5 = 'chorus'.

The acoustic samples were analyzed as instantaneous sound pressure level (SPL) by using the software SpectraPlus 5.0 (Pioneer Hill Software, Sequim, WA, USA; Hanning windows, 32768-pt FFT size, 75%FFT overlap, and averaging fast) previously calibrated with a signal of 100 mV RMS at 1 kHz and the hydrophone sensitivity value. This software utilizes the discrete fast Fourier transform algorithm to compute the frequency spectrum among 1/3 octave bands (center frequencies of 50–20,000 Hz) and the broadband SPL per each second. Per each sampling station, a single SPL value was further obtained (i) for the 1/3 octave bands and (ii) along the whole broadband by log-averaging in the dB domain the obtained values over the sample (300 s). Energy levels (hereafter called "ELs") were also calculated for the frequency range of 100–3150 Hz ($EL_{100–3150}$; range: 89–3548 Hz), which was consistent with the species' audiogram, and 200–630 Hz ($EL_{200–630}$; range: 178–708 Hz), which corresponded to the species' best hearing range [30]. Energy levels were calculated by summing the energy of the corresponding 1/3 octave bands by a log-sum in the dB domain.

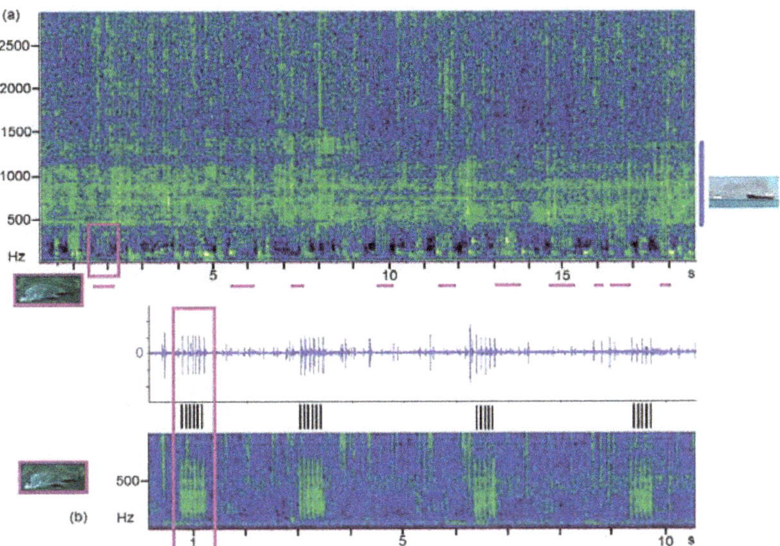

Figure 2. (**a**) Spectrogram (FS = 44.1 kHz, resampled at 6 kH, FFT = 512, and 50% overlap) of an acoustic sample containing both the noise of a distant cargo passing through the Venice inlet (most of the energy was located in the frequency range of 400–1100 Hz; blue line) and the sounds of the target species, *Sciaena umbra*, visible below 400 Hz (indicated by the purple square and lines). (**b**) Spectrogram (FS = 44.1 kHz, resampled at 6 kH, FFT = 512, and 50% overlap) and waveform of four *Sciaena umbra* sounds made by a series of low frequency pulses; the black bars indicate the number of pulses per sound, as scored in accordance with [11,26].

Median one-third octave band levels generated from all the 5-min SPL averages were compared to the *S. umbra* audiogram [30] to estimate which parts of underwater noise spectra might be audible to the species. For a quantitative evaluation, the average difference between the *S. umbra* audiogram and the median values in the sensitive frequency band ($EL_{200-630}$) was calculated for (i) the afternoon (n = 80) and (ii) night (n = 80) recordings, (iii) the night recordings characterized by the *S. umbra* chorus (n= 11) and (iv) the recordings containing more than three boat passages (n = 31); almost all the files containing more than three boat passages were recorded in the afternoon in the absence of fish sounds (see results). As a consequence, their ELs were representative of the anthropic contribution to the local background noise in case of high boating traffic.

In order to group the recording stations of the three inlets according to the local vocal activity of the target species, a hierarchical cluster analysis based on Ward's algorithm was applied to the listening points by using the *S. umbra* PC as a variable; for this analysis, the PC values were standardized as (x − min)/(max − min), with x being the average PC value over the 5-min-file recorded per single station and the min and max being the minimum and maximum average PC recorded along all the monitored stations, respectively. The cluster groups were calculated by using the Ward's minimum variance method using Euclidean distances. Groups were based on an *a priori* level of 70% of similarity.

To compare the collected data, statistical analyses were performed with non-parametric tests, with an alpha level of 0.05: (1) the Kruskal–Wallis Test was used for spatial comparisons between the three inlets or for comparisons of the *S. umbra* vocalization rate or noise levels between the groups of recording point, (2) the Mann–Whitney U Test was used for temporal comparisons between the two surveys (end of July/beginning of August vs. end of August), and (3) the Wilcoxon pair test was used for temporal comparisons between afternoon vs. night collected data per each recording point.

3. Results

3.1. Acoustic Characterisation of the Data Collected in the Venice Inlets

The acoustic data collected in the Venice inlets were characterized by the sounds produced by the waves and the water current (present in 53% of the collected acoustic samples), the irregular transit of different type of vessels (70%) or cargo (5%) contributing to both the high and low frequency bands, and the sounds produced by biological sources such as the snapping shrimps (100%) and Sciaenid fish vocalizations (50% of the collected nocturnal acoustic samples).

From a quantitative point of view, the background noise (broadband) varied from 107 to 137 dB re 1 µPa in the afternoon recordings and between 109 and 133 dB re 1 µPa in the night recordings. The broadband SPLs calculated per each listening point did not differ between these two periods (Wilcoxon signed rank test, $p = 0.38$), suggesting a temporal consistency of the noise levels within this broadband frequency spectrum. Furthermore, no difference was found in the broadband SPLs of samples collected during the first (end of July/beginning of August) vs. the second (end of August) surveys (Mann–Whitney U test: $p = 0.84$ for the afternoon samples and $p = 0.53$ for the nocturnal samples) nor when comparing the average SPLs recorded at the three inlets (Kruskal–Wallis Test: $p = 0.49$ for the afternoon samples and $p = 0.38$ for the nocturnal samples).

The local SPLs were highly influenced by the vessel passages, as clearly shown by Figure 3. A significantly higher presence of vessel passing along the inlets was detectable in the afternoon (2.4 ± 2.2 vessel passages per five-minute sample, corresponding to about 30 vessel passages per hour along the 17–19 period) compared to the night (0.8 ± 0.8 vessel passages per 5-min sample, corresponding to about 10 vessel passages per hour along the 19.30–23; Wilcoxon signed rank test, $p < 0.001$). As a result, 30 out of 31 acoustic files containing more than three boat passages were recorded in the afternoon in the absence of fish sounds.

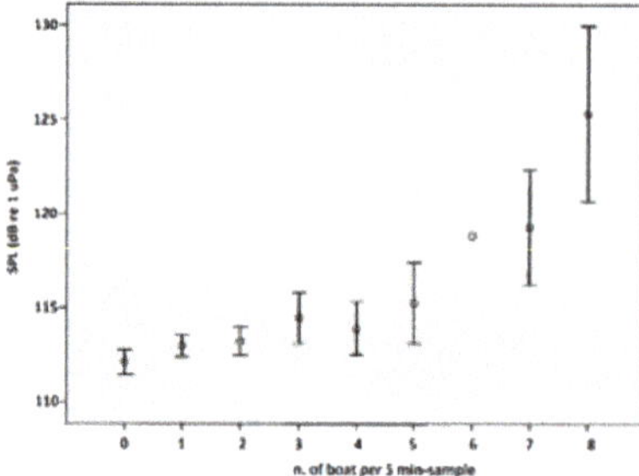

Figure 3. Variation of the mean (±standard error) sound pressure levels (SPLs) of recordings containing a variable number of boat passages.

The background noise, as described by the 1/3 octave band spectrum (Figure 4), showed consistently high SPLs at low frequency, which increased above 2 kHz with a peak around 4 kHz, which was mainly attributable to the activity of snapping shrimps (the *Alpheus* and *Synalpheus* genera). After comparing the recorded levels with the *Sciaena umbra* audiogram (Figure 4), an overlap of the background noise with the species hearing thresholds was evident.

An average $EL_{100-3150}$ value of 113 (±5 SD) dB re 1 µPa and an $EL_{200-630}$ value of 103 (±7 SD) dB re 1 µPa were found along the *S. umbra* hearing frequencies (100–3150 Hz) and the restricted range of 200–630 Hz, respectively. The average difference between the *S. umbra* audiogram and the median $EL_{200-630}$ in the sensitive frequency band (200–630 Hz) was equal to approximately 10 dB re 1 µPa for both the afternoon (10.7 dB re 1 µPa; n = 80) and night (10.3 dB re 1 µPa; n = 80) recordings. This increased to 16.6 dB re 1 µPa for the afternoon recordings containing more than three boat passages (n = 31) and to 14 dB re 1 µPa for the night recordings characterized by the *S. umbra* chorus (n = 11). It has to be noticed that out of eleven files with chorusing activity, only two contained the signals produced by one passing recreational boat.

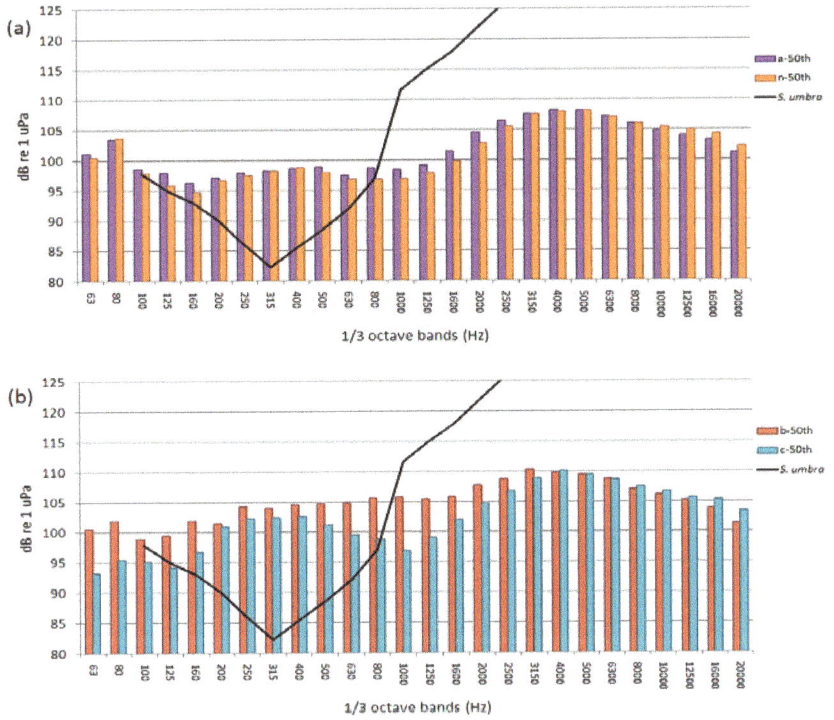

Figure 4. One-third octave band levels (median) related to (**a**) the afternoon (a-50th; 17–19, sample size of n = 80) and night (n-50th; 19.30–23, sample size of N = 80) recordings; (**b**) the afternoon recordings containing three or more boat (b-50th; sample size of N = 31) passages and the nocturnal *Sciaena umbra* chorus (c-50th; sample size of n = 11). The levels are compared to the *Sciaena umbra* audiogram; noise above the audiogram lines is expected to be audible.

The background noise for each recording point was slightly higher in the afternoon than in the night recordings when considering the 100–3150 Hz ($EL_{100-3150}$ = 114 ± 5 dB re 1 µPa vs. $EL_{100-3150}$ = 112 ± 4 dB re 1 µPa; Wilcoxon signed rank test, p = 0.027) but not the 200–630 Hz frequency ranges ($EL_{200-630}$ = 104 ± 7 vs. $EL_{200-630}$ = 103 ± 7 dB re 1 µPa, respectively; Wilcoxon signed rank test, p = 0.131). In the 200–630 Hz range, however, the *S. umbra* nocturnal chorus was responsible for increased values, like the ones resulting from the boat noise contributions observed in the afternoon (Figure 4) [31].

The afternoon SPL levels were similar across the three inlets for both the 100–3150 and 200–630 Hz frequency ranges (Kruskal–Wallis test: $EL_{100-3150}$, p = 0.07 and $EL_{200-630}$, p = 0.25, respectively), indicating a spatial homogeneity in accord with the case of the broadband SPLs.

3.2. Sciaena umbra Vocalizations

Sciaena umbra vocalizations were recorded at sea only after the sunset. About half of the collected nocturnal samples included these sounds (39 out of a total of 80). During the first and second surveys, *S. umbra* sounds were found in 21 and 18 out of 40 listening points (i.e., in 52% and 42% of points), respectively. The species' PR (i.e., the number of pulses per minute) ranged between 0 and 350 pulse min^{-1} (mean: 68 pulse min^{-1} ± 125 SD), not being consistent along the monitored area. In detail, the mean PR varied across the three tested inlets (Kruskal–Wallis test, $p = 0.04$), with a mean of 100 pulse min^{-1} ($\pm$21 SD) and 113 pulse min^{-1} ($\pm$23 SD) at the Lido and Malamocco inlets, respectively, whereas a mean of only 11 pulse min^{-1} ($\pm$4 SD) was recorded at the Chioggia inlet. Boat noises were present in about half of the samples containing *S. umbra* vocalizations (19 out of a total of 39; in six cases, one of the noise sources was a cargo ship).

A cluster analysis based on the *S. umbra* pulse code (PC) created a total of 14 possible groupings, one for each node at descending Euclidean distances (Figure 5). The grouping produced at a distance of 0.3 have three distinct assemblages that clearly differed for the species vocalizations activity: group "1" included seven points characterized by a higher vocalization rate (PC = 4.3 $\pm$ 0.3, i.e., characterized by a continuous sound production of >100 pulses min^{-1} and/or the chorus, following [11]) distributed mainly in the internal side of the Lido and Malamocco inlets (see also Figure 1) whereas group "3" contained nineteen points with a very low vocalization rate (PC = 0.1 $\pm$ 0.08, i.e., characterized by less than 30 pulses min^{-1}, following [11]), mostly facing the external sea side of the inlets and/or located in the Chioggia inlet. Group "2" consisted of an intermediate vocal activity of the target species (PC = 1.5 $\pm$ 0.3, i.e., characterized by 30–50 pulses min^{-1}, following [11]).

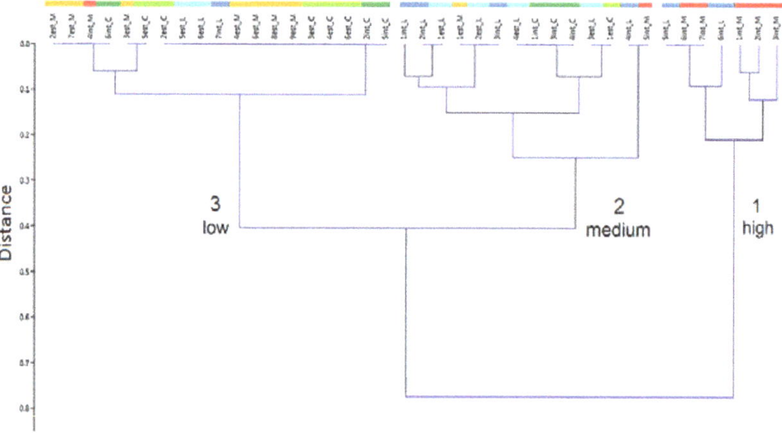

Figure 5. Tree diagram provided by cluster analysis using the pulse code to visualize which listening points were more similar in the Venice inlets; colored lines refer to points located at Malamocco internal side (red), Malamocco external sea side (orange), Lido internal side (blue), Lido external sea side (light blue), Chioggia internal side (green), and Chioggia external sea side (light green). The positions of the points in the inlets are highlighted in Figure 1 (black dots are used for group coded "1," dark grey dots are used for group coded "2," and white dots are used for group coded "3").

The three groups of locations produced by the cluster analysis were characterized by a slightly but significantly different background noise levels along the afternoon recordings (Kruskal–Wallis test, broadband $p = 0.001$; EL$_{100–3150}$ $p = 0.018$, EL$_{200–630}$ $p = 0.017$), with group 1 and 2 having higher values compared to group 3 for all considered frequency ranges. Figure 6a shows the case of EL$_{200–630}$. Consistently, groups 1 and 2 were characterized by a higher number of vessel passages compared to group 3 (Kruskal–Wallis test, $p = 0.021$; Figure 6b).

On the contrary, neither the SPLs and ELs (Kruskal–Wallis test, $p = 0.1$, 0.4, and 0.5 for the broadband, EL$_{100–3150}$, and EL$_{200–630}$, respectively) nor the number of vessel passages (Kruskal–Wallis test, $p = 0.602$) varied between the three groups of locations during the night recordings. Though not significant, a slightly higher value was evident while comparing the average EL$_{200–630}$ of group 1 vs. group 3 (Figure 7a). This was likely due to the contribution of the *S. umbra* vocalizations, given the overall low number of vessel passages after sunset (Figure 7b).

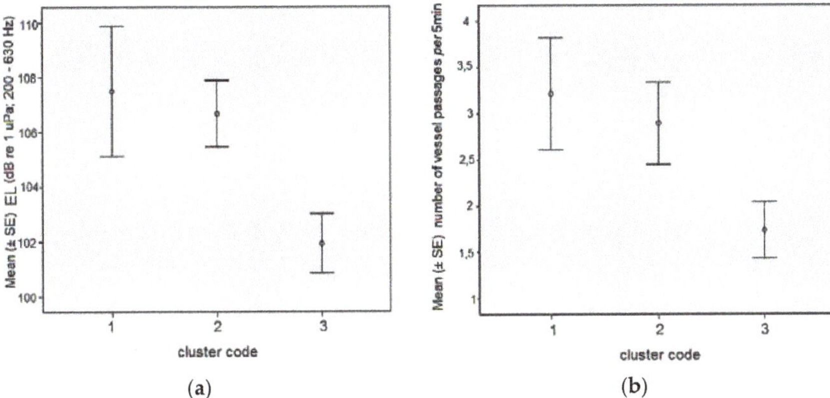

Figure 6. Mean ($\pm$standard error) (**a**) noise levels (energy level (EL)$_{200-630}$, dB re 1 μPa) and (**b**) number of vessel passages at the listening points belonging to the three groups highlighted by cluster analysis for the afternoon recordings.

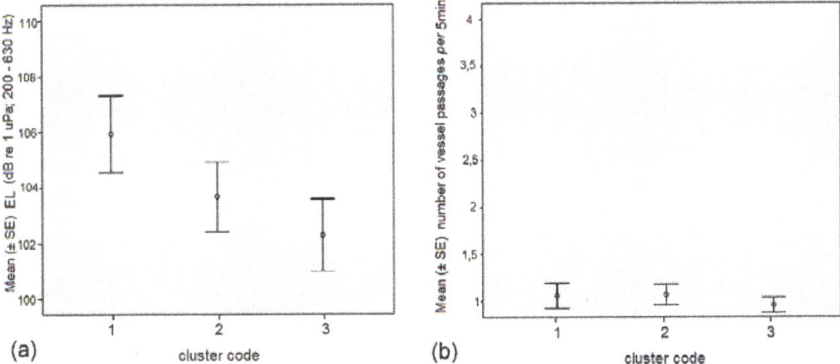

Figure 7. Mean ($\pm$standard error) (**a**) noise levels (EL$_{200-630}$, dB re 1 μPa) and (**b**) number of vessel passages at the listening points belonging to the three groups highlighted by cluster analysis for the night recordings.

4. Discussion

The Venice lagoon, located in the northern part of the Adriatic Sea, is one of the largest lagoons in Europe. Its port is characterized by a double function: a passenger port in the lagoon city (hosting two million passengers and providing services to 200 mega-yachts) and a commercial port on the mainland. It is therefore not surprising that such vessel traffic influences the underwater acoustics in the inlet areas, resulting in high underwater noise levels (overall recorded SPLs$_{50Hz-20kHz}$ ranging from 107 to 137 dB re 1 μ), with maximum values similar to those reported in other Italian and Portuguese areas (i.e., Gulf of Trieste: SPLs$_{50Hz-20kHz}$ 76–141 dB re 1 μPa; Gulf of Naples: SPLs$_{16Hz-40kHzH}$ 108–140 dB re 1 μPa; and Port of Civitavecchia: SPLs$_{12.5-16\ kHz}$ 45–158 dB re 1 μPa) [32–34]. Noise levels at the low-frequencies are mainly produced by vessel engines, electrical machinery, and propeller cavitation (<3 kHz [35,36]). They were found to be slightly, but significantly, higher during the afternoon than night-recordings, mirroring the number of operating vessels in the inlets. A similar pattern was recently found by the authors of [37] in a large Portuguese coastal lagoon: during summer, the underwater noise tended to remain sustained from 7 to 17 at levels that approached or exceeded 120 dB re 1 μPa, with a smooth and progressive variation at dawn and evening as a result of the reduced boat traffic nearby.

The present study could not fully define a continuous and extremely variable phenomenon such as marine background noise by using a non-continuous monitoring approach; nevertheless, the acquired data are indicative of an environment saturated by the anthropic presence at the lower frequencies (<2–3 kHz) during the afternoon period, whereas the vessel presence was found to be reduced at night. Vessels generated acoustic signals that were unpredictable in intensity, in a context of a persistent low intensity background. According to the present paper, the background noises were at similar levels to the *Sciaena umbra* audiogram. Further, if vessel activity expanded into evening hours, then the associated acoustic input associated to boating had the potential to exceed the *S. umbra* vocalizations even during chorusing activity. This likely led to a reduction in the detectability of signals of interest. This effect was supported by a laboratory study, where the playback of one single recreational boat noise (mean ambient noise—LLeq$_{1\ min}$ 132 dB, with a maximum instantaneous SPL of 138 dB re 1 μPa) induced an upward shift in the *S. umbra* auditory threshold by about 25 dB and a reduction of the species acoustic communication from 500 m under ambient noise to only about 1 m under the boat noise conditions [22].

Boat noise is known to affect the efficiency of fitness-related fish behaviors such as foraging and antipredator behaviors, risk assessment, nest-defense, and parental care [38–44]. Fish responses, however, depend on many variables, including boat and engine types, boat speed, distance from noise source, motivational and physiological fish state, and the social context [41,43,45]. Therefore, there can be a continuum of responses to disturbance in in situ conditions ranging from mild to more severe forms.

In the context of conservation, the critical factor is whether a disturbance results in lower population sizes, especially in case of an already vulnerable species, such as the brown meagre. Usually, these assessments rely on proximate measures as the behavioral responses to a stimulus. In this context, the in situ exposition to boat noises (the mean SPL ranged from 134 to 146 dB re 1 μPa) did not cause displacement or elicit any significant activity changes in *S. umbra* groups (for a total of 65 tested brown meagres) living in an Italian MPA besides a reduction in the duration of active swimming [19]. An individual variability in response was found by the authors: on average, one third of exposed fish in a group reacted to noise by flighting and hiding. They resumed behavior quickly after exposures.

Conversely, visual-based methods could not be applied to the Venice inlets due to the risks connected with high boating and cargo traffic. As a consequence, it was not possible to investigate the *S. umbra* behaviors in relation to the local traffic. Since *S. umbra* is a vocal fish, the passive acoustic method (PAM) was the best option for the species-specific recognition of its presence and distribution in the inlets.

Noise effects could also be evaluated by relating the presence of animals to varying rates of disturbance across a number of sites [46]. Along the Venice inlets, brown meagre vocalizations were found in about 50% of the investigated points; this was an intermediate value compared to the case of the old and younger MPAs investigated in the northwestern Mediterranean Sea, where *S. umbra* was found to be present in approximately 70–80% and 30% of the monitored stations, respectively [21]. Such a result is somehow surprising: since animals select a location based upon its perceived quality, areas degraded by anthropogenic (noise) disturbance are expected to be poorly occupied.

Following [19], brown meagre groups were reported to remain on site year by year despite high boat traffic (18.7 boat passages h^{-1} during the tourist season), showing a higher abundance than other surrounded less anthropized sites. The present study also indicated that the brown meagre did not avoid the Venice inlets despite a vessel traffic corresponding to an estimation of 10–30 boat passages h^{-1} on average due to motorboats, cargos, or cruises. Interestingly, most of the listening points characterized by a high *S. umbra* pulse rate (groups 1 and 2) were located in the internal side of the inlets. These areas represent the only water connections between the inner lagoon and the sea and are therefore characterized by a high boating traffic. Given the sedentary and site-related *S. umbra* attitude [14,18,19], the nocturnal distribution is expected to mirror the diurnal distribution.

Furthermore, only the busiest locations in the internal side of the inlets were characterized by chorusing (group 1). Since the latter is a reliable natural indicator of the *S. umbra* breeding sites [24], we conclude that the brown meagre does reproduce in these sites, despite the relatively high anthropogenic noise levels experienced mostly but not exclusively during the diurnal hours. This result was also unexpected if we consider that a reduction in the ability to detect conspecific signals due to boat noise [22,47] could potentially affect the courtship efficacy. It has to be stressed, however, that in the inlets, the vessel traffic was reduced during the sunset-nocturnal hours, when the brown meagre vocalized. In its turn, this could result in a lower masking effect on the species calls, thus diminishing the costs expected for the animals living in the area.

The animals could be following the best-of-a-bad-job strategy: if the resources found in the Venice inlets are unique in the local coastal area, the fish will not leave them. Exposed fish could increase tolerance by a declining response from learning that the stimulus does not have any detrimental consequences or through shifts in hearing sensitivity thresholds. Behavioral and physiological attenuation have been found in fish after the repeated playback of the same motorboat-noise [48]; accordingly, the responses to motorboat noise in wild endemic cichlids in Lake Malawi were lower in areas with higher levels of motorboat disturbance [49]. One way to potentially assess this effect would be to generate and compare audiograms of the fish living in the three groups of locations in the inlets.

The decision of whether or not to stay in disturbed areas is determined not only by the quality of the site but also by the distance to and quality of other suitable sites and/or their relative risk of predation, the availability of prey, the density of competitors, and the investment that an individual has made for establishing a territory, gaining dominance status, and so on [46]. The Venice inlets are a structurally homogeneous area made by artificial standard blocks that are matched by distance to the shore. Theoretically, there are plenty of suitable sites for *S. umbra* along the inlets where noise disturbance is minimal, such as the locations facing the open sea. Therefore, these sites were expected to be more exploited by the species. As this was not the case, other factors (current, water depth, salinity, bottom composition, and so on) seem to be crucial for the species, "forcing" it to remain in some areas regardless of whether or not noise represents a disturbance. A deep assessment of the habitat characteristics of those areas, where the vocal production and likely the reproductive activity was more intense, should be carried out in future studies. In other words, we need to address the structure of the reproductive habitat of the brown meagre in this highly anthropic ecological context. Such information could help to evaluate the suitability of surrounding locations for the reproduction of the target species. It could also be used to evaluate the real extent of disturbance on the local fish population: if there are relevant, less disturbed, but not exploited areas in the surroundings, we can conclude that *S. umbra* is not strongly affected by boat noise. On the contrary, the species is likely to be impacted by the anthropic pressure because it is constrained to stay and to tolerate the costs of disturbance [46]. Data are currently being acquired to confirm the observed spatial distribution of *S. umbra* in the Venice inlets and to evaluate the role of environmental factors leading the habitat selection of this species.

Author Contributions: M.P., S.M. and C.F. conceived and designed the study and data collection; M.P., R.F. and F.R. collected the data; M.P. analyzed the data; M.Z. and C.F. contributed equipment and analytical tools and software; M.P. wrote the original draft and all authors contributed to editing and final reviews of the paper. All authors have read and agreed to the published version of the manuscript.

Funding: Scientific activity performed with the contribution of the Provveditorato for the Public Works of Veneto, Trentino Alto Adige and Friuli Venezia Giulia, provided through the concessionary of State Consorzio Venezia Nuova and coordinated by CORILA.

Institutional Review Board Statement: Ethical review and approval were waived for this study, due to the use of non-invasive sampling methods. Animals were not collected, and data were achieved by means of underwater records of sounds that are naturally produced. Human presence for investigation purpose was comparable with the usual anthropogenic pressure in the areas.

Informed Consent Statement: Not applicable.

Data Availability Statement: The data presented in this study are available on request from the corresponding author. The data are not publicly available because the funding project is not ended. Data will be available at project end, in 2022.

Acknowledgments: We are very grateful to three anonymous referees for the fruitful and the critical revision of this manuscript.

Conflicts of Interest: The authors declare no conflict of interest. The funders had no role in the design of the study; in the collection, analyses, or interpretation of data; in the writing of the manuscript, or in the decision to publish the results.

References

1. Slabbekoorn, H.; Bouton, N.; van Opzeeland, I.; Coers, A.; ten Cate, C.; Popper, A.N. A noisy spring: The impact of globally rising underwater sound levels on fish. *Trends Ecol. Evol.* **2010**, *25*, 419–427. [CrossRef] [PubMed]
2. Radford, A.N.; Kerridge, E.; Simpson, S.D. Acoustic communication in a noisy world: Can fish compete with anthropogenic noise? *Behav. Ecol.* **2014**, *25*, 1022–1030. [CrossRef]
3. Kunc, H.P.; McLaughlin, K.E.; Schmidt, R. Aquatic noise pollution: Implications for individuals, populations, and ecosystems. *Proc. R. Soc. B* **2016**, *283*, 20160839. [CrossRef] [PubMed]
4. Cox, K.; Brennan, L.P.; Gerwing, T.G.; Dudas, S.E.; Juanes, F. Sound the alarm: A meta-analysis on the effect of aquatic noise on fish behavior and physiology. *Glob. Chang. Biol.* **2018**, *24*, 3105–3116. [CrossRef] [PubMed]
5. Nichols, T.A.; Anderson, T.W.; Širović, A. Intermittent Noise Induces Physiological Stress in a Coastal Marine Fish. *PLoS ONE* **2015**, *9*, e0139157. [CrossRef]
6. Di Franco, E.; Pierson, P.; Di Iorio, L.; Calò, A.; Cottalorda, J.M.; Derijard, B.; Di Franco, A.; Galvé, A.; Guibbolini, M.; Lebrun, J.; et al. Effects of marine noise pollution on Mediterranean fishes and invertebrates: A review. *Mar. Pollut. Bull.* **2020**, *159*, 111450. [CrossRef] [PubMed]
7. Bolgan, M.; Picciulin, M.; Codarin, A.; Fiorin, R.; Zucchetta, M.; Malavasi, S. Is the Venice Lagoon Noisy? First Passive Listening Monitoring of the Venice Lagoon: Possible Effects on the Typical Fish Community. In *Effects of Noise on Aquatic Life II*; Popper, A.N., Hawkins, A.D., Eds.; Springer: New York, NY, USA, 2016; Volume 875, pp. 83–90. [CrossRef]
8. Macura, B.; Byström, P.; Airoldi, L.; Eriksson, B.K.; Rudstam, L.; Støttrup, J. Impact of structural habitat modifications in coastal temperate systems on fish recruitment: A systematic review. *Environ. Evid.* **2019**, *8*, 1–22. [CrossRef]
9. Bonacito, C.; Costantini, M.; Picciulin, M.; Ferrero, E.A.; Hawkins, A.D. Passive hydrophone census of *Sciaena umbra* (Sciaenidae) in the Gulf of Trieste (Northern Adriatic Sea, Italy). *Bioacoustics* **2002**, *12*, 292–294. [CrossRef]
10. Fabi, G.; Panfili, M.; Spagnolo, A. Note on feeding of *Sciaena umbra* L. (Osteichthyes: Sciaenidae) in the central Adriatic sea. *Rapp. Comm. Int. Mer Médit.* **1998**, *35*, 426–427.
11. Picciulin, M.; Bolgan, M.; Codarin, A.; Fiorin, R.; Zucchetta, M.; Malavasi, S. Passive acoustic monitoring of *Sciaena umbra* on rocky habitats in the Venetian littoral zone. *Fish. Res.* **2013**, *145*, 76–81. [CrossRef]
12. Chater, I.; Romdhani-Dhahri, A.; Dufour, J.L.; Mahé, K.; Chakroun-Marzouk, N. Age, growth and mortality of Sciaena umbra (Sciaenidae) in the Gulf of Tunis. *Sci. Mar.* **2018**, *82*, 17–25. [CrossRef]
13. Aydın, M.; Bodur, B. Morphologic characteristics and length-weight relationships of *Sciaena umbra* (Linnaeus, 1758) in the Black Sea coast. *Mar. Sci. Technol. Bull.* **2021**, *10*, 8–15. [CrossRef]
14. La Mesa, M.; Coltella, S.; Riannetti, G.; Arnesi, E. Age and growth of brown meagre *Sciaena umbra* (Sciaenidae) in the Adriatic Sea. *Aquat. Living Resour.* **2008**, *21*, 153–161. [CrossRef]
15. Lloret, J.; Zaragoza, N.; Caballero, D.; Font, T.; Casadevall, M.; Riera, V. Spearfishing pressure on fish communities in rocky coastal habitats in a Mediterranean marine protected area. *Fish. Res.* **2008**, *98*, 84–91. [CrossRef]
16. Garcia-Rubies, A.; Hereu, B.; Zabalà, M. Long-term recovery patterns and limited spillover of large predatory fish in a Mediterranean MPA. *PLoS ONE* **2013**, *8*, e73922. [CrossRef]
17. Alos, J.; Cabanellas-Reboredo, M. Experimentalacoustic telemetry experiment reveals strong site fidelity during the sexual resting period of wild brown meagre, *Sciaena umbra*. *J. Appl. Ichthyol.* **2012**, *28*, 606–611. [CrossRef]
18. Harmelin-Vivien, M.; Cottalorda, M.; Dominici, J.M.; Harmelin, J.G.; Le Direach, L.; Ruitton, S. Effects of reserve protection level on the vulnerable fish species *Sciaena umbra* and implications for fishing management and policy. *Glob. Ecol. Conserv.* **2015**, *3*, 279–287. [CrossRef]
19. La Manna, G.; Manghi, M.; Perretti, F.; Sarà, G. Behavioural response of brown meagre (*Sciaena umbra*) to boat noise. *Mar. Pollut. Bull.* **2016**, *110*, 324–334. [CrossRef] [PubMed]
20. Grau, A.; Linde, M.; Grau, A.M. Reproductive biology of the vulnerable species *Sciaena umbra* Linnaeus, 1758 (Pisces: Sciaenidae). *Sci. Mar.* **2009**, *73*, 67–81. [CrossRef]

21. Di Iorio, L.; Bonhomme, P.; Michez, N.; Ferrari, B.; Gigou, A.; Panzalis, P.; Desiderà, E.; Navone, A.; Boissery, P.; Lossent, J.; et al. Spatio-temporal surveys of the brown meagre *Sciaena umbra* using passive acoustics for management and conservation. *bioRxiv* **2020**. [CrossRef]

22. Codarin, A.; Wysocki, L.E.; Ladich, F.; Picciulin, M. Effects of ambient and boat noise in hearing and communication in three fishes living in a marine protected area (Miramare, Italy). *Bull. Mar. Poll.* **2009**, *58*, 1880–1887. [CrossRef] [PubMed]

23. Picciulin, M.; Calcagno, G.; Sebastianutto, L.; Bonacito, C.; Codarin, A.; Costantini, M.; Ferrero, E.A. Diagnostics of nocturnal calls of *Sciaena umbra* (L., fam.,Sciaenidae) in a nearshore Mediterranean marine riserve. *Bioacoustics* **2012**, *22*, 109–120. [CrossRef]

24. Picciulin, M.; Fiorin, R.; Facca, C.; Malavasi, S. Sound features and vocal rhythms as a proxy for locating the spawning ground of *Sciaena umbra* in the wild. *Aquat. Conserv. Mar. Freshw. Ecosyst.* **2020**. [CrossRef]

25. Parmentier, E.; Di Iorio, L.; Picciulin, M.; Malavasi, S.; Lagardère, J.P.; Bertucci, F. Consistency of spatiotemporal sound features supports the use of passive acoustics for long-term monitoring. *Anim. Conserv.* **2018**, *21*, 211–220. [CrossRef]

26. Colla, S.; Pranovi, F.; Fiorin, R.; Malavasi, S.; Picciulin, M. Using passive acoustics to assess habitat selection by the brown meagre *Sciaena umbra* in a northern Adriatic Sea mussel farm. *J. Fish Biol.* **2018**, *92*, 1627–1634. [CrossRef] [PubMed]

27. Di Franco, A.; Bussotti, S.; Navone, A.; Panzalis, P.; Guidetti, P. Evaluating effects of total and partial restrictions to fishing on Mediterranean rocky-reef fish assemblages. *Mar. Ecol. Progr. Ser.* **2009**, *387*, 275–285. [CrossRef]

28. Guidetti, P.; Baiata, P.; Ballesteros, E.; Di Franco, A.; Hereu, B.; Macpherson, E.; Micheli, F.; Pais, A.; Panzalis, P.; Rosenberg, A.A.; et al. Large- scale assessment of Mediterranean Marine Protected Areas effects on fish assemblages. *PLoS ONE* **2014**, *9*, e91841. [CrossRef] [PubMed]

29. Picciulin, M.; Sebastianutto, L.; Codarin, A.; Calcagno, G.; Ferrero, E.A. Brown meagre vocalization rate increases during repetitive boat noise exposures: A possible case of vocal compensation. *J. Acoust. Soc. Am.* **2012**, *132*, 3118–3124. [CrossRef] [PubMed]

30. Wysocki, L.E.; Codarin, A.; Ladich, F.; Picciulin, M. Sound pressure and particle acceleration audiograms in three marine fish species from the Adriatic Sea. *J. Acoust. Soc. Am.* **2009**, *126*, 2100–2107. [CrossRef]

31. Picciulin, M.; Codarin, A.; Spoto, M. Characterization of small-boat noises compared with the chorus of *Sciaena umbra* (Sciaenidae). *Bioacoustics* **2008**, *17*, 210–212. [CrossRef]

32. Codarin, A.; Picciulin, M. Underwater noise assessment in the Gulf of Trieste (North Adriatic Sea, Italy) using and MSFD approach. *Mar. Pollut. Bull.* **2015**, *101*, 694–700. [CrossRef] [PubMed]

33. Pieretti, N.; Lo Martire, M.; Corinaldesi, C.; Musco, L.; Dell'Anno, A.; Danovaro, R. Anthropogenic noise and biological sounds in a heavily industrialized coastal area (Gulf of Naples, Mediterranean Sea). *Mar. Environ. Res.* **2020**, *159*, 105002. [CrossRef] [PubMed]

34. Cafaro, V.; Piazzolla, D.; Melchiorri, C.; Burgio, C.; Fersini, G.; Conversano, F.; Piermattei, V.; Marcelli, M. Underwater noise assessment outside harbor areas: The case of Port of Civitavecchia, northern Tyrrhenian Sea, Italy. *Mar. Pollut. Bull.* **2018**, *133*, 865–871. [CrossRef] [PubMed]

35. Arveson, P.; Vendittis, D. Radiated noise characteristics of a modern cargo ship. *J. Acoust. Soc. Am.* **2000**, *107*, 118–129. [CrossRef]

36. Hildebrand, J.A. Anthropogenic and natural sources of ambient noise in the ocean. *Mar. Ecol. Prog. Ser.* **2009**, *395*, 5–20. [CrossRef]

37. Soares, C.; Pacheco, A.; Zabel, F.; González-Goberña, E.; Sequeira, C. Baseline assessment of underwater noise in the Ria Formosa. *Mar. Pollut. Bull.* **2019**, *150*, 110731. [CrossRef]

38. Picciulin, M.; Sebastianutto, L.; Codarin, A.; Farina, A.; Ferrero, E.A. In situ behavioural responses to boat noise exposure of *Gobius cruentatus* (Gmelin, 1789; fam. Gobiidae) and *Chromis chromis* (Linnaeus, 1758; fam. Pomacentridae) living in a Marine Protected Area. *J. Exp. Mar. Biol. Ecol.* **2010**, *386*, 125–132. [CrossRef]

39. Voellmy, I.K.; Purser, J.; Flynn, D.; Kennedy, P.; Simpson, S.D.; Radford, A.N. Acoustic noise reduces foraging success in two sympatric fish species via different mechanisms. *Anim. Behav.* **2014**, *89*, 191–198. [CrossRef]

40. Simpson, S.D.; Purser, J.; Radford, A.N. Anthropogenic noise compromises antipredator behaviour in European eels. *Glob. Chang. Biol.* **2015**, *21*, 586–593. [CrossRef] [PubMed]

41. Magnhagen, C.; Johansson, K.; Sigray, P. Effects of motorboat noise on foraging behaviour in Eurasian perch and roach: A field experiment. *Mar. Ecol. Prog. Ser.* **2017**, *564*, 115–125. [CrossRef]

42. Nedelec, S.L.; Radford, A.N.; Pearl, L.; Nedelec, B.; McCormick, M.I.; Meekan, M.G.; Simpson, S.D. Motorboat noise impacts parental behaviour and offspring survival in a reef fish. *Proc. R. Soc. B Biol. Sci.* **2017**, *284*, 20170143. [CrossRef] [PubMed]

43. McCormick, M.I.; Allan, B.J.; Harding, H.; Simpson, S.D. Boat noise impacts risk assessment in a coral reef fish but effects depend on engine type. *Sci. Rep.* **2018**, *8*, 3847. [CrossRef]

44. de Jong, K.; Amorim, M.C.P.; Fonseca, P.J.; Heubel, K.U. Noise affects multimodal communication during courtship in a marine fish. *Front. Ecol. Evol.* **2018**, *6*, 113. [CrossRef]

45. Bruintjes, R.; Radford, A.N. Context-dependent impacts of anthropogenic noise on individual and social behaviour in a cooperatively breeding fish. *Anim. Behav.* **2013**, *85*, 1343–1349. [CrossRef]

46. Gill, J.A.; Sutherland, W.J.; Watkinson, A.R. A method to quantify the effects of human disturbance on animal populations. *J. Appl. Ecol.* **1996**, *33*, 786–792. [CrossRef]

47. Sprague, M.S.; Luczkovich, J.J. Measurement of an individual silver perch *Bairdiella chrysoura* sound pressure level in a field recording. *J. Acoust. Soc. Am.* **2004**, *116*, 3186–3191. [CrossRef]

48. Nedelec, S.L.; Mills, S.C.; Lecchini, D.; Nedelec, B.; Simpson, S.D.; Radford, A.N. Repeated exposure to noise increases tolerance in a coral reef fish. *Environ. Pollut.* **2016**, *216*, 428–436. [CrossRef]
49. Harding, H.R.; Gordon, T.A.C.; Hsuan, R.E.; Mackaness, A.C.E.; Radford, A.N.; Simpson, A.D. Fish in habitats with higher motorboat disturbance show reduced sensitivity to motorboat noise. *Biol. Lett.* **2018**, *14*, 20180441. [CrossRef]

Journal of
Marine Science and Engineering

Article

An Acoustic Treatment to Mitigate the Effects of the Apple Snail on Agriculture and Natural Ecosystems

Marta Solé [1,*], José-Manuel Fortuño [2], Mike van der Schaar [1] and Michel André [1,*]

1   Laboratory of Applied Bioacoustics, Technical University of Catalonia-BarcelonaTech,
    08800 Vilanova i la Geltrú, Spain; mike.vanderschaar@upc.edu
2   Institute of Marine Sciences, Spanish National Research Council, 08003 Barcelona, Spain; jmanuel@icm.csic.es
*   Correspondence: marta.sole@upc.edu (M.S.); michel.andre@upc.edu (M.A.)

Abstract: Global change is the origin of increased occurrence of disturbance events in natural communities, with biological invasions constituting a major threat to ecosystem integrity and functioning. The apple snail (*Pomacea maculata*) is a freshwater gastropod mollusk from South America. Considered one of the 100 most harmful invasive species in the world, due to its voracity, resistance, and high reproductive rate, it has become a global problem for wetland crops. In Catalonia, it has affected the rice fields of the Ebre Delta since 2010 with significant negative impact on the local economy. As a gastropod mollusc it possesses statocysts consisting of a pair of sacs, one located on each side of the foot, that contain multiple calcium carbonate statoconia. This study shows the first ultrastructural images of pathological changes in the sensory epithelium of the statocyst of apple snail adults with an increase in the severity of the lesions over time after exposure to low frequency sounds. Sound-induced damage to the statocyst could likely result in an inhibition of its vital functions resulting in a potential reduction in the survival ability of the apple snail and lead to an effective mitigation method for reducing damage to rice fields.

Keywords: apple snail; *Pomacea maculata*; acoustic trauma; scanning electron microscopy; invasive species; plague; mitigation method

Citation: Solé, M.; Fortuño, J.-M.; van der Schaar, M.; André, M. An Acoustic Treatment to Mitigate the Effects of the Apple Snail on Agriculture and Natural Ecosystems. *J. Mar. Sci. Eng.* **2021**, *9*, 969. https://doi.org/10.3390/jmse9090969

Academic Editor: Giuseppa Buscaino

Received: 12 June 2021
Accepted: 25 August 2021
Published: 6 September 2021

Publisher's Note: MDPI stays neutral with regard to jurisdictional claims in published maps and institutional affiliations.

1. Introduction

The introduction of invasive biological species as a result of globalization represents a worldwide threat to the integrity of ecosystems. Invasive species are one of the main causes of biodiversity loss, as they are the second main cause of species extinction, especially in ecosystems that are geographically and evolutionarily isolated, such as small islands [1,2]. To understand the factors that determine the success of the invasion and its effects on native species most studies have focused on individual species or taxonomic groups. As a result of trophic and non-trophic interactions (e.g., mutualism) invasions can profoundly alter the structure of the entire food web [3]. A holistic view of the problem that considers species interactions within trophic networks can contribute to a better understanding of the effects of invasions on complex communities. Aspects to be considered in this global view are the local biodiversity, diet amplitude, number of predators, or bioenergetic thresholds below which invasive and native species become extinct [3,4]. Simpler food webs are more vulnerable to invasions and relatively isolated mammals are amongst the most successful invaders. Invasive species modify the food web structure by decreasing biodiversity [4]. Resident species, on the other hand, evolve to try to tolerate or exploit invaders, a process that can lead to a more well-adjusted food web and may help to avoid extinctions [3]. A comprehensive approach identifies combinations of trophic factors that facilitate or prevent the introduction of new species and provides contrasting predictions about their effects on the structure and dynamics of ecosystems [4], enabling global problem management.

Another aspect to consider is climate change. Global warming allows alien species to settle in new ecosystems by locally increasing temperatures or eliminating the winter hypoxia that prevented their survival in the past [5]. In addition, global warming is increasing the competitive and predatory effects of invasive species on native ones and will enhance the virulence of some diseases [5]. Invasive species may constitute a health risk, such as the introduction of disease and parasite vectors that can lead to global pandemics, and/or an economic risk, as in the case of agricultural pests. Predation is the most dramatic damage to invaded ecosystems and often affects primary producers [3].

The apple snail (*Pomacea maculata*) is a freshwater gastropod mollusc (*Ampullariidae*) from the Amazon basin (Brazil, Bolivia and Argentina). Considered one of the 100 most harmful invasive species in the world, due to its voracity, resistance, and high reproductive rate it has become a global problem (USA, Southeast Asia (China, Japan, Taiwan, Cambodia, Malaysia, Pakistan, and the Philippines), Israel and Europe (Spain and Belgium)) for wetland crops [6–8] (Figure 1).

Figure 1. (**A**): *P. maculata* world distribution. Reproduced with permission from CABI 2021 [8] (**B**): Different images of *P. maculata specimens* used in the experiments.

There are two vectors that increase the risk of introduction of *P. maculata* in new habitats, the aquaculture industry and the aquarium trade. The necessity of replacing expensive sources of protein with cheaper alternatives at the global level has resulted in the deliberate introduction of apple snail into new areas (from South America to Asia) as a potential source of food [9]. But in Asia the market for *P. maculata* did not develop because it was not well liked as a food [10]. The deliberate release into agricultural or natural ecosystems and the abandonment of snail farms allowed *P. maculata* specimens to escape and become agricultural pests. The pet trade has been another cause of the introduction of apple snails. Pet stores receive freshwater snails from multiple sources. *P. maculata* was probably introduced in the southeastern USA, Belgium, Israel, and Spain via the aquarium trade [11–14]. Once introduced, *P. maculata* spreads naturally by floating in canals and rivers, during flooding or by attaching to birds, as has been reported in Hawaii [15]. People may also act as an introduction vector by accidentally transporting eggs on boats [12].

As most ampullariids, adult and juvenile *P. maculata* can be considered a particularly voracious generalist herbivore that feeds on diverse aquatic plants [16]. They can affect two aquatic crops, taro (*Colocasia esculenta*) and rice (*Oryza sativa*) and can feed on macroalgae, submerged plants or freely floating macrophytes. Typically, the *P. maculata* female size exceeds the male in size. They reproduce by internal fertilization and oviparous development [6]. Their female lays clutches of pink eggs out of the water on an emergent substrate. Their reproductive capacity is bigger than other *Pomacea* species [17]. Egg clutches take 10–14 days to hatch [6] and hatchlings fall into the water trying to adhere to some type of substrate. *P. maculata* can survive long periods without access to water [18] under muddy substrates by closing the shell with the operculum. They can breathe with their lung rather than their gill and survive brief periods out of the water (e.g., during egg laying).

Herbivory by apple snails impacts habitats and the biodiversity of ecosystems. As generalist herbivores they can opportunistically consume available resources very quickly (e.g., can quickly eliminate aquatic machrophytes). In populations with high densities this behaviour increases its ecological impact, for example changing the stable state of a lake from a clear to turbid condition [19]. The impact of *P. maculata* on biodiversity could occur through several mechanisms like competition [20–22] or hybridization with local species, or influencing the foraging behaviour of native species and consequently negatively influencing higher trophic levels [23]. In addition, this snail acts as a vector of various parasites including *Angiostrongyulus cantonensis*, a nematode which causes human eosinophillic meningitis [10].

Early detection and eradication together with transport regulatory legislation are the best measures to fight against the introduction of invasive species and to prevent their evolution to a pest. The first sign of *P. maculata* infestation is the presence of pink eggs highly visible above the water line. Eradication by removing the eggs manually or by trapping the snails is only possible when the population is small and restricted to one area. But when a bigger population is stablished this control treatment is ineffective [11].

Biological control of *P. maculata* has been tried in Alabama by introducing native redear sunfish (*Lepomis microlophus*) which eat the snail hatchlings [11]. The species eradication was unsuccessful but this trial suggested that some species of fishes could help in the control of the snail hatchlings. In Asia, fire ants (*Solenopsis geminate*) feed on eggs and juveniles of apple snails and have been suggested as possible biocontrol agents [24], but the introduction of a new invasive pest species would be inappropriate.

Limited success has been achieved by chemical control of *P. maculata*. Traditional pesticides failed because the apple snails close the operculum for long periods of time. The use of chelated copper or copper sulphate in the USA provoked high snail mortality but some large, adult snails still survived by filling their shells with air and floating away from the pesticide, and the eggs survived until the next year [11]. The copper treatments are very expensive and there is very little research on the indirect impacts of the pesticide on ecosystems, by poisoning not only the snails but also other invertebrates in the ecosystem, desirable or not, native or introduced.

In Catalonia, the species has affected the rice fields of the left hemi-delta of the Ebre Delta since 2010 [25]. *P. maculata* moves, actively, against the current on the bottom or, passively, closing the operculum and floating in the direction of the current. It is currently occupying part of the hydraulic network and could also affect the river itself in the near future. As a gastropod mollusc, the apple snail presents statocysts, the organs responsible for determining position with respect to gravity, as well as acoustic perception [26]. The ability of aquatic invertebrates, both larvae and adults, to perceive gravity and live under gravitational load is critical to their survival. In snails, statocysts consist of a pair of sacs, each located on a side of the foot, which contain multiple calcium carbonate statoconia in adulthood. Noteworthy in terms of their physiological relevance, they are the first neural structures to appear in larval development. A change in the snail position causes movement of the statoconia which in turn mechanically press the hair cells leading to a modification of the perception of gravity. These cilia act as transducers of the information that reaches the snail's nervous system, triggering a regulatory response based on its needs [27,28].

Research on the sensitivity of aquatic invertebrates to noise have determined ultra-structural effects of anthropogenic sound on cephalopods, cnidarians, and crustaceans even though they lack proper auditory receptors [29–33]. Only a few works reported the effects on behavior and development of gastropods and bivalves after sound exposure [34–37], but there are no studies reporting lesions on sensory epithelia of snail statocyst. Given our previous experience on the determination of the effects produced at ultrastructural and physiological levels in the statocyst of invertebrates exposed to artificial sounds [30–32], we hypothesized the possibility, by applying the same approach, of affecting snail sensory epithelia through exposure to sound. We present here the first images of *P. maculata* sensory epithelia after sound exposure. Presumably, ultrastructural effects, which cause behavioural effects, as observed in work on cephalopods—loss of appetite and decreased reproductive rate (see [38])—would be an effective and applicable measure in the fight against the apple snail plague in the Ebre Delta.

2. Methods

2.1. Sound Exposure Protocol

Sequential Controlled Exposure Experiments (CEE) were conducted on individuals of *P. maculata* (n = 40). An additional set of individuals (n = 25) was used as control. Individuals were maintained in the LAB maintenance tank system (see detailed system description and exposure protocol in [38]. The maintenance tanks consisted of a closed system where the two tanks (control and exposed animals) were connected. Controls and exposed animals were kept in the same conditions (natural freshwater at 20–25 °C and natural oxygen pressure). The exposure consisted of 50–400 Hz sinusoidal wave sweeps with 100% duty cycle and a 1 s sweep period for two hours. The sweep was produced and amplified through an in-air loudspeaker (QSC KW153) while the level received was measured by a calibrated B&K 8106 hydrophone. Figure 2 displays the power spectral density of the signal received by the hydrophone (RL). The average level calculated over the sweep was 157 ± 5 dB re 1 μ Pa^2/Hz with peak levels up to 175 dB re 1 μ Pa. The controls were placed in the exposure tank for as long as those exposed (2 h), without any sound playback (Figure 3A). The sequential sacrificing process after exposure was identical for controls and exposed animals. Five animals were sacrificed to obtain statocyst samples upon arrival at the LAB as initial controls. Following exposure, samples were obtained from individuals (exposed, n = 20; and controls, n = 10) at 48 h and 120 h after sound exposure (Figure 3B).

2.2. Removal of Statocysts

The animals were euthanized with anaesthesia (magnesium chloride 25 g/L), extracted from the shell and the anterior portion of the body was sectioned through a dorso-ventral cut that revealed the statocysts. This anterior body section was detached and chemically fixed for observation and analysis. The two statocysts were analysed for each snail, taking special care to prevent mechanical damage to the tissues and making sure to expose the sensory epithelia overlaid by multiple statoconias.

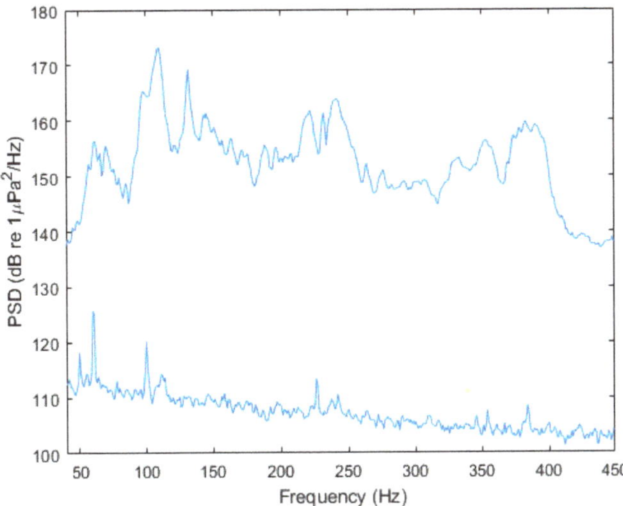

Figure 2. Power spectral density of the 50 to 400 Hz sweep as received by the hydrophone on top with below the background noise level recorded just before exposure.

2.3. Scanning Electron Microscopy

One hundred and thirty statocysts from sixty five *P. maculata* were used for this study. Fixation was performed in 2.5% glutaraldehyde for 24–48 h at 4 °C. Statocysts embedded on the snail muscular mass were dehydrated in graded ethanol solutions and critical-point dried with CO_2 in a Bal-Tec CPD 030 unit (Leica Mycrosystems, Austria). The dried samples were mounted on specimen stubs with double-sided tape. The mounted tissues were gold coated with a Quorum Q150R S sputter coated unit (Quorum Technologies, Ltd. Laughton, East Sussex, United Kingdom) and viewed with a variable pressure Hitachi S-3500N scanning electron microscope (Hitachi High-Technologies Co., Ltd., Tokyo, Japan) at an accelerating voltage of 5 kV at the Institute of Marine Sciences of the Spanish Research Council (CSIC) facilities.

2.4. Quantification and Data Analysis

We selected the sensory areas of the statocyst. The surface of these statocyst areas comprising hair cells was determined for each sample. Sampling squares of 400 μm^2 (20 μm × 20 μm) were placed in this area at 20%, 40%, 60%, and 80% along its centre axis (Figure 4). Hair cell damage was analysed by classifying them as intact (hair cell undamaged), damaged (kinocilium or surrounding stereocilia partially or entirely missing or fused), extruded (hair cell partially extruded of the epithelium) and missing (hole in the epithelium caused by the total extrusion of the hair cell). The severity of the lesions was quantified as the percentage of extruded and missing hair cells with respect to the total hair cell count of the sampling square. The damaged category encompassed a wide range of different types of lesions with different severities; this made direct comparison between animals more difficult. The extruded and missing categories were well-defined and easily compared, and the presence of extruded cells showed the limit of severe damage after sound exposure. Two statocysts of each exposed animal were analysed; the hair cell counts of the statocysts were combined to obtain a single measurement per region per animal.

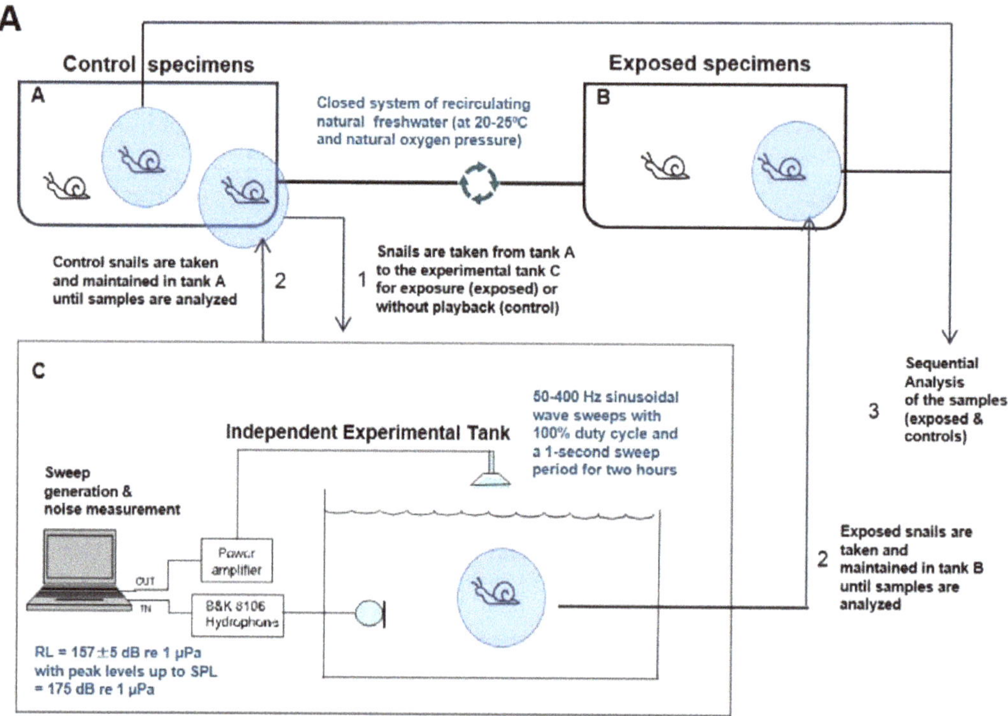

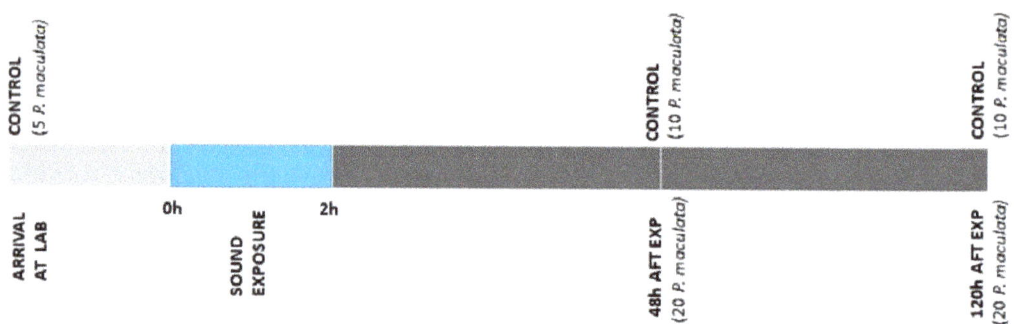

Figure 3. Sound exposure protocol, sampling collection, and analysis. (**A**): Sound exposure protocol. *P. maculata* were maintained in tank A until some were transferred to an independent experimental tank C where they were exposed to sound (1). At the end of the exposure experiments, the snails were transferred to tank B (2) that presented the same environmental conditions as tank A. Control specimens were transferred to tank C for 2 h without any playback and after that, they were taken back to tank A (2). Samples of control and exposed plants were sequentially taken for analysis (3). (Figure modified from 38). (**B**): Sampling collection and analysis. Before the sound exposure started, control specimens were taken and analyzed at the arrival of the laboratory facilities. Samples of control and exposed snails were sequentially analyzed at 48 h and 120 h after sound exposure.

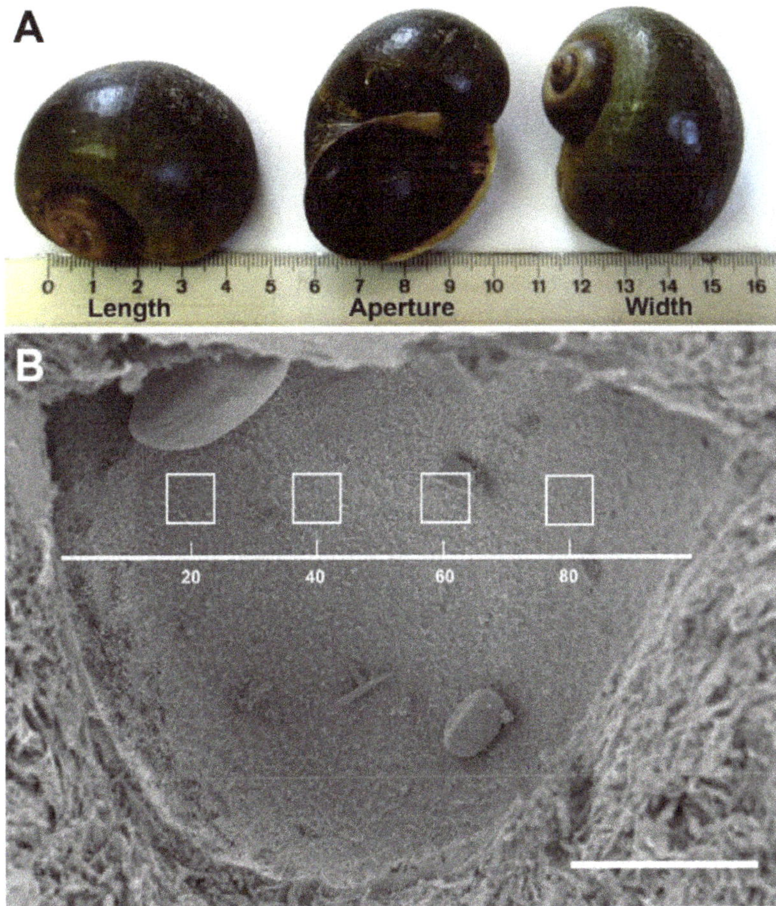

Figure 4. (**A**): Each individual of *P. maculata* was characterized through its shell dimensions (length, width and aperture length). (**B**): Hair cell bundle count locations (sampling squares) on the *P. maculata* statocyst. Hair cell counts were sampled at four predetermined locations: 20%, 40%, 60% and 80% of the total statocysts length. A 400 μm^2 box was placed at each sampling area and hair cells were counted within each box. Scale bar: (**B**) = 100 μm.

The influence of recovery time after sound exposure in groups of animals sacrificed 48 h vs. 120 h after sound exposure was tested comparing the mean damage count between groups using permutation tests repeated multiple times with N = 1000. To evaluate the difference in damage between regions, the median hair cell counts between all regions (for control and exposed animals at 48 h and 120 h separately) were compared using a Kruskal–Wallis test.

3. Results

3.1. P. maculata Statocyst Morphology

All the animals used in this study were adult snails with 3–5 cm shell length, 2–4 cm shell width and 4–5 cm shell aperture length (Figure 4A). *P. maculata* individuals presented two statocysts, 340–350 μm in diameter, located in the foot and linked ventrolaterally to each pleural ganglion (Figure 5A). Statocysts consisted essentially of a cavity, covered by an epithelium of mechanosensitive hair cells, which contained a fluid and a large number of highly morphologically dissimilar aragonite (calcium carbonate) crystals, statoconia (Figure 5B–D). Two types of epithelial cells were found on the statocyst: small supporting cells carrying microvilli and sensory hair cells which in addition of a crown of stereocilia bear one large kinocilia (Figure 5E). The hair cells projected the kinocilia into the lumen of the statocyst, which also contained endolymph and statoconia. The majority of statoconia were oval shaped with smoothened edges (Figure 5B,C). The individual statoconia filled all the statocyst cavity and were constantly moving independently of each other.

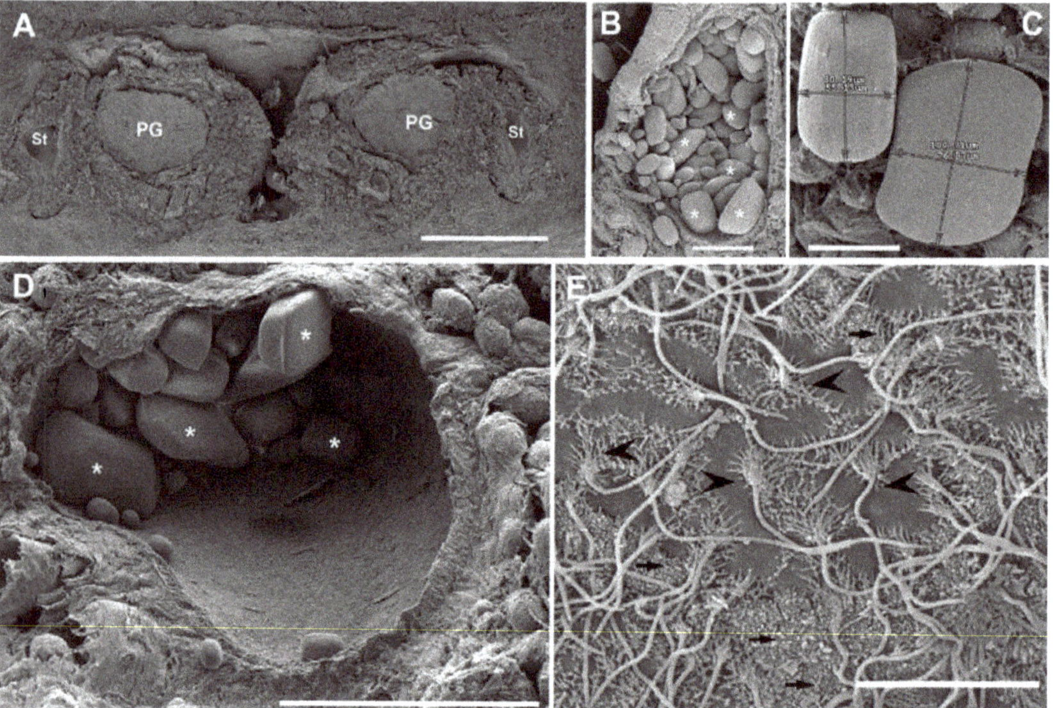

Figure 5. SEM. Internal morphology of *P. maculata statocyst*. Control animal. (**A**): Dorso-ventral cutting of the body anterior portion showing the statocyst location in the snail foot and linked ventrolaterally to each pleural ganglion (PG: pleural ganglion, St: statocyst). (**B**): Opened statocyst completely filled with statoconia (asterisk). (**C**): Differently sized statoconia. (**D**): Internal cavity of the statocyst covered by the sensory epithelium. Some of the aragonite crystals are visible (asterisk). (**E**): Inner statocyst sensory epithelia. Arrowheads point to the hair cells exhibiting their lonely kinocilia surrounded by a crown of stereocilia. Between them microvilli of the supporting cells are visible (arrows). Scale bar: (**A**) = 1 mm. (**D**) = 200 μm. (**B**) = 100 μm. (**C**) = 50 μm. (**E**) = 5 μm.

3.2. Ultrastructural Analysis of the Statocyst Sensory Epithelium

The exposed animals presented lesions in the statocyst sensory epithelium that incremented against time. In comparison with the control animals (Figures 5E and 6A), damaged hair cells on exposed animals presented swelling (Figure 6B) or missing (Figure 6C–E) kinocilia or the whole hair cell was missing or extruded (ejected) from the sensory epithelium into the statocysts cavity (Figure 6F–H). A considerable number of hair cells had totally lost the unique kinocilium (Figure 6E) and the crown of stereocilia surrounding the large kinocilium was totally fused (Figure 6D) or lost (Figure 6E). Some ejected hair cells presented spherical holes (Figure 6H), probably due to the cell swelling and the extrusion of the inner cellular material.

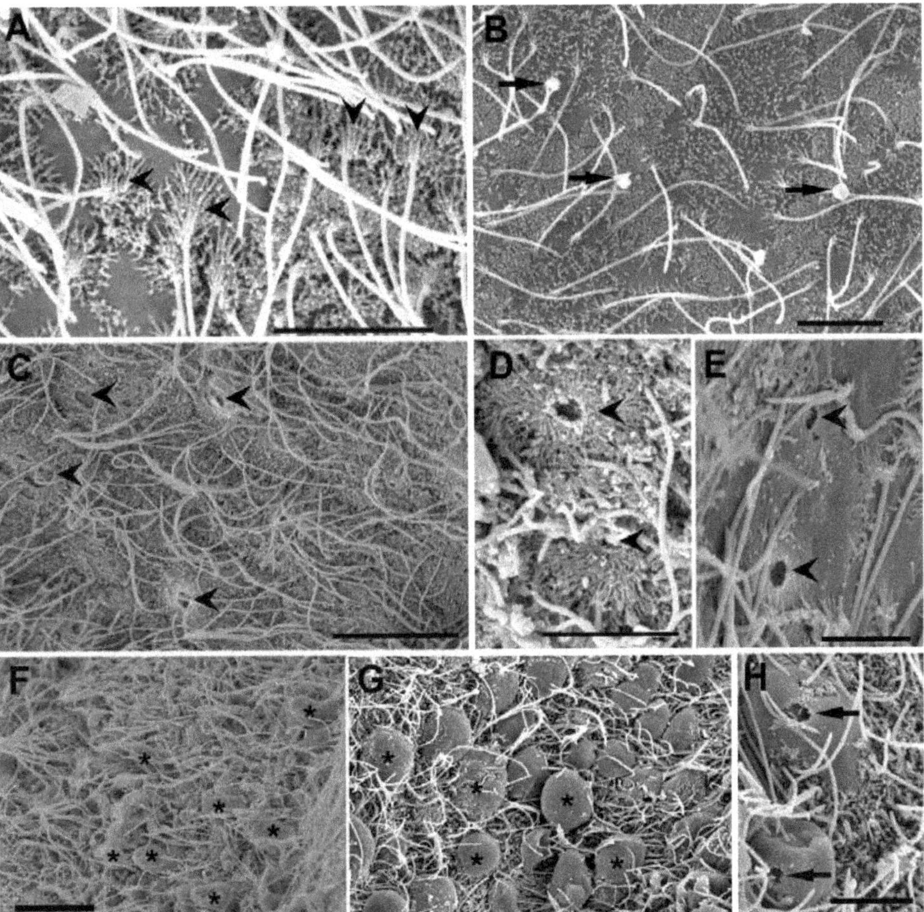

Figure 6. SEM. *P. maculata* statocyst sensory epithelium. (**A**). Control animal. (**B–D**): 48 h after sound exposure. (**F–H**): 120 h after sound exposure. (**A**): Healthy hair cells show the kinocilia surrounded by a crown of stereocilia (arrowheads). (**B**): Some hair cells present swollen kinocilia (arrows). (**C**): Arrowheads point to the holes left by the single kinocilia after their ejection. (**D**): Detail from (**C**). Surrounding the hole left by the ejected kinocilia a crown of stereocilia are fused in a compact structure. (**E**): Detail from (**C**). The hair cells have lost the entire kinocilia and the crown of stereocilia (arrowheads). (**F**): The hair cells show severe damage. Most hair cells are (asterisks) ejected from the sensory epithelium. (**G**): Detail from (**F**), shows the cellular body of the sensory cells extruded from the epithelium (asterisks). (**H**): Two extruded hair cells show spherical holes in the body due to the extrusion of the inner cellular material (arrows). Their kinocilia has lost the stereocilia crown. Scale bar: (**A,B,D,E,H**) = 5 μm. (**C,G**) = 10 μm. (**F**) = 20 μm.

3.3. Image and Data Analysis

The abnormal features we identified on the surface of sound exposed epithelia included damaged (kinocilium or surrounding stereocilia partially or entirely missing or fused), extruded (hair cell extruded from the epithelium) and missing hair cells (hole in the epithelium caused by the total extrusion of the hair cell). The number of damaged, extruded and missing hair cells was counted for each image. The severity of the lesions was chosen to be quantified as the percentage of extruded and missing hair cells because these were well-defined and easy-to-compare categories. The damaged category was not useful for this purpose as with increasing physical damage it was increasing and then decreasing again as a damaged hair cell was at an intermediate state. Therefore, in order to be able to quantify the magnitude of overall damage to the sensory organs "lesion severity" was based on the combination of extruded and missing hair cells. The presence of extruded cells showed the start of severe damage after sound exposure. None of the control animals showed extruded or missing hair cells on the sensory epithelia and all exposed animals had various degrees of damage.

First, we analysed if there was a difference in lesion severity between the different regions. No difference in hair cell counts would allow us to sum the counts over all the regions for each animal, without the need to consider weighting factors. Comparing the median counts between the regions of all control animals (in this case summing the counts across all categories for each region together as the categories considered for lesion severity have a 0 count), each region appeared to have a similar median count with $p = 0.074$. Performing the same test for the exposed animals, but now considering the extruded and missing categories, we found for the animals sacrificed at 48 h that median counts were dissimilar with $p = 0.027$ and at 120 h they were similar again with $p = 0.73$. Taking a closer look at 48 h, we performed the KW test with the regions 40% to 80%, leaving out the 20% region resulting in $p = 0.91$. Considering these outcomes, we concluded that there was no significant difference in hair cell count or hair cell damage between regions (Figure 7A–C; Figure 8B,C).

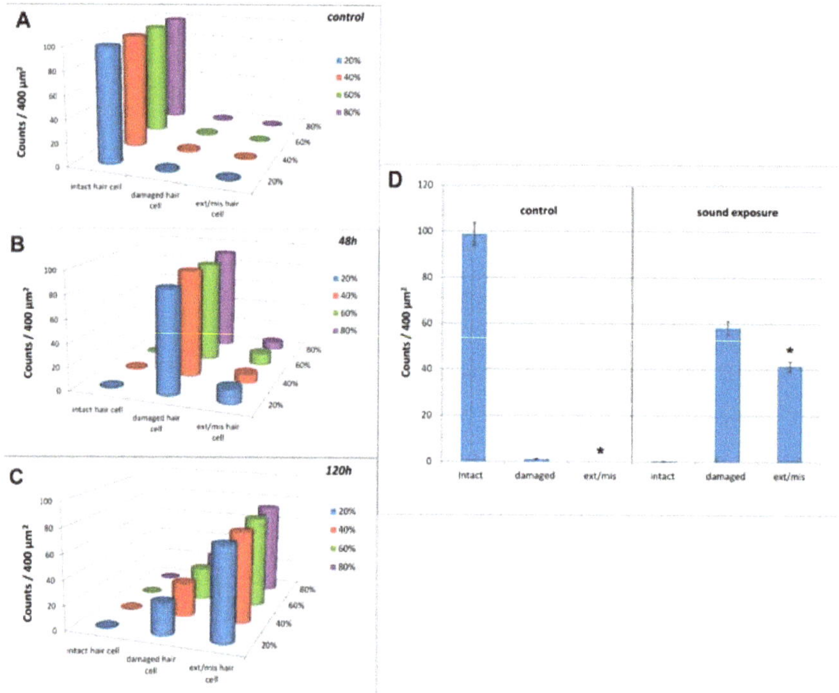

Figure 7. (**A–C**): Mean intact, damaged, and extruded/missing hair cells at 20%, 40%, 60%, and 80% of the total length of sensory epithelium on *P. maculata* statocyst (48 h and 120 h after sound exposure versus control animals). Note the increase of damaged, extruded, and missing cells versus controls with increase in time. (**D**): Mean (±SE) intact, damaged and extruded/missing hair cells after sound exposure ($n = 80$) versus control ($n = 50$) on statocyst sensory epithelium of *P. maculata*. Each bar is the average over the 4 regions with the line indicating the standard error. The percentage was computed by dividing by the total count for each individual sample. The lesion severity (see text for definition) of the sound exposure group was found to be significantly higher than that of the control group with $p = 9.99 \times 10^{-4}$ (* significant difference).

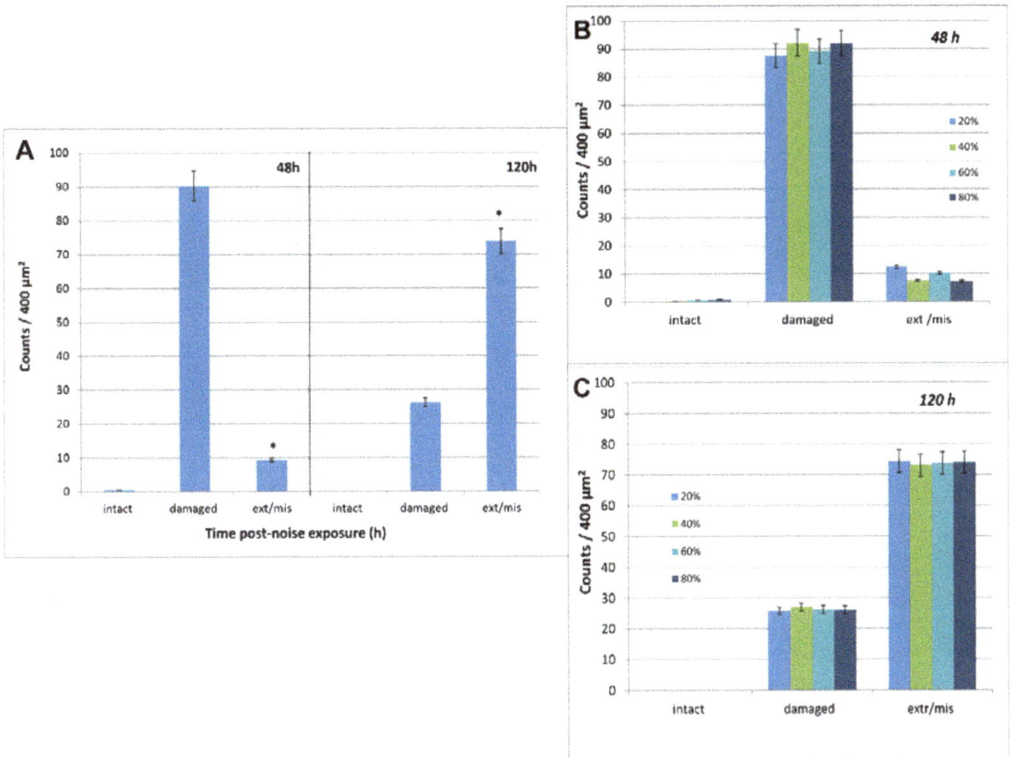

Figure 8. (**A**): Mean (+SE) intact, damaged, extruded/missing hair cells over the total length of sensory epithelium on *P. maculata* statocyst (48 h versus 120 h after sound exposure). The lesion severity was found to be significantly higher at 120 h than at 48 h ($p = 9.99 \times 10^{-4}$; $n = 40$). (* significant difference). (**B**): Hair cell counts between the different regions in animals 48 h after sound exposure. The lesion severity was found to be similar between 40%, 60%, and 80% regions ($p = 0.91$; $n = 40$; see text for 20% region consideration). (**C**): Hair cell count between the different regions in animals 120 h after sound exposure. Each bar is the average over the 4 regions with the line indicating the standard error. The percentage was computed by dividing with the total count for each individual sample. The lesion severity was found to be similar between all regions ($p = 0.73$; $n = 40$).

4. Discussion

Our study not only shows the first ultrastructural images of *P. maculata* statocyst sensory epithelium but also describes pathological changes after sound exposure in these tissues.

The apple snail sensory epithelia and statoconia presented a similar structure as in other pulmonate gastropods [39–41]. As in terrestrial pulmonate snails [42] the presence of small and very small statoconia together with large statoconia in the same statocyst suggest their state of generation and growth during the whole life of the animal.

The sensory epithelium hair cell death and damage in vertebrates after loud or prolonged noise exposure or ototoxic drugs can consist of a dysfunction of mechanotransduction complex, loss of ribbon synapses or hair cell death by apoptosis or necrosis [43]. Noise trauma produces an increase in calcium levels in hair cells, which leads to the stimulation of reactive oxygen species (ROS) production. In addition, higher metabolic demand in hair cells during exposure to sound could also lead to increased ROS levels, leading to various forms of damage in the cell (including apoptosis or necrosis) [44]. These dying cells are eliminated from the sensory epithelium by extrusion, a mechanism used to eliminate unfit, excess, or dying cells [45]. The apex of the hair cell or its entire body is expelled from the epithelium [44] and the lost hair cells are replaced by expansion of adjacent supporting cells, which form a scar. This process has been observed in vertebrates [44] and some invertebrates, specifically in cephalopods and cnidarians [32,38]. The lesions observed on *P. maculata* statocysts sensory epithelium showed similarities in their nature and intensity to those described in cephalopod statocysts (hair cells partially or totally ejected from the sensory epithelium into the statocyst cavity; swollen, bent, fused or lost kinocilia) after being exposed to the same low-frequency sounds [30,46], as well as in cnidarians [32] and crabs [47]. The consequences for the cephalopod's species exposed to low-frequency sounds were an immediate loss of the ability to orient, mate, and feed. Probably due the similarity of their sensory systems, because of their taxonomic proximity (both are molluscs) sound exposure in *P. maculata* would affect the physiology and functioning of the *P. maculata* statocyst. Further investigation is needed to determine threshold levels and to explain these effects in apple snail species.

In regards to pathology findings, there was an increase in the severity of the lesions over time after exposure to low frequency sounds. The increase of the damage severity with time after sound exposure in the sensory epithelia affected by acoustic trauma is a common process described in vertebrate and invertebrate species. In mammals this can provoke the entire destruction of the Corti organ [48]. In birds, hair cells which are affected by apoptosis within the basilar papilla after sound exposure continue degenerating after the beginning of the cell regeneration [44,49]. In invertebrates [30,32,38,50] and plants [33] similar degeneration processes of hair cells increases with time after the exposure.

The Ebre Delta is a particularly sensitive area that has experienced successive introductions of invasive species that have become real pests to their ecosystems. A non-exhaustive list of the most recently introduced species includes red swamp crayfish (*Procambarus clarkii*), which was initially introduced to the river in 1979 for commercial purposes. Artificial introduction of wels catfish (*Silurus glanis*), the largest-bodied European freshwater fish, ocurred in 1983. Since 1993, black fly (*Simuliidae*) have had periodic episodes in the area. Since 1997, piranha (*Serrasalmus* sp.) individuals have been detected in the Ebre River, probably originating from aquarofilia installations. To date, zebra mussel (*Dreissena polymorpha*) has been the most aggressive freshwater invader worldwide and is present in Ebre Delta since 2001. Since the beginning of the twentieth century red palm weevil (*Rhynchophorus ferrugineus*) has been reported as one of the most important pests of ornamental and oil palms and it represents a plague for the palms on the Delta since 2003. In 2008, the first apple snail (*P. maculata*) was detected. Since 2012 the American bullfrog *(Lithobates catesbeianus)* and American blue crab (*Callinectes sapidus*) have become new plagues in the Ebre Delta [8]. After some years of detection of the *P. maculata* pest in the Ebre Delta a new invasive species was established in the same area, the American blue crab *Callinectes sapidus*. Although this species has been present in Europe (Atlantic cost of France, North Sea, Mediterranean Sea, Baltic Sea, Black Sea) since the early years of the 20th century [51], where multiple independent introductions may have taken place through ballast water, it was not detected at the Ebre Delta until 2012 [52]. The future impact of this new pest on the Delta ecosystem is unclear, but at this moment the crab is acting as a biocontrol agent of *P. maculata*, reducing its population by feeding on them. In any case, the introduction into the ecosystem of a foreign species that is particularly voracious and has a high reproduction rate, such as *C. sapidus*, cannot be a long-term solution because of the problems that accompany such an invasion.

Different treatments have been used to fight the *P. maculata* rice crops invasion of Ebre Delta: sea water flooding of the rice fields, chemical treatments with saponin, winter drying, channel cleaning, manual removal of the eggs clutches, traps, and advance obstructions [53]. None of these treatments had an acceptable success, and due to the severe infestation of the species in the area a total eradication is non-viable. In this context, the use of an alternative treatment such as exposure to sound would be an innovative and ecological option. The results obtained in the present study are consistent with the hypothesis that exposure to sound may impair the sensory perception of *P. maculata*. Sound-induced damage to the statocyst could likely result, as in cephalopods [38], in an inhibition of its vital functions, which would result in a dramatic reduction of the survival ability of the apple snail. Cellular effects, which can trigger behavioral effects (e.g., in cephalopods, loss of appetite and a decrease in the reproductive rate) would be an effective measure to fight apple snail plagues over the world. However, further efforts should analyse the possible interactions of the method with other trophic levels in the ecosystem, in order to manage the problem holistically and avoid damage to other native species. This is essential to provide global benefits for complex communities. The choice of the sound parameters (frequencies, amplitudes, and exposure times) used during this study was made for the sole purpose of verifying the possible harmful effects of exposure to acoustic sources on apple snails and not to determine a precise threshold for inducing effective lesions. Further investigation is also needed to determine the levels that produce lesions in association with behavioural changes, which would result in a pest reduction, while at the same time being innocuous for the other species and the ecosystem as a whole.

This method is patented (WO 2018/167003 A1) and could become a valuable approach to mitigate the invasions of any aquatic alien species that are expected to interfere with ecological communities due to the increasing negative effects of climate change.

5. Patent

André M., Solé M., Van der Schaar, De Vreese S (International Patent WO 2018/167003 A1). 20-09-2018. A method for inducing lethal lesions in sensory organs of undesirable aquatic organisms by use of sound. Licensed to SEASEL SOLUTIONS AS [NO/NO]; P.O.BOX 93 N-6282 BRATTVÅG (NO).

Author Contributions: M.S. and M.A. planned the research and designed the study. M.S., M.A. and M.v.d.S. conducted experimental/lab work. M.S. and J.-M.F. performed SEM analysis. M.S. and M.v.d.S. analysed the data. M.S. prepared the figures. M.S. wrote the article and all authors reviewed the manuscript. All authors have read and agreed to the published version of the manuscript.

Funding: This research received no external funding.

Institutional Review Board Statement: Not applicable.

Informed Consent Statement: Not applicable.

Data Availability Statement: The authors declare that the data supporting the findings of this study are available within the paper.

Conflicts of Interest: The authors declare no conflict of interest.

References

1. Clarke, B.; Murray, J.; Johnson, M.S. The extinction of endemic species by a program of biological control. *Pac. Sci.* **1984**, *38*, 97–104.
2. Savidge, J.A. Extinction of an Island Forest Avifauna by an Introduced Snake. *Ecology* **1987**, *68*, 660–668. [CrossRef]
3. David, P.; Thébault, E.; Anneville, O.; Duyck, P.-F.; Chapuis, E.; Loeuille, N. *Impacts of Invasive Species on Food Webs*; Elsevier: Amsterdam, The Netherlands, 2017; pp. 1–60. [CrossRef]
4. Galiana, N.; Lurgi, M.; Montoya, J.M.; López, B.C. Invasions cause biodiversity loss and community simplification in vertebrate food webs. *Oikos* **2014**, *123*, 721–728. [CrossRef]
5. Rahel, F.J.; Olden, J. Assessing the Effects of Climate Change on Aquatic Invasive Species. *Conserv. Biol.* **2008**, *22*, 521–533. [CrossRef]
6. Barnes, M.A.; Fordham, R.K.; Burks, R.L.; Hand, J.J. Fecundity of the exotic applesnail, *Pomacea insularum. J. North Am. Benthol. Soc.* **2008**, *27*, 738–745. [CrossRef]
7. Rawlings, T.A.; Hayes, K.A.; Cowie, R.H.; Collins, T.M. The identity, distribution, and impacts of non-native apple snails in the continental United States. *BMC Evol. Biol.* **2007**, *7*, 97. [CrossRef] [PubMed]
8. CABI. Pomacea maculata. In *Invasive Species Compendium*; CAB International: Wallingford, UK, 2019. Available online: https://www.cabi.org/isc/datasheet/116486#toDistributionMaps (accessed on 24 August 2021).
9. Naylor, R. Invasions in agriculture: Assessing the cost of the golden apple snail in Asia. *Ambio* **1996**, *25*, 443–448.

10. Yang, T.-B.; Wu, Z.-D.; Lun, Z.-R. The Apple Snail *Pomacea canaliculata*, A Novel Vector of the Rat Lungworm, *Angiostrongylus cantonensis*: Its Introduction, Spread, and Control in China. *Hawai'i. J. Med. Public Health J. Asia Pac. Med. Public Health* **2013**, *72*, 23–25.

11. Martin, C.W.; Bayha, K.M.; Valentine, J.F. Establishment of the Invasive Island Apple Snail *Pomacea insularum* (*Gastropoda: Ampullaridae*) and Eradication Efforts in Mobile, Alabama, USA. *Gulf Mexico Sci.* **2012**, *1–2*, 30–38. [CrossRef]

12. Panel, E.; Plh, H.; Baker, R.; Candresse, T.; Simon, E.D.; Gilioli, G.; Grégoire, J.; Puglia, P.; Rafoss, T.; Rossi, V.; et al. Scientific Opinion on the evaluation of the pest risk analysis on *Pomacea insularum*, the island apple snail, prepared by the Spanish Ministry of Environment and Rural and Marine Affairs. *EFSA J.* **2012**, *10*. [CrossRef]

13. Hayes, K.A.; Joshi, R.C.; Thiengo, S.; Cowie, R.H. Out of South America: Multiple origins of non-native apple snails in Asia. *Divers. Distrib.* **2008**, *14*, 701–712. [CrossRef]

14. Roll, U.; Dayan, T.; Simberloff, D.; Mienis, H.K. Non-indigenous land and freshwater gastropods in Israel. *Biol. Invasions* **2008**, *11*, 1963–1972. [CrossRef]

15. Levin, P.; Cowie, R.; Taylor, J.; Hayes, K.; Burnett, K.; CA, F. Apple snail invasions and the slow road to control: Ecological, economic, agricultural, and cultural perspectives in Hawaii. In *Global Advances in Ecology and Management of Golden Apple Snails*; Joshi, R.C., Sebastian, L.S., Eds.; Philippine Rice Research Institute (PhilRice): Los Baños, Philippines, 2006; pp. 325–335.

16. Morrison, W.E.; Hay, M.E. Herbivore Preference for Native vs. Exotic Plants: Generalist Herbivores from Multiple Continents Prefer Exotic Plants That Are Evolutionarily Naïve. *PLoS ONE* **2011**, *6*, e17227. [CrossRef] [PubMed]

17. Cowie, R. Apple snails (*Ampullariidae*) as agricultural pests: Their biology, impacts and management. In *Molluscs as Crop Pests*; Barker, G.M., Ed.; CABI: Los Baños, Philippines, 2002; pp. 145–192.

18. Ramakrishnan, V. Salinity, pH, Temperature, Desiccation and Hypoxia Tolerance in the Invasive Freshwater Apple Snail *Pomacea insularum*. Ph.D. Thesis, University of Texas at Arlington, Arlington, TX, USA, 2007.

19. Carlsson, N.; Kestrup, Å.; Mårtensson, M.; Nyström, P. Lethal and non-lethal effects of multiple indigenous predators on the invasive golden apple snail (*Pomacea canaliculata*). *Freshw. Biol.* **2004**, *49*, 1269–1279. [CrossRef]

20. Warren GL Nonindigenous freshwater invertebrates. In *Strangers in Paradise*; Simberloff, D.; Schmitz, D.C.; Brown, T.C. (Eds.) Island Press: Washington, DC, USA, 1997; pp. 101–108.

21. Wood, T.S.; Anurakpongsatorn, P.; Chaichana, R.; Mahujchariyawong, J.; Satapanajaru, T. Heavy Predation on Freshwater Bryozoans by the Golden Apple Snail, *Pomacea canaliculata* Lamarck, 1822 (Ampullariidae). *Nat. Hist. J. Chulalongkorn Univ.* **2006**, *6*, 31–36.

22. Carlsson, N.O.L.; Brönmark, C.; Hansson, L.-A. Invading herbivory: The golden apple snail alters ecosystem functioning in asian wetlands. *Ecology* **2004**, *85*, 1575–1580. [CrossRef]

23. Cattau, C.E.; Martin, J.; Kitchens, W.M. Effects of an exotic prey species on a native specialist: Example of the snail kite. *Biol. Conserv.* **2010**, *143*, 513–520. [CrossRef]

24. Way, M.; Islam, Z.; Heong, K.; Joshi, R. Ants in tropical irrigated rice: Distribution and abundance, especially of *Solenopsis geminata* (*Hymenoptera: Formicidae*). *Bull. Èntomol. Res.* **1998**, *88*, 467–476. [CrossRef]

25. Departament D'agricultura Alimentació i Acció Rural. *Generalitat de Catalunya Ordre AAR/404/2010, de 27 de Juliol, per la Qual es Declara Oficialment l'existència d'un Focus del Cargol Poma (Pomacea sp.) A L'hemidelta Esquerra del Delta de l'Ebre*; DOGC: Catalonia, Spain, 2010. Available online: https://portaljuridic.gencat.cat/eli/es-ct/o/2010/07/27/aar404 (accessed on 24 August 2021).

26. Budelmann, B.U. Hearing in N onarthropod Invertebrates. In *The Evolutionary Biology of Hearing*; Webster, D.B., Fay, R.R., Eds.; Springer: New York, NY, USA, 1992; pp. 141–155.

27. Wiederhold, M.L.; Sharma, J.S.; Driscoll, B.P.; Harrison, J.L. Development of the statocyst in *Aplysia californica* I. Observations on statoconial development. *Hear. Res.* **1990**, *49*, 63–78. [CrossRef]

28. Balaban, P.M.; Malyshev, A.Y.; Ierusalimsky, V.; Aseyev, N.; Korshunova, T.A.; Bravarenko, N.I.; Lemak, M.S.; Roshchin, M.; Zakharov, I.S.; Popova, Y.; et al. Functional Changes in the Snail Statocyst System Elicited by Microgravity. *PLoS ONE* **2011**, *6*, e17710. [CrossRef]

29. Day, R.D.; McCauley, R.D.; Fitzgibbon, Q.P.; Hartmann, K.; Semmens, J.M. *Assessing the Impact of Marine Seismic Surveys on Southeast Australian Scallop and Lobster Fisheries*; University of Tasmania: Hobart, Tasmania, 2016; ISBN 9780646959108.

30. Solé, M.; Sigray, P.; Lenoir, M.; Van Der Schaar, M.; Lalander, E.; André, M. Offshore exposure experiments on cuttlefish indicate received sound pressure and particle motion levels associated with acoustic trauma. *Sci. Rep.* **2017**, *7*, 45899. [CrossRef]

31. Solé, M.; Monge, M.; André, M.; Quero, C. A proteomic analysis of the statocyst endolymph in common cuttlefish (*Sepia officinalis*): An assessment of acoustic trauma after exposure to sound. *Sci. Rep.* **2019**, *9*, 1–12. [CrossRef]

32. Solé, M.; Lenoir, M.; Fontuño, J.M.; Durfort, M.; Van Der Schaar, M.; André, M. Evidence of Cnidarians sensitivity to sound after exposure to low frequency noise underwater sources. *Sci. Rep.* **2016**, *6*, 37979. [CrossRef]

33. Solé, M.; Lenoir, M.; Durfort, M.; Fortuño, J.-M.; van der Schaar, M.; De Vreese, S.; André, M. Seagrass *Posidonia* is impaired by human-generated noise. *Commun. Biol.* **2021**, *4*, 1–11. [CrossRef]

34. La Bella, G.; Cannata, S.; Froglia, C.; Ratti, S.; Rivas, G. First assessment of effects of air-gun seismic shooting on marine resources in the Central Adriatic Sea. *Int. Conf. Heal. Saf. Environ. Oil Gas Explor. Prod.* **1996**, *1*, 227–238. [CrossRef]

35. Stocks, J.R. Response of Marine Invertebrate Larvae to Natural and Anthropogenic Sound: A Pilot Study. *Open Mar. Biol. J.* **2012**, *6*, 57–61. [CrossRef]

36. Nedelec, S.L.; Radford, A.N.; Simpson, S.D.; Nedelec, B.; Lecchini, D.; Mills, S.C. Anthropogenic noise playback impairs embryonic development and increases mortality in a marine invertebrate. *Sci. Rep.* **2014**, *4*, srep05891. [CrossRef] [PubMed]
37. De Soto, N.A.; Delorme, N.; Atkins, J.; Howard, S.; Williams, J.M.; Johnson, M.J.A. Anthropogenic noise causes body malformations and delays development in marine larvae. *Sci. Rep.* **2013**, *3*, 2831. [CrossRef] [PubMed]
38. Solé, M.; Lenoir, M.; Durfort, M.; López-Bejar, M.; Lombarte, A.; van der Schaar, M.; André, M. Does exposure to noise from human activities compromise sensory information from cephalopod statocysts? *Deep. Sea Res. Part II Top. Stud. Oceanogr.* **2013**, *95*, 160–181. [CrossRef]
39. Stahlschmidt, V.; Wolff, H.G. The fine structure of the statocyst of the prosobranch mollusc *Pomacea paludosa*. *Z. Mikroskop. Anat.* **1972**, *133*, 529–537. [CrossRef] [PubMed]
40. Zaitseva, O. Structural organization of the sensory systems of the snail. *Neurosci. Behav. Physiol.* **1994**, *24*, 47–57. [CrossRef]
41. Kononenko, N.L.; Kiss, T.; Elekes, K. The neuroanatomical and ultrastructural organization of statocyst hair cells in the pond snail, *Lymnaea stagnalis*. *Acta Biol. Hung.* **2012**, *63*, 99–113. [CrossRef] [PubMed]
42. Gorgiladze, G.I.; Bukia, R.D.; Davitashvili, M.T.; Taktakishvili, A.D.; Gelashvili, N.S.; Kalandarishvili, E.L.; Satdykova, G.P. Morphological peculiarities statoconia in statocysts of terrestrial pulmonary snail *Helix lucorum*. *Bull. Exp. Biol. Med.* **2010**, *149*, 269–272. [CrossRef]
43. Wagner, E.L.; Shin, J.-B. Mechanisms of Hair Cell Damage and Repair. *Trends Neurosci.* **2019**, *42*, 414–424. [CrossRef]
44. Leonova, E.V.; Raphael, Y. Organization of cell junctions and cytoskeleton in the reticular lamina in normal and ototoxically damaged organ of Corti. *Hear. Res.* **1997**, *113*, 14–28. [CrossRef]
45. Atieh, Y.; Wyatt, T.; Zaske, A.M.; Eisenhoffer, G.T. Pulsatile contractions promote apoptotic cell extrusion in epithelial tissues. *Curr. Biol.* **2021**, *31*, 1129–1140. [CrossRef] [PubMed]
46. Solé, M.; Lenoir, M.; Fortuño, J.-M.; van der Schaar, M.; André, M. A critical period of susceptibility to sound in the sensory cells of cephalopod hatchlings? *Biol. Open* **2018**, *7*, bio033860. [CrossRef] [PubMed]
47. Day, R.D.; McCauley, R.D.; Fitzgibbon, Q.; Hartmann, K.; Semmens, J.M. Exposure to seismic air gun signals causes physiological harm and alters behavior in the scallop Pecten fumatus. *Proc. Natl. Acad. Sci. USA* **2017**, *114*, E8537–E8546. [CrossRef] [PubMed]
48. Spoendlin, H. Primary Structural Changes in the Organ of Corti After Acoustic Overstimulation. *Acta Oto Laryngolog.* **1971**, *71*, 166–176. [CrossRef]
49. Corwin, J.; Cotanche, D. Regeneration of sensory hair cells after acoustic trauma. *Science* **1988**, *240*, 1772–1774. [CrossRef] [PubMed]
50. Solé, M.; Lenoir, M.; Fortuño, J.-M.; De Vreese, S.; van der Schaar, M.; André, M. Sea Lice Are Sensitive to Low Frequency Sounds. *J. Mar. Sci. Eng.* **2021**, *9*, 765. [CrossRef]
51. Nehring, S. Invasion History and Success of the American Blue Crab Callinectes sapidus in European and Adjacent Waters. In *the Wrong Place Alien Marine Crustaceans: Distribution, Biology and Impacts*; Springer: Amsterdam, The Netherlands, 2011; pp. 607–624.
52. Castejón, D.; Guerao, G. A new record of the American blue crab, *Callinectes sapidus* Rathbun, 1896 (Decapoda: Brachyura: Portunidae), from the Mediterranean coast of the Iberian Peninsula. *BioInvasions Rec.* **2013**, *2*, 141–143. [CrossRef]
53. Generalitat de Catalunya. Memòria D'actuacions de Control del Cargol Poma. Periode 2015–2018. 2018. Available online: http://mediambient.gencat.cat/web/.content/home/ambits_dactuacio/patrimoni_natural/especies_exotiques_medinatural/llista_sp_catalogades/invertebrat/cargol/Memoria-cargol-poma.pdf (accessed on 24 August 2021).

Journal of
*Marine Science
and Engineering*

Article

A Case Study of Whistle Detection and Localization for Humpback Dolphins in Taiwan

Ching-Tang Hung, Wei-Yen Chu, Wei-Lun Li, Yen-Hsiang Huang, Wei-Chun Hu and Chi-Fang Chen *

Department of Engineering Science and Ocean Engineer, National Taiwan University, Taipei City 10216, Taiwan;
d08525005@ntu.edu.tw (C.-T.H.); r06525012@ntu.edu.tw (W.-Y.C.); r05525004@ntu.edu.tw (W.-L.L.);
r05525002@ntu.edu.tw (Y.-H.H.); d03525001@ntu.edu.tw (W.-C.H.)
* Correspondence: chifang@ntu.edu.tw; Tel.: +886-2-3355-5735

Abstract: In recent years, Taiwan's government has focused on policies regarding offshore wind farming near the Indo-Pacific humpback dolphin habitat, where marine mammal observation is a critical consideration. The present research developed an algorithm called National Taiwan University Passive Acoustic Monitoring (NTU_PAM) to assist marine mammal observers (MMOs). The algorithm performs whistle detection processing and whistle localization. Whistle detection processing is based on image processing and whistle feature extraction; whistle localization is based on the time difference of arrival (TDOA) method. To test the whistle detection performance, we used the same data to compare NTU_PAM and the widely used software PAMGuard. To test whistle localization, we designed a real field experiment where a sound source projected simulated whistles, which were then recorded by several hydrophone stations. The data were analyzed to locate the moving path of the source. The results show that localization accuracy was higher when the sound source position was in the detection region composed of hydrophone stations. This paper provides a method for MMOs to conveniently observe the migration path and population dynamics of cetaceans without ecological disturbance.

Keywords: marine mammal; whistle detection; time difference of arrival; underwater acoustic; underwater sound sensing; ocean sound measurement

Citation: Hung, C.-T.; Chu, W.-Y.; Li, W.-L.; Huang, Y.-H.; Hu, W.-C.; Chen, C.-F. A Case Study of Whistle Detection and Localization for Humpback Dolphins in Taiwan. *J. Mar. Sci. Eng.* **2021**, 9, 725. https://doi.org/10.3390/jmse9070725

Academic Editors: Michel André, Christine Erbe and Giuseppa Buscaino

Received: 30 April 2021
Accepted: 28 June 2021
Published: 30 June 2021

Publisher's Note: MDPI stays neutral with regard to jurisdictional claims in published maps and institutional affiliations.

1. Introduction

Currently, most of Taiwan's raw materials for energy production, including coking coal, fuel coal, crude, and liquefied natural gas [1], are imported and have a large and immediate impact on the environment. Therefore, the government has actively developed green energy, including offshore wind farms [2], but most sites overlap with Indo-Pacific humpback dolphin reservation zones. The noise from pile driving during construction may impact marine mammals and cause auditory injury, ranging from temporary threshold shift (TTS) to permanent threshold shift (PTS) in hearing [3]. To minimize the noise-induced impact on cetaceans caused by construction and the operation of wind turbines, establishing a marine mammal detection mechanism is a priority. The traditional method to detect cetaceans is visual, whereby marine mammal Observers (MMOs) work from vehicles, using the naked eye to search for cetaceans, an operation that is expensive and offers only a low probability of success; moreover, it is limited to daylight hours. Underwater acoustics provide an alternative technique to detect marine mammals, and the cetacean call can be used as a specific characteristic of detection. We used passive acoustic monitoring (PAM) to develop an algorithm and NTU_PAM to monitor cetacean calls followed by motion tracking. In addition to overcoming the weaknesses of the visual method, NTU_PAM can show the correlation between the results of the visual method and PAM.

Cetaceans produce two major types of cetacean calls [4,5]: (1) the "whistle" is a continuous, narrow-band, and frequency-modulated signal that is thought to be a form of social communication; (2) the "click" is considered a bio-sonar and is a short, broadband, and directional impulse signal used to navigate, detect, and identify objects. In marine mammal research, PAM has proved a useful tool. For example, (1) Spaulding et al. [6] built a near-real-time buoy system to automatically detect North Atlantic right whale calls in Cape Cod Bay and near the Boston Harbor. When the buoy system detects a whale call, an alarm signal is transmitted and the call is recorded. (2) Linnenschmidt et al. [7] equipped an acoustic data logger on a porpoise to record clicks and determine the relationship among the click, movement, and diving behavior. (3) Akamatsu et al. [8] used an underwater pulse event recorder (A-TAG) to record clicks and analyze critical parameters such as interclick interval (ICI).

Previous studies of cetacean whistle detection have been vigorous. Gannier et al. [9] developed Seafox software to extract whistle characteristics (length, beginning frequency, ending frequency, maximum frequency, minimum frequency, etc.) on a time spectrogram and used a regression tree to classify five dolphin species. Lai [10] used the mel-frequency cepstral coefficient to simulate human auditory features, namely the critical band and auditory masking, and to extract the characteristics of the whistle. The whistle characteristics were then used in a support vector machine (SVM) to identify the cetacean species. Caldwell and Caldwell [11–13] hypothesized that signature whistle variations, which dolphins emit and which carry information, are required so distinctive whistles can be used to identify individual dolphins. Datta and Sturtivant [14] considered two whistle features on a spectrogram (overall contour shape and detailed contour structure difference) as parameters of the signature whistle and grouped whistles using the hidden Markov model (HMM) method. Bahoura and Simard [15] used an artificial neural network to classify blue whale calls. The above research is based on supervised machine learning methods requiring numerous sets of clean training data, manually labeling the calls, and building the model. These models are only suitable for specific or regional species.

To avoid the disadvantages of labeling, training, and specific targeting, Gillespie et al. [16] developed a whistle detector based on image processing on a spectrogram which is implemented as the Whistle and Moan Detector module in PAMGuard. PAMGuard software includes a user-friendly, human–machine interface and modules for data processing and marine mammal detection [17] and has been widely used for real-time marine mammal monitoring. Lin [18] devised a non-targeted algorithm on the MATLAB platform that helps users grasp the position of whistles across many audio files, making further processing convenient. Lin et al. [19] first denoised the spectrogram and then detected the whistle characteristics. Gillespie's and Lin's methods include four main steps: (1) spectrogram, (2) image processing, (3) whistle feature extraction, and (4) combination of the whistle data points. We applied the same pattern to develop the whistle detection algorithm. A similar concept is applied in steps 2–4, but the detailed methods are different. We also compared NTU_PAM and PAMGuard, which is regarded as a standard of whistle detection.

Tracking cetaceans is another recent primary research subject. Janik et al. [20] deployed three hydrophones to form a two-dimensional, triangular array in Beauly Firth, Northern Scotland, U.K. The interhydrophone distances were 208, 513, and 506 m. An artificial sound was then projected at a depth of 1 m. The time difference of signal arrival for each pair of hydrophones became markers to conduct localization of a sound source. Wang et al. [21] deployed a two-dimensional, cross-shaped array consisting of five hydrophones from the side of the boat at a depth of 1 m in Pearl River Estuary, China, and Beibu Gulf of Guangxi, China. The inter-hydrophone distances were 1.47, 1.54, 2.08, and 2.18 m. The boat followed the dolphin group at a close distance to receive the dolphin call, and they used the difference in arrival time of a sound at each hydrophone pair to localize the targets. Wiggins et al. [22] deployed a tracking high-frequency acoustic recording package (HARP) [23] consisting of four hydrophones at 3 m above the seafloor offshore of Southern California to track beached whales and dolphins. Wiggins et al. [24] also deployed four HARPs offshore of Southern California to track whistling dolphins. Both of Wiggins's methods used the TDOA method. Building on the demonstrated effectiveness of TDOA for tracking and localization, we utilized four hydrophone stations to form a kilometer-scale array for tracking the source based on TDOA.

We designed an experiment that simulated different whistle types in the real field and developed four PAM stations to track the artificial source. Four stations were deployed near Taichung Harbor to record the simulated calls. After processing the detected algorithm, finding the whistle time, and tracking the source, we compared the results from the algorithm and the moving path of the boat carrying the sound source. In this study, we developed an algorithm that does not require a trained model for the automatic detection of the whistle. The algorithm is based on the time length and frequency band of the whistle feature. Furthermore, the automatic detection algorithm and localization method were combined as NTU_PAM. NTU_PAM can work as an auxiliary tool for MMO during the daytime, and it can function as the main monitoring tool at night.

2. Whistle Detector Algorithm

Passive acoustic monitoring has been used widely in marine monitoring to amass longitudinal data and requires high-efficiency algorithms to assist researchers in finding the required file segments. We developed a whistle detector algorithm, which was then improved according to Li's prototype algorithm [25]. The algorithm can detect any creature producing a whistle and the whistle's detected frequency range, depending on the species. There are six main processes in the algorithm:

1. Transfer time-series data to a spectrogram by short-time Fourier transform (STFT);
2. Remove the noise on the time axis of the spectrogram;
3. Remove the salt and pepper noise in the spectrogram;
4. Find the data point that satisfies the condition of the power spectral density (PSD) and signal-to-noise ratio (SNR);
5. Extract data points using the features of whistles;
6. Cluster data points into different whistles.

A flow chart of the algorithm is shown in Figure 1. In order to present whistles clearly on the spectrogram, some processes are based on image processing. Each process will be described in detail. Figure 2 shows each step of the results.

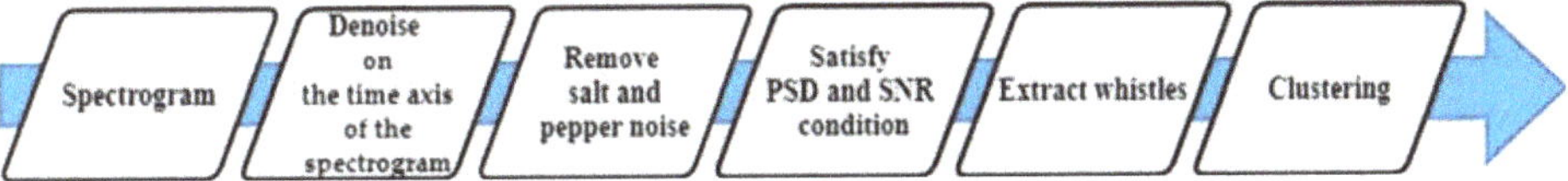

Figure 1. Whistle detector algorithm flow chart.

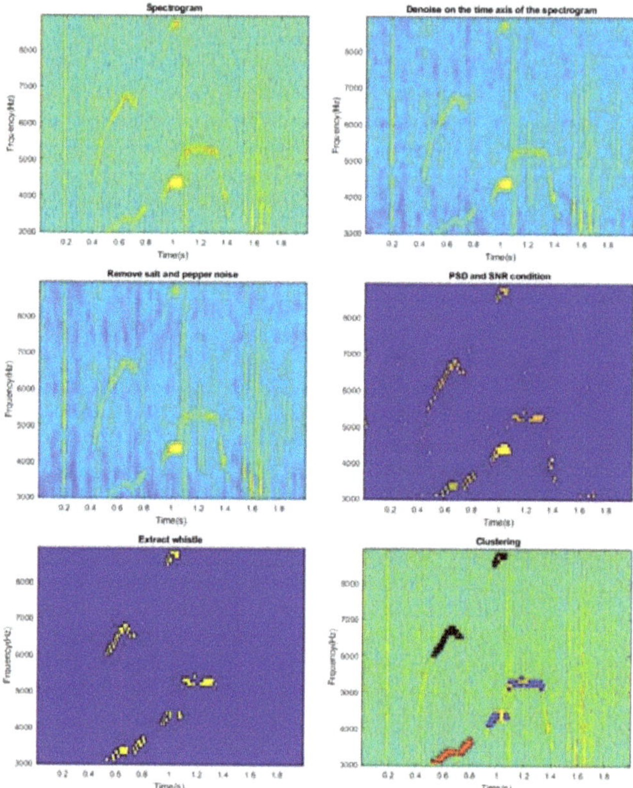

Figure 2. Each step of the results.

2.1. Spectrogram

We used the STFT [26], which adds a window function to obtain the frequency domain information changed by the time domain. This establishes a frame to slide on the time domain signal and extracts the signal in the frame, which convolves with the window function to perform the Fourier transform. This information is used to produce the spectrogram. The window function is the Hamming window [27], the frame length is 0.01 s, and the overlap is 90%. The STFT formula is shown in Equation (1), where $w(t)$ is window function and $x(t)$ is raw data.

$$S_{t,f} = \int_{-\infty}^{+\infty} w(t-\tau)x(t)e^{-ift}dt \tag{1}$$

2.2. Denoising on the Time Axis of the Spectrogram

Whistle length is long compared to impulse noise; therefore, we use the moving average method to remove impulse noise on the spectrogram. Every 20 points on the time axis of each single frequency band are averaged to build a new spectrogram; the formula is shown in Equation (2), where $S_{t,f}$ is the original spectrogram and $S'_{t,f}$ is the new spectrogram after denoising.

$$S'_{t,f} = \frac{1}{20} \sum_{n=0}^{19} S_{t-n,f} \tag{2}$$

2.3. Removing Salt and Pepper Noise

A median filter, often used in image processing and a technique for nonlinear signal processing, was used to remove salt and pepper noise [28]. The median of every 3-by-3 matrix on the spectrogram is calculated. The formula is shown in Equation (3), where $S'_{t,f}$ is the spectrogram after the denoising and $S''_{t,f}$ is the spectrogram after using the median filter.

$$S''_{t,f} = median\left(S'_{t+i,f+j}\right); \; i,j = -1,0,1 \tag{3}$$

2.4. Satisfying PSD and SNR Conditions

Since a whistle is a narrow frequency band signal, with the occurrence of a whistle, its PSD is much larger than that of the point whose frequency is very close to the whistle. The definition of SNR in this study is shown in Equation (4). If the PSD is larger than the PSD threshold and the SNR is larger than the SNR threshold simultaneously at a data point, the data point will be replaced by one. If this is not the case, the data point will be replaced by zero. The formula is shown in Equation (5). The new spectrogram $B_{t,f}$ is a binary image. The default value of the SNR threshold and the PSD threshold are 6 dB and 40 dB (re 1 µPa^2/Hz), respectively.

$$SNR = \frac{2S''_{t,f}}{\left(S''_{t,f+1} + S''_{t,f-1}\right)} \tag{4}$$

$$B_{t,f} = \begin{array}{l} 1, \; SNR_{t,f} \geq SNR_{threshold} \; \& \; S''_{t,f} \geq PSD_{threshold} \\ 0, \; otherwise \end{array} \tag{5}$$

2.5. Extracting the Whistle

As mentioned in Section 2.4, the whistle is a narrow frequency band and a continuous signal. In this method, the nearby data points whose value is one are connected and labeled as a segment. Next, two conditions are set: the frequency bandwidth threshold and the time length threshold. Lastly, the segments whose frequency bandwidth is smaller than the frequency bandwidth threshold and whose time length is longer than the time length threshold are retained. The binary image $B_{t,f}$ will be refreshed as a new image $B'_{t,f}$. The default values of frequency bandwidth threshold and time length threshold are 300 Hz and 0.06 seconds, respectively.

2.6. Clustering

The k-means method [29] is used to cluster the data points in $B'_{t,f}$. According to the difference of frequency and time, some of the whistle segments from Section 2.5 and above are merged. If the time interval of two segments is smaller than 0.3 seconds and the difference of frequency between two segments is smaller than 1 kHz simultaneously, two segments will be considered as one whistle segment. After merging, the k (number of clusters) is decided by the new number of segments. Each data point automatically combines into k whistles by calculating Euclidean distance of frequency and time index in $B'_{t,f}$. Each whistle's start time, end time, start frequency, and end frequency are recorded after k-means.

3. Localization Method

TDOA was used to track the whistle. We devised an experiment to track the moving path of the artificial source by a whistle detector algorithm and TDOA.

3.1. Time Difference of Arrival (TDOA)

TDOA is often used in signal source positioning [30]. It only requires the received signal time and the speed that the signal travels. Once the signal is received at the two receiving stations, the difference in arrival time can be used to draw the hyperbola of possible location by the equation shown in Equations (6) and (7). If we have three receiving stations, least two hyperbolas are produced, as shown in Figure 3, and their intersection will be the signal source location. To realize this hypothesis, the receiving stations must be time-synchronized.

$$\sqrt{(x-x_1)^2 + (y-y_1)^2} - \sqrt{(x-x_3)^2 + (y-y_3)^2} = c(t_1 - t_3) \qquad (6)$$

$$\sqrt{(x-x_2)^2 + (y-y_2)^2} - \sqrt{(x-x_3)^2 + (y-y_3)^2} = c(t_2 - t_3) \qquad (7)$$

where t_1, t_2, and t_3 are the times when the same signal arrives at different hydrophones; (x, y) is the position of the unknown signal source; and c is the sound speed from the local sound speed profile.

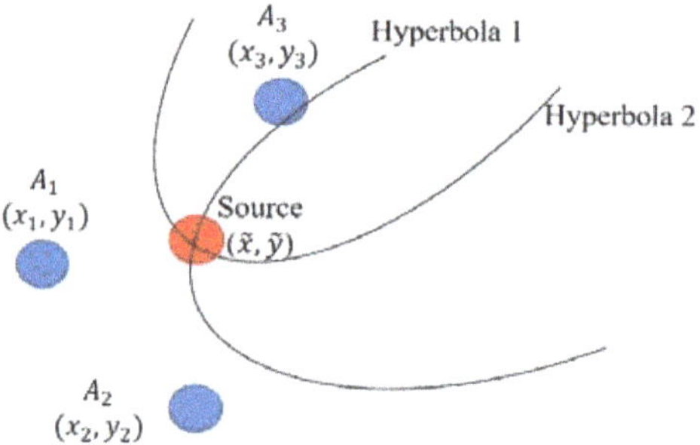

Figure 3. TDOA schematic.

3.2. Taichung Harbor TDOA Experimental Configuration

We deployed four hydrophone stations near Taichung Harbor, an area where Indo-Pacific humpback dolphins are extremely active [31,32]. The locations of the hydrophones are shown in Figure 4, and the exact latitude and longitude are shown in Table 1. The Beaufort Sea state was below 3, and the ambient noise is illustrated in Figure 5 as a percentile level. The highest PSD was around 95 dB (re 1 μPa^2/Hz) from 60–70 Hz on L50, possibly produced by shipping noise, and the PSD from 3 kHz–10 kHz was around 65 dB (re 1 μPa^2/Hz).

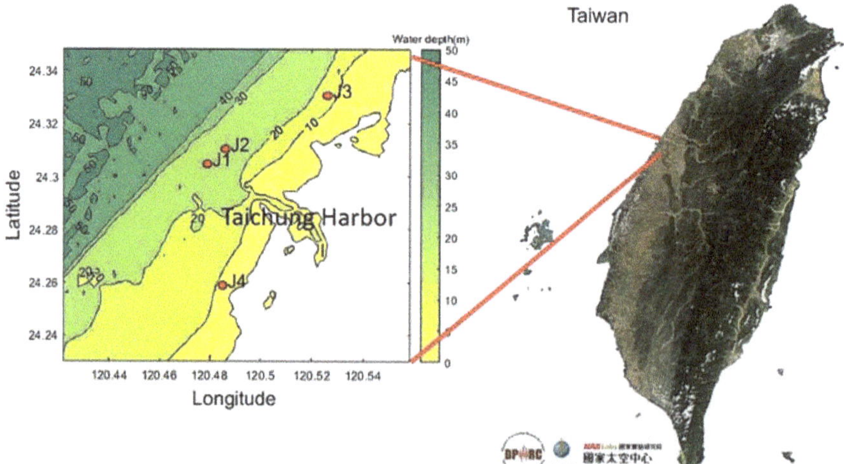

Figure 4. Hydrophone station locations.

Table 1. Latitude and longitude of hydrophone stations.

Station	Latitude (N)	Longitude (E)	Depth (m)
J1	24.3305°	120.4788°	29.1
J2	24.3101°	120.4861°	28.7
J3	24.3305°	120.5259°	8.0
J4	24.2588°	120.4851°	11.0

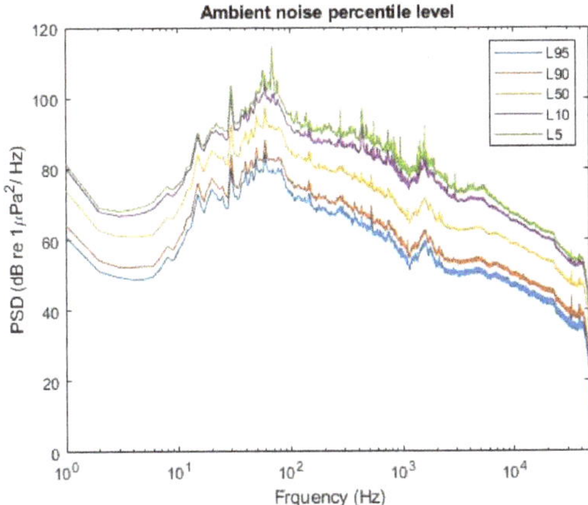

Figure 5. Ambient noise percentile level: Ln is the noise level exceeding *n*% of the measurement time, i.e., L50 is the noise level exceeding 50% of the measurement time.

The SoundTrap ST500 hydrophone recorder was used at point J3, and three Wildlife Acoustics SM3M hydrophone recorders were used at points J1, J2, and J4. They were deployed using the bottom-mounted method with sampling frequency set to 96 kHz. To achieve time synchronization for all recorders, we produced an impulse signal as a benchmark for correcting the time before deploying. To simulate the whistle of an actual Indo-Pacific humpback dolphin, which features a frequency range of 3–9 kHz, three kinds of artificial sound signals were employed: (a) rising frequency (5–9 kHz), (b) U-type (9–5–9 kHz), and (c) decreasing frequency (9 5 kHz), with a time length of one second, as shown in Figure 6. The source level (SL) was 160 dB (re 1 μ Pa at 1 m). The underwater acoustic projector SQS-23 was placed at a water depth of 5 m (Figure 7), since Indo-Pacific humpback dolphins often stay about 5 m below sea level [33]. Figure 8 shows where the artificial sound signals were played, every 10 seconds for 10 minutes, in the 15 spots (T1–T15) outside Taichung Harbor.

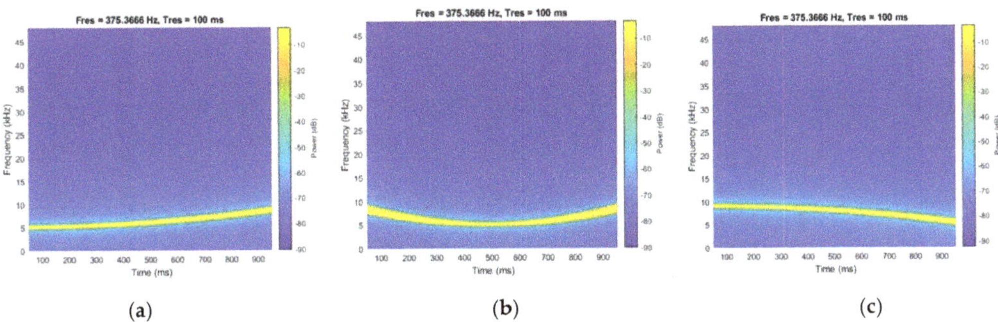

Figure 6. (**a**) Rising frequency; (**b**) U-type; (**c**) decreasing frequency.

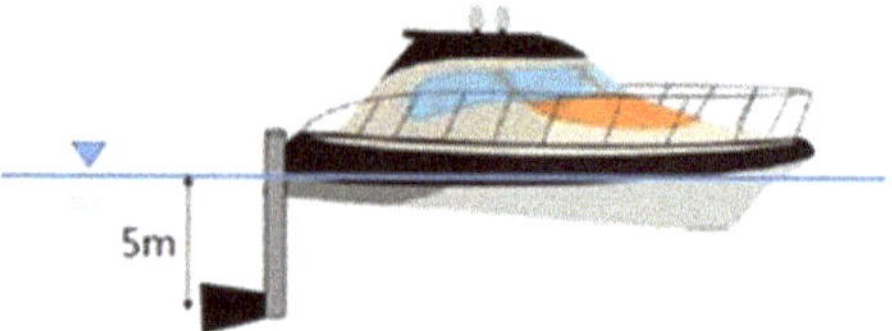

Figure 7. Schematic of the installation position of the projector.

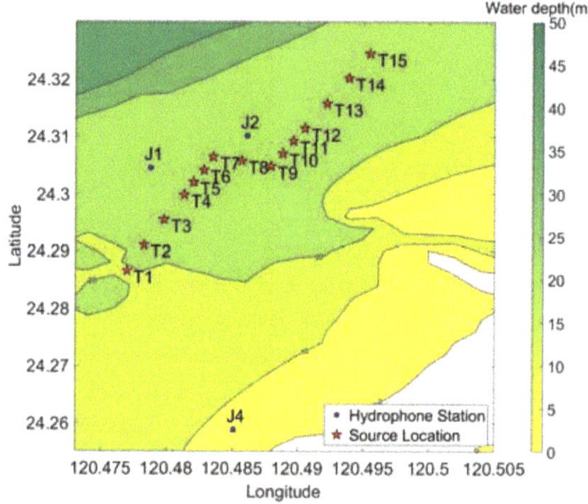

Figure 8. Locations of the signal sources.

3.3. Experimental Data Analysis Method

In this experiment, the SNR of the received signal was larger than 10 dB, exceeding the NTU_PAM-recommended SNR threshold of 6 dB. The signals recorded by each of the hydrophones at the four stations when the source was at point T10 are shown in Figure 9. To find the artificial whistle within the sound file, NTU_PAM was used to extract information, namely the start and end times from the raw data of the four hydrophones. However, the extracted time information was not precise enough for TDOA. For increased accuracy, the raw data of the start and end times of the whistle were directly analyzed without being processed by the algorithm. The time of the J2 station was considered as the central time, and cross-correlation analysis with the full frequency band raw data of the central station and three other stations was performed to determine the time difference, as shown in Equations (8) and (9), where X_2 is J2 station's whistle raw data; X_o is the three other stations' whistle raw data; R is the result of cross-correlation; and td is the time difference, which was used to obtain the location of the signal source by the TDOA method.

$$R(\tau) = \int X_2(t)X_o(t-\tau)dt \tag{8}$$

$$td = max(R(\tau)) \tag{9}$$

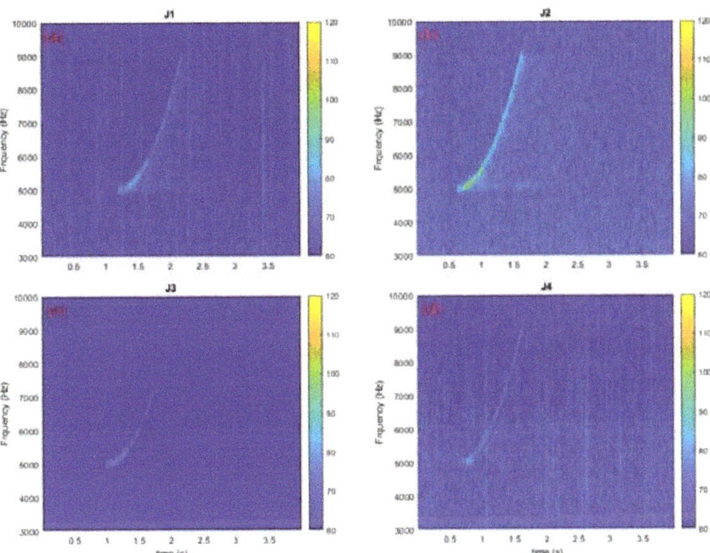

Figure 9. Received signal of each station when the source was at point T10. (**a**) J1 station; (**b**) J2 station; (**c**) J3 station; (**d**) J4 station.

4. Results

4.1. Comparison with PAMGuard

As mentioned, PAMGuard software is widely used in the field of marine mammal observation. In this research, the performance of NTU_PAM and the Whistle and Moan Detector module of PAMGuard were compared using the same hardware (an i9-9900 CPU from Intel Corporation with 64 GB of memory). The test audio is a two-minute sound file, rich in whistles and with a sampling frequency of 96 kHz, recorded near the sea area of Yunlin, Taiwan [34]. We manually confirmed that the file contained a total of 33 whistles.

When the PAMGuard Whistle and Moan Detector's parameters were set at a window length of 2048 data points (0.02 s) and 1024 data points (0.01 s), and when the overlap ratios were 50% and 90%, the NTU_PAM's recommended window length was 0.01 s with an overlap ratio of 90% and SNR set to 6 dB. As shown in Table 2, PAMGuard with settings of window length at 1024 data points, 90% overlap ratio, and 6 dB SNR shows the closest result of the 47 detected whistles to the manually confirmed 33 whistles. A total of 30 whistles were detected by NTU_PAM.

Table 2. Comparison of results.

Algorithm	Parameters	Detected Numbers
Manually confirmed	-	33
PAMGuard	Window = 2048 Overlap = 50% SNR = 6 dB	79
PAMGuard	Window = 2048 Overlap = 90% SNR = 6 dB	50

Table 2. *Cont.*

Algorithm	Parameters	Detected Numbers
PAMGuard	Window = 1024 Overlap = 50% SNR = 6 dB	91
PAMGuard	Window = 1024 Overlap = 90% SNR = 6 dB	47
NTU_PAM	Window = 0.01 s Overlap = 90% SNR = 6 dB	30

4.2. Experimental Results

At least three signal receiving stations were used to calculate TDOA. When the intersection of the hyperbolic curves is plural, the center point is taken as the final judgment location. To verify localization accuracy, GPS data from the experimental ship bearing the sound source were compared to results from TDOA.

In the series of graphs in Figure 10, the blue dot is the hydrophone station position (J1, J2, and J4), the red dot is the signal source position of the experimental ship's GPS record, and the yellow star is the TDOA positioning result. The results from the first experiment testing the rising frequency (5–9 kHz) signal are shown in Figure 10a. The positioning accuracy was higher when the sound source was nearer to the center positions J1 and J2 from the group of hydrophone stations. The nearest positioning points T4 to T11 showed an average positioning error of 24.7 m, and the overall positioning error was 143.5 m, which was affected by the lower accuracy of the outer point.

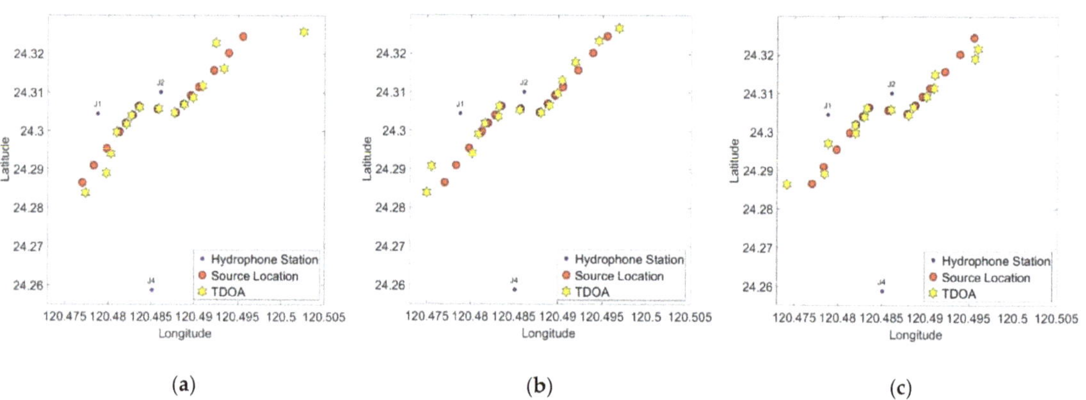

Figure 10. (a) Result of rising frequency signal; (b) result of decreasing frequency signal; (c) result of U-shaped signal.

The second experiment was the decreasing frequency (9–5 kHz) signal, and its positioning trend was similar to the rising frequency signal (Figure 10b). It also showed higher positioning accuracy when the signal source was close to the J1 and J2 stations. The average positioning error of T4 to T11 was 44.8 m, larger than that of the rising frequency signal, and the overall positioning error was 145.9 m. Finally, the U-shaped (9–5–9 kHz) signal displayed a similar trend as the aforementioned signals (Figure 10c). The average positioning error of T4 to T11 was 39.6 m, but the overall positioning error was the smallest of the three signals at 116.1 m.

5. Discussion

In the comparison between PAMGuard and NTU_PAM, the results were close to the number of whistles that was manually confirmed and showed that both performed well on whistle detection. The reason for the different numbers detected may be that PAMGuard is a real-time auxiliary tool mainly provided to visual method researchers for detecting the occurrence of a call; as such, it only needs a few window lengths of data to detect the whistle. As to the amount of audio data required, NTU_PAM needs one second or more of data to build a spectrogram and to initiate processing. However, PAMGuard may, at times, break one call into several calls, as shown in Figure 11. According to the results, NTU_PAM is suitable for to processing measurements captured over a longer duration, and it proves as robust as PAMGuard.

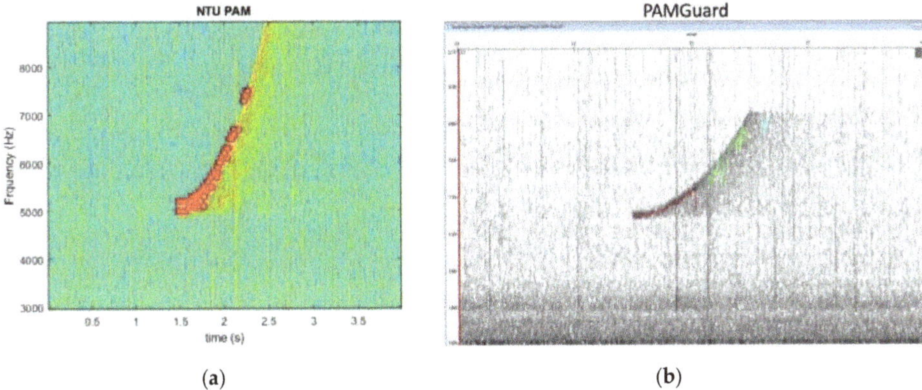

Figure 11. Output results of rising frequency signals: (**a**) NTU_PAM; (**b**) PAMGuard with 1024 data points window length, 90% overlap ratio and 6 dB SNR.

In the localization experiment, the TDOA method proved useful for localizing the whistle source. Figure 12 plots the errors of the three different types of signals at each spot and indicates that the error is small when the source is inside the region of the four hydrophone recorders (points T4–T11); when outside the region (points T1–T3 and T12–T15), location was only approximate (Figure 13). The results of this experiment indicate strengths in using the NTU_PAM for successful tracking of cetaceans.

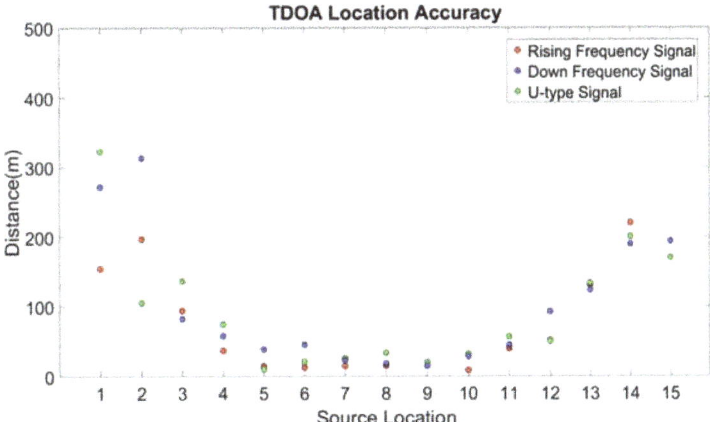

Figure 12. Distribution of localization errors.

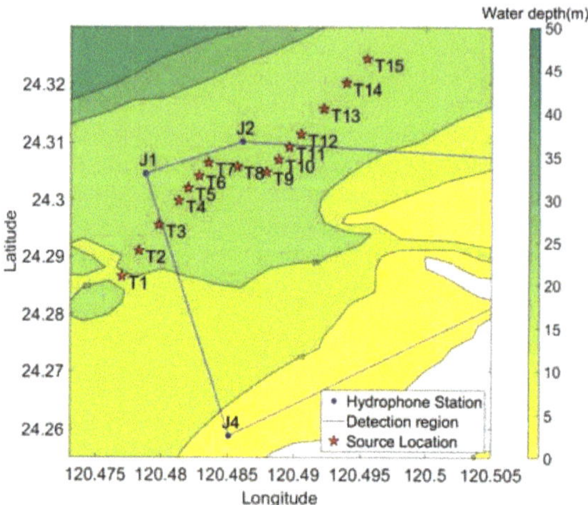

Figure 13. Detection region consisted of hydrophones.

6. Conclusions

In this research, we devised and developed the NTU_PAM algorithm, which performs whistle detection and whistle localization based on the TDOA method. The results showed NTU_PAM is able to localize and track the whistle sound source with high accuracy. In the future, MMOs can monitor the moving path of marine mammals via the visual method combined with NTU_PAM, making it possible to monitor cetaceans without being limited by daylight hours.

Author Contributions: Conceptualization, C.-T.H., W.-L.L., Y.-H.H., W.-Y.C. and C.-F.C.; methodology, C.-T.H., W.-L.L., Y.-H.H., W.-Y.C. and C.-F.C.; software, C.-T.H., W.-L.L. and W.-Y.C.; formal analysis, W.-Y.C. and W.-C.H.; writing—original draft preparation, C.-T.H. and W.-Y.C.; writing—review and editing, C.-T.H. and C.-F.C.; supervision, C.-F.C.; project administration, C.-F.C.; funding acquisition, C.-F.C. All authors have read and agreed to the published version of the manuscript.

Funding: This research was funded by the Ministry of Science and Technology, Taiwan (MOST 109-2221-E-002-198-MY3).

Institutional Review Board Statement: Not applicable.

Informed Consent Statement: Not applicable.

Data Availability Statement: Request from the corresponding author of this article.

Acknowledgments: The authors would like to thank the Formosa Plastics Group and the Formosa Petrochemical Corporation, 1141946-00, for the 2 minutes of data. The authors would like to thank the Ministry of Science and Technology, Taiwan, for the funding; and Professor Lien-Siang Chou for help with marine mammal knowledge.

Conflicts of Interest: The authors declare no conflict of interest.

Short Biography of Authors

Ching-Tang Hung received his B.S. degree from Department of Hydraulic and Ocean Engineering, National Cheng Kung University and M.S. degree from Department of Engineering Science and Ocean Engineering, National Taiwan University. He is currently a Ph.D. candidate in Department of Engineering Science and Ocean Engineering, National Taiwan University. His research interests include underwater acoustic, whistle detection and automated unmanned surface vehicle system.

Yen-Hsiang Huang received his B.S. degree from the Department of Mechanical Engineering, National Chun Hsing University and M.S. degree from Department of Engineering Science and Ocean Engineering, National Taiwan University. He is currently working as firmware engineer in Taiwan. His research interest including underwater acoustics, USV system integration and strong skill in C/C++ programming.

Wei-Yen Chu received his B.S. degree from Department of Marine Engineering, National Taiwan Ocean University and M.S. degree from Department of Engineering Science and Ocean Engineering, National Taiwan University. He is currently working in the semiconductor manufacturing area in Taiwan. His research interests include underwater acoustics, signal processing, whistle detection, localization and simulation.

Wei-Chun Hu is a Ph.D. candidate at the Department of Engineering Sciences and Ocean Engineering, National Taiwan University, Taiwan. His research focuses specifically on the soundscape and the propagation of underwater noise. Recent participated publications can be found in Ecological Indicators and Entropy journal.

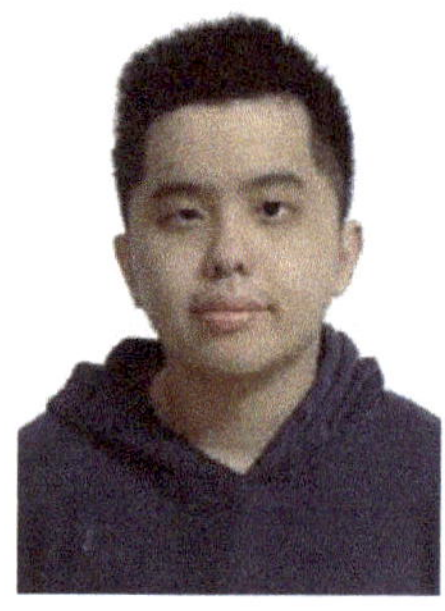

Wei-Lun Li received his B.S. degree from Department of Hydraulic and Ocean Engineering, National Cheng Kung University and M.S. degree from Department of Engineering Science and Ocean Engineering, National Taiwan University. He is currently working as research assistant, Institute of hydrobiology, Chinese Academy of science. His research interests include underwater acoustic and underwater detection of marine mammals.

Chi-Fang Chen received her Ph.D. in the Department of Ocean Engineering, Massachusetts Institute of Technology in 1991, and started her career as the faculty member of the Department of Naval Architecture of National Taiwan University from 1991 till now. (Department of Naval Architecture was renamed as Department of Engineering Science and Ocean Engineering in 2000). Her research expertise and interests are underwater acoustics and underwater acoustic propagation. She is 0conducting passive acoustic monitoring (PAM) in recognizing sounds from different species in the ocean which includes Sousa Chinensis in Taiwan waters. She also has interests in autonomous ocean sensing, and has supervised two master's theses in AUV, and is now supervising five graduate students in autonomous surface vehicle study.

References

1. Bureau of Energy, Ministry of Economic Affairs (Ed.) *Bureau of Energy Annual Report*; Bureau of Energy, Ministry of Economic Affairs: Taiwan, 2018.
2. Department of Information Services, Executive Yuan (Ed.) *Four-Year Wind Power Promotion Plan*; Department of Information Services, Executive Yuan: Taiwan, 2019.
3. Bailey, H.; Senior, B.; Simmons, D.; Rusin, J.; Picken, G.; Thompson, P. Assessing underwater noise levels during pile-driving at an offshore windfarm and its potential effects on marine mammals. *Mar. Pollut. Bull.* **2010**, *60*, 888–897. [CrossRef] [PubMed]
4. Au, W.W.L. Hearing in Whales and Dolphins: An Overview. In *Hearing by Whales and Dolphins*; Springer: Berlin/Heidelberg, Germany, 2000; pp. 1–42.
5. Oswald, J.N.; Rankin, S.; Barlow, J. To Whistle or Not to Whistle? Geographic Variation in the Whistling Behavior of Small Odontocetes. *Aquat. Mamm.* **2008**, *34*, 288–302. [CrossRef]
6. Spaulding, E.; Robbins, M.; Calupca, T.; Clark, C.W.; Tremblay, C.; Waack, A.; Warde, A.; Kemp, J.; Newhall, K. An autonomous, near-real-time buoy system for automatic detection of North Atlantic right whale calls. In Proceedings of the Meetings on Acoustics 157ASA, Portland, Oregon, 18–22 May 2009; Acoustical Society of America: New York, NY, USA, 2009; Volume 6.
7. Linnenschmidt, M.; Teilmann, J.; Akamatsu, T.; Dietz, R.; Miller, L.A. Biosonar, dive, and foraging activity of satellite tracked harbor porpoises (*Phocoena phocoena*). *Mar. Mammal Sci.* **2012**, *29*, E77–E97. [CrossRef]
8. Akamatsu, T.; Wang, D.; Wang, K.; Naito, Y. Biosonar behaviour of free-ranging porpoises. *Proc. Soc. B Biol. Sci.* **2005**, *272*, 797–801. [CrossRef] [PubMed]
9. Gannier, A.; Fuchs, S.; Quèbre, P.; Oswald, J.N. Performance of a contour-based classification method for whistles of Mediterranean delphinids. *Appl. Acoust.* **2010**, *71*, 1063–1069. [CrossRef]
10. Lai, Y.-C. The Analysis of Identification of Cetaceans' Whistle with Support Vector Machine. Master's Thesis, Department of Engineering Science and Ocean Engineering, National Taiwan University, Taipei, Taiwan, 2014; pp. 1–61.
11. Caldwell, M.C.; Caldwell, D.K. Individualized whistle contours in bottle-nosed dolphins (*Tursiops truncatus*). *Nature* **1965**, *207*, 434–435. [CrossRef]
12. Caldwell, M.C.; Caldwell, D.D. *Statistical Evidence for Individual Signature Whistles in the Pacific Whitesided Dolphin, Lagenorhynchus obliquidens*; Los Angeles County Museum CA: Los Angeles, CA, USA, 1970.
13. Caldwell, M.C.; Caldwell, D.K.; Miller, J.F. *Statistical Evidence for Individual Signature Whistles in the Spotted Dolphin, Stenella plagiodon*; Los Angeles County Museum Calif: Los Angeles, CA, USA, 1970.
14. Datta, S.; Sturtivant, C. Dolphin whistle classification for determining group identities. *Signal Process.* **2002**, *82*, 251–258. [CrossRef]

15. Bahoura, M.; Simard, Y. Blue whale calls classification using short-time Fourier and wavelet packet transforms and artificial neural network. *Digit. Signal Process.* **2010**, *20*, 1256–1263. [CrossRef]
16. Gillespie, D.; Caillat, M.; Gordon, J.; White, P. Automatic detection and classification of odontocete whistles. *J. Acoust. Soc. Am.* **2013**, *134*, 2427–2437. [CrossRef] [PubMed]
17. Gillespie, D.; Mellinger, D.; Gordon, J.; Mclaren, D.; Redmond, P.; McHugh, R.; Trinder, P.; Deng, X.; Thode, A. PAMGUARD: Semiautomated, open source software for real-time acoustic detection and localisation of cetaceans. *J. Acoust. Soc. Am.* **2008**, *30*, 54–62. [CrossRef]
18. Lin, T.-H. The Application of Passive Acoustic Monitoring for Studying Indo-Pacific Humpback Dolphin Behavior and Habitat Use off Western Taiwan. Master's Thesis, Institute of Ecology and Evolutionary Biology, National Taiwan University, Taipei, Taiwan, 2013; pp. 1–150.
19. Lin, T.-H.; Chou, L.-S.; Akamatsu, T.; Chan, H.-C.; Chen, C.-F. An automatic detection algorithm for extracting the representative frequency of cetacean tonal sounds. *J. Acoust. Soc. Am.* **2013**, *134*, 2477–2485. [CrossRef] [PubMed]
20. Janik, V.M.; Parijs, S.M.; Thompson, P.M. A two-dimensional acoustic localization system for marine mammals. *Mar. Mammal Sci.* **2006**, *16*, 437–447. [CrossRef]
21. Wang, Z.-T.; Au, W.W.; Rendell, L.; Wang, K.-X.; Wu, H.-P.; Wu, Y.-P.; Liu, J.-C.; Duan, G.-Q.; Cao, H.-J.; Wang, D. Apparent source levels and active communication space of whistles of free-ranging Indo-Pacific humpback dolphins (*Sousa chinensis*) in the Pearl River Estuary and Beibu Gulf, China. *PeerJ* **2016**, *4*, 1695. [CrossRef] [PubMed]
22. Wiggins, S.M.; McDonald, M.A.; Hildebrand, J.A. Beaked whale and dolphin tracking using a multichannel autonomous acoustic recorder. *J. Acoust. Soc. Am.* **2012**, *131*, 156–163. [CrossRef]
23. Wiggins, S.M.; Hildebrand, J.A. High-frequency Acoustic Recording Package (HARP) for broad-band, long-term marine mammal monitoring. In Proceedings of the 2007 Symposium on Underwater Technology and Workshop on Scientific Use of Submarine Cables and Related Technologies, Tokyo, Japan, 17–20 April 2007; Institute of Electrical and Electronics Engineers (IEEE): Piscataway, NJ, USA, 2007; pp. 551–557.
24. Wiggins, S.M.; Frasier, K.E.; Henderson, E.E.; Hildebrand, J.A. Tracking dolphin whistles using an autonomous acoustic recorder array. *J. Acoust. Soc. Am.* **2013**, *133*, 3813–3818. [CrossRef] [PubMed]
25. Li, W.-L. Study of Dolphin Whistle Detection. Master's Thesis, Department of Engineering Science and Ocean Engineering, National Taiwan University, Taipei, Taiwan, 2018; pp. 1–73.
26. Griffin, D.; Lim, J. Signal estimation from modified short-time Fourier transform. *IEEE Trans. Acoust. Speech, Signal Process.* **1984**, *32*, 236–243. [CrossRef]
27. Podder, P.; Khan, T.Z.; Khan, M.H.; Rahman, M.M. Comparative Performance Analysis of Hamming, Hanning and Blackman Window. *Int. J. Comput. Appl.* **2014**, *96*, 1–7. [CrossRef]
28. Chan, R.H.; Ho, C.-W.; Nikolova, M. Salt-and-pepper noise removal by median-type noise detectors and detail-preserving regularization. *IEEE Trans. Image Process.* **2005**, *14*, 1479–1485. [CrossRef] [PubMed]
29. Steinley, D. K-means clustering: A half-century synthesis. *Br. J. Math. Stat. Psychol.* **2006**, *59*, 1–34. [CrossRef] [PubMed]
30. Kaune, R. In Accuracy studies for TDOA and TOA localization. In Proceedings of the 2012 15th International Conference on Information Fusion, Singapore, 9–12 July 2012; IEEE: Piscataway, NJ, USA, 2012; pp. 408–415.
31. Chou, L.-S.; Ding, J.-J.; Lin, H.-J.; Suen, J.-P. *Population Ecology and Estuary Habitat Monitoring for Chinese White Dolphin (Sousa chinensis) (II)*; Forestry Bureau, Council of Agriculture, Executive Yuan: Taiwan, 2019.
32. Chou, L.-S.; Lin, H.-J., Suen, J.-P. *Population Ecology and Estuary Habitat Monitoring for Chinese white Dolphin (Sousa Chinensis)*; Forestry Bureau, Council of Agriculture, Executive Yuan: Taiwan, 2017.
33. Kuo, L.-H. Statistical Analysis of Whistle and Ambient Noise of Summer in the Habitat of Indo-Pacific Humpback Dolphin (*Sousa chinensis*). Master's Thesis, Department of Engineering Science and Ocean Engineering, National Taiwan University, Taipei, Taiwan, 2014; pp. 1–99.
34. Chi-Fang, C.; Wei-Chun, H.; Chi-Hung, L. *Population Ecology for Indo-Pacific Humpback Dolphins (Sousa chinensis) in Yunlin, Taiwan*; Forestry Bureau, Council of Agriculture, Executive Yuan: Taiwan, 2021.

Journal of
Marine Science and Engineering

Article

Marine Acoustic Zones of Australia

Christine Erbe [1,*], David Peel [2], Joshua N. Smith [3] and Renee P. Schoeman [1]

[1] Centre for Marine Science and Technology, Curtin University, Perth, WA 6102, Australia; renee.p.schoeman@gmail.com

[2] Data61, CSIRO, CSIRO Marine Laboratories, Hobart, TAS 7004, Australia; david.peel@data61.csiro.au

[3] Centre for Sustainable Aquatic Ecosystems, Harry Butler Institute, Murdoch University, Perth, WA 6150, Australia; joshua.smith@uqconnect.edu.au

* Correspondence: c.erbe@curtin.edu.au

Abstract: Underwater sound is modelled and mapped for purposes ranging from localised environmental impact assessments of individual offshore developments to large-scale marine spatial planning. As the area to be modelled increases, so does the computational effort. The effort is more easily handled if broken down into smaller regions that could be modelled separately and their results merged. The goal of our study was to split the Australian maritime Exclusive Economic Zone (EEZ) into a set of smaller acoustic zones, whereby each zone is characterised by a set of environmental parameters that vary more across than within zones. The environmental parameters chosen reflect the hydroacoustic (e.g., water column sound speed profile), geoacoustic (e.g., sound speeds and absorption coefficients for compressional and shear waves), and bathymetric (i.e., seafloor depth and slope) parameters that directly affect the way in which sound propagates. We present a multivariate Gaussian mixture model, modified to handle input vectors (sound speed profiles) of variable length, and fitted by an expectation-maximization algorithm, that clustered the environmental parameters into 20 maritime acoustic zones corresponding to 28 geographically separated locations. Mean zone parameters and shape files are available for download. The zones may be used to map, for example, underwater sound from commercial shipping within the entire Australian EEZ.

Keywords: underwater noise; sound propagation modelling; multivariate mixture model; acoustic zone; ship noise; Australian EEZ

Citation: Erbe, C.; Peel, D.; Smith, J.N.; Schoeman, R.P. Marine Acoustic Zones of Australia. *J. Mar. Sci. Eng.* **2021**, *9*, 340. https://doi.org/10.3390/jmse9030340

Academic Editor: Philippe Blondel

Received: 23 February 2021
Accepted: 16 March 2021
Published: 19 March 2021

Publisher's Note: MDPI stays neutral with regard to jurisdictional claims in published maps and institutional affiliations.

1. Introduction

Australia, continent and island country, has the third largest Exclusive Economic Zone (EEZ) after France and the United States of America. Australia's EEZ covers 8.5 Mkm^2, which is a rather large region to manage for a country with <25 M citizens. Management of an EEZ involves administration, surveillance, and enforcement, as well as research and planning [1].

In the era of blue economy, pressure on our ocean resources is increasing from industries that include fisheries and seafood, oil and gas, minerals, renewable energies, biotechnology, defence, tourism, and transport. Concerns for the marine ecosystem range from overexploitation to pollution. One aspect of pollution is marine noise. No blue industry can proceed without the generation of underwater noise; and so ocean management increasingly involves the assessment and regulation of marine noise emission [2].

The first step in the noise assessment and permitting process is typically the quantification and charting of noise (to be) emitted. In many cases, the activity to be permitted might be limited in time and space. For example, pile driving for bridge construction might last a few weeks to months and occur right at the location of the bridge, ensonifying an area of a few tens to hundreds of square-kilometres (e.g., [3]). Similarly, a seismic survey for oil or gas might last several weeks, during which an array of seismic airguns is discharged every few seconds and every few metres, covering tens to hundreds of kilometres of line transects and ensonifying an area of hundreds to thousands of square-kilometres (e.g., [4]).

In the case of shipping, the noise sources are distributed over the entire EEZ and each ship emits noise continuously while *en route*. Quantifying this noise and charting the ensonified areas is a big task. To be computationally practicable, the EEZ needs to be broken down into a number of smaller regions that can be more efficiently and accurately modelled. These smaller regions would ideally be defined based on their acoustic properties so that one sound propagation model may be set up and applied throughout an entire region and a different model set up for the neighbouring regions.

The propagation of sound in the ocean is affected by the hydroacoustic parameters of the water, the geoacoustic parameters of the seafloor, and the bathymetry [5]. From the sea surface to the seafloor, temperature and salinity vary, yielding site-characteristic sound speed profiles in the water column. Changes in the speed of sound evoke changes in the direction of sound propagation (via Snell's law). Minima in sound speed as a function of depth may give rise to sound channels within which sound may propagate with little loss over vast ranges, crossing entire oceans. The more pronounced this minimum in the sound speed profile is, the stronger the focussing of sound propagation paths about the channel axis (which is located at the depth of the minimum sound speed). Above the axis, the sound speed profile is downward refracting, below upward refracting. At the seafloor, some of the sound is reflected back into the water and some sound is transmitted into the seafloor. The amounts of acoustic energy that are reflected and transmitted depend on the density and sound speed (i.e., both compressional and shear sound speeds) of the seafloor and on the angle of incidence (hence, bathymetry). Regions with similar acoustic properties will conduct sound in similar ways. It is thus both sensible and desirable to partition the EEZ into a series of distinct acoustic zones, which was the goal of our study.

2. Materials and Methods

The idea was to cluster the hydroacoustic parameters of the water, the geoacoustic properties of the seafloor, and the bathymetry such that a manageable number of acoustic zones would emerge. As the upper water column parameters may vary over the course of the year, season might affect the zoning. We selected values representative of the austral winter.

2.1. Data

All the data were collated and projected to be on a common 10×10 km grid in UTM coordinates for the Australian EEZ. The parameters that affect acoustic propagation and the databases from which they were extracted are listed in Table 1.

The water column was characterised by sea surface temperature and salinity taken from [6,7] and sound speed profiles [8]—all for the month of July. The latter was computed with the formulae by Mackenzie [9] applied to monthly average temperature and salinity data (as a function of depth below the sea surface) from the Generalized Digital Environmental Model—Variable Resolution (GDEM-V) [10].

For the seafloor, compressional and shear sound speeds, compressional and shear wave attenuations, and density were computed as a weighted sum from the %clay, %silt, %sand, and %gravel measures [11], based on the geoacoustic properties of typical seafloor materials (Table 1.3 in [5]). The %mud provided in [11] was split into %clay and %silt by considering grain size from [12]. Sediment thickness was taken from [13]. It refers to the combined thickness of unconsolidated and consolidated sedimentary layers. The Australian continental shelf consists, where known, of a calcarenite seabed [14,15]. Bathymetry was obtained from the Geoscience Australia bathymetry and topography grid [16]. The slope is the gradient of the bathymetry and was calculated as the maximum rate of change of the surface from the centre cell to its eight neighbours.

Table 1. List of input and derived variables that affect sound propagation.

Group Variable	Derived Variable	Input Variable	Source
Water Column	sea surface temperature	sea surface temperature	[6]
	sea surface salinity	sea surface salinity	[7]
	sound speed gradient profile	sound speed profile	[8]
Seafloor	compressional sound speed, shear sound speed, compressional absorption coefficient, shear absorption coefficient, density	% clay	
		% silt	[11]
		% sand	
		% gravel	
	sediment thickness	sediment thickness	[13]
	bedrock type	bedrock type	[14,15]
Bathymetry	water depth	water depth	[16]

2.2. Clustering Model

The input variables to the clustering model were derived from the variables affecting sound propagation (Table 1). The input variables representing the water column were sea surface temperature and salinity, and parameterised sound speed gradients. The salinity data had significant gaps in coastal waters. However, the speed of sound is a function of salinity. Therefore, we estimated sea surface salinity from the speed of sound at 0 m depth, by fitting a spatial generalised additive model as a tensor of sound speed at 0 m and longitude, using the *mgcv* library in *R* [17]. With regards to the speed of sound at deeper depths, the absolute speed at which sound arrives at a receiver does not matter for cumulative noise mapping. What matters is how sound travelling along different paths converges and diverges. The direction in which sound travels and the changes in direction depend on the sound speed gradient (a vector quantity) over the water column. We therefore parameterised the first derivative of sound speed at seven depth points (20, 40, 100, 200, 400, 1000, and 2000 m), and used these scalar quantities, together with sea surface temperature and salinity, in the clustering model.

Representing the seafloor, compressional and shear sound speeds, compressional and shear absorption coefficients, and density of the seafloor surface layer were highly correlated, and so only the least correlated variables were used in the clustering model: the compressional sound speed and the shear absorption coefficient. The sediment layer (combined unconsolidated and consolidated) was thick across much of the EEZ. While only the geoacoustic properties of the upper seafloor (<~200 m) matter much in underwater sound propagation, there is rather little detail available on the ocean seafloor in general (compared to information on ground geoacoustics on land). Hence, we decided to keep sediment thickness in the clustering despite its relatively low variability across most zones. The bedrock was assumed to not vary across the Australian EEZ and was, therefore, excluded from clustering. It should, however, be used for later sound propagation modelling. Water depth (i.e., bathymetry) was the final input into the clustering model. Seafloor slope was excluded, because it was large only along short and thin lines corresponding to the edges of subsea canyons and had limited spatial variability everywhere else.

The clustering model was a multivariate Gaussian mixture model [18]. Let the *p*-dimensional data of size n be denoted by $y = \{y, \dots y_n,\}$, then each data point is modelled by the probability density function

$$f(y; \theta) = \sum_{i=1}^{g} \pi_i N(y; \mu_i, \Sigma_i),$$

where the mixing proportions π_i are ≥ 0 and sum to 1, and where N is the multivariate normal probability density function with mean μ_i and covariance Σ_i. Thus, the unknown parameters are $\theta = \{\pi_i, \mu_i, \Sigma_i\}$ for $i = 1, \ldots, g$. The model was fitted via the Expectation-Maximization (EM) algorithm [19], implemented in $R/C{+}{+}$. Mixture models provide a flexible framework allowing any distribution to be used in the model. After some preliminary analysis, we decided on one of the simplest parameterisations of a mixture: normal distributions with equal diagonal covariance matrix Σ. This will favour elliptical clusters of similar shape and size that have axes parallel to the variable axes. Although this application is obviously highly spatial in nature, for simplicity, we chose to ignore the spatial context of the data and cluster only on data values. The spatial correlation of the underlying data meant clusters were generally spatially continuous.

We modified the mixture model to deal with the absence of deep sound speed gradient values in shallow water. The water column sound speed vector is only as long as the water is deep, and so at locations <2000 m deep, the deep sound speed gradient values were absent. The likelihood at each location was calculated on the sound speed gradient values that were defined at that location, and the number of dimensions of the normal density was adjusted accordingly.

An important and difficult question in clustering is how many clusters there are in the data. However, in this application, the number of clusters is not as important an issue, because the goal was to discretise variables. That is, we are not necessarily looking for clumps of denseness in the variable space, but rather wish to simply segment or discretise the variables such that the data points within a segment are distinct from other segments in multi-dimensional space. For example, if data were spread perfectly uniformly in space, from a clustering perspective, this would be seen as not having clusters, whereas for our purpose, we would hope to divide the data into a specified number of bins equally sized in the parameter space. If we think of this analysis as discretising the variables, the number of clusters is more a question of how accurately we wish to represent the data. From this viewpoint, the answer is to use as many clusters (i.e., acoustic zones) as we can practically manage in any subsequent acoustic mapping analysis. From the perspective of future sound propagation modelling, the fewer zones there are to handle, the simpler the modelling exercise. However, matrix sizes relative to typical computer random access memory and computation time (which is a function of grid size and vessel density) limit how large any one zone can be. In this application, we settled for a trade-off aiming at 20–30 acoustic zones for future acoustic modelling.

An overview of the clustering process is given in Figure 1. The EM algorithm requires starting values. We used a combination of 100 random starts and 100 k-means to initialise the algorithm. Each start was run for a set of 100 initial iterations and the start that gave the best fit (highest likelihood) was chosen, and 1200 further iterations were done to refine the result. The number of clusters was fixed at 20. This model was fitted using the 10×10 km data and was then used to predict/allocate cluster labels for a finer 5×5 km grid. This allowed for efficient and fast model fitting while producing a fine scale over a large area. In post-processing, spatially disjoint areas were identified using the image analysis function *connectedComponents* from the *Rvision* library, which is based on OpenCV [20] and disjoint areas were then assigned different zone labels. Finally, any spatial zone less than 30,000 km^2 in area was combined with the neighbouring or surrounding zone that was predicted most likely by the mixture model's posterior probabilities. This size limit corresponds to a radius of just under 100 km, which is often the maximum range over which sound is propagated in noise maps (e.g., [21,22]).

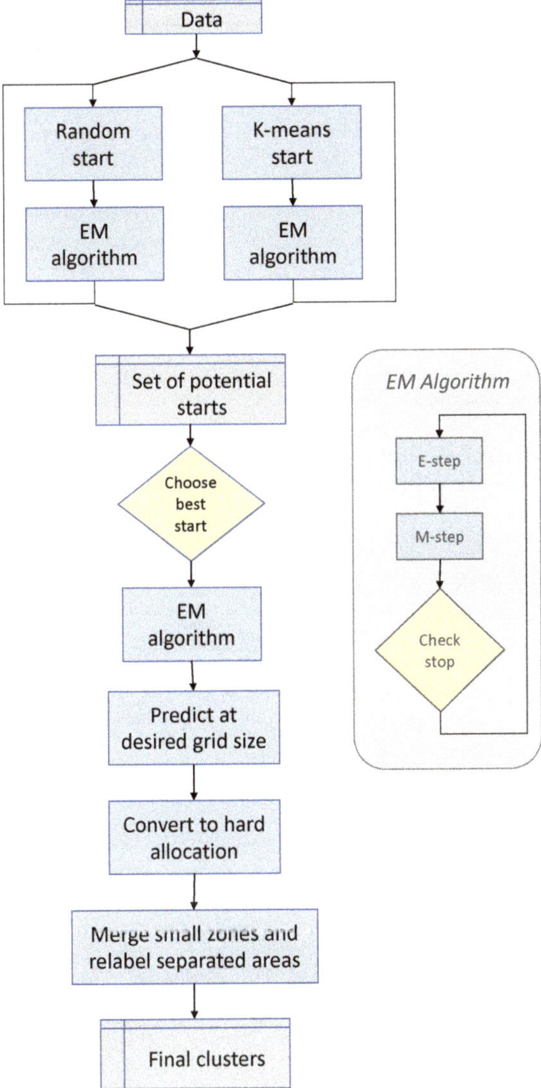

Figure 1. Flow chart of the clustering process.

3. Results

The clustering model resulted in 20 acoustic zones, eight of which were represented twice in geographically separated locations within the EEZ and so consequently there were 28 geographically separated zones. The 20 acoustic zones resulting from the clustering model are mapped in Figure 2, while the distributions of input variables amongst zones can be examined from the box-and-whisker plots in Figures 3 and 4.

When mapping some of the environmental variables across the zones (Figure 5), a natural division of the clusters appears at the edge of the continental shelf (~250 m bathymetric contour; Figure 5, Depth). The zone separation along the continental shelf is not completely surprising given the abrupt change in depth along the continental slope. It is further acoustically meaningful, given the importance of water depth in sound propagation.

The inshore continental shelf was characterised by shallow water with a sandy seafloor, and included five of the twenty zones—1, 4, 6, 10, and 20 (Figure 5, %Sand). Among these zones, there was an expected latitudinal difference in the more dynamic variable of sea surface temperature, and to a lesser extent sea surface salinity. Sea surface temperature ranged from hot (tropical) in the North to cold in the South (Figure 5, SST). Salinity was lowest in the Gulf of Carpentaria and highest at temperate latitudes (Figure 5, Salinity).

Zones 2 and 15 straddled shallow shelf and deep waters. Zone 2 occurred once within the Gulf of Carpentaria and once offshore in the Northwest. The specifics of Zone 2 include a high mud percentage (Figure 5, %Mud) and a very steep downward refracting sound speed profile (Figure 6). Sediment thickness was the driving parameter in Zone 15, being significantly greater than elsewhere (Figure 5, Sediment).

The offshore zones (3, 5, 7–9, 11–14, and 16–19) were characterised by deep water with a mostly muddy seafloor. There was a latitudinal separation driven by gradual changes in sea surface temperature, sea surface salinity, and water column sound speed gradients. The southernmost and coldest zones (11 and 12) exhibited the least sound-focussing sound speed profiles.

Table 2 summarises the features of the 20 marine acoustic zones of Australia. Table 3 presents the means and standard deviations of the acoustic zone parameters. As some of the zones occur multiple times in separate locations (e.g., on the East and West coasts), we also offer the acoustic zone parameters for 28 renumbered zones, so that geographically split zones may be modelled separately with slightly better accuracy (see Appendix A).

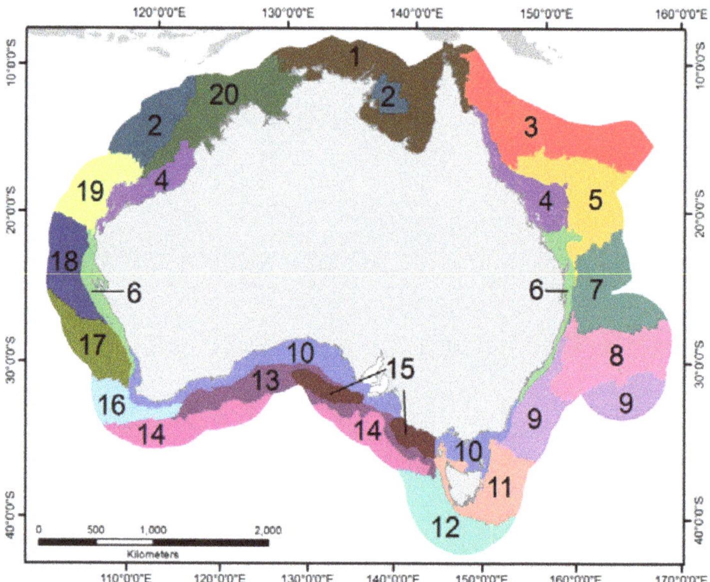

Figure 2. Map of the 20 acoustic zones of the Australian Exclusive Economic Zone (EEZ) for austral winter.

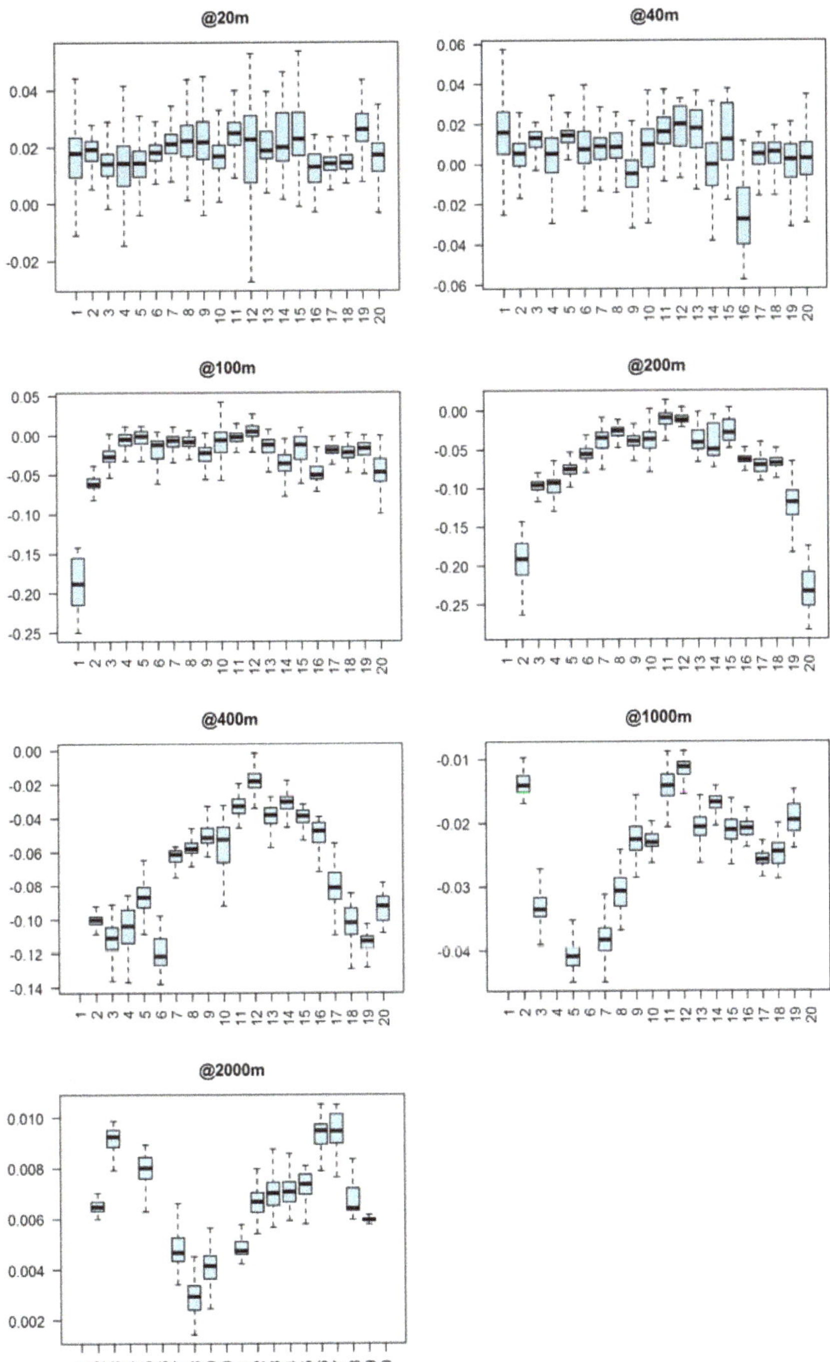

Figure 3. Box-and-whisker plots of July sound speed profile gradients (in units of (m/s)/m) at selected depths, by zone (x-axes show zone number).

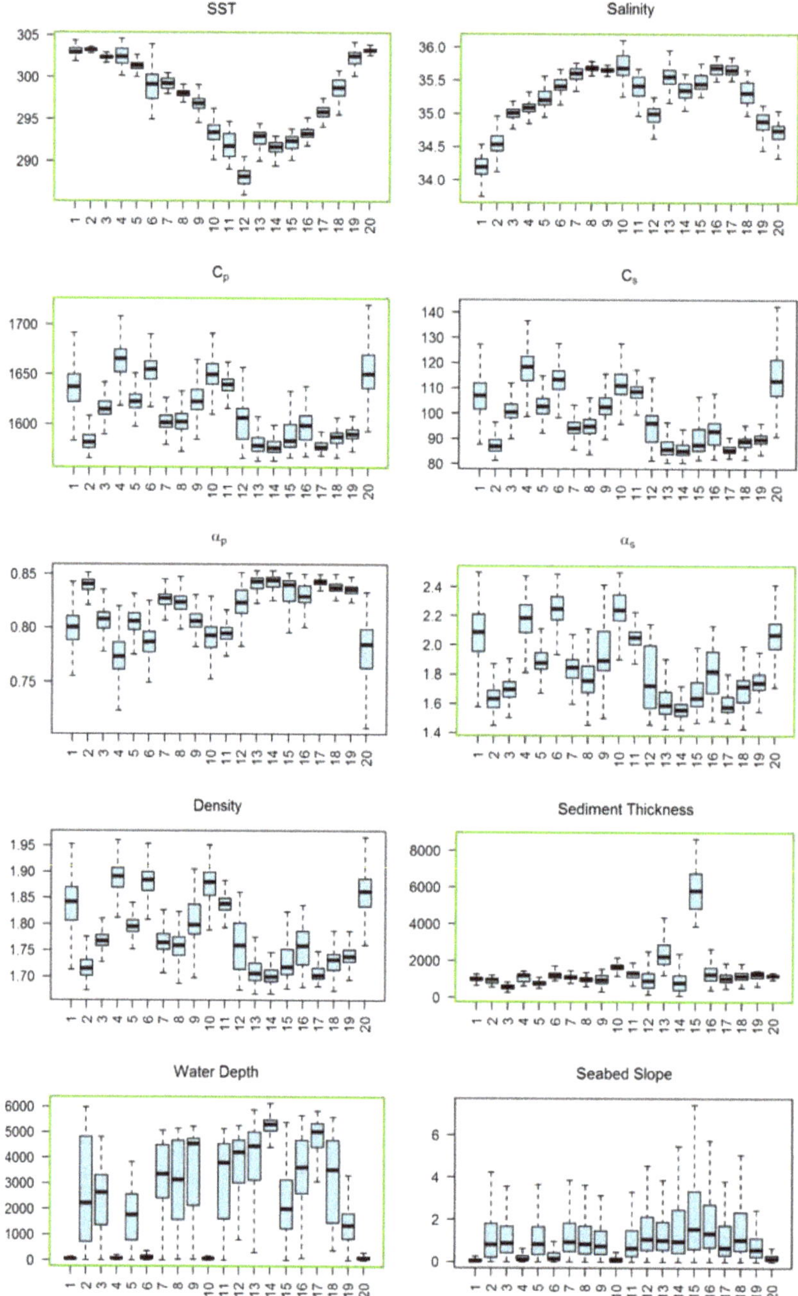

Figure 4. Box-and-whisker plots of various variables by zone (x-axes show zone number): Sea surface temperature (SST) [kelvin], sea surface salinity [psu], compressional sound speed at the seafloor (Cp) [m/s], shear sound speed at the seafloor (Cs) [m/s], compressional absorption coefficient at the seafloor (α_p) [dB/λ], shear absorption coefficient at the seafloor (α_p) [dB/λ], density of the top seafloor layer [1000 kg/m^3], thickness of the unconsolidated and consolidated sediment [m], water depth [m], and seabed slope [degrees]. Variables used directly in the clustering model are in green boxes.

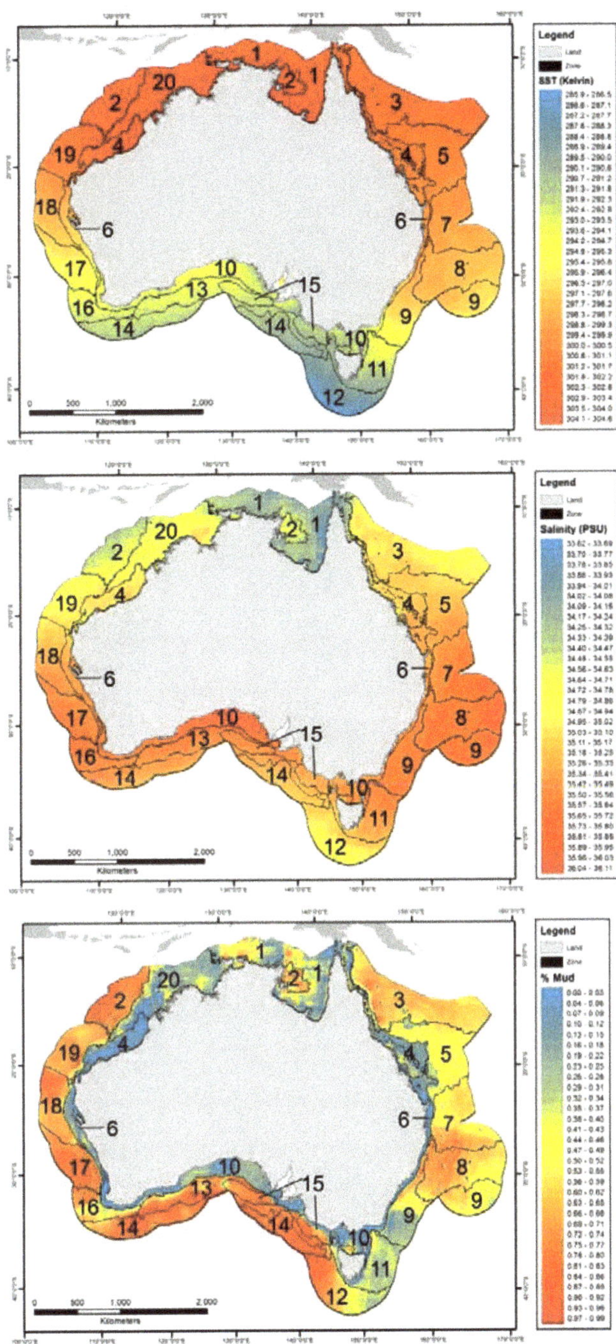

Figure 5. *Cont.*

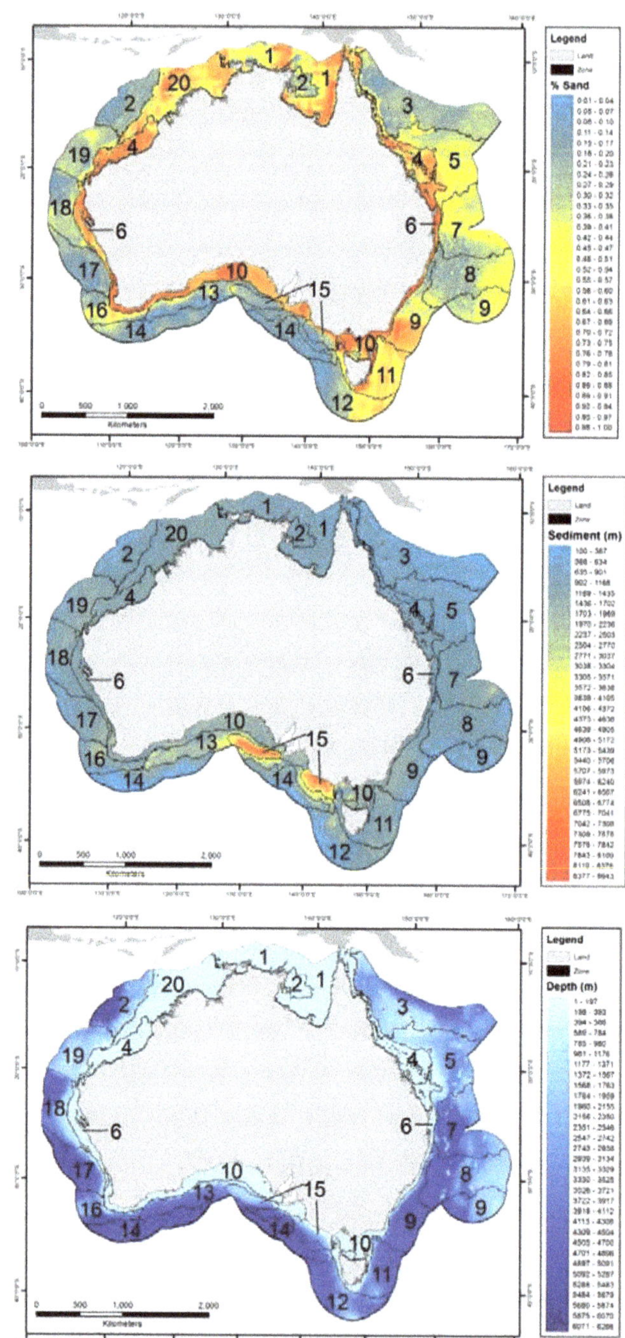

Figure 5. Maps of selected physical variables with acoustic zones overlaid: Sea surface temperature (SST), sea surface salinity, seafloor %mud, seafloor %sand, sediment thickness, and water depth.

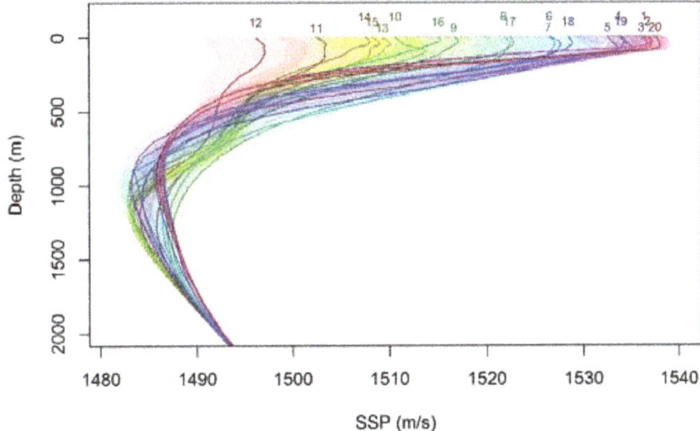

Figure 6. Sound speed profiles representative of the month of July, coloured by zone. The mean profile for each zone is drawn as a line on top of all profiles within that zone, which are merely shaded.

Table 2. Interpretation of zones given by the clustering.

Zone	Name	Description
1	Northern Tropical Shelf	Hot northern zone, shallow water with low salinity. Borders appear to align with salinity change.
2	Muddy Tropical Shallow	Two areas of northern Australia, one offshore in the West and one in shallow water in the Gulf of Carpentaria. Their predominant features are a muddy seabed, hot temperature, and low salinity.
3	Eastern Tropical 1	Excludes coastal waters but spans a range of depths. Separate from neighbouring Zone 5 to the South by water column sound speed profile. Zones 1–3 and 20 have the steepest downward refracting gradients. An abrupt change in salinity defines the border with Zone 1.
4	Tropical Shelf	Exists on both coasts. Mostly shallow, sandy, coastal water. Hot temperature, medium salinity.
5	Eastern Tropical 2	Geoacoustic parameters drive the border with Zone 4, hydroacoustic parameters (i.e., sound speed profiles) drive the borders with Zones 3 and 7.
6	Sub-Tropical to Temperate Shelf	Exists on both coasts. Warm, shallow, sandy.
7	Eastern Sub-Tropical Deep	Inner bound at continental shelf. Hydroacoustic parameters drive the separation from Zone 8.
8	Eastern Temperate Deep 1	Inner bound at continental shelf. High sea surface salinity. Less downward refracting than Zone 7.
9	Eastern Temperate Deep 2	Inner bound at continental shelf. High salinity. Less downward refracting than Zone 8.
10	Southern Temperate Shelf	Shallow water bounded by the continental slope. High salinity. Sandy seafloor.
11	Southern Cold	Covers shallow coastal Tasmania but also extends to deep water South-East of Tasmania. Cold water with a strong surface duct in July.
12	Southern Cold Deep	Most southerly zone. Cold. Less saline surface water than Zone 11. The least sound-focussing sound speed profiles.
13	Great Australian Bight Temperate Deep 1	Southern, cold, deep. Inner bound at continental shelf. Shallower yet thicker sediment than Zone 14.
14	Great Australian Bight Temperate Deep 2	Southern, cold, deep.
15	Southern Thick Sediment	An area (2 locations) between Zones 10 and 13 with very high sediment thickness.
16	Western Temperate Deep 1	Offshore with a sandy, shallower band resulting in different seabed acoustic properties from neighbouring Zones 14 and 17. Sound speed gradients at 40–1000 m different from neighbouring zones. Inner bound at continental shelf.
17	Western Temperate Deep 2	Inner bound at continental shelf. Cooler and more saline than Zone 18 to the North.
18	Western Sub-Tropical Deep	Offshore, warm. Inner bound at continental shelf.
19	Western Tropical Offshore	Warm-hot. Downward refracting July profile. Inner bound at continental shelf. Shallower than Zone 2 to the North and Zone 18 to the South.
20	Western Tropical Shelf	Shallow water. Wide, sandy continental shelf. Hot sea surface; strongly downward refracting.

Table 3. Average (±SD) parameters of the 20 acoustic zones produced by the clustering model: Water depth, seafloor slope, sediment thickness, compressional (c_p) and shear (c_s) speeds, compressional (α_p) and shear (α_s) absorption coefficients, and density. λ is the acoustic wavelength.

Zone	Depth [m]	Slope [Degrees]	Sediment Thickness [m]	C_p [m/s]	C_s [m/s]	α_p [dB/λ]	α_s [dB/λ]	Density [kg/m^3]
1	51 ± 49	0.1 ± 0.5	990 ± 139	1635 ± 23	107 ± 9	0.80 ± 0.02	2.05 ± 0.18	1833 ± 45
2	2583 ± 2047	1.5 ± 2.4	917 ± 160	1584 ± 12	88 ± 4	0.84 ± 0.01	1.65 ± 0.10	1720 ± 26
3	2460 ± 1230	1.4 ± 1.8	580 ± 132	1617 ± 13	101 ± 5	0.81 ± 0.01	1.70 ± 0.09	1770 ± 24
4	94 ± 131	0.4 ± 1.0	1076 ± 230	1662 ± 23	117 ± 10	0.78 ± 0.03	2.17 ± 0.13	1882 ± 37
5	1773 ± 966	1.6 ± 2.2	764 ± 114	1624 ± 13	103 ± 5	0.80 ± 0.01	1.89 ± 0.09	1800 ± 23
6	146 ± 200	0.5 ± 1.2	1225 ± 194	1654 ± 16	113 ± 7	0.79 ± 0.02	2.24 ± 0.12	1878 ± 29
7	3354 ± 1085	1.8 ± 2.7	1163 ± 231	1603 ± 11	95 ± 4	0.83 ± 0.01	1.83 ± 0.10	1765 ± 24
8	3151 ± 1438	1.6 ± 2.5	976 ± 154	1603 ± 13	95 ± 5	0.82 ± 0.01	1.76 ± 0.11	1757 ± 26
9	3588 ± 1452	1.6 ± 2.6	1008 ± 287	1623 ± 15	102 ± 5	0.81 ± 0.01	1.93 ± 0.17	1803 ± 37
10	126 ± 265	0.5 ± 1.9	1727 ± 386	1650 ± 17	112 ± 7	0.79 ± 0.02	2.25 ± 0.12	1874 ± 32
11	3032 ± 1758	1.4 ± 2.4	1278 ± 316	1638 ± 13	108 ± 5	0.80 ± 0.01	2.06 ± 0.11	1836 ± 26
12	3821 ± 1057	1.9 ± 2.6	988 ± 490	1603 ± 22	95 ± 8	0.82 ± 0.02	1.77 ± 0.21	1758 ± 51
13	3899 ± 1505	2.0 ± 3.1	2401 ± 730	1580 ± 11	87 ± 4	0.84 ± 0.01	1.62 ± 0.12	1711 ± 27
14	5240 ± 364	2.3 ± 3.5	840 ± 498	1578 ± 9	86 ± 3	0.84 ± 0.01	1.57 ± 0.08	1704 ± 20
15	2157 ± 1413	2.6 ± 3.2	5848 ± 1120	1596 ± 30	92 ± 11	0.83 ± 0.02	1.73 ± 0.24	1745 ± 65
16	3616 ± 1252	2.5 ± 3.5	1339 ± 553	1597 ± 14	92 ± 5	0.83 ± 0.01	1.81 ± 0.16	1755 ± 36
17	4478 ± 1388	1.8 ± 3.4	1094 ± 317	1579 ± 6	86 ± 2	0.84 ± 0.00	1.60 ± 0.08	1708 ± 16
18	3180 ± 1587	1.9 ± 2.6	1163 ± 225	1586 ± 8	89 ± 3	0.84 ± 0.01	1.70 ± 0.10	1728 ± 22
19	1515 ± 692	1.0 ± 1.1	1269 ± 171	1591 ± 9	90 ± 3	0.84 ± 0.01	1.75 ± 0.10	1740 ± 22
20	135 ± 0134	0.3 ± 0.7	1200 ± 49	1655 ± 28	115 ± 12	0.78 ± 0.03	2.07 ± 0.13	1861 ± 40

4. Discussion

The acoustic zones are based on the clustering of hydroacoustic, geoacoustic, and bathymetric variables. These three types of variables have different drivers and their spatial patterns are quite different. Therefore, the clustering process is a compromise, which results in different variables dominating the separation of different acoustic zones.

The water column parameters, on which the zones are based, vary with season (i.e., sea surface temperature, sea surface salinity, and sound speed). While we used parameters from the austral winter, the clusters might be different in the austral summer.

The benefit of this clustering model is that we can predict the zones of new data; for example, if we wish to increase the resolution. It is a useful feature of the clustering approach that once the model has been developed, it can be used to predict zoning at any grid resolution—as long as the environmental covariate data are available at that resolution. However, when increasing the resolution, computational effort and time will also increase. For example, modelling of acoustic propagation from ships in split zone 17, a medium-sized zone with 16 340 source cells, took 55 h to complete, using full capacity on a MacBook Pro laptop with a 2.9 GHz 6-Core Intel Core i9 processor and 32 GB 2400 MHz DDR4 memory (5 km resolution). Increasing to 2.5 km resolution added 20 h of computation time.

The geographically separated (split) zones (Figure A1 and Table 1) are perhaps more manageable in size than some of the unsplit zones (Figure 2). In particular, if sound propagation modelling is required on only one of the coasts, using the split zones avoids handling unnecessary regions. Having recomputed the means of the acoustic zone parameters, Table 1 offers slightly improved accuracy over Table 3, given the parameters are presented in 28 instead of 20 clusters. When modelling small regions, accuracy may be improved even further, by using raw, gridded acoustic parameters instead of their zone means. However, most of the parameters are not available on a spatially fine resolution—except for depth. Thus, while depth played an important role in the clustering, sound propagation modellers will most likely use finer bathymetry data within each zone.

Having these zones, a sound propagation model can be set up for each zone separately and sound sources (such as ships, seismic surveys, sonars, whales, dolphins, and fishes, covering a frequency range from a few hertz to tens of kilohertz) can be modelled within each zone. Noise maps can then be merged into an EEZ-wide map. At the zone boundaries, the sound from sources within one zone needs to be modelled into the neighbouring zones, adding to the sound map of each of the neighbouring zones. Care must be taken not to model any source locations near boundaries multiple times; in other words, source locations must be uniquely mapped to a zone. The resulting area-wide maps can ultimately inform the management of marine noise, identifying "hot-spots". If noise maps are overlain with habitat and species distribution maps, areas of concern (i.e., high biodiversity and animal density, and high noise) and areas of opportunity (i.e., high biodiversity and animal density, and low noise) will emerge, potentially informing the establishment of marine protected areas [21,23,24].

Author Contributions: Conceptualization, C.E., D.P. and J.N.S.; methodology, D.P., C.E., J.N.S. and R.P.S.; software, D.P.; formal analysis, D.P.; data curation, D.P., J.N.S. and R.P.S.; writing—original draft preparation, C.E. and D.P.; writing—review and editing, J.N.S. and R.P.S.; visualization, D.P.; project administration, D.P.; funding acquisition, D.P., J.N.S. and C.E. All authors have read and agreed to the published version of the manuscript.

Funding: This work was funded as part of the Australian National Environmental Science Programme, Marine Biodiversity Hub, under Project E2.

Institutional Review Board Statement: Not applicable.

Informed Consent Statement: Not applicable.

Data Availability Statement: A spreadsheet with the acoustic parameters that characterise each zone (i.e., mean winter sound speed profile; mean water depth and slope; sediment thickness; and compressional sound speed, shear sound speed, compressional absorption coefficient, shear absorption coefficient, and mean seafloor density) is available for download, as is a shape file of the spatially separated 28 acoustic zones (see https://catalogue.aodn.org.au/geonetwork/srv/eng/metadata.show?uuid=0c1bb667-29b2-4848-ade7-a98417121a66, accessed on 14 March 2021).

Conflicts of Interest: The authors declare no conflict of interest.

Appendix A

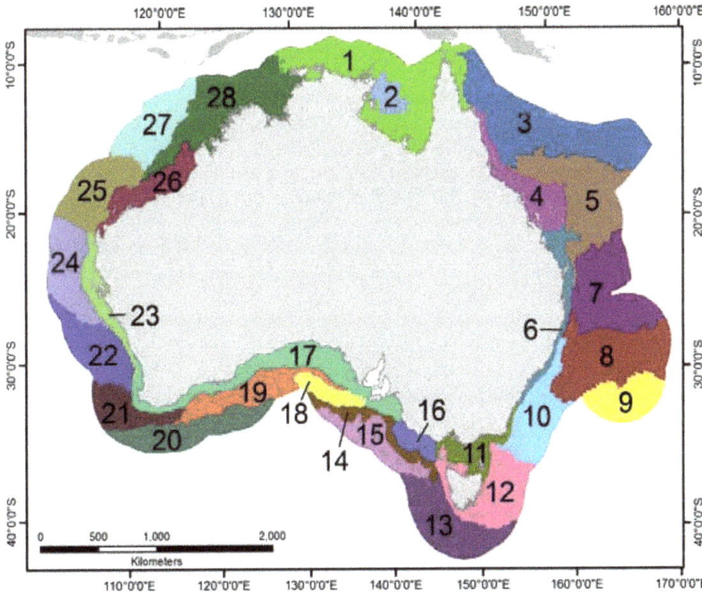

Figure A1. Acoustic zones after geographically separated zones were split and given a new zone label.

Table 1. Average parameters ($\pm$SD) of the 28 spatially separated acoustic zones: Water depth, seafloor slope, sediment thickness, compressional (c_p) and shear (c_s) speeds, compressional (α_p) and shear (α_s) absorption coefficients, and density. λ is the acoustic wavelength.

Split Zone	Zone	Depth [m]	Slope [Degrees]	Sediment Thickness [m]	c_p [m/s]	c_s [m/s]	α_p [dB/λ]	α_s [dB/λ]	Density [kg/m^3]
1	1	51 $\pm$ 49	0.1 $\pm$ 0.5	990 $\pm$ 139	1635 $\pm$ 23	107 $\pm$ 9	0.8 $\pm$ 0.02	2.05 $\pm$ 0.18	1833 $\pm$ 45
2	2	54 $\pm$ 9	0 $\pm$ 0.1	1013 $\pm$ 26	1599 $\pm$ 12	93 $\pm$ 5	0.83 $\pm$ 0.01	1.76 $\pm$ 0.10	1752 $\pm$ 26
3	3	2460 $\pm$ 1230	1.4 $\pm$ 1.8	580 $\pm$ 132	1617 $\pm$ 13	101 $\pm$ 5	0.81 $\pm$ 0.01	1.70 $\pm$ 0.09	1770 $\pm$ 24
4	4	118 $\pm$ 166	0.5 $\pm$ 1.3	876 $\pm$ 53	1654 $\pm$ 23	114 $\pm$ 10	0.78 $\pm$ 0.02	2.12 $\pm$ 0.13	1866 $\pm$ 37
5	5	1773 $\pm$ 966	1.5 $\pm$ 2.2	764 $\pm$ 114	1624 $\pm$ 13	103 $\pm$ 5	0.8 $\pm$ 0.01	1.88 $\pm$ 0.09	1800 $\pm$ 23
6	6	136 $\pm$ 166	0.6 $\pm$ 1.5	1100 $\pm$ 131	1659 $\pm$ 13	115 $\pm$ 6	0.78 $\pm$ 0.02	2.27 $\pm$ 0.11	1888 $\pm$ 21
7	7	3354 $\pm$ 1085	1.8 $\pm$ 2.7	1163 $\pm$ 231	1603 $\pm$ 11	95 $\pm$ 4	0.83 $\pm$ 0.01	1.83 $\pm$ 0.10	1765 $\pm$ 24
8	8	3151 $\pm$ 1438	1.6 $\pm$ 2.5	976 $\pm$ 154	1603 $\pm$ 13	95 $\pm$ 5	0.82 $\pm$ 0.01	1.76 $\pm$ 0.11	1757 $\pm$ 26
9	9	2590 $\pm$ 1398	1.1 $\pm$ 1.2	808 $\pm$ 138	1619 $\pm$ 7	101 $\pm$ 2	0.81 $\pm$ 0.01	1.85 $\pm$ 0.08	1789 $\pm$ 17
10	9	4375 $\pm$ 909	1.9 $\pm$ 3.3	1165 $\pm$ 275	1626 $\pm$ 18	103 $\pm$ 6	0.81 $\pm$ 0.01	2.00 $\pm$ 0.19	1814 $\pm$ 43
11	10	106 $\pm$ 176	0.5 $\pm$ 1.7	1741 $\pm$ 509	1655 $\pm$ 16	114 $\pm$ 7	0.79 $\pm$ 0.02	2.28 $\pm$ 0.12	1884 $\pm$ 29
12	11	3032 $\pm$ 1758	1.4 $\pm$ 2.4	1278 $\pm$ 316	1638 $\pm$ 13	108 $\pm$ 5	0.8 $\pm$ 0.01	2.06 $\pm$ 0.11	1836 $\pm$ 26
13	12	3821 $\pm$ 1057	1.9 $\pm$ 2.6	988 $\pm$ 490	1603 $\pm$ 22	95 $\pm$ 8	0.82 $\pm$ 0.02	1.77 $\pm$ 0.21	1758 $\pm$ 51
14	13	4389 $\pm$ 1310	2.4 $\pm$ 3.5	2768 $\pm$ 768	1579 $\pm$ 8	86 $\pm$ 3	0.84 $\pm$ 0.01	1.59 $\pm$ 0.07	1706 $\pm$ 17
15	14	5258 $\pm$ 380	0.9 $\pm$ 1.6	1099 $\pm$ 471	1575 $\pm$ 5	85 $\pm$ 2	0.84 $\pm$ 0.00	1.54 $\pm$ 0.04	1696 $\pm$ 10
16	15	1867 $\pm$ 1527	2.4 $\pm$ 2.8	5704 $\pm$ 1083	1606 $\pm$ 34	96 $\pm$ 12	0.82 $\pm$ 0.03	1.83 $\pm$ 0.30	1768 $\pm$ 77
17	10	133 $\pm$ 289	0.6 $\pm$ 2.0	1723 $\pm$ 333	1649 $\pm$ 17	111 $\pm$ 7	0.79 $\pm$ 0.02	2.24 $\pm$ 0.12	1870 $\pm$ 32
18	15	2406 $\pm$ 1256	2.8 $\pm$ 3.6	5972 $\pm$ 1137	1588 $\pm$ 23	90 $\pm$ 9	0.83 $\pm$ 0.02	1.65 $\pm$ 0.13	1726 $\pm$ 45
19	13	3674 $\pm$ 1535	1.8 $\pm$ 2.8	2232 $\pm$ 645	1581 $\pm$ 12	87 $\pm$ 4	0.84 $\pm$ 0.01	1.63 $\pm$ 0.13	1714 $\pm$ 30
20	14	5229 $\pm$ 353	3.2 $\pm$ 4.1	672 $\pm$ 439	1580 $\pm$ 10	87 $\pm$ 4	0.84 $\pm$ 0.01	1.60 $\pm$ 0.09	1709 $\pm$ 23
21	16	3616 $\pm$ 1252	2.5 $\pm$ 3.5	1339 $\pm$ 553	1597 $\pm$ 14	92 $\pm$ 5	0.83 $\pm$ 0.01	1.81 $\pm$ 0.16	1755 $\pm$ 36
22	17	4478 $\pm$ 1388	1.8 $\pm$ 3.4	1094 $\pm$ 317	1579 $\pm$ 6	86 $\pm$ 2	0.84 $\pm$ 0.00	1.60 $\pm$ 0.08	1708 $\pm$ 16
23	6	156 $\pm$ 229	0.4 $\pm$ 0.8	1349 $\pm$ 164	1648 $\pm$ 17	111 $\pm$ 7	0.79 $\pm$ 0.02	2.21 $\pm$ 0.12	1867 $\pm$ 32
24	18	3180 $\pm$ 1587	1.9 $\pm$ 2.6	1163 $\pm$ 225	1586 $\pm$ 8	89 $\pm$ 3	0.84 $\pm$ 0.01	1.69 $\pm$ 0.10	1728 $\pm$ 22
25	19	1515 $\pm$ 692	1.0 $\pm$ 1.1	1269 $\pm$ 171	1591 $\pm$ 9	90 $\pm$ 3	0.84 $\pm$ 0.01	1.75 $\pm$ 0.10	1740 $\pm$ 22
26	4	65 $\pm$ 54	0.1 $\pm$ 0.2	1319 $\pm$ 81	1671 $\pm$ 20	121 $\pm$ 9	0.77 $\pm$ 0.02	2.24 $\pm$ 0.11	1901 $\pm$ 24
27	2	3244 $\pm$ 1782	1.9 $\pm$ 2.6	891 $\pm$ 170	1580 $\pm$ 8	87 $\pm$ 3	0.84 $\pm$ 0.01	1.62 $\pm$ 0.08	1712 $\pm$ 19
28	20	135 $\pm$ 134	0.3 $\pm$ 0.6	1200 $\pm$ 49	1655 $\pm$ 28	115 $\pm$ 12	0.78 $\pm$ 0.03	2.07 $\pm$ 0.13	1861 $\pm$ 40

References

1. Juda, L. The exclusive economic zone and ocean management. *Ocean Dev. Int. Law* **1987**, *18*, 305–331. [CrossRef]
2. Erbe, C. International regulation of underwater noise. *Acoust. Aust.* **2013**, *41*, 12–19.
3. Erbe, C. Underwater noise from pile driving in Moreton Bay, Qld. *Acoust. Aust.* **2009**, *37*, 87–92.
4. Erbe, C.; King, A.R. Modeling cumulative sound exposure around marine seismic surveys. *J. Acoust. Soc. Am.* **2009**, *125*, 2443–2451. [CrossRef] [PubMed]
5. Jensen, F.B.; Kuperman, W.A.; Porter, M.B.; Schmidt, H. *Computational Ocean Acoustics*, 2nd ed.; Springer: New York, NY, USA, 2011.
6. Integrated Marine Observing System (IMOS). IMOS—SRS—SST—L4—GAMSSA—Australia Database. Available online: https://portal.aodn.org.au (accessed on 9 October 2019).
7. Boutin, J.; Vergely, J.-L.; Koehler, J.; Rouffi, F.; Reul, N. ESA Sea Surface Salinity Climate Change Initiative (Sea_Surface_Salinity_cci): Version 1.8 Data Collection. *Centre Environ. Data Anal.* **2019**. [CrossRef]
8. Barlow, J. *Global Ocean Sound Speed Profile Library (GOSSPL), An RData Resource for Studies of Ocean Sound Propagation (NOAA Technical Memorandum No. NMFS-SWFSC-612)*; U.S. Department of Commerce: La Jolla, CA, USA, 2019. [CrossRef]
9. MacKenzie, K.V. Nine-term equation for sound speed in the oceans. *J. Acoust. Soc. Am.* **1981**, *70*, 807–812. [CrossRef]
10. Carnes, M.R. *Description and Evaluation of GDEM-V 3.0 (Memorandum Report No. NRL/MR/7330–09-9165)*; Naval Research Laboratory, United States Navy: Stennis Space Center, MS, USA, 2009.
11. Li, J.; Heap, A.D.; Potter, A.; Daniell, J.J. *Predicting Seabed Mud Content Across the Australian Margin II: The Performance of Machine Learning Methods and Their Combinations with Ordinary Kriging and Inverse Distance Squared*; Record 2011/07, GeoCat No. 71407; Geoscience Australia: Canberra, Australia, 2011.
12. Geoscience Australia. Marine Sediments (MARS) Database. Available online: http://dbforms.ga.gov.au/pls/www/npm.mars.search (accessed on 21 December 2020).
13. Straume, E.O.; Gaina, C.; Medvedev, S.; Hochmuth, K.; Gohl, K.; Whittaker, J.M.; Fattah, R.A.; Doornenbal, J.C.; Hopper, J.R. GlobSed: Updated Total Sediment Thickness in the World's Oceans. *Geochem. Geophys. Geosyst.* **2019**, *20*, 1756–1772. [CrossRef]
14. Duncan, A.J.; Gavrilov, A.; Fan, L. Acoustic propagation over limestone seabeds. In Proceedings of the Acoustics 2009, Adelaide, Australia, 23–25 November 2009.
15. Santosh, M.; Maruyama, S.; Komiya, T.; Yamamoto, S. Orogens in the evolving Earth: From surface continents to 'lost continents' at the core–mantle boundary. *Geol. Soc. Lond. Spéc. Publ.* **2010**, *338*, 77–116. [CrossRef]

16. Whiteway, T.G. *Australian Bathymetry and Topography Grid, June 2009*; Record 2009/21, GeoCat No. 67703; Geoscience Australia: Canberra, Australia, 2009.
17. Wood, S.N. *Generalized Additive Models: An Introduction with R*, 2nd ed.; CRC Press: Boca Raton, FL, USA, 2017.
18. McLachlan, G.J.; Peel, D. *Finite Mixture Models*; John Wiley & Sons: New York, NY, USA, 2000.
19. Dempster, A.P.; Laird, N.M.; Rubin, D.B. Maximum likelihood from incomplete data via the EM algorithm. *J. R. Stat. Soc. Ser. B* **1977**, *39*, 1–22.
20. Bradski, G. The OpenCV Library. *Dr. Dobb's J. Softw. Tools* **2000**, *25*, 120–125.
21. Erbe, C.; MacGillivray, A.; Williams, R. Mapping cumulative noise from shipping to inform marine spatial planning. *J. Acoust. Soc. Am.* **2012**, *132*, EL423–EL428. [CrossRef] [PubMed]
22. Erbe, C.; Farmer, D.M. Zones of impact around icebreakers affecting beluga whales in the Beaufort Sea. *J. Acoust. Soc. Am.* **2000**, *108*, 1332–1340. [CrossRef] [PubMed]
23. Erbe, C.; Williams, R.; Sandilands, D.; Ashe, E. Identifying Modeled Ship Noise Hotspots for Marine Mammals of Canada's Pacific Region. *PLoS ONE* **2014**, *9*, e89820. [CrossRef]
24. Williams, R.; Erbe, C.; Ashe, E.; Clark, C.W. Quiet(er) marine protected areas. *Mar. Pollut. Bull.* **2015**, *100*, 154–161. [CrossRef]

Journal of
*Marine Science
and Engineering*

Article

Noise Waveforms within Seabed Vibrations and Their Associated Evanescent Sound Fields

Richard Hazelwood [1,*] **and Patrick Macey** [2]

1   R & V Hazelwood Associates LLP, Guildford GU 8UT, UK
2   PACSYS Ltd., Strelley Hall, Nottingham NG8 6PE, UK; patrickm@pafecfe.com
*   Correspondence: acoudick@gmail.com; Tel.: +44-1483-568646

Abstract: While the effects of sound pressures in water have been studied extensively, very much less work has been done on seabed vibrations. Our previous work used finite element modeling to interpret the results of field trials, studying propagation through graded seabeds as excited by impulsive energy applied to a point. A new simulation has successfully replicated further features of the original observations, and more field work has addressed other questions. We have concentrated on the water-particle motion near the seabed, as this is well known to be critical for benthic species. The evanescent pressure sound fields set up as the impulsive vibration energy passes are expected to be important for the local species, such as crabs and flatfish. By comparison with effects occurring away from boundaries, these seismic interface waves create vigorous water-particle motion but proportionately less sound pressure. This comparative increase ratio exceeds 12 for unconsolidated sediment areas, as typically used for piling operations.

Keywords: seismic interface waves; dispersion; water-particle velocity; seabed vibration

Citation: Hazelwood, R.; Macey, P. Noise Waveforms within Seabed Vibrations and Their Associated Evanescent Sound Fields. *J. Mar. Sci. Eng.* **2021**, *9*, 733. https://doi.org/10.3390/jmse9070733

Academic Editors: Christine Erbe, Michel André and Chun Feng Li

Received: 30 March 2021
Accepted: 15 June 2021
Published: 2 July 2021

Publisher's Note: MDPI stays neutral with regard to jurisdictional claims in published maps and institutional affiliations.

1. Introduction

1.1. The Need for More Research into Seabed Vibration

Sound pressure waves within the marine environment are regularly monitored and regulated to minimize adverse effects on marine life [1–6]. In contrast, few authors have raised concern over the potential effects of seabed vibration. The topic is more complex for many reasons. First, there are a large range of seabed types. Many measurements have been made, and the reviews by E.L. Hamilton [7] have been found useful. This work has concentrated on flat areas of saturated sediments.

Where seabed structures have been considered in acoustic modeling, most consider the seabed to be a fluid, as did Heaney et al. [8]. This reduces the complexity of the calculation, but also misses many of the effects discussed here. There have been other approaches to modeling the seabed structure that will be discussed later. Hawkins et al. [9] have called for a better approach.

The initial work was driven by a desire to understand measurements made at various sites, as well as effects reported in the literature. The relatively simple propagation in the open water environments inhabited by the more charismatic species, such as marine mammals, is here contrasted with highly selective and delayed propagation through the sediment structure. There are some useful analogies with the ducted pathways found in the ocean, such as the mixed-layer water-surface duct described by Urick [10]. If the surface waters are isothermal, the refraction of sound is controlled by a gradual increase of the speed of sound with depth as the pressure rises. This causes energy to be returned to the sea surface following circular pathways, and reduces the geometric attenuation with distance near the surface. However, attempts to use this increase for communication are affected by unpredictable converging paths.

1.2. The Issues Addressed in the Paper

The paper is structured as follows: We discuss the choice of physical models and the care required to obtain valid results with the transient finite element (FE) modeling technique. We then present summary results from the ongoing research with some new conclusions regarding the predicted effects on those creatures inhabiting the benthic region in and on the seabed. We compare results with the mathematics of continuous waves. A separate section will discuss some additional measurements made to study the water motion induced by seismic interface waves in a reservoir.

2. Using Models to Understand the Observations

2.1. Modeling the Physical Environment

A quantitative study of seabed propagation requires the adoption of a physical model. One proposed by Lord Rayleigh [11] used an infinite interface, such as a horizontal division of all space. Rayleigh assumed a uniform solid under a vacuum. He showed how the waves that travel along this idealized interface show many properties of earthquakes. Such seismic interface waves create coupled horizontal and vertical motions with a characteristic difference in phase. Later, Scholte [12] added a fluid upper half-space with similar results.

A notable flaw in these models is the lack of any dispersion. Observed seismic interface waves are seen as wave packets (see Stein and Wysession [13]), wherein various frequency components travel at different speeds. Typically, a short packet becomes extended in its duration, so that the peak intensity is reduced as the energy is distributed in time.

As pointed out by Shearer [14] and Achenbach [15], dispersion does not occur in either of the half-space models of Rayleigh or Scholte. We used a more realistic half-space model from our earlier work [16] with gradually increasing material stiffness with depth. This is seen to support vibration wave modes that travel at different speeds. However, in this work we also compared these modes with those created by a sharply layered substrate. Results from this layered model contrasted strongly with those of the graded sediment model. A layer of much stiffer material, with its second distinct interface, created a defined depth with a strong effect on the modal structure of the response.

2.2. The Smoothly Graded Sediment Model

The graded half-space model is based on the Scholte model [12], but with a steady increase of the solid material stiffness with depth below the seabed, rather than a uniform material. This stiffness can be described by the speed of shear waves, V_s, as propagated within a uniform material. Material properties can be defined by this and the speed of compressive waves, V_p, plus the solid density, ρ_s. Published data was reviewed by Hamilton [7] in this format to provide a range of measured seabed properties. We set a constant solid density of 2250 kg/m$^{3'}$, which was not found critical. V_p, the value of the solid compressive wave speed at the interface, was set at 1520 m/s. The water density was set at 1000 kg/m^3, while the speed of waves in water, c_w, was set at 1500 m/s. Again, these are not critical properties.

Unlike the model proposed by Godin and Chapman [17] we specified a step change of material shear speed, V_s, at the interface, followed by a linear increase with sediment depth, rather than their power-law increase, with its imperceptible initial change at the interface. Our choice of parameters was based on Hamilton's data, mainly lying within the considerable variation he described. The initial fit by Hazelwood and Macey in 2016 [18] was simplified for the 2018 paper [16] to a linear increase of V_s with depth d. The function $V_s(d)$ then has a gradient, g_r.

$$V_s\,(d) \;=\; V_s(0) \;+\; g_r\,d \tag{1}$$

Typical gradients of around 4 (m/s)/m were also discussed in Jensen et al. [19]. Specifying this parameter creates an energy duct with some similarities to the surface mixed water duct described by Urick [10]. The linear gradient gives rise to refraction of wave energy back towards the interface, where it combines to form seismic interface waves. These trap the energy as an elastic wave travelling close to the interface. Our studies used transient finite element techniques (FE) to show the nature of the coupling of the solid motion with that of the adjacent sea water. Note that we did not specify any absorption, or consider the effects of gravity. We considered the material within each element as isotropic without the internal structure described by the Biot theory [20]. We are only studying wavelengths much bigger than the element size.

We can define the graded physical model with one additional condition, that of Poisson's ratio at great depth. In our model, this tended towards a value of 0.25, creating a solid similar to granite, a type known as a Poisson solid. This controlled the increase in *Vp* with depth as set out by Hazelwood and Macey 2018 [16]. These few parameters were judged to be reasonably realistic, and their consequences were used to make useful predictions, some of which remain to be tested.

2.3. The Sharply Layered Model

The smoothly graded model can be contrasted with a sharply layered model, wherein different adjacent layers have quite different material properties, rather than a gradual change. Such models are much used by seismologists, as the reflected waves yield useful information on the location of minerals, but we are more interested in the seabed interface and its substrate.

We investigated this using the mathematics given by Fahy [21] for a flexural wave excited in a steel plate, which then excites an adjacent fluid volume. This would apply equally to a rock layer, but we have kept the steel layer, with its well-defined material properties, for our computer analysis and comparison of the FE results with continuous wave theory.

3. The Mathematics of a Continuous Evanescent Pressure Field

Most available mathematics use continuous waves with no start time that extend throughout all space, but yield convenient mathematical forms such as the sine wave. They can be compared to excitation by an impulse, such as that used for our computer modeling. The mathematical nature of these pressure fields was discussed by Fahy [21]. He showed that when the speed of the solid wave is less than that of the pressure wave in the adjacent fluid, there will be no radiation of energy from the plate into the fluid. The resultant exponential decay of the acoustic pressure with increasing distance from the plate is described as an evanescent field.

The decay can be described by a field thickness δ. While both particle velocities and pressures decay at the same rate, it is convenient to consider the variation of the acoustic pressure, p:

$$p = p_0\, e^{-\frac{z}{\delta}} \tag{2}$$

The pressure at the interface p_0 will have dropped to a fraction $1/e$ when the distance from the interface z is equal to δ. This field thickness will vary with the oscillatory frequency and other parameters. This exponential decay is similar to the "skin effect" of electromagnetic theory.

The evanescent field thickness is proportional to the wavelength of the interface wave, λ:

$$\delta = \frac{\lambda}{2\pi\sqrt{\left(1 - \left(\frac{c_d}{c_w}\right)^2\right)}} \approx \frac{\lambda}{2\pi} \quad \text{when } \frac{c_d}{c_w} \text{ is small} \tag{3}$$

where c_w is the compressive water wave speed and c_d is the particle displacement interface wave speed. This equation has been reconfigured to use the wavelength, λ, from Fahy's original, which used the wavenumber. When $c_w \gg c_d$, δ becomes close to $\lambda/2\pi$, the inverse of the wavenumber, k, for the wave being driven along the solid interface, is given by:

$$k = \frac{2\pi}{\lambda} = \frac{2\pi f}{c} \tag{4}$$

In Equation (3), Fahy's wavenumber ratio has been replaced by a ratio of wave speeds. Since both waves have the same frequency, f, their ratio is the inverse of the wavenumber ratio.

4. Results from Transient Finite Element (FE) Modeling

4.1. The FE Graded Model

Unlike an infinite physical model, the FE mesh is finite. We created a set of elements by defining a data file of their properties for the computation. These small contiguous blocks had specified isotropic material properties, which were then subjected to small elastic deformations by the imposed force. These deformations were transmitted from element to element by a large number of elastic responses between the elements. The FE process is linear, in that if the force is doubled, all responses are also doubled. We thus used an arbitrary 1 MN peak force to generate realistic responses. A benefit of the transient analysis is the ability to use an impulsive excitation force. This has to be kept short in both time and space to avoid unwanted errors, but will track changes with time. We used a more flexible impulsive force for this work than those we used previously.

The FE model was a pair of stacked discs, shown in cross-section in Figure 1. The lower disc was solid, with a depth usually specified as 128 m, of which just over 2 m is shown, but the disc radius varied, often set to 256 m. Many thousands of elements were defined, with the smaller elements shown by the grid in this figure being 0.125 m square. These needed to be kept small compared with the wavelength. By doing so, the discrete divisions between the elements had little effect, and the FE model could simulate the continuous variation in material properties of the physical model.

The displacement waves are shown by an exaggerated deformation of the interface. Here, five troughs or dips are shown with a wavelength of approximately 12 elements or 1.5 m. The upper disc is water, with a depth of 16 m, of which the lower 1.7 m is shown. Note that this model assumed axisymmetry, with no circumferential variation.

Defined points in the FE model are called nodes, corresponding to particles of solid or water. Node 700 is marked on the interface. This is halfway out from the centre at 128 m radius. Node 920 is within the water at 0.5 m above the interface. More effort was placed on the analysis of such nodes in this paper than in our previous work. The primary calculation for these acoustic elements is of the acoustic pressure, but postprocessing can be done to recover the water-particle motion.

4.2. The FE Layered Model

The layered physical model is more easily rendered by the FE technique, as the substrate layer is finite. Steel was used, with its well-defined material properties, but the mathematics will be similar for a rock layer.

The diameter was made large enough to be able to avoid confusion between the outgoing wave and the return after a reflection at the circumference. This FE model will be referred to as the tank model, representing a 16 m deep water tank of diameter 512 m with a 0.125 m thick steel base. Excitation forces were applied downward to the tank centre, using low frequencies (e.g.,16 Hz) to create an evanescent pressure field in the water near the bottom.

Although the motion of the steel plate was primarily vertical, the adjacent water motions retained the substantial horizontal velocities of the graded model. This indicated that these water-particle velocities were a consequence of the intrinsic fluid mechanics within the evanescent field, and were independent of the horizontal motion of the solid below. The fluid as modeled had no viscosity.

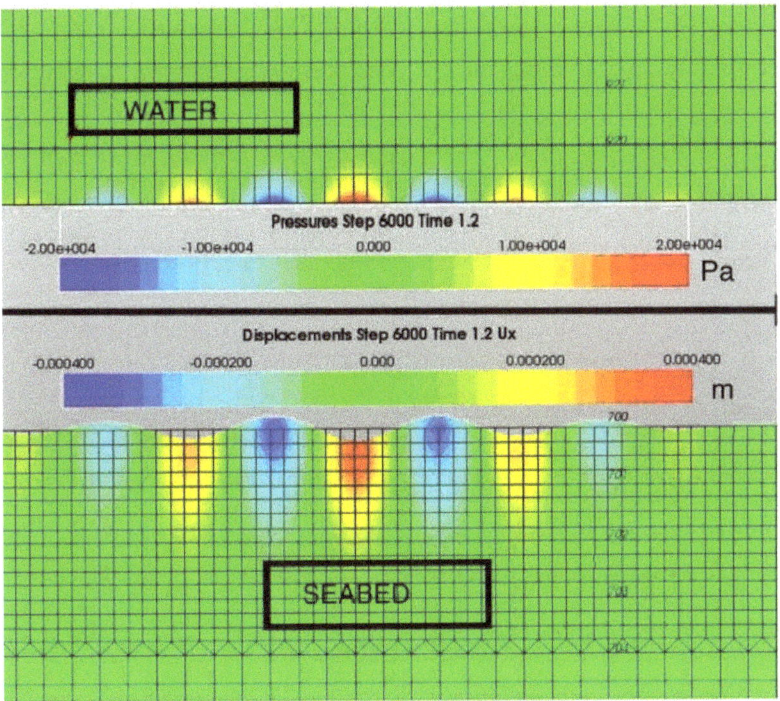

Figure 1. An enlarged view of a section through the FE model shows two parts. The upper section has been processed as a fluid (water) to give a distribution of acoustic pressures. The lower section has been processed to show the deformation of the solid. A colour barcode shows the range of the vertical displacements, from +0.4 mm to −0.4 mm. The upper section was connected by computation to the lower section to create the acoustic pressure range ±20 kPa. This snapshot view was at a time of 1.2 s, after 6000 computational time steps had been completed.

4.3. The Nature and Influence of the Energy Source

Both the graded and layered FE models were excited by a force applied as a downward thrust at the centre of the substrate disc. This was chosen to represent the actions of a pile driver or dredger. However, there was limited realism in this choice. The FE method had difficulty in modeling a cutting action. Both piling and dredging require the rupture of the seabed material, and this was outside the scope of this work. However, the elastic waves as modeled showed most features of the observed seabed vibrations, so we studied the propagation from the idealized thrust to the consequent motion.

The displacements of the particles as shown in Figure 1 created two particle velocity components, described henceforth as upward and outward. In contrast, the applied thrust was only vertical, and applied to a point at the centre of the model. The horizontal velocities that were integral to the seismic interface wave structure were thus generated within the model, as the energy was converted from that of a vertical thrust into a two-dimensional velocity vector. There was no circumferential velocity in this axisymmetric model. This conversion process may have some similarity to the process that converts a fluctuating volume source (a "simple source" as described by Kinsler et al. [22]) into a plane pressure wave within open water. This is usually described by comparing the simple physics of the near field to that of the far field, with a more complicated structure as the conversion proceeds.

4.4. The Energy-Source Time Profile

Our 2016 work used a polynomial variation of thrust with time over a defined period, designed to limit the acceleration values. For the 2018 work, this was changed to a Ricker pulse, an infinite pulse profile, with very low values at times well away from the peak. The profile was truncated at a time when the level was extremely low.

Both these pulse forms have been replaced in this work by a truncated windowed cosine form. The cosine form was windowed by a Gaussian "bell" curve, as seen in Figure 2. In addition to the cosine frequency f, set here to 40 Hz, the bell curve was specified by a time width, t_w, and a peak time, t_0. The peak amplitude force F was set at an arbitrary $K = 1$ MN. Here, the window width $t_w = 0.0226$ s.

$$F = K\, e^{-\left(\frac{t-t_0}{t_w}\right)^2} \cos 2\pi f \tag{5}$$

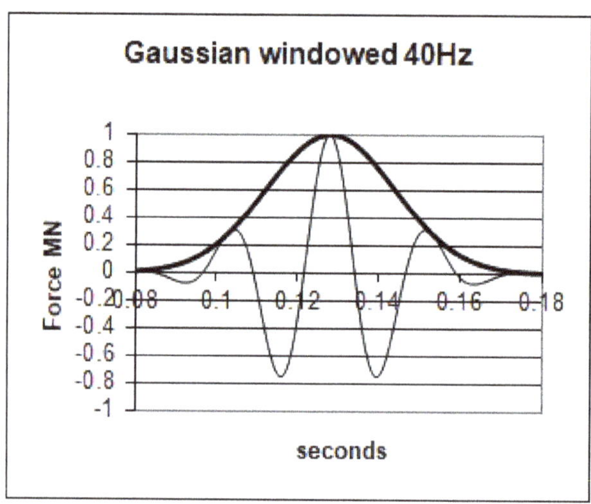

Figure 2. An example of a time profile used in recent FE work.

The envelope form is shown in Figure 2 by the bold black line, while the applied force is shown by the lighter grey line. This improved impulse driver allowed the frequency of the propagated displacement wave to be set. However, it was not based on any measurements of piling action, but rather adopted to replicate features of the observed waveforms.

4.5. The Time Profile of the Response

The form of the time response is shown in Figure 3 at two positions, 64 m and 128 m from the centre of the model. The response at a radius of 128 m was taken at node 920, which is labeled in Figure 1. The water-particle kinetic energy depended on the particle speed, a scalar found by combining the upward and outward velocity components.

Notably, it was the time profile of the particle speed that followed the form of the applied impulse window function, the envelope shape of the wave packet. This bell shape was seen to propagate unchanged from the node at radius 64 m to that at 128 m. In contrast, the individual velocity components of the displacement wave followed the more rapid oscillations of the applied force. At different positions, the peaks and troughs of the two components occurred at different times within the envelope. This was a consequence of the displacement wave speed being faster than the speed of the envelope. This behaviour is well documented for other branches of physics, in which they are referred to as the "phase velocity" and "group velocity", respectively. These include ocean surface waves, controlled by gravity, where the wave structure is highly dispersive, with a characteristic ensuing chaotic motion.

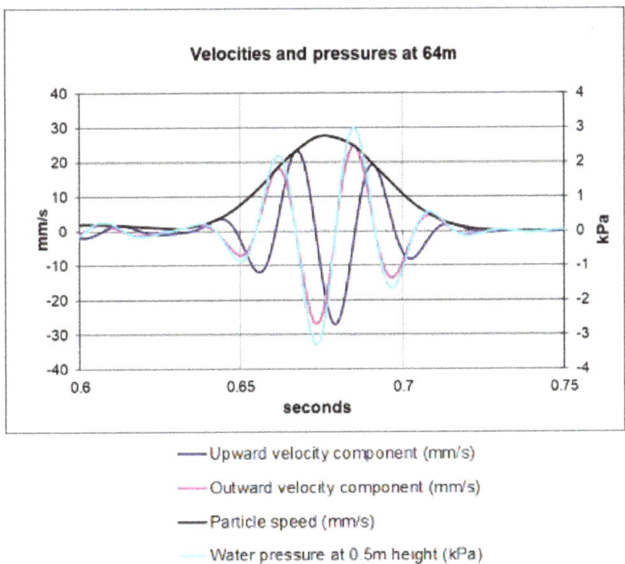

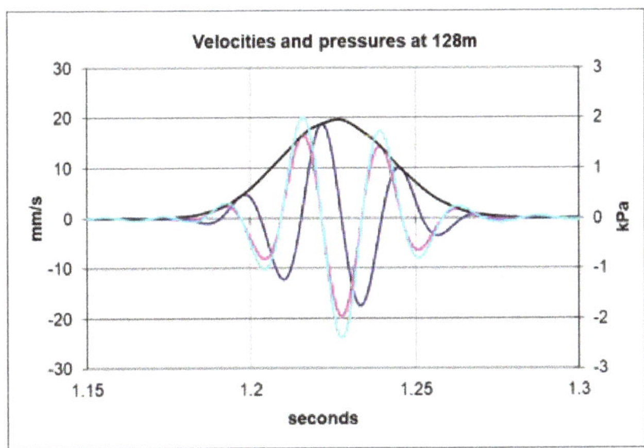

Figure 3. The response of the FE graded model to a Gaussian windowed cosine impulse is shown at two positions, each plotted against time. The two velocity components (upward (dark blue) and outward (mauve)) are combined to show the vector magnitude, the particle speed (black). The pressures (light blue) at this height (0.5 m above the seabed) matched the shape of the outward velocity component.

4.6. Dispersive and Nondispersive Groups: A Comparison with a Dispersive Substrate

Dispersive mechanisms are best displayed using animations, and good examples are available online from ISVR [23], and also on Wikipedia as "phase velocity". As shown in Figure 4, these wave packets can propagate without dispersion, but with the packet continuously changing in form as the peaks apparently move through the envelope. The stability of the bell shape was tracked in our models for more than a 500 m range, showing that there was no dispersion within the packet.

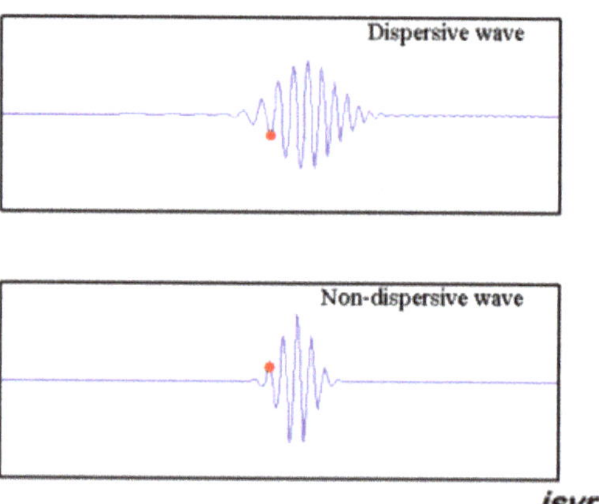

Figure 4. The difference between a dispersive wave packet and a nondispersive wave packet. The latter shows a uniform frequency content, and remains compact as it travels.

The observed structure of the groups or modes generated within the graded sediment substrate could be compared with a dispersive substrate, that of a steel tank base. Transverse (here, "upward") vibrations within such plates will radiate sound into an adjacent fluid space if the wavelength within the plate is larger than the corresponding wavelength within the fluid. However, if the wave speed within the vibrating solid is less than that of the compression waves within the fluid, an evanescent pressure field is set up in the fluid near the interface.

The steel baseplate of the tank model was made 0.125 m thick to give a suitably slow speed for 16 Hz flexural waves, a frequency commensurate with results for the graded seabed model. The solid motion was predominantly vertical, but the water particle response shown in Figure 5 had both horizontal and vertical motion.

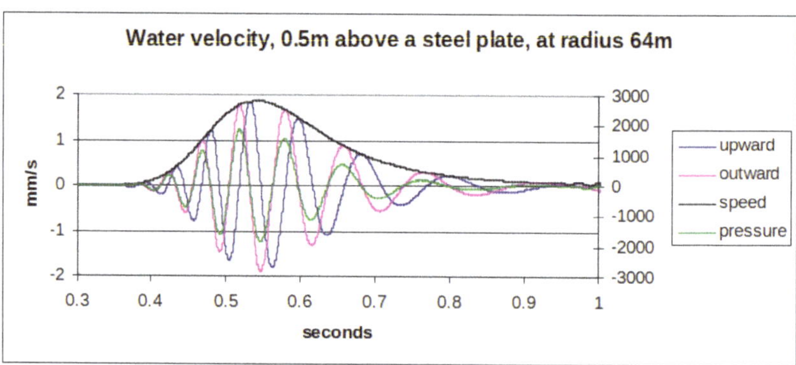

Figure 5. A water-particle response in a dispersive-tank model. The envelope is no longer symmetrical with time, being stretched on its trailing edge by dispersion in the flexural waves of the tank baseplate.

The underlying dispersion is described by the mathematics of the flexural (bending) waves being driven in the baseplate, as given by Fahy [21]:

$$c_b^2 = 2\pi f \sqrt{\frac{D}{m}} \tag{6}$$

The propagation speed, c_b, for these bending waves is dependent on the frequency, f, as well as the material properties, including the mass per unit area, m. The bending stiffness D is given by the Young's modulus, E; the Poisson ratio, v; as well as the plate thickness, h. (Timoshenko et al.) [24].

$$D = \frac{E\,h^3}{12\,(1 - v^2)} \tag{7}$$

The dispersive character could be clearly seen in snapshot views of the impulsive wave train as it propagates.

At 0.35 s, the wave train was well established, seen in Figure 6 as a snapshot including node 600 at radius 64 m. The acoustic pressures in the water were well correlated with the deformation of the adjacent steel plate.

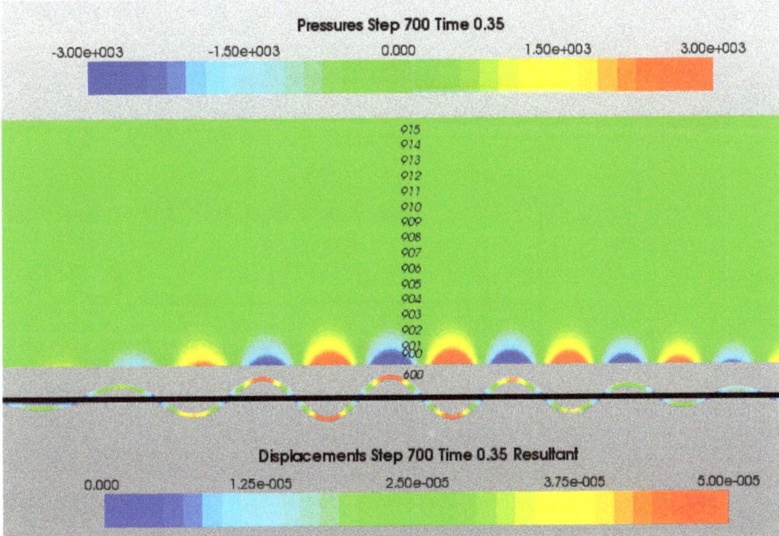

Figure 6. A snapshot view of the flexural wave as an exaggerated displacement of the baseplate. This is shown as a horizontal black line between its exaggerated displacements, in colour, peaking at 50 μm. The water finite element edges are not shown here, but a set of monitor nodes are numbered in the water column up to the water surface (16 m deep). The oscillating pressures covered a range ±3 kPa.

In Figure 7, at 0.8 s the pulse has moved out from node 600 to include node 800 at a 192 m radius. The strong dispersion dramatically increased the length of the wave train. This view had a reduced magnification, with the extended wave train almost extended to the edge of the FE model at a 256 m radius. Note the lack of significant reverberation at this time.

4.7. A Comparison of Different Modes in the Graded Model

The bell envelope shown in Figure 3 occurred again (Figure 8), showing two separate bell-shaped modal envelopes in water. The variation with time of the water particle velocity components were plotted, as was the vector magnitude, the particle speed.

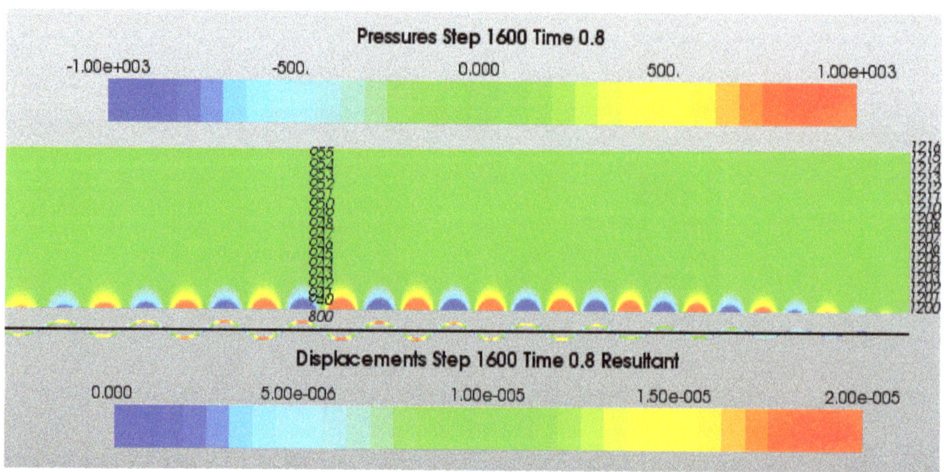

Figure 7. A second snapshot at 0.8 s also shows the numbered nodes at the edge of the model at a 256 m radius.

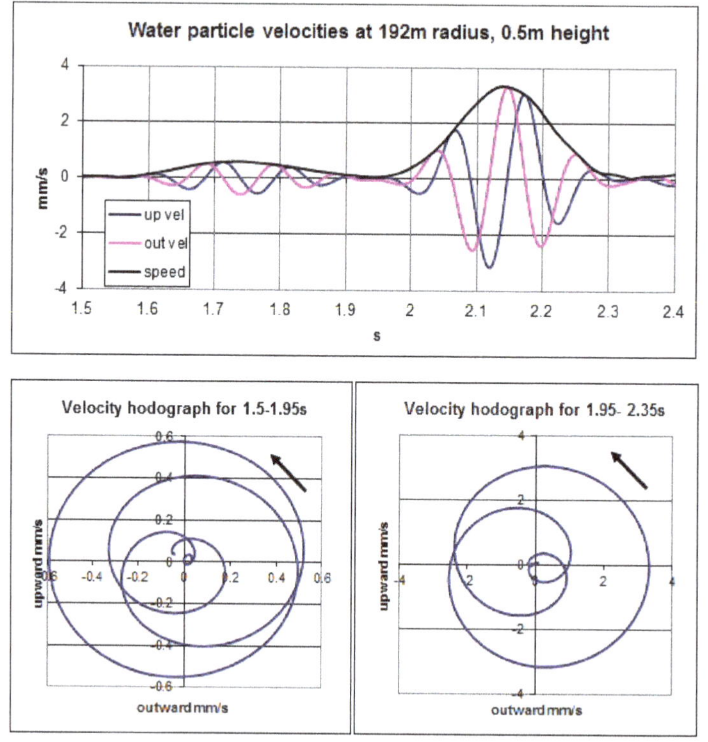

Figure 8. A time series from 1.5 s to 2.4 s shows an early mode that is less intense.

The early mode data in Figure 8 was selected (1.5 s to 1.95 s) and plotted as a hodograph below the time series. This shows the locus of the motion of the water particle as the early mode passed. A second hodograph shows the data for the period 1.95 to 2.35 s, after a brief moment of inactivity.

This hodograph data display was so named by W.R. Hamilton [25], and provided additional insight into the motion. Unlike the solid motion, the water particles followed a nearly circular path. As more cycles were included within the envelope, more circuits were observed. For a continuous wave, a single circle would be followed repeatedly, but for an impulse, the motion had to start and stop at the origin.

The arrows show the direction of circulation, which is described as retrograde, with the horizontal velocity being at a maximum in the direction of the source at the moment the upward displacement was at a maximum.

In Figure 9, the motion of a solid particle at the interface is shown. Now the hodographs are highly contrasting both in shape and direction. The less-intense early mode travelled faster (~140 m/s) than the more intense mode, which was slower (~117 m/s).

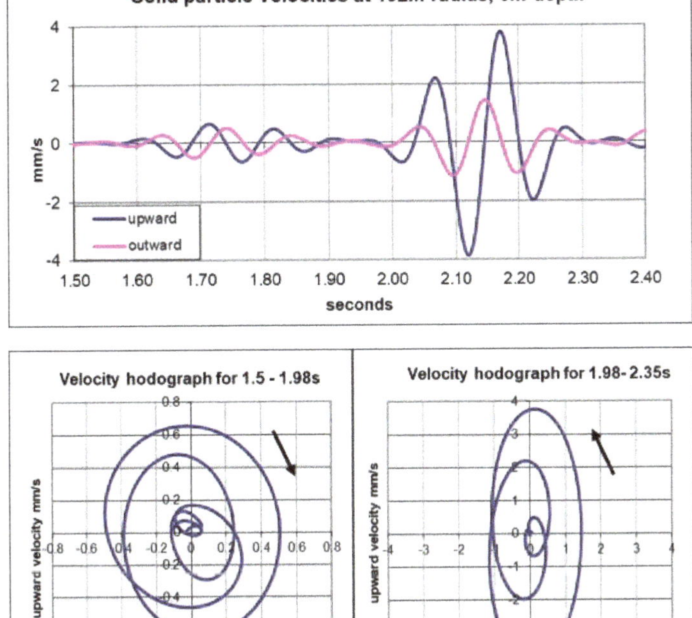

Figure 9. The time series for a solid particle at the interface is shown for the same period of 1.5–2.4 s. Two periods were selected for the hodographs shown below, which differ in both shape and circulation direction.

The more intense retrograde mode is that which we previously studied, but we intended to further investigate the prograde mode, which cannot exist in the Rayleigh half-space model, but propagates stably in the graded half-space.

In Figure 10, both hodographs are prograde, but with contrasting shapes. These examples show how the graded model dispersed energy between modes, but we found no evidence of dispersion within individual modes. In 2018, we showed this by tracking their peak intensities against travel time. There was a good fit for the cylindrical spreading of peak energy, as the wave circumference increased without dispersion.

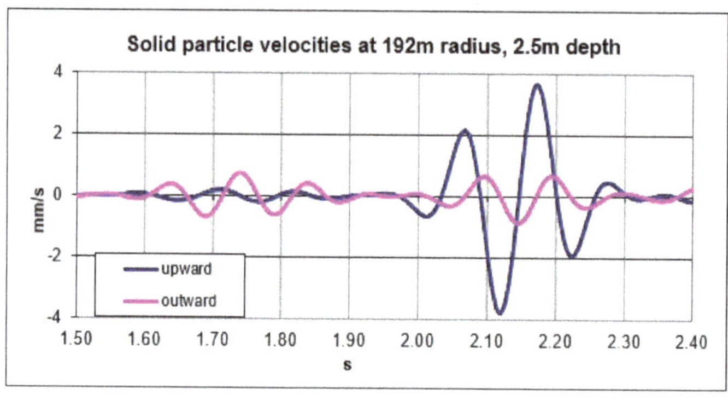

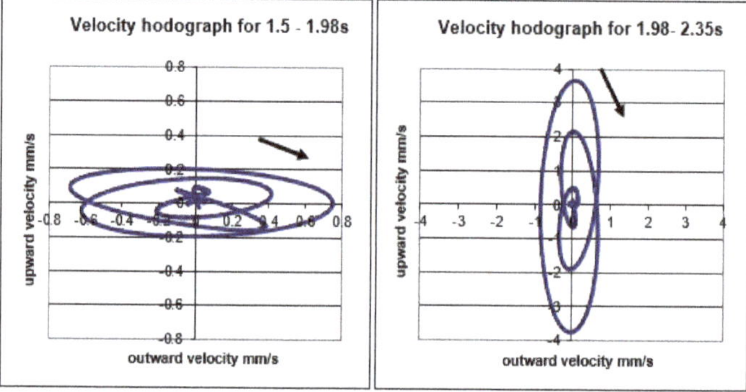

Figure 10. At greater depths, such as 2.5 m below the interface, both modes became prograde.

For the later mode, a change in direction of the solid motion occurred at a critical depth (~1.1 m) below the interface. Here, there was no horizontal motion. This phenomenon was described by Rayleigh for the interface waves in a solid half-space, and this similarity showed how the later mode in the graded seabed model was a version of the "Rayleigh wave". The existence of prograde waves at depth was thus expected, but the Rayleigh theory only predicted retrograde motions at the surface with no dispersion.

In contrast, the energy in the graded model was dispersed into at least two modes that travelled at different speeds, and thus reduced the intensity to be found at a distance, but this dispersion was still much less than that to be expected over a layered seabed.

4.8. The Contrasting Group and Phase Velocities

Time profiles at two positions were used to measure the travel speed. The Gaussian envelope bell form helped by allowing a precise fit to the data extracted from the FE analysis. This speed was found to be 117 m/s when the material shear speed, $V_s(0)$, at the interface was specified as 128 m/s. This ratio of 91.4% can be compared with the interface wave speed for the Rayleigh model given by Achenbach [15] as varying from 86.7% to 95.5% dependent on the Poisson ratio of the uniform solid in the half-space. We estimated the precision of our measurement at ±0.5 m/s, but recognize the uncertainty of the FE procedure, which depends on the accuracy of many successive calculations.

A mathematical analysis of the impulse clearly would be better, if one could be found to give a closed-form solution. However, a discussion with Godin (private communication) suggested that his published method [17] for a power-law depth profile cannot cope with the step change at the interface of our model.

4.9. The Influence of the Depth Profile Gradient on the Speed of the Displacement Wave

The gradient of the material shear wave speed with depth has been found to influence the speed of the displacement wave, c_d. We know that the greater this depth profile gradient, the more rapidly the wave energies are refracted back to the interface.

Figure 11 shows a snapshot view that was used to measure the wavelength. With a specified frequency of 80 Hz, this determined the displacement (or "phase") wave speed. The smallest elements in this fine-mesh density model were only 0.0625 m wide.

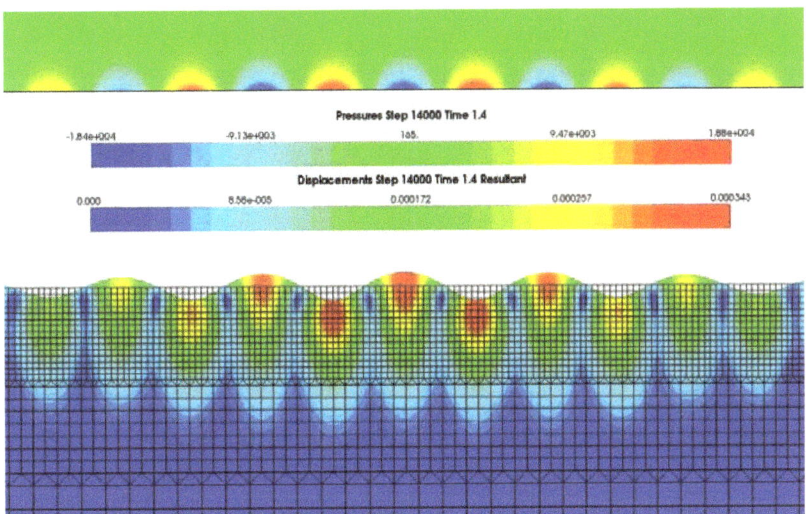

Figure 11. By increasing the mesh density and frequency, the wavelength can be better measured.

Five whole waves were selected, and the corresponding number of elements was found to be 119. This gave a wavelength of 1.4875 m and a wave speed of 119 m/s. The gradient in this model was 4 (m/s)/m. FE models in a sequence were run, each with different values of the gradient. The results are shown in Table 1. They were of limited precision, estimated at ±0.5 m/s, but showed a clear trend.

Table 1. The variation of wave speed with depth profile.

Shear Speed Profile Gradient, g_r, (m/s)/m	2	4	6	8	10
Displacement wave speed, c_d, m/s	118	119	120	120.7	121.7

Improved methods for using the FE methodology to determine c_d are being considered, including better postprocessing. The precision was increased by using a finer mesh density, but at the expense of much longer run times and large data files.

Making these changes also helped to show that the results were not an artifact of the modeling process. Other checks on the validity of the FE process include changing the boundary conditions of the FE model. A change from a rigid boundary (no motion) to a soft boundary (no forces) would change the phase of any reflected waves, so that the extent of any such reverberation can be monitored and shown not to affect the conclusions.

4.10. The Evanescent Form of the Associated Pressure Fields in the FE Modelling

Figure 1 shows the distribution of acoustic pressures within the water as the wave passed. A rapid reduction in amplitude is shown by the green coloration well away from the interface. This distribution was typical of an evanescent field. No energy was radiated away from the interface, with the stored energy being supplied by the travelling seismic interface wave, and recovered for onward transmission, assuming there were no losses. In practice, there will be absorption, as the acoustic energy is converted to heat, but that will depend on the local circumstances, and was not modeled here.

Figure 12 shows the acoustic water pressures as a function of the height above the interface. Notably, the waveforms of these time profiles were all in alignment, with peaks occurring simultaneously.

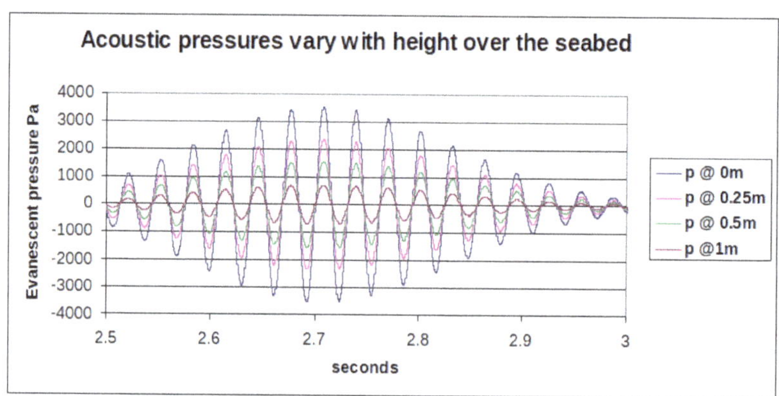

Figure 12. The pressures created at a distance of 256 m from the source are shown at different heights above the seabed. They oscillated simultaneously, independent of height.

The good fit to the straight line of the continuous wave theory in Figure 13 shows that the evanescent water field followed an exponential decay within this FE model. The field thickness was 0.6 m for the fitted line, shown for a frequency of 32 Hz. The deviations at greater heights were due to reverberation in the model, but were small compared with the pressure levels at the interface.

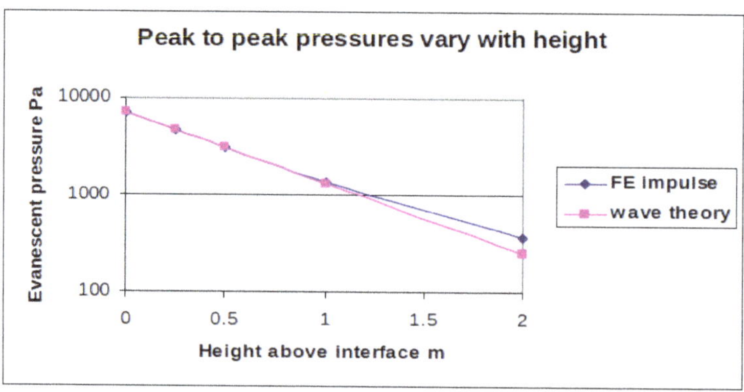

Figure 13. The pressure amplitudes (peak to peak) plotted on a log scale against height.

The displacements below the interface were also shown to be evanescent, with an eventual exponential decay occurring at depths >10 m, but these fields were more complicated than those in the water above. Different modes travelled at different speeds, with changes in the direction of the motion at defined depths, as can be seen in contrasting Figures 9 and 10. However, the simpler evanescent water pressure field was independent of the horizontal motion of the solid.

4.11. Linking the Acoustic Pressure and the Horizontal Water-Particle Velocity

For continuous plane waves in open water, well away from boundaries, the ratio of acoustic pressure to the water particle speed is called the specific acoustic impedance (Kinsler et al. [22]). This is a scalar, dependent on the water density, ρ_w. Only compression waves are propagated in a fluid, and the particle motion lies along the direction of energy propagation, so that the speed is just this velocity component. However, in an evanescent field, the water particles move with velocities that have components in multiple directions.

It is then appropriate to use the inverse of the impedance, the admittance, as a complex vector, which is the water-particle velocity vector, u, divided by the scalar pressure, *p*. The complex vertical admittance component is imaginary, the pressure waveform is out of phase with the vertical velocity, and there is then no work done on the fluid by the motion of the solid (Fahy [21]). We denoted the vertical velocity component as u_v and the horizontal component as u_h.

Then, the vertical admittance becomes:

$$\frac{u_v}{p} = \frac{j}{\rho_w c_d}\sqrt{\left(1 - \left(\frac{c_d}{c_w}\right)^2\right)} \tag{8}$$

The horizontal admittance is a real quantity and derived in the same way as that for continuous plane waves (Kinsler et al. [22]), but is dependent on the displacement wave speed, c_d.

$$\frac{u_h}{p} = \frac{1}{\rho_w c_d} \tag{9}$$

The square root term in Equations (3) and (8) can now be considered as a "shape factor", since it will control the shape of the water-particle hodographs. When $c_w \gg c_d$, this factor is close to unity, and the hodograph is circular, but as the ratio decreases, the hodograph becomes elliptical with comparatively reduced vertical motion.

4.12. A Comparison of Water-Particle Motions in the Evanescent Field to Those in Plane Waves

With the knowledge that it is the water-particle motion that is sensed by many subsea creatures, it is common to infer this parameter from a measured acoustic pressure, using the specific acoustic impedance, $\rho_w c_w$. However, this inference is invalid when the acoustic pressure is created by a seismic interface wave and the displacement wave speed, c_d, should be used in place of c_w.

It is convenient to use a unitless velocity-to-pressure ratio (VPR). This is done by comparing the evanescent field horizontal admittance to the value for acoustic plane waves in open water. From the data shown in Figure 3, we found a VPR just over 12.0 for the windowed wave packet used.

In contrast, if we used Equation (9) for a continuous wave, with the value for c_d of 119 m/s for a value of gr = 4 (Table 1) and a value of c_w = 1500 m/s, we calculated a VPR of 12.6. More work is planned to investigate how this difference was influenced by other factors, such as the properties of different finite-length wave packets. However, it was clear that the VPR was substantial in all realistic cases, and a value of 12 provided a fair working estimate. This showed how much more significant the water-particle motions were in comparison to the acoustic pressures in the evanescent field.

It is also interesting to measure this ratio in practice by comparing the values from a hydrophone with a water-particle velocity sensor.

5. Additional Field Trials

5.1. Methods

The prediction of the strong linkage between the local evanescent pressures and the horizontal water particle velocity led to a trial in the still waters of Wraysbury reservoir, where a test facility is operated by the National Physical Laboratory (NPL) Teddington, London.

Figure 14 shows the stainless-steel sledge fitted with various instruments on cables to allow their deployment onto the clay bed 20 m below the raft. The central housing contained a set of three orthogonally mounted geophones. This is the same instrument that was deployed into the Rhine estuary at Kinderdijk and reported by Hazelwood and Macey 2016 [18]. As before, the deployment used an extra line attached to the sledge prow (Figure 14) to tow the sledge a short distance, thus aligning it with respect to the raft.

Figure 14. The test rig ready for deployment at Wraysbury in March 2017.

Two additional instruments were added, each with their own cables. A hydrophone was supplied by NPL, and an experimental glass buoy was held in a holster, which allowed it to float clear once deployed into the water. This was then held down by a light coiled spring, which had been previously tested to give a vertical resonance with a low natural frequency <5 Hz.

One of the set of geophones in the buoy can be seen through the glass. The use of more geophones allowed the same low-noise, low-impedance preamplifiers to be used as developed for the Kinderdijk trial. These geophones were well adapted to sense velocity over the frequency range 10–200 Hz, with a flat sensitivity of 20 V/(m/s). It was thus useful to compare their outputs with that of the hydrophone, which had a pressure sensitivity that was flat over the same frequency range.

The seismic interface wave was generated by the impact of a 20 kg bronze cylinder with the bed of the reservoir, after being dropped from the raft. The cylinder was streamlined with a cylindrical rear fin to maintain its attitude during the descent, and recovered by winch using a rope, flaked out on the raft deck before the drop to minimize drag. Several drops were made with similar results.

5.2. Results

The principal finding shown in Figure 15 was that the shape and phase of the hydrophone time series (dark blue) largely matched that of the horizontal y velocity component (green) given by one of the geophones. The data shown was used to estimate the VPR. The peak-to-peak (p-p) amplitude of the y component data was 43.6 μm/s. The amplitude of hydrophone data was 16.5 Pa. The ratio of pressure to velocity was 378 Pa/(mm/s).

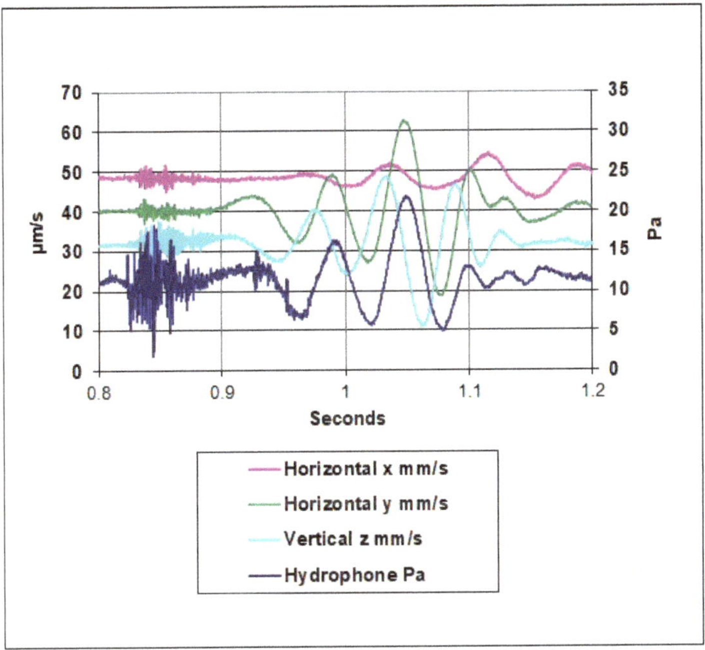

Figure 15. Test results from the trial showed three velocity components and the acoustic pressure. The right side vertical scale applies to the hydrophone pressure data, the left side to the geophone water-motion data. The plots are offset for clarity. The recorded values returned to zero after the wave had passed.

Comparing this with a typical open water impedance of 1500 Pa/(mm/s) gave a VPR of 4.0, much less than the continuous wave theoretical value of 12.6 or the FE model VPR of 12.0. This was disappointing, and suggested the need to further investigate the performance of the experimental glass geophone buoy. We also recognized the risk that the heavy sledge may have affected the nearby water motion, and that we did not have details of the properties of the sediment. However, the confirmation of the wave-packet shape was encouraging.

6. Conclusions

The change of the forcing impulse in the FE model provided more results for higher frequencies than the previous papers. The Gaussian windowed cosine form allowed comparisons with analytic mathematical theory for continuous waves, showing similar effects, but with some changes in magnitude.

Additional analysis of the water-particle motions generated by the seismic interface waves showed a Gaussian bell-shaped time series for the water-particle kinetic energies in various modes. This was used in a fit procedure for the wave envelope to allow more precise timing to give better data for the envelope speed.

The displacement wave speed was shown to be controlled by the rate of increase with depth of the material shear wave speed. It was faster than the wave envelope speed, creating apparent movement of the displacement wave peaks through the envelope window. This explained the morphing effect reported for the pulses seen in earlier work.

The retention of energy within the most intense mode, as reported before, was confirmed by the contrast in the dispersion between the modes in the graded model, as well as that for modes generated by a layered model. Other modes were seen to have a quite different structure in the solid, without a change from retrograde to prograde predicted by Rayleigh [11] for the uniform half-space.

Evanescent pressure fields in the water have been analysed both by flexural plate theory and by the use of FE models, both for the graded seabed and a single-layer substrate. The water-particle motions were shown to be large by comparison to the acoustic pressures, with a large velocity pressure ratio when comparing this effect to that found in open water.

The study has also raised many questions that it is hoped can be answered by future work. More measurements and theory could help further explain the features outlined here.

Author Contributions: R.H. conceived and designed the experiments; P.M. provided support for the work. All authors have read and agreed to the published version of the manuscript.

Funding: This research received no external funding.

Acknowledgments: This work has received no external funding, but the assistance of S.P. Robinson and the staff at the National Physical Laboratory was important for the work at Wraysbury. The authors are grateful for the assistance of colleagues John King and Jim Delderfield, who helped with the figures, and many others who also provided comments on the text. Tim Swann made especially useful comments.

Conflicts of Interest: The authors declare no conflict of interest.

References

1. Thomsen, F.; Gill, A.; Kosecka, M.; Andersson, M.; Andre, M.; Degraer, S.; Folegot, T.; Gabriel, J.; Judd, A.; Neumann, T.; et al. *MaRVEN—Environmental Impacts of Noise, Vibrations and Electromagnetic Emissions from Marine Renewable Energy*; Final Study Report of Project RTD-KI-NA-27-738-EN-N; EUR 27738; European Commission: Brussels, Belgium, 2015.
2. Hawkins, A.D.; Johnston, C.; Popper, A.N. How to set sound exposure criteria for fishes. *J. Acoust. Soc. Am.* **2020**, *147*, 1762–1777. [CrossRef] [PubMed]
3. Popper, A.N.; Hawkins, A.D. The importance of particle motion to fishes and invertebrates. *J. Acoust. Soc. Am.* **2018**, *143*, 470–486. [CrossRef] [PubMed]
4. Madsen, P.T.; Wahlberg, M.; Tougaard, J.; Lucke, K.; Tyack, P. Wind Turbine Underwater Noise and Marine Mammals: Implications of Current Knowledge and Data Needs. *Mar. Ecol. Prog. Ser.* **2006**, *309*, 279–295. [CrossRef]
5. Halvorsen, M.B.; Casper, B.M.; Woodley, C.M.; Carlson, T.J.; Popper, A.N. *Predicting and Mitigating Hydroacoustic Impacts on Fish from Pile Installations*; NCHRP Research Results Digest 363, Project 25–28; National Cooperative Highway Research Program; Transportation Research Board, National Academy of Sciences: Washington, DC, USA, 2011.
6. Hawkins, A.D.; Popper, A.N. Directional hearing and sound source localization by fishes. *J. Acoust. Soc. Am.* **2018**, *144*, 3329–3350. [CrossRef] [PubMed]
7. Hamilton, E.L. Vp/Vs and Poisson's ratio in marine sediments and rocks. *J. Acoust. Soc. Am.* **1979**, *66*, 1093–1101. [CrossRef]
8. Heaney, K.; Ainslie, M.; Halvorsen, M.; Seger, K.; Müller, R.; Nijhof, M.; Lippert, T. *A Parametric Analysis and Sensitivity Study of the Acoustic Propagation for Renewable Energy*; US Department of the Interior Bureau of Ocean Energy Management Office of Renewal Energy Program: Stuart, FL, USA, 2020.
9. Hawkins, A.D.; Hazelwood, R.A.; Popper, A.N.; Macey, P.C. Substrate vibrations and their potential effects upon Fishes and Invertebrates. *J. Acoust. Soc. Am* **2021**, *149*, 2782. [CrossRef] [PubMed]
10. Urick, R.J. *Principles of Underwater Sound*, 3rd ed.; McGraw Hill: New York, NY, USA, 1983.
11. Rayleigh, L. Interface waves. *Proc. Lond. Math. Soc* **1887**, *17*, 4–11.
12. Scholte, J. On true and pseudo Rayleigh waves. *Proc. K. Ned. Akad. Wet.* **1949**, *52*, 652–653.
13. Stein, S.; Wysession, M. *An introduction to Seismology Eathquakes and Earth Structure*; Blackwell: Oxford, UK, 2003.
14. Shearer, P.M. *Introduction to Seismology*; Cambridge University Press: Cambridge, UK, 1999.
15. Achenbach, J.D. *Wave Propagation in Elastic Solids*; Elsevier Science Publishers: Amsterdam, The Netherlands, 1975.
16. Hazelwood, R.A.; Macey, P.C.; Robinson, S.P.; Wang, L.S. Optimal transmission of interface vibration wavelets—A simulation of seabed seismic responses. *J. Mar. Sci. Eng.* **2018**, *6*, 61. [CrossRef]
17. Godin, O.A.; Chapman, D.M. Dispersion of interface waves in sediments with power-law shear speed profiles. I. Exact and approximate analytical results. *J. Acoust. Soc. Am.* **2001**, *110*, 1890–1907. [CrossRef] [PubMed]
18. Hazelwood, R.A.; Macey, P.C. Modeling water motion near seismic waves propagating across a graded seabed, as generated by man-made impacts. *J. Mar. Sci. Eng.* **2016**, *4*, 47. [CrossRef]
19. Jensen, F.B.; Kuperman, W.A.; Porter, M.B.; Schmidt, H. *Computational Ocean Acoustics*; Springer Science & Business Media: New York, NY, USA, 2011.
20. Biot, M.A. Theory of propagation of elastic waves in a fluid saturated porous solid. *J. Acoust. Soc. Am.* **1956**, *28*, 168–178. [CrossRef]

21. Fahy, F.J. *Sound and Structural Vibration*; Academic Press: Cambridge, MA, USA, 1985.
22. Kinsler, L.E.; Frey, A.R.; Coppens, A.B.; Sanders, J.V. *Fundamentals of Acoustics*; Wiley: New York, NY, USA, 1999.
23. Institute of Sound and Vibration Research, Southampton UK, Accessed in June 2021. Available online: https://www.southampton.ac.uk/engineering/research/centres/isvr.page (accessed on 1 March 2020).
24. Timoshenko, S.P.; Woinowsky-Krieger, S. *Theory of Plates and Shells*; McGraw-Hill: New York, NY, USA, 1959.
25. Hamilton, W.R. The Hodograph. *Proc. R. Ir. Acad.* **1847**, *3*, 344–353.

Journal of
Marine Science and Engineering

Article

A Modeling Comparison of the Potential Effects on Marine Mammals from Sounds Produced by Marine Vibroseis and Air Gun Seismic Sources

Marie-Noël R. Matthews [1,*], Darren S. Ireland [2], David G. Zeddies [3], Robert H. Brune [4] and Cynthia D. Pyć [5]

[1]   JASCO Applied Sciences (Canada) Ltd., Halifax, NS B3B 1Z1, Canada
[2]   LGL Ecological Research Associates Inc., Bryan, TX 77802, USA; DIreland@lglalaska.com
[3]   JASCO Applied Sciences (USA) Inc., Silver Spring, MD 20910, USA; david.zeddies@jasco.com
[4]   Robert Brune, LLC., Evergreen, CO 80439, USA; rhbrune@aol.com
[5]   JASCO Applied Sciences (Canada) Ltd., Victoria, BC V8Z 7X8, Canada; cynthia.pyc@jasco.com
*   Correspondence: marie-noel.matthews@jasco.com

Abstract: Concerns about the potential environmental impacts of geophysical surveys using air gun sources, coupled with advances in geophysical surveying technology and data processing, are driving research and development of commercially viable alternative technologies such as marine vibroseis (MV). MV systems produce controllable acoustic signals through volume displacement of water using a vibrating plate or shell. MV sources generally produce lower acoustic pressure and reduced bandwidth (spectral content) compared to air gun sources, but to be effective sources for geophysical surveys they typically produce longer duration signals with short inter-signal periods. Few studies have evaluated the potential effects of MV system use on marine fauna. In this desktop study, potential acoustic exposure of marine mammals was estimated for MV and air gun arrays by modeling the source signal, sound propagation, and animal movement in representative survey scenarios. In the scenarios, few marine mammals could be expected to be exposed to potentially injurious sound levels for either source type, but fewer were predicted for MV arrays than air gun arrays. The estimated number of marine mammals exposed to sound levels associated with behavioral disturbance depended on the selection of evaluation criteria. More behavioral disturbance was predicted for MV arrays compared to air gun arrays using a single threshold sound pressure level (SPL), while the opposite result was found when using frequency-weighted sound fields and a multiple-step, probabilistic, threshold function.

Keywords: animat; air gun; impact assessment; marine vibroseis; marine mammal; sound propagation; ocean noise; underwater noise modeling and mapping; underwater noise effects; ocean noise regulations

Citation: Matthews, M.R.; Ireland, D.S.; Zeddies, D.G.; Brune, R.H.; Pyć, C.D. A Modeling Comparison of the Potential Effects on Marine Mammals from Sounds Produced by Marine Vibroseis and Air Gun Seismic Sources. *J. Mar. Sci. Eng.* **2021**, *9*, 12. https://dx.doi.org/10.3390/jmse9010012

Received: 6 November 2020
Accepted: 16 December 2020
Published: 24 December 2020

Publisher's Note: MDPI stays neutral with regard to jurisdictional claims in published maps and institutional affiliations.

1. Introduction

Compressed air sources, commonly referred to as seismic air guns, are the most common marine geophysical survey sound source for oil and gas exploration [1] and are also used for construction siting and studying subsea geomorphology. Substantial technological and geophysical data processing advancements, combined with concerns about potential effects of seismic air gun sources on marine fauna, are driving a resurgence in research and commercial development of alternate technologies. Vibroseis technology, used extensively on land, has been investigated for marine use since the 1970s, but it had limited success until recently.

Marine vibroseis (MV) systems produce controllable acoustic signals through volume displacement of water using a vibrating plate or shell that, unlike air guns, produce broadband acoustic signals when an ejected air bubble collapses in water. The output signals of MV are often narrowband tones or swept frequency signals, but they can be tuned to produce a variety of waveforms (e.g., pseudorandom noise, upsweeps, and downsweeps). Compared to air gun signals, MV signals are generally lower in zero-to-peak sound pressure level (PK) and root-mean-square sound pressure level (SPL), and they have reduced bandwidth [2–4]. To obtain enough return of acoustic energy for seismic data processing, MV signals are typically longer waveforms with shorter inter-signal intervals than air guns.

This study builds on previous modeling assessments of MV sources [5], comparing signal characteristics and estimated sound exposures to assess the potential effects on marine mammals using MV versus air gun sources. Here, modeled results for a generic MV array configuration are compared to a realistic operational air gun array, in three acoustically different hydrocarbon-producing basins. Except where otherwise indicated, acoustical terminology follows ISO 18405:2017 [3]. Underwater acoustic metrics used in this study are listed in Table 1.

Table 1. List of underwater acoustic metrics. dB: decibel; µPa: micropascal.

Metric	Symbol	Units	Definition
Sound pressure level (SPL)	L_p	dB re 1 µPa	The root-mean-square (rms) pressure level in a stated frequency band over a specified time window.
Peak sound pressure level (PK)	L_{PK}	dB re 1 µPa	The maximum instantaneous sound pressure level, in a stated frequency band, within a specified period.
Sound exposure level (SEL)	L_E	dB re 1 µPa²·s	A cumulative measure related to the sound energy in one or more pulses or sweeps, within a specified period.
Zero-to-peak source level (SLPK)	-	dB re 1 µPa m	The sound level measured in the far field and scaled
Source level (SL)	-	dB re 1 µPa m	back to a standard reference distance of 1 metre from
Evergy source level (ESL)	-	dB re 1 µPa²·s m²	the acoustic centre of the source.

2. Methods

2.1. Scenarios

Marine geophysical surveys are performed globally in widely different ocean environments where differing biota may be present. To address this variability, three survey environments were defined (Table 2) based on water depth regimes, the likely occurrence of oil and gas exploration, and the presence of marine mammals: a Transition Zone offshore Indonesia (10–25 m), a Shallow Water area in the northern North Sea (110–130 m), and a Deep Water area in the Gulf of Mexico (>1000 m). In each study area, the source array, survey design, and month were selected from the most common survey parameters and operational seasons. In total, four scenarios were evaluated: one in the Transition Zone, two in the Shallow Water area, and one in the Deep Water area (Table 2).

Table 2. Summary list of scenarios.

Scenario	Area	Month	Water Depth (m)	Survey Line Spacing (m)	Survey Line Length (m)	Air Gun Array Chamber Volume	MV Configurations
1	Transition Zone (Java Sea)	July	10–25	100	25,000	12,290 cm³ (750 in³)	18 elements, 15 × 16 m
2	Shallow Water (North Sea)	August	110–130	100	25,000	67,680 cm³ (4130 in³)	18 elements, 15 × 16 m
3				500			
4	Deep Water (Gulf of Mexico)	February	1500–1600	500	50,000	67,680 cm³ (4130 in³)	18 elements, 15 × 16 m

MV and air gun arrays were configured to produce a similar broadband energy source level (ESL) in the vertical direction: 223 dB re 1 µPa²·s m² per sweep for the MV array and 218–233 dB re $\mu Pa^2 \cdot s\ m^2$ per pulse for the air gun array. Two survey patterns were evaluated: a tight-spaced (100 m) pattern and a wide-spaced (500 m) pattern. A tow speed of 2.3 m/s (4.5 knots) and acquisition spacing of 25 m was used for all surveys.

2.2. Comparison

To compare potential effects on marine mammals from sounds produced by MV and air guns, we compared the characteristics of modeled received signals and estimated marine mammal exposure to injury and disturbance thresholds. The received signal characteristics studied were PK, sound exposure level (SEL), pulse duration, duty cycle, and the time and frequency domain representations of the received signal. To calculate the received pulse duration, the SPL was calculated using a 0.125 s sliding window and compared to the estimated ambient level in each region (see Sections 2.3.2 and 2.3.4). The duty cycle was defined as the percentage of time when the received signal was more than 6 dB above ambient level (see Section 2.3.4).

The probability of detecting and discriminating a signal in the presence of noise increases as a signal becomes louder relative to the noise [6]. The signal-to-noise ratio (SNR) is often used in signal detection theory to estimate when a signal can be reliably detected. While humans can detect speech in noise when the signal is about the same level as the noise (SNR $\approx$ 0 dB), many factors contribute to this performance such as signal redundancy and the direction of the signal compared to the direction of the noise (e.g., [7]). Given that in this study we did not know the amount of redundant information in the signal nor did we know the direction of the signal relative to the noise, we assumed that reliable detection would occur when the signal had a sound pressure level 6 dB higher than the ambient noise. The critical ratio level is the difference between the SPL of a barely audible pure tone in the presence of a continuous noise of constant spectral density and the spectral density level (SDL) for that noise [3]; it depends on frequency and duration of the signal. In other words, it is the minimal ratio between a signal and noise levels at which the signal can be perceived. For marine mammals, the critical ratio level is ~10 to 35 dB re 1 Hz [8]. The chosen SNR threshold of 6 dB, for a noise bandwidth of 25 kHz, corresponds to a ratio of 50 dB re 1 Hz, which exceeds the largest critical ratio level for marine mammals, indicating that the chosen SNR threshold is conservative. In cases where detection requires a greater SNR (e.g., the signal is relatively new or non-redundant, or the direction of the signal is the same as that of the noise), the reported pulse duration and duty cycle would be overestimates. In cases where animals could detect signals closer to ambient levels (i.e., for smaller SNR), the reported values would be underestimates.

These received signal characteristics were compared at multiple distances from the sources. This comparison highlights the differences between the air gun array impulsive sounds (typically <<1 s) and the MV array non-impulsive sounds (>>1 s) [9,10]. Impulsive and non-impulsive sounds affect marine life differently, especially in terms of their potential to cause injury [9,11]. Thus, the inherent difference between the sources required the application of different criteria when predicting and comparing their potential effect on marine mammals. The relative novelty of the MV technology and lack of research on its effect on marine mammals means, however, that no effect criteria or guidelines are specific to this type of seismic source. The comparison of received signal characteristics may be useful in the development of MV-specific effect criteria.

Estimated marine mammal exposure to injury and disturbance thresholds were also compared. These exposure estimates were calculated using an agent-based model (see Section 2.3.5) and the current effect criteria that are most likely to be applied, based on the best available science today.

The injury criteria recommended by Southall et al. [12] and the US National Marine Fisheries Service (NMFS) [13] were used to calculate exposure estimates. They are the most commonly used criteria for estimating the potential for injury to exposure to air gun sounds, and both criteria provide acoustic thresholds for the onset of permanent threshold shift (PTS) for impulsive (air gun array) and non-impulsive (MV array) sounds. They recommend using dual criteria for assessing exposures to potentially injurious sound levels: a frequency-weighted SEL metric, which accumulates over a set exposure duration (i.e., 24 h), and a PK metric, which is a measure of acute exposure to high-amplitude sounds. These criteria also divide marine mammals into functional hearing groups (low-, high-, and very high-frequency cetaceans, phocids, and otariids [12]), each with different threshold levels and weighting functions applied to the sound field to account for the hearing sensitivity as a function of frequency for the hearing group. The auditory weighting functions associated with the Southall et al. [12] set of criteria are similar to those recommended by the NMFS [13], only the name of the hearing groups differ. Here, the NMFS [13] names have been used; Table 3 shows the PTS onset thresholds used in this study.

Table 3. Summary of PTS onset acoustic thresholds from NMFS [13]. L_{pk}, flat: unweighted zero-to-peak sound pressure level referenced to 1 µPa; L_E: frequency-weighted sound exposure level accumulated over a 24 h period referenced to 1 µPa^2·s; the subscript associated with L_E thresholds indicates the designated hearing group and the associated frequency-weighting function.

HEARING GROUP	Impulsive Source	Non-Impulsive Source
Low-frequency (LF) cetaceans	$L_{pk,flat}$: 219 dB $L_{E,LF,24h}$: 183 dB	$L_{E,LF,24h}$: 199 dB
Mid-frequency (MF) cetaceans	$L_{pk,flat}$: 230 dB $L_{E,MF,24h}$: 185 dB	$L_{E,MF,24h}$: 198 dB
High-frequency (HF) cetaceans	$L_{pk,flat}$: 202 dB $L_{E,HF,24h}$: 155 dB	$L_{E,HF,24h}$: 173 dB
Phocid pinnipeds (in water; PW)	$L_{pk,flat}$: 218 dB $L_{E,PW,24h}$: 185 dB	$L_{E,PW,24h}$: 201 dB
Otariid pinnipeds (in water; OW)	$L_{pk,flat}$: 232 dB $L_{E,OW,24h}$: 203 dB	$L_{E,OW,24h}$: 219 dB

Despite numerous studies on behavioral responses of marine mammals to sound, there is little consensus among the scientific community on the most appropriate approach for assessing noise as an impact producing factor. Behavioral responses to sound are highly variable and depend on the context in which a sound is received. Many factors may contribute to the likelihood of a response in addition to the received sound level [9,14]. Here, two different criteria were used to evaluate potential behavioral disturbance. The first set of criteria involves thresholds of unweighted SPL as currently used by NMFS [15]: 160 dB re 1 µPa for impulsive sounds and 120 dB re 1 µPa for non-impulsive sounds. These thresholds have been used for many years by NMFS and other agencies around the world to assess behavioral impacts from various sound-producing offshore projects (seismic or other). Alternative methods to these simplistic thresholds that are based on observations of mysticetes alone [16–19], have been proposed (e.g., [14,20]). Alternative criteria used in this study include frequency-weighted SPL step functions proposed by Wood et al. [21] for impulsive sounds, and frequency-weighted SPL values adopted by the US Department of the Navy (DoN) [22] for non-impulsive sounds. These criteria associate a series of increasing sound level thresholds with increasing probability of behavioral response. The DoN criterion uses a smooth continuous function, adapted from Feller [23]. We discretized the DoN function at 10%, 50% and 90% probabilities of response, corresponding to the Wood et al. [21] step functions, giving weighted SPL of 155, 165, and 180 dB re 1 µPa, respectively.

2.3. Computation

2.3.1. Sources Levels

Marine Vibroseis (MV)

Few measured output signals from marine vibrators were publicly available at the time of this study; none showed spectral levels at more than ~100 Hz. The MV array layout and synthetic source waveform [3] in this study were based on publicly available information for General Dynamic's CoHerent Acoustic Modulation Projector (CHAMP) system [24]. This system consists of three strings of six identical vibrators units (18 elements in total) towed at 6 m depth (Figure 1). It was designed to produce signals at frequencies between 5 and 100 Hz and includes an adaptive, in-line, compensation filter that uses acceleration data at the piston assembly for gain control (along with other controller technology) to reduce sound levels at frequencies outside the desired frequency band [25].

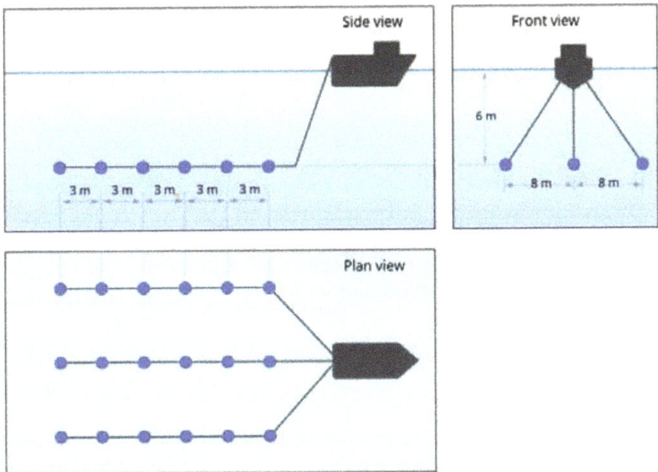

Figure 1. Illustration of MV layout B. Port and starboard arrays are assumed to be identical.

In general, three signal types are possible from MV elements: linear frequency sweeps, logarithmic frequency sweeps [26], and pseudorandom noise (PRN) [27–31]. Here, we studied the most common signal type, a linear frequency upsweep, with specifications within the Marine Vibrator Joint Industry Project (MVJIP) guidelines [32,33]. This input signal is defined as:

$$p(t) = A(t) \sin\left(2\pi t \left[f_0 + \left(\frac{df}{dt} \right) t \right] \right),$$ (1)

where df/dt is positive (upsweep), f_0 is the starting frequency, and $A(t)$ is the amplitude of the signal. The synthetic signal is a 5 s linear upsweep from 5 to 100 Hz, with a 10 dB amplitude ramp up from 5 to 10 Hz and cosine taper of 0.1 s at the start and end of the signal to reduce transients produced in real systems during abrupt signal changes. Harmonics will arise from distortion of signals in marine vibrator systems, but their nature is not empirically characterized. Harmonics were added to the synthetic MV signal based on the progression of a damped harmonic oscillator whose amplitude decreased linearly with time and exponentially with harmonic order. For this study, the amplitude of the source waveform of one MV element was set to 5 kPa (zero-to-peak source level (SLPK) [34] of 194 dB re 1 µPa m), resulting in an energy source spectral density level (ESSL) [35] of 178 dB re 1 µPa²·s m²/Hz from 10 to 100 Hz. Harmonics were assumed to be at least 20 dB below the main signal spectral level for frequencies ≤150 Hz (i.e., with an ESSL of 158 dB re 1 µPa²·s/Hz at 5 Hz, decreasing linearly with time and exponentially with harmonic order) and at least 40 dB less than the main signal spectral level for frequencies ≥150 Hz, as prescribed by the MVJIP guidelines [32,33].

The array source waveform was modeled by summing the contributions of the 18 individual MV elements using linear superposition of signals from individual vibrator elements. While the signal parameters were well informed by industry experts involved in developing the technology, the modeled output signals have not been compared to measured MV signals. Mutual impedance (the load on one piston due to the pressure waveform from the other elements in the array) may be important for some arrays [36–38]. This physical effect is not currently implemented in the MV source waveform model because of the lack of available output signal measurements that would allow proper benchmarking of the model. The studied MV array synthetic sound waveforms and ESSL are presented in Figure 2; the corresponding ESSL is shown for two frequency ranges (1 Hz to 25 kHz; Figure 2, left). The array signal outputs (source levels) are summarized in Table 4.

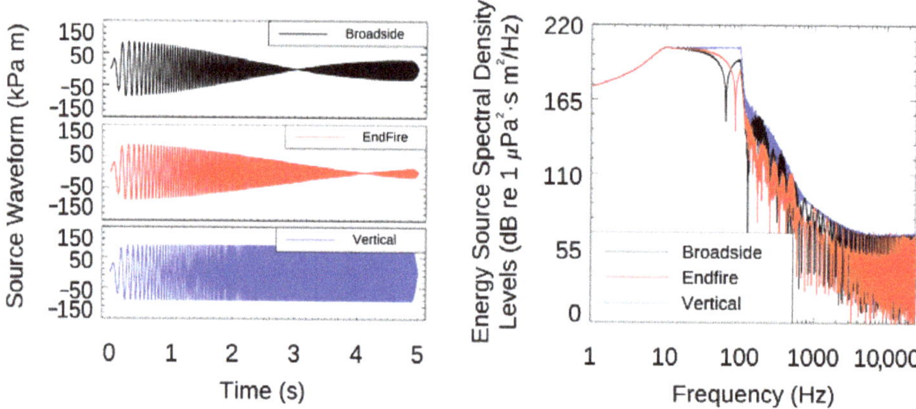

Figure 2. MV array 5–100 Hz linear upsweep (**left**) source waveform and (**right**) energy source spectral density levels (ESSL) for broadside (perpendicular to tow direction), endfire (directly aft of the array), and vertical directions. The ESSL is shown on two frequency scales (1 Hz to 25 kHz). Tow depth 6 m.

Table 4. Marine vibroseis source levels. SLPK: zero-to-peak source level [34]; SL: source level [3]; ESL: energy source level [3].

Specifications	Broadside	Endfire	Vertical
SLPK (dB re 1 µPa m)	218.6	218.8	219.1
SL (dB re 1 µPa m)	210.6	213.7	216.1
ESL (dB re 1 µPa²·s m²) 0.05–0.1 kHz	217.2	218.2	222.9
ESL (dB re 1 µPa²·s m²) 0.05–1 kHz	217.2	218.2	222.9
ESL (dB re 1 µPa²·s m²) 1–25 kHz	101.7	99.9	107.0

Air Gun Arrays

Seismic air gun arrays comprise individual air guns with different air chamber volumes. Air gun arrays are configured to generate a desired dominant low-frequency beam pattern by placement of the individual air guns. Two hypothetical air gun arrays were modeled in this study: a relatively small volume, 12,290 cm³ (750 in³) array operating in the Transition Zone (scenario 1), and a larger 67,680 cm³ (4130 in³) array operating in the Shallow Water and Deep Water areas (scenarios 2–4). The smaller volume array consisted of one triangular cluster towed at 6 m depth (at the cluster center; Figure 3). The cluster was made of three identical 4100 cm³ (250 in³) air guns with a 0.9 m spacing separation between elements. The larger volume array consisted of 31 elements on three strings, towed at 6 m depth (Figure 4), with individual volumes ranging from 660 to 4100 cm³ (40 to 250 in³). In each array, the air guns simultaneously released compressed air pulses at a firing pressure (chamber pressure relative to atmospheric pressure) of 13.8 MPa (2000 lbf/in²).

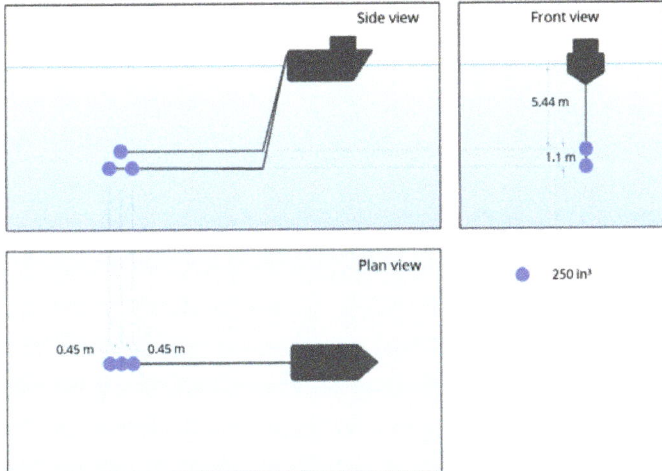

Figure 3. A 750 in^3 air gun array.

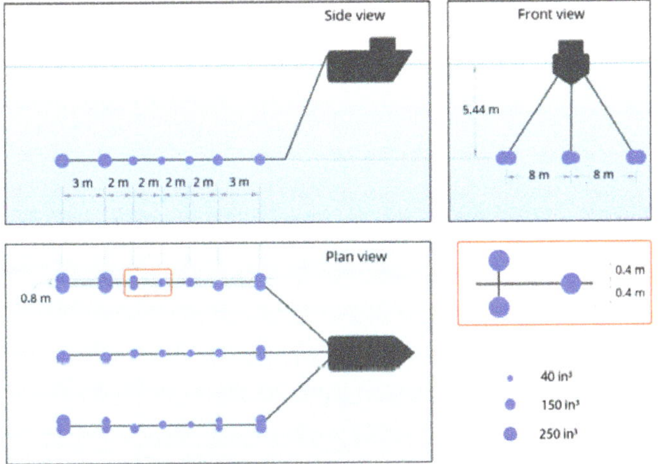

Figure 4. A 4130 in^3 air gun array.

The source waveforms of the air gun arrays were predicted using JASCO's Air gun Array Source Model (AASM) [39], which is based on the physics of oscillation and radiation of air gun bubbles [40]. Physical effects accounted for in the simulation include pressure interactions between air gun array elements, port throttling, bubble damping, and generator-injector element [41–43]. The high-frequency module of AASM uses a stochastic simulation to predict the sound emissions of individual elements above 800 Hz using a multivariate statistical model. A global optimization algorithm tunes free parameters in the model to a large library of air gun source waveforms. AASM has been tuned to fit a large library of high-quality, air gun source, waveform data [44].

The horizontal source waveforms and corresponding ESSL for each air gun array were computed at frequencies up to 25 kHz (Figure 5; Table 5). Figure 5 (right) presents the ESSL over the same frequency range (1 Hz to 25 kHz) as for the MV array results (Figure 2); levels above 2 kHz decreased at an average rate of 23 dB per decade. The source waveforms consisted of a strong primary peak related to the initial pulse followed by a series of pulses associated with bubble oscillations. Most acoustic energy was produced at frequencies below 500 Hz, and the spectra showed peaks and nulls resulting from interactions between the array elements (Figure 5 right).

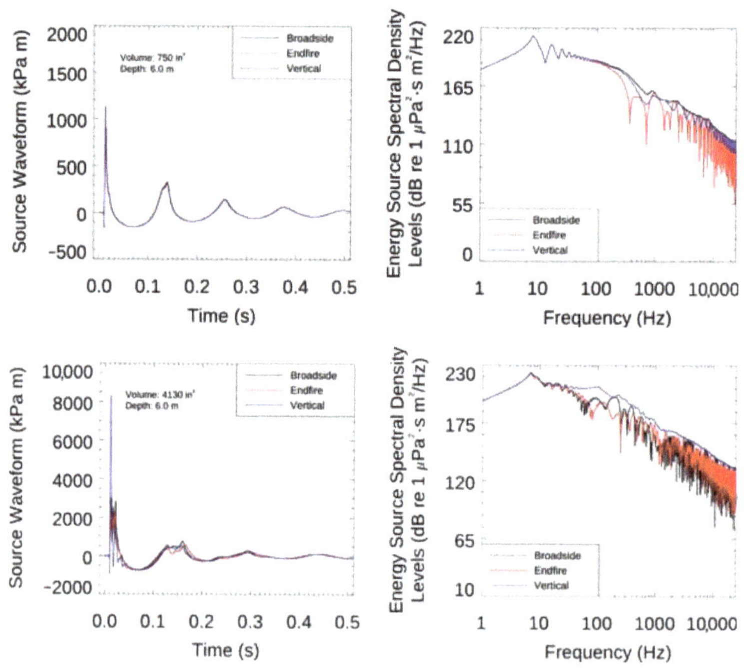

Figure 5. (**Top**) 12,290 cm^3 (750 in^3) and (**bottom**) 676,780 cm^3 (4130 in^3) air gun array (**left**) source waveform and (**right**) energy source spectral density levels (ESSL) for broadside (perpendicular to tow direction), endfire (directly aft of the array), and vertical directions. The ESSL is shown on two frequency scales (1 Hz to 25 kHz). Surface ghost effects are not included in the presented source waveform and spectra.

Table 5. Air Gun array source levels. SLPK: zero-to-peak source level [34]; SL: source level [3]; ESL: energy source level [3].

Specifications	12,290 cm^3 (750 in^3) Air Gun Array			67,680 cm^3 (4130 in^3) Air Gun Array		
	Broadside	Endfire	Vertical	Broadside	Endfire	Vertical
SLPK (dB re 1 μPa m)	241.1	239.3	240.5	249.6	247.3	258.4
SL (dB re 1 μPa m)	224.1	223.4	223.8	236.3	235.7	240.9
ESL (dB re 1 μPa^2·s m^2) 0.05–0.1 kHz	218.2	218.1	218.1	229.1	228.8	231.4
ESL (dB re 1 μPa^2·s m^2) 0.05–1 kHz	218.6	218.3	218.5	229.7	229.0	232.7
ESL (dB re 1 μPa^2·s m^2) 1–25 kHz	182.9	179.6	181.3	187.2	191.1	199.9

2.3.2. Sound Propagation

Sound Propagation Models

Sound propagation was modeled using JASCO's Marine Operations Noise Model (MONM), which combines a wide-angle parabolic equation model [45–48] and a ray-tracing model [49], and JASCO's Full Waveform Range-dependent Acoustic Model (FWRAM; [50]). Both models have been extensively validated with experimental data from several underwater acoustic measurement programs [48,51–58]. MONM and FWRAM incorporate the following site-specific environmental properties: a bathymetric grid of the modeled area, underwater sound speed as a function of depth, and a geoacoustic profile based on the overall stratified composition of the seafloor. The models account for the reflection loss at the seabed, which results from partial conversion of incident compressional waves to shear waves at the seabed and sub-bottom interfaces, and they include wave attenuations in all layers.

Per-pulse SEL sound fields over the frequency range 4 Hz–25 kHz, in decidecade frequency bands, were predicted using MONM. (A decidecade is one tenth of a decade. One tenth of a decade is approximately equal to one third of an octave. For this reason, a decidecade is sometimes incorrectly referred to as a "one-third octave".) Three-dimensional (3-D) acoustic fields were modeled by computing the propagation loss along 36 two-dimensional (2-D) vertical planes radiating from the source with a $10°$ angular step and adding the directional source levels (presented in Section 2.3.1). The full-wave modeling approach (FWRAM) was used to calculate time-domain waveforms along eight radials to then calculate a range-dependent conversion factor. This factor was applied to convert the predicted sound fields in terms of per-pulse SEL to SPL. Sound fields in terms of PK were calculated directly from the time-domain waveforms. All sound fields were modeled to a distance of up to 30 km in the Transition Zone and 50 km in the Shallow Water and Deep Water areas. The computation range step decreased from 30 to 1 m with increasing frequency. The output field was sampled at depths spanning the entire water column, with step sizes ranging from 1 to 100 m, increasing with depth.

Environmental Parameters

To address the variability in ocean environments where geophysical surveys are performed, three sound propagation environments were modeled (Table 2). Profiles of geoacoustic properties characterizing each survey area (Table 6) were developed using an empirical relationship between the sediments physical and acoustic properties [59,60]. The area-specific sediment properties (i.e., grain sizes and sediment porosity) were gathered from publicly available borehole data and borehole logs. The seabed in the Transition Zone (Java Sea) consisted of young sediments made of sand and muddy swamp deposits, reaching a thickness of ~20 m [61]. This surficial sediment layer overlaid claystone with sandstone, conglomerates, and limestone [62]. The Shallow Water area (Northern North Sea) was modeled with a hard to very hard sandy clay layer, 700 m thick [63]. The Deep Water area (Gulf of Mexico) was modeled with a layer of unconsolidated sediments at least several hundred meters thick [64].

One representative sound speed profile in each area was derived from the US Naval Oceanographic Office's Generalized Digital Environmental Model (GDEM) V 3.0 [65,66]. Monthly averaged temperature–salinity profiles were converted to sound speed profiles as described by Coppens [67]. The final profiles represent the monthly climatic conditions in the water column.

In the Transition Zone, the sound speed profile for July was used, representing a time of the year suitable climate wise (June to August) for geophysical surveys for oil and gas exploration (known as seismic surveys) and the most suitable for long-distance propagation. In the Shallow Water area, the profile for August was assumed to best represent a time of the year suitable climate wise for survey operations. In the Deep Water area, the profile for February was used, representing the time of the year most suitable for long-distance propagation, due to the weak surface sound channel to a depth of 100 m, a strong downward refracting environment to 800 m depth, and a weak upward refracting environment below that depth. Figure 6 presents the profiles associated with each survey area.

Table 6. Estimated geoacoustic properties of the sub-bottom sediments as a function of depth below the seafloor. P wave: compressional wave; S wave: shear wave.

Depth Regime	Depth (m)	Density (g/cm^3)	P-Wave Speed (m/s)	P-Wave Attenuation(dB/λ)	S-Wave Speed (m/s)	S-Wave Attenuation(dB/λ)
Transition Zone	0–5	1.5–1.6	1500–1535	0.16–0.26		
	5–10	1.6–1.7	1535–1565	0.26–0.32		
	10–20	1.7–1.8	1565–1630	0.32–0.37		
	20–50	2.2	1800–1900	0.8–1.1	150	3.6
	50–100	2.2	1900–2150	1.1–1.6		
	100–250	2.2	2150–2300	1.6–2.0		
	250–500	2.2	2300–2625	2.0–2.3		
	>500	2.2	2625	2.3		
Shallow Water	0–50	1.70–1.95	1550–1650	0.60–0.82		
	50–700	1.95–2.20	1650–1800	0.82–0.8	150	3.0
	>700	2.40	2200	0.2		
Deep Water	0–50	1.5–1.6	1500–1550	0.14–0.42		
	50–180	1.6–1.8	1550–1670	0.42–0.50		
	180–250	1.8–2.0	1670–1900	0.50–1.70	100	0.1
	250–1000	2.0–2.5	1900–2200	1.70		
	>1000	2.5	3000	0.20		

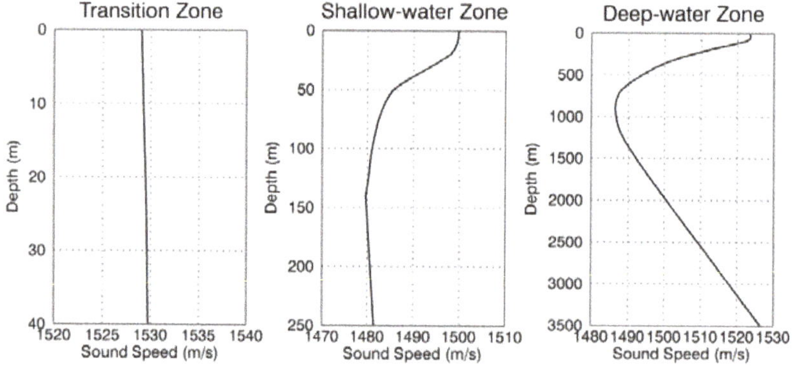

Figure 6. Sound speed profile for (**left**) July in the Transition Zone, (**middle**) August in Shallow Water, and (**right**) February in Deep Water.

Ambient Sound Levels

Ambient spectral levels (Figure 7) were estimated based on published data near the modeled areas or in similar environments. Broadband levels were used to compute the received signal duration and duty cycle (as defined in Sections 2.2 and 2.3.3) and spectral levels (integrated over the period of the modeled received signal) were compared to the modeled received spectral levels (presented in Section 3.1).

Published data on ambient sound levels in the Java Sea were inadequate for this study; representative ambient levels in the Transition Zone area were based on measurements from the shallow waters of the southwestern Bay of Bengal [68]. The measured levels during the transition period between monsoon seasons were used to derive spectral levels at frequencies between 150 and 5000 Hz. For frequencies 20–150 Hz, the spectral levels were assumed to be equal to the level at 150 Hz. Ambient levels measured in four decidecade frequency bands, centered at 63, 125, 250, and 500 Hz [69,70], were used to estimate spectral levels for the Shallow Water survey area. The spectral levels were interpolated between 56 and 560 Hz based on the published decidecade-band SPL values. For frequencies

20–56 Hz, the levels were assumed to be equal to the level at 56 Hz. Ambient spectral levels measured at the Green Canyon monitoring station between 10 and 1000 Hz [71] were used for the Deep Water survey area. In all areas, the spectral levels were extrapolated below 20 Hz by assuming an 8 dB per decade increase with decreasing frequency, and above the maximum available spectral level, assuming a 22 dB per decades decrease with increasing frequency [72].

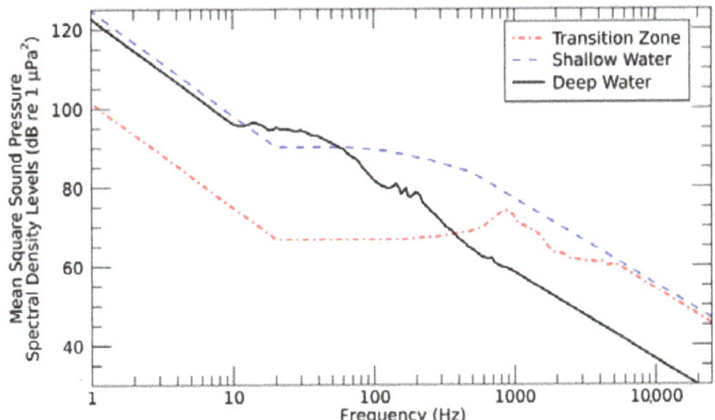

Figure 7. Estimated ambient sound spectral density levels in the survey areas (Transition Zone, Shallow Water, and Deep Water).

2.3.3. Surveys Characteristics

In all scenarios, both sources were simulated with a 2.3 m/s (4.5 knots) tow speed and emitting one pulse every 25 m along the survey lines. Therefore, the duty cycle of the MV array was 45% (5 s sweep every 11 s) and the duty cycle of the air gun array was <9%. Marine animal exposure was calculated over a 24 h of survey and averaged over 7 days (Section 2.3.5). To do so, per-pulse sound fields were computed at three locations in each survey area, with different water depths, and transposed geographically along the survey lines to simulate a moving source. Because of the low variability in water depth within the Shallow Water survey area (North Sea), the per-pulse sound field was calculated at one location, in the center of the survey area. Two survey types were modeled: tight surveys (scenarios 1 and 2; Table 2) and wide surveys (scenarios 3 and 4; Table 2). The adjacent lines in the tight survey were surveyed consecutively using half "dog-bone" turns. A "racetrack" pattern was used for the wide survey with half-circle turns with a radius of 2500 m (5 survey lines). For both survey types, the arrays were fully active during the turns (as opposed to using one mitigation element or ramp up).

2.3.4. Received Signal Characteristics

PK, SEL, duration of the signal, duty cycle, and time and frequency domain representations of the received signal were calculated as a function of distance from the sources from the synthetic pulses modeled by FWRAM. Broadband SPL (4 Hz to 25 kHz) was calculated over a 0.125 s sliding time window (as an approximate integration time of the mammalian ear [73], which is believed to be similar for land and aquatic mammals [74,75]). The time interval over which the SPL exceeded the estimated broadband (5–25,000 Hz) ambient level by more than 6 dB was used as a proxy for the audible duration of the signal. This approach was not species specific; it did not consider frequency-dependent hearing sensitivities of various species. The duty cycle was defined as the percentage of time, over a time interval common to all sources, when the broadband SPL of the received signal was more than 6 dB above ambient level; a common period of 11 s was chosen.

The per-pulse sound field modeled at the center location of each survey area was used to calculate the received signal characteristics. The signal characteristics were sampled at up to 10 distances between 50 m and 50 km from the source along one representative azimuth.

2.3.5. Agent-Based Model

The acoustic effects criteria (Section 2.2) indicate the lowest received sound levels that could result in injury or behavioral disturbance. To use the criteria in a realistic context, the sound levels received by animals must be estimated considering the time-evolving distance of animals relative to a source. An agent-based model was used to predict the probability that animals could be exposed to sound levels exceeding sound threshold criteria. In this approach, simulated animals (animats) moving in a realistic way are used to sample the sound fields. The combined received levels of many animats generates a probability density function (PDF) predicting animal exposure in the evolving sound field. This repeated random sampling (Monte Carlo method) yields an estimate of the probability that animals are exposed above effects thresholds.

The JASCO Animal Simulation Model Including Noise Exposure (JASMINE) used in this study is based on the open-source Marine Mammal Movement and Behavior Model (3MB) [76], one of several animal movement models [76–78], and is integrated with MONM and FWRAM acoustic propagation models (Section 2.3.2). JASMINE includes survey source track inputs and allows animats to change behavioral states based on time and space dependent variables, such as received sound level history.

JASMINE's behavioral input parameters include travel rate and direction, dive details and surface intervals. Animats were randomly seeded within the simulated environment at an animat density of 0.5 km^{-2}. The simulation area for potential behavioral responses was limited in this analysis to a maximum distance of 30 km from the simulated survey tracks in the Transition Zone (scenario 1) and 50 km in other areas (scenarios 2–4). Seven-day simulations were modeled for each scenario (Section 2.3.3) to achieve multiple examples of exposures and to mimic the approximate scale of the animal movement data [76]. The average number of exposures above threshold levels per 24 h period was calculated over the duration of the simulation. The thresholds used (discussed in Section 2.2) were based on the recommendation by NMFS [13].

To obtain the number of real-world animals predicted to be exposed to levels at or exceeding threshold values, the output PDF was scaled by the ratio of the estimated real-world density to simulation density. Each survey area contains more than 20 species of marine mammals. We selected a representative species from each functional hearing group (as recommended by Southall et al. [12] and NMFS [13]) in each area. Other behaviorally sensitive species were added to represent different behavioral categories, such as Cuvier's beaked whales in the Deep Water survey area. The species considered are listed in Table 7.

Density estimates in Indonesia (Transition Zone) were from a variety of sources, including best estimates for humpback whale using habitat-based models [79]. Bottlenose dolphin density estimates were from Kreb and Budiono [80], finless porpoise density estimates were based on Shirakihara et al. [81], and sea turtle density estimates were best estimates from Reyne et al. [82]. Density estimates for most marine mammal species in the North Sea were obtained from Hammond et al. [83], with the assumption that densities were constant across seasons. Harbor seal density estimates were from the US Navy OPAREA Density Estimate NODE; NODE; [84] model. The marine mammal density data used in this assessment for the Gulf of Mexico region were from the Duke University Marine Geospatial Ecological Laboratory model [85]. The sources of animal density information in Indonesia are limited relative to the compiled data and models for the Gulf of Mexico [85], so their uncertainties were higher. Because the same marine mammal and sea turtle densities were used in each area, the same level of uncertainty relating to animal density was expected for both source types.

Table 7. Representative species used in exposure models for each survey area, classified by hearing group, as defined by NMFS [13].

Hearing Group [13]	Transition Zone	Shallow Water	Deep Water
Low-frequency (LF) cetaceans	Humpback whale (*Megaptera novaeangliae*)	Minke whale (*Balaenoptera acutorostrata*)	Bryde's whale (*Balaenoptera edeni*)
Mid-frequency (MF) cetaceans	Bottlenose dolphin (*Tursiops truncates*)	Bottlenose dolphin White-beaked dolphin (*Lagenorhynchus albirostris*)	Bottlenose dolphin Sperm whale (*Physeter macrocephalus*)Cuvier's beaked whale (*Ziphius cavirostris*)
High-frequency (HF) cetaceans	Finless porpoise (*Neophocaena phocaenoides*)	Harbor porpoise (*Phocoena phocoena*)	Pygmy sperm whale (*Kogia breviceps*)
Phocid pinnipeds (in water; PW)	N/A	Harbor seal (*Phoca vitulina*)	N/A

3. Results

3.1. Signal Characteristics with Distance

The received signal characteristics were compared to highlight the differences between the air gun array impulsive sounds and the MV array non-impulsive sounds. The modeled sound fields were sampled at 10 distances along one direction in each survey area; the direction was selected to be representative of the entire (360°) field. At each sampling distance, the PK, SEL, duration of the signal, and duty cycle were calculated as the maximum broadband (4 Hz to 25 kHz) value over all modeled depths. The results are presented in Figure 8.

Because of propagation loss effects, the difference in received PK (Figure 8, top left) between MV and air gun sounds decreased somewhat as the distance from the sources increased. Our results indicated, however, that large differences in PK remained over several kilometers in all modeled environments. In the Transition Zone, PK values remained at least 10 dB greater for the air gun array (black line; Figure 8) than the MV array (grey line; Figure 8), up to 30 km (maximum modeled distance). In the Shallow Water and Deep Water environments, the air gun array PK values (blue and red lines; Figure 8) remained at least 20 and 17 dB greater than those for the MV sources (cyan and yellow lines; Figure 8), respectively, up to 50 km (maximum modeled distance).

Because MV arrays have lower source levels (SL; compare Tables 4 and 5), the sweeps produced must have a longer duration to result in a signal with similar SEL as air gun array pulses (Figure 8, top right and bottom left). The longer duration means shorter periods of quiet time between MV sweeps than between air gun pulses (~55% quiet time per period (the complement of duty cycle) for MV array at less than 1 km, compared to 90% for air gun array; see bottom right plot in Figure 8). However, because the SL for the MV array was lower than for the air gun arrays, its received SPL decreased below detectable levels (considered here as 6 dB above ambient levels) at a shorter distance. In the studied environments, this resulted in the signal duration for the MV array becoming shorter than that of the air gun array, and the duty cycle becoming shorter, between 2 and 5 km from the sources.

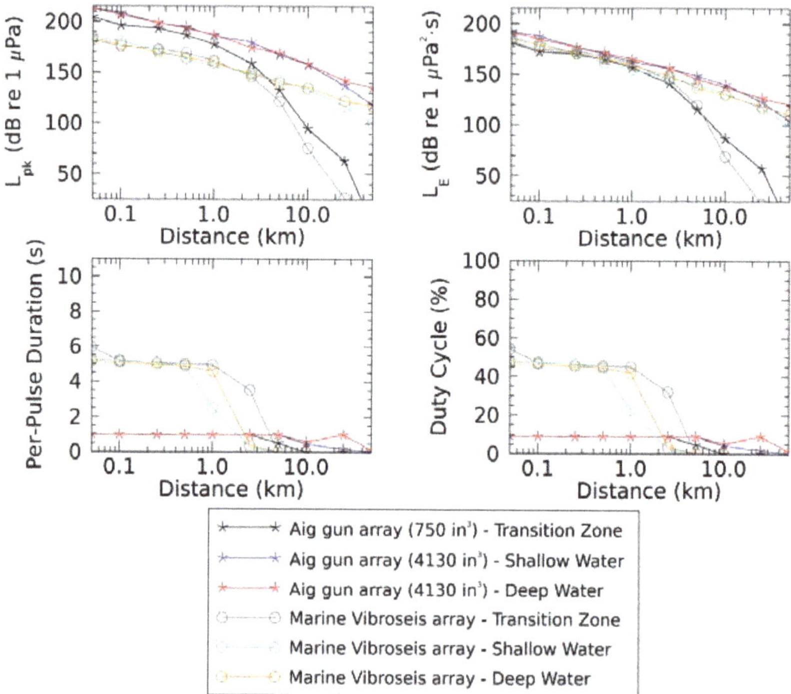

Figure 8. Received peak sound pressure levels (L_{pk}; **top left**), sound exposure levels (L_E; **top right**), pulse duration (**bottom left**) and duty cycle (**bottom right**) as a function of distance. L_E, pulse duration and percentage of quiet time are calculated over a common time interval of 11 s.

Figure 9 presents the received signals as sound pressure (left) and spectral density levels (right). For conciseness, only the results in the Shallow Water area, sampled in approximately the middle of the water column (50 m), are presented here; results were similar for all scenarios. The dash line on the spectral density levels plots represent the ambient levels integrated over the same period as the modeled received signal (10 s for the MV array and 1 s for the air gun array). Note that the same scale was used to present the sources spectral density levels (left), but different scales had to be used to present the sound pressure (right).

Figure 9 shows that at all modeled distances both sources produced similar spectral levels below 100 Hz, while the spectral levels at higher frequencies were much higher for the air gun array. Here, results show MV sound above 200 Hz was below ambient levels at distances as short as 100 m, while spectral levels for the air gun array in the same frequency band remained above ambient levels for several (>10) kilometers.

Per-pulse frequency-weighted SEL sound fields over the frequency range 4 Hz–25 kHz (in decidecade frequency bands) were predicted using MONM and then input to JASMINE. Figure 10 presents an example of received frequency-weighted SEL for the MV array (left) and the air gun array (right), at 1000 m from the arrays in the middle of the water column (50 m; scenario 2). This example illustrates the effect of frequency weighting on the received sound field, following the NMFS guidance [13].

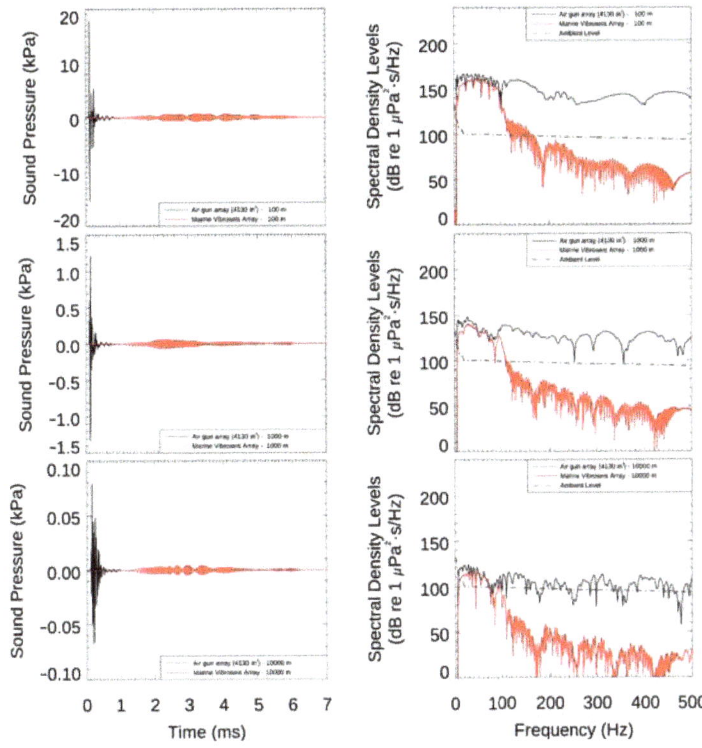

Figure 9. MV array (red) and 4130 in^3 air gun (black). Sound pressure (**left**) and energy spectral density levels (**right**) received at three sampling locations: 100 m (**top**), 1000 m (**middle**), and 10 km (**bottom**). The ambient level (dashed lines, right) was based on the integration period of 10 s. Results are shown along the tow direction (0°) at a depth of 50 m in the Shallow Water area (scenario 2; Table 2).

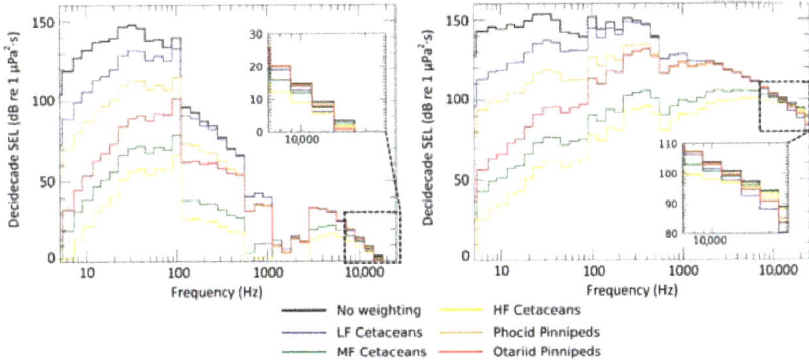

Figure 10. Example of received frequency-weighted sound exposure levels (SEL) in decidecade bands for the MV array (**left**) and the air gun array (**right**). Results are shown at a distance of 1000 m from the arrays, along the tow direction (0°), at a depth of 50 m in the Shallow Water area (scenario 2; Table 2). The frequency weighting was applied according to the designated NMFS [13] hearing group and the associated frequency-weighting functions.

3.2. Exposure Estimates from Agent-Based Model

The agent-based model was used to estimate marine mammal exposure to injury and behavioral thresholds for each scenario. The mean number of marine mammals exposed to sound levels above injury thresholds based on NMFS guidelines [13] due to MV and air gun sounds are presented in Table 8. The mean number of marine mammals exposed to sound levels above behavioral thresholds based on frequency-weighted step functions and on single unweighted values (discussed in Section 2.2) are shown in Table 9.

Table 8. Mean number of marine mammals expected to be exposed to sound levels at or above injury thresholds during a 24 h survey period. Peak sound pressure level (PK) and sound exposure level (SEL) thresholds based on NMFS [13]. SEL sound fields were frequency-weighted and accumulated over a 24 h survey period according to the NMFS guidelines [13].

Scenario	Species Name	MV Array	Air Gun Array	
		SEL	PK	SEL
1	Humpback whale	<0.01	<0.01	0.03
	Bottlenose dolphin	<0.01	0.07	<0.01
	Finless porpoise	<0.01	24.07	<0.01
2	Minke whale	<0.01	0.26	0.45
	Bottlenose dolphin	<0.01	<0.01	<0.01
	White-beaked dolphin	<0.01	0.01	<0.01
	Harbor porpoise	<0.01	12.9	0.10
	Harbor seal	<0.01	0.08	0.08
3	Minke whale	<0.01	0.27	0.47
	Bottlenose dolphin	<0.01	<0.01	<0.01
	White-beaked dolphin	<0.01	0.01	<0.01
	Harbor porpoise	<0.01	13.6	0.06
	Harbor seal	<0.01	0.13	0.06
4	Bryde's whale	<0.01	<0.01	<0.01
	Sperm whale	<0.01	<0.01	<0.01
	Bottlenose dolphin	<0.01	<0.01	<0.01
	Cuvier's beaked whale	<0.01	<0.01	<0.01
	Pygmy sperm whale	<0.01	<0.01	<0.01

Although the ESL of MV and air gun sounds were similar, applying the frequency-weighting functions and the SEL-based injury criteria resulted in fewer predicted exposures to levels at or above injury thresholds from the MV array than the air gun array (Table 8) because the MV energy is concentrated at lower frequencies than the air gun array (Figure 10). In this study, however, SEL for both sources quickly decreased to below injury thresholds, such that few marine mammals were predicted to be exposed to injurious acoustic levels in any of the scenarios (Table 8).

Increasing the width between survey lines (scenarios 2 and 3; Table 2), increased the area ensonified at or above the injury thresholds for air gun array sounds. This resulted in a small increase (≤0.7) in the number of marine mammals potentially exposed to injurious acoustic levels from the air gun array. The variation in the survey line spacing did not affect the results for the MV array.

A change in behavioral criteria resulted in the reversal of conclusions regarding which source was likely to cause more behavioral disturbance. Results showed a higher number of animals exposed to acoustic levels at or above the regulatory-defined thresholds for behavioral response (single unweighted values) when exposed to MV sounds compared to air gun sounds. In contrast, when frequency-weighted step functions were used, a higher number of animals were predicted to be exposed to sound levels resulting in behavioral responses when exposed to air gun sounds versus MV sounds.

Table 9. Mean number of marine mammals expected to be exposed to sound levels at or above behavioral thresholds during a 24 h survey period. Step function thresholds are based on DoN [22] for MV array and Wood et al. [21] for the air gun arrays; single-value thresholds are based on NMFS [86].

Scenario	Species Name	Step Function Thresholds		Single-Value Thresholds	
		MV Array	Air Gun Array	MV Array	Air Gun Array
1	Humpback whale	0.06	0.08	0.09	0.09
	Bottlenose dolphin	2.95	8.89	23.1	13.9
	Finless porpoise	107	239	266	172
2	Minke whale	1.76	15.3	44.2	23.6
	Bottlenose dolphin	0.01	0.08	0.24	0.12
	White-beaked dolphin	0.20	3.05	7.82	5.33
	Harbor porpoise	26.4	175	113	64.3
	Harbor seal	0.23	1.32	3.84	1.95
3	Minke whale	1.75	15.6	45.8	23.9
	Bottlenose dolphin	0.02	0.14	0.38	0.22
	White-beaked dolphin	0.17	3.21	8.40	5.72
	Harbor porpoise	28.7	183	120	68.9
	Harbor seal	0.24	1.50	4.23	2.30
4	Bryde's whale	<0.01	0.05	0.28	0.05
	Sperm whale	0.03	0.70	6.89	1.24
	Bottlenose dolphin	<0.01	<0.01	0.20	<0.01
	Cuvier's beaked whale	0.05	0.62	0.81	0.12
	Pygmy sperm whale	<0.01	<0.01	0.92	<0.01

4. Discussion

From a biological point of view, one advantage of MV arrays is their lower SLPK; MV arrays are expected to have a lower potential than air gun arrays to cause onset of permanent or temporary threshold shift in marine mammals or to cause mortality and injury in fish and other animals. In the current study, SLPK values for the air gun arrays were 21 to 39 dB higher than for the MV array (see Table 5). Our results indicated that a large difference in PK persist over several kilometers (>30 km). Thus, the likelihood of injury due to high PK values was much lower for an MV array than for an air gun array.

To acquire the desired geophysical data, a certain level of energy must be produced by the seismic source. Therefore, for similar SEL and lower PK levels, the MV array must produce a much longer signal than the air gun array. The longer duration means there is less opportunity for "dip-listening" between MV sounds than between air gun pulses. This would appear to be a clear advantage of air guns over MVs in terms of a lower risk of masking effects, but the lower SPL of MV array sounds means the distances within which this masking may occur (considered here as the maximum distance where per-pulse duration was greater than zero) is shorter than for air guns (~5 km for the MV array versus 10 to 50 km for the air gun arrays, for the modeled scenarios). Additionally, if the harmonic content of MV array sounds above ~100 Hz is kept low (i.e., the MV array follows the specifications required by the MVJIP), then potential masking of mid- and high-frequency cetaceans may be negligible and greatly reduced for low-frequency cetaceans.

Responses to disturbance include a variety of effects, ranging from subtle to conspicuous changes in behavior, movement, and displacement. Available detailed data on reactions of marine mammals to air gun sounds (and other anthropogenic sounds) are limited to relatively few species and situations (see reviews in [9,87–92]). Behavioral reactions to sound are highly variable and context specific, and they may differ for species, state of maturity, experience, current activity, reproductive state, time of day, and many other factors [9,87,90,91,93,94].

Given the many uncertainties in predicting behavioral responses in marine mammals, it is common practice to assume that behavioral responses will occur if an animal is within a predicted distance, as is done with the single-value criteria currently used by NMFS [86]. Even though MV arrays have lower source amplitudes, the distances to the estimated behavioral response SPL isopleths are longer than for the MV arrays. Therefore, when these single-step threshold criteria were used in exposure modeling simulations, results showed a higher number of animals exposed to MV sounds than air gun sound (Table 9). This was due to the substantially lower NMFS SPL threshold for non-impulsive sounds compared to that for impulsive sounds (120 vs. 160 dB re 1 μPa). It is therefore imperative to either have clear rules for determining the impulsiveness of the sound produced or to develop criteria that do not depend so critically on this distinction. While the duration of a signal at its source has historically been used to separate sources into two groups (impulsive vs. non-impulsive), other metrics such as kurtosis [95], crest factor, and the Harris impulse factor are being proposed to quantify the impulsiveness of ocean sounds [10].

As an alternative to the single-value criteria currently used by NMFS [86], animal exposure to sound potentially resulting in behavioral disturbance was calculated using weighted sound fields to account for the hearing range of the animals, and a 2- or 3-step (depending on the species and activity group) probability of response to impulsive sounds, as proposed by Wood et al. [21], or a continuous probability of response function, for non-impulsive sounds, as used by DoN [22]; the two probability of response functions are similar. Because of the effect of frequency weighting, distances to behavioral disturbance thresholds are much shorter for MV sounds than air gun sounds. Therefore, exposure modeling using the graded probability thresholds resulted in a smaller estimated number of exposures to the MV array than the air gun array (Table 9).

The reversal of conclusions regarding which source was likely to cause more behavioral disturbance depending on the assessment criteria used was notable, although perhaps not surprising given the differences in the criteria (with vs. without frequency weighting; step function vs. single value; difference in thresholds for impulsive vs. non-impulsive sound). If the graded probability of response functions [21,22] better reflects actual behavioral responses when frequency-weighting functions are applied, then MV arrays may elicit substantially fewer behavioral disturbances than air gun arrays. However, the variability in observed behavioral response to anthropogenic sounds [9] and the importance of context [90] means that we do not know how animals might respond to sounds from MV arrays.

Because MV is a new technology still largely under development, there are no data available documenting injury or behavioral responses of marine mammals to sounds produced by this source. Some naval sonar sources produce sounds of similar durations but at somewhat higher frequencies (100–500 Hz for low-frequency active sonar, 1–8 kHz for mid-frequency active sonar). A recent review of behavioral responses to naval sonar source (from 1–8 kHz) shows similarly high levels of variability among species and individuals as observed for other sources [91]. Systematic well-controlled studies of animal responses to MV sounds are necessary before the relative behavioral responses from MV and air gun sources can be meaningfully compared.

5. Conclusions

The goal of this study was to quantify the potential effects of MV array sounds by comparing the signal characteristics and estimate marine mammal exposures associated with geophysical surveys conducted using MV arrays versus air gun arrays. Various survey designs scenarios were selected to allow meaningful comparisons among representative survey operations in three different environmental settings. Because of the difference in signal type (impulsive versus non-impulsive), different acoustic thresholds were necessarily applied for each seismic source, complicating the comparison. Nonetheless, the results of this study provide important insights into the relative potential for the two sources to cause injury and behavioral disturbance.

The lower source amplitudes of MV array sounds mean that they are less likely than air gun arrays to exceed the currently prescribed marine mammal injury thresholds based on PK levels. For arrays with similar energy source levels (ESL), the frequency weighting used in estimating distances to injurious thresholds for most marine mammal hearing groups has a greater filtering effect on the sound field of the MV than the air gun array. This is due to differences in acoustic energy propagating at frequencies outside the main frequency band of interest for seismic surveys; the spectral levels for the MV array are expected to be much lower than that of the air gun array above 100 to 200 Hz. Therefore, MV arrays with well-suppressed harmonics are expected to exceed the studied SEL injury thresholds at shorter distances than air gun arrays with similar ESL. However, sounds from air gun arrays typically decrease to below current injurious thresholds at relatively short distances (tens of meters in this study, depending on the hearing group). None of the scenarios in this modeling study, regardless of source, resulted in more than 25 regulatory-defined injurious exposures (Table 8).

The single-step unweighted SPL thresholds currently used by NMFS [13] resulted in higher estimates of behavioral effects from MV arrays than air gun arrays, primarily as a result of the lower SPL threshold (120 dB re 1 µPa) used for non-impulsive sounds. Because it is unlikely that sounds outside of the hearing range of an animal will result in behavioral response, it can be argued that frequency weighting should be applied and that a more statistical approach should be used when assessing behavioral effects on marine mammals. When the frequency-weighted and multiple-step functions proposed by Wood et al. [21] and DoN [22] were used, the modeled air gun arrays are predicted to cause more behavioral disturbance than the MV array. This is primarily caused by the higher frequency-weighted source pressure levels of air gun arrays resulting in longer distances to nearly equivalent behavioral response threshold levels for the two source types.

Author Contributions: Conceptualization: D.S.I., D.G.Z., R.H.B. and C.D.P.; Formal analysis: M.-N.R.M., D.S.I. and D.G.Z.; Funding acquisition: C.D.P.; Investigation: M.-N.R.M. and R.H.B.; Methodology: M.-N.R.M., D.S.I., D.G.Z. and R.H.B.; Project administration: C.D.P.; Resources: C.D.P.; Software: M.-N.R.M.; Validation: D.G.Z.; Visualization: M.-N.R.M. and D.G.Z.; Writing—original draft: M.-N.R.M., D.S.I., D.G.Z. and and C.D.P. All authors have read and agreed to the published version of the manuscript.

Funding: This research was funded by the E&P Sound and Marine Life Joint Industry Programme (SML JIP).

Acknowledgments: Thanks to Kirsty Speirs and Andrew Feltham with Total, and Michael Jenkerson with ExxonMobil, for their invaluable guidance and information on the various source configurations used in the modeling. Several modelers and scientists contributed to this work, including Z. Alavizadeh, J. Christian, S. Denes, T.J. Deveau, H. Frouin-Mouy, V. Moulton, G. Warner, J. Richardson, M.A. Ainsley, K. Lucke, and D.E. Hannay.

Conflicts of Interest: The authors declare no conflict of interest.

References

1. Parkes, G.E.; Hatton, L. *The Marine Seismic Source*, 1st ed.; Springer Science & Business Media: Dordrecht, The Netherlands, 1986.
2. Carroll, P. A note on synthetic vibroseis sweep generation. *Can. J. Explor. Geophys.* **1971**, *7*, 80–82.
3. International Organization for Standardization. *Underwater Acoustics—Terminology*; ISO 18405:2017; ISO: Geneva, Switzerland, 2017; p. 51.
4. Spence, J.H. Seismic survey noise under examination. In *Offshore: World Trends and Technology for Offshore Oil and Gas Operations*; Endeavor Business Media LCC.: Nashville, TN, USA, 2009; Volume 69.
5. LGL. *Environmental Assessment of Marine Vibroseis*; Report for the Joint Industry Programme on E&P Sound and Marine Life; LGL: King City, CA, USA; Marine Acoustics Inc.: Middletown, RI, USA, 2011; p. 207.
6. Green, D.M.; Swets, J.A. *Signal Detection Theory and Psychophysics*; Krieger: Huntington, WV, USA, 1974.
7. Moore, B.C.J. *An Introduction to the Psychology of Hearing*, 6th ed.; Emerald Group Publishing: Bingley, UK, 2012.
8. Erbe, C.; Reichmuth, C.; Cunningham, K.; Lucke, K.; Dooling, R. Communication Masking in Marine Mammals: A Review and Research Strategy. *Mar. Pollut. Bull.* **2016**, *103*, 15–38. [CrossRef]

9.   Southall, B.L.; Bowles, A.E.; Ellison, W.T.; Finneran, J.J.; Gentry, R.L.; Greene, C.R., Jr.; Kastak, D.; Ketten, D.R.; Miller, J.H.; Nachtigall, P.E.; et al. Marine Mammal Noise Exposure Criteria: Initial Scientific Recommendations. *Aquat. Mamm.* **2007**, *33*, 411–521. [CrossRef]

10.   Martin, S.B.; Lucke, K.; Barclay, D.R. Techniques for Distinguishing between Impulsive and Non-Impulsive Sound in the Context of Regulating Sound Exposure for Marine Mammals. *J. Acoust. Soc. Am.* **2020**, *147*, 2159–2176. [CrossRef]

11.   Popper, A.N.; Hawkins, A.D.; Fay, R.R.; Mann, D.A.; Bartol, S.; Carlson, T.J.; Coombs, S.; Ellison, W.T.; Gentry, R.L.; Halvorsen, M.B.; et al. *Sound Exposure Guidelines for Fishes and Sea Turtles: A Technical Report Prepared by ANSI-Accredited Standards Committee S3/SC1 and Registered with ANSI*; ASA S3/SC1.4 TR-2014; ASA Press: Washington, DC, USA; Springer: Berlin/Heidelberg, Germany, 2014.

12.   Southall, B.L.; Finneran, J.J.; Reichmuth, C.J.; Nachtigall, P.E.; Ketten, D.R.; Bowles, A.E.; Ellison, W.T.; Nowacek, D.P.; Tyack, P.L. Marine Mammal Noise Exposure Criteria: Updated Scientific Recommendations for Residual Hearing Effects. *Aquat. Mamm.* **2019**, *45*, 125–232. [CrossRef]

13.   National Marine Fisheries Service. *2018 Revision to: Technical Guidance for Assessing the Effects of Anthropogenic Sound on Marine Mammal Hearing (Version 2.0): Underwater Thresholds for Onset of Permanent and Temporary Threshold Shifts*; US Department of Commerce: Washington, DC, USA; NOAA: Washington, DC, USA, 2018; p. 167.

14.   Ellison, W.T.; Frankel, A.S. A common sense approach to source metrics. In *The Effects of Noise on Aquatic Life*; Popper, A.N., Hawkins, A.D., Eds.; Springer: New York, NY, USA, 2012; pp. 433–438.

15.   National Oceanic and Atmospheric Administration. *Notice of Public Scoping and Intent to Prepare an Environmental Impact Statement*; NOAA: Washington, DC, USA, 2005.

16.   Malme, C.I.; Miles, P.R.; Clark, C.W.; Tyack, P.L.; Bird, J.E. *Investigations of the Potential Effects of Underwater Noise from Petroleum Industry Activities on Migrating Gray Whale Behavior. Phase II: January 1984 migration*; US Department of the Interior. Minerals Management Service: Cambridge, MA, USA, 1984.

17.   Richardson, W.J.; Würsig, B.; Greene, C.R., Jr. Reactions of bowhead whales, Balaena mysticetus, to drilling and dredging noise in the Canadian Beaufort Sea. *Mar. Environ. Res.* **1990**, *29*, 135–160. [CrossRef]

18.   Richardson, W.J.; Würsig, B.; Greene, C.R., Jr. Reactions of bowhead whales, Balaena mysticetus, to seismic exploration in the Canadian Beaufort Sea. *J. Acoust. Soc. Am.* **1986**, *79*, 1117–1128. [CrossRef]

19.   Malme, C.I.; Miles, P.R.; Clark, C.W.; Tyack, P.L.; Bird, J.E. *Investigations of the Potential Effects of Underwater Noise from Petroleum Industry Activities on Migrating Gray Whale Behavior*; US Department of the Interior: Washington, DC, USA; Minerals Management Service: Cambridge, MA, USA, 1983.

20.   Nedwell, J.R.; Turnpenny, A.W.; Lovell, J.; Parvin, S.J.; Workman, R.; Spinks, J.A.L.; Howell, D. *A Validation of the dB$_{ht}$ as a Measure of the Behavioural and Auditory Effects of Underwater Noise*; Department for Business, Enterprise and Regulatory Reform: London, UK, 2007; p. 74.

21.   Wood, J.D.; Southall, B.L.; Tollit, D.J. *PG&E offshore 3-D Seismic Survey Project Environmental Impact Report–Marine Mammal Technical Draft Report*; SMRU Ltd.: St Andrews, UK, 2012; p. 121.

22.   Department of the Navy (US). *Final Supplemental Environmental Impact Statement/Supplemental Overseas Environmental Impact Statement for Surveillance Towed Array Sensor System Low Frequency Active (SURTASS LFA) Sonar*; Department of the Navy: Washington, DC, USA, 2012.

23.   Feller, W. *An Introduction to Probability Theory and Its Applications*, 3rd ed.; John Wiley & Sons: New York, NY, USA, 1968; Volume 1.

24.   McConnell, J.A.; Berkman, E.F.; Murray, B.S.; Abraham, B.M. Coherent Sound Source for Marine Seismic Surveys. U.S. Patent 9,625,598, 18 April 2017.

25.   McConnell, J.A.; Berkman, E.F.; Murray, B.S.; Abraham, B.M.; Roy, D.A. Coherent Sound Source for Marine Seismic Surveys. U.S. Patent 9,562,982, 18 April 2017.

26.   Tellier, N.; Ollivrin, G. Low-frequency Vibroseis: Current achievements and the road ahead? *First Break* **2019**, *37*, 49–54. [CrossRef]

27.   Sallas, J.; Gibson, J.; Maxwell, P.; Lin, F. Pseudorandom sweeps for simultaneous sourcing and low-frequency generation. *Lead. Edge* **2011**, *30*, 1162–1172. [CrossRef]

28.   Scholtz, P. Pseudo-random sweeps for built-up area seismic surveys. *Lead. Edge* **2013**, *32*, 276–282. [CrossRef]

29.   Dean, T. Establishing the limits of vibrator performance-experiments with pseudorandom sweeps. In *SEG Technical Program Expanded Abstracts 2012*; Society of Exploration Geophysicists: Tulsa, OK, USA, 2012; pp. 1–5.

30.   Dean, T. The use of pseudorandom sweeps for vibroseis surveys. *Geophys. Prospect.* **2014**, *62*, 50–74. [CrossRef]

31.   Dean, T.; Tulett, J.; Lane, D. The use of pseudorandom sweeps for vibroseis acquisition. *First Break* **2017**, *35*, 107–112.

32.   Feltham, A.; Girard, M.; Jenkerson, M.; Nechayuk, V.; Griswold, S.; Henderson, N.; Johnson, G. The Marine Vibrator Joint Industry Project: Four years on. *Explor. Geophys.* **2017**, *49*, 675–687. [CrossRef]

33.   Schostak, B.; Jenkerson, M. The Marine Vibrator Joint Industry Project. In *SEG Technical Program Expanded Abstracts*; Society of Exploration Geophysicists: Tulsa, OK, USA, 2015; pp. 4961–4962.

34.   Ainslie, M.A. *Principles of Sonar Performance Modeling*; Praxis Books; Springer: Berlin/Heidelberg, Germany, 2010.

35.   Ainslie, M.A.; de Jong, C.A.F.; Martin, S.B.; Miksis-Olds, J.L.; Warren, J.D.; Heaney, K.D.; Hillis, C.A.; MacGillivray, A.O. *ADEON Project Dictionary: Terminology Standard*; JASCO: Oklahoma City, OK, USA, 2020.

36.   Sallas, J.; Teyssandier, B. Vibrator Source Array Load-Balancing Method and System. U.S. Patent 9482765, 1 November 2016.

37.   Teyssandier, B.; Sallas, J.J. The shape of things to come—Development and testing of a new marine vibrator source. *Lead. Edge* **2019**, *38*, 680–690. [CrossRef]

38. Scandrett, C.L.; Baker, S.R. *Pritchard's Approximation in Array Modeling*; Naval Postgraduate School: Monterey, CA, USA, 1999.
39. MacGillivray, A.O. *An Acoustic Modelling Study of Seismic Airgun Noise in Queen Charlotte Basin*; University of Victoria: Victoria, BC, Canada, 2006; p. 98.
40. Ziolkowski, A.M. A method for calculating the output pressure waveform from an air gun. *Geophys. J. Int.* **1970**, *21*, 137–161. [CrossRef]
41. Dragoset, W.H. A comprehensive method for evaluating the design of airguns and airgun arrays. In Proceedings of the 16th Annual Offshore Technology Conference, Houston, TX, USA, 7–9 May 1984; pp. 75–84.
42. Laws, R.M.; Hatton, L.; Haartsen, M. Computer modeling of clustered airguns. *First Break* **1990**, *8*, 331–338. [CrossRef]
43. Landro, M. Modeling of GI gun signatures. *Geophys. Prospect.* **1992**, *40*, 721–747. [CrossRef]
44. Mattsson, A.; Jenkerson, M. Single Airgun and Cluster Measurement Project. In *Joint Industry Programme (JIP) on Exploration and Production Sound and Marine Life Proramme Review*; International Association of Oil and Gas Producers: Houston, TX, USA, 2008.
45. Zhang, Z.Y.; Tindle, C.T. Improved equivalent fluid approximations for a low shear speed ocean bottom. *J. Acoust. Soc. Am.* **1995**, *98*, 3391–3396. [CrossRef]
46. Collins, M.D.; Cederberg, R.J.; King, D.B.; Chin-Bing, S. Comparison of algorithms for solving parabolic wave equations. *J. Acoust. Soc. Am.* **1996**, *100*, 178–182. [CrossRef]
47. Collins, M.D. A split-step Padé solution for the parabolic equation method. *J. Acoust. Soc. Am.* **1993**, *93*, 1736–1742. [CrossRef]
48. Hannay, D.E.; Racca, R.G. *Acoustic Model Validation*; JASCO Research Ltd.: Oklahoma City, OK, USA, 2005; p. 34.
49. Porter, M.B.; Liu, Y.C. Finite-element ray tracing. In Proceedings of the International Conference on Theoretical and Computational, Manchester, UK, 13–16 December 1994; pp. 947–956.
50. MacGillivray, A.O.; Chapman, N.R. Modeling underwater sound propagation from an airgun array using the parabolic equation method. *Can. Acoust.* **2012**, *40*, 19–25. [CrossRef]
51. Aerts, L.A.M.; Blees, M.; Blackwell, S.B.; Greene, C.R., Jr.; Kim, K.H.; Hannay, D.E.; Austin, M.E. *Marine Mammal Monitoring and Mitigation during BP Liberty OBC Seismic Survey in Foggy Island Bay, Beaufort Sea, July-August 2008: 90-Day Report*; NOAA: Washington, DC, USA, 2008; p. 199.
52. Funk, D.W.; Hannay, D.E.; Ireland, D.S.; Rodrigues, R.; Koski, W.R. *Marine Mammal Monitoring and Mitigation during Open Water Seismic Exploration by Shell Offshore Inc. in the Chukchi and Beaufort Seas, July–November 2007: 90-Day Report*; NOAA: Washington, DC, USA, 2008; p. 218.
53. Funk, D.W.; Hannay, D.E.; Ireland, D.S.; Rodrigues, R.; Koski, W.R. *Marine Mammal Monitoring and Mitigation during Open Water Seismic Exploration by Shell Offshore Inc. in the Chukchi and Beaufort Seas, July–October 2008: 90-Day Report*; NOAA: Washington, DC, USA, 2009; p. 277.
54. O'Neill, C.; Leary, D.; McCrodan, A. Sound Source Verification. In *Marine Mammal Monitoring and Mitigation during open water seismic exploration by Statoil USA E&P Inc. in the Chukchi Sea, August-October 2010: 90-Day Report*; Blees, M.K., Ed.; LGL Report P1119; NOAA: Washington, DC, USA, 2010; pp. 1–34.
55. Warner, G.A.; Erbe, C.; Hannay, D.E. Underwater Sound Measurements. In *Marine Mammal Monitoring and Mitigation during Open Water Shallow Hazards and Site Clearance Surveys by Shell Offshore Inc. in the Alaskan Chukchi Sea, July-October 2009: 90-Day Report*; Reiser, C.M., Ed.; NOAA: Washington, DC, USA, 2010; pp. 1–54.
56. Racca, R.G.; Rutenko, A.N.; Bröker, K.; Gailey, G. *Model Based Sound Level Estimation and in-Field Adjustment for Real-Time Mitigation of Behavioural Impacts from a Seismic Survey and Post-Event Evaluation of Sound Exposure for Individual Whales*; Acoustics: Fremantle, Australia, 2012.
57. Racca, R.G.; Rutenko, A.N.; Bröker, K.; Austin, M.E. A Line in the Water—Design and Enactment of a Closed Loop, Model Based Sound Level Boundary Estimation Strategy for Mitigation of Behavioural Impacts from a Seismic Survey. In Proceedings of the 11th European Conference on Underwater Acoustics, Edinburgh, UK, 2–6 July 2012.
58. Martin, B.; Bröker, K.; Matthews, M.-N.R.; MacDonnell, J.T.; Bailey, L. Comparison of measured and modeled air-gun array sound levels in Baffin Bay, West Greenland. In Proceedings of the OceanNoise, Barcelona, Spain, 11–15 May 2015.
59. Hamilton, E.L. Geoacoustic modeling of the sea floor. *J. Acoust. Soc. Am.* **1980**, *68*, 1313–1340. [CrossRef]
60. Buckingham, M.J. Compressional and Shear Wave Properties of Marine Sediments: Comparisons between Theory and data. *J. Acoust. Soc. Am.* **2005**, *117*, 137–152. [CrossRef] [PubMed]
61. Hanebuth, T.J.J.; Voris, H.K.; Yokoyama, Y.; Saito, Y.; Okuno, J. Formation and fate of sedimentary depocentres on Southeast Asia's Sunda Shelf over the past sea-level cycle and biogeographic implications. *Earth-Sci. Rev.* **2011**, *104*, 92–110. [CrossRef]
62. Bishop, M.G. *Petroleum Systems of the Northwest Java Province, Java and offshore Southeast Sumatra, Indonesia, in Open-File Report*; US Department of the Interior: Washington, DC, USA; US Geological Survey: Reston, VA, USA, 2000; p. 31.
63. [NPD] Oljedirektoratet–Norwegian Petroleum Directorate. *The NPD's Fact-pages-Hordaland Group*. 2011. Available online: http://www.npd.no/engelsk/cwi/pbl/en/su/all/67.htm (accessed on 21 December 2020).
64. Shipboard Scientific Party. Site 96, Leg 619. In *Deep Sea Drilling Projects Initial Reports*; US Government Printing Office: Washington, DC, USA, 1987.
65. Teague, W.J.; Carron, M.J.; Hogan, P.J. A comparison between the Generalized Digital Environmental Model and Levitus climatologies. *J. Geophys. Res.* **1990**, *95*, 7167–7183. [CrossRef]
66. Carnes, M.R. *Description and Evaluation of GDEM-V 3.0*; MS. NRL Memorandum Report 7330-09-9165; US Naval Research Laboratory, Stennis Space Center: New Orleans, MS, USA, 2009; p. 21.

67. Coppens, A.B. Simple equations for the speed of sound in Neptunian waters. *J. Acoust. Soc. Am.* **1981**, *69*, 862–863. [CrossRef]
68. Mahanty, M.M.; Sanjana, M.C.; Latha, G.; Raguraman, G. *An Investigation on the Fluctuation and Variability of Ambient Noise in Shallow Waters of South West Bay of Bengal*; National Institute of Science Communication and Information Resources: New Delhi, India, 2014; Volume 43, pp. 747–753.
69. International Electrotechnical Commission. *IEC 61260-1:2014 Electroacoustics—Octave-Band and Fractional-Octave-Band Filters—Part 1: Specifications*; IEC: London, UK, 2014; p. 88.
70. Merchant, N.D.; Brookes, K.L.; Faulkner, R.C.; Bicknell, A.W.J.; Godley, B.J.; Witt, M.J. Underwater noise levels in UK waters. *Sci. Rep.* **2016**, *6*, 1–10. [CrossRef]
71. Wiggins, S.M.; Hall, J.M.; Thayre, B.J.; Hildebrand, J.A. Gulf of Mexico low-frequency ocean soundscape impacted by airguns. *J. Acoust. Soc. Am.* **2016**, *140*, 176–183. [CrossRef]
72. Urick, R.J. *Principles of Underwater Sound*, 3rd ed.; McGraw-Hill: New York, NY, USA; London, UK, 1983; p. 423.
73. Martin, B.; MacDonnell, J.T.; Bröker, K. Cumulative sound exposure levels—Insights from seismic survey measurements. *J. Acoust. Soc. Am.* **2017**, *141*, 3603. [CrossRef]
74. Plomp, R.; Bouman, M.A. Relation between Hearing Threshold and Duration for Tone Pulses. *J. Acoust. Soc. Am.* **1959**, *31*, 749–758. [CrossRef]
75. MacGillivray, A.O.; Racca, R.G.; Li, Z. Marine mammal audibility of selected shallow-water survey sources. *J. Acoust. Soc. Am.* **2014**, *135*, EL35–EL40. [CrossRef] [PubMed]
76. Houser, D.S. A method for modeling marine mammal movement and behavior for environmental impact assessment. *IEEE J. Ocean. Eng.* **2006**, *31*, 76–81. [CrossRef]
77. Ellison, W.T.; Clark, C.W.; Bishop, G.C. Potential use of surface reverberation by bowhead whales, Balaena mysticetus, in under-ice navigation: Preliminary considerations. In *Report of the International Whaling Commission*; International Whaling Commission: Cambridge, UK, 1987; pp. 329–332.
78. Frankel, A.S.; Ellison, W.T.; Buchanan, J. Application of the acoustic integration model (AIM) to predict and minimize environmental impacts. In *OCEANS'02 MTS/IEEE*; IEEE: Biloxi, MI, USA, 2002; pp. 1438–1443.
79. Kaschner, K.; Rius-Barile, J.; Kesner-Reyes, K.; Garilao, C.; Kullander, S.O.; Rees, T.; Froese, R. AquaMaps: Predicted Range Maps for Aquatic Species; World Wide Web Electronic Publication. 2016. Available online: www.aquamaps.org (accessed on 21 December 2020).
80. Kreb, D. Cetacean diversity and habitat preferences in tropical waters of East Kalimantan, Indonesia. *Raffles Bull. Zool.* **2005**, *53*, 149–155.
81. Shirakihara, K.; Shirakihara, M.; Yamamoto, Y. Distribution and abundance of finless porpoise in the Inland Sea of Japan. *Mar. Biol.* **2007**, *150*, 1025–1032. [CrossRef]
82. Reyne, M.; Webster, I.; Huggins, A. A Preliminary Study on the Sea Turtle Density in Mauritius. *Mar. Turt. Newsl.* **2017**, *152*, 5–8.
83. Hammond, P.S.; Macleod, K.; Berggren, P.; Borchers, D.L.; Burt, L.; Cañadas, A.; Desportes, G.; Donovan, G.P.; Gilles, A.; Gillespie, D. Cetacean abundance and distribution in European Atlantic shelf waters to inform conservation and management. *Biol. Conserv.* **2013**, *164*, 107–122. [CrossRef]
84. [DoN] Department of the Navy (US). *Navy OPAREA Density Estimate (NODE) for the Gulf of Mexico*; Contract #N62470-02 D-9997, CTO 0030; Department of the Navy: Washington, DC, USA, 2007.
85. Roberts, J.J.; Best, B.D.; Mannocci, L.; Fujioka, E.; Halpin, P.N.; Palka, D.L.; Garrison, L.P.; Mullin, K.D.; Cole, T.V.N.; Khan, C.B.; et al. Habitat-based cetacean density models for the U.S. Atlantic and Gulf of Mexico. *Sci. Rep.* **2016**, *6*, 22615. [CrossRef]
86. [NOAA] National Oceanic and Atmospheric Administration (US). Notice of Public Scoping and Intent to Prepare an Environmental Impact Statement. *Fed. Regist.* **2005**, *70*, 1871–1875.
87. Richardson, W.J.; Greene, C.R., Jr.; Malme, C.I.; Thomson, D.H. *Marine Mammals and Noise*; Academic Press: San Diego, CA, USA, 1995; p. 576.
88. Gordon, J.; Gillespie, D.; Potter, J.R.; Frantzis, A.; Simmonds, M.P.; Swift, R.; Thompson, D. A Review of the Effects of Seismic Surveys on Marine Mammals. *Mar. Technol. Soc. J.* **2003**, *37*, 16–34. [CrossRef]
89. Nowacek, D.P.; Thorne, L.H.; Johnston, D.W.; Tyack, P.L. Responses of cetaceans to anthropogenic noise. *Mammal Rev.* **2007**, *37*, 81–115. [CrossRef]
90. Ellison, W.T.; Southall, B.L.; Clark, C.W.; Frankel, A.S. A New Context-Based Approach to Assess Marine Mammal Behavioral Responses to Anthropogenic Sounds. *Conserv. Biol.* **2012**, *26*, 21–28. [CrossRef] [PubMed]
91. Southall, B.L.; Nowaceck, D.P.; Miller, P.J.O.; Tyack, P.L. Experimental field studies to measure behavioral responses of cetaceans to sonar. *Endanger. Species Res.* **2016**, *31*, 293–315. [CrossRef]
92. Dunlop, R.A.; Noad, M.J.; McCauley, R.D.; Scott-Hayward, L.; Kniest, E.; Slade, R.; Paton, D.; Cato, D.H. Determining the behavioural dose–response relationship of marine mammals to air gun noise and source proximity. *J. Exp. Biol.* **2017**, *220*, 2878–2886. [CrossRef] [PubMed]
93. Wartzok, D.; Popper, A.N.; Gordon, J.; Merrill, J. Factors affecting the responses of marine mammals to acoustic disturbance. *Mar. Technol. Soc. J.* **2003**, *37*, 6–15. [CrossRef]
94. Weilgart, L.S. A Brief Review of Known Effects of Noise on Marine Mammals. *Int. J. Comp. Psychol.* **2007**, *20*, 159–168.
95. Müller, R.A.J.; von Benda-Beckmann, A.M.; Halvorsen, M.B.; Ainslie, M.A. Application of kurtosis to underwater sound. *J. Acoust. Soc. Am.* **2020**, *148*, 780–792. [CrossRef]

Journal of
Marine Science and Engineering

Review

When Is Temporary Threshold Shift Injurious to Marine Mammals?

Dorian S. Houser

National Marine Mammal Foundation, 2240 Shelter Island Drive, Suite 200, San Diego, CA 92106, USA; dorian.houser@nmmf.org

Abstract: Evidence for synaptopathy, the acute loss of afferent auditory nerve terminals, and degeneration of spiral ganglion cells associated with temporary threshold shift (TTS) in traditional laboratory animal models (e.g., mice, guinea pigs) has brought into question whether TTS should be considered a non-injurious form of noise impact in marine mammals. Laboratory animal studies also demonstrate that both neuropathic and non-neuropathic forms of TTS exist, with synaptopathy and neural degeneration beginning over a narrow range of noise exposures differing by ~6–9 dB, all of which result in significant TTS. Most TTS studies in marine mammals characterize TTS within minutes of noise exposure cessation, and TTS generally does not achieve the levels measured in neuropathic laboratory animals, which have had initial TTS measurements made 6–24 h post-exposure. Given the recovery of the ear following the cessation of noise exposure, it seems unlikely that the magnitude of nearly all shifts studied in marine mammals to date would be sufficient to induce neuropathy. Although no empirical evidence in marine mammals exists to support this proposition, the regulatory application of impact thresholds based on the onset of TTS (6 dB) is certain to capture the onset of recoverable fatigue without tissue destruction.

Keywords: permanent threshold shift; synaptopathy; neuropathy; auditory brainstem response; behavioral thresholds

Citation: Houser, D.S. When Is Temporary Threshold Shift Injurious to Marine Mammals?. *J. Mar. Sci. Eng.* **2021**, *9*, 757. https://doi.org/10.3390/jmse9070757

Academic Editors: Michel André and Christine Erbe

Received: 31 May 2021
Accepted: 6 July 2021
Published: 9 July 2021

Publisher's Note: MDPI stays neutral with regard to jurisdictional claims in published maps and institutional affiliations.

1. Introduction

The last two decades have seen an explosion in the scientific literature related to the impact of anthropogenic noise on marine mammals. Arguably, the topic also dominates the distribution of research dollars related to marine mammal science. Research investment has sought to provide insight on the types of responses that marine mammals exhibit when exposed to anthropogenic noise, and more importantly, the short- and long-term consequences of such exposures. Relationships between animal responses and signal frequency, level, duration, and duty cycle, as well as the importance of novel to repeated exposures, have been the focus of many studies, the results of which have informed regulatory practices of countries actively engaged in the marine mammals and noise issue.

The avoidance of injury to marine mammals incidentally exposed to anthropogenic noise is a common goal of regulators. However, there is a lack of legal and regulatory consensus among countries with environmental management frameworks as to what defines an 'injury'. This contributes to differences in the noise exposure thresholds at which impacts are regulated, an issue that has particular relevance when considering direct, physiological impacts to the auditory system. For example, under the authority of the Marine Mammal Protection Act (MMPA), the National Marine Fisheries Service (NMFS) regulates noise producers in US territorial waters that have the potential to impact marine mammals. As part of its regulatory framework, NMFS adopted a definition of injury that involved the destruction of tissue [1,2]. This definition formed the basis for a legal distinction under the MMPA between two forms of noise-induced hearing loss (NIHL)–a temporary elevation of hearing threshold (temporary threshold shift, or TTS), which was believed to be a fully recoverable form of auditory fatigue, and permanent threshold shift

(PTS), which was a permanent loss of hearing sensitivity believed to arise from damage to the auditory system tissues (e.g., disarticulation of the middle ear bones, loss of hair cells). The occurrence of TTS was not considered injury under this regulatory interpretation, whereas PTS was. In contrast, all forms of hearing impairment caused by exposure to anthropogenic noise are considered injury under German regulation: "An injury within the meaning of the prohibition on taking under species protection law is an impairment of an animal's physical welfare or damage to its health. This encompasses any impairment of its physical integrity [3]." Thus, a TTS is considered an injury under German law once the threshold for TTS has been exceeded.

There has been a greater focus in recent years on whether TTS is truly non-injurious under an injury definition that incorporates the destruction of tissue [4]. Indeed, since TTS magnitude grows with the degree of noise exposure and eventually becomes PTS, there is obviously some threshold of noise exposure beyond which tissue damage occurs. What was historically less apparent was whether the onset of PTS and the onset of tissue damage were due to equivalent exposures. In 2009, Kujawa and Liberman [5] provided evidence that fully-recoverable threshold shifts in mice could be associated with the permanent loss of tissues within the auditory system. The question that followed for the marine mammal community was a natural extension of the findings–if TTS can be associated with the destruction of tissue, then at what magnitude of TTS can injury be present in marine mammals?

2. TTS and the Loss of Auditory System Tissues

There are relatively few studies demonstrating that TTS can be associated with the destruction of tissue. To date, relevant studies have only been performed in terrestrial laboratory animals. Kujawa and Liberman [5] exposed mice (*Mus musculus*) to octave-band noise (8–16 kHz) for two hours at a sound pressure level (SPL) of 100 dB re 20 µPa. Utilizing measurements of the auditory brainstem response (ABR), they measured a ~40 dB TTS in the mice 24 h after the noise exposure. (Note that a 40 dB shift reflects a several order of magnitude reduction in hearing sensitivity relative to the 6 dB of shift used to define the onset of TTS in some marine mammal regulations.) Kujawa and Liberman demonstrated that the mice suffered an acute loss of afferent nerve terminals, termed synaptopathy, while cochlear sensory (hair) cells remained intact. Degeneration of the cochlear nerve (loss of spiral ganglion cells) was also observed, although it occurred slowly over the course of one to two years. The magnitude and cochlear placement of the syanptopathy and nerve degeneration were spatially related to the corresponding frequency at which the greatest threshold shift was observed. Both phenomena were noted concomitant with hearing thresholds that returned to baseline days to weeks after the exposure, suggesting that conventional threshold testing alone was insufficient to determine pathologies associated with noise over-exposure.

Subsequent work has reinforced these findings. Lin and colleagues [6] performed a similar experiment to that of Kujawa and Liberman [5] but utilized guinea pigs (*Cavia porcellus*) in order to address concerns that neurons of the mouse ear might be particularly susceptible to noise over-exposure. Subjects were exposed to octave-band noise (4–8 kHz) for two hours at SPLs of 106 and 109 dB re 20 µPa. Again, utilizing ABR measurements, the magnitude of TTS 24 h after the noise exposure was found to be ~50 dB, but returned to normal by ten days after the exposure. Significant synaptopathy was noted at this time, although there was no loss of either inner or outer hair cells (IHCs and OHCs, respectively). Long-term monitoring subsequently showed the slow loss of spiral ganglion cells over the course of a two-year period, with losses closely associated with the sites of synaptopathy.

Wang and Ren [7] performed a noise exposure experiment in mice similar to that of Kujawa and Liberman [5] but utilized a repeated exposure paradigm in which a subset of mice were exposed to fatiguing stimuli, either two or three times following recovery of the initial ABR threshold shift. Mice were exposed to octave band noise centered at 12 kHz for two hours at ~100 dB re 20 µPa, and as in previous studies, ABR thresholds were measured prior to and 24 h after the exposure. For mice receiving a single noise exposure, threshold shifts were initially found to be about the same as observed in previous experiments (30–40 dB), recovering to normal in approximately the same time. Similarly, the occurrence of synaptopathy without the loss of IHCs or OHCs was observed, although the loss of spiral ganglion cells was not investigated.

Collectively, these studies provide a small but sufficient amount of evidence to suggest that a fully recoverable TTS can occur despite permanent auditory system tissue damage. Further, progressive loss of auditory system tissues, specifically spiral ganglion cells, can occur long after recovery of hearing thresholds. How this applies to marine mammals warrants discussion, and the relevance of laboratory animal work conducted to date requires consideration in the context of marine mammal TTS studies.

3. Relevance of Laboratory Animal TTS Findings to Marine Mammals

To reconcile the findings of the TTS literature demonstrating tissue damage with the TTS work performed in marine mammals, there must first be an understanding of the magnitude of threshold shifts achieved in traditional laboratory animal models and the time courses at which shifts were measured. Threshold shifts for which tissue damage has been associated in laboratory animals range from ~30 to 50 dB of TTS [5–7]. Threshold shifts in these studies were measured 24 h post-exposure, and the measurements were made using ABR threshold procedures. The majority of marine mammal TTS studies have behaviorly measured smaller amounts of TTS (<20 dB) within minutes of noise exposure (for review, see [8]; for studies after 2015, see [9–14]), though initial threshold shifts measured behaviorally have occasionally been greater than 40 dB [15]. A smaller number of studies have measured ABR threshold shifts and found initial shifts as high as 63 dB when measured within a couple of minutes of the cessation of noise exposure [16]. Behavioral and ABR measurements are not equivalent, however, and reconciling threshold shifts measured with the two approaches requires an understanding of the differences between them [8].

Behavioral measurements of hearing sensitivity require an animal to act in response to hearing a sound (e.g., paddle push, produce a whistle), thus providing an integrated response that includes the animal's perception of the sound and its decision to respond. ABR measurements of hearing sensitivity do not reflect this integrated animal response but measure only voltages generated by portions of the ascending auditory system. Temporary threshold shifts determined from ABR measurements generally demonstrate an earlier onset of TTS, generally characterized as 6 dB of threshold shift, larger shifts than those observed with behavioral methods and longer recovery times than observed with behavioral methods. This suggests that some mechanism accommodates the restoration of the hearing threshold even though the auditory system has not fully recovered from the fatiguing noise exposure [8,17]. Finneran et al. [17] showed that TTS measured with ABRs could be 19–33 dB greater than those measured behaviorally and that ABR threshold shifts of ~10 dB could be found in the absence of a behavioral shift. The time courses of recovery measured with ABRs were always longer than those measured behaviorally. In a subsequent study of TTS induced by exposure to air gun impulses, no behavioral threshold shifts were observed, whereas a small amount of TTS was detected by measuring ABR thresholds [14]–in one dolphin, a 9-dB TTS was measured at a test frequency of 8 kHz. Thus, caution should be exercised in making comparisons between studies that used behavioral or ABR threshold measurement methods, and the synthesis of findings across studies should account for these differences.

If the difference between the magnitude of ABR and behavioral threshold shifts measured following noise exposure is consistent across mammals in general, then the modest behaviorally measured threshold shifts from marine mammal studies could appear similar to the ABR threshold shifts observed in laboratory animals that have been associated with tissue damage (e.g., a 20 dB behavioral TTS could potentially correspond to a 50 dB TTS measured with ABRs). However, the initial TTS measurements in marine mammal studies are typically made within minutes of noise exposure, not 24 h after the exposure (as described for traditional laboratory animal models). Recovery from TTS induced by narrowband or tonal noise can crudely be described as a function of the logarithm of time with recovery rates increasing in variability as recovery time increases. In marine mammals, measured recovery rates range from ~4 to 23 dB/decade of time [12,17–27] and generally demonstrate a positive relationship with the magnitude of the initial threshold shift. Thus, TTS measured within minutes of the noise exposure would be much higher than that measured 24-h after the exposure. Comparisons between laboratory animal studies with 24-h post-exposure TTS measures and marine mammal studies made within minutes of exposure cessation must keep this difference in mind, particularly since marine mammal TTS studies often recover to baseline thresholds within 24 h of noise exposure, even when TTS measured behaviorally and immediately following the noise exposure was as high as ~30 dB. It is important to note that in the one marine mammal study in which PTS was observed, the behavioral threshold shift was ~30 dB 24 h after the exposure, which could equate to an ABR threshold shift as high as 60 dB [24].

Little information exists on the relationship between the growth of TTS and quantifiable tissue damage in terrestrial mammals, and none exists in marine mammals. However, some limited work in mice demonstrates that there exist both neuropathic and non-neuropathic levels of TTS. Mice exposed to octave-band (8–16 kHz) noise exposures ranging from 91 to 100 dB re 20 µPa for periods of two hours demonstrated significant synaptopathy at exposures >97 dB re 20 µPa, but not at exposures <94 dB re 20 µPa [28–30]. The degree of synaptopathy appeared progressive and frequency-dependent, i.e., the degree of synaptopathy varied as a function of the cochlear frequency-place map, as previously observed. The magnitude of TTS measured after noise exposure ranged from up to 55 dB measured 6 h after exposure to ~35–40 dB measured 24 h after exposure in non-neuropathic mice, showing substantial TTS could occur without the presence of synaptopapthy. However, the observance of synaptopathy onset at noise exposures that differed by as little to 6–9 dB from those that were non-neuropathic suggested a narrow range over which the onset and growth of synaptopathy occurs. Thus, the limited evidence that is available suggests that relatively large TTS (>30 dB, 24-h post-exposure) can occur without tissue damage, but that damage begins to occur along some noise exposure continuum as noise exposures (and TTS) increase.

4. Discussion

A limited amount of evidence from terrestrial laboratory animals suggests that both neuropathic and non-neuropathic TTS are feasible, with the onset of neuropathology occurring at noise exposures well exceeding those corresponding to the onset of TTS. Given this evidence, it is probable that threshold shifts in marine mammals can occur with noise exposures that also range in magnitude and effect from fully recoverable TTS without tissue damage, through fully recoverable TTS with tissue damage, to the destruction of tissue producing PTS. In other words, TTS is a graded phenomenon that is fully recoverable at low levels but can lead to tissue damage as it becomes more extreme–not all TTS results in the destruction of tissue. The threshold of exposure at which neuropathy would occur is unknown and likely varies between marine mammal species, as does the noise exposure required for the onset of TTS [8]. Based on laboratory animal studies, the onset of neuropathic TTS would appear to occur at only more extreme threshold shifts, exceeding the magnitude of TTS commonly induced in the marine mammal studies conducted thus far. Nevertheless, if a legal definition of injury includes the destruction of tissue, then synaptopathy qualifies as injury and must be considered in the framework of potential acoustic impacts to marine mammals.

Countries actively regulating the potential impact of ocean noise to marine mammals often employ thresholds for the onset of injury that would be conservative relative to the findings related to neuropathic TTS, regardless of whether following a broad definition of injury that encompasses impacts to behavior or one that more narrowly relies on a definition involving the destruction of tissue. In the least conservative case, such as is employed by US regulators [31], the use of an initial (i.e., measured minutes after exposure) 40 dB of TTS as the onset of injury falls below the magnitude and time scale of TTS associated with neuropathic TTS (i.e., 30–50 dB of TTS measured 24 h after noise exposure) observed in conventional laboratory animal models. Therefore, even though it has been demonstrated that a fully-recoverable TTS of sufficient magnitude can result in underlying tissue damage [5], the implementation of regulatory thresholds based on TTS onset should encompass recoverable auditory fatigue without the occurrence of tissue damage [32].

Funding: This research received no external funding.

Institutional Review Board Statement: Not applicable.

Acknowledgments: The author wishes to thank J. Mulsow for his thoughtful review of a draft of this manuscript. This is scientific contribution #308 of the National Marine Mammal Foundation.

Conflicts of Interest: The author declares no conflict of interest.

References

1. National Marine Fisheries Service. Final Rule SURTASS LFA Sonar. In *Federal Register*; U.S. Department of Commerce: Washington, DC, USA, 2002; Volume 67, pp. 46712–46789.
2. National Marine Fisheries Service (NMFS). Final rule for the shock trial of the USS Mesa Verde, (LPD-19). In *Federal Register*; Department of Commerce: Washington, DC, USA, 2008; Volume 73, pp. 43727–43730.
3. Bundesministerium für Umwelt. Concept for the protection of harbour porpoises from sound exposures during the construction of offshore wind farms in the German North Sea. In Proceedings of the 21st ASCOBANS Advisory Committee Meeting, Gothenberg, Sweden, 24 December 2014.
4. Tougaard, J.; Wright, A.J.; Madsen, P.T. Cetacean noise criteria revisited in the light of proposed exposure limits for harbour porpoises. *Mar. Pollut. Bull.* **2015**, *90*, 196–208. [CrossRef] [PubMed]
5. Kujawa, S.G.; Liberman, M.C. Adding insult to injury: Cochlear nerve degeneration after "temporary" noise-induced hearing loss. *J. Neurosci.* **2009**, *29*, 14077–14085. [CrossRef] [PubMed]
6. Lin, H.W.; Furman, A.C.; Kujawa, S.G.; Liberman, M.C. Primary neural degeneration in the guinea pig cochlea after reversible noise-induced threshold shift. *J. Assoc. Res. Otolaryngol.* **2011**, *12*, 605–616. [CrossRef] [PubMed]
7. Wang, Y.; Ren, C. Effects of repeated "benign" noise exposures in young cba mice: Shedding light on age-related hearing loss. *J. Assoc. Res. Otolaryngol.* **2012**, *13*, 505–515. [CrossRef] [PubMed]
8. Finneran, J.J. Noise-induced hearing loss in marine mammals: A review of temporary threshold shift studies from 1996 to 2015. *J. Acoust. Soc. Am.* **2015**, *138*, 1702–1726. [CrossRef]
9. Kastelein, R.A.; Helder-Hoek, L.; Cornelisse, S.A.; Defillet, L.N.; Huijser, L.A.E.; Terhune, J.M. Temporary hearing threshold shift in harbor seals (*Phoca vitulina*) due to one-sixth-octave noise bands centered at 0.5, 1, and 2 kHz. *J. Acoust. Soc. Am.* **2020**, *148*, 3873–3885. [CrossRef]
10. Kastelein, R.A.; Helder-Hoek, L.; Cornelisse, S.A.; Huijser, L.A.E.; Gransier, R. Temporary hearing threshold shift at ecololgically relevant frequencies in a harbor porpoise (*Phocoena phocoena*) due to exposure to a noise band centered at 88.4 kHz. *Aquat. Mammal.* **2020**, *46*, 444–453. [CrossRef]
11. Kastelein, R.A.; Helder-Hoek, L.; Cornelisse, S.A.; Defiller, L.N.; Huijser, L.A.E. Temporary threshold shift in a second harbor porpoise (*Phocoena phocoena*) after exposure to a one-sixth-octave noise band at 1.5 kHz and a 6.5 kHz continuous wave. *Aquat. Mammal.* **2020**, *46*, 431–443. [CrossRef]
12. Kastelein, R.A.; Helder-Hoek, L.; Cornelisse, S.A.; Huijser, L.A.E.; Terhune, J.M. Temporary hearing threshold shift in harbor seals (*Phoca vitulina*) due to a one-sixth-octave noise band centered at 32 kHz. *J. Acoust. Soc. Am.* **2020**, *147*, 1885–1896. [CrossRef]
13. Sills, J.M.; Ruscher, B.; Nichols, R.; Southall, B.L.; Reichmuth, C. Evaluating temporary threshold shift onset levels for impulsive noise in seals. *J. Acoust. Soc. Am.* **2020**, *148*, 2973–2986. [CrossRef]
14. Finneran, J.J.; Schlundt, C.E.; Branstetter, B.K.; Trickey, J.; Bowman, V.; Jenkins, K. Effects of multiple impulses from a seismic air gun on bottlenose dolphin hearing and behavior. *J. Acoust. Soc. Am.* **2015**, *137*, 1634–1646. [CrossRef]
15. Kastelein, R.A.; Gransier, R.; Hoek, L. Comparative temporary threshold shifts in a harbor porpoise and harbor seal, and severe shift in a seal. *J. Acoust. Soc. Am.* **2013**, *134*, 13–16. [CrossRef] [PubMed]
16. Popov, V.V.; Supin, A.Y.; Rozhnov, V.V.; Nechaev, D.I.; Sysuyeva, E.V.; Klishin, V.O.; Pletenko, M.G.; Tarakanov, M.B. Hearing threshold shifts and recovery after noise exposure in beluga whales *Delphinapterus Leucas*. *J. Exp. Biol.* **2013**, *216*, 1587–1596. [CrossRef] [PubMed]
17. Finneran, J.J.; Schlundt, C.E.; Branstetter, B.; Dear, R.L. Assessing temporary threshold shift in a bottlenose dolphin (*Tursiops truncatus*) using multiple simultaneous auditory evoked potentials. *J. Acoust. Soc. Am.* **2007**, *122*, 1249–1264. [CrossRef] [PubMed]
18. Nachtigall, P.E.; Supin, A.Y.; Pawloski, J.; Au, W.W.L. Temporary threshold shifts after noise exposure in the bottlenose dolphin (*Tursiops truncatus*) measured using evoked auditory potentials. *Mar. Mammal Sci.* **2004**, *20*, 673–687. [CrossRef]

19. Mooney, T.A.; Nachtigall, P.E.; Vlachos, S. Sonar-induced temporary hearing loss in dolphins. *Biol. Lett.* **2009**, *5*, 565–567. [CrossRef]

20. Kastelein, R.A.; Gransier, R.; Hoek, L.; Macleod, A.; Terhune, J.M. Hearing threshold shifts and recovery in harbor seals (*Phoca vitulina*) after octave-band noise exposure at 4 kHz. *J. Acoust. Soc. Am.* **2012**, *132*, 2745–2761. [CrossRef]

21. Kastelein, R.A.; Gransier, R.; Hoek, L.; Olthuis, J. Temporary threshold shifts and recovery in a harbor porpoise (*Phocoena phocoena*) after octave-band noise at 4 kHz. *J. Acoust. Soc. Am.* **2012**, *132*, 3525–3537. [CrossRef]

22. Kastelein, R.A.; Gransier, R.; Hoek, L.; Rambags, M. Hearing frequency thresholds of a harbor porpoise (*Phocoena phocoena*) temporarily affected by a continuous 1.5 kHz tone. *J. Acoust. Soc. Am.* **2013**, *134*, 2286–2292. [CrossRef]

23. Popov, V.V.; Supin, A.Y.; Rozhnov, V.V.; Nechaev, D.I.; Sysueva, E.V. The limits of applicability of the sound exposure level (SEL) metric to temporal threshold shifts (TTS) in beluga whales, *Delphinapterus Leucas*. *J. Exp. Biol.* **2014**, *217*, 1804–1810. [CrossRef]

24. Reichmuth, C.; Sills, J.M.; Mulsow, J.; Ghoul, A. Long-term evidence of noise-induced permanent threshold shift in a harbor seal (*Phoca vitulina*). *J. Acoust. Soc. Am.* **2019**, *146*, 2552–2561. [CrossRef]

25. Kastelein, R.A.; Helder-Hoek, L.; Cornelisse, S.; Huijser, L.A.E.; Terhune, J.M. Temporary hearing threshold shift in harbor seals (*Phoca vitulina*) due to a one-sixth-octave noise band centered at 16 kHz. *J. Acoust. Soc. Am.* **2019**, *146*, 3113–3122. [CrossRef]

26. Kastelein, R.A.; Helder-Hoek, L.; Gransier, R. Frequency of greatest temporary hearing threshold shift in harbor seals (*Phoca vitulina*) depends on fatiguing sound level. *J. Acoust. Soc. Am.* **2019**, *145*, 1353–1362. [CrossRef] [PubMed]

27. Kastelein, R.A.; Parlog, C.; Helder-Hoek, L.; Cornelisse, S.A.; Huijser, L.A.E.; Terhune, J.M. Temporary hearing threshold shift in harbor seals (*Phoca vitulina*) due to a one-sixth-octave noise band centered at 40 kHz. *J. Acoust. Soc. Am.* **2020**, *147*, 1966–1976. [CrossRef] [PubMed]

28. Hickox, A.E.; Liberman, M.C. Is noise-induced cochlear neuropathy key to the generation of hyperacusis or tinnitus? *J. Neurophysiol.* **2014**, *111*, 552–564. [CrossRef] [PubMed]

29. Fernandez, K.A.; Jeffers, P.W.C.; Lall, K.; Liberman, M.C.; Kujawa, S.G. Aging after noise exposure: Acceleration of cochlear synaptopathy in "recovered" ears. *J. Neurosci.* **2015**, *35*, 7509–7520. [CrossRef]

30. Jensen, J.B.; Lysaght, A.C.; Liberman, M.C.; Qvortrup, K.; Stankovic, K.M. Immediate and delayed cochlear neuropathy after noise exposure in pubescent mice. *PLoS ONE* **2015**, *10*, e0125160. [CrossRef]

31. National Marine Fisheries Service. *Revision to: Technical Guidance for Assessing the Effects of Anthropogenic Sound on Marine Mammal Hearing (Version 2.0)—Underwater Acoustic Thresholds for Onset of Permanent and Temporary Threshold Shifts*; National Oceanic and Atmospheric Administration: Silver Springs, MD, USA, 2018; p. 167.

32. Le Prell, C.G. Effects of noise exposure on auditory brainstem response and speech-in-noise tasks: A review of the literature. *Int. J. Audiol.* **2019**, *58*, S3–S32. [CrossRef] [PubMed]

Journal of
Marine Science and Engineering

Review

The Use of Psychoacoustics in Marine Mammal Conservation in the United States: From Science to Management and Policy

Shane Guan [1,*] and Tiffini Brookens [2,*]

1   Department of Mechanical Engineering, The Catholic University of America, Washington, DC 20064, USA
2   U.S. Marine Mammal Commission, Bethesda, MD 20814, USA
*   Correspondence: guan@cua.edu (S.G.); tbrookens@mmc.gov (T.B.)

Abstract: Underwater sound generated from human activities has been long recognized to cause adverse effects on marine mammals, ranging from auditory masking to behavioral disturbance to hearing impairment. In certain instances, underwater sound has led to physical injuries and mortalities. Research efforts to assess these impacts began approximately four decades ago with behavioral observations of large whales exposed to seismic surveys and rapidly progressed into the diverse field that today includes studies of behavioral, auditory, and physiological responses of marine mammals exposed to anthropogenic sound. Findings from those studies have informed the manner in which impact assessments have been and currently are conducted by regulatory agencies in the United States. They also have led to additional questions and identified information needed to understand more holistically the impacts of underwater sound, such as population- and species-level effects, long-term, chronic, and cumulative effects, and effects on taxa for which little or no information is known. Despite progress, the regulatory community has been slow to incorporate the best available science in marine mammal management and policy and often has relied on outdated and overly simplified methods in its impact assessments. To implement conservation measures effectively, regulatory agencies must be willing to adapt their regulatory scheme to ensure that the best available scientific information is incorporated accordingly.

Keywords: underwater sound impacts; marine mammal conservation; impact assessment; behavioral disturbance; hearing impairment; auditory masking

Citation: Guan, S.; Brookens, T. The Use of Psychoacoustics in Marine Mammal Conservation in the United States: From Science to Management and Policy. *J. Mar. Sci. Eng.* **2021**, *9*, 507. https://doi.org/10.3390/jmse9050507

Academic Editor: Giuseppa Buscaino

Received: 2 April 2021
Accepted: 5 May 2021
Published: 8 May 2021

Publisher's Note: MDPI stays neutral with regard to jurisdictional claims in published maps and institutional affiliations.

1. Introduction

Since the Industrial Revolution, with the mechanization and expansion of human activities into the sea, humans have been introducing pervasive anthropogenic sound into the marine environment [1]. Given that many marine species rely on acoustic cues for their life functions, such as communication, sensing the environment, migration, and detecting predators and prey, elevated anthropogenic sound can have detrimental effects on them [2,3]. To address these concerns and to provide sound scientific information for the conservation of marine mammal species, interdisciplinary studies have been conducted over the past four decades to support impact assessments by various regulatory entities [4,5].

Those assessments typically follow the source–path–receiver model, where the "source" is the anthropogenic sound source, "path" is the underwater sound propagation, and "receiver" is the marine mammal that is exposed to the sound [6]. The former two aspects lie within the field of underwater acoustics, while the latter is addressed in animal bio- and psychoacoustics. An overview of the application of underwater acoustics in marine conservation, which focuses on knowledge gained regarding anthropogenic sound sources and sound propagation relevant to impact assessments is discussed in a companion review paper [7]. Here, we provide a review of the application of psychoacoustics in marine mammal impact assessments that support management and policy in the United States.

2. An Early History of Research Involving the Effects of Sound on Marine Mammals and Its Application in Impact Assessments: Up to 2000

The invention of an underwater acoustic transducer to aid navigation and to engage anti-submarine warfare at the turn of the 20th century opened a new field of research in underwater acoustics [8,9]. However, it was not until almost half a century later that researchers first documented underwater sound production and communication by marine mammals [10–13]. In the early 1970s, a group of researchers began to recognize the potential adverse effects of anthropogenic sound on marine species, particularly marine mammals. Payne and Webb were the first to hypothesize that sound emitted from modern ships significantly reduced the communication range of 20 Hz fin whale calls by up to 3000 nautical miles [14]. Using audiograms of the bottlenose dolphin (*Tursiops truncatus*), harbor seal (*Phoca vitulina*), and California sea lion (*Zalophus californianus*), Myrberg proposed that auditory masking of these species could occur from vessel traffic and industrial activities [15].

With the passage of the National Environmental Policy Act (NEPA) in 1969, the Marine Mammal Protection Act (MMPA) in 1972, and the Endangered Species Act in 1973 in the United States [16–19], government agencies faced the challenge of properly regulating various human activities that had been overlooked previously. The lack of science-based information to implement regulatory requirements and to assess environmental impacts from those activities prompted numerous studies investigating the effects of underwater sound on marine mammals [4,5,20].

Some of the initial pressing questions were focused on the impacts that underwater detonations, typically used during naval shock trials and other exercises, may have on marine mammals [21]. Underwater detonation experiments were conducted using live animals such as sheep, dogs, and monkeys submerged in water to extrapolate the potential physical injuries and mortalities to marine mammals [21–23]. Those extrapolations, along with additional theoretical analysis on potential hearing impairment [24] and field observation of behavioral disturbance from sound exposure [25], were the basis for the U.S. Navy's (the Navy) environmental impact statement (EIS) for the shock trial of the Seawolf submarine [26].

In addition, from the early 1980s through the 1990s, the U.S. Minerals Management Service (MMS, the predecessor of the current Bureau of Ocean Energy Management (BOEM)) funded a number of studies to investigate adverse effects of oil and gas development activities on marine mammals in the Arctic. Most of those studies were based on aerial or vessel observations of marine mammal behavioral responses and movements when exposed to industrial activities, such as oil and gas exploration using seismic airgun arrays [27–39]. Many of those studies were reviewed by Myrberg [40] and summarized in a landmark book "Marine Mammals and Noise" by Richardson et al. [6].

A relatively simple scheme for assessing the impacts of sound on marine mammals was proposed by Richardson et al. [6] using the source–path–receiver model. In the model, potential adverse impacts from a given intense sound source are based on the distance of the source from the receiver (animal) (Figure 1). When the animal is in close proximity to the source, the impacts would be expected to be "severe" and physical injury could occur. When the animal is farther away from the source, the expected impacts would gradually decrease, until a distance at which the impacts would be negligible [6].

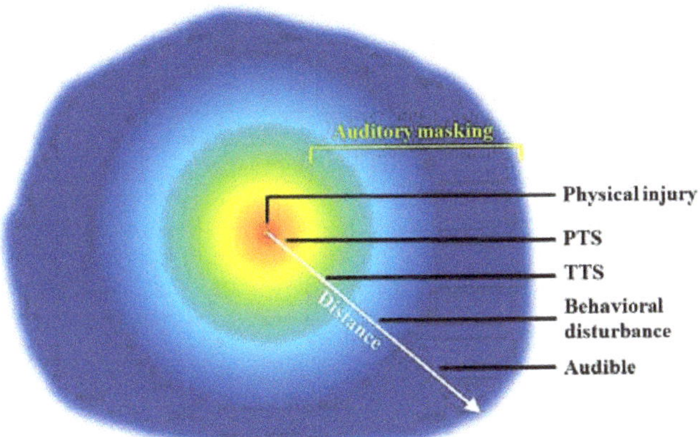

Figure 1. A simple acoustic impact model based on the distance of the source from the receiver (animal). PTS = permanent threshold shift; TTS = temporary threshold shift. See text for explanation.

Contrary to the simplicity of the model, the relationship between a given sound exposure level and the severity of the impact expected is not well defined. To delineate and assess impacts in a more meaningful manner, the 1994 amendments to the MMPA included definitions for both Level A and Level B harassment [41]. Based on the 1994 amendments, Level A harassment "has the potential to injure a marine mammal or marine mammal stock in the wild," while Level B harassment "has the potential to disturb a marine mammal or marine mammal stock in the wild by causing disruption of behavioral patterns, including, but not limited to, migration, breathing, nursing, breeding, feeding, or sheltering." A simple interpretation of Level A vs. Level B harassment is that Level A harassment could cause physical injury to a marine mammal, while Level B harassment could result in behavioral disturbance or displacement.

Given the lack of a standard for each type of impact, a number of different criteria, thresholds[1], and metrics have been proposed and used in the impact assessment and review processes over the years (see [42]). For example, the Navy used the 50 percent tympanic membrane rupture criterion as onset of auditory injury for Level A harassment based on an energy flux density (EFD) of 1.17 in-lb/in^2 (equivalent to 205 dB re 1 μPa^2-s) in its EIS for the Seawolf shock trial [20,26,42]. In the same EIS, the Navy used dual criteria (based on two acoustic metrics) for Level B harassment: (1) an energy-based temporary threshold shift (TTS) threshold[2] of 182 dB re 1 μPa^2-s and (2) a peak pressure-based TTS threshold of 12 lb/in^2 (or psi) [26,42]. At the time, the U.S. National Marine Fisheries Service (NMFS) used interim acoustic criteria for marine mammal harassment: 70 dB over a hearing threshold defined behavioral harassment, 80–100 dB over the threshold defined TTS, 133 dB over the threshold defined pain, and 155 dB over the threshold defined permanent threshold shift (PTS) [42,43].

Since information was unavailable concerning the onset of hearing injury in the form of PTS for marine mammals, an expert working group, the High Energy Seismic Survey Team (HESS Team) was convened by MMS in 1996 to provide a "roadmap" for applicants that used airgun arrays to acquire geophysical data and needed to comply with the various environmental statutes and regulations [43]. In its 1997 workshop, the HESS Team reached a consensus that exposures to root-mean-square sound pressure levels (SPL$_{rms}$ or $L_{p,rms}$) greater than 180 dB re 1 μPa$_{rms}$ were likely to have the potential to cause serious behavioral,

---

1 Only in-water thresholds are described in detail herein. Details regarding in-air thresholds are beyond the scope of the paper.
2 The metric considered the total energy of all exposures based on the greatest EFD in any one-third octave band for frequencies greater than 100 Hz for odontocetes and 10 Hz for mysticetes.

physiological, and hearing effects [43]. Thus, 180 dB re 1 μPa_{rms} became the Level A harassment threshold for cetaceans, and 190 dB re 1 μPa_{rms} was used for pinnipeds [44].

Malme et al. [29] documented that migrating gray whales (*Eschrichtius robustus*) avoided areas when exposed to continuous low-frequency sound levels above 120 dB re 1 μPa_{rms} but female-calf pairs exhibited behavioral disturbances when exposed to impulsive sound levels above 160 dB re 1 μPa_{rms} [28,29]. Richardson et al. [32,35,45] documented similar responses in migrating bowhead whales. The 120 and 160 dB re 1 μPa_{rms} sound levels were subsequently used as the onset Level B harassment thresholds for marine mammals when exposed to continuous and impulsive sound, respectively [44]. These Level B harassment thresholds continue to be used by the regulatory agencies for certain sound sources (e.g., seismic and high-resolution geophysical surveys, vibratory and impact pile driving, drilling).

Although hearing impairment, including TTS and PTS, had been widely documented in psychoacoustic studies on humans [46,47] and some terrestrial mammal species [48–50], the effects on marine mammals were only hypothesized until the mid-1990s (e.g., [4,6]). As a result, many of the impact assessments, with the exception of a few high-profile cases (e.g., Navy shock trials [26], military sonar [51], and the Acoustic Thermometry of Ocean Climate (ATOC) experiment [52]), did not consider the effects of TTS or PTS.

3. Advances in Marine Mammal Psychoacoustic Research and the Consideration of Frequency-Based Auditory Responses in Impact Assessments: Since 2000

The last 20 years have seen great progress in marine mammal psychoacoustic research, especially in the field of noise-induced threshold shift (NITS) studies. The first instance of NITS in a marine mammal was documented by Kastak and Schusterman [53], when a harbor seal was inadvertently exposed to intense broadband construction sound for 6 days, which resulted in a TTS of 8 dB at 100 Hz. Soon afterwards, the U.S. Office of Naval Research funded a number of NITS studies to investigate marine mammal TTS by exposing the animals to various types of intense sounds (e.g., [54–59]). Many of the earlier studies were conducted on animals that were trained to respond to acoustic stimuli (behavioral or psychophysical methods), but more recently auditory evoked potentials (AEP) measurements (electrophysiological methods) have been used to study NITS (e.g., [60–64]). A comprehensive review of marine mammal NITS studies up to 2015 was provided by Finneran [65]. The results of those studies, coupled with behavioral and electrophysiological audiograms of approximately 20 marine mammal species, basic knowledge of their hearing capabilities (e.g., [6,66,67]), and known noise-induced PTS in humans, provided the foundation for assessing auditory injuries in marine mammals.

Around the same time in the early 2000s, naval exercises involving mid-frequency active sonar[3] (MFAS) were purportedly linked to mass strandings of various deep-diving cetaceans, particularly beaked whales [68–72]. One of the hypotheses suggested that mortalities associated with sonar exposure-linked strandings were the result of decompression sickness (i.e., "the bends") due to rapid ascension to the sea surface by the mammals when exposed to MFAS sounds [73–75]. To fully understand whether behavioral responses from sonar exposure could be the cause of cetacean strandings, the Navy funded several controlled-exposure experiments (CEEs) beginning in the late-2000s (e.g., [76]). The researchers attached dataloggers equipped with various sensors (e.g., accelerometers, acoustic and pressure sensors) to individual whales and conducted playbacks of simulated MFAS or other anthropogenic sounds to the tagged animals. Behavioral responses of the exposed animals were analyzed from the recovered dataloggers [76–78]. More recently, actual naval sonar has been used to determine whether and how marine mammals respond to MFAS, e.g., [79]. Despite several large-scale CEEs that have been conducted around the world to address the effects of naval sonar on marine mammals, results have shown that

---

[3] Which is termed by the Navy as sonar within the 1–10 kHz range. Other navies, particularly those in Europe, define a portion of the MFAS range as low-frequency active sonar (LFAS), whereas LFAS in the United States is any sonar that operates below 1 kHz.

responses are highly variable and may not be fully predictable with only sound level-based thresholds, e.g., [79–87].

Based on comprehensive review of the then up-to-date best available information, Southall et al. recommended a set of criteria for assessing sound exposure in marine mammals [88]. In their recommendations, marine mammals were grouped into five functional hearing groups based on their generalized auditory frequency responses and, for pinnipeds, the medium in which they listen. The functional hearing groups were: low-frequency (LF) cetaceans (all baleen whales), mid-frequency (MF) cetaceans (delphinids, beaked whales, and the sperm whale), high-frequency (HF) cetaceans (porpoises, river dolphins, and *Kogia* spp.), pinnipeds in water, and pinnipeds in air. An "M-weighting" function representing a generic auditory frequency response was established for each of the five functional hearing groups using the paradigm similar to the human 100-phon equal-loudness function (or "C-weighting") [88].

Three sound types were identified in Southall et al.'s [88] criteria: (1) a single pulse, (2) multiple pulses, and (3) nonpulses. The distinction between pulse and nonpulse was determined based on a 3 dB difference between the continuous and impulse setting of a sound level meter (SLM). Specifically, if the SLM measurement from the impulse setting (35 ms) was 3 dB or greater than the continuous setting (1 s) for the sound, that sound would be classified as a pulse, otherwise it would be deemed a nonpulse [88,89].

For assessing onset of auditory injury (defined as PTS), Southall et al. [88] recommended dual criteria for impact assessments, similar to the TTS criteria for the Seawolf shock trial EIS. The instantaneous pressure criteria were based on certain received peak sound pressure levels (SPL$_{peak}$ or L_{pk}) above which PTS could occur, while the total energy criteria were based on received cumulative sound exposure levels (SEL$_{cum}$ or L_E) above which PTS could occur. The SEL$_{cum}$ criteria also incorporated the M-weighting functions such that the sound source's frequency content, as well as its broadband sound levels, were considered, while the SPL$_{peak}$ criteria only considered the overall broadband sound levels absent frequency weighting. The authors further recommended that whichever criterion is exceeded first (e.g., that which resulted in the largest impact zones) should be used as the operative injury criterion. The specific PTS thresholds were established by interpreting the available marine mammal TTS data and adding 6 dB to the TTS thresholds for the peak pressure metrics and 15 dB for the energy metrics [88]. For marine mammals where TTS data were lacking, such as LF cetaceans, their auditory anatomy (e.g., [66]) was considered along with extrapolations from MF cetacean TTS values [88].

For the onset of behavioral disturbance thresholds for single pulses, similar dual criteria were recommended based on marine mammal TTS-onset thresholds for cetaceans, pinnipeds in water, and pinnipeds in air [88]. For sound exposure from multiple pulses and nonpulses, Southall et al. [88] were not able to recommend a single value for each functional hearing group and acoustic metric. Instead, the authors conducted an extensive literature review and developed a "severity scale" system to rank the observed behavioral responses of marine mammals (both free-ranging and captive) exposed to a variety of anthropogenic sound with differing received SPL$_{rms}$ values [88]. At the time, the authors argued against using a single value for onset of behavioral disturbance from multiple pulses and nonpulses due to the vast variability and context-specific nature of animal responses to those types of sound.

The concepts from Southall et al.'s [88] recommendations soon became widely accepted and served as the basis for a renewed approach for conducting marine mammal impact assessments. In 2012[4], the Navy released its first comprehensive compilation of the various criteria and thresholds for assessing impacts on marine mammals from acoustic and explosive sources [92]. Those criteria further updated the Southall et al. [88] recom-

---

[4] At the same time, Wood et al. [90] developed simple probabilistic dose response functions for seismic surveys based on a given received level at 90, 50, and 10 percent response rates for porpoises and beaked whales, migrating mysticetes, and all other species—the dose response functions also were intended to be used with the M-weighting functions. Variations of the Wood et al. [90] dose response functions have been used only once by the U.S. regulatory community [91].

mendations by modifying the M-weighting functions for cetaceans. Specifically, the Navy developed Type II weighting functions to account for both Type I weighting functions that were the same as, or similar to, M-weighting functions from Southall et al. [88] and an equal loudness weighting function that incorporated the increased susceptibility to sound observed by Finneran and Schlundt [93] in bottlenose dolphins. The Navy's 2012 criteria also divided pinnipeds into two functional hearing groups: phocids (eared seals or "true seals") and otariids (sea lions and fur seals) and added sirenians (manatees and dugong) to the phocid functional hearing group and odobenids (walruses), mustelids (sea otters and marine otters), and ursids (polar bear) to the otariid functional hearing group. In addition, the Navy's 2012 criteria consolidated various criteria into a set of six categories that were specific to underwater explosive sources: mortality, slight lung injury, gastrointestinal (GI) tract injury, PTS, TTS, and behavioral disturbance [92][5]. Instead of using NMFS's generic 120 and 160 dB re 1 μPa_{rms} thresholds for onset of behavioral disturbance for acoustic sources, the Navy's 2012 criteria relied on behavioral response functions (BRFs)[6] that were based on probabilistic Feller [94] functions it developed for its TAP I EISs[7] for all functional hearing groups except beaked whales and harbor porpoises. The Navy used unweighted 140 and 120 dB re 1 μPa_{rms} thresholds for assessing impacts on beaked whales and harbor porpoises, respectively, from acoustic sources. For explosive sources, the Navy used behavior thresholds that were 5 dB less than the TTS thresholds for each functional hearing group[8]. The Navy's 2012 criteria and thresholds were used in multiple Phase II EISs for the Navy's training and testing activities.

In 2016, the Navy drafted another technical report that provided updated marine mammal auditory weighting functions based on the human 40-phon equal-loudness function (or "A-weighting") and revised TTS and PTS thresholds [96]. Besides modifying various variables, methods, and functions for determining onset of TTS and PTS, the report also established sirenians as a separate functional hearing group [96]. The Navy also provided a singular equation to determine the weighting function amplitude, $W(f)$ in dB, at a given frequency, f in kHz:

$$W(f) = C + 10 \log_{10} \left\{ \frac{(f/f_1)^{2a}}{\left[1 + (f/f_1)^2\right]^a \left[1 + (f/f_2)^2\right]^b} \right\} \tag{1}$$

where C, f_1, f_2, a and b are constants that define the shape of the filter for each functional hearing groups [96].

In the same year, the Navy's 2016 in-water criteria were incorporated into NMFS's technical guidance for assessing the effects of underwater sound on species under its jurisdiction (i.e., cetaceans and pinnipeds except odobenids) [97]. In 2018, NMFS released a revision to its 2016 technical guidance [98]. However, there were no substantial differences in the weighting functions or thresholds between the 2018 revision[9] and the 2016 version. These criteria and thresholds currently are used by U.S. regulatory agencies for marine mammal impact assessments, primarily PTS for Level A harassment under the MMPA.

---

5 The Navy used many of the same criteria in its EISs for previous shock trials and for training and testing activities analyzed under the Tactical Training Theater Assessment and Planning (TAP I) documents.

6 Along with Type I weighting functions.

7 For more than two decades, the Navy also has used a metric it has termed "single ping equivalent" (SPE) to estimate behavioral responses of marine mammals to Surveillance Towed Array Sensor System Low Frequency Active (SURTASS LFA) sonar. SPE is a quasi-metric that the Navy has used to apply its SPE-based behavioral risk function, even though the metric is not based on any sort of physical quantity nor is it recognized by either the American National Standards Institute or the International Organization for Standardization. The U.S. Marine Mammal Commission, an independent oversight agency, has reviewed the deficiencies and inappropriatness of SPE and the underlying behavioral risk function [95].

8 The Navy, and thus NMFS, maintained and continues to maintain that the behavior thresholds only apply to multiple underwater detonations, not single detonations regardless of the net explosive weight.

9 Southall et al. [99] also recommended that the same weighting functions and TTS and PTS thresholds be used. However, the authors termed MF cetaceans as HF cetaceans and HF cetaceans as very high frequency (VHF) cetaceans. Southall et al. [99] developed a modified nomenclature that accounted for additional subdivisions within the LF and HF cetacean functional hearing groups but acknowledged that there were insufficient data to define further the exposure criteria within those subdivisions.

TTS thresholds are considered Level B harassment under the MMPA and are used to assess impacts only from acoustic sources used in military readiness activities and from explosive sources used both in military readiness and construction activities. A summary of the current onset TTS and PTS thresholds for marine mammals, which originated from the Navy's 2016 technical report, is provided in Table 1.

Table 1. Summary of current marine mammal onset TTS and PTS thresholds (data from [96]). SEL_{cum} thresholds in dB re 1 μPa^2-s, SPL_{peak} thresholds in dB re 1 μPa_{peak}.

Functional Hearing Group	Impulsive				Non-Impulsive	
	TTS		PTS		TTS	PTS
	SEL_{cum}	SPL_{peak}	SEL_{cum}	SPL_{peak}	SEL_{cum}	SEL_{cum}
LF	168	213	183	219	179	199
MF	170	224	185	230	178	198
HF	140	196	155	202	153	173
SI	175	220	190	226	186	206
OW	188	226	203	232	199	219
PW	170	212	185	218	181	201

Notation: LF = low-frequency cetaceans; MF = mid-frequency cetaceans; HF = high-frequency cetaceans; SI = sirenians; OW = otariids in water (also includes odobenids, mustelids, and ursids); PW = phocids in water.

In 2017, the Navy released another round of updates to its marine mammal criteria and thresholds for the Phase III EISs for its training and testing activities [100]. The 2017 technical report retained all previous auditory weighting functions and the TTS and PTS thresholds but revised the Navy's behavior thresholds for acoustic sources[10] and the mortality and slight lung and GI tract injury criteria and thresholds for explosive sources. Most notably, "cut-off distances" were introduced and defined as the distances beyond which significant behavioral responses to acoustic sources are unlikely to occur and harassment under the MMPA would not occur. The cut-off distances were used in conjunction with the unweighted 120 dB re 1 μPa_{rms} threshold that the Navy continues to use for harbor porpoises and the Navy's revised BRFs for all other species. The Bayesian biphasic BRFs explicitly were intended to describe both level- and context-based responses as proposed by Ellison et al. [101]. At higher amplitudes, a level-based response relates the received sound level to the probability of a behavioral response, whereas, at lower amplitudes, sound can cue the presence, proximity, and approach of a sound source and stimulate a context-based response based on factors other than received sound level (e.g., the animal's previous experience, separation distance between sound source and animal, and behavioral state including feeding, traveling) [101].

4. Application of Psychoacoustics in the Marine Mammal Regulatory Scheme: Successes and Deficiencies

Decades of marine mammal psychoacoustic research has provided the much needed scientific basis for marine mammal management and policy. However, incorporation of scientific information into the regulatory scheme has proven to be a protracted process.

Among one of the biggest achievements in the application of best available science was the adoption of marine mammal auditory weighting functions for assessing auditory impacts, particularly for Level A harassment under the MMPA. This was a significant improvement from the previous generic Level A harassment thresholds of 180 and 190 dB re 1 μPa_{rms}. Since the weighting function equation can be easily solved analytically to derive the weighted SEL_{cum} thresholds at a given frequency or frequencies, it is possible to estimate distances at which Level A harassment could occur using a spreadsheet and simplified assumptions [102]. However, such a simple approach for estimating injury

---

[10] For explosive sources, the Navy again used behavior thresholds that were 5 dB less than the TTS thresholds for each functional hearing group and assumed that the behavior thresholds only apply to multiple underwater detonations.

zones may not always be accurate. First, the simplified spreadsheet approach only uses geometrical spreading for sound propagation [7], and under certain circumstances (e.g., a moving source), only spherical spreading is incorporated [103]. Second, the estimation of SEL_{cum} for transient sounds does not consider potential recovery between exposures—but neither does sophisticated modeling. Third, and the most serious shortcoming, the simplified spreadsheet approach assumes a stationary receiver that is exposed for the entire duration of the sound exposure, which can last as long as 24 h. Often times, the distances at which Level A harassment can occur exceed the distances at which Level B harassment, specifically behavioral responses, occur [104–108]. To address this issue, as well as to calculate more realistic marine mammal take estimates, several models have been developed to incorporate simulations of marine mammal movements [109–114]. Those models include: Acoustic Integration Model (AIM) [109], Marine Mammal Movement and Behavior (3MB)[11], Navy Acoustic Effects Model (NAEMO) [110,111], and JASCO Animal Simulation Model Including Noise Exposure (JASMINE) [115]. The parameters that seed the virtual marine mammals, or animats, are based on known animal distribution, behavior, and movement under various conditions [116–118]. Animat modeling can be embedded within simple spreadsheet tools to better inform the distances at which Level A harassment can occur and to avoid results that are contrary to common sense.

Regarding behavioral harassment and despite the large volume of studies conducted in the past few decades on marine mammal behavioral response to anthropogenic noise exposure, regulatory agencies continue to use the existing rudimentary thresholds of 120 and 160 dB re 1 μPa_{rms} for onset of behavioral disturbance or Level B harassment under the MMPA for all activities except military readiness and underwater or confined detonations. The overly simplified threshold of 160 dB re 1 μPa_{rms} that originally was intended to apply only to impulsive sounds has been systematically applied to all intermittent sounds, both impulsive and non-impulsive, by one regulatory agency, e.g., [119,120] and to all sounds—continuous, intermittent, impulsive, and non-impulsive[12]—by another agency, e.g., [121–123]. The rudimentary thresholds are not only outdated and not based on best available science, but they are being used in a manner that was never intended.

The Navy has made considerable progress in updating its behavior thresholds by way of its Bayesian BRFs for acoustic sources. The Bayesian BRFs were based on more applicable data and were a much needed update to the Navy's original Feller [94] function BRFs. However, use of cut-off distances by the Navy and subsequently NMFS has completely undermined the Bayesian BRFs' original intent. Specifically, inclusion of additional cut-off distances[13] contradicts the data underlying the Bayesian BRFs, negates the intent of the functions themselves, and ultimately underestimates the numbers of takes, e.g., [124]. To investigate this issue further, Tyack and Thomas [125] compared results between setting a threshold where 50 percent of the animals respond and using the actual Bayesian BRF. Setting the threshold at a 50 percent response led to an underestimation of effect by a factor of 280 [125]. Given that the arbitrary cut-off distances were synonymous with an up to 45 percent response, the behavioral impacts and numbers of takes of the various species would have been underestimated as well [124]. As noted by Tyack and Thomas [125], given the shape of the dose–response function and how efficiently sound propagates in the ocean, the number of animals that are predicted to have a low probability of response may in fact represent the dominant impact from a given sound source.

For explosive sources, the behavior thresholds have yet to be updated and continue to be based on a value that was derived from observed onset behavioral responses of captive bottlenose dolphins during non-impulsive TTS testing of 1 sec tones [57]. In addition, the regulatory community only applies the behavior thresholds to multiple detonations and assumes that marine mammals would not be impacted by single detonations, including

---

[11] 3MB is available at http://oalib.hlsresearch.com/Sound%20and%20Marine%20Mammals/3MB%20HTML.htm (accessed on 29 March 2021).

[12] Including drilling, vibratory pile driving and removal, dynamic positioning systems and other vessel sounds.

[13] Furthermore, the distances themselves are unsubstantiated.

those with high net explosive weights. The justification for such an assumption is lacking. Moreover, the derivation of mortality and serious injury (GI tract and slight lung) thresholds continue to be based on terrestrial and domestic animals. While it is clear that directed research on these types of adverse impacts likely would not be permitted for marine mammals, the applicability of the previous studies to marine mammals is questionable. Such deficiencies highlight the importance of prompt retrieval, imaging, and necropsy of animals unintentionally killed by underwater detonations, i.e., common dolphins killed at Silver Strand [126].

Despite the availability of tools to assess the impacts of sound on marine mammals, the regulatory community has yet to incorporate the latest scientific information or use the tools (e.g., animats) and models (e.g., sound propagation models) in-house for its impact assessments [7]. This has prevented regulatory agencies from determining more realistic estimates when calculating harassment takes. Rather, the agencies generally use the simplest arithmetic method of calculating marine mammal takes (N) using the equation [127]:

$$N = D \times A \times T \tag{2}$$

where D is the animal density (number/km^2) in the area, A is the ensonified area (km^2), and T is the project duration (days).

The failure to incorporate information on marine mammal behavior, distribution, and densities to ground-truth arithmetic results has often yielded unrealistic take estimates. For example, when assessing the potential behavioral disturbance of harbor seals from a small mooring float construction project at Sentinel Island in Alaska that involved installing up to six 24 in piles on up to six days, one regulatory agency estimated that a total of 36,180 harbor seal takes could occur, among a presumed local population of approximately 134 animals [128]. Since the same regulatory agency assumes that a marine mammal can be taken only once on a given day, the maximum number of takes for the project would have been less than 1000 seals. Although determining whether and how to use pinniped haul-out counts appropriately, when such counts should be adjusted based on haul-out correction factors or time spent at sea (e.g., [124,129–132]), and whether certain animal movement models (e.g., [133]) are appropriate may be challenging, such evaluations are necessary for ensuring that take estimates are informed by the basic biology of the species.

5. Future Needs in Marine Mammal Psychoacoustic Research

Despite the achievements made in the first 20 years of the 21st century toward understanding marine mammal NITS, auditory effects on many taxa (or functional hearing groups) have still not been studied empirically. Specifically, there are no audiograms or TTS data from mysticetes (LF cetaceans). Hearing sensitivities of those species only have been derived theoretically based on anatomical observation and modeling [134–139].

Similarly, for many large odontocetes, such as the sperm whale (*Physeter macrocephalus*) and beaked whales (Ziphiidae), there are very few measured audiograms [140–142]. Therefore, hearing sensitivity and TTS thresholds for those species are largely extrapolated from a few species of delphinids (MF cetaceans). The same is true for HF cetaceans, which include a number of species from different taxa but the composite audiogram was derived from the harbor porpoise (*Phocoena phocoena*), finless porpoise (*Neophocoena pocaenoides*), and three species of river dolphins [88]. Data similarly are scant for otariids, walruses, sea otters, and sirenians and non-existent for polar bears [88]. Further research on audiograms and TTS for data-poor marine mammal species will provide further insights needed to validate or revise the existing weighting functions used by the regulatory community [99].

While it has been recognized that impulsive sounds generally are more harmful than non-impulsive sounds in terms of auditory effects [143–148], there are certain situations where both impulsive and non-impulsive sound occur simultaneously. For example, down-the-hole (DTH) pile installation is one such source that generates complex sounds [149]. The authors are unaware of any NITS study that has been conducted on marine mammals exposed to complex sounds, despite the fact that studies on human and terrestrial species have shown that exposure to such sounds is more detrimental than to pure non-impulsive continuous sounds [150–154].

Additionally, neither sufficient nor appropriate data currently exist to develop sophisticated BRFs for sources other than MFAS [100,155,156]. For example, data are lacking regarding near-field behavioral response effects from seismic surveys and in general for behavioral responses of species other than mysticetes and sperm whales, e.g., [157–160]. Similarly, near-field behavioral response data are lacking for pile-driving and -removal activities for all species, while far-field data generally exist only for harbor porpoises and harbor seals, primarily from studies at large-scale wind farms in Europe where larger diameter piles are installed[14] than for coastal construction projects, e.g., [162,163]. A plethora of monitoring data are available for coastal construction projects in the United States[15]. However, much of the data have not been collected in a manner or with sufficient specificity to determine what, if any, behavioral response may have occurred at a given received level. Moreover, behavioral response data are lacking for LFAS sources, such as SURTASS LFA sonar, and underwater and confined detonations. Further research is needed to inform source-specific[16] BRFs for the various groups of marine mammals.

The interpretation of behavioral response data in relation to biologically significant impacts remains a challenge. Many factors, such as age and sex, behavioral context, motivation, and naivety of the animal need to be considered when interpreting behavioral response, or the lack thereof [101]. Many of these factors cannot be easily quantified (see [88]). Regardless, the current step function thresholds for the onset of behavioral disturbance for two generalized sound types are overly simplistic. At a minimum and until such time that more sophisticated BRFs are developed, dose response functions similar to those from Wood et al. [90] should be adopted to better assess behavioral disturbance of marine mammals. Researchers developing BRFs also should provide quantile response data at given received levels (see Table IV in Miller et al. [156]) to allow regulators that are unable to implement dose response functions to implement the quantiles using simple spreadsheets [125].

Quantifying how sub-lethal impacts, particularly behavioral disturbance, translate to population-level impacts has been challenging for the regulatory community. In more recent years, frameworks have been developed and various case studies have been investigated to link quantitatively behavioral responses of individuals to impacts to the population as a whole, see [164–170]. For many marine mammal populations, baseline data necessary to inform such analyses are lacking [171]. As such, basic demographic, physiological, and health data should be collected to inform future population-level impact assessments.

Compared to behavioral disturbance, auditory masking is a relatively less studied area in marine mammal psychoacoustic research [172,173]. Auditory masking of biologically important acoustic signals can lead to a reduced communication space [14,174,175] and, for species that largely rely on acoustic cues for their life functions, should be considered a significant issue in marine mammal conservation [176,177]. This is particularly true in regard to vessel traffic, which is not currently regulated by U.S. regulatory agencies.

Marine mammals exposed to elevated background sound have been observed to change their vocal behaviors to compensate, including increasing the length or repetition rates of the calls [160,178–181], shifting call frequencies [182], or increasing call intensities [181,183–189] (which also is known as Lombard effect [190]). It is likely that some of these vocal behavioral changes could be energetically costly [191–193].

---

[14] Acoustic deterrent devices also are used to deter marine mammals from close ranges, which add a confounding factor to any behavioral response data collected [161].

[15] https://www.fisheries.noaa.gov/national/marine-mammal-protection/incidental-take-authorizations-construction-activities (accessed on 29 March 2021).

[16] Or groups of similar sources. For example, BRFs could be developed for vibratory pile driving and drilling combined or BRFs for seismic surveys could apply to underwater and confined detonations if data are lacking.

While there are many studies on the cause–effect relationship between chronic sound exposure and physiological stress and overall health in humans [194–199], fewer studies have been conducted on marine mammals. Based on evidence from terrestrial mammals and humans, it has long been suspected chronic sound exposure is a potential source of stress for marine mammals [200]. Although numerous techniques are widely used to assess marine mammal health status by examining stress-related hormone biomarkers or gene expression index [201–205], the cause and effect that links sound exposure to physiological impacts or health status are often difficult to ascertain due to the compounding nature of the stressors.

It also is worth noting that anthropogenic sound is just one of the many stressors that may affect marine mammals. As the marine environment is undergoing unprecedented changes, cumulative effects arising from a wide range of stressors, such as climate change, overfishing, pollution, habitat loss, and ocean acidification, must be considered [206–209]. The specific topic of cumulative impacts is beyond the scope of this paper. Nevertheless, additional research is needed to incorporate and evaluate multiple stressors and assess the cumulative impact of those stressors on marine mammals, similar to what has been described by the National Academies of Sciences [207].

Last but not the least, all the scientific and technological advances in marine mammal psychoacoustics will not improve management and implementation of appropriate conservation measures if regulatory agencies do not have the expertise and skills to translate science into policy. It is encouraging that several government agencies, such as the Navy [210,211] and BOEM[17], have taken tremendous steps in building their internal capacity by funding underwater acoustic studies and conducting modeling for environmental impact assessments, while other agencies have yet to do so. All agencies with regulatory responsibilities for evaluating the potential impacts of sound on marine mammals must make it a priority to bolster their internal capacity, in terms of both determining whether the analyses provided are scientifically sound and incorporating the most recent and best available science into their marine mammal take assessments. A lack of science-based management can have legal and, more importantly, conservation implications for marine mammals, including for those populations that are declining.

Author Contributions: Conceptualization, S.G. and T.B.; writing—original draft, S.G. and T.B.; review and revisions, S.G. and T.B. All authors have read and agreed to the published version of the manuscript.

Funding: This research received no external funding.

Acknowledgments: The authors thank Peter Thomas for a thorough review of and comments on this manuscript. The authors also appreciate the constructive comments from two anonymous reviewers.

Conflicts of Interest: The authors declare no conflict of interest.

References

1. Duarte, C.M.; Chapuis, L.; Collin, S.P.; Costa, D.P.; Devassy, R.P.; Eguiluz, V.M.; Erbe, C.; Gordon, T.A.C.; Halpern, B.S.; Harding, H.R.; et al. The soundscape of the anthropocene ocean. *Science* **2021**, *371*. [CrossRef] [PubMed]
2. National Research Council. *Marine Mammal. Populations and Ocean. Noise: Determining When Noise Causes Biologically Significant Effects*; National Academies Press: Washington, DC, USA, 2005.
3. Nowacek, D.P.; Thorne, L.H.; Johnston, D.W.; Tyack, P.L. Responses of cetaceans to anthropogenic noise. *Mamm. Rev.* **2007**, *37*, 81–115. [CrossRef]
4. National Research Council. *Low-Frequency Sound and Marine Mammals: Current Knowledge and Research Needs*; The National Academies Press: Washington, DC, USA, 1994.
5. Shannon, G.; McKenna, M.F.; Angeloni, L.M.; Crooks, K.R.; Fristrup, K.M.; Brown, E.; Warner, K.A.; Nelson, M.D.; White, C.; Briggs, J.; et al. A synthesis of two decades of research documenting the effects of noise on wildlife. *Biol. Rev.* **2016**, *91*, 982–1005. [CrossRef]

---

[17] BOEM recently established the Center for Marine Acoustics that is staffed with experts to conduct underwater acoustic and animat modeling in support of its environmental impact assessments. https://www.boem.gov/center-marine-acoustics (accessed on 21 March 2021).

6. Richardson, W.J.; Malme, C.I.; Thomson, D.H.; Greene, C.R. *Marine Mammals and Noise*; Academic Press: San Diego, CA, USA, 1995.
7. Guan, S.; Brookens, T.; Vignola, J. Use of underwater acoustics in marine conservation and policy: Previous advances, current status, and future needs. *J. Mar. Sci. Eng.* **2021**, *9*, 173. [CrossRef]
8. Lasky, M. Review of undersea acoustics to 1950. *J. Acoust. Soc. Am.* **1977**, *61*, 283–297. [CrossRef]
9. Bjørnø, L. Features of underwater acoustics from Aristotle to our time. *Acoust. Phys.* **2003**, *49*, 24–30. [CrossRef]
10. Schevill, W.E.; Lawrence, B. Underwater listening to the white porpoise (*Delphinapterus leucas*). *Science* **1949**, *109*, 143–144. [CrossRef] [PubMed]
11. Kellogg, W.N.; Kohler, R.; Morris, H.N. Porpoise sounds as sonar signals. *Science* **1953**, *117*, 239–243. [CrossRef] [PubMed]
12. Ray, C.; Watkins, W.A.; Burns, J.J. The underwater song of *Erignathus* (bearded seal). *Zool. N. Y. Zool. Soc.* **1969**, *54*, 79–83.
13. Payne, R.S.; McVay, S. Songs of humpback whales. *Science* **1971**, *173*, 585–597. [CrossRef]
14. Payne, R.; Webb, D. Orientation by means of long range acoustic signaling in baleen whales. *Ann. N. Y. Acad. Sci.* **1971**, *188*, 110–141. [CrossRef] [PubMed]
15. Myrberg, A.A., Jr. Ocean noise and the behavior of marine animals: Relationships and implications. In *Effects of Noise on Wildlife*; Academic Press: New York, NY, USA, 1978; pp. 169–208.
16. Llewellyn, L.G.; Peiser, C. *NEPA and the Environmental Movement: A Brief. History*; National Bureau of Standards: Washington, DC, USA, 1973.
17. Doub, J.P. *The Endangered Species Act.: History, Implementation, Successes, and Controversies*, 1st ed.; CRC Press: Boca Raton, FL, USA, 2016.
18. Roman, J.; Altman, I.; Dunphy-Daly, M.M.; Campbell, C.; Jasny, M.; Read, A.J. The Marine Mammal Protection Act at 40: Status, recovery, and future of U.S. marine mammals. *Ann. N. Y. Acad. Sci.* **2013**, *1286*, 29–49. [CrossRef] [PubMed]
19. Greenwald, N.; Suckling, K.F.; Hartl, B.; Mehrhoff, L.A. Extinction and the U.S. Endangered Species Act. *PeerJ* **2019**, *7*. [CrossRef] [PubMed]
20. Gisiner, R.C. *Workshop on the Effects of Anthropogenic Noise in the Marine Environment*; Office of Naval Research: Arlington, VA, USA, 1998.
21. Goertner, J.F. *Prediction of Underwater Explosion Safe. Range for Sea Mammals*; Naval Surface Weapons Center (CODE R15); White Oak: Silver Spring, MD, USA, 1982.
22. Richmond, D.; Yelverton, J.; Fletcher, E. *Far-Field Underwater-Blast Injuries Produced by Small Charges*; Defense Nuclear Agency: Washington, DC, USA, 1973.
23. Yelverton, J.T.; Richmond, D.R.; Fletcher, E.R.; Jones, R.K. *Safe. Distances from Underwater Explosions for Mammals and Birds*; Lovelace Foundation for Medical Education and Research: Albuquerque, NM, USA, 1973.
24. Ketten, D.R. Estimates of blast injury and acoustic trauma zones for marine mammals from underwater explosions. In *Sensory Systems of Aquatic Mammals*; Kastelein, R.A., Thomas, J.A., Nachtigall, P.E., Eds.; De Spil Publishers: Woerden, The Netherlands, 1995; pp. 391–407; ISBN 90-72743-05-9.
25. Todd, S.; Lien, J.; Marques, F.; Stevick, P.; Ketten, D. Behavioural effects of exposure to underwater explosions in humpback whales (*Megaptera novaeangliae*). *Can. J. Zool.* **1996**, *74*, 1661–1672. [CrossRef]
26. U.S. Navy. *Final Environmental Impact Statement: Shock Testing the Seawolf Submarine*; Naval Facilities Engineering Command, Southern Division: North Charleston, SC, USA, 1998.
27. Richardson, W.J. *Behavior, Disturbance Responses and Distribution of Bowhead Whales Balaena mysticetus in the Eastern Beaufort Sea, 1982*; U.S. Department of the Interior, Minerals Management Service: Reston, VA, USA, 1983.
28. Malme, C.I.; Miles, P.R.; Clark, C.W.; Tyack, P.; Bird, J.E. *Investigations of the Potential Effects of Underwater Noise from Petroleum Industry Activities on Migrating Gray Whale Behavior*; U.S. Department of the Interior, Minerals Management Service: Anchorage, AK, USA, 1983.
29. Malme, C.I.; Miles, P.R.; Clark, C.W.; Tyack, P.; Bird, J.E. *Investigations of the Potential Effects of Underwater Noise from Petroleum Industry Activities on Migrating Gray Whale Behavior—Phase II: January 1984 Migration*; U.S. Department of the Interior, Minerals Management Service: Anchorage, AK, USA, 1984.
30. Ljungblad, D.K.; Würsig, B.; Reeves, R.R.; Clarke, J.; Greene, C.R., Jr. *Fall 1983 Beaufort Sea Seismic Monitoring and Bowhead Whale Behavior Studies*; Minerals Management Service: Anchorage, AK, USA, 1984.
31. Malme, C.I.; Miles, P.; Tyack, P.; Clark, C.; Bird, J.E. *Investigations of the Potential Effects of Underwater Noise from Petroleum Industry Activities on Feeding Humpback Whale Behavior*; U.S. Department of the Interior, Minerals Management Service: Anchorage, AK, USA, 1985.
32. Richardson, W.J.; Fraker, M.A.; Würsig, B.; Wells, R.S. Behaviour of bowhead whales *Balaena mysticetus* summering in the Beaufort Sea: Reactions to industrial activities. *Biol. Conserv.* **1985**, *32*, 195–230. [CrossRef]
33. Fraker, M.A.; Ljungblad, D.K.; Richardson, W.J.; Van Schoik, D.R. *Bowhead Whale Behavior in Relation to Seismic Exploration, Alaska Beaufort Sea, Autumn 1981*; U.S. Department of the Interior, Minerals Management Service: Reston, VA, USA, 1985.
34. Richardson, W.J. *Behavior, Disturbance Responses and Distribution of Bowhead Whales Balaena mysticetus in the Eastern Beaufort Sea. 1980–1984*; LGL Ecological Research Associates, Inc.: Bryan, TX, USA, 1985.
35. Richardson, W.J.; Würsig, B.; Greene, C.R. Reactions of bowhead whales, *Balaena mysticetus*, to seismic exploration in the Canadian Beaufort Sea. *J. Acoust. Soc. Am.* **1986**, *79*, 1117–1128. [CrossRef] [PubMed]

36. Richardson, W.J.; Davis, R.A.; Evans, C.R.; Ljungblad, D.K.; Norton, P. Summer distribution of bowhead whales, *Balaena mysticetus*, relative to oil industry activities in the Canadian Beaufort Sea, 1980–1984. *Arctic* **1987**, *40*, 93–104. [CrossRef]
37. Ljungblad, D.K.; Würsig, B.; Swartz, S.L.; Keene, J.M. Observations on the behavioral responses of bowhead whales (*Balaena mysticetus*) to active geophysical vessels in the Alaskan Beaufort Sea. *Arctic* **1988**, *41*, 183–194. [CrossRef]
38. Johnson, S.R.; Burns, J.J.; Malme, C.I.; Davis, R.A. *Synthesis of Information on the Effects of Noise and Disturbance on Major Haulout Concentrations of Bering Sea Pinnipeds*; U.S. Department of the Interior, Minerals Management Service: Anchorage, AK, USA, 1989.
39. Richardson, W.J.; Greene, C.R., Jr.; Hanna, J.S.; Koski, W.R.; Miller, G.W.; Patenaude, N.J.; Smultea, M.A. *Acoustic Effects of Oil Production Activities on Bowhead and White Whales Visible during Spring Migration Near Pt. Barrow, Alaska—1991 and 1994 Phases: Sound Propagation and Whale Responses to Playbacks of Icebreaker Noise*; Mineral Management Service: Herndon, VA, USA, 1995.
40. Myrberg, A.A. The effects of man-made noise on the behavior of marine animals. *Environ. Int.* **1990**, *16*, 575–586. [CrossRef]
41. National Research Council. *Marine Mammals and Low-Frequency Sound: Progress Since 1994*; National Academies Press: Washington, DC, USA, 2000.
42. Cavanagh, R.C. *Criteria and Thresholds for Adverse Effects of Underwater Noise on Marine Animals*; Air Force Research Laboratory: Dayton, OH, USA, 2000.
43. High Energy Seismic Survey. *High Energy Seismic Survey Review Process and Interim Operational Guidelines for Marine Surveys Offshore Southern California*; The High Energy Seismic Survey Team: Santa Barbara, CA, USA, 1999.
44. Scholik-Schlomer, A.R. Where the decibels hit the water: Perspectives on the application of science to real-world underwater noise and marine protected species issues. *Acoust. Today* **2015**, *11*, 36–44.
45. Richardson, W.J.; Würsig, B.; Greene, C.R. Reactions of bowhead whales, *Balaena mysticetus*, to drilling and dredging noise in the Canadian Beaufort Sea. *Mar. Environ. Res.* **1990**, *29*, 135–160. [CrossRef]
46. Kryter, K.D. *Effects of Noise on Man*; Academic Press: New York, NY, USA, 1970.
47. Nixon, C.W.; Krantz, D.W.; Johnson, D.L. *Human Temporary Threshold Shift and Recovery from 24 Hour Acoustic Exposures*; Defense Technical Information Center: Fort Belvoir, VA, USA, 1975.
48. Davis, H. Acoustic Trauma in the Guinea Pig. *J. Acoust. Soc. Am.* **1953**, *25*, 1180–1189. [CrossRef]
49. Peters, E.N. Temporary Shifts in auditory thresholds of chinchilla after exposure to noise. *J. Acoust. Soc. Am.* **1965**, *37*, 831–833. [CrossRef] [PubMed]
50. Benitez, L.D.; Eldredge, D.H.; Templer, J.W. Temporary threshold shifts in chinchilla: Electrophysiological correlates. *J. Acoust. Soc. Am.* **1972**, *52*, 1115–1123. [CrossRef]
51. U.S. Navy. *Draft Overseas Environmental Impact Statement and Environmental Impact Statement for Surveillance Towed Array Sensor System Low Frequency Active (SURTASS LFA) Sonar*; Department of the Navy, Chief of Naval Operations: Arlington, VA, USA, 1999.
52. Advanced Research Projects Agency. *Final Environmental Impact Statement for the Kauai Acoustic Thermometry of Ocean. Climate Project and Its Associated Marine Mammal. Research Program. (Scientific Research Permit Application [P557E])*; Advanced Research Projects Agency: Arlington, VA, USA, 1995; Volume 1.
53. Kastak, D.; Schusterman, R.J. Temporary threshold shift in a harbor seal (*Phoca vitulina*). *J. Acoust. Soc. Am.* **1996**, *100*, 1905–1908. [CrossRef] [PubMed]
54. Ridgway, S.H.; Carter, D.A.; Smith, R.R.; Kamolnick, T.; Schlundt, C.E. *Behavioral Responses and Temporary Shift in Masked Hearing Threshold of Bottlenose Dolphins, Tursiops truncatus, to 1-Second Tones of 141 to 201 dB re 1 μPa*; Naval Command, Control, and Ocean Surveillance Center, RDT&E Division: San Diego, CA, USA, 1997.
55. Kastak, D.; Schusterman, R.J.; Southall, B.L.; Reichmuth, C.J. Underwater temporary threshold shift induced by octave-band noise in three species of pinniped. *J. Acoust. Soc. Am.* **1999**, *106*, 1142–1148. [CrossRef]
56. Finneran, J.J.; Schlundt, C.E.; Carder, D.A.; Clark, J.A.; Young, J.A.; Gaspin, J.B.; Ridgway, S.H. Auditory and behavioral responses of bottlenose dolphins (*Tursiops truncatus*) and a beluga whale (*Delphinapterus leucas*) to impulsive sounds resembling distant signatures of underwater explosions. *J. Acoust. Soc. Am.* **2000**, *108*, 417–431. [CrossRef]
57. Schlundt, C.E.; Finneran, J.J.; Carder, D.A.; Ridgway, S.H. Temporary shift in masked hearing thresholds of bottlenose dolphins, *Tursiops truncatus*, and white whales, *Delphinapterus leucas*, after exposure to intense tones. *J. Acoust. Soc. Am.* **2000**, *107*, 3496–3508. [CrossRef]
58. Finneran, J.J.; Schlundt, C.E.; Dear, R.; Carder, D.A.; Ridgway, S.H. Temporary shift in masked hearing thresholds in odontocetes after exposure to single underwater impulses from a seismic watergun. *J. Acoust. Soc. Am.* **2002**, *111*, 2929–2940. [CrossRef]
59. Nachtigall, P.E.; Pawloski, J.L.; Au, W.W.L. Temporary threshold shifts and recovery following noise exposure in the Atlantic bottlenosed dolphin (*Tursiops truncatus*). *J. Acoust. Soc. Am.* **2003**, *113*, 3425–3429. [CrossRef]
60. Nachtigall, P.E.; Supin, A.Y.; Pawloski, J.; Au, W.W.L. Temporary threshold shifts after noise exposure in the bottlenose dolphin (*Tursiops truncatus*) measured using evoked auditory potentials. *Mar. Mamm. Sci.* **2004**, *20*, 673–687. [CrossRef]
61. Finneran, J.J.; Schlundt, C.E.; Branstetter, B.; Dear, R.L. Assessing temporary threshold shift in a bottlenose dolphin (*Tursiops truncatus*) using multiple simultaneous auditory evoked potentials. *J. Acoust. Soc. Am.* **2007**, *122*, 1249–1264. [CrossRef]
62. Lucke, K.; Siebert, U.; Lepper, P.A.; Blanchet, M.-A. Temporary shift in masked hearing thresholds in a harbor porpoise (*Phocoena phocoena*) after exposure to seismic airgun stimuli. *J. Acoust. Soc. Am.* **2009**, *125*, 4060–4070. [CrossRef] [PubMed]
63. Popov, V.V.; Klishin, V.O.; Nechaev, D.I.; Pletenko, M.G.; Rozhnov, V.V.; Supin, A.Y.; Sysueva, E.V.; Tarakanov, M.B. Influence of acoustic noises on the white whale hearing thresholds. *Dokl. Biol. Sci.* **2011**, *440*, 332–334. [CrossRef] [PubMed]

64. Popov, V.V.; Supin, A.Y.; Wang, D.; Wang, K.; Dong, L.; Wang, S. Noise-induced temporary threshold shift and recovery in Yangtze finless porpoises *Neophocaena phocaenoides asiaeorientalis*. *J. Acoust. Soc. Am.* **2011**, *130*, 574–584. [CrossRef]

65. Finneran, J.J. Noise-induced hearing loss in marine mammals: A review of temporary threshold shift studies from 1996 to 2015. *J. Acoust. Soc. Am.* **2015**, *138*, 1702–1726. [CrossRef]

66. Wartzok, D.; Ketten, D. Marine Mammal Sensory Systems. In *Biology of Marine Mammals*; Smithsonian Institute Press: Washington, DC, USA, 1999; pp. 117–175.

67. Houser, D.S.; Helweg, D.A.; Moore, P.W.B. A bandpass filter-bank model of auditory sensitivity in the humpback whale. *Aquat. Mamm.* **2001**, *27*, 82–92.

68. Balcomb, K.C., III; Claridge, D.E. A mass stranding of cetaceans caused by naval sonar in the Bahamas. *Bahamas J. Sci.* **2001**, *8*, 1–12.

69. Evans, D.L.; England, G.R. *Joint Interim Report Bahamas Marine Mammal. Stranding Event of 15–16 March 2000*; National Oceanic and Atmospheric Administration: Silver Spring, MD, USA, 2001.

70. Schrope, M. Whale deaths caused by US Navy's sonar. *Nature* **2002**, *415*, 106. [CrossRef] [PubMed]

71. D'Amico, A.; Gisiner, R.C.; Ketten, D.R.; Hammock, J.A.; Johnson, C.; Tyack, P.L.; Mead, J. Beaked whale strandings and naval exercises. *Aquat. Mamm.* **2009**, *35*, 452–472. [CrossRef]

72. Filadelfo, R.; Pinelis, Y.K.; Davis, S.; Chase, R.; Mintz, J.; Wolfanger, J.; Tyack, P.L.; Ketten, D.; D'Amico, A. Correlating military sonar use with beaked whale mass strandings: What do the historical data show? *Aquat. Mamm.* **2009**, *35*, 445–451. [CrossRef]

73. Jepson, P.D.; Arbelo, M.; Deaville, R.; Patterson, I.A.P.; Castro, P.; Baker, J.R.; Degollada, E.; Ross, H.M.; Herráez, P.; Pocknell, A.M.; et al. Gas-bubble lesions in stranded cetaceans. *Nature* **2003**, *425*, 575–576. [CrossRef] [PubMed]

74. Jepson, P.D.; Deaville, R.; Patterson, I.A.P.; Pocknell, A.M.; Ross, H.M.; Baker, J.R.; Howie, F.E.; Reid, R.J.; Colloff, A.; Cunningham, A.A. Acute and chronic gas bubble lesions in cetaceans stranded in the United Kingdom. *Vet. Pathol.* **2005**, *42*, 291–305. [CrossRef] [PubMed]

75. Fernández, A.; Edwards, J.F.; Rodríguez, F.; de los Monteros, A.E.; Herráez, P.; Castro, P.; Jaber, J.R.; Martín, V.; Arbelo, M. "Gas and fat embolic syndrome" involving a mass stranding of beaked whales (family Ziphiidae) exposed to anthropogenic sonar signals. *Vet. Pathol.* **2005**, *42*, 446–457. [CrossRef]

76. Southall, B. *Biological and Behavioral Response Studies of Marine Mammals in Southern California, 2011 ("SOCAL-11") Final Project Report*; Naval Postgraduate School: Monterey, CA, USA, 2012.

77. Tyack, P.; Gordon, J.; Thompson, D. Controlled-exposure experiments to determine the effects of noise on marine mammals. *Mar. Technol. Soc. J.* **2003**, *37*, 39–51. [CrossRef]

78. Tyack, P.L.; Zimmer, W.M.X.; Moretti, D.; Southall, B.L.; Claridge, D.E.; Durban, J.W.; Clark, C.W.; D'Amico, A.; DiMarzio, N.; Jarvis, S.; et al. Beaked whales respond to simulated and actual Navy sonar. *PLoS ONE* **2011**, *6*, e17009. [CrossRef]

79. Falcone, E.A.; Schorr, G.S.; Watwood, S.L.; DeRuiter, S.L.; Zerbini, A.N.; Andrews, R.D.; Morrissey, R.P.; Moretti, D.J. Diving behaviour of Cuvier's beaked whales exposed to two types of military sonar. *R. Soc. Open Sci.* **2017**, *4*, 170629. [CrossRef]

80. Sivle, L.D.; Kvadsheim, P.H.; Fahlman, A.; Lam, F.P.A.; Tyack, P.L.; Miller, P.J.O. Changes in dive behavior during naval sonar exposure in killer whales, long-finned pilot whales, and sperm whales. *Front. Physiol.* **2012**, *3*. [CrossRef]

81. DeRuiter, S.L.; Southall, B.L.; Calambokidis, J.; Zimmer, W.M.X.; Sadykova, D.; Falcone, E.A.; Friedlaender, A.S.; Joseph, J.E.; Moretti, D.; Schorr, G.S.; et al. First direct measurements of behavioural responses by Cuvier's beaked whales to mid-frequency active sonar. *Biol. Lett.* **2013**, *9*, 20130223. [CrossRef]

82. Goldbogen, J.A.; Southall, B.L.; DeRuiter, S.L.; Calambokidis, J.; Friedlaender, A.S.; Hazen, E.L.; Falcone, E.A.; Schorr, G.S.; Douglas, A.; Moretti, D.J.; et al. Blue whales respond to simulated mid-frequency military sonar. *Proc. R. Soc. B Biol. Sci.* **2013**, *280*, 20130657. [CrossRef]

83. Stimpert, A.K.; DeRuiter, S.L.; Southall, B.L.; Moretti, D.J.; Falcone, E.A.; Goldbogen, J.A.; Friedlaender, A.; Schorr, G.S.; Calambokidis, J. Acoustic and foraging behavior of a Baird's beaked whale, *Berardius bairdii*, exposed to simulated sonar. *Sci. Rep.* **2014**, *4*, 7031. [CrossRef] [PubMed]

84. Sivle, L.D.; Kvadsheim, P.H.; Curé, C.; Isojunno, S.; Wensveen, P.J.; Lam, F.-P.A.; Visser, F.; Kleivane, L.; Tyack, P.L.; Harris, C.M.; et al. Severity of expert-identified behavioural responses of humpback whale, minke whale, and northern bottlenose whale to naval sonar. *Aquat. Mamm.* **2015**, *41*, 469–502. [CrossRef]

85. Southall, B.; Nowacek, D.; Miller, P.; Tyack, P. Experimental field studies to measure behavioral responses of cetaceans to sonar. *Endang. Species. Res.* **2016**, *31*, 293–315. [CrossRef]

86. Curé, C.; Isojunno, S.; Visser, F.; Wensveen, P.; Sivle, L.; Kvadsheim, P.; Lam, F.; Miller, P. Biological significance of sperm whale responses to sonar: Comparison with anti-predator responses. *Endang. Species. Res.* **2016**, *31*, 89–102. [CrossRef]

87. Southall, B.L.; DeRuiter, S.L.; Friedlaender, A.; Stimpert, A.K.; Goldbogen, J.A.; Hazen, E.; Casey, C.; Fregosi, S.; Cade, D.E.; Allen, A.N.; et al. Behavioral responses of individual blue whales (*Balaenoptera musculus*) to mid-frequency military sonar. *J. Exp. Biol.* **2019**, *222*. [CrossRef] [PubMed]

88. Southall, B.L.; Bowles, A.E.; Ellison, W.T.; Finneran, J.J.; Gentry, R.L.; Greene, C.R., Jr.; Kastak, D.K.; Ketten, D.R.; Miller, J.H.; Nachtigall, P.E.; et al. Marine mammal noise exposure criteria: Initial scientific recommendations. *Aquat. Mamm.* **2007**, *33*, 411–521. [CrossRef]

89. Harris, C.M. *Handbook of Acoustical Measurements and Noise Control*, 3rd ed.; American Institute of Physics: Woodbury, NY, USA, 1998; ISBN 978-1-56396-774-0.

90. Wood, J.; Southall, B.; Tollit, D. *PG&E Offshore 3-D Seismic Survey Project EIR—Marine Mammal. Technical Report*; SMRU Ltd.: Fife, Scotland, UK, 2012.

91. National Marine Fisheries Service. Taking and importing marine mammals; Taking marine mammals incidental to geophysical surveys related to oil and gas activities in the Gulf of Mexico. *Fed. Reg.* **2021**, *86*, 5322–5450.

92. Finneran, J.J.; Jenkins, A.K. *Criteria and Thresholds for U.S. Navy Acoustic and Explosive Effects Analysis*; Space and Naval Warfare Systems Center Pacific: San Diego, CA, USA, 2012.

93. Finneran, J.J.; Schlundt, C.E. Subjective loudness level measurements and equal loudness contours in a bottlenose dolphin (*Tursiops truncatus*). *J. Acoust. Soc. Am.* **2011**, *130*, 3124–3136. [CrossRef]

94. Feller, W. *An Introduction to Probability Theory and Its Applications*, 3rd ed.; John Wiley & Sons, Inc.: New York, NY, USA, 1968; Volume 1.

95. Marine Mammal Commission. *Comments and Recommendations from the U.S. Marine Mammal. Commission on National Marine Fisheries Service's Proposed Issuance of Regulations to the U.S. Navy to Take Marine Mammals Incidental to Conducting Training, Testing, and Routine Military Operations that Use Surveillance Towed Array Sensor System Low Frequency Active Sonar*; Marine Mammal Commission: Bethesda, MD, USA, 2019. Available online: https://www.mmc.gov/wp-content/uploads/19-04-01-Harrison-Navy-SURTASS-LFA-PR.pdf (accessed on 20 March 2021).

96. Finneran, J.J. *Auditory Weighting Functions and TTS/PTS Exposure Functions for Marine Mammals Exposed to Underwater Noise*; Space and Naval Warfare Systems Center Pacific: San Diego, CA, USA, 2016.

97. National Marine Fisheries Service. *Technical Guidance for Assessing the Effects of Anthropogenic Sound on Marine Mammal. Hearing: Underwater Acoustic Thresholds for Onset of Permanent and Temporary Threshold Shifts*; National Oceanic and Atmospheric Administration: Silver Spring, MD, USA, 2016; p. 178.

98. National Marine Fisheries Service. *2018 Revision to: Technical Guidance for Assessing the Effects of Anthropogenic Sound on Marine Mammal. Hearing (Version 2.0): Underwater Thresholds for Onset of Permanent and Temporary Threshold Shifts*; National Oceanic and Atmospheric Administration: Silver Spring, MD, USA, 2018; p. 167. Available online: https://www.fisheries.noaa.gov/national/marinemammal-protection/marine-mammal-acoustic-technical-guidance (accessed on 29 January 2021).

99. Southall, B.L.; Finneran, J.J.; Reichmuth, C.; Nachtigall, P.E.; Ketten, D.R.; Bowles, A.E.; Ellison, W.T.; Nowacek, D.P.; Tyack, P.L. Marine mammal noise exposure criteria: Updated scientific recommendations for residual hearing effects. *Aquat. Mamm.* **2019**, *45*, 125–232. [CrossRef]

100. U.S. Navy. *Criteria and Thresholds for U.S. Navy Acoustic and Explosive Effects Analysis (Phase III)*; SSC Pacific: San Diego, CA, USA, 2017.

101. Ellison, W.T.; Southall, B.L.; Clark, C.W.; Frankel, A.S. A new context-based approach to assess marine mammal behavioral responses to anthropogenic sounds. *Conserv. Biol.* **2012**, *26*, 21–28. [CrossRef]

102. National Marine Fisheries Service. *Manual for Optional USER SPREADSHEET TOOL (Version 2.2, December) for 2018 Revision to: Technical Guidance for Assessing the Effects of Anthropogenic Sound on Marine Mammal. Hearing (Version 2.0): Underwater Thresholds for Onset of Permanent and Temporary Threshold Shifts*; National Oceanic and Atmospheric Administration: Silver Spring, MD, USA, 2020. Available online: https://media.fisheries.noaa.gov/2020-12/User_Manual%20_DEC_2020_508.pdf?null (accessed on 13 February 2021).

103. Sivle, L.D.; Kvadsheim, P.H.; Ainslie, M.A. Potential for population-level disturbance by active sonar in herring. *ICES J. Mar. Sci.* **2015**, *72*, 558 567. [CrossRef]

104. Marine Mammal Commission. *Comments and Recommendations from the U.S. Marine Mammal. Commission on National Marine Fisheries Service's Proposed Issuance of the Technical Guidance for Assessing the Effects of Anthropogenic Sound on Marine Mammal. Hearing: Underwater Acoustic Thresholds for Onset of Permanent and Temporary Threshold Shifts*; Marine Mammal Commission: Bethesda, MD, USA, 2017. Available online: https://www.mmc.gov/wp-content/uploads/17-07-11-Bettridge-NMFS-Technical-Guidance.pdf (accessed on 20 March 2021).

105. Marine Mammal Commission. *Comments and Recommendations from the U.S. Marine Mammal. Commission on National Marine Fisheries Service's Proposed Issuance of an Incidental Harassment Authorization to Vineyard Wind, LLC, to Take Marine Mammals Incidental to Construction of Commercial Wind Energy Turbines and Associated Facilities off Massachusetts*; Marine Mammal Commission: Bethesda, MD, USA, 2019. Available online: https://www.mmc.gov/wp-content/uploads/19-06-03-Harrison-NMFS-Vineyard-Wind-wind-farm-construction-IHA.pdf (accessed on 20 March 2021).

106. Marine Mammal Commission. *Comments and Recommendations from the U.S. Marine Mammal. Commission on National Marine Fisheries Service's Proposed Issuance of an Incidental Harassment Authorization to Chesapeake Tunnel Joint Venture to take marine mammals incidental to conducting construction activities for the Parallel Thimble Shoal Tunnel Bridge. Project in Virginia*; Marine Mammal Commission: Bethesda, MD, USA, 2019. Available online: https://www.mmc.gov/wp-content/uploads/19-12-26-Harrison-Chesapeake-Tunnel-Joint-Venture-IHA.pdf (accessed on 20 March 2021).

107. Marine Mammal Commission. *Comments and Recommendations from the U.S. Marine Mammal. Commission on National Marine Fisheries Service's Proposed Issuance of an Incidental Harassment Authorization to Virginia Electric and Power Company d/b/a/ Dominion Energy Virginia to Take Marine Mammals Incidental to Construction of Wind Energy Turbines off the Coast. of Virginia*; Marine Mammal Commission: Bethesda, MD, USA, 2020. Available online: https://www.mmc.gov/wp-content/uploads/20-04-15-Harrison-Dominion-wind-energy-construction-IHA.pdf (accessed on 20 March 2021).

108. Marine Mammal Commission. *Comments and Recommendations from the U.S. Marine Mammal. Commission on National Marine Fisheries Service's Proposed Incidental Harassment Authorization Renewal to Port. of Kalama to Renew. its Authorization to Take Marine Mammals Incidental to Construction of the Kalama Manufacturing and Marine Export Facility on the Columbia River in Washington*; Marine Mammal Commission: Bethesda, MD, USA, 2020. Available online: https://www.mmc.gov/wp-content/uploads/20-11-05-Harrison-Port-of-Kalama-IHA-renewal.pdf (accessed on 20 March 2021).

109. Frankel, A.S.; Ellison, W.T.; Buchanan, J. Application of the Acoustic Integration Model (AIM) to predict and minimize environmental impacts. In Proceedings of the Oceans '02 MTS/IEEE, Biloxi, MI, USA, 29–31 October 2002; pp. 1438–1443.

110. Houser, D.S. A method for modeling marine mammal movement and behavior for environmental impact assessment. *IEEE J. Ocean. Eng.* **2006**, *31*, 76–81. [CrossRef]

111. U.S. Navy. *Determination of Acoustic Effects on Marine Mammals and Sea Turtles for the Hawaii-Southern California Training and Testing Environmental Impact Statement/Overseas Environmental Impact Statement*; Naval Undersea Warfare Center Division: Newport, RI, USA, 2012.

112. Ellison, W.; Racca, R.; Clark, C.; Streever, B.; Frankel, A.; Fleishman, E.; Angliss, R.; Berger, J.; Ketten, D.; Guerra, M.; et al. Modeling the aggregated exposure and responses of bowhead whales *Balaena mysticetus* to multiple sources of anthropogenic underwater sound. *Endang. Species. Res.* **2016**, *30*, 95–108. [CrossRef]

113. Blackstock, S.A.; Fayton, J.O.; Hulton, P.H.; Moll, T.E.; Jenkins, K.K.; Kotecki, S.; Henderson, E.; Rider, S.; Martin, C.; Bowman, V. *Quantifying Acoustic Impacts on Marine Mammals and Sea Turtles: Methods and Analytical Approach for Phase III Training and Testing*; Naval Undersea Warfare Center Division: Newport, RI, USA, 2017.

114. Denes, S.L.; Zeddies, D.G.; Weirathmueller, M.J. *Turbine Foundation and Cable Installation at South. Fork Wind Farm: Underwater Acoustic Modeling of Construction Noise*; JASCO Applied Sciences: Silver Spring, MD, USA, 2018.

115. Matthews, M.-N.R.; Ireland, D.S.; Zeddies, D.G.; Brune, R.H.; Pyć, C.D. A modeling comparison of the potential effects on marine mammals from sounds produced by marine vibroseis and air gun seismic sources. *J. Mar. Sci. Eng.* **2021**, *9*, 12. [CrossRef]

116. Shaffer, S.A.; Costa, D.P. A database for the study of marine mammal behavior: Gap analysis, data standardization, and future directions. *IEEE J. Ocean. Eng.* **2006**, *31*, 82–86. [CrossRef]

117. Frankel, A.S.; Vigness-Raposa, K.; Giard, J.; White, A.; Ellison, W.T. Exposures v. individuals: Effects of varying movement patterns and animal behavior on long-term animat model exposure predictions. *Proc. Meet. Acoust.* **2016**, *27*, 10038. [CrossRef]

118. Borcuk, J.R.; Mitchell, G.H.; Watwood, S.L.; Moll, T.E.; Oliveira, E.M.; Robinson, E.R. *Dive Distribution and Group Size Parameters for Marine Species Occurring in the U.S. Navy's Atlantic and Hawaii-Southern California Training and Testing Study Areas*; NUWC-NPT Technical Report; Naval Undersea Warfare Center Division: Newport, RI, USA, 2017.

119. National Marine Fisheries Service. Taking and importing marine mammals; taking marine mammals incidental to Southwest Fisheries Science Center Fisheries Research. *Fed. Reg.* **2020**, *85*, 53606–53640.

120. National Marine Fisheries Service. Takes of marine mammals incidental to specified activities; taking marine mammals incidental to site characterization surveys off the coast of Massachusetts. *Fed. Reg.* **2021**, *86*, 11930–11947.

121. U.S. Fish and Wildlife Service. Marine mammals; incidental take during specified activities; proposed incidental harassment authorization for Pacific walruses and polar bears in Alaska and associated Federal waters. *Fed. Reg.* **2017**, *82*, 25304–25322.

122. U.S. Fish and Wildlife Service. Marine mammals; incidental take during specified activities: Cook Inlet, Alaska. *Fed. Reg.* **2019**, *84*, 10224–10251.

123. U.S. Fish and Wildlife Service. Marine Mammals; Incidental Take During Specified Activities; Proposed Incidental Harassment Authorizations for Northern Sea Otters in Southeast Alaska. *Fed. Reg.* **2019**, *84*, 32932–32945.

124. Marine Mammal Commission. *Comments and Recommendations from the U.S. Marine Mammal. Commission on U.S. Navy's Draft Supplemental Environmental Impact Statement/Overseas Environmental Impact Statement for Training Activities Conducted within the Temporary Maritime Activities Area in the Gulf of Alaska*; Marine Mammal Commission: Bethesda, MD, USA, 2021. Available online: https://www.mmc.gov/wp-content/uploads/21-01-04-Naval-Facilities-Engineering-Command-Northwest-GOA-Phase-III-DSEIS.pdf (accessed on 20 March 2021).

125. Tyack, P.L.; Thomas, L. Using dose–response functions to improve calculations of the impact of anthropogenic noise. *Aquat. Conser. Mar. Freshwater Ecosys.* **2019**, *29*, 242–253. [CrossRef]

126. Danil, K.; St. Leger, J.A. Seabird and dolphin mortality associated with underwater detonation exercises. *Mar. Technol. Soc. J.* **2011**, *45*, 89–95. [CrossRef]

127. National Marine Fisheries Service. Takes of marine mammals incidental to specified activities; taking marine mammals incidental to Washington State Department of Transportation Purdy Bridge Rehabilitation Project, Pierce County, WA. *Fed. Reg.* **2020**, *85*, 81886–81904.

128. National Marine Fisheries Service. Takes of marine mammals incidental to specified activities; taking marine mammals incidental to Gastineau Channel Historical Society Sentinel Island moorage float project, Juneau, Alaska. *Fed. Reg.* **2020**, *85*, 18196–18213.

129. Marine Mammal Commission. *Comments and Recommendations from the U.S. Marine Mammal. Commission on National Marine Fisheries Service's Proposed Issuance for an Incidental Harassment Authorization to Lamont-Doherty Earth Observatory to Take Marine Mammals Incidental to Conducting Two Marine Geophysical Surveys in the North. Pacific Ocean*; Marine Mammal Commission: Bethesda, MD, USA, 2018. Available online: https://www.mmc.gov/wp-content/uploads/18-07-20-Harrison-LDEO-HI-IHA.pdf (accessed on 20 March 2021).

130. Marine Mammal Commission. *Comments and Recommendations from the U.S. Marine Mammal. Commission on National Marine Fisheries Service's Proposed Issuance of Regulations to U.S. Navy to take Marine Mammals Incidental to Conducting Training and Testing Activities in the Hawaii-Southern California Training and Testing Study Area*; Marine Mammal Commission: Bethesda, MD, USA, 2018. Available online: https://www.mmc.gov/wp-content/uploads/18-07-13-Harrison-Navy-HSTT-PR-Phase-III.pdf (accessed on 20 March 2021).
131. Marine Mammal Commission. *Comments and Recommendations from the U.S. Marine Mammal. Commission on U.S. Navy's Draft Supplemental Environmental Impact Statement/Overseas Environmental Impact Statement to Conduct Training and Research, Development, Testing, and Evaluation Activities within the Northwest. Training and Testing Study Area*; Marine Mammal Commission: Bethesda, MD, USA, 2019. Available online: https://www.mmc.gov/wp-content/uploads/19-04-15-Naval-Facilities-Engineering-Command-Northwest-NWTT-DSEIS.pdf (accessed on 20 March 2021).
132. Marine Mammal Commission. *Comments and Recommendations from the U.S. Marine Mammal. Commission on National Marine Fisheries Service's Proposed Issuance of an Incidental Harassment Authorization to U.S. Army Corps of Engineers, Portland District to Take Marine Mammals Incidental to Replacing Dike Markers in the Columbia River. 11 September 2019*; Marine Mammal Commission: Bethesda, MD, USA, 2019. Available online: https://www.mmc.gov/wp-content/uploads/19-09-11-Harrision-USACE-CR-marker-IHA.pdf (accessed on 20 March 2021).
133. Marine Mammal Commission. *Comments and Recommendations from the U.S. Marine Mammal. Commission on National Marine Fisheries Service's Proposed Issuance of an Incidental Harassment Authorization to Jordan Cove Energy Project, LP to Take Marine Mammals Incidental to Construction of the Jordan Cove Liquefied Natural Gas. (LNG) Facility and Ancillary Activities in Coos Bay, Oregon*; Marine Mammal Commission: Bethesda, MD, USA, 2019. Available online: https://www.mmc.gov/wp-content/uploads/19-12-18-Harrison-NMFS-proposed-IHA-Jordan-Cove-LNG_corrected.pdf (accessed on 20 March 2021).
134. Ketten, D.R. Structure and function in whale ears. *Bioacoustics* **1997**, *8*, 103–135. [CrossRef]
135. Parks, S.E.; Ketten, D.R.; O'Malley, J.T.; Arruda, J. Anatomical predictions of hearing in the North Atlantic right whale. *Anat. Rec.* **2007**, *290*, 734–744. [CrossRef]
136. Mooney, T.A.; Yamato, M.; Branstetter, B.K. Chapter Four—Hearing in Cetaceans: From Natural History to Experimental Biology. In *Advances in Marine Biology*; Lesser, M., Ed.; Academic Press: New York, NY, USA, 2012; Volume 63, pp. 197–246.
137. Cranford, T.W.; Krysl, P. Fin whale sound reception mechanisms: Skull vibration enables low-frequency hearing. *PLoS ONE* **2015**, *10*, e0116222. [CrossRef]
138. Tubelli, A.; Zosuls, A.; Ketten, D.; Mountain, D.C. Prediction of a mysticete audiogram via finite element analysis of the middle ear. In *The Effects of Noise on Aquatic Life II*; Advances in Experimental Medicine and Biology; Springer: New York, NY, USA, 2015; Volume 875, pp. 57–59.
139. Tubelli, A.A.; Zosuls, A.; Ketten, D.R.; Mountain, D.C. A model and experimental approach to the middle ear transfer function related to hearing in the humpback whale (*Megaptera novaeangliae*). *J. Acoust. Soc. Am.* **2018**, *144*, 525–535. [CrossRef] [PubMed]
140. Cook, M.L.H.; Varela, R.A.; Goldstein, J.D.; McCulloch, S.D.; Bossart, G.D.; Finneran, J.J.; Houser, D.; Mann, D.A. Beaked whale auditory evoked potential hearing measurements. *J. Comp. Physiol. A* **2006**, *192*, 489–495. [CrossRef] [PubMed]
141. Finneran, J.J.; Houser, D.S.; Mase-Guthrie, B.; Ewing, R.Y.; Lingenfelser, R.G. Auditory evoked potentials in a stranded Gervais' beaked whale (*Mesoplodon europaeus*). *J. Acoust. Soc. Am.* **2009**, *126*, 484–490. [CrossRef]
142. Pacini, A.F.; Nachtigall, P.E.; Quintos, C.T.; Schofield, T.D.; Look, D.A.; Levine, G.A.; Turner, J.P. Audiogram of a stranded Blainville's beaked whale (*Mesoplodon densirostris*) measured using auditory evoked potentials. *J. Exp. Biol.* **2011**, *214*, 2409–2415. [CrossRef] [PubMed]
143. Ward, W.D.; Glorig, A.; Sklar, D.L. Temporary threshold shift from octave-band noise: Applications to damage-risk criteria. *J. Acoust. Soc. Am.* **1959**, *31*, 522–528. [CrossRef]
144. Luz, G.A.; Hodge, D.C. Recovery from impulse-noise induced TTS in monkeys and men: A descriptive model. *J. Acoust. Soc. Am.* **1971**, *49*, 1770–1777. [CrossRef] [PubMed]
145. Sulkowski, W.J.; Lipowczan, A. Impulse noise-induced hearing loss in drop forge operators and the energy concept. *Noise Control. Eng.* **1982**, *18*, 24–29. [CrossRef]
146. Hamernik, R.P.; Patterson, J.H.; Salvi, R.J. The Effect of impulse intensity and the number of impulses on hearing and cochlear pathology in the chinchilla. *J. Acoust. Soc. Am.* **1987**, *81*, 1118–1129. [CrossRef]
147. Dunn, D.E.; Davis, R.R.; Merry, C.J.; Franks, J.R. Hearing loss in the chinchilla from impact and continuous noise exposure. *J. Acoust. Soc. Am.* **1991**, *90*, 1979–1985. [CrossRef]
148. Hamernik, R.P.; Qiu, W. Energy-independent factors influencing noise-induced hearing loss in the chinchilla model. *J. Acoust. Soc. Am.* **2001**, *110*, 3163–3168. [CrossRef]
149. Guan, S.; Miner, R. Underwater noise characterization of down-the-hole pile driving activities off Biorka Island, Alaska. *Mar. Pollut. Bull.* **2020**, *160*, 111664. [CrossRef]
150. Qiu, W.; Davis, B.; Hamernik, R.P. Hearing loss from interrupted, intermittent, and time varying Gaussian noise exposures: The applicability of the equal energy hypothesis. *J. Acoust. Soc. Am.* **2007**, *121*, 1613–1620. [CrossRef]
151. Hamernik, R.P.; Qiu, W.; Davis, B. Hearing loss from interrupted, intermittent, and time Varying non-Gaussian noise exposure: The applicability of the equal energy hypothesis. *J. Acoust. Soc. Am.* **2007**, *122*, 2245–2254. [CrossRef]
152. Zhao, Y.; Qiu, W.; Zeng, L.; Chen, S.; Cheng, X.; Davis, R.I.; Hamernik, R.P. Application of the kurtosis statistic to the evaluation of the risk of hearing loss in workers exposed to high-level complex noise. *Ear Hear.* **2010**, *31*, 527–532. [CrossRef]

153. Qiu, W.; Hamernik, R.P.; Davis, R.I. The value of a kurtosis metric in estimating the hazard to hearing of complex industrial noise exposures. *J. Acoust. Soc. Am.* **2013**, *133*, 2856–2866. [CrossRef] [PubMed]

154. Xie, H.; Qiu, W.; Heyer, N.J.; Zhang, M.; Zhang, P.; Zhao, Y.; Hamernik, R.P. The use of the kurtosis-adjusted cumulative noise exposure metric in evaluating the hearing loss risk for complex noise. *Ear Hear.* **2016**, *37*, 312–323. [CrossRef] [PubMed]

155. Moretti, D.; Thomas, L.; Marques, T.; Harwood, J.; Dilley, A.; Neales, B.; Shaffer, J.; McCarthy, E.; New, L.; Jarvis, S.; et al. A risk function for behavioral disruption of Blainville's beaked whales (*Mesoplodon densirostris*) from mid-frequency active sonar. *PLoS ONE* **2014**, *9*, e85064. [CrossRef]

156. Miller, P.J.O.; Antunes, R.N.; Wensveen, P.J.; Samarra, F.I.P.; Catarina Alves, A.; Tyack, P.L.; Kvadsheim, P.H.; Kleivane, L.; Lam, F.-P.A.; Ainslie, M.A.; et al. Dose-response relationships for the onset of avoidance of sonar by free-ranging killer whales. *J. Acoust. Soc. Am.* **2014**, *135*, 975–993. [CrossRef] [PubMed]

157. Richardson, W.J.; Miller, G.W.; Greene, C.R. Displacement of migrating bowhead whales by sounds from seismic surveys in shallow waters of the Beaufort Sea. *J. Acoust. Soc. Am.* **1999**, *106*, 2281. [CrossRef]

158. Miller, G.W.; Moulton, V.D.; Davis, R.A.; Holst, M.; Millman, P.; MacGillivray, A.; Hannay, D. Monitoring seismic effects on marine mammals—southeastern Beaufort Sea, 2001–2002. In *Offshore Oil and Gas Environmental Effects Monitoring: Approaches and Technologies*; Armsworthy, S.L., Cranford, P.J., Lee, K., Eds.; Battelle Press: Columbus, OH, USA, 2005; pp. 511–542.

159. Miller, P.J.O.; Johnson, M.P.; Madsen, P.T.; Biassoni, N.; Quero, M.; Tyack, P.L. Using at-sea experiments to study the effects of airguns on the foraging behavior of sperm whales in the Gulf of Mexico. *Deep Sea Res. Part. I* **2009**, *56*, 1168–1181. [CrossRef]

160. Blackwell, S.B.; Nations, C.S.; McDonald, T.L.; Thode, A.M.; Mathias, D.; Kim, K.H.; Greene, C.R., Jr.; Macrander, A.M. Effects of airgun sounds on bowhead whale calling rates: Evidence for two behavioral thresholds. *PLoS ONE* **2015**, *10*, e0125720. [CrossRef] [PubMed]

161. Graham, I.M.; Merchant, N.D.; Farcas, A.; Barton, T.R.; Cheney, B.; Bono, S.; Thompson, P.M. Harbour porpoise responses to pile-driving diminish over time. *R. Soc. Open Sci.* **2019**, *6*. [CrossRef]

162. Russell, D.J.F.; Hastie, G.D.; Thompson, D.; Janik, V.M.; Hammond, P.S.; Scott-Hayward, L.A.S.; Matthiopoulos, J.; Jones, E.L.; McConnell, B.J. Avoidance of wind farms by harbour seals is limited to pile driving activities. *J. Appl. Ecol.* **2016**, *53*, 1642–1652. [CrossRef]

163. Brandt, M.J.; Dragon, A.-C.; Diederichs, A.; Bellmann, M.A.; Wahl, V.; Piper, W.; Nabe-Nielsen, J.; Nehls, G. Disturbance of harbour porpoises during construction of the first seven offshore wind farms in Germany. *Mar. Ecol. Prog. Ser.* **2018**, *596*, 213–232. [CrossRef]

164. New, L.F.; Moretti, D.J.; Hooker, S.K.; Costa, D.P.; Simmons, S.E. Using energetic models to investigate the survival and reproduction of beaked whales (family Ziphiidae). *PLoS ONE* **2013**, *8*, e68725. [CrossRef] [PubMed]

165. Villegas-Amtmann, S.; Schwarz, L.K.; Sumich, J.L.; Costa, D.P. A bioenergetics model to evaluate demographic consequences of disturbance in marine mammals applied to gray whales. *Ecosphere* **2015**, *6*, 1–19. [CrossRef]

166. Schwarz, L.K.; McHuron, E.; Mangel, M.; Wells, R.S.; Costa, D.P. Stochastic dynamic programming: An approach for modelling the population consequences of disturbance due to lost foraging opportunities. *Proc. Meet. Acoust.* **2016**, *27*, 040004. [CrossRef]

167. McHuron, E.A.; Peterson, S.H.; Hückstädt, L.A.; Melin, S.R.; Harris, J.D.; Costa, D.P. The energetic consequences of behavioral variation in a marine carnivore. *Ecol. Evol.* **2018**, *8*, 4340–4351. [CrossRef]

168. Pirotta, E.; Booth, C.G.; Costa, D.P.; Fleishman, E.; Kraus, S.D.; Lusseau, D.; Moretti, D.; New, L.F.; Schick, R.S.; Schwarz, L.K.; et al. Understanding the population consequences of disturbance. *Ecol. Evol.* **2018**, *8*, 9934–9946. [CrossRef]

169. Wilson, L.J.; Harwood, J.; Booth, C.G.; Joy, R.; Harris, C.M. A decision framework to identify populations that are most vulnerable to the population level effects of disturbance. *Conserv. Sci. Pract.* **2020**, *2*, e149. [CrossRef]

170. Pirotta, E.; Booth, C.G.; Cade, D.E.; Calambokidis, J.; Costa, D.P.; Fahlbusch, J.A.; Friedlaender, A.S.; Goldbogen, J.A.; Harwood, J.; Hazen, E.L.; et al. Context-dependent variability in the predicted daily energetic costs of disturbance for blue whales. *Conserv. Physiol.* **2021**, *9*. [CrossRef]

171. Booth, C.G.; Sinclair, R.R.; Harwood, J. Methods for monitoring for the population consequences of disturbance in marine mammals: A review. *Front. Mar. Sci.* **2020**, *7*. [CrossRef]

172. Reichmuth, C. Psychophysical Studies of Auditory Masking in Marine Mammals: Key Concepts and New Directions. In *The Effects of Noise on Aquatic Life*; Popper, A.N., Hawkins, A., Eds.; Advances in Experimental Medicine and Biology; Springer: New York, NY, USA, 2012; Volume 730, pp. 23–27.

173. Erbe, C.; Reichmuth, C.; Cunningham, K.; Lucke, K.; Dooling, R. Communication masking in marine mammals: A review and research strategy. *Mar. Pollut. Bull.* **2016**, *103*, 15–38. [CrossRef]

174. Clark, C.; Ellison, W.; Southall, B.; Hatch, L.; Van Parijs, S.; Frankel, A.; Ponirakis, D. Acoustic masking in marine ecosystems: Intuitions, analysis, and implication. *Mar. Ecol. Prog. Ser.* **2009**, *395*, 201–222. [CrossRef]

175. Guan, S.; Southall, B.L.; Barlow, J.; Vignola, J.F.; Judge, J.A.; Turo, D. Inter-ping sound field from a simulated mid-frequency active sonar, and its implication to marine mammal tonal masking. *Proc. Meet. Acoust.* **2016**, *27*, 070023. [CrossRef]

176. Pine, M.K.; Hannay, D.E.; Insley, S.J.; Halliday, W.D.; Juanes, F. Assessing vessel slowdown for reducing auditory masking for marine mammals and fish of the western Canadian Arctic. *Mar. Pollut. Bull.* **2018**, *135*, 290–302. [CrossRef] [PubMed]

177. Pine, M.K.; Nikolich, K.; Martin, B.; Morris, C.; Juanes, F. Assessing auditory masking for management of underwater anthropogenic noise. *J. Acoust. Soc. Am.* **2020**, *147*, 3408–3417. [CrossRef] [PubMed]

178. Miller, P.J.O.; Biassoni, N.; Samuels, A.; Tyack, P.L. Whale songs lengthen in response to sonar. *Nature* **2000**, *405*, 903. [CrossRef] [PubMed]

179. Foote, A.D.; Osborne, R.W.; Hoelzel, A.R. Whale-call response to masking boat noise. *Nature* **2004**, *428*, 910. [CrossRef]

180. Di Iorio, L.; Clark, C.W. Exposure to seismic survey alters blue whale acoustic communication. *Biol. Lett.* **2010**, *6*, 51–54. [CrossRef] [PubMed]

181. Thode, A.M.; Blackwell, S.B.; Conrad, A.S.; Kim, K.H.; Marques, T.; Thomas, L.; Oedekoven, C.S.; Harris, D.; Bröker, K. Roaring and repetition: How bowhead whales adjust their call density and source level (Lombard effect) in the presence of natural and seismic airgun survey noise. *J. Acoust. Soc. Am.* **2020**, *147*, 2061–2080. [CrossRef]

182. Parks, S.E.; Clark, C.W.; Tyack, P.L. Short- and long-term changes in right whale calling behavior: The potential effects of noise on acoustic communication. *J. Acoust. Soc. Am.* **2007**, *122*, 3725–3731. [CrossRef]

183. Scheifele, P.M.; Andrew, S.; Cooper, R.A.; Darre, M.; Musiek, F.E.; Max, L. Indication of a Lombard vocal response in the St. Lawrence River beluga. *J. Acoust. Soc. Am.* **2005**, *117*, 1486–1492. [CrossRef]

184. Holt, M.M.; Noren, D.P.; Veirs, V.; Emmons, C.K.; Veirs, S. Speaking up: Killer whales (*Orcinus orca*) increase their call amplitude in response to vessel noise. *J. Acoust. Soc. Am.* **2009**, *125*, EL27–EL32. [CrossRef] [PubMed]

185. Parks, S.E.; Johnson, M.; Nowacek, D.; Tyack, P.L. Individual right whales call louder in increased environmental noise. *Biol. Lett.* **2011**, *7*, 33–35. [CrossRef] [PubMed]

186. Dunlop, R.A.; Cato, D.H.; Noad, M.J. Evidence of a Lombard response in migrating humpback whales (*Megaptera novaeangliae*). *J. Acoust. Soc. Am.* **2014**, *136*, 430–437. [CrossRef] [PubMed]

187. Fournet, M.E.H.; Matthews, L.P.; Gabriele, C.M.; Haver, S.; Mellinger, D.K.; Klinck, H. Humpback whales *Megaptera novaeangliae* alter calling behavior in response to natural sounds and vessel noise. *Mar. Ecol. Prog. Ser.* **2018**, *607*, 251–268. [CrossRef]

188. Helble, T.A.; Guazzo, R.A.; Martin, C.R.; Durbach, I.N.; Alongi, G.C.; Martin, S.W.; Boyle, J.K.; Henderson, E.E. Lombard effect: Minke whale boing call source levels vary with natural variations in ocean noise. *J. Acoust. Soc. Am.* **2020**, *147*, 698–712. [CrossRef]

189. Guazzo, R.A.; Helble, T.A.; Alongi, G.C.; Durbach, I.N.; Martin, C.R.; Martin, S.W.; Henderson, E.E. The Lombard effect in singing humpback whales: Source levels increase as ambient ocean noise levels increase. *J. Acoust. Soc. Am.* **2020**, *148*, 542–555. [CrossRef]

190. Brumm, H.; Zollinger, S.A. The evolution of the Lombard effect: 100 years of psychoacoustic research. *Behaviour* **2011**, *148*, 1173–1198. [CrossRef]

191. Noren, D.P.; Holt, M.M.; Dunkin, R.C.; Williams, T.M. The metabolic cost of communicative sound production in bottlenose dolphins (*Tursiops truncatus*). *J. Exp. Biol.* **2013**, *216*, 1624–1629. [CrossRef]

192. Holt, M.M.; Noren, D.P.; Dunkin, R.C.; Williams, T.M. Vocal performance affects metabolic rate in dolphins: Implications for animals communicating in noisy environments. *J. Exp. Biol.* **2015**, *218*, 1647–1654. [CrossRef]

193. Noren, D.P.; Holt, M.M.; Dunkin, R.C.; Williams, T.M. The metabolic cost of whistling is low but measurable in dolphins. *J. Exp. Biol.* **2020**, *223*. [CrossRef] [PubMed]

194. Ising, H.; Kruppa, B. Health effects caused by noise: Evidence in the literature from the past 25 years. *Noise Health* **2004**, *6*, 5–13. [PubMed]

195. Skogstad, M.; Johannessen, H.A.; Tynes, T.; Mehlum, I.S.; Nordby, K.-C.; Lie, A. Systematic review of the cardiovascular effects of occupational noise. *Occup. Med.* **2016**, *66*, 10–16. [CrossRef] [PubMed]

196. Park, T.; Kim, M.; Jang, C.; Choung, T.; Sim, K.-A.; Seo, D.; Chang, S.I. The public health impact of road-traffic noise in a highly populated city, Republic of Korea: Annoyance and sleep disturbance. *Sustainability* **2018**, *10*, 2947. [CrossRef]

197. Bolm-Audorff, U.; Hegewald, J.; Pretzsch, A.; Freiberg, A.; Nienhaus, A.; Seidler, A. Occupational noise and hypertension risk: A systematic review and meta-analysis. *Int. J. Environ. Res. Public Health* **2020**, *17*, 6281. [CrossRef]

198. Shin, S.; Bai, L.; Oiamo, T.H.; Burnett, R.T.; Weichenthal, S.; Jerrett, M.; Kwong, J.C.; Goldberg, M.S.; Copes, R.; Kopp, A.; et al. Association between road traffic noise and incidence of diabetes mellitus and hypertension in Toronto, Canada: A population-based cohort study. *J. Am. Heart Ass.* **2020**, *9*, e013021. [CrossRef]

199. Liu, J.; Zhu, B.; Xia, Q.; Ji, X.; Pan, L.; Bao, Y.; Lin, Y.; Zhang, R. The effects of occupational noise exposure on the cardiovascular system: A review. *J. Public Health Emerg.* **2020**, *4*. [CrossRef]

200. Wright, A.J.; Soto, N.A.; Baldwin, A.L.; Bateson, M.; Beale, C.M.; Clark, C.; Deak, T.; Edwards, E.F.; Fernández, A.; Godinho, A.; et al. Do marine mammals experience stress related to anthropogenic noise? *Int. J. Comp. Psychol.* **2007**, *20*, 274–316.

201. Kellar, N.M.; Catelani, K.N.; Robbins, M.N.; Trego, M.L.; Allen, C.D.; Danil, K.; Chivers, S.J. Blubber cortisol: A potential tool for assessing stress response in free-ranging dolphins without effects due to sampling. *PLoS ONE* **2015**, *10*, e0115257. [CrossRef]

202. Bechshoft, T.; Wright, A.J.; Weisser, J.J.; Teilmann, J.; Dietz, R.; Hansen, M.; Björklund, E.; Styrishave, B. Developing a new research tool for use in free-ranging cetaceans: Recovering cortisol from harbour porpoise skin. *Conserv. Physiol.* **2015**, *3*. [CrossRef]

203. Keogh, M.J.; Gastaldi, A.; Charapata, P.; Melin, S.; Fadely, B.S. Stress-related and reproductive hormones in hair from three North Pacific otariid species: Steller sea lions, California sea lions and northern fur seals. *Conserv. Physiol.* **2020**, *8*. [CrossRef]

204. Pujade Busqueta, L.; Crocker, D.E.; Champagne, C.D.; McCormley, M.C.; Deyarmin, J.S.; Houser, D.S.; Khudyakov, J.I. A blubber gene expression index for evaluating stress in marine mammals. *Conserv. Physiol.* **2020**, *8*, coaa082. [CrossRef] [PubMed]

205. Bechshoft, T.; Wright, A.J.; Styrishave, B.; Houser, D. Measuring and validating concentrations of steroid hormones in the skin of bottlenose dolphins (*Tursiops truncatus*). *Conserv. Physiol.* **2020**, *8*. [CrossRef] [PubMed]

206. Moore, S.E.; Reeves, R.R.; Southall, B.L.; Ragen, T.J.; Suydam, R.S.; Clark, C.W. A new framework for assessing the effects of anthropogenic sound on marine mammals in a rapidly changing Arctic. *BioScience* **2012**, *62*, 289–295. [CrossRef]

207. National Academies of Sciences, Engineering, and Medicine. *Approaches to Understanding the Cumulative Effects of Stressors on Marine Mammals*; National Academies Press: Washington, DC, USA, 2017; ISBN 978-0-309-44048-6.
208. Faulkner, R.C.; Farcas, A.; Merchant, N.D. Guiding principles for assessing the impact of underwater noise. *J. Appl. Ecol.* **2018**, *55*, 2531–2536. [CrossRef]
209. Pirotta, E.; Mangel, M.; Costa, D.P.; Goldbogen, J.; Harwood, J.; Hin, V.; Irvine, L.M.; Mate, B.R.; McHuron, E.A.; Palacios, D.M.; et al. Anthropogenic disturbance in a changing environment: Modelling lifetime reproductive success to predict the consequences of multiple stressors on a migratory population. *Oikos* **2019**, *128*, 1340–1357. [CrossRef]
210. Houser, D.S.; Finneran, J.J.; Ridgway, S.H. Research with Navy marine mammals benefits animal care, conservation and biology. *Int. J. Comp. Psychol.* **2010**, *23*, 249–268.
211. Fetherston, T. Environment in a (high-tech) box: Navy model simulates undersea sound fields & marine mammal locations to plan training & testing activities. *Currents* **2011**, 42–43.

Journal of
*Marine Science
and Engineering*

Review

Use of Underwater Acoustics in Marine Conservation and Policy: Previous Advances, Current Status, and Future Needs

Shane Guan [1,*], Tiffini Brookens [2,*] and Joseph Vignola [1]

[1] Department of Mechanical Engineering, The Catholic University of America, Washington, DC 20064, USA; vignola@cua.edu
[2] U.S. Marine Mammal Commission, Bethesda, MD 20814, USA
* Correspondence: guan@cua.edu (S.G.); tbrookens@mmc.gov (T.B.)

Abstract: The interdisciplinary field of assessing the impacts of sound on marine life has benefited largely from the advancement of underwater acoustics that occurred after World War II. Acoustic parameters widely used in underwater acoustics were redefined to quantify sound levels relevant to animal audiometric variables, both at the source and receiver. The fundamental approach for assessing the impacts of sound uses a source-pathway-receiver model based on the one-way sonar equation, and most numerical sound propagation models can be used to predict received levels at marine animals that are potentially exposed. However, significant information gaps still exist in terms of sound source characterization and propagation that are strongly coupled with the type and layering of the underlying substrate(s). Additional challenges include the lack of easy-to-use propagation models and animal-specific statistical detection models, as well as a lack of adequate training of regulatory entities in underwater acoustics.

Keywords: underwater acoustics; underwater sound impacts; marine conservation; impact assessment

Citation: Guan, S.; Brookens, T.; Vignola, J. Use of Underwater Acoustics in Marine Conservation and Policy: Previous Advances, Current Status, and Future Needs. *J. Mar. Sci. Eng.* **2021**, *9*, 173. https://doi.org/10.3390/jmse9020173

Academic Editors: Philippe Blondel and Michel André
Received: 19 December 2020
Accepted: 2 February 2021
Published: 9 February 2021

Publisher's Note: MDPI stays neutral with regard to jurisdictional claims in published maps and institutional affiliations.

1. Introduction—Historical Perspective

As a visually oriented species, it is not surprising that our knowledge of the world has been based largely on reasoning and experimentation through visual means. This includes research on marine organisms whose natural habitats beneath the ocean surface are mostly beyond visual observation, e.g., [1,2]. Moreover, the lack of an auditory system that functions efficiently underwater led humankind to consider the marine environment as a "silent world" for eons, e.g., [3].

The need for navigational safety, especially after the sinking of the HMS Titanic as a result of striking an iceberg in the North Atlantic Ocean, and to detect enemy submarines (i.e., antisubmarine warfare or ASW) and warships during the two world wars prompted tremendous advancements in the field of underwater acoustics between the 1910s and 1950s [4,5]. However, it was not until the discovery of underwater sound production and orientation [6–8] and communication [9,10] in several cetacean species that marine biologists began to investigate the potential effects of underwater sound on marine mammals.

While it was widely recognized before the 1970s that high sound levels, either from elevated ambient sound or from sonar ping reverberation, could adversely affect signal detection in the naval sonar community [4,5], Payne and Webb [11] were the first to document that long-distance acoustic communication ranges could be greatly reduced as a result of increased background sound levels from ships. By using simple sonar equations, Payne and Webb [11] determined that, with the advent of propeller-driven ships, the transmission range of 20-Hz fin whale calls was reduced by nearly 100 and 3000 nmi using spherical and cylindrical spreading models, respectively.

Some of the greatest achievements in environmental conservation in the United States occurred in the 1970s with the enactment of various laws, including passage of the Marine Mammal Protection Act (MMPA) in 1972 and the Endangered Species Act (ESA) in 1973 [12–14]. Implementation of the MMPA led to strict regulations by U.S. Federal agencies to reduce the incidental taking of marine mammals, initially in regard to fishery bycatches [15,16].

The need to implement measures to mitigate impacts on marine mammals from human activities beyond commercial fisheries led to many government-sponsored studies and workshops. For example, in the early 1980s through the 1990s, the U.S. Minerals Management Service (MMS, the predecessor of the current Bureau of Ocean Energy Management (BOEM)) funded several studies investigating disturbance of marine mammals from oil and gas development activities in the Arctic [17]. Many of those studies provided novel information on how underwater sound from various industrial activities affected marine mammals [18–21].

In the early to mid-1990s, two global oceanographic experiments became controversial based on the concern that the intense underwater sound used for climate research would harm whales [22]. The Heard Island Feasibility Test (HIFT) and the subsequent Acoustic Thermometry of Ocean Climate (ATOC) used sufficiently intense low-frequency signals to measure large-scale and long-term temperature changes in the upper ocean layers [23–25]. The acoustic signal used in HIFT had a source level of 221 dB re 1 µPa at 1 m with a center frequency of 57 Hz [24], while the signal used in ATOC operated at 420 W (or 197 dB re 1 µPa at 1 m) and was centered at 75 Hz [26]. To address the concerns of potential impacts from the sound emitted, extensive field studies were conducted on marine mammals and other marine organisms, e.g., [27–32].

Besides industry and academia, the military—particularly, the naval community—produces intense underwater sound for various purposes. The sound sources include naval sonars, live-fire munitions, and underwater detonations used during training and testing activities and ship shock trials. However, the acoustic impacts from those sources were not broadly known until the early 1990s [33]. The situation changed dramatically in the late-1990s to early 2000s when several marine mammal mass-stranding events occurred in areas where the U.S. and North Atlantic Treaty Organization (NATO) navies had conducted exercises involving the use of active sonars, e.g., [34–38]. Those stranding events received considerable attention from environmental organizations, academia, and the public, which led to a surge in field and laboratory studies that have greatly increased our knowledge regarding the impacts of sound on marine mammals, as well as other marine life, including fish and invertebrates, in the past two decades [39–43].

Although our understanding of the impacts of sound on marine organisms has increased, the regulatory community has struggled to evaluate and incorporate new findings and data into impact assessments and environmental policies in general [44]. Since assessing acoustic impacts on marine life is an interdisciplinary field, it requires that regulators and policymakers have knowledge and education in both underwater acoustics and marine biology [45]. Therefore, a solid understanding of physical principles in acoustics is imperative for assessing the impacts of underwater sound.

This paper addresses many of the physical principles in underwater acoustics that have been and currently are applied to the regulation and management of underwater sound and what information needs to be obtained in the future.

2. Application of Underwater Acoustic Principles in Marine Conservation and Policy—Current Status

Impact assessments of various anthropogenic sound-generating activities involve the evaluation of the physical characteristics of the sound sources and the propagation of sound in the marine environment. Most of the concepts used in these assessments are based on the field of underwater acoustics. These include acoustic parameters, the characterization of underwater sources (measurements and modeling), the application of the sonar equation, and sound propagation modeling.

2.1. Acoustic Parameters

A number of physical quantities can be used to describe underwater sound in terms of acoustic energy (in joules), power (in watts), intensity (in watts per unit area), and pressure (in pascals or micropascals or μPa). However, more commonly, the quantity of sound is expressed in a relative unit of decibels (dB), which is a logarithm (base 10) ratio of a physical quantity to a reference quantity, as expressed in the following equation:

$$\mathrm{SPL} = 10 \log_{10} \left(\frac{p}{p_0} \right)^2 \tag{1}$$

where SPL is the sound pressure level, p is the acoustic pressure, and p_0 is the reference acoustic pressure.

This expression converts the physical unit to a "level". For example, the sound pressure (in μPa) can be expressed in the sound pressure level (SPL) in dB in reference to 1 μPa (dB re 1 μPa), which is the standard reference unit in underwater acoustics. For airborne sound, the standard referenced sound pressure level is 20 μPa. The mismatch of acoustic impedance between water and air due to the differences in the sound speeds of these two media results in different acoustic pressures from a source with the same acoustic intensity. Specifically, for a plane wave with a far-field intensity of I, the underwater acoustic pressure p_w and airborne acoustic pressure p_a are

$$\begin{aligned} p_w &= \sqrt{I \rho_w c_w} \\ p_a &= \sqrt{I \rho_a c_a} \end{aligned} \tag{2}$$

where c_w and c_a are the sound speed in water and air, and ρ_w and ρ_a are the density of water and air, respectively. The product ρc is referred to as a characteristic acoustic impedance. Given that the nominal sound speed in water is 1500 m/s, the nominal sound speed in air is 340 m/s, the density of water is about 1000 kg/m^3, and the density of air is about 1.225 kg/m^3, the underwater acoustic pressure from a sound source with the same intensity would be approximately 60 times great than that in the air.

For example, for a sound source with an intensity of 1 W/m^2, the underwater and airborne acoustic pressures would be approximately 1225 Pa and 20 Pa, respectively, based on Equation (2), and the underwater and airborne SPLs would be 182 dB re 1 μPa and 120 dB re 20 μPa, respectively. These issues often create confusion among lay persons and regulators who may not be well-versed in physical acoustics [46].

Notwithstanding the simple definition of SPL provided herein, several variations of broadband "sound levels" are tailored to address different types of source characteristics that are pertinent to various marine organisms that have different vibroacoustic sensitivities[1] and exhibit varying responses [47,48]. Some of the commonly used sound levels are the peak sound pressure level (L_{pk}, $L_{0\text{-}pk}$, or SPL_{pk}); root mean square (rms) sound pressure level ($L_{p,rms}$ or SPL_{rms}); sound exposure level (L_E or SEL)l single-strike (single-shot or single-ping) sound exposure level ($L_{E,ss}$, $L_{E,sp}$, or SEL_{ss}); and cumulative sound exposure level ($L_{E,cum}$ or SEL_{cum}). The usage of these sound level metrics is summarized in Table 1.

---

[1] The authors acknowledge the importance of particle motion. However, particle motion, velocity, and acceleration are beyond the scope of this paper. Please see [47,48] for more details regarding particle motion.

Table 1. Summary of the sound level metrics commonly used in assessing the impacts of underwater sound on marine life.

Metric & Notation	Equation for Derivation	Usage in Impact Assessment
Peak sound pressure level (L_{pk}, $L_{0\text{-}pk}$, or SPL_{pk})	$L_{pk} = 10\log_{10}\left(\frac{p_{pk}}{p_0}\right)^2$	The maximum instantaneous sound pressure, which is used to assess a potential permanent threshold shift (PTS) and temporary threshold shift (TTS) in the hearing of marine mammals [49–51], gastrointestinal tract injury in marine mammals [52], and mortality and injury in fish and sea turtles [53] exposed to impulsive sound.
Root-mean-square sound pressure level ($L_{p,rms}$ or SPL_{rms})	$L_{p,rms} = 10\log_{10}\left[\frac{1}{T}\int_T \frac{p^2(t)}{p_0^2}dt\right]$	The square root of the average of the sound pressure squared over a given duration, which is used to assess potential behavioral disturbance in marine mammals [54] from impulsive and non-impulsive sound exposure—a time window that consists of 90% of the acoustic energy is used to calculate $L_{p,rms}$ for impulsive sound [55]. It also is used to assess the potential mortality, injury, or TTS in fish and sea turtles exposed to non-impulsive sound [53].
Sound exposure level (L_E or SEL)	$L_E = 10\log_{10}\left[\frac{1}{T_0}\int_{T_{100}} \frac{p^2(t)}{p_0^2}dt\right]$	A 1-s normalized L_E is used to characterize the source level for non-impulsive sound [56].
Single-strike, single-shot, single-ping sound exposure level ($L_{E,ss}$, $L_{E,sp}$, or SEL_{ss})	$L_{E,ss} = 10\log_{10}\left[\int_{T_{100}} \frac{p^2(t)}{p_0^2}dt\right]$	For impulsive or non-impulsive intermittent sounds, this is the L_E for a single hammer strike for pile driving [56,57], a single air gun shot for a seismic survey, or a single ping for sonar.
Cumulative sound exposure level ($L_{E,cum}$ or SEL_{cum})	$L_{E,cum} = 10\log_{10}\left[\int_{T_{cum}} \frac{p^2(t)}{p_0^2}dt\right]$	This is the L_E for the entire duration of sound exposure. It is used to assess potential PTS and TTS in marine mammals when exposed to impulsive or non-impulsive sounds [49–51] and the mortality or injury of fish and sea turtles exposed to impulsive sound [53].

Notation: p_{pk} = peak acoustic pressure in a time series, $p(t)$ = time varying acoustic pressure in a waveform, p_0 = referenced acoustic pressure, which is 1 µPa, T = duration of the time series, T_{100} = the entire (100%) time duration of the time series, T_0 = a referenced time interval of 1 s, and T_{cum} = the entire duration of sound exposure.

2.2. Source Characterization

In acoustics, a source is a physical device or object that generates acoustic disturbance(s) in a medium. A simple point source can be viewed as a pulsating sphere with its radius varying sinusoidally with time. The acoustic pressure generated by such a sphere is time-varying and contains one or more frequencies.

Similar to almost all real-world sources, very few anthropogenic sources can be treated as a simple point source. Sound sources that have routinely been evaluated for adverse impacts on marine mammals include seismic air guns, military sonars, various types of in-water pile driving, underwater detonations, drilling, and, to some extent, civilian sonars and high-resolution geophysical (HRG) devices. Although it is well-recognized that vessel noise is the most pervasive source of anthropogenic sound both in terms of temporal and spatial extents in the marine environment [58,59], its potential impacts are not well-addressed, nor is it currently regulated in most countries. Additionally, with the exception of certain military and civilian sonars, the majority of these sound sources are considered broadband.

Based on the temporal characteristics and the types of impacts[2], underwater sound sources are classified by the following categories: impulsive, non-impulsive, continuous,

---

[2] In the U.S. regulatory framework, impacts on marine mammals are classified into two categories: Level A harassment, which has the potential to cause injury, and Level B harassment, which has the potential to cause behavioral disturbance, as well as temporary threshold shifts (TTS).

and intermittent. It should be noted that the definitions of these four categories within the regulatory community generally are qualitative, although quantitative methods have been proposed in a few cases when clear-cut distinctions between categories are evident. For example, when differentiating between impulsive and non-impulsive sources, a 3-dB difference in measurements between the continuous and impulse settings of a sound level meter (SLM) has been used [57]. Specifically, if the SLM measurement from the impulse setting (a 35-ms window) is 3 dB or greater than the continuous setting (a 1-s window), the sound should be classified as impulsive [60]. A recent study by Martin et al. [61] used the kurtosis of a 1-min time window to determine whether a sound was impulsive or non-impulsive.

However, not all of these categories are mutually exclusive. For example, a source that is impulsive is typically intermittent (e.g., impact pile driving), but a source that is non-impulsive can be either continuous (e.g., vibratory pile driving and removal) or intermittent (e.g., sonar). In addition, not all sources fit into a single category. For example, down-the-hole (DTH) pile installation produces both percussive hammering and continuous drilling sounds, while HRG devices can emit impulsive or non-impulsive intermittent sounds. Some common examples of the source categories used by the U.S. regulatory community are provided in Table 2. An explanation of how these different categories of sound sources should be analyzed under the MMPA is provided in a User Spreadsheet Tool by the National Marine Fisheries Service [62][3].

Table 2. Examples of common categories of sound sources regulated by the U.S. regulatory community.

Source Type	For Assessing PTS and TTS	For Assessing Behavioral Disturbance
Seismic air gun	Impulsive	Intermittent
Impact pile driving	Impulsive	Intermittent
Underwater detonations	Impulsive	Intermittent
Vibratory pile driving and removal	Non-impulsive	Continuous
DTH pile installation	Impulsive and non-impulsive	Intermittent and continuous
Sonar	Non-impulsive	Intermittent
HRG devices	Non-impulsive and impulsive	Intermittent
Drilling	Non impulsive	Continuous
Icebreaking	Non-impulsive	Continuous

Notation: DTH = down-the-hole and HRG = high-resolution geophysical.

For the most part, source levels are based on broadband sound levels measured at given locations back-calculated to 1 m from the source or modeled (in which case, the spectra also are considered). For in-water pile driving for construction activities, the term "source level" used by the regulatory community in the United States typically refers to the broadband sound level (L_{pk}, $L_{p,rms}$, or $L_{E,ss}$) measured at or normalized to 10 m as opposed to the more conventional 1 m from the pile, e.g., [56]. For seismic air guns, source levels are obtained from in situ measurements at various distances back-calculated to 1 m from the source, e.g., [63–67]. For many sources for which measurements are not available, source models (e.g., Gundalf, Nucleus, and Airgun Array Source Model (AASM)) are used to estimate the source levels that then are fed into sound propagation models, e.g., [68].

---

3   The User Spreadsheet Tool is available at https://media.fisheries.noaa.gov/2020-12/2020_BLANK_USER_SPREADSHEET_-508_DEC.xlsx?null (accessed on 2 February 2021).

2.3. Sound Propagation

As with all underwater acoustic analyses, the basic sonar equation with only the geometric spreading loss term is most commonly used by conservation biologists and regulators to estimate received sound levels. That equation is expressed in dB as

$$\text{TL} = F \log_{10}(R) \tag{3}$$

where TL is the transmission (or propagation)[4] loss [4,45,69–72], F is a coefficient for the TL, and R is the distance from the source to receiver (i.e., the animal). For a simple point source within a lossless infinite medium, F is 20, which implies the "spherical spreading" of acoustic energy. In a shallow-water environment, the boundary condition dictates that the acoustic energy predominantly follows a "cylindrical spreading" model, where the transmission loss would be expressed as 10log(R) [4,69]. In addition, there is a "combined spreading loss" model that calculates transmission loss using spherical spreading to a certain range H where the sound reaches the sea floor—after which, cylindrical spreading is assumed [73].

The combined spreading loss is expressed as

$$\text{TL} = 20 \log_{10}(H) + 10 \log_{10}\left(\frac{R}{H}\right) \tag{4}$$

at a range (R) greater than the water depth (H). Although additional loss mechanisms such as absorption and scattering (i.e., volume and boundary scattering) also contribute to the decay of acoustic intensity over range, models that incorporate absorption and scattering terms are seldom used by regulators, mainly due to the fact that such models cannot be solved analytically. Similarly, transmission loss models that incorporate low-frequency cutoffs or leakages in shallow water also are rarely used by regulators.

To account for the additional losses due to absorption and scattering, and to partially account for acoustic energy that is confined within the boundaries, regulatory agencies often use 15 (i.e., the arithmetic mean between 20 and 10) as the transmission loss coefficient and define it as "practical spreading". The practical spreading model primarily is used to assess the impacts from pile-driving activities, e.g., [74]. Other transmission loss coefficients that have been used include the derivation of decay slopes from linear fit models of field measurements at varying distances, e.g., [75]. However, transmission loss coefficients obtained using field measurements are location- and season-specific, because received sound levels at distances from the source are products of multiple attenuation mechanisms. Factors such as sediment type, bathymetry, and temperature/salinity profiles of the water column often dictate far-field sound level measurements.

However, sophisticated numerical sound propagation models (such as ray theory, wavenumber integration, normal mode, the parabolic equation, etc.; see [69]) generally are not used by the regulatory community. Regulatory agencies typically rely on results provided by applicants or their contractors who have those modeling capabilities, e.g., [76,77]. In those cases, it sometimes is unclear whether the regulatory agencies adequately evaluated or validated the modeling results.

2.4. Impact Assessment Analyses

In general, the underlying approach for assessing the impacts of underwater sound on marine life uses a source–path–receiver model, where the source is the anthropogenic sound emitted, the path describes the assumed sound propagation, and the receiver is the animal(s) that detects the sound.

---

[4] The authors recognize the difference between "transmission" and "propagation" under certain circumstances, where "transmission" could mean traveling of the acoustic wave from one medium to another. However, these two terms are used interchangeable herein, because the term "TL" is more widely used for "transmission loss" than "PL" for "propagation loss" in the underwater acoustics literature; see [4,45,70–73].

This model can be best presented in the form of a simple passive sonar equation:

$$RL = SL - TL \tag{5}$$

where RL[5] is echo level or received level, SL is source level, and TL is transmission loss [4,71].

The metrics used for received levels mirror those of acoustic parameters described herein; however, the numerical values of the levels, or thresholds, have been revised, as more studies have been conducted investigating the acoustic impacts of underwater sound. For example, the auditory injury thresholds (defined as permanent threshold shift (PTS)) were revised from the $L_{p,rms}$ thresholds of 180- and 190-dB re 1 μPa for cetaceans and pinnipeds, respectively [54], to the dual criteria of $L_{E,cum}$ and $L_{p,pk}$, with the incorporation of frequency-based, auditory weighting functions for the $L_{E,cum}$ metric [49,50].

The receivers that are pertinent to impact assessment analyses include all aquatic organisms that are sensitive to underwater sound and vibroacoustic disturbance. The levels upon which adverse impacts occur depend on the taxonomy, physiology, and behavioral ecology of specific species or individual animals, which is not within the scope of this paper. Interested readers are referred to several research, review, and guidance papers for the relevant information, e.g., [45,49–53,60,78].

The statistical detection theory at the receiver (i.e., the animal) is not currently considered in impact assessments of underwater sound. Such considerations would include quantitative studies of detection thresholds, the minimum signal-to-noise ratio needed to perceive the signal, the frequency spectrum and bandwidth of the signal, and the ambient sound, as well as receiver operating characteristic (ROC) curves, which describe the probability of detection at the receiver given a detection threshold and the signal-to-noise ratio [5].

Additionally, receiver (animal) movement modeling can be used to better inform an impact assessment by estimating the number of animals, in the form of "animats" that could be affected (taken). Animal movement modeling falls within the field of behavioral ecology; therefore, it is not discussed further. However, multiple animal movement models do exist; see [79] for information on the Marine Mammal Movement and Behavior (3MB)[6] and [80] for information on the Navy Acoustic Effects Model (NAEMO).

2.5. Chronic Impact Assessment and Soundscape Analyses

Over the past decade or so, there has been increased interest in addressing potential chronic and cumulative impacts from low-intensity sound sources (e.g., commercial ships and smaller vessels) that are not typically regulated [81,82]. Many of these studies have shown that chronic exposure to low-intensity sound can cause various adverse effects, such as communication masking, changes in vocalizations and echolocation, and increased stress levels [83].

With the recent advances in underwater acoustic sensing technology available to nonmilitary researchers, the accessibility of large acoustic datasets from global sensor networks, and the enhanced computational resources for signal processing of large acoustic datasets, the large-scale, long-term monitoring of the underwater acoustic environment is feasible. These new opportunities have created considerable possibilities for studying the relationship between underwater acoustic and biological phenomena [84].

Many of these studies build on earlier research on ambient sound by analyzing spectral contents of long-term acoustic recordings. A frequency–time analysis has been used to investigate the inter-relationships of three sound types—biophony, geophony, and anthrophony—within an ecosystem. This relatively new subfield, ecoacoustics and

---

5 In most underwater acoustics literature, "EL" (echo level) is typically used to indicate the received (echo) level at the receiving transducer, and "RL" is reserved for the reverberation level in the sonar equation (e.g., [5,60,70,73]. However, this paper uses "RL" to indicate "received level", which is a more common practice within the ocean sound community (e.g., [39]).

6 3MB is available at http://oalib.hlsresearch.com/Sound%20and%20Marine%20Mammals/3MB%20HTML.htm (accessed on 8 February 2021).

soundscape ecology [85,86], takes a holistic approach for studying underwater sound and their relationship to marine life. Ecoacoustics and soundscape ecology allow for the assessment of the overall health of the ecosystem by including the acoustic component, a very important element that has been long overlooked.

2.6. Knowledge and Expertise of Regulatory Community

The regulatory community that oversees the implementation of marine conservation and policy measures concerning the impacts of underwater sound primarily are composed of conservation biologists and environmental policy specialists, many of whom lack a formal educational background in physics, mathematics, or underwater acoustics. Staff analysts and managers who conduct impact assessments and make regulatory decisions may receive on-the-job training through seminars and web-based tutorials, such as those on the Discovery of Sound in the Sea website (https://dosits.org/, accessed on 2 February 2021). However, such ad hoc training is inadequate to bring analysts within the U.S. regulatory community beyond the level of performing simple analytical calculations of sound propagation using scripted spreadsheets. Few are able to evaluate sophisticated acoustic models or sound source measurements. The regulatory agencies have yet to prioritize the knowledge and skills of physical acoustics that are necessary to conduct impact assessments of underwater sound. These shortcomings have resulted in frequent errors, the omission of pertinent information, and inconsistencies in agency decision-making documents, as documented in multiple comment letters from the U.S. Marine Mammal Commission, an independent oversight agency, e.g., [87–96].

For example, when addressing the potential impacts of the relatively novel DTH pile installation method, one regulatory agency repeatedly mischaracterized the source, used inappropriate thresholds, and underestimated the source levels, which resulted in much smaller impact zones [89,92–94]. In another example, the same agency fabricated a method termed "log average of the sources"—taking a log average of log-based sound levels to derive a source level for DTH pile installation—which is not rooted in the principles of underwater acoustics [89]. The agencies also have routinely used inappropriate and inconsistent source levels for pile driving and removal, as well as inappropriate thresholds in general [87,88] and inappropriate assumptions and inputs for estimating the extents of the various impact zones [89,91–93]. The aforementioned issues result in inaccurate and often underestimated impact zones, which are used to determine whether and how an animal may be affected and to inform the mitigation measures necessary to minimize those impacts.

3. Needs for Using Underwater Acoustics in Marine Conservation

While underwater acoustic concepts are well-understood, information gaps exist, and training is necessary to improve the accuracy of and consistency among impact assessments. Data and information needs include source characteristics of novel sound sources; robust and easy-to-use sound propagation models; and statistical detection models, as well as quantitative exposure models that evaluate the acute, chronic, and cumulative impacts on marine organisms. While the last topic is addressed primarily by the field of bioacoustics, the knowledge of behavioral ecology and physiology needs to be incorporated into any impact assessment [38] and is beyond the scope of this paper. The research needs regarding source characterization, propagation modeling, and detection theory are provided herein.

3.1. Source Characterization

Despite the numerous technical reports that involve measurements of underwater sound generated by various sound sources, robust data are still lacking concerning some of the sound sources. Those deficiencies include: (1) novel sound sources, (2) uncommon sources for which few data exist, and (3) a lack of scientific rigor in measurements. Large variations in source levels also are evident among the same source type, which adds another complicating factor.

Some sound sources are novel, which either have not been used in the marine environment until recently or have not been well-documented. For example, in-water pile installation using DTH pile installation is a relatively new application in the marine environment and uses a combination of percussive and drilling mechanisms [57]. During DTH pile installation, a percussive hammer acts directly upon the bedrock to create a hole for the pile to enter, while the drill cuttings and debris at the rock surface are removed by an airlift exhaust through the inside of a pile. Therefore, the sound generated from DTH pile installation contains both impulsive, intermittent components (from percussive hammer strikes) and non-impulsive, continuous components (from drilling actions and airlifts of debris). Currently, only a few studies have conducted measurements of DTH pile installations [57,97–101]. Additionally, in situ measurements of DTH pile installations have been limited to piles with diameters of only 0.20 m, 0.46 m, 0.61 m, and 1.07 m. Those data are scant and inadequate, particularly for larger-sized piles. Piles used in coastal construction projects can be larger and are generally much larger for offshore wind turbine structures. In addition, the substrates associated with the measured sound levels often are not specified.

Other sound sources that lack the full complement of the relevant acoustic information include various nonmilitary shipboard or towed sonars, transducers, other HRG sources, and acoustic deterrent devices. While potential effects from exposure to these sources are still under debate due to their generally lower intensity and high-frequency components (which are subject to greater absorption losses) [102–105], the source levels of these devices have not been well-documented beyond the manufacturer specifications, e.g., [106].

There also are new sources that are being developed. One example is marine vibroseis, a source that is being developed to replace a conventional air gun array with much reduced SPLs [107–109]. Source characteristics of marine vibroseis are mostly modeled; there are few in situ measurements of sound levels of marine vibroseis to date [110].

While a number of models were developed and countless measurements were made for open-water underwater detonations years ago (as one of the sound sources for underwater acoustics research was small charges), e.g., [4,111–114], few studies are available on sound characteristics and propagation from detonations that are embedded in bedrock or other structures, e.g., [115,116]. The lack of measurements from confined underwater detonations presents significant challenges when assessing environmental impacts for projects that use such methods, particularly for shipping channel deepening or structure removal.

Conversely, sound generated from marine seismic surveys using air guns have been well-documented since the 1980s [45]. Numerous measurements of seismic air gun arrays have been acquired in the Arctic for the purpose of environmental compliance from the mid-2000s to early 2010s, e.g., [63–67,117–119]. Nevertheless, due to differences in the volumes of air gun arrays and their deployment configurations, those measurements were only pertinent to those specific surveys. Industrial standard models such as Gundalf [120,121], Nucleus [122], and AASM [123] have been used to predict air gun array sound levels to form the sound propagation modeling used in impact assessments; however, none of the models are available to the regulatory community for use at this time. In addition, the accuracy of these models is still being evaluated by the underwater acoustic modeling community, e.g., [124,125].

Similar issues exist for in-water pile-driving data. Despite the fact that large quantities of pile-driving sound level measurements exist, e.g., [56], the regulatory community has struggled to use representative source levels consistently for specific pile materials and dimensions. Some of the inconsistencies are due to differences in bathymetry, substrate type, hammer energy, and other environmental parameters at the locations where measurements have been collected. An attempt is underway by one of the authors to review and analyze all available pile-driving measurement data and to recommend a set of "generic" 10-m normalized source levels based on the various pile types and diameters. Similar to air gun source models, models for pile-driving sound sources also exist, e.g., [126–128]. None of these models are readily available for use by the regulatory community to form its environmental impact assessments.

In addition to the need to characterize broadband levels from many of the aforementioned sound sources, spectral information regarding these sources is in demand, as impact assessments of auditory effects are often frequency-dependent, especially for marine mammals [50].

Finally, most sound level measurements have been conducted for the purpose of environmental compliance and were collected and analyzed by contractors of regulated entities with varying professional experience and/or knowledge in underwater acoustics. Most of these measurements exist in the form of gray literature, e.g., [56], and few of them have been peer-reviewed or published in peer-reviewed journals. Therefore, the quality of some sound source measurements is questionable and should be evaluated further.

3.2. Sound Propagation Models

Although numerous models exist for underwater sound propagation [69], the majority of these models assume that the source is in the open water. However, sources from some of the regulated activities occur within sediment or structures (e.g., confined underwater detonations) or are coupled with the sediment (e.g., DTH pile installation). The authors are unaware of an available propagation model for these sources, despite the increasing use of these sources in recent years.

For the majority of sources that are used in a water column, the existing propagation model commonly used by the regulatory community is a simple spreading model with a transmission loss coefficient of 20 or 15, depending on the source. Given that underwater sound propagation is almost always a complex process that involves bathymetry and topography of the location, substrate layers and types, temperature and salinity profiles of the water column, sea surface conditions, and the frequency spectrum of the source, sophisticated numerical modeling typically is required to obtain the reasonably accurate results needed for impact assessments.

Although many of the sophisticated numerical propagation models are derived from well-established propagation theories [129], the implementation of these models is beyond the expertise of the regulatory community due to the lack of necessary technical skills within the agencies. For example, high-level programming languages such as MATLAB or Octave are not among the standard software used by the U.S. regulatory community for conducting impact assessments.

Given these resource and technical limitations, it is beneficial to develop relatively simple numerical models that can be incorporated into an Excel spreadsheet format. For example, the U.S. National Marine Fisheries Service has developed a simple spreadsheet tool that incorporates the frequency of absorption and beam width for determining sound propagation and estimating the distances at which behavioral harassment could occur in marine mammals [130].

In another example, a damped cylindrical spreading (DCS) model-based spreadsheet, or DCSiE, was recently developed with funding from the BOEM to estimate the distances of certain received sound levels from impact pile driving for offshore wind turbine installations [131]. This spreadsheet tool incorporates information related to bathymetry and the substrate type, in addition to the measured sound level at a reference distance (typically, no less than three times the water depth at the source). It is based on a reasonably simple but more accurate DCS model [132–134]. To implement DCSiE properly for estimating impact zones, one must have an understanding of the sediment composition and the layering of those sediments (including the sediment porosity and particle size) in the project area. Unfortunately, these data are lacking in most regions and are not routinely described when pile-driving measurements are collected. In addition, a comparable sound propagation model for vibratory pile driving and removal and DTH pile installation currently does not exist.

Although these spreadsheet tools can perform simple propagation modeling to a certain degree, they cannot replace sophisticated numerical models that are commonly used by the underwater acoustics community. To address such deficiencies, a standalone software package that does not require programming skills needs to be developed for the regulatory community.

3.3. Statistical Detection Models

None of the impact assessment tools for underwater sound currently address statistical signal detection at the receiver. The received level at the animal is therefore considered the level of exposure. Such an approach generally is acceptable when assessing the PTS or temporary threshold shift (TTS), as most data on those effects are based on direct measurements at the animals. However, they may not be accurate for quantitatively addressing behavioral disturbances and acoustic masking. Most research on marine animal audiograms and hearing thresholds is conducted in the absence of background sound or at very low ambient conditions with a higher signal-to-noise ratio than is typical in the marine environment. Only a few studies have addressed signal detection in the presence of noise, which could elucidate detection thresholds of some marine mammal species, e.g., [135–137] and review [138]. The authors are not aware of any such studies in species other than marine mammals. Although the detection theory falls within the fields of auditory physiology and behavioral psychology, the information from such studies is critical in the application of underwater acoustics to impact the assessments of sound. The lack of information on the auditory detection thresholds under various noise conditions by many marine species makes it impossible to conduct assessments of masking using the well-established statistical detection theory with ROC curves [5].

3.4. Needs for Chronic Impact Assessment and Soundscape Analyses

Despite the recent progress in understanding the impacts of chronic low-intensity sound on marine life (e.g., [83]), most of the regulatory community has been slow to implement the associated analyses. For example, shipping noises receive relatively little consideration during conservation planning and regulatory management. One of the reasons appears to be that the adverse effects from the low-intensity sound are difficult to quantify on a project and area basis, which is the main mechanism underpinning the various regulations. Therefore, models that can quantify fine-scale and project-specific impacts from low-intensity sound exposure should be developed. These models would be able to assess the energetic cost to marine life from sound exposure in the form of behavioral modification, changes in vocalizations and echolocation, communication masking, habitat displacement, and increased stress levels.

Another "low-level" impact that has received little consideration is reverberation—specifically, the reverberation field between intense intermittent sounds due to multipath propagation [139–141]. It has been suggested that the elevated background sound levels from reverberation have the potential to mask vital marine mammal acoustic cues [142]. However, there are few studies that provide quantitative data on the threshold level associated with auditory masking [143].

In addition, further studies are needed to investigate how the soundscape changes as a result of long- and short-term habitat modifications, which may affect certain species and, in turn, set off a cascade through various trophic levels and affect the ecosystem as a whole. While most of the questions being addressed lie in the field of ecoacoustics, the technical capability required to analyze large amounts of acoustic data that are being collected continuously by many global observation networks is a critical need to be addressed [144–146].

3.5. Needs for Expertise and Knowledge within the Regulatory Community

Last, but not least, advancements in the knowledge of underwater acoustics and its applications are not possible without a regulatory community that is well-versed in underwater acoustics. Since assessing acoustic impacts on marine life is an interdisciplinary field that involves physical acoustics, oceanography, and biology, scientifically sound environmental policy and conservation measures can only be developed through a solid understanding of the scientific principles within these fields. Specifically, the knowledge, skills, and expertise in performing and evaluating numerical source and propagation models; acoustic measurements; and exposure and impact models are key areas where gaps currently exist. It is imperative for regulatory agencies to integrate professional physical acousticians into their hierarchy, rather than relying on policy experts that lack formal education in and an understanding of physics, mathematics, or engineering. As the renowned acoustician Dr. Allan D. Pierce stated, "a deep understanding of acoustical principles is not acquired by superficial efforts" [147].

4. Conclusions

The field of the environmental conservation that addresses the impacts of underwater sound on marine life has advanced considerably over the past half-century. Although one of the initial concerns involved ever-increasing ocean ambient impacts on the communication space of baleen whales [11], the field advanced most readily when acute impacts from ATOC sources, seismic air guns, and military sonar were investigated [17,27,34]. The environmental impact assessments of these sound sources were assisted largely by knowledge within the field of underwater acoustics, e.g., [29,39,45].

Over the years, assessing the impacts of underwater sound has gradually evolved into a research area with its own unique definition of acoustic parameters, sound sources, propagation modeling, and measures of biological impacts, e.g., [51,52,60]. Besides addressing the direct and acute impacts of sound, which is largely under the purview of natural resource agencies that implement various regulatory statutes and measures [49–52,54], recent developments in this field include research that addresses the overall acoustic environment. These new studies have broadened the scope of the relatively narrow-focused field that only addressed acute impacts into the emergent subfield of ecoacoustics and soundscape ecology [85,86]. Holistic approaches for studying underwater sound in relation to marine life allow for the assessment of the overall ecosystem health, which includes an acoustic component, and of certain elements that have long-term and chronic adverse effects on marine life (e.g., low-intensity but pervasive sound from ships).

As with all emerging scientific fields, many information gaps still exist and likely will exist for the foreseeable future. Sound characteristics of many known and novel emerging anthropogenic sound sources have yet to be assessed and validated. Data on sediment composition and associated layering are lacking in many regions, which compromises the integrity and accuracy of any sophisticated sound propagation model. Sound propagation modeling, though firmly established within the underwater acoustics field, needs to be made accessible to the regulatory community, which is largely composed of conservation biologists and environmental policy specialists. Simple analytical computational methods and/or standalone software that does not require programming skills are desirable and need to be developed for regulators to conduct impact assessments. Finally, prioritizing the hiring of scientists who have formal educations in physics, mathematics, or engineering to co-lead or co-manage environmental impact assessments is essential to forming scientifically sound policy and conservation measures that minimize the impacts of underwater sound on marine life.

Author Contributions: Conceptualization: S.G. and T.B.; writing—original draft: S.G. and T.B.; and review and revisions: S.G., T.B., and J.V. All authors have read and agreed to the published version of the manuscript.

Funding: This research received no external funding.

Institutional Review Board Statement: Not applicable.

Informed Consent Statement: Not applicable.

J. Mar. Sci. Eng. **2021**, 9, 173

Acknowledgments: The authors thank Peter Thomas for a thorough review of and comments on this manuscript. The authors also appreciate the constructive comments from the three anonymous reviewers.

Conflicts of Interest: The authors declare no conflict of interest.

References

1. Scammon, C.M. *The Marine Mammals of the North-Western Coast of North America, Described and Illustrated: Together with an Account of the American Whale-Fishery*; J H. Carmany and Co.: San Francisco, CA, USA, 1874.
2. Great Britain Challenger Office; Wyville, T.C.; Sie Murray, J.; Nares, G.S.; Tourle, T.F. *Report on the Scientific Results of the Voyage of H. M. S. Challenger during the Years 1873-76 under the Command of Captain George, S. Nares. v.32, pt.82 (1889)*; Neill and Company: Edinburgh, UK, 1880. Available online: https://www.biodiversitylibrary.org/bibliography/6513#/summary (accessed on 28 September 2020).
3. Cousteau, J.Y. *The Silent World*; Harper & Brother Publishers: New York, NY, USA, 1953.
4. Urick, R.J. *Principles of Underwater Sound*, 3rd ed.; McGraw-Hill Book Company: New York, NY, USA, 1983.
5. Ainslie, M.A. *Principles of Sonar Performance Modelling*; Springer: Berlin/Heidelberg, Germany, 2010.
6. Schevill, W.E.; Lawrence, B. Underwater listening to the white porpoise (*Delphinapterus leucas*). *Science* **1949**, *109*, 143–144. [CrossRef]
7. Kellogg, W.N.; Kohler, R.; Morris, H.N. Porpoise sounds as sonar signals. *Science* **1953**, *117*, 239–243. [CrossRef]
8. Norris, K.N.; Prescott, J.H.; Asa-Dorian, P.V.; Perkins, P. An experimental demonstration of echo-location behavior in the porpoise, *Tursiops truncatus* (Montagu). *Biol. Bull.* **1961**, *120*, 163–176. [CrossRef]
9. Ray, C.; Watkins, W.A.; Burns, J.J. The underwater song of *Erignathus* (bearded seal). *N. Y. Zool. Soc.* **1969**, *54*, 79–83.
10. Payne, R.S.; McVay, S. Songs of humpback whales. *Science* **1971**, *173*, 585–597. [CrossRef] [PubMed]
11. Payne, R.; Webb, D. Orientation by means of long range acoustic signaling in baleen whales. *Ann. N. Y. Acad. Sci.* **1971**, *188*, 110–141. [CrossRef] [PubMed]
12. Doub, J.P. *The Endangered Species Act: History, Implementation, Successes, and Controversies*; Taylor & Francis Group: Boca Raton, FL, USA, 2013.
13. Roman, J.; Altman, I.; Campbell, C.J.; Jasny, M.; Read, A.J.; Dunphy-Daly, M.M. The marine mammal protection act at 40: Status, recovery, and future of U.S. marine mammals. *Ann. N. Y. Acad. Sci.* **2013**, *1286*, 29–49. [CrossRef]
14. Greenwald, N.; Suckling, K.F.; Hartl, B.; Mehrhoff, L.A. Extinction and the U.S. endangered species act. *PeerJ* **2019**, *7*, e6803. [CrossRef] [PubMed]
15. National Research Council. *Dolphins and the Tuna Industry*; National Academies Press: Washington, DC, USA, 1992.
16. Moore, J.E.; Wallace, B.P.; Lewison, R.L.; Žydelis, R.; Cox, T.M.; Crowder, L.B. A review of marine mammal, sea turtle and seabird bycatch in USA fisheries and the role of policy in shaping management. *Mar. Policy* **2009**, *33*, 435–451. [CrossRef]
17. Richardson, W.J.; Wells, R.S.; Würsig, B. Project rationale, design and summary. In *Disturbance Responses and Distribution of Bowhead Whales Balaena Mysticetus in the Eastern Beaufort Sea*; Richardson, W.J., Ed.; Minerals Management. Service: Reston, VA, USA, 1984.
18. Richardson, W.J.; Wells, R.S.; Würsig, B. Disturbance responses of bowheads. In *Behavior, Disturbance Responses and Distribution of Bowhead Whales Balaena Mysticetus in the Eastern Beaufort Sea*; Richardson, W.J., Ed.; Minerals Management Service: Reston, VA, USA, 1984.
19. Richardson, W.J.; Fraker, M.A.; Würsig, B.; Wells, R.S. Behaviour of bowhead whales *Balaena mysticetus* summering in the Beaufort Sea: Reactions to industrial activities. *Biol. Conserv.* **1985**, *32*, 195–230. [CrossRef]
20. Richardson, W.J.; Würsig, B.; Greene, C.R., Jr. Reactions of bowhead whales, *Balaena mysticetus*, to seismic exploration in the Canadian Beaufort Sea. *J. Acoust. Soc. Am.* **1986**, *79*, 1117–1128. [CrossRef] [PubMed]
21. Richardson, W.J.; Greene, C.R., Jr.; Hanna, J.S.; Koski, W.R.; Miller, G.W.; Patenaude, N.J.; Smultea, M.A. *Acoustic Effects of Oil Production Activities on Bowhead and White Whales Visible during Spring Migration Near Pt. Barrow, Alaska–1991 and 1994 Phases: Sound Propagation and Whale Responses to Playbacks of Icebreaker Noise*; Minerals Management Service: Herndon, VA, USA, 1995.
22. Cohen, J. Was underwater "shot" harmful to the whales? *Science* **1991**, *252*, 912–914. [CrossRef] [PubMed]
23. Munk, W.H.; Spindel, R.C.; Baggeroer, A.; Birdsall, T.G. The Heard Island feasibility test. *J. Acoust. Soc. Am.* **1994**, *96*, 2330–2342. [CrossRef]
24. Munk, W.H.; Worcester, P. *Ocean Acoustic Tomography*; Cambridge University Press: Cambridge, UK, 1995.
25. Dushaw, B.; Worcester, P.F.; Munk, W.H.; Spindel, R.C.; Mercer, J.A.; Howe, B.M.; Metzger, K.; Birdsall, T.G.; Andrew, R.K.; Dzieciuch, M.A.; et al. A decade of acoustic thermometry in the North Pacific Ocean. *J. Geophys. Res. Space Phys.* **2009**, *114*, C07021. [CrossRef]
26. Howe, B.M.; Anderson, S.G.; Baggeroer, A.; Colosi, J.A.; Hardy, K.R.; Horwitt, D.; Karig, F.W.; Leach, S.; Mercer, J.A.; Metzger, K., Jr. Instrumentation for the Acoustic Thermometry of Ocean Climate (ATOC) prototype Pacific Ocean network. In *Challenges of our Changing Global Environment, Proceedings of the OCEANS 1995 MTS/IEEE, San Diego, CA, USA, 9–12 October 1995*; IEEE: New York, NY, USA, 1995; Volume 3, pp. 1483–1500.
27. Bowles, A.E.; Smultea, M.; Würsig, B.; DeMaster, D.P.; Palka, D. Relative abundance and behavior of marine mammals exposed to transmissions from the Heard Island Feasibility test. *J. Acoust. Soc. Am.* **1994**, *96*, 2469–2484. [CrossRef]

28. National Research Council. *Low-Frequency Sound and Marine Mammals: Current Knowledge and Research Needs*; National Academies Press: Washington, DC, USA, 1994.

29. National Research Council. *Marine Mammals and Low-Frequency Sound: Progress Since 1994*; National Academies Press: Washington, DC, USA, 2000.

30. Au, W.W.L.; Nachtigall, P.E.; Pawloski, J.L. Acoustic effects of the ATOC signal (75 Hz, 195 dB) on dolphins and whales. *J. Acoust. Soc. Am.* **1997**, *101*, 2973–2977. [CrossRef]

31. Klimley, A.P.; Beavers, S.C. Playback of acoustic thermometry of ocean climate (ATOC) -like signal to bony fishes to evaluate phonotaxis. *J. Acoust. Soc. Am.* **1998**, *104*, 2506–2510. [CrossRef]

32. Costa, D.P.; Crocker, D.E.; Gedamke, J.; Webb, P.M.; Houser, D.S.; Blackwell, S.B.; Waples, D.; Hayes, S.A.; Le Boeuf, B.J. The effect of a low-frequency sound source (acoustic thermometry of the ocean climate) on the diving behavior of juvenile northern elephant seals *Mirounga angustirostris*. *J. Acoust. Soc. Am.* **2003**, *113*, 1155–1165. [CrossRef]

33. Simmonds, M.P.; López-Jurado, L.F. Whales and the military. *Nature* **1991**, *351*, 448. [CrossRef]

34. Frantzis, A. Does acoustic testing strand whales? *Nature* **1998**, *392*, 29. [CrossRef]

35. Evans, D.L.; England, G.R. *Joint Interim Report Bahamas Marine Mammal Stranding Event of 15–16 March 2000*; National Oceanic and Atmospheric Administration: Silver Spring, MD, USA, 2001.

36. D'Amico, A.; Gisiner, R.C.; Ketten, D.R.; Hammock, J.A.; Johnson, C.; Tyack, P.L.; Mead, J. Beaked Whale strandings and naval exercises. *Aquat. Mamm.* **2009**, *35*, 452–472. [CrossRef]

37. Filadelfo, R.; Mintz, J.; Michlovich, E.; D'Amico, A.; Tyack, P.L.; Ketten, D.R. Correlating military sonar use with beaked whale mass strandings: What do the historical data show? *Aquat. Mamm.* **2009**, *35*, 435–444. [CrossRef]

38. Fernández, A.; Sierra, E.; Martín, V.; Méndez, M.; Sacchinni, S.; de Quirós, Y.B.; Andrada, M.; Rivero, M.; Quesada, O.; Tejedor, M.; et al. Last "atypical" beaked whales mass stranding in the Canary Islands (July 2004). *J. Mar. Sci. Res. Deve.* **2012**, *2*, 3. [CrossRef]

39. National Research Council. *Ocean Noise and Marine Mammals*; The National Academies Press: Washington, DC, USA, 2003.

40. National Research Council. *Marine Mammal Population and Ocean Noise: Determining When Noise Causes Biologically Significant Effects*; The National Academies Press: Washington, DC, USA, 2005.

41. National Research Council. *Approaches to Understanding the Cumulative Effects of Stressors on Marine Mammals*; National Academies Press: Washington, DC, USA, 2017.

42. Williams, R.; Wright, A.J.; Ashe, E.; Blight, L.K.; Bruintjes, R.; Canessa, R.; Clark, C.W.; Cullis-Suzuki, S.; Dakin, D.; Erbe, C.; et al. Impacts of anthropogenic noise on marine life: Publication patterns, new discoveries, and future directions in research and management. *Ocean Coast. Manag.* **2015**, *115*, 17–24. [CrossRef]

43. Hawkins, A.D.; Popper, A.N. A sound approach to assessing the impact of underwater noise on marine fishes and inverte-brates. *ICES J. Mar. Sci.* **2016**, *74*, 635–651. [CrossRef]

44. Faulkner, R.C.; Farcas, A.; Merchant, N.D. Guiding principles for assessing the impact of underwater noise. *J. Appl. Ecol.* **2018**, *55*, 2531–2536. [CrossRef]

45. Richardson, W.J.; Greene, C.R., Jr.; Malme, C.I.; Thomson, D.H. *Marine Mammals and Noise*; Academic Press: San Diego, CA, USA, 1995.

46. Chapman, D.M.F.; Ellis, D.D. The elusive decibel: Thoughts on sonars and marine mammals. *Can. Acoust.* **1998**, *26*, 29–31.

47. Nedelec, S.L.; Campbell, J.; Radford, A.N.; Simpson, S.D.; Merchant, N.D. Particle motion: The missing link in underwater acoustic ecology. *Methods Ecol. Evol.* **2016**, *7*, 836–842. [CrossRef]

48. Popper, A.N.; Hawkins, A.D. The importance of particle motion to fishes and invertebrates. *J. Acoust. Soc. Am.* **2018**, *143*, 470–488. [CrossRef]

49. National Marine Fisheries Service. *Technical Guidance for Assessing the Effects of Anthropogenic Sound on Marine Mammal Hearing: Underwater Acoustic Thresholds for Onset of Permanent and Temporary Threshold Shifts*; National Oceanic and Atmospheric Administration: Silver Spring, MD, USA, 2016; p. 178.

50. National Marine Fisheries Service. *Revisions to: Technical Guidance for Assessing the Effects of Anthropogenic Sound on Marine Mammal Hearing Underwater Thresholds for Onset of Permanent and Temporary Threshold Shifts*; National Oceanic and Atmospheric Administration: Silver Spring, MD, USA, 2018; p. 167. Available online: https://www.fisheries.noaa.gov/national/marine-mammal-protection/marine-mammal-acoustic-technical-guidance (accessed on 17 November 2020).

51. Southall, B.L.; Finneran, J.J.; Reichmuth, C.; Nachtigall, P.E.; Ketten, D.R.; Bowles, A.E.; Ellison, W.T.; Nowacek, D.P.; Tyack, P.L. Marine mammal noise exposure criteria: Updated scientific recommendations for residual hearing effects. *Aquat. Mamm.* **2019**, *45*, 125–232, Erratum in **2019**, *45*, 569–572. [CrossRef]

52. Finneran, J.J.; Jenkins, A.K. *Criteria and Thresholds for U.S. Navy Acoustic and Explosive Effects Analysis (Phase III)*; SSC Pacific: San Diego, CA, USA, 2017.

53. Popper, A.N.; Hawkins, A.D.; Fay, R.R.; Mann, D.A.; Bartol, S.; Carlson, T.J.; Coombs, S.; Ellison, W.T.; Gentry, R.L.; Halvorsen, M.B.; et al. *Sound Exposure Guidelines for Fishes and Sea Turtles: A Technical Report Prepared by ANSI-Accredited Standards Committee S3/SC1 and Registered with ANSI*; Springer Nature: Berlin/Heidelberg, Germany, 2014.

54. Scholik-Schlomer, A.R. Where the decibels hit the water: Perspectives on the application of science to real-world underwater noise and marine protected species issues. *Acoust. Today* **2015**, *11*, 36–44.

55. Madsen, P.T. Marine mammals and noise: Problems with root mean square sound pressure levels for transients. *J. Acoust. Soc. Am.* **2005**, *117*, 3952–3957. [CrossRef]

56. Buehler, D.; Oestman, R.; Reyff, J.; Pommerenck, K.; Mitchell, B. *Technical Guidance for Assessment and Mitigation of the Hydroacoustic Effects of Pile Driving on Fish*; California Department of Transportation: Sacramento, CA, USA, 2015. Available online: http://website.dot.ca.gov/env/bio/docs/biotech-guidance-hydroacoustic-effects-110215.pdf (accessed on 28 September 2020).

57. Guan, S.; Miner, R. Underwater noise characterization of down-the-hole pile driving activities off Biorka Island, Alaska. *Mar. Pollut. Bull.* **2020**, *160*, 111664. [CrossRef]

58. Wenz, G.M. Acoustic ambient noise in the ocean: Spectra and sources. *J. Acoust. Soc. Am.* **1962**, *34*, 1936–1956. [CrossRef]

59. Urick, R.J. *Ambient Noise in the Sea*; Naval Sea Systems Command, Department of the Navy: Washington, DC, USA, 1984.

60. Southall, B.L.; Bowles, A.E.; Ellison, W.T.; Finneran, J.J.; Gentry, R.L.; Greene, C.R., Jr.; Kastak, D.K.; Ketten, D.R.; Miller, J.H.; Nachtigall, P.E.; et al. Marine mammal noise-exposure criteria: Initial scientific recommendations. *Aquat. Mamm.* **2007**, *33*, 411–521. [CrossRef]

61. Martin, S.B.; Lucke, K.; Barclay, D.R. Techniques for distinguishing between impulsive and non-impulsive sound in the context of regulating sound exposure for marine mammals. *J. Acoust. Soc. Am.* **2020**, *147*, 2159–2176. [CrossRef] [PubMed]

62. National Marine Fisheries Service. *Manual for Optional USER SPREADSHEET Tool (Version 2.2, December) for: 2018 Revision to: Technical Guidance for Assessing the Effects of Anthropogenic Sound on Marine Mammal Hearing (Version 2.0): Underwater Thresholds for Onset of Permanent and Temporary Threshold Shifts*; National Oceanic and Atmospheric Administration: Silver Spring, MD, USA, 2020. Available online: https://media.fisheries.noaa.gov/2020-12/User_Manual%20_DEC_2020_508.pdf?null (accessed on 22 December 2020).

63. Ireland, D.; Rodrigues, R.; Hannay, D.; Jankowski, M.; Hunter, A.; Patterson, H.; Haley, B.; Funk, D.W. *Marine Mammal Monitoring and Mitigation during Open Water Seismic Exploration by ConocoPhillips Alaska Inc. in the Chukchi Sea, July–October 2006: 90-Day Report*; Technical Report Prepared for Shell Offshore, Inc.; National Marine Fisheries Service; U.S. Fish and Wildlife Service; LGL Research Associates, Inc.: Anchorage, AK, USA, 2007.

64. Ireland, D.S.; Rodrigues, R.; Funk, D.; Koski, W.; Hannay, D. *Marine Mammal Monitoring and Mitigation during Open Water Seismic Exploration by Shell Offshore Inc. in the Chukchi and Beaufort Seas, July–October 2008: 90-Day Report*; Technical Report Prepared for Shell Offshore Inc.; National Marine Fisheries Service; U.S. Fish and Wildlife Service; LGL Research Associates, Inc.: Anchorage, AK, USA, 2009.

65. Funk, D.; Hannay, D.; Ireland, D.; Rodrigues, R.; Koski, W. *Marine Mammal Monitoring and Mitigation during Open Water Seism Offshore Inc. In the Chukchi and Beaufort Seas, July–November 2007: 90-Day Report*; Technical Report Prepared for Shell Offshore Inc.; National Marine Fisheries Service; U.S. Fish and Wildlife Service; LGL Research Associates, Inc.: Anchorage, AK, USA, 2008.

66. Reiser, C.M.; Funk, D.W.; Rodrigues, R.; Hannay, D. *Marine Mammal Monitoring and Mitigation during Open Water Seismic Exploration by Shell Offshore, Inc. in the Alaskan Chukchi Sea, July–October 2009: 90-Day Report*; Technical Report Prepared for Shell Offshore Inc.; National Marine Fisheries Service; U.S. Fish and Wildlife Service; LGL Research Associates, Inc.: Anchorage, AK, USA, 2010.

67. Reiser, C.M.; Funk, D.W.; Rodrigues, R.; Hannay, D. *Marine Mammal Monitoring and Mitigation during Marine Geophysical Surveys by Shell Offshore, Inc. In the Alaskan Chukchi and Beaufort seas, July–October 2010: 90-Day Report*; Technical Report Prepared for Shell Offshore Inc.; National Marine Fisheries Service; U.S. Fish and Wildlife Service; LGL Research Associates, Inc.: Anchorage, AK, USA, 2011.

68. Denes, S.L.; Zeddies, D.G.; Weirathmueller, M.M. *Turbine Foundation and Cable Installation at South Fork Wind Farm: Underwater Acoustic Modeling of Construction Noise*; Technical Report for Jacobs Engineering Group Inc. JASCO Applied Sciences (USA) Inc.: Silver Spring, MD, USA, 2020.

69. Etter, P.C. *Underwater Acoustic Modeling and Simulation*, 4th ed.; CRC Press: Boca Raton, FL, USA, 2013.

70. Medwin, H.; Clay, C.S. *Fundamentals of Acoustical Oceanography*; Academic Press: San Diego, CA, USA, 2008.

71. Au, W.W.L.; Hastings, M.C. *Principles of Marine Bioacoustics*; Springer: New York, NY, USA, 2008.

72. Lurton, X. *An Introduction to Underwater Acoustics—Principles and Applications*, 2nd ed.; Springer: Chichester, UK, 2010.

73. Erbe, C. Underwater Acoustics: Noise and the effects on marine mammals. *Pocket Handb.* **2011**, *164*, 9–10.

74. National Marine Fisheries Service. Takes of marine mammals incidental to specified activities; Taking marine mammals incidental to the transit protection program pier and support facilities project at Naval Base Kitsap Bangor, Washington. *Fed. Reg.* **2020**, *85*, 48206–48225.

75. Austin, M.; Denes, S.; MacDonnell, J.; Warner, G. *Hydroacoustic Monitoring Report: Anchorage Port Modernization Project Test Pile Program. Version 3.0*; Technical Report for Kiewit Infrastructure West Co. JASCO Applied Sciences (Alaska) Inc.: Anchorage, AK, USA, 2016.

76. National Marine Fisheries Service. Taking and importing marine mammals; Taking marine mammals incidental to the U.S. Navy training and testing activities in the Hawaii-Southern California training and testing study area. *Fed. Reg.* **2018**, *83*, 66846–67031.

77. National Marine Fisheries Service. Takes of marine mammals incidental to specified activities; Taking marine mammals incidental to construction of the Vineyard Wind Offshore Wind Project. *Fed. Reg.* **2019**, *84*, 18346–18381.

78. Finneran, J.J. Noise-induced hearing loss in marine mammals: A review of temporary threshold shift studies from 1996 to 2015. *J. Acoust. Soc. Am.* **2015**, *138*, 1702–1726. [CrossRef] [PubMed]

79. Houser, D.S. A method for modeling marine mammal movement and behavior for environmental impact assessment. *IEEE J. Ocean. Eng.* **2006**, *31*, 76–81. [CrossRef]

80. Blackstock, S.A.; Fayton, J.O.; Hulton, P.H.; Moll, T.E.; Jenkins, K.; Kotecki, S.; Henderson, E.; Rider, S.; Martin, C.; Bowman, V. *Quantifying Acoustic Impacts on Marine Mammals and Sea Turtles: Methods and Analytical Approach for Phase III Training and Testing*; Space and Naval Warfare Systems Center Pacific: San Diego, CA, USA, 2017.

81. Bassett, C.; Polagye, B.; Holt, M.; Thomson, J. A vessel noise budget for Admiralty Inlet, Puget Sound, Washington (USA). *J. Acoust. Soc. Am.* **2012**, *132*, 3706–3719. [CrossRef] [PubMed]

82. McKenna, M.F.; Ross, D.; Wiggins, S.M.; Hildebrand, J.A. Underwater radiated noise from modern commercial ships. *J. Acoust. Soc. Am.* **2012**, *131*, 92–103. [CrossRef]

83. Erbe, C.; Marley, S.A.; Schoeman, R.P.; Smith, J.N.; Trigg, L.E.; Embling, C.B. The effects of ship noise on marine mammals—A Review. *Front. Mar. Sci.* **2019**, *6*, 606. [CrossRef]

84. Howe, B.M.; Miksis-Olds, J.; Rehm, E.; Sagen, H.; Worcester, P.F.; Haralabus, G. Observing the Oceans Acoustically. *Front. Mar. Sci.* **2019**, *6*, 426. [CrossRef]

85. Farina, A. *Soundscape Ecology: Principles, Patterns, Methods and Applications*; Springer: New York, NY, USA, 2014.

86. Farina, A.; Gage, S.H. *Ecoacoustics: The Ecological Role of Sound*; John Wiley and Sons: Hoboken, NJ, USA, 2017.

87. Marine Mammal Commission. *Comments and Recommendations from the U.S. Marine Mammal Commission on Fish and Wildlife Service's Proposed Issuance of a Regulation to Hilcorp Alaska, Harvest Alaska, and the Alaska Gasline Development Corporation*; Marine Mammal Commission: Bethesda, MD, USA, 2019. Available online: https://www.mmc.gov/wp-content/uploads/19-04-18-Putnam-Cook-Inlet-OG-activities-proposed-ITR.pdf (accessed on 10 January 2021).

88. Marine Mammal Commission. *Comments and Recommendations from the U.S. Marine Mammal Commission on Fish and Wildlife Service's Proposed Issuance of Incidental Harassment Authorizations to the City and Borough of Sitka and Duck Point Development II, LLC.*; Marine Mammal Commission: Bethesda, MD, USA, 2019. Available online: https://www.mmc.gov/wp-content/uploads/19-07-24-Lemons-and-Putnam-CBS-and-DPD-IHAs.pdf (accessed on 10 January 2021).

89. Marine Mammal Commission. *Comments and Recommendations from the U.S. Marine Mammal Commission on National Marine Fisheries Service's Proposed Issuance of an Incidental Harassment Authorization to Alaska Marine Lines*; Marine Mammal Commission: Bethesda, MD, USA, 2020. Available online: https://www.mmc.gov/wp-content/uploads/20-01-09-Harrison-Alaska-Marines-Line-IHA-003.pdf (accessed on 10 January 2021).

90. Marine Mammal Commission. *Comments and Recommendations from the U.S. Marine Mammal Commission on National Marine Fisheries Service's Proposed Issuance of an Incidental Harassment Authorization to the National Science Foundation's Office of Polar Programs to Take Marine Mammals Incidental to Conducting a Marine Geophysical Survey in the Amundsen Sea in February 2020. 21 January 2020*; Marine Mammal Commission: Bethesda, MD, USA, 2020. Available online: https://www.mmc.gov/wp-content/uploads/20-01-21-Harrison-NSF-Amundsen-Sea-IHA.pdf (accessed on 10 January 2021).

91. Marine Mammal Commission. *Comments and Recommendations from the U.S. Marine Mammal Commission on National Marine Fisheries Service's Proposed Issuance of an Incidental Harassment Authorization to Port of Alaska to take Marine Mammals Incidental to Construction of a New Petroleum and Cement Terminal in Anchorage, Alaska. 23 January 2020*; Marine Mammal Commission: Bethesda, MD, USA, 2020. Available online: https://www.mmc.gov/wp-content/uploads/20-01-23-Harrison-POA-IHAs.pdf (accessed on 10 January 2021).

92. Marine Mammal Commission. *Comments and Recommendations from the U.S. Marine Mammal Commission on National Marine Fisheries Service's Proposed Issuance of an Incidental Harassment Authorization to the Power Systems & Supplies of Alaska to Take Marine Mammals Incidental to Dock Construction in Ketchikan, Alaska. 23 March 2020*; Marine Mammal Commission: Bethesda, MD, USA, 2020. Available online: https://www.mmc.gov/wp-content/uploads/20-03-23-Harrison-PSSA-IHA.pdf (accessed on 10 January 2021).

93. Marine Mammal Commission. *Comments and Recommendations from the U.S. Marine Mammal Commission on National Marine Fisheries Service's Proposed Issuance of an Incidental Harassment Authorization to Hampton Roads Connector Partners to Take Marine Mammals Incidental to Conducting Construction Activities for the Hampton Roads Bridge-Tunnel Expansion Project in Virginia. 20 April 2020*; Marine Mammal Commission: Bethesda, MD, USA, 2020. Available online: https://www.mmc.gov/wp-content/uploads/20-04-20-Harrison-HRCP-IHA.pdf (accessed on 10 January 2021).

94. Marine Mammal Commission. *Comments and Recommendations from the U.S. Marine Mammal Commission on National Marine Fisheries Service's Proposed Issuance of an Incidental Harassment Authorization to the Gastineau Channel Historical Society to Take Marine Mammals Incidental to Constructing a Mooring Float Near Juneau, Alaska. 24 April 2020*; Marine Mammal Commission: Bethesda, MD, USA, 2020. Available online: https://www.mmc.gov/wp-content/uploads/20-04-24-Harrison-GCHS-IHA.pdf (accessed on 10 January 2021).

95. Marine Mammal Commission. *Comments and Recommendations from the U.S. Marine Mammal Commission on National Marine Fisheries Service's Proposed Issuance of an Incidental Harassment Authorization to the City and County of San Francisco to Take Marine Mammals Incidental to Conducting Various Construction Activities on Treasure Island in San Francisco Bay, California. 29 June 2020*; Marine Mammal Commission: Bethesda, MD, USA, 2020. Available online: https://www.mmc.gov/wp-content/uploads/20-06-29-Harrison-City-of-SF-IHA.pdf (accessed on 10 January 2021).

96. Marine Mammal Commission. *Comments and Recommendations from the U.S. Marine Mammal Commission on National Marine Fisheries Service's Proposed Issuance of an Incidental Harassment Authorization to Lamont-Doherty Earth Observatory to Take Marine Mammals Incidental to Conducting a Marine Geophysical Survey in the Bering Sea and North Pacific Ocean. 13 August 2020*; Marine Mammal Commission: Bethesda, MD, USA, 2020. Available online: https://www.mmc.gov/wp-content/uploads/20-08-13-Harrison-LDEO-Aleutian-IHA.pdf (accessed on 10 January 2021).

97. Dazey, E.; McIntosh, B.; Brown, S.; Dudzinski, K.M. Assessment of underwater anthropogenic noise associated with construction activities in Bechers Bay, Santa Rosa Island, California. *J. Environ. Prot.* **2012**, *3*, 1286–1294. [CrossRef]

98. Denes, S.L.; Warner, G.J.; Austin, M.E.; MacGillivray, A.O. *Hydroacoustic Pile Driving Noise Study—Comprehensive Report*; JASCO Applied Sciences Inc.: Anchorage, AK, USA, 2016; Available online: http://www.dot.state.ak.us/stwddes/research/assets/pdf/4000-135.pdf (accessed on 10 January 2021).

99. Denes, S.; Vallarta, J.; Zeddies, D. *Sound Source Characterization of Down-the-Hole Hammering: Thimble Shoal, Virginia*; JASCO Applied Sciences (USA) Inc.: Silver Spring, MD, USA, 2019. Available online: https://www.fisheries.noaa.gov/webdam/download/105110147 (accessed on 10 January 2021).

100. Reyff, J. *Review of Down-the-Hole Rock Socket Drilling Acoustic Data Measured for White Pass and Yukon Route (WP&YR) Mooring Dolphins*; Illingworth & Rodkin, Inc.: Cotati, CA, USA, 2020; p. 8.

101. Reyff, J.; Heyvaert, C. *White Pass and Yukon Railroad Mooring Dolphin Installation: Pile Driving and Drilling Sound Source Verifica-tion*; Illingworth & Rodkin, Inc.: Cotati, CA, USA, 2019. Available online: https://www.fisheries.noaa.gov/webdam/download/104795528 (accessed on 28 September 2020).

102. Deng, Z.D.; Southall, B.L.; Carlson, T.J.; Xu, J.; Martinez, J.J.; Weiland, M.A.; Ingraham, J.M. 200 kHz commercial sonar systems generate lower frequency side lobes audible to some marine mammals. *PLoS ONE* **2014**, *9*, e95315. [CrossRef]

103. Hastie, G.D.; Donovan, C.; Götz, T.; Janik, V.M. Behavioral responses by grey seals (*Halichoerus grypus*) to high frequency sonar. *Mar. Pollut. Bull.* **2014**, *79*, 205–210. [CrossRef]

104. Cholewiak, D.M.; DeAngelis, A.I.; Palka, D.; Corkeron, P.J.; Van Parijs, S.M. Beaked whales demonstrate a marked acoustic response to the use of shipboard echosounders. *R. Soc. Open Sci.* **2017**, *4*, 170940. [CrossRef]

105. Varghese, H.K.; Miksis-Olds, J.L.; DiMarzio, N.; Lowell, K.; Linder, E.; Mayer, L.; Moretti, D. The effect of two 12 kHz multibeam mapping surveys on the foraging behavior of Cuvier's beaked whales off of southern California. *J. Acoust. Soc. Am.* **2020**, *147*, 3849–3858. [CrossRef]

106. Crocker, S.E.; Fratantonio, F.D. *Characteristics of Sounds Emitted during High-Resolution Marine Geophysical Surveys*; Naval Undersea Warfare Center Division: Newport, RI, USA, 2016.

107. LGL Ecological Research Associates Inc. *Marine Acoustics Inc. Environmental Assessment of Marine Vibroseis*; Technical Report for Joint Industry Programme, E & P Sound and Marine Life; International Association of Oil & Gas Producers: London, UK, 2011; p. 207.

108. CSA. *Quieting Technologies for Reducing Noise during Seismic Surveying and Pile Driving Workshop*; Bureau of Ocean Energy Management: Herdon, VA, USA, 2014.

109. Long, A.; Tenghamn, R. Marine vibrator concepts for modern seismic challenges. *ASEG Ext. Abstr.* **2018**, *2018*, 1–4. [CrossRef]

110. Orji, O.; Oscarsson-Nagel, M.D.C.; Söllner, W.; Trætten, Ø.; Armstrong, B.; Nams, D.; Yeatman, P. Marine vibrator source: Modular projector system. In *SEG Technical Program Expanded Abstracts 2019*; Bevc, D., Nedorub, O., Eds.; Society of Exploration Geophysicists: Tulsa, OK, USA, 2019. [CrossRef]

111. Arons, A.B. Secondary pressure pulses due to gas globe oscillation in underwater explosions. II. Selection of adiabatic parameters in the theory of oscillation. *J. Acoust. Soc. Am.* **1948**, *20*, 277–282. [CrossRef]

112. Cole, R.H.; Weller, R. *Underwater Explosions*; Princeton University Press: Preston, NJ, USA, 1948.

113. Weston, D.E. Underwater Explosions as Acoustic Sources. *Proc. Phys. Soc.* **1960**, *76*, 233–249. [CrossRef]

114. Rogers, P.H. Weak-shock solution for underwater explosive shock waves. *J. Acoust. Soc. Am.* **1977**, *62*, 1412–1419. [CrossRef]

115. Nedwell, J.R.; Thandavamoorthy, T.S. The waterborne pressure wave from buried explosive charges: An experimental investigation. *App. Acoust.* **1992**, *37*, 1–14. [CrossRef]

116. Hall, M.V. Underwater signals from confined explosions in very shallow water. In Proceedings of the 20th International Congress Acoustics ICA 2010, Sydney, Australia, 3–27 August 2010; p. 5.

117. Beland, J.A.; Ireland, D.S.; Bisson, L.N.; Hannay, D. *Marine Mammal Monitoring and Mitigation during a Marine Seismic Survey by ION Geophysical in the Arctic Ocean, October–November 2012: 90-Day Report*; Technical Report for ION, Geophysical; Nature Marine Fisheries Service, and U.S. Fish and Wildfire Service; LGL Research Associates, Inc.: Anchorage, AK, USA, 2013.

118. Cate, J.R.; Smultea, M.; Blees, M.; Larson, M.; Simpson, S.; Jefferson, T.; Steckler, D. *90-Day Report of Marine Mammal Monitoring and Mitigation during a 2D Seismic Survey by TGS in the Chukchi Sea, August through October 2013*; Technical Report for TGS-NOPEC Geophysical Company; National Marine Fisheries Service and U.S. Fish and Wildlife Service; ASRC Energy Services: Anchorage, AK, USA, 2014.

119. Cate, J.R.; Blees, M.; Larson, M.; Simpson, S.; Mills, R.; Cooper, R. *90-Day Report of Marine Mammal Monitoring and Mitigation during a Shallow Geohazard Survey Hilcorp in Foggy Island Bay, Alaska, July 2015*; Technical Report for Hilcorp Alaska, LLC; National Marine Fisheries Service and U.S. Fish and Wildlife Service; ASRC Energy Services: Anchorage, AK, USA, 2015.

120. Laws, R.M. The Interaction of Marine Seismic Sources. Ph.D. Thesis, University of London, London, UK, 1991.

121. Laws, R.M.; Hatton, L.; Haartsen, M. Computer Modelling of Clustered Airguns. *First Break* **1990**, *8*, 331–338. [CrossRef]

122. PGS. Nucleus+: Survey Design and Modeling. 2020. Available online: https://www.pgs.com/marine-acquisition/tools-and-techniques/acquisition-solutions/technology/nucleus/ (accessed on 27 November 2020).

123. MacGillivray, A.O. Acoustic Modelling Study of Seismic Airgun Noise in Queen Charlotte Basin. Master's Thesis, University of Victoria, Victoria, BC, USA, 2006.

124. Ainslie, M.A.; Halvorsen, M.B.; Dekeling, R.; Laws, R.M.; Duncan, A.J.; Frankel, A.S.; Heaney, K.D.; Küsel, E.T.; MacGillivray, A.; Prior, M.K.; et al. Verification of airgun sound field models for environmental impact assessment. *Proc. Meet. Acoust.* **2016**, *27*, 070018. [CrossRef]

125. Ainslie, M.A.; Laws, R.M.; Sertlek, H.Ö. International Airgun Modeling Workshop: Validation of source signature and sound propagation models—Dublin (Ireland), 16 July 2016—Problem description. *IEEE J. Ocean. Eng.* **2019**, *44*, 565–574. [CrossRef]

126. MacGillivray, A. A model for underwater sound levels generated by marine impact pile driving. *Proc. Meet. Acoust.* **2014**, *20*, 045008.

127. Lippert, S.; Nijhof, M.; Lippert, T.; Wilkes, D.; Gavrilov, A.; Heitmann, K.; Ruhnau, M.; Von Estorff, O.; Schafke, A.; Schafer, I.; et al. COMPILE—A generic benchmark case for predictions of marine pile-driving noise. *IEEE J. Ocean. Eng.* **2016**, *41*, 1061–1071. [CrossRef]

128. Jeong, C.; Manalaysay, A.; Gharti, H.N.; Guan, S.; Vignola, J. Applicability of 3D Spectral Element Method for Computing Close-Range Underwater Piling Noises. *J. Theor. Comput. Acoust.* **2019**, *27*, 1950012. [CrossRef]

129. Jensen, F.B.; Kuperman, W.A.; Porter, M.B.; Schmidt, H. *Computational Ocean Acoustics*, 2nd ed.; Springer: New York, NY, USA, 2010.

130. National Marine Fisheries Service. *Interim Recommendation for Sound Source Level and Propagation Analysis for High Resolution Geophysical (HRG) Sources*; National Marine Fisheries Service: Silver Spring, MD, USA, 2020; p. 5.

131. Heaney, K.D.; Ainslie, M.A.; Halvorsen, M.B.; Seger, K.D.; Müller, R.A.J.; Nijhof, M.J.J.; Lippert, T. *A Parametric Analysis and Sensitivity Study of the Acoustic Propagation for Renewable Energy Sources*; U.S. Department of the Interior Bureau of Ocean Energy Management: Sterling, VA, USA, 2020; p. 165. Available online: https://espis.boem.gov/final%20reports/BOEM_2020-011.pdf (accessed on 28 November 2020).

132. Lippert, T.; Ainslie, M.A.; Von Estorff, O. Pile driving acoustics made simple: Damped cylindrical spreading model. *J. Acoust. Soc. Am.* **2018**, *143*, 310–317. [CrossRef]

133. Müller, R.A.; Ainslie, M.A.; Halvorsen, M.B.; Lippert, T. Empirical modelling for derived metrics as function of sound exposure level in marine pile driving. *J. Acoust. Soc. Am.* **2018**, *144*, 1809. [CrossRef]

134. Ainslie, M.A.; Halvorsen, M.B.; Müller, R.A.J.; Lippert, T. Application of damped cylindrical spreading to assess range to injury threshold for fishes from impact pile driving. *J. Acoust. Soc. Am.* **2020**, *148*, 108–121. [CrossRef] [PubMed]

135. Erbe, C.; Farmer, D. Masked hearing thresholds of a beluga whale (*Delphinapterus leucas*) in icebreaker noise. *Deep Sea Res.* **1998**, *45*, 1373–1388. [CrossRef]

136. Branstetter, B.K.; Trickey, J.S.; Bakhtiari, K.; Black, A.; Aihara, H.; Finneran, J.J. Auditory masking patterns in bottlenose dolphins (*Tursiops truncatus*) with natural, anthropogenic, and synthesized noise. *J. Acoust. Soc. Am.* **2013**, *133*, 1811–1818. [CrossRef] [PubMed]

137. Cunningham, K.C.; Southall, B.L.; Reichmuth, C. Auditory sensitivity in complex listening scenarios. *J. Acoust. Soc. Am.* **2014**, *136*, 3410–3421. [CrossRef] [PubMed]

138. Erbe, C.; Reichmuth, C.; Cunningham, K.; Lucke, K.; Dooling, R. Communication masking in marine mammals: A review and research strategy. *Mar. Pollut. Bull.* **2016**, *103*, 15–38. [CrossRef]

139. Guerra, M.; Thode, A.M.; Blackwell, S.B.; Macrander, A.M. Quantifying seismic survey reverberation off the Alaskan North Slope. *J. Acoust. Soc. Am.* **2011**, *130*, 3046–3058. [CrossRef]

140. Guan, S.; Vignola, J.; Judge, J.; Turo, D. Airgun inter-pulse noise field during a seismic survey in an Arctic ultra shallow marine environment. *J. Acoust. Soc. Am.* **2015**, *138*, 3447–3457. [CrossRef] [PubMed]

141. Guan, S.; Southall, B.L.; Vignola, J.F.; Judge, J.A.; Turo, D. Sonar inter-ping noise field characterization during cetacean be-havioral response studies off southern California. *Acoust. Phys.* **2017**, *63*, 204–215. [CrossRef]

142. Clark, C.; Ellison, W.; Southall, B.; Hatch, L.; Van Parijs, S.; Frankel, A.; Ponirakis, D. Acoustic masking in marine ecosystems: Intuitions, analysis, and implication. *Mar. Ecol. Prog. Ser.* **2009**, *395*, 201–222. [CrossRef]

143. Sills, J.M.; Southall, B.L.; Reichmuth, C. The influence of temporally varying noise from seismic air guns on the detection of underwater sounds by seals. *J. Acoust. Soc. Am.* **2017**, *141*, 996–1008. [CrossRef]

144. Dugan, P.; Pourhomayoun, M.; Shiu, Y.; Paradis, R.; Rice, A.N.; Clark, C. Using high performance computing to explore large complex bioacoustic soundscapes: Case study for right whale acoustics. *Procedia Comput. Sci.* **2013**, *20*, 156–162. [CrossRef]

145. Sethi, S.S.; Jones, N.S.; Fulcher, B.D.; Picinali, L.; Clink, D.J.; Klinck, H.; Orme, C.D.L.; Wrege, P.H.; Ewers, R.M. Characterizing soundscapes across diverse ecosystems using a universal acoustic feature set. *Proc. Natl. Acad. Sci. USA* **2020**, *117*, 17049–17055. [CrossRef]

146. Shiu, Y.; Palmer, K.J.; Roch, M.A.; Fleishman, E.; Liu, X.; Nosal, E.-M.; Helble, T.; Cholewiak, D.; Gillespie, D.; Klinck, H. Deep neural networks for automated detection of marine mammal species. *Sci. Rep.* **2020**, *10*, 1–12. [CrossRef]

147. Pierce, A.D. *Acoustics: An Introduction to Its Physical Principles and Applications*; Acoustical Society of America: Woodbury, NW, USA, 1991.

Journal of
*Marine Science
and Engineering*

Article

Turning Scientific Knowledge into Regulation: Effective Measures for Noise Mitigation of Pile Driving

Carina Juretzek, Ben Schmidt and Maria Boethling *

Bundesamt für Seeschifffahrt und Hydrographie (BSH), Bernhard-Nocht Straße 78, 20359 Hamburg, Germany;
carina.juretzek@bsh.de (C.J.); ben.schmidt@bsh.de (B.S.)
* Correspondence: maria.Boethling@bsh.de

Abstract: Pile driving is one of the most intense anthropogenic noise sources in the marine environment. Each foundation pile may require up to a several thousand strokes of high hammer energy to be driven to the embedded depth. Scientific evidence shows that effects on the marine environment have to be anticipated if mitigation measures are not applied. Effective mitigation measures to prevent and reduce the impact of pile driving noise should therefore be part of regulation. The role of regulators is to demonstrate and assess the applicability, efficiency and effectiveness of noise mitigation measures. This requires both, scientific knowledge on noise impacts and the consideration of normative aspects of noise mitigation. The establishment of mitigation procedures in plans and approvals granted by regulatory agencies includes several stages. Here, we outline a step-wise approach in which most of the actions described may be performed simultaneously. Potential measures include the appropriate maritime spatial planning to avoid conflicts with nature conservation, site development for offshore wind farms to avoid undesirable activities in time and space, coordination of activities to avoid cumulative effects, and the application of technical noise abatement systems to reduce noise at the source. To increase the acceptance of noise mitigation applications, technical measures should fulfil a number of requirements: (a) they are applicable and affordable, (b) they are state-of-the-art or at least advanced in development, (c) their efficiency can be assessed with standardised procedures. In this study, the efficiency of noise mitigation applied recently in offshore wind farm construction projects in the German North Sea is explained and discussed with regard to the regulation framework, including the technical abatement of impulsive pile driving noise.

Keywords: ocean noise mitigation; ocean noise regulation; ocean sound measurement

Citation: Juretzek, C.; Schmidt, B.; Boethling, M. Turning Scientific Knowledge into Regulation: Effective Measures for Noise Mitigation of Pile Driving. *J. Mar. Sci. Eng.* **2021**, *9*, 819. https://doi.org/10.3390/jmse9080819

Academic Editors: Michel André and Christine Erbe

Received: 14 June 2021
Accepted: 21 July 2021
Published: 29 July 2021

Publisher's Note: MDPI stays neutral with regard to jurisdictional claims in published maps and institutional affiliations.

1. Introduction

An essential characteristic of the human being is the ability to reach ever new horizons. The development of new technologies is increasingly expanding the reach of human influence on the sea. For a decade now, this has also included the first-time and large-scale exploration of nearshore and offshore sea regions for the generation of green energy as part of the growing Blue Economy. Today's generations of people are shaping and experiencing this new type of development at sea. However, we should bear in mind that marine species also experience the effects of these human activities. The reconciliation of both needs, those of humans and those of marine species, is therefore a process that must accompany these new developments. The most common installation method for foundations of offshore structures in the (southern) North and Baltic Sea is pile driving [1]. A major effort has been made to better understand the risks for adverse effects on the marine environment due to underwater noise during foundation installations by pile driving. These risks have been found to include very relevant adverse effects, such as impaired hearing and temporary habitat loss. Effects of pile driving on marine life and the need for environmental monitoring and noise mitigation measures that invariably accompany impact pile driving, particularly in biologically sensitive subsea habitats, are well described [2]. Alternative installation methods such as gravity foundations, suction buckets or vibratory pile driving are either still under development or to date only applicable to specific site conditions. Due to local site conditions, including geological conditions and technical developments, pile driving is expected to remain one of the most common installation methods for the next decade [2–4].

The presence of underwater noise in the ocean and its impacts on marine life have received increasing attention in recent years [5–10]. Science has made progress on both related aspects, the physics of underwater noise and the effects of noise on marine organisms. At the same time, work is underway to research or develop underwater noise-reducing technologies for various anthropogenic activities [11,12].

Since 2011, guidance documents for underwater noise measurements during construction of offshore wind farms have been published in Germany, the Netherlands and the UK to provide advice on good practice and to make data comparable [13–15]. This was followed by an international effort to establish a common terminology for underwater acoustics that defines terms and expressions used in the field of underwater acoustics, including natural, biological, and anthropogenic sound by ISO [16]. ISO 18406:2017 [17] describes the methods, procedures and measurement systems for measuring radiated underwater sound generated by pile driving operations using hammer strokes. Furthermore, many research projects have developed models to describe sound generation in the near field and sound propagation during pile driving activities [18–21].

The description of noise criteria and frequency weighting functions for marine mammals grouped according to their hearing abilities [22] has supported noise regulation and mitigation efforts worldwide. In 2019, the proposed noise criteria were revised and evaluated in the light of new scientific findings. Dual exposure metrics are provided for impulsive noise criteria, including the frequency-weighted sound exposure level (SEL) and the unweighted peak sound pressure level (SPL) (expressed in units relative to 1 μPa for water). Exposures exceeding the respective criterion level for each exposure metric are interpreted as causing a predicted temporary threshold shift (TTS) or permanent threshold shift (PTS) onset [22].

The critical role of exposure context in addition to received noise levels has been extensively studied. One review of studies on noise criteria found that behavioural responses in cetaceans (measured using a linear severity scale) are best described by the interaction between sound source (continuous sound, sonar or seismic/explosion) and functional auditory group (a proxy for hearing capabilities) [10]. Importantly, more severe behavioural responses were not consistently associated with higher received noise levels and vice versa. The authors recommend replacing the severity score of behavioural response with a dichotomous approach (response/no response) that can provide a measure of impact in terms of habitat loss and degradation [10].

Based on scientific results, efforts were made to describe regulatory frameworks for mitigating noise from impulsive sources. Further, a frequency-dependent weighting function and numerical thresholds for the onset of temporary threshold shift (TTS) and permanent threshold shift (PTS) were derived for each group of marine mammals from the available data describing the hearing ability and the effects of noise on marine mammals [23].

Acoustic thresholds based on TTS and PTS onset levels for marine mammal auditory groups for assessing the effects of anthropogenic sound were updated and included in a technical guidance published by NMFS [24] for regulatory purposes. The acoustic thresholds according to NMFS [24] are based on dual metrics of cumulative sound exposure level (SEL_{cum}) and peak sound level (L_{peak}) for impulsive sounds and on SEL_{cum} for non-impulsive sounds.

Policy options to mitigate underwater noise and employ noise abatement systems to reduce the impact on marine life have recently been reviewed [25]. Today, noise emissions can be reduced by placing acoustic barriers around the pile driving operation. Various designs have been developed, using air bubbles, solid barriers, or combinations of these. However, regulatory responses to offshore wind farm construction noise have differed.

As noted in [25], in Germany and several other Western European countries (e.g., the Netherlands and Denmark), regulations are routinely applied which de facto mandate the use of noise abatement measures such as bubble curtains (see below). On the other hand, noise abatement has not yet been implemented in other countries, including two of the three largest offshore wind energy producers by wattage. Looking at the economic factors, the author [25] concludes that the German approach, based on a command-and-control regulation, has substantially reduced noise pollution from pile driving while allowing renewable energy development to continue, and is therefore the most effective noise management model currently available.

Here, we outline a possible way to build a mitigation strategy and detail a step-wise approach for implementing noise mitigation measures in practice. Furthermore, we present and discuss recent examples of such a regulation scheme in regard of noise mitigation of impulsive pile driving noise by considering representative examples of monitoring data from acoustic measurements with regard to current conditions at offshore construction sites. Moreover, we analyse the noise emissions of mitigated piling in the German Bight from an offshore wind farm cluster at different distances. Additionally, we discuss implications for future monitoring requirements in terms of the variety of processes at the construction site.

2. Materials and Methods

As a strictly protected species according to the Habitat directive (92/43/EWG) and the Federal Nature Conservation Act in Germany, the harbour porpoise represents the key species in German waters of the North Sea and Baltic Sea and temporary hearing threshold shift (TTS) in harbour porpoises is classified as injury. Since 2008, the com-pliance with a dual sound pressure threshold criterion for pile driving activities has been mandatory for all wind farm construction projects in the German exclusive eco-nomic zone (EEZ) approved by BSH in order to prevent TTS in harbour porpoises. Fol-lowing the precautionary principle for the protection of harbour porpoises and the marine environment, noise emissions from pile driving at a measuring distance of 750 m from the piling location must not exceed the dual criterion given by

- an unweighted sound exposure level SEL (L_E) of 160 dB re 1 $\mu Pa^2 s$,
- a zero-to-peak sound pressure level (L_{peak}) of 190 dB re 1 μPa.

Frequency weighting has not been included intentionally. Since several thousand blows are required to reach the final penetration depth, a percentile statistic was chosen to account for multiple blows. In this sense, the SEL_{05} describes the sound exposure level exceeded by 5% of the total number of SEL measurements.

The sound exposure E is defined as

$$E = \int_{T_1}^{T_2} p(t)^2 dt \tag{1}$$

and has the unit $Pa^2 s$. $T1$ and $T2$ indicate the start and end of the evaluated time span respectively, which includes exactly one hammer stroke here. $p(t)$ is the time-variant sound pressure. For the calculation of the dimensionless level, this ex-pression is divided by the reference E_0

$$E_0 = p_0^2 T_0 \tag{2}$$

where T_0 is defined as 1 s and p_0 is defined as 1 μPa. The SEL is thus given by

$$SEL = 10 \log_{10} \left[\frac{\int_{T_1}^{T_2} p(t)^2 dt}{p_0^2 T_0} \right] \left[dB \text{ re } 1 \, \mu Pa^2 s \right] \tag{3}$$

The L_{peak} is given by

$$L_{peak} = 20 \, \log_{10}\left[\frac{\max(|p\,(t)|)}{p_0}\right] [\text{dB re 1 } \mu\text{Pa}] \tag{4}$$

The definition of the dual criterion was based on research results on the onset of TTS in a harbour porpoise as a results of a single impulsive sound exposure [12,15].

All acoustic measurements of underwater noise radiated during pile driving requested by competent agencies in Germany follow the specifications of ISO 18406:2017 and the terminology of ISO 18405:2017. Accreditation according to DIN EN ISO/IEC 17025 [26] for ISO 18406:2017 [16] and DIN SPEC 45653:2017 [27] is required from the measuring institutes.

In comparison to SEL_{cum}, as proposed by NMFS (2016), using percentile statistics of measured SEL to monitor compliance with target values and the performance of noise abatement systems provides a practical and convenient option to monitor and steer the ramming process and the noise abatement systems on a tight schedule. Especially as each pile driving operation is regularly limited to a maximum of three hours, including any deterrence measures and soft-start procedures.

These noise threshold values were purposely defined as unweighted levels which, on one hand, provide a multi-species framework, including the key species, for the development of technical noise mitigation for offshore construction sites. On the other hand, they allow for a robust and standardised monitoring of compliance. In this way, the achievements of the technical targets for noise reduction at the source and the associated reduction of habitat loss due to avoidance and disturbance can be quantified. The compliance with these thresholds requires the application of technical noise abatement measures. Since 2011, the application of such technical noise abatement systems is mandatory at all offshore construction sites in the German EEZ. For a standardised evaluation and documentation of the compliance with the threshold criteria and of the noise exposure of habitats, the noise emissions are monitored and evaluated according at a distance of 750 m and 1500 m to the source and in the nearest nature conservation site or alternatively at a distance of a few kilometres to the wind farm according to the Measuring Instruction by BSH [17] and ISO 18406:2017 [18]. Measurements of sound pressure must cover a frequency range of at least 10 Hz to 20 kHz. All acoustic monitoring data and corresponding technical data on noise abatement systems and the construction process during pile driving activities must be delivered to BSH via the e-reporting portal of an expert information system for underwater noise (https://MarinEARS.bsh.de (28 July 2021)) [28], where the data is undergoing a quality assurance before being made available for internal evaluations.

To evaluate the effectiveness of noise abatement systems and the source level of impact hammers for different site and project characteristics, an instruction was specified in Germany and further developed into a national DIN Specification [26]. According to DIN SPEC 45653:2017 [26], reference and test measurements following a standard design are mandatory for every construction project. Standards for assessment procedures, such as the DIN SPEC 45653:2017 [27], are the prerequisite for the development of abatement systems to reduce underwater noise from pile driving and for an effective protection of marine ecosystems.

In this study, we evaluated acoustic measurements for 23 pile driving events from recent offshore wind farm constructions in the German Bight. Acoustic data were recorded by a high quality, temporary monitoring station with a fixed location, corresponding to distances between 1.7 km and 14.6 km from the pile driving locations. Acoustic data were also recorded at distances of 750 and 1500 m to the source, according to the measuring instructions and standards detailed above [17,18]. The measurement height above seabed was 1.5 m for the temporary fixed monitoring station and 2 m above seabed for the remaining monitoring locations.

All hydrophone data correspond to single-channel recordings (MPEG1 Audio-Layer 3 at the fixed station and PCM Wave-data at 750 and 1500 m distance) and were recorded by autonomous remote measurement systems with a sampling frequency of fs = 44.1 kHz. Due to the higher sound pressure levels expected at 750 and 1500 m distance to the source, the sensitivity of hydrophones at these locations is reduced in comparison to the distant monitoring location, limiting the sensitivity for the higher frequency range above a few kilohertz. Since most of the radiated acoustic energy is well below 1 kHz, this is not a limitation for the evaluation of pile driving noise emissions.

Pile driving events analysed here, occurred during the foundation construction phase in a wind farm cluster with installation positions located in water depths between approximately 38 and 40 m. All foundations correspond to monopiles of the same diameter. With respect to the geographic location of the temporary fixed monitoring stations, pile driving event locations can be attributed to a narrow range of source azimuths.

We perform an in-depth analysis regarding certain aspects of interest, in order to study the effects of various noise emissions at the increasingly active construction sites.

All evaluations of the acoustic metrics were performed using BSH's own acoustic evaluation tool BSoundH, which was developed by Fraunhofer IDMT as part of the research and development project SOUND Mapping (BSH project number 10044386) and which is specifically designed for the evaluation of monitoring data in the context of the regulation framework in the German EEZ. The maximum zero-to-peak level L_{peak} and the sound exposure level SEL were calculated for each hammer stroke detected. Furthermore, we calculated the equivalent continuous sound pressure level LEQ for the period during pile driving and a few hours before via

$$\text{LEQ} = 10 \, \log_{10} \left[\frac{1}{T} \int_0^T \frac{p(t)^2}{p_0^2} dt \right] \left[\text{dB re 1 } \mu\text{Pa}^2 \right] \tag{5}$$

where 0 and T indicate the start and end of the evaluated time span respectively and p_0 is defined as 1 μPa. Here, the duration of each evaluation window T is 5 s. For all quantities, SEL, LEQ and L_{peak}, we determined the exceedance levels of 5%, 50%, 95% and the maximum. E.g. the 5% exceedance level of the L_{peak} corresponds to the sound pressure level which is exceeded in 5% of the hammer strokes of a pile driving event. Broadband levels of SEL and LEQ were calculated over the entire measured frequency range without time or frequency weighting. For the SEL and LEQ we further calculated 1/3-octave band levels between 10 Hz and 20 kHz (IEC 61260).

3. Results

In this section, we present recent examples from acoustic monitoring under the noise mitigation framework, which is further detailed in Section 4. In Section 3.1 we analyse the acoustic measurements at 750 and 1500 m distance and in Section 3.2 we evaluate the measurements from the temporary fixed measurement position, which is at greater distances.

3.1. Results of Pile Driving Noise Monitored at 750 m and 1500 m Distance to the Source during a Typical Recent Construction Project in the German EEZ

For more than six years, several noise abatement systems have been successfully applied for ensuring the reliable compliance with the threshold criteria for construction projects in the German EEZ. Since emitted sound pressure levels increase with the pile surface area below the sea surface, and therefore with increasing water depths and pile diameters [1], an ongoing development and adaption of noise abatement systems to these changing conditions is required for maintaining the reliable compliance with threshold values as observed during recent years.

Figure 1 shows resulting values of SEL_{05} and L_{peak} evaluated from acoustic monitoring data recorded at a distance of 750 m to the source for the 23 pile driving events were considered in this study. Horizontal lines indicate the dual threshold criterion, which indicates a compliance with the regulatory requirements. All but one pile in our data pool were installed using a combination of two noise abatement systems: a noise mitigation screen (IHC-NMS) and a double big bubble curtain (DBBC). For pile number 17, no DBBC was used and the IHC-NMS was removed before the second half of the pile driving process.

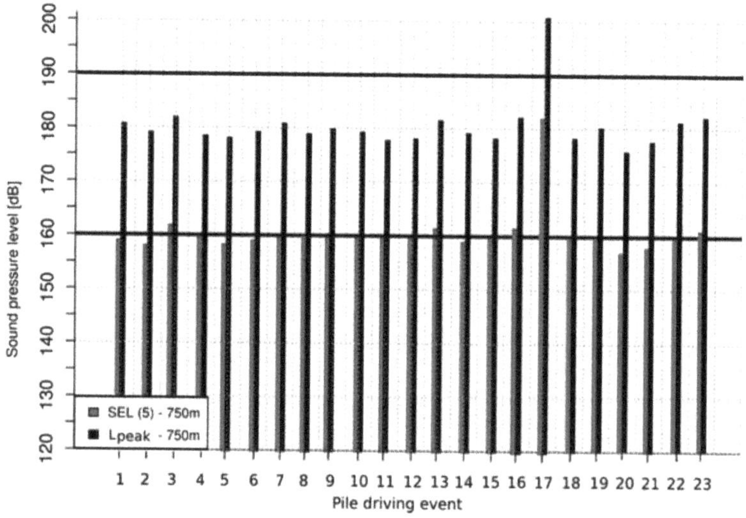

Figure 1. Bar-diagram showing SEL_{05} (blue, in dB re 1 $\mu Pa^2 s$) and L_{peak} (black, in dB re 1 μPa) values evaluated from acoustic monitoring data recorded in a distance of 750 m to the pile driving location in comparison to the dual threshold criterion. Pile number 17 corresponds to the test and reference measurement as detailed in the text.

Resulting SEL_{05} in 750 m distance to the source do not exceed a value of 160 dB for all but five piling events, and none of the mitigated pile driving activities yields a SEL_{05} above 161 dB. Observed L_{peak} values remain well below the threshold of 190 dB re 1 μPa. In contrast, significantly higher sound pressure levels for both metrics are observed for the unmitigated pile driving event, which exceeds both of the threshold criteria with more than 20 dB for the SEL_{05} and more than 12 dB for the L_{peak}. Such obligatory test and reference measurements are granted by BSH purposely for the evaluation of the offshore effectiveness of noise abatement systems applied. When construction projects face difficult site conditions, such as increasing water depths and/or new technical characteristics, such as increasing pile diameters, these evaluations are crucial for the assessment of necessary measures and technical development in further projects. In particular, the reduction target to be achieved can be estimated, which yields implications on the handling and optimisation of technical abatement systems. All results of measurements at 750 m and 1500 m distance to each pile driving site within the framework of the officially ordered construction monitoring by the regulatory agency are included in technical reports provided to the approval and nature conservation authorities.

For a comparison between the spectra of three different noise abatement system configurations, we consider unmitigated pile driving at pile number 17, mitigated pile driving at pile number 17 using an IHC-NMS and mitigated pile driving using an IHC-NMS and DBBC at pile number 23. For all three cases, we selected four hammer strokes during time spans, which yield a similar hammer energy within a narrow energy range between 2850 and 2890 kJ. Figure 2 shows the resulting spectrograms for the three different configurations evaluated for the acoustic measurements at 1500 m distance to the pile driving location.

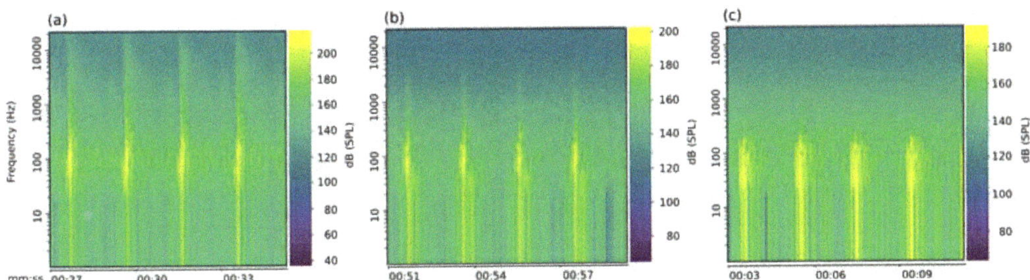

Figure 2. Narrow-band analysis (colour range in dB re 1 µPa) showing a few hammer strokes for different noise abatement system configurations but comparable hammer energies, evaluated at 1500 m distance to the pile driving location. (**a**) Pile driving event 17 without noise abatement systems and a hammer energy of 2870 kJ; (**b**) pile driving event 17 with IHC-NMS but without DBBC and a hammer energy of 2890 kJ; and (**c**) pile driving event 23 with IHC-NMS and DBBC and a hammer energy of 2850 kJ.

Figure 2a shows the unabated spectra of noise emissions for a sequence of four hammer strokes. The spectra for hammer strokes in presence of a single noise abatement system (IHC-NMS) and the combination of two systems (IHC-NMS and DBBC) is depicted in Figure 2b,c, respectively. Already with one noise abatement system, the sound emission of pile driving noise, measured at a distance of 1500 m, is significantly reduced over the entire spectral range considered here (cf. colour bars in figure insets (a) and (b)), With the combination of two systems, the sound pressure level is reduced even further (cf. color bars in figure insets (b) and (c)). Only the combination of both noise abatement systems leads to compliance with the dual threshold values. Importantly, both configurations also yield a particularly strong decrease in measured sound pressure levels for the higher frequency range. The most significant energy reduction is observed at frequencies above approximately 250 Hz when both noise abatement systems are used (see figure inset (c)). In this case, There the acoustic emission of pile-driving noise at high frequencies can no longer be distinguished from the overall noise background.

For a quantification of the ratio between sound energy contained within a low-frequency range (<250 Hz) and the entire frequency range, we consider a subset of pile driving events analysed in this study. Histograms presented in Figure 3 show the distribution of the energy ratio between the low-frequency range versus the entire frequency range for all strokes of six different pile driving events recorded at 1500 m distance. Consistently for the vast majority of mitigated hammer strokes during these pile driving events, more than 95% of the acoustic energy is confined to the frequency range below 250 Hz. The few exceptions can be attributed to background noise at the measurement position and correspond to the first pile driving strokes with very low pile driving energy. Even for unmitigated pile driving, at least 85% of the stroke noise energy is found below 250 Hz for almost all hammer strokes (see pile 17). For pile 17, most of the unmitigated pile driving strokes yield between 87% and 97% of low-frequent energy, while strokes during the application of one noise abatement system mostly yield more than 97% of low-frequent energy.

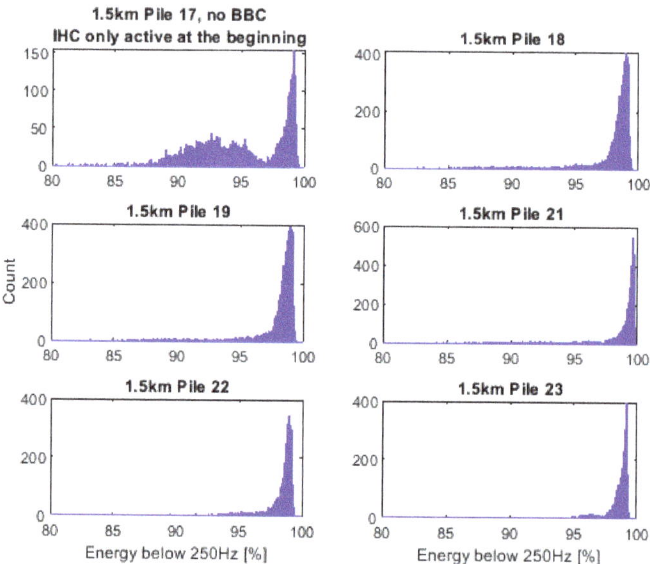

Figure 3. Histogram of the proportion of acoustic energy within the low-frequency range (<250 Hz) relative to the entire frequency range for all hammer strokes evaluated at 1500 m distance for selected pile driving events according to labels.

3.2. Acoustic Pressure at Larger Distances to Piling Site as Revealed from Standardized Measurements at the Same Construction Project

Here, data from a measuring station outside the construction site is primarily analysed in order to highlight effects related to the effectiveness of noise abatement systems used but also cumulative effects with noise from ship traffic in connection to the construction activities.

For the evaluation of the acoustic impact on the marine environment at larger distances to the piling locations, Figure 4 (top) depicts SEL_{05} and L_{peak} values at the temporary fixed monitoring station for the same pile driving events as shown in Figure 1. In Figure 4 (bottom) the corresponding distances between the evaluated pile driving location and the temporary fixed monitoring station are shown. For pile 17, corresponding to a distance of 11.3 km, only the first part of the pile driving process was evaluated at the fixed station. This first part of the ramming includes all hammer strokes for which the IHC-NMS system was used for noise abatement. In contrast to the evaluation at 750 m (in Figure 1) and 1500 m, which included the whole piling process, the unmitigated second part of the pile driving process could not be evaluated at the fixed measuring station. Due to the significantly higher sound pressure levels caused by the non-existing noise abatement system, the hydrophone was overdriven here (at a distance of 11.3 km), so that the measured values were clipped.

Figure 4 shows that measured SEL_{05} and L_{peak} do not ony depend on the distance to the pile, but also on other factors (site and propagation path dependence), i.e., there is no monotonic relationship between measurements and distance. However, an overall decreasing trend between SEL_{05} or L_{peak} and distance can still be observed. As expected, results for pile driving event 17 (evaluated period with IHC-NMS system as single noise abatement system) yield higher values compared to other piling events in similar distances. Generally, measurement results do not only depend on e.g., dispersion and attenuation during sound propagation, but also on background noise at the measurement position. These factors contribute to the observed difference between L_{peak} and SEL_{05} at 750 m distance in comparison to the fixed measurement position. If measured levels were only dependent on the source, the difference between L_{peak} and SEL_{05} for piles in Figures 1 and 4 would be identical.

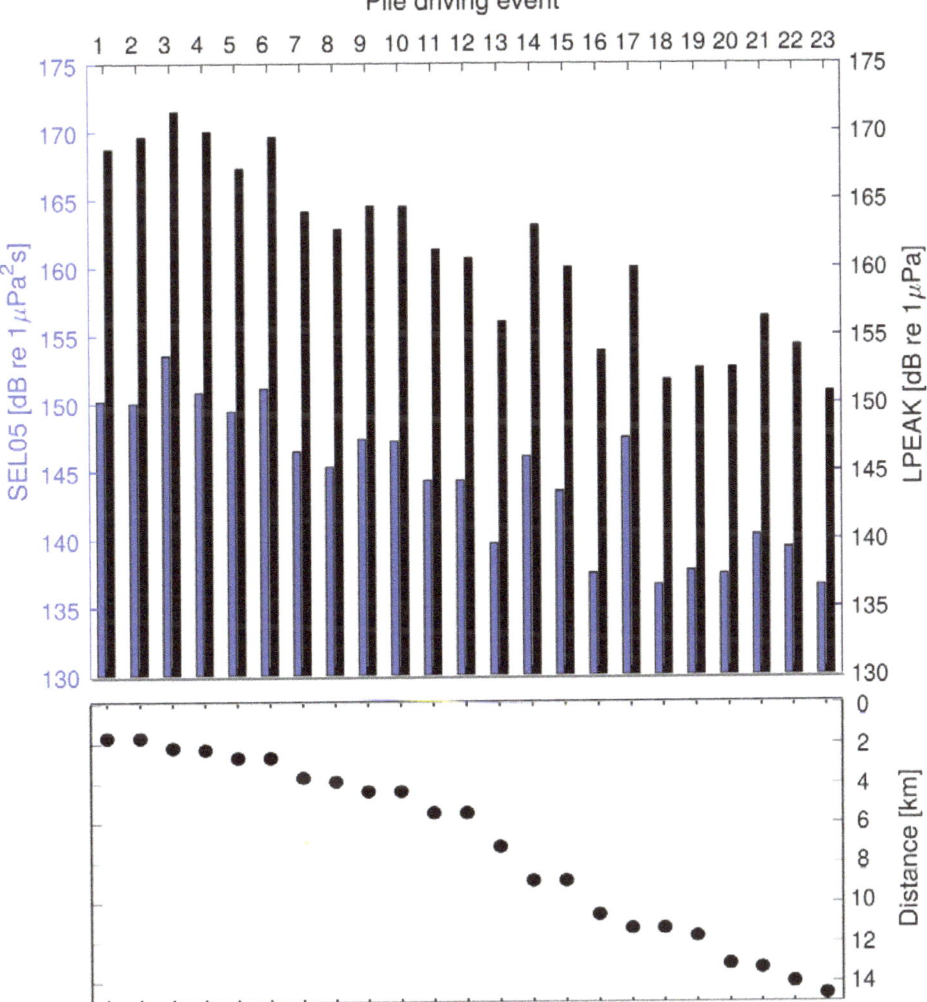

Figure 4. (**top**) SEL_{05} and L_{peak} values evaluated from acoustic monitoring data recorded at the temporary fixed monitoring station evaluated in this study. (**bottom**) Corresponding distances between the fixed monitoring station and the individual pile driving locations. Pile driving event number 17 corresponds to a test and reference measurement as detailed in the text.

We consider the dual threshold criterion, as described in Section 2, in terms of the distance dependent risk of injury due to the pile driving activity. Figure 5 shows the difference between measured SEL_{05} and L_{peak} values at our temporary fixed monitoring station and the dual threshold values defined for a distance of 750 m to the pile driving location. The risk for a harbour porpoise to experience injury due to a single exposure to a $_{peak}$ sound pressure level of 190 dB re 1 μPa or due to a cumulative exposure with a SEL_{05} of 160 dB re 1 μPa²s may be excluded at all distance ranges to the piling location considered here (i.e., >1.7 km).

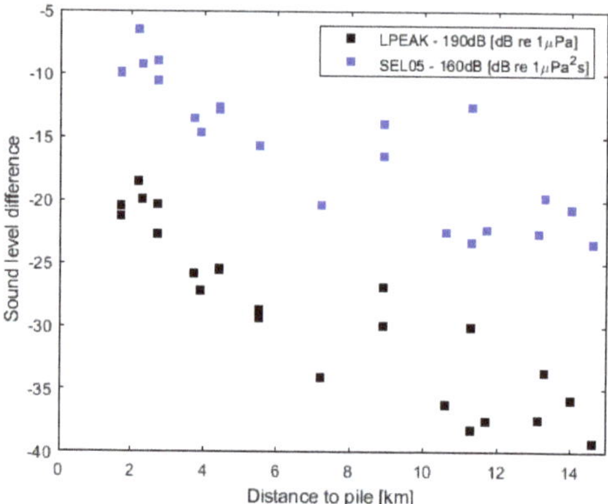

Figure 5. Difference between the measured SEL_{05} and L_{peak} at the temporary fixed monitoring station and the dual threshold values defined for a distance of 750 m to the pile driving location.

As mentioned above, we further observe a decreasing gap between L_{peak} values and SEL_{05} values with distance. In agreement with the 3D propagation of the sound wave field, the signal waveform of the peak will be quickly altered with distance, while the evaluation of the SEL over an integration duration of 1 s still captures the signal energy of the impact. The SEL_{05} thus represents the stricter and more robust condition of the dual threshold criteria over a larger distance range.

For the uppermost end of our distance ranges, but especially for pile driving event 17 at a distance of 11.3 km, we further noticed a decrease in the reliability of our automatic stroke detection algorithm, despite higher values of sound pressure due to the absence of a DBBC. The waveform of corresponding strokes differed from other pile driving events with both noise abatement systems, in agreement with the moderately higher frequency content for signals at pile 17 (cf. Figure 2b). Due to the remaining energy at higher frequencies, dispersion causes the broadening of wave packets for individual impacts at these greater distances. This characteristic implies that automatic signal detectors must be optimized and adapted to the desired distance ranges of monitoring. Furthermore, the determination of signal-to-noise ratios may be biased when evaluated locally around a stroke and the individual impacts recorded may start to overlap temporally before noise levels decay back to the background level.

For a distance range exceeding approximately 10 km to the pile driving location, signal-to-noise ratios for certain pile driving events were not sufficiently high to reliably detect hammer strokes. The lowermost signal-to-noise ratios were retrieved for two of the piling events in particular, events 18 and 23. Both piling events yield the lowest results for L_{peak} and SEL_{05} observed at the fixed monitoring station. However, these values fall not more than 1–2 dB below the next higher values observed for three other pile driving events (16, 19, and 20). However, it is not the absolute sound level that is decisive for the success of an automatic detection, but the amplitude difference between the signal and the background noise. Therefore, we next examine the background sound level at the fixed measurement position a few hours before the onset of pile driving.

For this we consider the LEQ evaluated over time windows of 5 s length. Figure 6a shows the resulting LEQ during these time spans preceding the pile driving activity. Relatively high background noise levels are observed during all of the considered time spans. However, while background noise levels remain below a value of 130 dB on average for most cases considered here, they exceed this threshold during the majority of the 7 h before pile driving event 18 and surpass this threshold during the hour directly before the onset of pile driving event 23. In combination with the low L_{peak} and SEL_{05} levels, this explains the low SNR observed. An analysis of the significant wave height retrieved as hourly time series measured at the FINO 1 station in the southern North

Sea [20], as depicted in Figure 6b, does not indicate a significant simultaneous increase in natural background noise due to an elevated sea state in the North Sea. Thus, it is reasonable to assume that these SPL fluctuations in the background sound level are an anthropogenic effect.

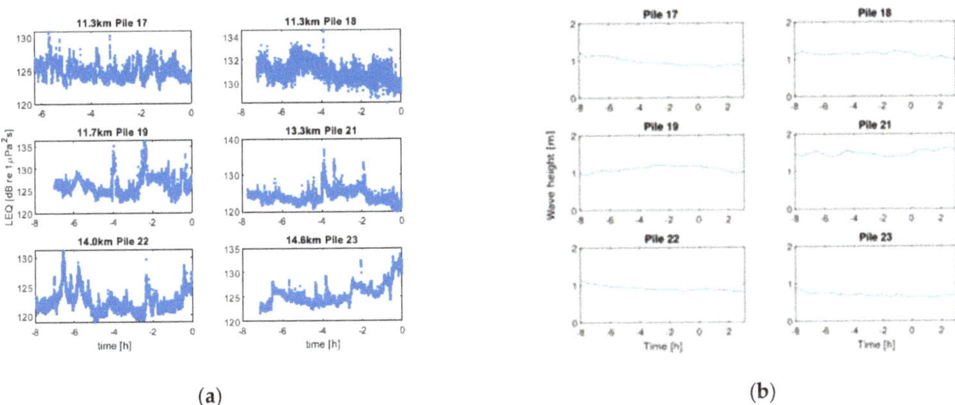

(a)   (b)

Figure 6. (a) LEQ evaluated over time windows of 5 s length during a few hours before the onset of pile driving at a subset of pile driving events. (b) Hourly time series of significant wave height at the BSH research platform FINO 1 retrieved via MarinEARS.

Furthermore, we do not observe a comparably disturbing and continuous background noise for the monitoring stations at closer distance ranges to the pile driving location. We therefore evaluate the spectral statistics of LEQ exceedance values of 5%, 50%, and 95% at the fixed monitoring station during the same time span before the onset of pile driving. The corresponding spectral statistics are depicted in Figure 7b. For comparison, the broadband LEQ values are shown in Figure 6a. All of these spectral results yield a strikingly low variability, indicating the predominance of a relatively continuous spectral characteristic of the background soundscape, and which likely masks piling noise from the distant source locations. By examining the spectral statistics in Figure 7b, the background soundscape yields higher frequency components above 10 kHz, unlike the characteristic emissions of abated pile driving.

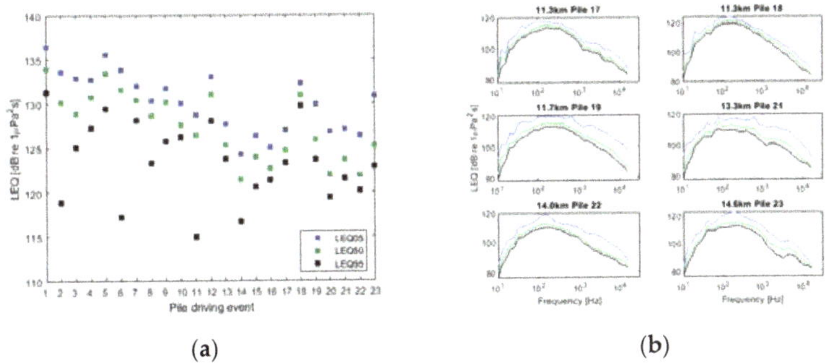

(a)   (b)

Figure 7. (a) Broadband LEQ exceedance values of 5% (blue), 50% (green), and 95% (black) at the fixed monitoring station during the same time span before the onset of pile driving as shown in Figure 5. (b) Spectral statistics of the 1/3 octave band LEQ exceedance values of 5% (blue), 50% (green), and 95% (black) at the fixed monitoring station during the same time span before the onset of pile driving as shown in Figure 5.

We analysed the simultaneous processes and operations carried out in addition to the installation of the foundations in the construction project under study, as well as in immediately adjacent construction sites. This revealed that a total of 15 vessels were present within a radius of up to 10 km from the construction site. On average, three vessels were involved in the pile driving procedure directly: one construction vessel in jack up mode and two support vessels in DP mode: one for the operation of the bubble curtain and one for securing the construction site. Jacking up of construction vessels is a procedure observed to take one to two hours depending on vessel and site [28]. It is assumed, that jacking up and down activities may add to overall background noise substantially. The same is true for vessels with dynamic positioning (DP) systems. In addition, at distances of a few kilometres to our temporary monitoring station, installation work on turbines took place involving jack up vessels accompanied by smaller vessels for the transport of components. Furthermore, a construction vessel for cable laying as well as additional smaller vessels were present in the study area. Most likely, these additional noise sources dominate the background soundscape not only before the onset of pile driving, but also decrease the detectability of the impulsive pile driving signals.

4. Discussion

With the introduction of a noise emission threshold for impulsive noise due to pile driving, and the associated requirement for the application of technical noise abatement systems, one of the strictest regulations for this noise emitting activity has been established in Germany. Similar regulations and requirements were at the same time also developed in Belgium, in the Netherlands and in Denmark.

Establishing a clear protection target as described by the German Environment Protection Agency [29] and in the Concept for the Protection of harbour porpoise from underwater noise emissions from Offshore Pile driving [30] and a comprehensive monitoring program for the observation of the compliance and effectiveness of technological solutions, were the fuel and guide for innovation in Germany. Today, a number of commercially accessible noise abatement systems have been applied successfully on a standard basis in numerous wind farm projects [1,11].

Based on internationally available scientific knowledge and building on dedicated offshore research on options to reduce noise radiated into the environment more than 10 year ago, noise abatement and mitigation for pile driving procedures has been subject to a steep learning curve up to today's standard application of state-of-the-art offshore usable noise abatement technologies. In recent years, however, installation practices have further been adapted, including more efficient, temporally optimised and faster work processes, which implies a different intensity of activities at the con-struction sites as 10 years ago.

In the offshore wind energy development in Europe over the past decade, scientists, regulators and industry have successfully worked hand in hand to advance knowledge of the most relevant adverse effects of noise emissions from pile driving construction. Moreover, the cooperation between these groups has been crucial for the development of a mitigation strategy for relevant adverse effects and, especially, for the implementation of practical approaches to reduce the risk of impaired hearing and habitat loss to marine mammals, which were identified to be most vulnerable to the impulsive noise emissions from pile driving. This includes the harbour porpoise in particular, which is considered as an indicator species for impulsive noise [12]. At the same time, the use of noise abatement measures can substantially reduce or even avoid impacts on marine life (fish, invertebrates).

Due to the data availability through the comprehensive monitoring program of off-shore wind farms under the regulation framework in Germany, recent studies have been able to analyse the effective impact ranges of harbour porpoise displacement based on the joint analysis of both, monitoring data of acoustics and harbour porpoise vocalisation for offshore wind farm (OWF) projects with and without noise mitigation in the German EEZ [4,13]. However, a most recent large scale study concludes that the effective displacement radii of harbour porpoise have not further decrease despite improvements in noise mitigation techniques [14].

In the following, we provide an overview of the prerequisites for establishing a command-and-control regulatory framework to reduce the impact of pile driving noise on marine ecosystems. These prerequisites for the introduction of a sustainable mitigation procedure include: (a) the description of the physical characteristics of the sound source to be mitigated, including knowledge of parameters such as intensity, frequency spectrum and duration of the event/s; (b) the identification of target species and knowledge on abundance and distribution patterns; (c) the evidence of adverse effects on target species by means of reliable and reproducible methods and (d) the consideration of potential mitigation measures.

A step-by-step approach may be followed for establishing an effective and practicable mitigation strategy based on the precautionary principle, covering all steps of the approval procedures for activities where significant impulsive noise emissions are expected. In this section, we outline eight steps for establishing mitigation procedures for pile driving activities in plans and approvals granted by regulatory agencies based on experiences gained in Germany:

Step (1) Develop scientific knowledge on species and habitats in the management area that may be affected: This step can be effectively initiated by the responsible agencies for the approval, environmental protection and nature conservation. A profound research basis may include ship- or aircraft-based surveys or acoustic investigations of priority species and may be strategically integrated into national monitoring and research programmes or specific site surveys. Another key component of the scientific knowledge base is targeted experimental work on noise sensitivity and hearing thresholds of species.

Step (2) Develop scientific knowledge on the physical characteristics of anthropogenic noise sources and on the transmission pathways into habitats: This step includes comprehensive research projects funded by the state and carried out by universities and research institutes. Importantly, this step needs to be repeated when construction site processes change or new technologies emerge.

Step (3) Adopt a noise mitigation strategy: This crucial step is led by the responsible agencies under consideration of: (a) the scientific knowledge developed in step 1 (species inventory, abundance and distribution); (b) the noise source and propagation loss characteristics investigated in step 2; (c) the information regarding threats to species and conservation status; (d) the requirements of European and national legislation and (e) the technical solutions (available or under development) to manage noise emissions and the applicability of measures in practice. The adopted strategy can be sustainable once it is robustly developed and only needs to be adapted or extended as necessary.

Step (4) Communicate the strategy and priority targets to stakeholders: Of particular importance is the timely communication of frameworks, targets, requirements and practical application issues to wind farm operators. Ongoing communication is necessary. However, the most intense communication phase takes place during the initial implementation phase.

Step (5) Develop technical noise abatement techniques: This step requires the joint cooperation and action of scientists, engineers and wind farm operators to develop noise abatement technologies. Research results (steps 1 and 2), the adopted noise mitigation strategy and the application framework defined in coordination with industry form the basis for technological developments and progress in field applications. The concerted action of all involved parties is the backbone for successful development.

Step (6) Develop normative rules to monitor the efficiency and effectiveness of measures: The adoption of standards for determining the performance of measures is A necessary step to take by approval and monitoring agencie. Standards for underwater noise measurements, for noise prognosis, for the quantitative determination of reduction potential of noise abatement systems and standards for the investigation of species abundance and distribution have to be in place to ensure the quality and comparability of monitoring data and assessments.

Step (7) Establish an administrative basis: This step provides a basis for environmental impact assessments (EIA's), including risk-based assessments of noise emissions, participation processes, approval procedures, incidental conditions with noise emission thresholds at activity level and thresholds to prevent cumulative effects on habitats.

Step (8) Implementation in plans and approval procedures: Targets and objectives of the adopted noise mitigation strategy are integrated in all planning steps including maritime spatial planning, the site development plan for offshore wind energy and the site suitability assessment. Incidental conditions on noise mitigation are part of approvals granted and construction releases include detailed noise mitigation measures and monitoring requirements.

The noise mitigation strategy and the so called command and control regulation [25] on pile driving noise implemented in German waters but also similar regulations in other countries (e.g., Belgium, the Netherlands, Denmark) are based on the application of noise abatement and have promoted rapid development of the technology to the state-of-the-art for at least three abatement systems. In recent years, it has become apparent that noise reduction to protect marine ecosystems from impacts due to driving of jacket piles or large monopiles can be effective, applicable and affordable for the industry.

Since the foundation of structures requires several thousand hammer strokes, cumulative effects due to multiple impacts were acknowledged by setting a threshold as the 5% exceedance level (SEL_{05}) at 160 dB re 1 $\mu Pa^2 s$, corresponding to a value 4 dB below the level which experimentally evoked TTS due to a single impulsive exposure. Cumulative effects on the key species harbour porpoise are further restricted according to the noise mitigation concept for the North Sea of the BMU [16,30], which limits the accepted acoustic pressure on habitats in terms of specific percentage-of-area thresholds applicable to the entire German EEZ and to certain nature conservation sites.

In our view, the application of the dual criterion based on unweighted broadband pressure level offer some strong advantages. Such criteria allow for broad development and application, and aim at reducing noise input in the marine ecosystem, and thus protect more marine species than frequency weighting for single species. What appears to be overregulation at a first glance, as described in [5], turns out to be an efficient measure for the marine ecosystem. Reduction of disturbance effects have so far been studied for the key species harbour porpoise. This is primarily due to availability of standard methods for such investigations. Even though a quantification of benefits for other marine species is lacking so far, it can be assumed that any substantial reduction of noise input in the marine environment is a good protective measure for the ecosystem.

Experiences so far shows how crucial it is from a regulatory perspective to implement norms and standards, since these are prerequisites for (a) a comparative evaluation between construction sites and their noise emissions; (b) the possibility of setting well defined thresholds and requirements and especially (c) for monitoring their compliance. In the context of pile driving and establishing the requirement of a field monitoring for noise reduction at the source, following aspects must be considered in the form of standardised procedures:

- the effectiveness of the noise-abatement and mitigation (e.g., deterrence) measures must be monitored and documented by means of measurements;
- a monitoring concept for assessing the effectiveness of measures must be submitted along with a noise mitigation concept and should be further concretised within the context of an implementation plan;
- concept and implementation plan for monitoring of underwater noise must be in accordance with standards, documented in a suitable manner and address construction-related noise due to vessels and due to pile-driving. During the execution of the noise-intensive works, underwater noise measurements should be performed at standardised distances (i.e., 750 m and 1500 m in Germany) to the source and in the nearest nature conservation area or an equivalent position for the purpose of habitat protection;

- the effectiveness of measures must additionally be monitored by means of temporarily deployed harbor porpoise acoustic detectors. These monitoring stations should preferably be co-located with acoustic monitoring stations.

Here, we demonstrated the effectiveness of noise abatement systems applied during recent construction projects in the German Bight in a water depth of approximately 40 m. We were able to confirm their significant reduction potential of sound pressure levels in comparison to unmitigated pile driving based on standardised monitoring data.

Building on recent results on effect ranges of mitigated and unmitigated pile driving [1,31–34], but also from the point of view of further construction sites in deeper waters with larger piles and heavy installation equipment, we have analysed potential noise sources that may have an impact on effect ranges despite the application of noise abatement systems. The study under [1] has shown that, according to recent underwater noise measurements, not only the number of vessels in and around the construction sites is important for background noise level, but also the type of drives, such as vessels with dynamic positioning systems (DP-system), as well as the use of underwater communication devices, such as echo sounders or sonars.

The analysis of potential sources of background noise during this study further revealed, that in addition to the pile driving, numerous other construction work processes took place, which increased the overall noise level. This includes turbine and rotor-blades installation, cable laying and preparatory processes ahead to operation. They may add substantially to noise levels. These sources are usually not included in noise emission and transmission loss modelling, which underlines the importance of sound measurements in the construction field. As shown in Section 3, pile driving was commonly conducted up to a significant wave height of 1.5 m. This implies that increased background noise levels recorded at our fixed monitoring station cannot be attributed to increased ambient noise levels. Most likely, the cumulative emission of a number construction processes in the vicinity to the station may be responsible for the background noise levels observed. This is further supported by the fact that calm weather conditions allow for diverse installation works at the site.

The number of installation vessels present at a time at the construction site has varied from two up to four at a time. Including accompanying vessels for crew and material transfers, safety vessels, vessels for the deployment of bubble curtain and for monitoring activities. As observed from measurements in the immediate vicinity of pile driving sites noise levels already increase a few hours before piling starts. Jack-up operations of construction vessel and operation in DP-modus of accompanying ships further contribute to the recorded increased noise levels. Measurements at greater distance to the piling sites, on the other hand, revealed variably high noise levels, which could be related to jacking-up processes or DP-modus operations of construction vessels, cable laying and ship transfers. Consequently, the overall sound-scape at construction sites is highly variable in terms of sound intensities, so that mitigated piling noise may not be detectable at all due to the low signal-to-ratio. This may also have an impact on the perception by harbour porpoises. The avoidance observed even at well-mitigated piling sites could therefore be more related to the overall noise levels at construction sites where multiple installation and cable laying activities occur simultaneously. As the construction phase of wind farm projects becomes more efficient, meaning that installation work is conducted following a tighter schedule and a rather large number of ships are in use at the construction sites, it is likely that avoidance solely due to these activities will take place to some distance range. On the other hand, the increasing efficiency of installation procedures means that the overall duration of construction noise due to special vessels and activities decreases, which also shortens the duration of animal disturbance. These trade-offs will be subject to a future in depth analysis and need to be taken into account for a comprehensive interpretation of potential effects on marine species.

Author Contributions: Conceptualisation, C.J.; software, B.S.; validation, B.S. and C.J.; formal analysis, B.S.; investigation, B.S. and C.J.; writing—original draft preparation, C.J.; writing—review and editing, B.S. and M.B.; visualisation, B.S.; supervision, C.J.; project administration, C.J. All authors have read and agreed to the published version of the manuscript.

Funding: This research received no external funding.

Institutional Review Board Statement: Not applicable.

Informed Consent Statement: Not applicable.

Data Availability Statement: The majority of data presented in this study are not publicly available due to restricted data sharing rules in the interest of national security and 3rd party data. Aggregated datasets presented in this study are publicly available. This data can be found here: https://marinears.bsh.de (accessed on 29 July 2021).

Conflicts of Interest: The authors declare no conflict of interest.

References

1. Bellmann, M.A.; Brinkmann, J.; May, A.; Wendt, T.; Gerlach, S.; Remmers, P. Underwater Noise during the Impulse Pile-Driving Procedure: Influencing Factors on Pile-Driving Noise and Technical Possibilities to Comply with Noise Mitigation Values. 2020; Supported by the Federal Ministry for the Environment, Nature Conservation and Nuclear Safety (Bundesministerium für Umwelt, Naturschutz und Nukleare Sicherheit (BMU)), FKZ UM16 881500. Commissioned and Managed by the Federal Maritime and Hydrographic Agency (Bundesamt für Seeschifffahrt und Hydrographie (BSH)), Order No. 10036866. Available online: https://www.itap.de/media/experience_report_underwater_era-report.pdf (accessed on 22 July 2021).
2. Dahl, P.H.; de Jong, C.A.; Popper, A.N. The underwater sound field from impact pile driving and its potential effects on marine life. *Acoust. Today* **2015**, *11*, 18–25.
3. Brandt, M.J.; Diederichs, A.; Betke, K.; Nehls, G. Responses of harbour porpoises to pile driving at the Horns Rev II offshore wind farm in the Danish North Sea. *Mar. Ecol. Prog. Ser.* **2011**, *421*, 205–216. [CrossRef]
4. Brandt, M.; Höschle, C.; Diederichs, A.; Betke, K.; Matuschek, R.; Nehls, G. Seal scarers as a tool to deter harbour porpoises from offshore construction sites. *Mar. Ecol. Prog. Ser.* **2013**, *475*, 291–302. [CrossRef]
5. Dähne, M.; Tougaard, J.; Carstensen, J.; Rose, A.; Nabe-Nielsen, J. Bubble curtains attenuate noise from offshore wind farm construction and reduce temporary habitat loss for harbour porpoises. *Mar. Ecol. Prog. Ser.* **2017**, *580*, 221–237. [CrossRef]
6. Boyd, I.; Brownell, B.; Cato, D.; Clarke, C.; Costa, D.; Evans, P.; Gedamke, J.; Gentry, R.; Gisiner, B.; Gordon, J.; et al. The Effects of Anthropogenic Sound on Marine Mammals. In *EMB Position Paper 13*; Connolly, N., Calewaert, J.-B., Eds.; Available online: http://marineboard.eu/publication/effects-anthropogenic-sound-marine-mammal (accessed on 29 July 2021).
7. Finneran, J.J. Noise-induced hearing loss in marine mammals: A review of temporary threshold shift studies from 1996 to 2015. *J. Acoust. Soc. Am.* **2015**, *138*, 1702–1726. [CrossRef] [PubMed]
8. Effects of Anthropogenic Noise on Animals. In *Springer Handbook of Auditory Research*; Springer Science and Business Media LLC, 2018; pp. 277–309. Available online: https://www.springer.com/gp/book/9781493985722 (accessed on 22 July 2021).
9. Popper, A.N.; Hawkins, A.D.; Thomsen, F. Taking the Animals' Perspective Regarding Anthropogenic Underwater Sound. *Trends Ecol. Evol.* **2020**, *35*, 787–794. [CrossRef] [PubMed]
10. Gomez, C.; Lawson, J.; Wright, A.; Buren, A.; Tollit, D.; Lesage, V. A systematic review on the behavioural responses of wild marine mammals to noise: The disparity between science and policy. *Can. J. Zoöl.* **2016**, *94*, 801–819. [CrossRef]
11. Southall, B.L.; Finneran, J.J.; Reichmuth, C.; Nachtigall, P.E.; Ketten, D.R.; Bowles, A.E.; Ellison, W.T.; Nowacek, D.P.; Tyack, P. Marine Mammal Noise Exposure Criteria: Updated Scientific Recommendations for Residual Hearing Effects. *Aquat. Mamm.* **2019**, *45*, 125–232. [CrossRef]
12. Koschinski, S.; Lüdemann, K. Noise Mitigation for the Construction of Increasingly Large Offshore Wind Turbines. In *Technical Options for Complying with Noise Limits*; The Federal Agency for Nature Conservation: Isle of Vilm, Germany, 2020; Available online: https://www.bfn.de/fileadmin/BfN/meeresundkuestenschutz/Dokumente/Noise-mitigation-for-the-construction-of-increasingly-large-offshore-wind-turbines.pdf (accessed on 29 July 2021).
13. De Jong, C.A.F.; Ainslie, M.A.; Blacquiere, G. Standard for Measurement and Monitoring of Underwater Noise, Part II: Procedures Formeasuring Underwater Noise in Connectionwith Offshore Wind Farm Licensing TNO Report, TNO-DV 2011 C251. 2011. Available online: https://tethys.pnnl.gov/sites/default/files/publications/TNO-Report-2011.pdf (accessed on 22 July 2021).
14. BSH Offshore Wind Farms: Measuring Instruction for Underwater Sound Monitoring. Current Approach with Annotations, Application Instructions. 2011. Available online: https://www.bsh.de/DE/PUBLIKATIONEN/_Anlagen/Downloads_Suchausschluss/Offshore/Anlagen-EN/Measuring-instruction-for-underwater-sound-monitoring.pdf?__blob=publicationFile&v=4 (accessed on 22 July 2021).
15. NPL. Good Practice Guide for Underwater Noise Measurement; No. 133; 2014; ISSN 1368-6550. Available online: https://www.npl.co.uk/special-pages/guides/gpg133underwater (accessed on 22 July 2021).

16. ISO 18405:2017. Underwater Acoustics—Terminology. Available online: https://www.iso.org/standard/62406.html (accessed on 22 July 2021).

17. ISO 18406:2017-04: Underwater Acoustics—Measurement of Radiated Underwater Sound From Percussive Pile Driving. Available online: https://www.iso.org/standard/62407.html (accessed on 22 July 2021).

18. Reinhall, P.G.; Dahl, P.H. Underwater Mach wave radiation from impact pile driving: Theory and observation. *J. Acoust. Soc. Am.* **2011**, *130*, 1209–1216. [CrossRef] [PubMed]

19. Zampolli, M.; Nijhof, M.J.J.; De Jong, C.A.F.; Ainslie, M.A.; Jansen, E.H.W.; Quesson, B. Validation of finite element computations for the quantitative prediction of underwater noise from impact pile driving. *J. Acoust. Soc. Am.* **2013**, *133*, 72–81. [CrossRef] [PubMed]

20. Lippert, T.; Ainslie, M.A.; Von Estorff, O. Pile driving acoustics made simple: Damped cylindrical spreading model. *J. Acoust. Soc. Am.* **2018**, *143*, 310–317. [CrossRef] [PubMed]

21. Ainslie, M.A.; Halvorsen, M.B.; Müller, R.A.J.; Lippert, T. Application of damped cylindrical spreading to assess range to injury threshold for fishes from impact pile driving. *J. Acoust. Soc. Am.* **2020**, *148*, 108–121. [CrossRef] [PubMed]

22. Southall, B.L.; Bowles, A.E.; Ellison, W.T.; Finneran, J.J.; Gentry, R.L.; Green, C.R.J.R.; Kastak, D.; Ketten, D.R.; Miller, J.H.; Nachtigall, P.E.; et al. Marine mammal noise exposure criteria: Initial scientific recommendations. *Aquat. Mamm.* **2007**, *33*, 411–521. [CrossRef]

23. Finneran, J.J. Auditory Weighting Functions and TTS/PTS Exposure Functions for Marine Mammals Exposed to Underwater Noise. TECHNICAL REPORT 3026. 2016. Available online: https://apps.dtic.mil/sti/citations/AD1026445 (accessed on 22 July 2021).

24. National Marine Fisheries Service. *Technical Guidance for Assessing the Effects of Anthropogenic Sound on Marine Mammal Hearing: Underwater Acoustic Thresholds for Onset of Permanent and Temporary Threshold Shifts*; NOAA Technical Memorandum NMFS-OPR-55; United States Department of Commerce: Washington, DC, USA, 2018; 178p.

25. Merchant, N.D. Underwater noise abatement: Economic factors and policy options. *Environ. Sci. Policy* **2019**, *92*, 116–123. [CrossRef]

26. DIN EN ISO/IEC 17025:2018-03, Allgemeine Anforderungen an die Kompetenz von Prüf-und Kalibrierlaboratorien (ISO/IEC_17025:2017); Deutsche und Englische Fassung EN_ISO/IEC_17025:2017. *EN ISO/IEC* 2019, Volume 17025. Available online: https://opus4.kobv.de/opus4-bam/frontdoor/index/index/docId/42383 (accessed on 28 July 2021).

27. DIN SPEC 45653:2017. Offshore Wind Farms—In-Situ Determination of the Insertion Loss of Control Measures Underwater. DIN/VDI-Normenausschuss Akustik, Lärmminderung und Schwingungstechnik (NALS). DIN Deutsches Institut für Normung e. V. Available online: https://webstore.ansi.org/standards/din/dinspec456532017 (accessed on 22 July 2021).

28. MarinEARS. Available online: https://MarinEARS.bsh.de (accessed on 12 June 2021).

29. Umweltbundesamt, 2011. Empfehlung von Lärmschutzwerten bei der Errichtung von Offshore-Windenergieanlagen (OWEA). Umweltbundesamt. Available online: www.umweltbundesamt.de (accessed on 28 July 2021).

30. BMU. Konzept für den Schutz der Schweinswale vor Schallbelastungen bei der Errichtung von Offshore-Windparks in der deutschen Nordsee (Schallschutzkonzept), Bundesministerium für Umwelt, Naturschutz und Reaktorsicherheit. 2013. Available online: https://www.bfn.de/fileadmin/BfN/awz/Dokumente/schallschutzkonzept_BMU.pdf (accessed on 22 July 2021).

31. Lucke, K.; Lepper, P.A.; Blanchet, M.A.; Siebert, U. Temporary shift in masked hearing thresholds in a harbor por-poise (*Phocoena phocoena*) after exposure to seismic airgun stimuli. *J. Acoust. Soc. Am.* **2009**, *125*, 4060–4070. [CrossRef] [PubMed]

32. Brandt, M.; Dragon, A.; Diederichs, A.; Bellmann, M.; Wahl, V.; Piper, W.; Nabe-Nielsen, J.; Nehls, G. Disturbance of harbour porpoises during construction of the first seven offshore wind farms in Germany. *Mar. Ecol. Prog. Ser.* **2018**, *596*, 213–232. [CrossRef]

33. Rose, A.; Brandt, M.J.; Vilela, R.; Diederichs, A.; Schubert, A.; Kosarev, V.; Nehls, G.; Volkenandt, M.; Wahl, V.; Michalik, A.; et al. Effects of Noise-Mitigated Offshore Pile Driving on Harbour Porpoise Abundance in the German Bight 2014–2016 (Gescha 2)—Assessment of Noise Effects, Technischer Abschlussbericht im Auftrag des Arbeitskreis Schallschutz" des Offshore Forums Windenergie, Erstellt von BioConsult SH GmbH & Co KG, Husum; IBL Umweltplanung GmbH, Oldenburg; Institut für Angewandte Ökosystemforschung GmbH, Hamburg. 2019. Available online: https//bwo-offshorewind.de/wp-content/uploads/2019/07/study-on-the-effects-of-noise-mitigated-construction-works-on-the-harbour-porpoise-population-in-the-german-north-sea.pdf (accessed on 28 July 2021).

34. Lucke, K.; Lepper, P.A.; Blanchet, M.; Siebert, U. Testing the acoustic tolerance of harbour porpoise hearing for impulsive sounds. *J. Acoust. Soc. Am.* **2008**, *123*, 3780. [CrossRef]

MDPI

St. Alban-Anlage 66

4052 Basel

Switzerland

Tel. +41 61 683 77 34

Fax +41 61 302 89 18

www.mdpi.com

Journal of Marine Science and Engineering Editorial Office

E-mail: jmse@mdpi.com

www.mdpi.com/journal/jmse